T0180804

Lecture Notes in Computer Science 12454

More information about this series at http://www.springer.com/series/7407

Meikang Qiu (Ed.)

Algorithms and Architectures for Parallel Processing

20th International Conference, ICA3PP 2020
New York City, NY, USA, October 2–4, 2020
Proceedings, Part III

Editor
Meikang Qiu
Columbia University
New York, NY, USA

ISSN 0302-9743 ISSN 1611-3349 (electronic)
Lecture Notes in Computer Science
ISBN 978-3-030-60247-5 ISBN 978-3-030-60248-2 (eBook)
https://doi.org/10.1007/978-3-030-60248-2

LNCS Sublibrary: SL1 – Theoretical Computer Science and General Issues

This Springer imprint is published by the registered company Springer Nature Switzerland AG
The registered company address is: Gewerbestrasse 11, 6330 Cham, Switzerland

Preface

This three-volume set contains the papers presented at the 20th International Conference on Algorithms and Architectures for Parallel Processing (ICA3PP 2020), held during October 2–4, 2020, in New York, USA.

There were 495 submissions. Each submission was reviewed by at least 3 reviewers, and on the average 3.5 Program Committee members. The committee decided to accept 147 papers. We will separate the proceeding into three volumes: LNCS 12452, 12453, and 12454. Yielding an acceptance rate of 29%.

ICA3PP 2020 was the 20th in this series of conferences started in 1995 that are devoted to algorithms and architectures for parallel processing. ICA3PP is now recognized as the main regular event of the world that is covering the many dimensions of parallel algorithms and architectures, encompassing fundamental theoretical approaches, practical experimental projects, and commercial components and systems. As applications of computing systems have permeated in every aspect of daily life, the power of computing systems has become increasingly critical. This conference provides a forum for academics and practitioners from countries around the world to exchange ideas for improving the efficiency, performance, reliability, security, and interoperability of computing systems and applications.

Following the traditions of the previous successful ICA3PP conferences held in Hangzhou, Brisbane, Singapore, Melbourne, Hong Kong, Beijing, Cyprus, Taipei, Busan, Melbourne, Fukuoka, Vietri sul Mare, Dalian, Japan, Zhangjiajie, Granada, Helsinki, Guangzhou, and Melbourne, ICA3PP 2020 was held in New York, USA. The objective of ICA3PP 2020 is to bring together researchers and practitioners from academia, industry, and governments to advance the theories and technologies in parallel and distributed computing. ICA3PP 2020 will focus on three broad areas of parallel and distributed computing, i.e., Parallel Architectures and Algorithms (PAA), Parallel computing with AI and Big Data (PAB), and Parallel computing with Cyberseucrity and Blockchain (PCB).

We would like to thank the conference sponsors: Springer LNCS, Columbia University, North America Chinese Talents Association, and Longxiang High Tech Group Inc.

October 2020

Meikang Qiu

Organization

Honorary Chairs

Sun-Yuan Kung	Princeton University, USA
Gerard Memmi	Télécom Paris, France

General Chair

Meikang Qiu	Columbia University, USA

Program Chairs

Yongxin Zhu	Shanghai Advanced Research Institute, China
Bhavani Thuraisingham	The University of Texas at Dallas, USA
Zhongming Fei	University of Kentucky, USA
Linghe Kong	Shanghai Jiao Tong University, China

Local Chair

Xiangyu Gao	New York University, USA

Workshop Chairs

Laizhong Cui	Shenzhen University, China
Xuyun Zhang	The University of Auckland, New Zealand

Publicity Chair

Peng Zhang	Stony Brook SUNY, USA

Finance Chair

Hui Zhao	Henan University, China

Web Chair

Han Qiu	Télécom-ParisTech, France

Steering Committee

Yang Xiang (Chair)	Swinburne University of Technology, Australia
Weijia Jia	Shanghai Jiao Tong University, China

Yi Pan	Georgia State University, USA
Laurence T. Yang	St. Francis Xavier University, Canada
Wanlei Zhou	University of Technology Sydney, Australia

Technical Committee

Dean Anderso	Bank of America Merrill Lynch, USA
Prem Chhetri	RMIT, Australia
Angus Macaulay	The University of Melbourne, Australia
Paul Rad	Rackspace, USA
Syed Rizvi	Penn State University, USA
Wei Cai	Chinese University of Hong Kong, Hong Kong, China
Abdul Razaque	University of Bridgeport, USA
Katie Cover	Penn State University, USA
Yongxin Zhao	East China Normal University, China
Sanket Desai	San Jose State University, USA
Weipeng Cao	Shenzhen University, China
Suman Kumar	Troy University, USA
Qiang Wang	Southern University of Science and Technology, China
Wenhui Hu	Peking University, China
Kan Zhang	Tsinghua University, China
Mohan Muppidi	UTSA, USA
Wenting Wei	Xidian University, China
Younghee Park	San Jose State University, USA
Sang-Yoon Chang	Advanced Digital Science Center, Singapore
Jin Cheol Kim	KEPCO KDN, South Korea
William de Souza	University of London, UK
Malik Awan	Cardiff University, UK
Mehdi Javanmard	Rutgers University, USA
Allan Tomlinson	University of London, UK
Weiwei Shan	Southeast University, China
Tianzhu Zhang	Télécom Paris, France
Chao Feng	National University of Defense Technology, China
Zhong Ming	Shenzhen University, China
Hiroyuki Sato	The University of Tokyo, Japan
Shuangyin Ren	Chinese Academy of Military Science, China
Thomas Austin	San Jose State University, USA
Zehua Guo	Beijing Institute of Technology, China
Wei Yu	Towson University, USA
Yulin He	Shenzhen University, China
Zhiqiang Lin	The University of Texas at Dallas, USA
Xingfu Wu	Texas A&M University, USA
Wenbo Zhu	Google Inc., USA
Weidong Zou	Beijing Institute of Technology, China
Hwajung Lee	Radford University, USA

Yuxuan Jiang	The Hong Kong University of Science and Technology, Hong Kong, China
Yong Guan	Iowa State University, USA
Chao-Tung Yang	Tunghai University Taiwan, China
Zonghua Gu	Zhejiang University, China
Gang Zeng	Nagoya University, Japan
Hui Zhao	Henan University, China
Yong Zhang	The University of Hong Kong, Hong Kong, China
Hanpin Wang	Peking University, China
Yu Hua	Huazhong University of Science and Technology, China
Yan Zhang	University of Oslo, Norway
Haibo Zhang	University of Otago, New Zealand
Hao Hu	Nanjing University, China
Zhihui Du	Tsinghua University, China
Jiahai Yang	Tsinghua University, China
Fuji Ren	Tokushima University, Japan
Long Fei	Google Inc., USA
Tianwei Zhang	Nanyang Technological University, Singapore
Ming Xu	Hangzhou Dianzi University, China
Golden Richard	Louisiana State University, USA
Virginia Franqueira	University of Derby, UK
Haoxiang Wang	Cornell University, USA
Jun Zhang	Shenzhen University, China
Xinyi Huang	Fujian Normal University, China
Debiao He	Wuhan University, China
Vijayan Sugumaran	Oakland University, USA
Ximeng Liu	Singapore Management University, Singapore
Zhan Qin	The University of Texas at San Antonio, USA
Dalei Wu	The University of Tennessee at Chattanooga, USA
Kathryn Seigfried-Spellar	Purdue University, USA
Jun Zheng	New Mexico Tech, USA
Paolo Trunfio	University of Calabria, Italy
Kewei Sha	University of Houston - Clear Lake, USA
David Dampier	The University of Texas at San Antonio, USA
Richard Hill	University of Huddersfield, UK
William Glisson	University of South Alabama, USA
Petr Matousek	Brno University of Technology, Czech Republic
Javier Lopez	University of Malaga, Spain
Dong Dai	Texas Tech University, USA
Ben Martini	University of South Australia, Australia
Ding Wang	Peking University, China
Xu Zheng	Shanghai University, China
Nhien An Le Khac	University College Dublin, Ireland
Shadi Ibrahim	Inria Rennes – Bretagne Atlantique, France
Neetesh Saxena	Bournemouth University, UK

Contents – Part III

Parallel Computing with Cybersecurity and Blockchain (PCB)

Understanding Privacy-Preserving Techniques in Digital Cryptocurrencies . . . 3
Yue Zhang, Keke Gai, Meikang Qiu, and Kai Ding

LNBFSM: A Food Safety Management System Using Blockchain and Lightning Network . . . 19
Zhengkang Fang, Keke Gai, Liehuang Zhu, and Lei Xu

Reputation-Based Trustworthy Supply Chain Management Using Smart Contract . . . 35
Haochen Li, Keke Gai, Liehuang Zhu, Peng Jiang, and Meikang Qiu

Content-Aware Anomaly Detection with Network Representation Learning . . . 50
Zhong Li, Xiaolong Jin, Chuanzhi Zhuang, and Zhi Sun

Blockchain Based Data Integrity Verification for Cloud Storage with T-Merkle Tree . . . 65
Kai He, Jiaoli Shi, Chunxiao Huang , and Xinrong Hu

IM-ACS: An Access Control Scheme Supporting Informal Non-malleable Security in Mobile Media Sharing System . . . 81
Anyuan Deng, Jiaoli Shi, Kai He, and Fang Xu

Blockchain-Based Secure and Privacy-Preserving Clinical Data Sharing and Integration . . . 93
Hao Jin, Chen Xu, Yan Luo, and Peilong Li

Blockchain Meets DAG: A BlockDAG Consensus Mechanism . . . 110
Keke Gai, Ziyue Hu, Liehuang Zhu, Ruili Wang, and Zijian Zhang

Dynamic Co-located VM Detection and Membership Update for Residency Aware Inter-VM Communication in Virtualized Clouds . . . 126
Zhe Wang, Yi Ren, Jianbo Guan, Ziqi You, Saqing Yang, and Yusong Tan

An Attack-Immune Trusted Architecture for Supervisory Intelligent Terminal . . . 144
Dongxu Cheng, Jianwei Liu, Zhenyu Guan, and Jiale Hu

A Simulation Study on Block Generation Algorithm Based on TPS Model 155
Shubin Cai, Huaifeng Zhou, NingSheng Yang, and Zhong Ming

Improving iForest for Hydrological Time Series Anomaly Detection 170
Pengpeng Shao, Feng Ye, Zihao Liu, Xiwen Wang, Ming Lu, and Yupeng Mao

Machine Learning-Based Attack Detection Method in Hadoop 184
Ningwei Li, Hang Gao, Liang Liu, and Jianfei Peng

Profiling-Based Big Data Workflow Optimization in a Cross-layer Coupled Design Framework 197
Qianwen Ye, Chase Q. Wu, Wuji Liu, Aiqin Hou, and Wei Shen

Detection of Loose Tracking Behavior over Trajectory Data 218
Jiawei Li, Hua Dai, Yiyang Liu, Jianqiu Xu, Jie Sun, and Geng Yang

Behavioral Fault Modelling and Analysis with BIP: A Wheel Brake System Case Study 231
Xudong Tang , Qiang Wang , and Weikai Miao

H2P: A Novel Model to Study the Propagation of Modern Hybrid Worm in Hierarchical Networks 251
Tianbo Wang and Chunhe Xia

Design of Six-Rotor Drone Based on Target Detection for Intelligent Agriculture 270
Chenyang Liao, Jiahao Huang, Fangkai Zhou, and Yang Lin

Consensus in Lens of Consortium Blockchain: An Empirical Study 282
Hao Yin, Yihang Wei, Yuwen Li, Liehuang Zhu, Jiakang Shi, and Keke Gai

Collaborative Design Service System Based on Ceramic Cloud Service Platform 297
Yu Nie, Yu Liu, Chao Li, Hua Huang, Fubao He, and Meikang Qiu

Towards NoC Protection of HT-Greyhole Attack 309
Soultana Ellinidou, Gaurav Sharma, Olivier Markowitch, Jean-Michel Dricot, and Guy Gogniat

Cross-shard Transaction Processing in Sharding Blockchains 324
Yizhong Liu, Jianwei Liu, Jiayuan Yin, Geng Li, Hui Yu, and Qianhong Wu

Lexicon-Enhanced Transformer with Pointing for Domains Specific Generative Question Answering 340
Jingying Yang, Xianghua Fu, Shuxin Wang, and Wenhao Xie

Design of Smart Home System Based on Collaborative Edge Computing and Cloud Computing 355
Qiangfei Ma, Hua Huang, Wentao Zhang, and Meikang Qiu

Classification of Depression Based on Local Binary Pattern and Singular Spectrum Analysis 367
Lijuan Duan, Hongli Liu, Huifeng Duan, Yuanhua Qiao, and Changming Wang

Cloud Allocation and Consolidation Based on a Scalability Metric 381
Tarek Menouer, Amina Khedimi, Christophe Cérin, and Congfeng Jiang

Adversarial Attacks on Deep Learning Models of Computer Vision: A Survey 396
Jia Ding and Zhiwu Xu

FleetChain: A Secure Scalable and Responsive Blockchain Achieving Optimal Sharding 409
Yizhong Liu, Jianwei Liu, Dawei Li, Hui Yu, and Qianhong Wu

DSBFT: A Delegation Based Scalable Byzantine False Tolerance Consensus Mechanism. 426
Yuan Liu, Zhengpeng Ai, Mengmeng Tian, Guibing Guo, and Linying Jiang

A Privacy-Preserving Approach for Continuous Data Publication 441
Mengjie Zhang, Xingsheng Zhang, Zhijun Chen, and Dunhui Yu

Web Attack Detection Based on User Behaviour Semantics 459
Yunyi Zhang, Jintian Lu, and Shuyuan Jin

A Supervised Anonymous Issuance Scheme of Central Bank Digital Currency Based on Blockchain 475
Wenhao Dai, Xiaozhuo Gu, and Yajun Teng

IncreAIBMF: Incremental Learning for Encrypted Mobile Application Identification 494
Yafei Sang, Mao Tian, Yongzheng Zhang, Peng Chang, and Shuyuan Zhao

A Multi-level Features Fusion Network for Detecting Obstructive Sleep Apnea Hypopnea Syndrome 509
Xingfeng Lv and Jinbao Li

BIMP: Blockchain-Based Incentive Mechanism with Privacy Preserving in Location Proof 520
Zhen Lin, Yuchuan Luo, Shaojing Fu, and Tao Xie

Indoor Positioning and Prediction in Smart Elderly Care: Model, System and Applications 537
Yufei Liu, Xuqi Fang, Fengyuan Lu, Xuxin Chen, and Xinli Huang

Research on Stylization Algorithm of Ceramic Decorative Pattern Based on Ceramic Cloud Design Service Platform 549
Xinxin Liu, Hua Huang, Meikang Qiu, and Meiqin Liu

Blockchain Consensus Mechanisms and Their Applications in IoT: A Literature Survey 564
Yujuan Wen, Fengyuan Lu, Yufei Liu, Peijin Cong, and Xinli Huang

Towards a Secure Communication of Data in IoT Networks: A Technical Research Report 580
Bismark Tei Asare, Kester Quist-Aphetsi, and Laurent Nana

Efficient Thermography Guided Learning for Breast Cancer Detection 592
Vishwas Rajashekar, Ishaan Lagwankar, Durga Prasad S N, and Rahul Nagpal

PTangle: A Parallel Detector for Unverified Blockchain Transactions 601
Ashish Christopher Victor, Akhilarka Jayanthi, Atul Anand Gopalakrishnan, and Rahul Nagpal

DOS-GAN: A Distributed Over-Sampling Method Based on Generative Adversarial Networks for Distributed Class-Imbalance Learning 609
Hongtao Guan, Xingkong Ma, and Siqi Shen

Effective Sentiment Analysis for Multimodal Review Data on the Web 623
Peiquan Jin, Jianchuan Li, Lin Mu, Jingren Zhou, and Jie Zhao

An Energy-Efficient AES Encryption Algorithm Based on Memristor Switch 639
Danghui Wang, Chen Yue, Ze Tian, Ru Han, and Lu Zhang

Digital Currency Investment Strategy Framework Based on Ranking 654
Chuangchuang Dai, Xueying Yang, Meikang Qiu, Xiaobing Guo, Zhonghua Lu, and Beifang Niu

Authentication Study for Brain-Based Computer Interfaces Using Music Stimulations 663
Sukun Li and Meikang Qiu

An Ensemble Learning Approach to Detect Malwares Based on Static Information 676
Lin Chen, Huahui Lv, Kai Fan, Hang Yang, Xiaoyun Kuang, Aidong Xu, and Siliang Suo

Automatic Medical Image Report Generation with Multi-view and Multi-modal Attention Mechanism 687
Shaokang Yang, Jianwei Niu, Jiyan Wu, and Xuefeng Liu

Poster Paper

PLRS: Personalized Literature Hybrid Recommendation System with Paper Influence 703
Fanghan Liu, Wenzheng Cai, and Kun Ma

Author Index 707

Parallel Computing with Cybersecurity and Blockchain (PCB)

Understanding Privacy-Preserving Techniques in Digital Cryptocurrencies

Yue Zhang[1], Keke Gai[2,3(✉)], Meikang Qiu[4], and Kai Ding[1]

[1] School of Computer Science and Technology, Beijing Institute of Technology, Beijing 100081, China
cheong.yue@hotmail.com, dingkai@bit.edu.cn

[2] School of Cyberspace Security, Beijing Institute of Technology, Beijing 100081, China
gaikeke@bit.edu.cn

[3] Henan Key Laboratory of Network Cryptography Technology, Zhengzhou, Henan, China

[4] College of Computer Science and Software Engineering, Shenzhen University, Shenzhen 518061, China
mqiu@szu.edu.cn

Abstract. Asset security and transaction efficiency have always been concerned, so that digital cryptocurrencies based on blockchain are increasingly favored. In the transaction, it is necessary to effectively protect private information such as identity and transaction content through technical means. This survey investigates privacy-preserving techniques commonly used in cryptocurrencies, and comparatively analyzes the advantages and disadvantages of different methods. The work mainly covers three technical dimensions. First, this paper investigates and summarizes privacy issues in cryptocurrencies. Second, we investigate mainstream privacy preserving techniques, such as Mixer, Homomorphic Encryption, and State Channel. Specially, we focus on the development of Zero-Knowledge Proof schemes and analyze current defects. Finally, the future direction of blockchain privacy protection technology is prospected.

Keywords: Blockchain · Cryptocurrencies · Privacy preserving · Zero-Knowledge Proofs · Mixer · Homomorphic Encryption · State Channel

1 Introduction

With the advent of global digital economy, the popularity of digital cryptocurrencies continues to grow. After Bitcoin, a large number of digital cryptocurrencies based on blockchain spring up [5]. Privacy leakage is threatening autonomous transaction systems based on digital cryptocurrencies [20]. Privacy protection has gradually become an important topic in the cryptocurrency field [9].

The protection of users' private information has been considered a challenging study for many years [49]. According to our investigations, we find that

M. Qiu (Ed.): ICA3PP 2020, LNCS 12454, pp. 3–18, 2020.
https://doi.org/10.1007/978-3-030-60248-2_1

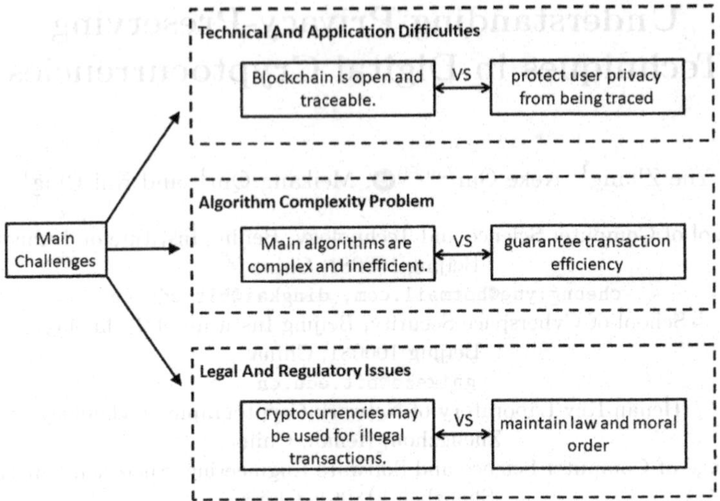

Fig. 1. Main challenges of applying privacy-preserving techniques to cryptocurrencies.

there are typical challenges in effectively applying privacy-preserving techniques to digital cryptocurrencies, as shown in Fig. 1. One of the challenges is that introducing privacy protection into digital currencies will encounter technical difficulties. Transaction records in blockchain need to be disclosed to all nodes according to the agreement. On one side, the anti-attack capabilities of each node are uneven, making weak nodes vulnerable to attack. On the other side, malicious nodes existing in the network can retrieve all historical transactions and may analyze hidden information through big data technology [17]. In addition, efficiency issues should be considered when designing the privacy-preserving algorithm [24]. People usually do not have a high tolerance for delay. Extra latency is introduced by security schemes and security related calculations [46]. The encryption and decryption process should not be too complicated, and the amount of calculation should be controlled within an acceptable range. The mismatch of CPU and memory speeds results in low processing efficiency. Latency can also be minimized by increasing memory access speed [48]. Other challenges come from legal and regulatory. While protecting users' privacy, we should be alert to illegal transactions and money laundering.

We have previously researched and proposed a series of privacy-preserving approaches, such as privacy protection in energy transactions [25,26], privacy protection in resource-constrained devices [19], and privacy protection in fog computing [21,27]. Since there is an urgent demand for introducing effective privacy-preserving method into cryptocurrencies, we investigate the effectiveness of key approaches and analyze their defects. We also hope to provide help in proposing new privacy-preserving mechanisms in cryptocurrencies.

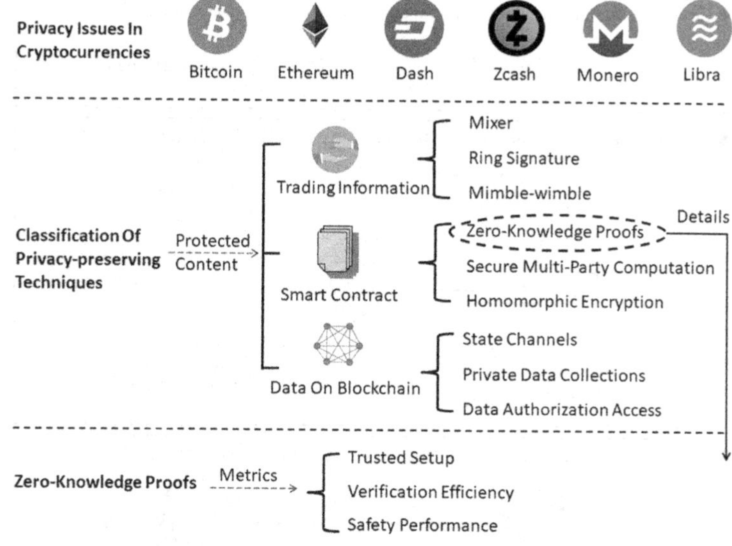

Fig. 2. Organization of this paper.

Figure 2 illustrates the organization of this paper. We take Bitcoin, Litecoin and Libra as examples to explore privacy issues in cryptocurrencies. Further research is conducted on cryptocurrencies that are committed to privacy protection, such as Monero, Zcash, and Dash. According to the protection content, we divide privacy-preserving techniques commonly used in cryptocurrencies into three categories, and analyze characteristics of different technologies in each category. The first category mainly includes Ring Signature, Range Proof, Mimblewimble and Mixer. Privacy protection is achieved by protecting information such as user identity and amount during the transaction. The second category mainly includes *Zero-Knowledge Proofs* (ZKP) [56], Trusted Execution Environment, *Secure Multi-Party Computation* (SMPC) and *Homomorphic Encryption* (HE) [1]. They are used to protect the privacy of data involved in complex computing scenarios in smart contracts. The third category mainly includes State Channels, private data collections and data authorization access. Privacy protection is achieved by protecting transaction records that have been put on the chain. In addition, we conduct a detailed study of existing ZKP schemes from three aspects: trusted setup, verification efficiency and safety performance. Finally, this survey prospects the development direction of privacy-preserving techniques.

Main contributions of this paper are twofold: (i) We analyze privacy issues in digital cryptocurrencies and investigate the key privacy-preserving techniques by category. (ii) This paper combines current research hotspots of ZKP to explore more effective encryption schemes in digital cryptocurrencies.

The remainder of this paper is organized as follows. We analyze privacy issues and summarize major privacy-preserving techniques in cryptocurrencies

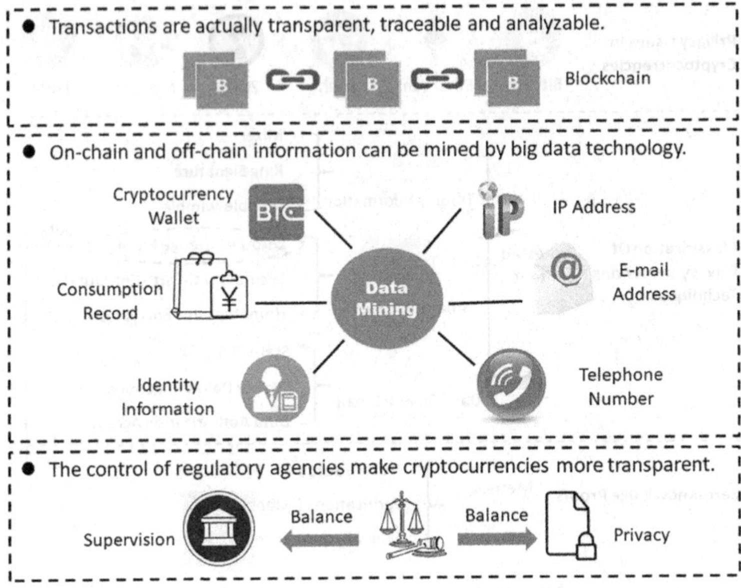

Fig. 3. Main privacy issues in digital cryptocurrencies.

in Sects. 2 and 3, respectively. Section 4 reviews ZKP in details. We discuss future privacy protection techniques and draw conclusions in Sects. 5 and 6.

2 Privacy Issues in Digital Cryptocurrencies

Figure 3 shows main privacy issues in digital cryptocurrencies currently. First, it is significant to realize true anonymous transactions, which require both pseudonymous and unconnectable conditions [42]. Cryptocurrencies such as Bitcoin, Ethereum and Libra are pseudonymous. Transactions are actually public, traceable and analyzable, so that participants can query the amount, time, sender and receiver of the transaction through blockchain browser. Once the true identity of the wallet address is revealed, all transactions at that address will be disclosed. To avoid exposing the true identity, some users create multiple wallet addresses for transactions. When conducting a larger transaction, it may be necessary to integrate many decentralized transactions. By analyzing the associated addresses and amounts of such multi-input transactions, the attacker can divide these source addresses into the same account, and then further track the account to obtain more connectable information. Other users try to use multiple accounts for transactions. Attackers can analyze different accounts belonging to the same user through address clustering technology.

Second, as blockchain transactions are gradually applied to the daily payment field, attackers may use data mining techniques to jointly analyze on-chain and off-chain information to infer account identities. When we conduct transactions, we will leave some traces on the network, such as IP addresses, shopping

Table 1. Comparison of privacy-preserving techniques and performance

Cryptocurrencies	Privacy-preserving Techniques	Transaction Address	Transaction Amount	Privacy	Anonymity Set	Traceability
Bitcoin	Pseudonym	Transparent	Transparent	Low	Small	Traceable
Ethereum	Pseudonym	Transparent	Transparent	Low	Small	Traceable
Dash	Mixer	Hidden	Transparent	Low	Medium	Partially Traceable
Monero	Ring Signature&Stealth Address&Ring CT	Hidden	Hidden	Medium-High	Large	Rarely Traceable
Zcash	zk-SNARKs	Transparent/Hidden	Hidden	High	Small	Mostly Traceable
Grin	Mimble-wimble	Hidden	Hidden	Medium	Large	Partially Traceable

records and contact information. Attackers may use various traces left by users inadvertently for big data analysis. For example, attackers can analyze account identities by comparing shopping records with Bitcoin account payment records. Chain Analysis (an American digital currency analysis company) has analyzed more than 80% of Bitcoin traders' information relying on big data mining, and it is monitoring more than 10 digital cryptocurrencies including Bitcoin, Ethereum and Litecoin. Libra officials promise not to mix financial data with social data. They also claim that most transactions are settled on blockchain, and the verification node cannot see the financial data on *Calibra* (the user's wallet) [3]. Facebook has already faced the confidence crisis due to several significant data security and privacy infringement events [13]. Facebook cannot convince the public if it just makes a promise on Libra's privacy issue.

Finally, the control of regulatory agencies has made some cryptocurrencies more transparent. In the past three years, the *OFAC* has successively blacklisted 25 Bitcoin and Litecoin addresses. At the request of U.S. law enforcement agency, *CENTRE*, the issuer of the USD Coin, has also blacklisted multiple Ethereum addresses and frozen account assets. Libra adopts an "incomplete decentralized" collective decision-making model and uses permissioned blockchain with a trusted third party. Each single transaction needs to be admitted and approved by Libra association [51]. The *WIRED* points that 15 founding members of Libra are directly or indirectly related to Facebook. Whether such a third party is eligible as a "trustworthy issuer" remains to be determined. Cryptocurrency transactions are not as private and free as people think. Third parties may track personal assets and hinder the freedom of transactions.

Privacy protection in cryptocurrencies has become a strong demand. Privacy-centric cryptocurrencies such as Zcoin, Monero, Grin and Dash are emerging. Cryptocurrencies with weak anonymity try to add privacy-preserving techniques to original projects. For instance, Litecoin has invested a lot of effort in Mimble-wimble, hoping to add confidential transaction. Although the privacy of these cryptocurrencies has been greatly improved, they are still not completely anonymous. Table 1 depicts the comparison of privacy-preserving techniques and privacy performance of different cryptocurrencies. Blockchain experts and cryptographers are making unremitting efforts to realize the real "anonymous currency".

3 Classification of Privacy-Preserving Techniques

In general, current privacy-preserving techniques fall into three categories according to the protection content, including transaction privacy protection, smart contract privacy protection and on-chain data privacy protection. The purpose of transaction privacy protection is to protect the privacy of traders' identity, the transaction amount and the knowledge behind transaction records. Main protection schemes include Mixer, Ring Signature, Range Proof and Mimblewimble. The purpose of smart contract privacy protection is to protect the privacy of data involved in complex computing scenarios. Main protection schemes include ZKP, SMPC and HE. The purpose of on-chain data privacy protection is to protect the privacy of the content on the chain. Main protection schemes include State Channels, private data collections and data authorization access.

3.1 Privacy-Preserving Techniques of Transaction

In transaction privacy protection, Bitcoin Fog [30] is an example of the centralized Mixer technology, which the private ledger is managed by a trusted third party. The third party knows the source of each fund, and there is a risk of fund theft and privacy leakage. Even though the audit mechanism can be introduced to supervise the third party [7], user privacy still cannot be guaranteed. CoinJoin [38] uses decentralized Mixer technology, which solves the problem of trusted third party. The disadvantages are complex structure, low efficiency, and weak resistance to Sybil and DoS attacks [55].

Monero is a privacy-centric cryptocurrency that has proposed privacy protection algorithms for transaction traders and amounts. The core technology of Monero is based on Ring Signatures [2]. Ring signatures become more complex as the ring grows, resulting in the need for larger storage space and increased communication overhead. There is the possibility that a malicious signer may discredit other members of the ring. Monero obscure transactions by mixing chaff coins (mixins) along with the coins actually spent. The work [41] states that Monero is not as difficult to track as it seems and evaluates two weaknesses in the mixin sampling strategy. Actual input can be deduced by elimination, and mixins can be easily differentiated from the actual coins by age distribution.

Jedusor [32] proposes Mimble-wimble protocol and a new transaction method to improve the privacy of public blockchain. Poelstra [45] makes a submission and proves the system security. Han [29] gives a presentation about the Mimble-wimble protocol at CSIRO-Data61. The protocol uses blind factors and Pedersen mode to obfuscate transaction values, so that only the sender and receiver in this transaction know the exact amount. The Mimble-wimble protocol does not hide the "transaction graph" well. "Eavesdroppers" in the network can still associate transactions by observing "transaction graph" before the transaction ends. In addition, Mimble-wimble relies on *Elliptic Curve Cryptography* (ECC), which can be theoretically destroyed by quantum computing.

3.2 Privacy-Preserving Techniques of Smart Contract

The biggest feature of Blockchain 2.0 is the introduction of smart contracts, which can develop various blockchain applications based on its architecture [20]. As for the smart contract privacy preserving, the most common approach is to combine multiple cryptographic techniques to ensure the confidentiality and security of the original data and calculation process. Hawk [34], which combines ZKP and SMPC, strives to build a privacy-protected smart contract system. Contract privacy preserving protects the consensus among participants, which includes the following four aspects.

- Input privacy. Each auction participant's quote is independent of others. Even if the auction executors collude, the participants cannot know others' quotations before entrusting their own quotations.
- Post-transaction privacy. If the executor does not disclose the information voluntarily, the quotation information will remain confidential after auction.
- Economic equity. Users, who suspend auctions to avoid payment, will have financial penalties. Other participants in the contract will be compensated.
- Punish dishonest performers. Malicious acts will be punished, such as suspending the auction and conspiring with other participants.

The ZKP technology is introduced in detail in Sect. 4. With characteristics of input privacy, calculation accuracy, and decentralization, SMPC can achieve controllable data sharing and collaborative computing under the premise of protecting privacy. Economic fairness cannot be achieved when most participants in the SMPC are dishonest. To solve this problem, the work [31] has proposed a fair SMPC protocol by constructing a penalty mechanism based on blockchain smart contract. The protocol uses a timeout mechanism to discriminate malicious premature termination and to impose financial penalties on malicious participants. SMPC may have multiple rounds of interaction in one calculation, and all participants must always stay online, which leads to inefficiency.

HE schemes can be divided into *Partial Homomorphic Encryption* (PHE) [35], *Somewhat Homomorphic Encryption* (SHE) [16], and *Fully Homomorphic Encryption* (FHE) [10,22,23]. The PHE can implement one kind of algebraic operation (addition or multiplication). The SHE can support a finite number of addition and multiplication homomorphic operations in ciphertext domain. The FHE can implement any number of addition and multiplication operations.

The FHE scheme basically follows two main research lines. One is based on the ideal lattice coset problem and the other is based on the *Learning With Errors* (LWE) and *Ring Learning With Errors* (RLWE) assumptions. These two approaches are considered to be quantum-resistant. The ideal lattice scheme has a large key size and is inefficient. The ciphertext based on the LWE and RLWE schemes is a vector, so that tensor product operation will cause the ciphertext dimension to expand rapidly. The BV scheme uses dimension-modulus reduction technology to control the dimensionality of the ciphertext. Key conversion technology is utilized in BGV to control the ciphertext dimension. The ciphertext of the GWS scheme is composed of a matrix, which can freely perform addition and multiplication operations, and also avoids the dimensional expansion.

The use of bootstrapping is the main reason that makes HE inefficient. In addition, the technology has to solve some application obstacles, one of which is the massive computing requirements of HE.

3.3 Privacy-Preserving Techniques of On-Chain Data

The principle of on-chain privacy preserving is not to disclose transaction details to others. One way is to use Block-DEF framework [52]. The evidence information is stored on the blockchain, and the evidence is stored on a trusted storage platform. Each participant can randomly generate multiple key pairs to hide their identity. In addition, the use of multi-signature technology in the submission and retrieval of evidence balances privacy and traceability. Block-DEF is constructed by one or multiple authority organizations, so that it relies on third-party institutions. Another method is to isolate the data of different business ledgers through State Channel [15]. Main processes are as follows.

1) The state of blockchain is locked through multi-signature or smart contracts, and can only be updated with the consent of all participants.

2) Participants update status by constructing and signing transactions. Temporarily reserved transactions may be submitted to blockchain.

3) Participants submit the final status to blockchain, and blockchain closes the State Channel and unlocks the status.

Anything that happens in the channel is not recorded on the chain nor broadcasted publicly. Only the final state is submitted to blockchain. The data on the chain is used to verify the validity of transaction, while actual transaction details are stored outside the chain, which is for reference when there is a dispute. Time lock and economic penalty are introduced into the State Channel to ensure that no party maliciously commits an earlier state [55].

Since each channel opening and closing requires an on-chain transaction, the strategy is not suitable for low-frequency operations and transactions that require frequent changes to participants. In addition, participants need to be always online and synchronized with blockchain in order to prevent asset loss. Pisa [39] helps participants resolve disputes by hiring third-party agents. Third-party agents only receive a hash of the state, not the state itself. When it is submitted to blockchain for dispute processing, the agent knows which status to check. The disadvantage of this method is that third-party agents may be irresponsible or bribed by others.

When there are multiple ledgers of participating organizations in the same channel, some private data needs to be shared among a small number of specific organizations, and private data scheme can solve this problem. Another way to protect on-chain data privacy is to encrypt data for authorized access. It means that sensitive data is encrypted before being uploaded to the chain, and only authorized users can view it.

Table 2. Comparison of several zero-knowledge proof schemes

Comparison	Trusted setup	Verification speed	Proof size	Interactive	Quantum-resistant
zk-SNARK	Yes	Fast	Super small (<1 kB)	No	No
zk-STARK	No	Fast	Large (>100 kB)	No	Yes
Groth16	Yes	Fast	Super small	No	No
Bulletproofs	No	Slow	Small	No	No
Supersonic	No	Fast	Small (<10 kB)	No	No

4 Zero-Knowledge Proofs

ZKP [8] balances the relationship between privacy and transparency by convincing the verifier that a statement is correct without revealing any other information. In recent years, a large amount of ZKP protocols with different characteristics have appeared. Some protocols need to re-establish the trusted setup for each application; others only need to do the trusted settings once or even completely cancel the trusted setup. The proof size of some protocols is related to the problem size. The more complex the problem is, the larger the proof size is. While other protocols have the same proof size no matter how complicated the problem is. Some protocols are interactive, requiring the prover and verifier to send many rounds of messages; others are non-interactive, and prover only needs to send a message to verifier according to the protocol. The most important goal of each new protocol is to solve problems that cannot be solved by existing protocols. A high-efficiency, constant-size, non-interactive ZKP protocol that does not require trusted setup has always been the goal of cryptographic researchers. We summarize current research hotspots of ZKP technology from perspectives of trusted setup, verification efficiency and security performance. Table 2 shows the performance comparison of several common ZKP schemes.

4.1 Trusted Setup

When designing encryption algorithms, we should minimize the cost and satisfy the timing constraint while guarantee the confidence probability [47,50]. As we all know, Zcash is the first typical application of *Zero-Knowledge Succinct Non-Interactive Argument of Knowledge* (zk-SNARK) [6]. zk-SNARK has super small proof size and efficient verification speed. The problem of this technology is that it needs an initialized trusted setup between prover and verifier. It requires a trusted third party to provide a set of public parameters, which is a very expensive and complicated process. In certain circumstance, failing to build trust successfully can introduce bugs, and it's complexed to make trusted setup work in the real world. Besides, new trusted Setups need to be re-established for each application. Ben Fisch, Findora's chief technology officer, points out that "trusted setup" is the core problem that new technologies strive to overcome, and it is clear that the future development trend is "no trusted setup".

STARK proposed by Sasson et al. [4] does not need a trusted setup, since it rely on a more streamlined symmetric encryption method by hash function collisions. The verification cost of this method is very high, and the proof size usually requires several hundred kilobytes. Similarly, Bulletproof [11,12] also does not require trusted setup and is more commonly used in low-complexity transactions. For instance, Monero reducedits cost by 97% by using Bulletproof to hide transaction amounts. In the Bulletproof protocol, you need to convert secret numbers to binary numbers first, then compute the inner product of two polynomials, and finally you can get a ZKP. For those complex transactions, the verification process can indeed take a long time. In addition, the scope of the protocol need to be a fixed form, that is $[0, ..., 2^n - 1]$ (where n is a bit number), which will undoubtedly be limited in certain applications.

Both PLONK [18] and Sonic [37] use the "generic and updateable" trusted setup, which means that a setting can be reused by other applications with the same complexity. While you deal with a more complicated dissertation, you need to reset it. DARK Proofs can help Sonic and PLONK remove the trusted setup. The Supersonic, combined with Sonic and Dark Proof, does not require any trusted setup and is efficient in high-complexity transactions. For many specific application scenarios, general tools may not be so efficient, so that we need to spend time exploring and optimizing efficient algorithms in specific scenarios.

4.2 Verification Efficiency

Parno et al. [44] have proposed the Pinocchio protocol, where customers can create a public evaluation key to describe their calculations. The worker then evaluates the computation on a particular input and uses the evaluation key to produce a proof of correctness. The proof size is 288 bytes, regardless of the computation performed or the size of the inputs and outputs. Besides, the Pinocchio protocol implements minute-level proofs and millisecond-level verifications, which brings ZKP from theory to applications. The well-known zk-SNARKS are also improvements based on Pinocchio. Ma et al. [36] propose an efficiently non-interactive Zero-Knowledge scheme at the cost of greater proof size.

Another method to improve verification efficiency is the *Zero-Knowledge Range Proof* (ZKRP) [33], which allows prover to convince verifier that a secret value is actually in an interval without revealing any secret information. For instance, it can prove that payment amount is within the limit and does not reveal the exact amount. Similarly, it can prove that someone's age is within a certain range and does not expose the specific age. This protocol can meet the actual needs while greatly reducing the time cost of transactions, which claims to be 10 times faster than ZKP. The zero-knowledge set membership proof have been proposed, so that the ZKRP can be applied not only to numbers but also other types of data, which will add more applications to corporate privacy. For example, you can prove that you are in a country and you do not need to give the exact location. The zero-knowledge set membership protocol is vulnerable and we will cover it in Sect. 4.3.

4.3 Safety Performance

ZKP schemes have certain security and privacy issues in blockchain applications, such as implementation vulnerabilities, trust risks, and unshielded transaction information leakage. For instance, the implementation of the LibSNARK project has a statute loophole in the R1CS to QAP protocol that fails to meet requirements of polynomial linear independence in QAP, which may cause soundness to be unsatisfactory [43]. The work [53] points out that Zcash is vulnerable to Ping and Reject attacks. When a node processes transactions related to itself in blockchain network, additional information leakage will occur during decryption, so that an attacker can send malicious forged transactions to determine which node an address belongs to, thereby breaking Zcash's unassociable feature.

Even a mathematically proven perfect ZKP scheme, such as Groth16 [28], will still have vulnerabilities in practical applications. There is no zero-knowledge when there is cipher text, and perfect zero-knowledge does not equal perfect privacy protection. An attacker can construct a new proof based on existing Groth16 proofs without obtaining the witness of prover.

Zero-Knowledge Conditional Payment (ZKCP) is a transaction protocol that allows buyers to trade with sellers in a private, scalable, and secure way. It does not require buyer and seller to trust each other, nor does it need to be arbitrated by a third party. Campanelli et al. [14] reveal that experiments in ZKCP are unfair to sellers. By certain means, buyers can take (part of) Sudoku answers without paying. Hence, they propose ZKCSP scheme on the basis of ZKCP, but they still have not solved the practical problem.

Although Koens et al. [33,40] have successfully implemented ZKRP, their methods also have defects that cannot be ignored. Lately, a potential security threat has been discovered by MIT's Madars Virza. Since this distance proof scheme has a fixed bit length parameter in practical applications, this allows the opponent to obtain a more accurate range of secret values. Since this distance proof scheme has a fixed bit length parameter in practical applications, which will enable the opponent to obtain a more accurate range of the secret value. The work [54] proposes an improved ZKRP for decentralized applications. They utilize the Fujisaki-Okamoto commitment, non-interactive Zero-Knowledge and computational bindingness through proof of knowledge in the cyclic group with secret order technique in the protocol. Comparing to [11], they provide an arbitrary range form for the non-interactive ZKRP scheme, which is more flexible and suitable for being applied in a real environment. The protocol is based on *Integer Factorization Problem* (IFP), which means that it cannot resist quantum attacks. Although STARK [4] can resist quantum attacks, it comes at a large cost. Proof size has increased from 288B to hundreds of bytes, and the cost is not worth it in certain circumstance.

5 Discussions and Future Work

We investigate privacy issues in digital cryptocurrencies, point out flaws in current schemes and make a summary as follows.

Table 3. Major flaws in current privacy-preserving techniques

Privacy-preserving techniques	Major flaws
Centralized mixer	Rely on trusted third party
Decentralized mixer	Low efficiency. Vulnerable to Sybil and DoS attacks
Ring signature	Become more complicated as the ring grows
Secure multi-party computation	Low efficiency. Partially applicable only to honest majority model
Homomorphic encryption	Low efficiency due to the use of bootstrap. Massive computing requirement
State channel	Not suitable for low frequency operation. Need to stay online all the time
Mimble-wimble	Not hide the "transaction graph" well. Cannot resist quantum attacks
zk-SNARK	Need trusted setup. Cannot resist quantum attacks
zk-STARK	Proof size is much larger than other protocols
Groth16	Need trusted setup. Cannot resist quantum attacks
Bulletproofs	Verification is slow. Cannot resist quantum attacks

First, certain cryptocurrencies use a pseudonym specification and transactions are actually linkable. By analyzing the correlation between transactions, the attacker may infer the trader's true identity and the wallet balance. Even though the Mimble-wimble protocol protects transaction addresses and amounts well, it still has limitations in terms of inputs and outputs linking. Mixer technology can break the linking between inputs and outputs, making it almost impossible to track the originator of a transaction. Centralized mixer technology relies on trusted third parties, while decentralized mixer technology is inefficient. Second, inputs, outputs, calculation processes, and results can be leaked in complex computing scenarios. SMPC can realize data controllable sharing and collaborative computing in complex scenarios. A prerequisite for SMPC to work properly is that the majority of participants are honest. Cryptologists are committed to creating an efficient, concise, non-interactive ZKP protocol that does not require trusted settings. Third, blockchain is a decentralized database and each user can store data as a single node. There may be situations where weak nodes are easily attacked or even malicious nodes steal data. The principle of privacy protection is trying not to disclose transaction details to other nodes, and the State Channel can achieve this. Participants in the State Channel need to be continuously online, since the representative may be attacked or bribed. Finally, most privacy preserving techniques are based on ECC and cannot resist quantum attacks. Even though zk-STARK can resist quantum attacks, the proof size is much larger than other protocols.

Privacy protection in cryptocurrencies is an ongoing battle and there is currently no perfect solution. Major flaws of the above privacy-preserving techniques are illustrated in Table 3. Future research includes continuous improvement and innovation of privacy-preserving schemes, such as improving efficiency, reducing communication overhead, and resisting quantum attacks. On the other hand, Blockchain has changed the production relationship on the basis of technology, which may lead to the reconstruction of supervision and legal rules in the future. We also intend to explore ways to integrate privacy protection and regulation.

6 Conclusions

This paper focuses on privacy-preserving issues in digital cryptocurrencies and investigates key techniques for the proposed problem. We categorize characteristics and applications of existing privacy protection strategies, and point out defects in current methods. In particular, the survey details current research hotspots and analyzes improvement directions in ZKP. Finally, we look forward to the development of privacy-preserving techniques in digital cryptocurrencies from the technical and regulatory perspective.

Acknowledgments. This work is partially supported by Natural Science Foundation of Beijing Municipality (Grant No. 4202068), National Natural Science Foundation of China (Grant No. 61972034, 61902025), Natural Science Foundation of Shandong Province (Grant No. ZR2019ZD10), Guangxi Key Laboratory of Cryptography and Information Security (No. GCIS201803), Henan Key Laboratory of Network Cryptography Technology (Grant No. LNCT2019-A08), Beijing Institute of Technology Research Fund Program for Young Scholars (Dr. Keke Gai).

References

1. Acar, A., Aksu, H., Uluagac, A., Conti, M.: A survey on homomorphic encryption schemes: theory and implementation. ACM CSUR **51**(4), 1–35 (2018)
2. Alahmari, F., Alqarni, T., Kar, J.: A research survey of ring signature scheme and implementation issues. IJSIA **10**(6), 295–304 (2016)
3. Association, T.L.: Security and privacy on the libranetwork (2019). https://libra.org/en-US/security-privacy/#overview
4. Ben-Sasson, E., et al.: Fast Reed-Solomon interactive oracle proofs of proximity. In: 45th ICALP. SDLZFI, Prague, Czech Republic (2018)
5. Biryukov, A., Tikhomirov, S.: Security and privacy of mobile wallet users in Bitcoin, Dash, Monero, and Zcash. Pervasive Mob. Comput. **59**, 101030 (2019)
6. Bitansky, N., Chiesa, A., Ishai, Y., Paneth, O., Ostrovsky, R.: Succinct non-interactive arguments via linear interactive proofs. In: Sahai, A. (ed.) TCC 2013. LNCS, vol. 7785, pp. 315–333. Springer, Heidelberg (2013). https://doi.org/10.1007/978-3-642-36594-2_18
7. Bonneau, J., Narayanan, A., Miller, A., Clark, J., Kroll, J.A., Felten, E.W.: Mixcoin: anonymity for bitcoin with accountable mixes. In: Christin, N., Safavi-Naini, R. (eds.) FC 2014. LNCS, vol. 8437, pp. 486–504. Springer, Heidelberg (2014). https://doi.org/10.1007/978-3-662-45472-5_31
8. Bootle, J., Cerulli, A., Chaidos, P., Groth, J.: Efficient zero-knowledge proof systems. In: Aldini, A., Lopez, J., Martinelli, F. (eds.) FOSAD 2015-2016. LNCS, vol. 9808, pp. 1–31. Springer, Cham (2016). https://doi.org/10.1007/978-3-319-43005-8_1

9. Borgonovo, E., Caselli, S., Cillo, A., et al.: Privacy and money: it matters. BAFFI CAREFIN Centre Research Paper, pp. 108–2019 (2019)
10. Brakerski, Z.: Fundamentals of fully homomorphic encryption. In: Providing Sound Foundations for Cryptography, pp. 543–563 (2019)
11. Bünz, B., Bootle, J., Boneh, D., Poelstra, A., Wuille, P., Maxwell, G.: Bulletproofs: efficient range proofs for confidential transactions. IACR Cryptology ePrint Archive **2017**, 1066 (2017)
12. Bünz, B., et al.: Bulletproofs: short proofs for confidential transactions and more. In: IEEE Symposium on SP, San Francisco, CA, USA, pp. 315–334. IEEE (2018)
13. Cadwalladr, C., Graham-Harrison, E.: Revealed: 50 million Facebook profiles harvested for Cambridge analytica in major data breach. Guardian **17**, 22 (2018)
14. Campanelli, M., Gennaro, R., Goldfeder, S., Nizzardo, L.: Zero-knowledge contingent payments revisited: attacks and payments for services. In: CCS, Dallas, Texas, USA, pp. 229–243 (2017)
15. Coleman, J., Horne, L., Xuanji, L.: Counterfactual: Generalized state channels (2018). https://ethresear.ch/t/counterfactual-generalized-state-channels/2223
16. Dowerah, U., Krishnaswamy, S.: A somewhat homomorphic encryption scheme based on multivariate polynomial evaluation. In: 29th International Conference Radioelektronika, CZ, pp. 1–6. IEEE (2019)
17. De Filippi, P.: The interplay between decentralization and privacy: the case of blockchain technologies. J. Peer Prod. Issue **PP**(7), 19 (2016)
18. Gabizon, A., Williamson, Z., Ciobotaru, O.: PLONK: permutations over lagrange-bases for oecumenical noninteractive arguments of knowledge. Technical report, Cryptology ePrint Archive (2019)
19. Gai, K., Choo, K., Qiu, M., Zhu, L.: Privacy-preserving content-oriented wireless communication in internet-of-things. IEEE IoT-J **5**(4), 3059–3067 (2018)
20. Gai, K., Guo, J., Zhu, L., Yu, S.: Blockchain meets cloud computing: a survey. IEEE Commun. Surv. Tutor. **PP**(99), 1 (2020)
21. Gai, K., Qin, X., Zhu, L.: An energy-aware high performance task allocation strategy in heterogeneous fog computing environments. IEEE Trans. Comput. **PP**(99), 1 (2020)
22. Gai, K., Qiu, M.: An optimal fully homomorphic encryption scheme. In: IEEE BigDataSecurity, Beijing, China (2017)
23. Gai, K., Qiu, M., Li, Y., Liu, X.: Advanced fully homomorphic encryption scheme over real numbers. In: IEEE CSCloud, New York, USA (2017)
24. Gai, K., Qiu, M., Zhao, H.: Privacy-preserving data encryption strategy for big data in mobile cloud computing. IEEE Trans. Big Data **PP**, 1 (2017)
25. Gai, K., Wu, Y., Zhu, L., Qiu, M., Shen, M.: Privacy-preserving energy trading using consortium blockchain in smart grid. IEEE Trans. Ind. Inform. **15**(6), 3548–3558 (2019)
26. Gai, K., Wu, Y., Zhu, L., Xu, L., Zhang, Y.: Permissioned blockchain and edge computing empowered privacy-preserving smart grid networks. IEEE IoT-J **6**(5), 7992–8004 (2019)
27. Gai, K., Zhu, L., Qiu, M., Xu, K., Choo, K.: Multi-access filtering for privacy-preserving fog computing. IEEE Trans. Cloud Comput. **PP**, 1 (2019)
28. Groth, J.: On the size of pairing-based non-interactive arguments. In: Fischlin, M., Coron, J.-S. (eds.) EUROCRYPT 2016. LNCS, vol. 9666, pp. 305–326. Springer, Heidelberg (2016). https://doi.org/10.1007/978-3-662-49896-5_11
29. Han, R.: The mimblewimble protocol (2019). https://www.researchgate.net/publication/334112184_The_Mimblewimble_Protocol

30. Hong, Y., Kwon, H., Lee, J., Hur, J.: A practical de-mixing algorithm for bitcoin mixing services. In: Proceedings of the 2nd ACM Workshop on BCC, Incheon, Republic of Korea, pp. 15–20 (2018)
31. Huang, J., Jiang, Y., Li, Z.: Constructing fair secure multi-party computation based on blockchain. Appl. Res. Comput. **37**(1), 225–230 (2020)
32. Jedusor, T.: Mimblewimble (2016). https://download.wpsoftware.net/bitcoin/wizardry/mimblewimble.txt
33. Koens, T., Ramaekers, C., Van Wijk, C.: Efficient zero-knowledge range proofs in ethereum. ING (2018)
34. Kosba, A., Miller, A., Shi, E., Wen, Z., Papamanthou, C.: Hawk: the blockchain model of cryptography and privacy-preserving smart contracts. In: IEEE symposium on SP, San Jose, USA, pp. 839–858. IEEE (2016)
35. Lu, Y., Zhu, M.: Privacy preserving distributed optimization using homomorphic encryption. Automatica **96**, 314–325 (2018)
36. Ma, S., et al.: An efficient NIZK scheme for privacy-preserving transactions over account-model blockchain. IACR Cryptology ePrint Archive **2017**, 1239 (2017)
37. Maller, M., et al.: Sonic: zero-knowledge snarks from linear-size universal and updatable structured reference strings. In: CCS, London, UK, pp. 2111–2128 (2019)
38. Maxwell, G.: Coinjoin: bitcoin privacy for the real world. In: Bitcoin forum (2013)
39. McCorry, P., Bakshi, S., Bentov, I., Meiklejohn, S., Miller, A.: Pisa: arbitration outsourcing for state channels. In: AFT, Zurich, Switzerland, pp. 16–30 (2019)
40. Morais, E., van Wijk, C., Koens, T.: Zero knowledge set membership (2018)
41. Möser, M., et al.: An empirical analysis of traceability in the monero blockchain. PoPETs **2018**(3), 143–163 (2018)
42. Narayanan, A., Bonneau, J., Felten, E., Miller, A., Goldfeder, S.: Bitcoin and cryptocurrency technologies. Curso elaborado pela (2015)
43. Parno, B.: A note on the unsoundness of vnTinyRAM's snark. IACR Cryptology ePrint Archive **2015**, 437 (2015)
44. Parno, B., Howell, J., Gentry, C., Raykova, M.: Pinocchio: nearly practical verifiable computation. In: IEEE Symposium on SP, Berkeley, CA, USA, pp. 238–252. IEEE (2013)
45. Poelstra, A.: Mimblewimble (2016). https://download.wpsoftware.net/bitcoin/wizardry/mimblewimble.pdf
46. Qiu, H., Qiu, M., Lu, Z., Memmi, G.: An efficient key distribution system for data fusion in V2X heterogeneous networks. Inf. Fusion **50**, 212–220 (2019)
47. Qiu, M., Sha, H.M.: Cost minimization while satisfying hard/soft timing constraints for heterogeneous embedded systems. ACM TODAES **14**(2), 1–30 (2009)
48. Qiu, M., et al.: Data allocation for hybrid memory with genetic algorithm. IEEE Trans. Emerg. Top. Comput. **3**(4), 544–555 (2015)
49. Qiu, M., et al.: Proactive user-centric secure data scheme using attribute-based semantic access controls for mobile clouds in financial industry. FGCS **80**, 421–429 (2016)
50. Shao, Z., Xue, C., Zhuge, Q., Qiu, M., Xiao, B., Sha, H.M.: Security protection and checking for embedded system integration against buffer overflow attacks via hardware/software. IEEE Trans. Comput. **55**(4), 443–453 (2006)
51. Taskinsoy, J.: Facebook's libra: big bang or big crunch? A technical perspective and challenges for cryptocurrencies. SSRN Electron. J. **PP**, 1–21 (2019)
52. Tian, Z., Li, M., Qiu, M., Sun, Y., Su, S.: Block-DEF: a secure digital evidence framework using blockchain. Inf. Sci. **491**, 151–165 (2019)
53. Tramer, F., Boneh, D., Paterson, K.: PING and REJECT: the impact of side-channels on Zcash privacy (2019)

54. Tsai, Y., Tso, R., Liu, Z., Chen, K.: An improved non-interactive zero-knowledge range proof for decentralized applications. In: IEEE DAPPCON, Newark, California, USA, pp. 129–134. IEEE (2019)
55. Wahab, J.: Privacy in blockchain systems. arXiv:1809.10642 (2018)
56. Zhao, C., Zhao, S., Zhao, M., Chen, Z., Gao, C., Li, H., Tan, Y.: Secure multi-party computation: theory, practice and applications. Inf. Sci. **476**, 357–372 (2019)

LNBFSM: A Food Safety Management System Using Blockchain and Lightning Network

Zhengkang Fang[1], Keke Gai[2,3](✉), Liehuang Zhu[2], and Lei Xu[2]

[1] School of Computer Science and Technology, Beijing Institute of Technology, Beijing 100081, China
1120161858@bit.edu.cn

[2] School of Cyberspace Security, Beijing Institute of Technology, Beijing 100081, China
{gaikeke,liehuangz,6120180029}@bit.edu.cn

[3] Henan Key Laboratory of Network Cryptography Technology, Zhengzhou, Henan, China

Abstract. Food safety management (FSM) has drawn attention recently, specifically after COVID-19 outbreak, due to lack of supervision in food products delivery chain. To address this issue, we propose a lightning network-blockchain FSM system. Our approach guarantees data authenticity by utilizing the technical characteristics of blockchain (e.g., tamper-resistant and traceability). We use lightning network to solve the throughput bottleneck issue in order to enable the proposed scheme adoptable in practice. The prototype of our approach has been evaluated and examined in our study.

Keywords: Food Safety Management · Life cycle record · Throughput optimization · Blockchain · Lightning network · Transaction verification

1 Introduction

Even though food safety is related to all individuals' daily lives, establishing an effective *Food Safety Management* (FSM) still encounters a variety of challenges due to different groups of participants (e.g., producers, logistics, dealers, consumers and management agencies) and potential complicated business processes. For instance, there may be multiple suppliers offering similar items with different qualities, such that multiple supply chain routes (interconnecting upstream and downstream service providers) will formulate a high-complex computation problem while considering both food safety controls and efficiency. Contemporary FSM systems are deployed in a centralized manner, which causes few issues, such as lack of supervision and transparency. We observe that the key of FSM lies in the trustworthy record of whole process (verification and traceability) in food supply chain [1,10].

M. Qiu (Ed.): ICA3PP 2020, LNCS 12454, pp. 19–34, 2020.
https://doi.org/10.1007/978-3-030-60248-2_2

The decentralized architecture of blockchain has an inherent advantage for information sharing and data reliability among multiple process and multiple parties [6,11,16]. As a distributed ledger technology, data on blockchain are recorded in all ledgers after a consensus is constructed. The feature of tamper-resistant can be utilized to protect food transaction information in FSM systems. When food suppliers/producers join the blockchain networks as nodes, their operations/actions will be kept in the blockchain ledger as consistent records.

To address the scenario of FSM systems, the whole process of food transaction needs to be recorded and supervised. In general, supervising operations on food transactions result in a high frequency so that the quantity of transactions is mostly large. Our investigation have shown that directly recording food transactions on blockchain is infeasible in practice. For instance, the throughput of bitcoin transaction system is only 7 transactions per second (TPS). This paper concentrates on enhancing throughput capacity of blockchain.

Many prior studies have demonstrated that lightning network is a potential solution to the enhancement of the blockchain throughput [8,9,13]. In essence, lightning network builds an off-chain transaction channel in which conducts real-time transactions. A single transaction in the channel will not be directly uploaded to blockchain for consensus process even though the transaction is completed, because it conducts multiple intra channel transactions before the transaction channel is closed. All transactions executed in the channel are integrated into a total transaction when the channel is closed and finally transactions are uploaded blockchain. Such a mechanism can effectively improve the throughput of blockchain system by reducing the frequency of transactions processed by consensus mechanism.

Prior studies [2,5] have explored blockchain-based applications in FSM; however, most of them focused on emphasizing advantages of blockchain rather than paying attention to solving low throughput issues. Latency generally had a positive relationship with the frequency of transactions [3,7,12]. Zhang *et al.* proposed a system architecture for grains supply chain, which combined multimode storage mechanism with the chain storage [15]. Even though foods supply chain could be supervised, the performance of system was not discussed. Another work developed a video surveillance system based on permissioned blockchain and edge computing [14]. Similarly, technical bottleneck caused by blockchain was not addressed, albeit edge computing somehow could contribute to data gathering and processing. Different from most prior attempts, our work used lightning network to optimize efficiency of transactions. We also proposed a verification scheme for securing transaction authenticity.

Our solution aims to support high-frequency food trading data uploads and monitors the whole life cycle of data via blockchain networks, for the purpose of foods data traceability. Briefly, our work intends to solve following problems. (i) Actions in foods supply chain shall be properly combined with blockchain. A convergence scheme seamlessly linking actions in foods supply chain with blockchain operations is needed. (ii) Lightning network shall be properly combined with blockchain. Establishing a set of connection mechanism between lightning network and blockchain is needed. (iii) Records in blocks shall be strictly

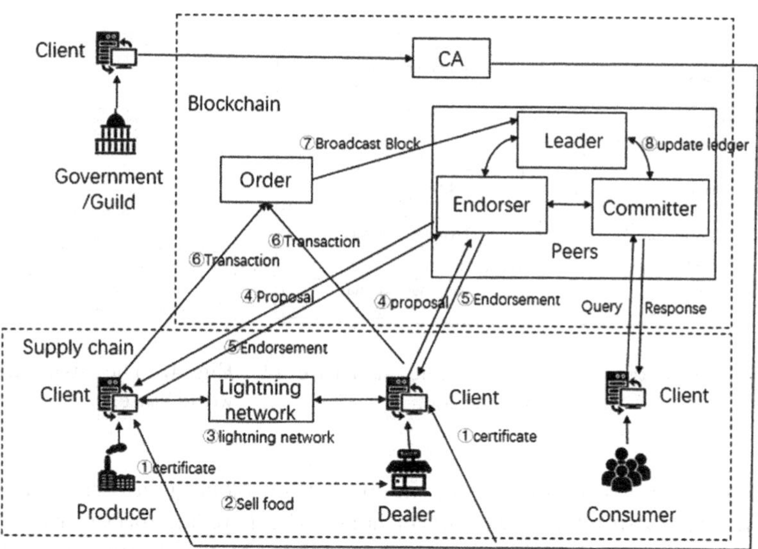

Fig. 1. The structure of LNBFSM system.

attached to actual operations in supply chain so that a transaction validation mechanism is needed.

To solve the issues above, we propose a lightning network-blockchain-based FSM system (LNBFSM). The system uses lightning network to conduct intra channel transactions and uses edge sensors to obtain food state information so that transactions are constructed for lightning network. Smart contract verifies transactions before lightning network processes. When the channel is settled, the transaction channel is closed and the a total transaction is uploaded to blockchain for consensus. After the consensus is conducted, transactions are recorded by distributed ledgers for users to query transaction information.

Our contributions mainly have cover two aspects. i) Our approach has re-constructed both block structures and transaction structures to make the blockchain system match the needs of foods supply chain. A trade between two participants in foods supply chain is considered a transaction between blockchain nodes. ii) The protocol of lightning network has been improved to improve the throughput.

Section 2 presents our specific system design. Next, Sects. 3 and 4 describe our algorithms in details and evaluation results, respectively. Finally, Sect. 5 summarizes the paper.

2 System Design

2.1 Overall Structure

We adopt a partially decentralized consortium blockchain due to the need of legal blockchain network access standards and the efficiency of the actual application,

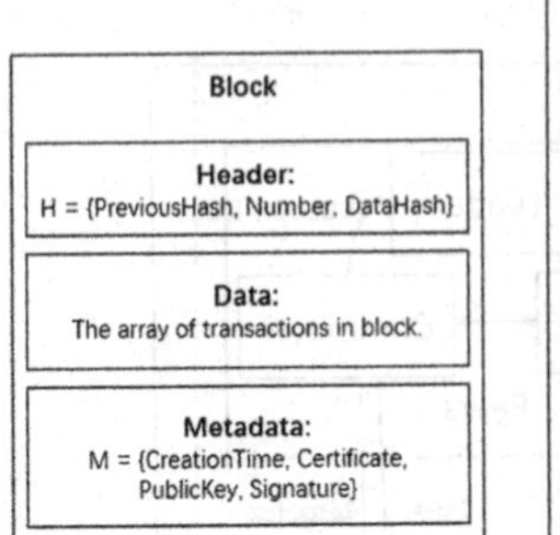

Fig. 2. The structure of block.

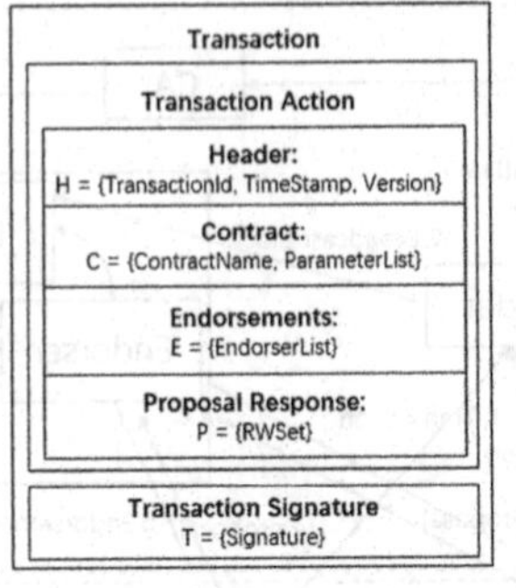

Fig. 3. The structure of transaction.

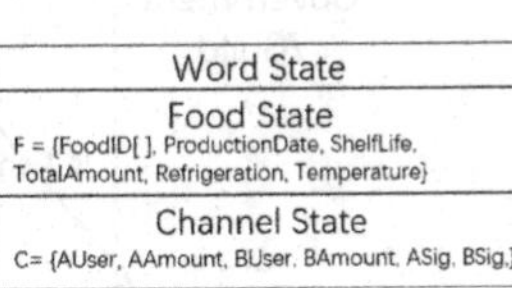

Fig. 4. The structure of world state.

As shown in Fig. 1, each participant in consortium blockchain corresponds to a client node that can interact with other nodes. Participants initiate transactions through the client node. We consider the government or industry association a trusted third party that can act as CA nodes for issuing certificates for other nodes in the blockchain. This certificate serves are evidences for transactions in blockchain network.

Lightning network implements transactions between clients and generates total transaction. When a transaction is generated in supply chain, the transaction initiator sends a transaction request to endorser nodes for signature and endorsing. Client then sends the transaction to order nodes. The order nodes pack the transaction results into a block and broadcast the block to peer nodes. After receiving the block, the leader node synchronizes it in peer nodes organization. Finally, the committer nodes update their own ledger and a consensus is made. At the endorsement stage, all nodes reach a consensus based on implementation results; all nodes reach a consensus based on the global transactions order at the sorting stage.

2.2 Data Structure

Block Structure: We refer to the block structure of Hyperledger fabric, which includes block header, block data and block metadata. The specific block structure is shown in Fig. 2. Block number is a natural number identifying the sorting position of blocks in whole blockchain. The block metadata includes the information when the block is created. It includes `CreationTime`, `Certificate`, `PublicKey`, and `Signature`. `CreationTime` records the block creation time. `Certificate` records the certification obtained by the block creator. `PublicKey` and `Signature` are the public key and digital signature of block creator. They are generated by asymmetric encryption algorithm and digital signature algorithm, which verify the block data source and the integrity of block data easily. This guarantees the authenticity of block data. The meaning of each filed is shown in Table 1.

Table 1. The main fields in block structure.

Field name	Meaning
Block header	
PreviousHash	The hash value of previous block
Number	The number of current block
DataHash	The hash value of current block
Block metadata	
CreationTime	The time of creating current block
Certificate	The certificate of current block creator
PublicKey	The public key of current block creator
Signature	The signature of current block creator

Table 2. The main fields in transaction record.

Field name	Meaning
Transaction action	
TransactionID	The ID of each transaction
TimeStamp	The time of transaction submission
Version	The version of the world state
ContractName	The smart contracts used in current transaction
ParameterList	The parameters input into the smart contract
EndorserList	The array of endorser's ID
RwSet	The transaction's read-write set of world state
Transaction signature	
Signature	The signature of transaction submitter

Table 3. The main fields in world state.

Field name	Meaning
Food state	
FoodID	The IDs of foods
ProductionDate	The production date of food
ShelfLife	Maximum save time of food
TotalAmount	The amount of food owned
Refrigeration	The flag marking whether refrigeration is required
Temperature	The current temperature of food
Channel state	
AUser	The user ID of channel
BUser	The user ID of channel
AAmount	The amount of A client pre-stored in channel
BAmount	The amount of B client pre-stored in channel
ASig	The signature of A client
BSig	The signature of B client

Transaction Structure: The array of transaction records is the specific storage content of block data in block. The transaction structure stores the detailed information of each transaction, which consists of transaction action and transaction signature. The transaction signature is the signature of transaction submitter. The transaction structure is shown in Fig. 3.

Data items in transaction structure record basic information and specific actions of the transaction. The transaction information has ID of the transaction, `timestamp` of transaction (submitted to the order nodes), the submitted food state version of transaction and signature of submitter. `ContractName` and `ParameterList` are set because the processing of a transaction is controlled by smart contract. When calling contract, the parameter list of the transaction will be inputted. In the process of consensus endorsement, multiple endorsement nodes are needed to simulate the execution of the submitted transaction proposal. The signature of simulated execution result is fed back to transaction submitter. Therefore, in the transaction, the `EndorserList` records the endorser nodes participating in the endorsement, and `RWSet` records the read-write set of the simulation execution result to the food state. The meaning of each field is shown in Table 2.

World State: World state records the information of food state in food supply chain and channel state updated by building or deleting transaction channel. The world state is updated by the read-write set generated after the transaction executed by smart contract. The version number increases automatically once it is updated. World state directly queries the state of food and channel recorded in current blockchain, such that the cost of FSM can be lowered down. The specific data structure is shown in Fig. 4.

Food state records mainly include food basic information and unique identification code of food marked by RFID chips. The fields related to food quality include `ProductionDate` (The Date of producing this food), `ShelfLife` (Time limit time to ensure food quality), `TotalAmount` (The owned food amount), `Refrigeration` (The sign of refrigeration or not), `Temperature` (Current temperature of stored food) and `Version` (current state version number). Channel state records the users of channel, the food amount pre-stored in channel and the signature of users. the fields are recorded in ledger in order to use lightning network conveniently. The meanings of main fields are shown in Table 3.

2.3 Smart Contract

Lightning network workflow of our FSM system is shown in Fig. 5. Our smart contract mainly includes `QueryValidation` contract, `TransactionChannelUp` contract, and `TransactionChannelDown` contract.

The `Query-Validation` contract is used to verify the trade in supply chain, which determines whether trade meets the security standards.

`TransactionChannelUp` contract is used to establish an off-chain transaction channel between two clients. When establishing a channel, the trade initiator launches a channel establishment request. To ensure both parties agree to create

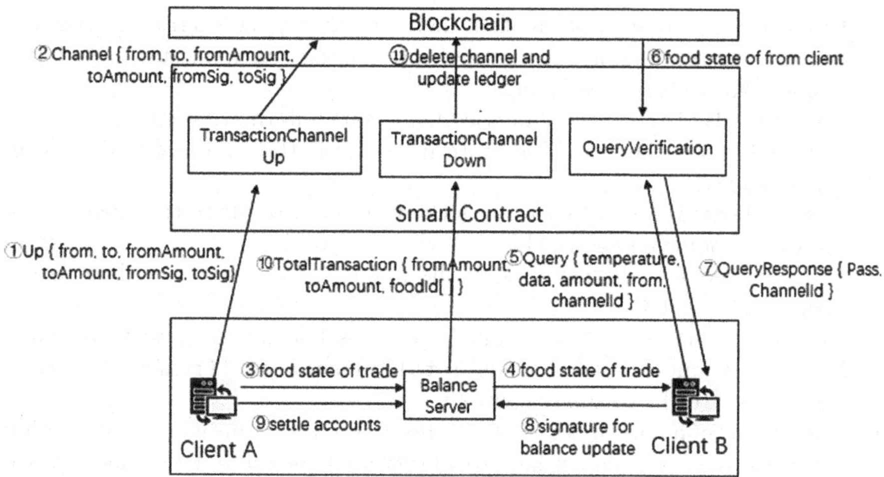

Fig. 5. The workflow of lightning network.

a trading channel, it is necessary to contain signature confirmation of both sides in channel establishment request. Two clients, which need to trade, submit the pre-stored food to the channel and upload channel information to the blockchain.

`TransactionChannelDown` contract is used to close transaction channel that settles the trading result and upload result to the blockchain. When closing channel, `TransactionChannelDown` contract must check signatures of both parties to balance contract first. After confirming that there is no objection, the contract modifies the pre-stored food quantity of both sides of trading, based on the balance state in the channel. Finally, the contract uploads the total trading result to blockchain.

The `Balance` contract works off-chain that is used for real-time transaction within the channel. By modifying the balance state attribute, `Balance` contract save the balance distribution state in channel temporarily. Two parties agree on the modified balance state by signing, which means that the transaction state is recognized. If one or both parties fail to sign, it is deemed that there is objection to transaction. The workflow is as follows:

1. Client sends channel opening message `Up (from, to, fromAmount, toAmount, fromSig, toSig)` to establish the channel. `from` and `to` are two clients establishing the channel. `fromAmount` and `toAmount` represent the quantity of food of two clients that can be traded in the channel. `fromSig` and `toSig` are the signatures of transaction parties to the channel opening message.
2. `TransactionChannel Up` contract handle the establishing channel request and uploads the channel information to blockchain in the structure of `Channel (from, to, fromAmount, toAmount, fromSig, toSig)`.
3. `from` client sends the state information of traded food to `Balance Server`.
4. `Balance` forwards the state information of food to `to` client.

5. to client sends query information including Query (temperature, data, amount, from, channelId) of multiple transaction food state attributes to QueryValidation contract.
6. QueryValidation contract queries food state information saved in ledger and verifies whether the food state in ledger is the same as food state in query request.
7. QueryValidation contract returns the validation result to the client in the form of a bool type Pass field.
8. to client decides whether to sign the update of balance according to the verification results.
9. Any client participating in channel sends a settlement request to Balance Server. Balance Server sends TotalTransaction (fromAmount, toAm ount, foodId[]) to smart contract.
10. TransactionChannelDown contract processes the transaction information and sends request to blockchain to modify the ledger and delete the channel.

2.4 Consensus Mechanism

The consensus mechanism ensures the consistency of each ledger's copies of FSM system. It is an important guarantee of data tamper proof. The consensus mechanism designed in our paper consists of three stages: transaction endorsement, sorting service and verification.

Transaction endorsement 1) In transaction endorsement stage, the client first submits the proposal (operation, source, policies, metadata) for simulated execution. The operation in proposal refers to chaincode operation function and parameter list. The source refers to chaincode source address. The metadata refers to relevant attributes of call and chaincode application. And policies refers to relevant policies for peer nodes to access chaincode. 2) After receiving the proposal from client, the endorser node verifies the signature of client. After verification, based on the endorsement strategy, endorser node uses chaincode and input parameters specified the transaction to simulate the execution of the transaction. Then endorser nodes send the endorsement signature of simulated results to client endorsement(sig, chaincode_info, rwset). sig is the signature of the endorsement node. chaincode_info is the chaincode information, and rwset represents the reading-writing set of the food state after the execution of the transaction simulation. 3) When the trading client receives enough endorsement() signature messages, it can judge whether the trading proposal has completed the endorsement. After this process, all nodes have reached consensus on transaction information.

Sorting Service. After binding received transaction endorsement and transaction information on client, the client sends transaction (header, contract, endorsements, proposed_response, tx_sig) to order nodes. header stores basic information of the transaction; contract records the contract name and parameter list of transaction call; endorsements records a list of endorsements node ID endorsing the transaction; proposal_response records the reading-writing set of the food state database of the ledger returned after endorsement;

`tx_sig` is to submit client's signature to the transaction; `transaction ()` message is broadcasted to orders network. There maybe multiple clients submitting multiple transactions. The order network needs to sort all submitted transactions globally for constructing a consensus.

PBFT algorithm is used in our system for sorting service that is a consensus algorithm supporting Byzantine fault tolerance [4]. We consider the safety requirement to be the priority in our scheme. When the number of Byzantine nodes is `f`, a consensus can be made only when the total number of nodes in the consensus is greater than `(3f + 1)`. There are one primary node and multiple backup nodes in the orders network. We assume that there is no possibility of forgery in primary node. Our consensus process consists of three stages, namely, pre-prepare, prepare, and commit stages, which are described in the following.

1. **Pre-prepare** After receiving the request, primary node sends `pe_prepare (n, d, m)` to all backup nodes. The `m` refers to request message; `d` refers to the hash value of the message; `n` refers to the sequence number assigned to `m`. After receiving the `pre_prepare ()` message, each backup node verifies it, mainly to verify whether there are other messages numbered `n`. When the validation is completed, the node enters the prepare phase.
2. **Prepare**: Nodes receive `prepare ()` and verify message `m` and `pre_prepare ()` message of `m` exist in the log. If it exists, nodes broadcast validation results to other nodes. When `2f` validation messages with the same validation results are received, the node's prepared state is recorded as `prepared (m, n, i)`.
3. **Commit**: When the nodes receive `2f commit (commit, n, d, i)`, `i` is the node number that reaches the prepared state and sends the commit message. After receiving `2f` messages (add itself is `2f+1`), the nodes achieve the `commit_local (m, n, i)` state.

Verification. After transactions are sorted by order nodes, they are packed into a block that is broadcasted to peer nodes. When leader node receives the sorted transactions, they are distributed to the committer nodes for verifications. Validate transactions within the block to ensure endorsement policies are executed. The ledger state of the read-write set variable has not been changed since the read-write set is generated by transaction execution. After the verification, it is marked as either a valid or an invalid transaction for ledger updates.

3 Algorithm

3.1 QueryVerification Algorithm

The `QueryVerification` in smart contract verifies whether the submitted transaction information conforms the real food state in ledgers. We obtain the food state of the trade initiator from the ledger. The `ProductionData` and `ShelfLife` in food state information determine whether the transaction date is within the food shelf life. The `Refrigeration` and `MaxTemperature` determines whether the food in the trade meets the refrigeration standard. The verification result

Algorithm 1. QueryVerification Algorithm

```
Require: Query
Ensure: Pass, channelID
fromKey ← stub.CreateCompositeKey("FoodState", Query.from)
fromAsBytes ← stub.GetState(fromKey)
Define fromFoodState as FoodState data struct
fromFoodState ← fromAsBytes
if (fromFoodState.ProductionData + fromFoodState.ShelfLife) >= Query.Data then
    if fromFoodState.TotalAmount >= Query.Amount then
        if fromFoodState.Refigeration = true then
            if fromFoodState.MaxTemperature >= Query.Temperature then
                Pass ← true
            else
                Pass ← false
            end if
        else
            Pass ← true
        end if
    else
        Pass ← false
    end if
else
    Pass ← false
end if
if Pass = true then
    channelAsBytes ← stub.GetState(Query.from)
    get channelID from channelAsBytes
else
    ChannelID ← NULL
end if
return Pass, channelID
```

is returned by `Pass` field. Query channel state is in the ledger. The channel ID will be returned through `channelId` field when there is a transaction channel between two parties; a NULL will be returned when there is no channel. Pseudo codes of QueryVerification algorithm are given in Algorithm (1) and a step-by-step description about the algorithm is presented below.

(1) Generate compound key `fromKey` for `FoodState`. Obtain the from food state value saved in world state of ledger according to `fromkey`.
(2) Assign the value of `from` food state to `fromFoodState`. Judge whether it is within the shelf life. If it is not within the shelf life, set `Pass` to false.
(3) Judge whether the quantity of food is sufficient within the shelf life. Set a `Pass` to false when insufficiency is detected.
(4) Continue to judge whether refrigeration is needed (Assign `Pass` to true if no refrigeration is needed). If refrigeration is needed, continue to judge whether

current food storage temperature meets the requirement. Assign a `Pass` to false if failing in meeting requirement; otherwise, assign a `Pass` to true.
(5) If `Pass` is true, obtain the transaction channel ID from the world state of the ledger. If `Pass` is a false, set the channel ID to NULL.

Time Complexity Analysis: The algorithm performs the same judgment process for each trade. And the judgment process is a selection structure. The number of trades processed by the algorithm is set to N, the average number of execution judgments for each trade is j, and the number of operations for each trade to read the ledger information and process is o, so the time complexity of Algorithm (1) is $T(n) = O((j + j)N)$.

Another algorithm running in smart contracts is `TransactionChannelUp`. This algorithm is executed when smart contract receives the channel open message from client. The main process is to build a data structure storing channel attributes and submit the structure to blockchain ledger in the form of transaction. A few major steps are described as follows. (i) Define `channelUp` as channel data structure. (ii) `channelUp` is assigned values related to the build channel with the input `Up` message. (iii) Convert `channelUp` to bytes `channelUpAsBytes`. (iv) Update `channelUpAsBytes` to channel state in ledger world state. This process only needs to perform fixed operations. If the number of algorithm steps is s, the time complexity can be expressed as $T(N) = O(s)$.

3.2 Balance Algorithm

`Balance` algorithm is to process high-frequency trades between clients and update the state attributes of food distribution in the channel. Operation mechanism of the algorithm is that the food distribution state in the channel will be updated when the trade message is submitted from client and the signature of trade confirmation is confirmed. The contract builds the `Total` structure when the signature of balance state is received at the settlement stage; settlement transaction, then, is sent to smart contract. We show pseudo codes of the algorithm in Algorithm (2) and provide a phase-by-phase description below.

(1) Assign the pre-stored food quantity value of both parties in the transaction channel to `fromAmount` and `toAmount`.
(2) The loop waits for receiving the trade if no settlement message is received. Once the trade is received, wait for the trade receiver to send `SigForUpdate`
(3) Update `fromAmount` and `toAmount` by pre-stored food quantity in the channel after `SigForUpdate` is received. Add `trade.FoodId` to `foodId [i]`.
(4) `Total` message is constructed and sent to the smart contract when a settlement message is received,

Time Complexity Analysis: Algorithm (2) is composed of two parts, namely, processing `Trade` and sending settlement transaction constructed. Set the number of trades to be processed in this channel as N, and the process of trade processing as P operations. Construct and send settlement information as s operations. Therefore, the time complexity is as follows: $T(N) = O(pN + s)$.

Algorithm 2. Balance Algorithm

```
Require: Trade, SigForUpdate, SigForBalance, SettleAccount
fromAmount←Trade.FromInitialAmount
toAmount←Trade.ToInitialAmount
while Not received SettleAccount do
    i←0
    if Receive Trade then
        Relay the Trade message to receive client
        Wait for SigForUpdate
        if SigForUpdate != NULL then
            fromAmount -= Trade.Amount
            toAmount += Trade.Amount
            foodId[i]←Trade.FoodId
            i++
        end if
    end if
end while
if Received SigForBalance then
    Define Total as TotalTransaction data struct
    Total.From←Trade.From
    Total.To←Trade.To
    Total.FromAmount←fromAmount
    Total.ToAmount←toAmount
    Total.FoodId[]←foodId[]
    Send total struct to smart contract
end if
```

3.3 TransactionChannelDown Algorithm

`TransactionChannelDown` contract is to process settlement transactions submitted at application layer. The algorithm firstly retrieves the food state structure from the ledger, then, modifies the food state information through parameters contained in settlement transaction message. Food state structure on the ledger and the channel state information will be modified and deleted, respectively, by running the algorithm. Phase-by-phase of the description about this algorithm is given as follows. Pseudo codes of the `TransactionChannelDown` algorithm are shown in Algorithm (3).

(1) Read the food state `fromFoodState` and `toFoodState` and to respectively from the ledger.
(2) Update `fromFoodFtate` and `toFoodState` based on the received *Total*.
(3) Accordingly update `fromFoodState` and `toFoodState` to the ledger, based on retrievals from prior steps.
(4) Delete the channel state value whose key is `Total.ChannelId` from the ledger (Table 4).

Algorithm 3. TransactionChannelDown Algorithm

```
Require: Total
Define fromFoodState, toFoodState as FoodState data struct
fromAsBytes ← stub.GetState(Total.from)
fromFoodState ← fromAsBytes
toAsBytes ← stub.GetState(Total.to)
toFoodState ← toAsBytes
fromFoodState.TotalAmount←Total.FromAmount
toFoodState.TotalAmount←Total.ToAmount
for ∀ FoodId[i] in Total.FoodId[] do
    Add FoodId[i] to toFoodState.FoodId[]
    Remove FoodId[i] from fromFoodState.FoodId[]
end for
Transform fromFoodState into bytes fromFoodStateAsBytes
stub.PutState(Total.From, fromFoodStateAsBytes)
Transform toFoodState into bytes toFoodStateAsBytes
stub.PutState(Total.To, toFoodStateAsBytes)
stub.PutState(Total.ChannelId, NULL)
```

Table 4. Channel creation message

ChannelID	AUser	BUser	AAmount		BAmount		ASig	BSig
0001	A	B	0101	100	0110	150	RSA(A)	RSA(B)

Table 5. Input trades message

Trade	Data	Amount
Trade1	April 15th	100
Trade2	April 15th	−120
Trade	Temperature	FoodId
trade1	−1	0101
trade2	10	0110

Table 6. Food state in ledger

	FoodId	TotalAmount	Refigeration
A	0101	150	1
B	0110	200	0
	MaxTemperature	ProductionData	ShelfLife
A	0	March 10th	60
B	30	March 20th	180

Table 7. Feedback verification results

Trade	Pass	ChannelId
Trade1	True	0001
Trade2	False	0001

Table 8. Modified channel balance allocation

From	To	FromAmount			
A	B	0101	0	0110	0
ToAmount				FoodId[0]	
0101	100	0110	150	0101	

Time Complexity Analysis: Assume that it is required to read and modify the ledger information except specific food ID information in s operations. It is set to complete $\bar{n}$ trades in each channel on average. Therefore, the time complexity is expressed as $T(N) = O(s + \bar{n})$.

4 Evaluations and Findings

The evaluation was based on an assumption of a set of trades submitted between clients. Trades included a trade (fully conforms to the real food state recorded in ledger) and a trade (forges the information of food state). A brief evaluation process was described as follows. We performed `Balance` contract to update channel state after trades passed validation. The settlement request was submitted after the state of all trades in channel was updated. The trades, then, was integrated in the channel into a single transaction to update the food state of the blockchain ledger. Our evaluation verified the adoption of the proposed scheme, which focused on examining the integration of blockchain and lightning network. Experiment was implemented on the Hyperledger Fabric platform; simulation was implemented to simulate outside environment (supply chain side). The system's inputs and processed outputs are presented in following tables.

Specifically, we create a transaction channel at `A` and `B`. `A` and `B` sign the message of creating a transaction channel. Channel information includes channel ID, channel creations (both sides), and the quantity of temporarily stored food in the channel. Table 6 shows the channel creation message sent to the contract. Meanwhile, the message also is recorded in channel state of ledger. Clients A and B submits `trade1` and `trade2`, respectively. The default amount is a positive number (means a trade from `A` to `B`). When the value becomes a negative number, it refers to a trade from `B` to `A`. Table 5 provides input trades message.

Table 9. The updated ledger

Food state						
	FoodId	TotalAmount	Refigeration	MaxTemperature	ProductionData	ShelfLife
A	0101	50	1	0	March 10th	60
B	0110	200	0	30	March 20th	180
B	0101	100	1	0	March 10th	60
Channel state						
NULL						

Client A forwards `trade1` to smart contract; client B forwards `trade2` to the smart contract for request validation. As shown in Table 6, `QueryVerification` contract obtains the food state information of two trade initiators from ledger. Table 7 shows feedback verification results. Client B afterward sends the confirmation signature of `trade1` to balance contract; client A will not send confirmation signature of `trade2` when receiving the verification result of a failed `Pass`. `Balance` updates the balance allocation state after receiving the confirmation signature. Table 8 displayed a modified channel balance allocation.

The updated ledger information after client A initiates the settlement request is shown in Table 9. From the running results of the system, it shows that proposed FSM system successfully uploads the trade information between clients to blockchain ledger through lightning network.

5 Conclusions

This paper put forward a method of dealing with food supply chain trades. Through the application of blockchain and lightning network technology, the decentralized FSM is realized. The transaction information can not be tampered, and the problem of insufficient block chain throughput in the high-frequency trading scenario is solved. Our proposed FSM system focuses on the seamless combination of blockchain technology, lightning network technology and FSM application scenarios. In order to achieve this goal, we have designed the data structure of blockchain, designed four algorithms to verify the trade before transaction, and then uploaded the blockchain through the lightning network after verification. Finally, we verify the feasibility of the system.

Acknowledgement. This work is partially supported by Natural Science Foundation of Beijing Municipality (Grant No. 4202068), National Natural Science Foundation of China (Grant No. 61871037, 61972034), Natural Science Foundation of Shandong Province (Grant No. ZR2019ZD10), Guangxi Key Laboratory of Cryptography and Information Security (No. GCIS201803), Henan Key Laboratory of Network Cryptography Technology (Grant No. LNCT2019-A08), Beijing Institute of Technology Research Fund Program for Young Scholars (Lei Xu, Keke Gai).

References

1. Aiello, D., De Luca, D., et al.: Review: multistage mass spectrometry in quality, safety and origin of foods. Eur. J. Mass Spectrom. **17**(1), 1 (2011)
2. Basnayake, B.M.A.L., Rajapakse, C.: A blockchain-based decentralized system to ensure the transparency of organic food supply chain. In: International Research Conference on SCSE, pp. 103–107 (2019)
3. Bonneau, J., Miller, A., Clark, J., Narayanan, A., Kroll, J.A., Felten, E.W.: Sok: research perspectives and challenges for bitcoin and cryptocurrencies. In: 2015 IEEE Symposium on Security and Privacy, pp. 104–121 (2015)
4. Castro, M., Liskov, B., et al.: Practical byzantine fault tolerance. In: OSDI, pp. 173–186 (1999)
5. Chandra, G.R., Liaqat, I.A., Sharma, B.: Blockchain redefining: the halal food sector. In: AICAI, pp. 349–354 (2019)
6. Gai, K., Guo, J., Zhu, L., Yu, S.: Blockchain meets cloud computing: a survey. Eur. J. Mass Spectrom. **PP**(99), 1 (2020)
7. Gai, K., Wu, Y., Zhu, L., Qiu, M., Shen, M.: Privacy-preserving energy trading using consortium blockchain in smart grid. IEEE Trans. Ind. Inform. **15**(6), 3548–3558 (2019)
8. Gai, K., Wu, Y., Zhu, L., Xu, L., Zhang, Y.: Permissioned blockchain and edge computing empowered privacy-preserving smart grid networks. IEEE Internet Things J. **6**(5), 7992–8004 (2019)
9. Guo, J., Gai, K., Zhu, L., Zhang, Z.: An approach of secure two-way-pegged multi-sidechain. In: Wen, S., Zomaya, A., Yang, L.T. (eds.) ICA3PP 2019. LNCS, vol. 11945, pp. 551–564. Springer, Cham (2020). https://doi.org/10.1007/978-3-030-38961-1_47

10. Ketney, O.: Food safety legislation regarding of aflatoxins contamination. Acta Universitatis Cibiniensis **67**(1), 71–76 (2015)
11. Nakamoto, S.: Bitcoin: a peer-to-peer electronic cash system. Technical report, Manubot (2019)
12. Pappalardo, G., Matteo, T.D., Caldarelli, G., Aste, T.: Blockchain inefficiency in the bitcoin peers network. EPJ Data Sci. **7**(30), (2017)
13. Poon, J., Dryja, T.: The bitcoin lightning network: scalable off-chain instant payments (2016)
14. Wang, R., Tsai, W., et al.: A video surveillance system based on permissioned blockchains and edge computing. In: IEEE International Conference on BigComp, pp. 1–6 (2019)
15. Zhang, X., Sun, P., Xu, J., Wang, X., et al.: Blockchain-based safety management system for the grain supply chain. IEEE Access **8**, 36398–36410 (2020)
16. Zhu, L., Wu, Y., Gai, K., Choo, K.: Controllable and trustworthy blockchain-based cloud data management. Future Gener. Comput. Syst. **91**, 527–535 (2019)

Reputation-Based Trustworthy Supply Chain Management Using Smart Contract

Haochen Li[1], Keke Gai[2,3](✉), Liehuang Zhu[2], Peng Jiang[2], and Meikang Qiu[4]

[1] School of Computer Science and Technology, Beijing Institute of Technology, Beijing 100081, China
1120161867@bit.edu.cn

[2] School of Cyberspace Security, Beijing Institute of Technology, Beijing 100081, China
{gaikeke,liehuangz,pengjiang}@bit.edu.cn

[3] Henan Key Laboratory of Network Cryptography Technology, Zhengzhou, Henan, China

[4] Department of Computer Science, Texas A&M University-Commerce, Commerce, TX 75428, USA
Meikang.Qiu@tamuc.edu

Abstract. *Supply Chain Management* (SCM) has been deemed a promising method to obtain higher-level profits, while traditional SCM methods are limited by its deficiency of trust system. Blockchain is an emerging technology to achieve decentralized trusted transactions with high level automation and security, which is adoptable to address the limitation of the trust deficiency. In this paper, a blockchain-enabled reputation-based trustworthy SCM model is proposed in order to achieve trusted SCM. Besides, we propose a smart contract-based reputation rating model for achieving incentive SCM, which uses tokens to assess reputation and realize reputation reward and punishment.

Keywords: Blockchain · Smart contract · Supply Chain Management · Reputation management · Trustworthiness

1 Introduction

Modern massive business cooperations have widely adopted *Supply Chain* (SC) for achieving higher-level business efficiency. In order to obtain high profits and reduce costs, companies/stakeholders consider implementing *Supply Chain Management* (SCM), in which includes all processes to optimize operations with low cost by coordinating the business process. We observe that various information technologies have been utilized in SCM, e.g. cloud computing [3] and *Internet of Things* (IoT) [8,15,18,19,21], etc. Such technologies have improved SCM in many aspects, such as decision making and automatic management [7].

M. Qiu (Ed.): ICA3PP 2020, LNCS 12454, pp. 35–49, 2020.
https://doi.org/10.1007/978-3-030-60248-2_3

Despite the improvements brought by traditional information technologies, current SCM systems are still encountering the issue of trust deficiency as data are mostly stored and processed in a central server [1,17,22]. This centralized architecture in current SCM system result in the risk of data modification, and possible data modification will challenge data trustworthiness. Moreover, most current SCM systems use the users' ratings to assess stakeholders' reputations, such that the falsification threatens the authenticity of reputation. Hence, a distributed trustworthy environment fitting in SCM is needed.

Blockchain is a promising technology to solve the trust deficiency issue, because it is a distributed ledger-based database with the tamper-resistant feature and consensus-based data storage [5,6,9,11]. Auto-running smart contracts facilitate a flexible auto controls based on pre-configured rules. In addition, a token mechanism offers a *reward/punishment* (R&P) approach that can be used for constructing reputation management systems. In fact, many recent studies have explored the implementation of blockchain in SCM. One of research directions focused on auto controls of SCM by using blockchain. For example, Chen *et al.* [2] presented a multi-layer model of blockchain system to realize the automation of commodity quality SCM. Smart contract in this model was used to realize automatic quality control and logistics planning. However, this work only focused on the framework design of blockchain-based SCM without offering specific function design of smart contract. Similarly, Qu *et al.* [16] presented a smart contract-enabled *Business-to-Consumer* (B2C) SCM to create and process customer orders, which were demonstrated by experimental simulations.

Our investigations also found that some other researches focused on blockchain-based trust management [23]. Feng *et al.* [4] proposed an RFID-blockchain-based SC traceability system for agri-food safety management. Authentic data were gathered by RFID technology for tracing information in a blockchain-based SCM. Malik *et al.* [14] proposed a trust management model in IoT-blockchain-based SCM, in which reputations were assessed to represent the users' credibilities. However, the reputation score setting was not combined with blockchain.

Differing from most prior researches, in this paper, we propose a blockchain-based SCM system that assess and update the reputation information of stakeholders based on authentic quality status of transactions. Representative samples of smart contract designs will be provided in this work with experimental results. For instance, our work pays attention to the application scenario of trust management in SCM and smart contract mechanisms are different from prior explorations (in order management). Meanwhile, rather than remaining on traceability, our approach assesses each stakeholder's reputation (overall credibility) via a proposed blockchain network, so that evaluations on the quality status of transactions can be given by running smart contracts.

Two challenges are addressed by this work. The first challenge is to find out a method that can determine the quality status of the transaction. It requires the authentic commodity information of the transaction in SC. Blockchain clients shall be able to access/process commodity data in SC and detect transaction quality issues. The difficulty is to construct the reputation system of stakeholders by establishing rules rewards or punishments based on the evaluation of the transaction quality. The other challenge is to investigate the assessment approach for examining stakeholders' reputations on a blockchain system. The assessment needs to be operated through deploying smart contracts with a pool of rules (rewards/punishments) for the purpose of the reputation updates. Presenting objects' reputations by offering accurate evaluation results is challenging.

To address challenges above, we propose a blockchain-based SCM improvement model, in which tokens are used to evaluate the reputation of stakeholders. Each stakeholder corresponds to a blockchain user account and the reputation is assessed by the amount of tokens in the account. Transaction records with authentic commodity information gathered by sensors are used to generate quality status of transactions. The quality status of a transaction is the input of smart contract, and nodes execute smart contract for rewarding/deducting tokens of the seller corresponding to quality status. The proposed blockchain-based SCM blockchain builds up a trustworthy reputation system among all SC participants.

Main contributions of this paper address two aspects. (i) This work proposes an automatic approach for estimating commodity quality status within an SC. Majority quality issues (e.g., falsification) can be avoided by being alarmed by implementing a blockchain-based quality status assessment system. (ii) We propose a blockchain-based reputation evaluation mechanism for SCM, which assesses stakeholders' reputations by using blockchain tokens. A smart contract approach is proposed for implementing token-based incentive management of reputation.

The remainder of the paper are organized as follows. Section 2 and 3 describe model designs and core algorithms, respectively. Evaluation findings and conclusions are given in Sects. 4 and 5.

2 Proposed Model

Our model is called *Reputation-based Trustworthy Blockchain Supply Chain Management* (RTB-SCM), which is based on the consortium blockchain as an SC consists of a limited number of permissioned stakeholders. Only authorized nodes maintain distributed shared ledgers [12]. Like other consortium blockchain systems, different nodes are kept by various stakeholders and each transaction needs a global consistency among all nodes. Specifically, consortium blockchain ensures decentralization among all permissioned stakeholders in SC with relative high transaction throughput, which has a higher decentralization-degree than private blockchain and a higher transaction throughput than public blockchain.

2.1 Key Components and Model Design

The RTB-SCM model mainly consists of five key components, namely, sensors, SCM database, clients, *Certificate Authority* (CA), and *Reputation Assessment*

Blockchain (RAB). Specifically, sensors gather the authentic commodity information from trades and SCM database stores trade records and the commodity information. "Clients" receive requests from users and query trade records from SCM database to generate corresponding quality status. CA encrypts data during the query process of SCM database. RAB stores/updates the reputation of stakeholders based on quality status generated by clients. Figure 1 shows the high level architecture of RTB-SCM model.

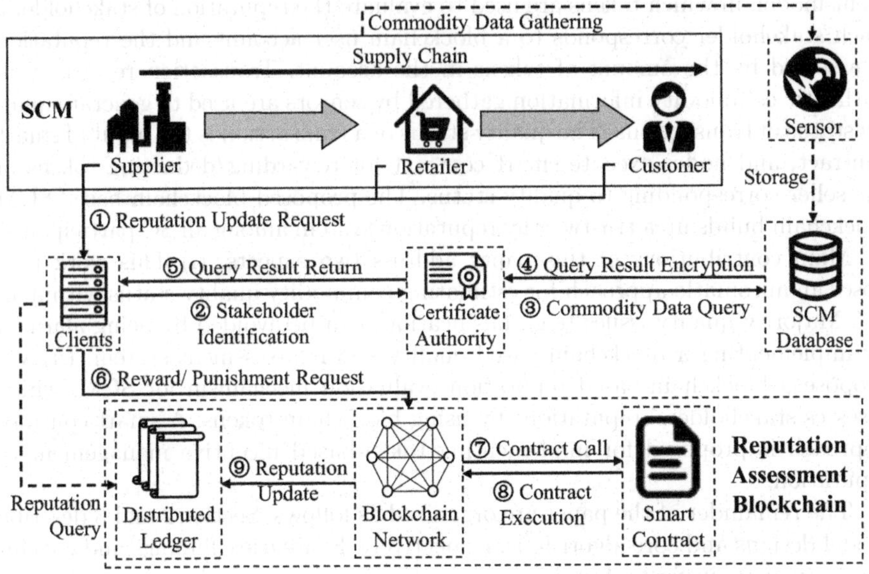

Fig. 1. The high level architecture of RTB-SCM.

In RTB-SCM model, SCM database stores each trade's information collected from SC, off the RAB. The buyer uses sensors to automatically gather commodity's information for each trade, in order to avoid artificial falsification. The collected commodity information will be combined with the order information to generate a transaction record, which is sent to SCM database for storage.

The structure of the transaction record in SCM database is shown in Fig. 2. As shown in the figure, each trade record consists of two data fields, including the order information and commodity information list. Order information field stores the ID number of orders ($OrderID$), ID number of seller stakeholder ($Seller$), ID number of buyer stakeholder ($Buyer$), type of commodity ($Type$), and the amount of the commodity ($Amount$). Commodity information list stores the information records of each commodity in this trade. Each commodity information record $C_i = \{Attr_1, Attr_2, \dots\}$ consists of commodity attributes which are strictly required in the corresponding commodity quality standard.

Clients are connected to both RAB and SCM database. After receiving a reputation update request of a trade from a stakeholder, a client will query corresponding trade record in SCM database. CA verifies the stakeholder and encrypts query results based on *Secure Electronic Transaction* (SET) protocol, which ensures the safety and authenticity of SCM database query. After verifying the user's identity, the query request will be sent to SCM database to obtain the corresponding trade record. The query result (recording trade record) is, then, encrypted by CA and is sent back to client.

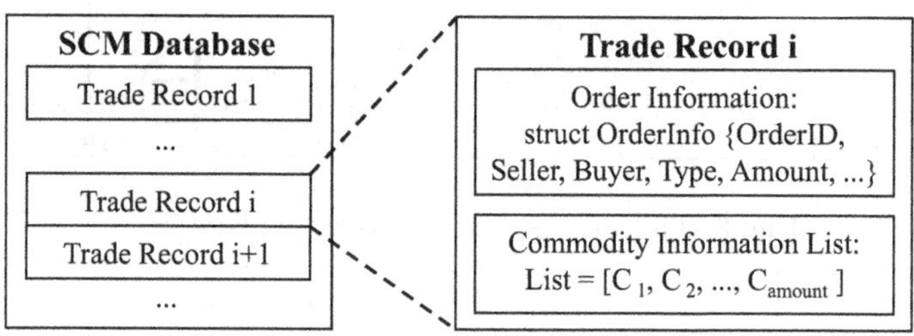

Fig. 2. The structure of trade records in SCM database.

After receiving and decrypting the query results of trade records from SCM database, client will use corresponding commodity information in this transaction to generate corresponding quality status. The quality standard of a certain type of commodity records the minimum value of each attribute for qualified commodities as corresponding attribute standard. Quality standards are stored in clients that use corresponding quality standard of a certain commodity to determine its quality status. Based on the quality status, a client sends the R&P request to RAB network for the reputation update, after We present detailed description about the proposed algorithm for commodity quality status generation in Sect. 3.

RAB is a crucial component in our model that will be detailedly explained in Sect. 2.2.

2.2 Reputation Assessment Blockchain (RAB)

RAB governs the assessment of reputation of SC stakeholders, in which the reputation of a stakeholder is assessed by the amount of tokens in its account. Each copy of the distributed ledgers is stored in a node in the blockchain network. Hence, a few key components of RAB include blockchain network, smart contract, and distributed ledgers. Blockchain network updates reputation information by running smart contract calls. Quality status information sent by clients triggers corresponding reputation R&P rule in smart contract to update reputation of stakeholders. Figure 3 shows the architecture of RAB in RTB-SCM.

Blockchain network in RAB consists of blockchain nodes connected by *Point to Point* (P2P) network. Each node in RAB is maintained by a trusted organization (e.g., an enterprise). Nodes in blockchain network of RAB, as shown in Fig. 3, are categorized into two types, based on specific function, which include

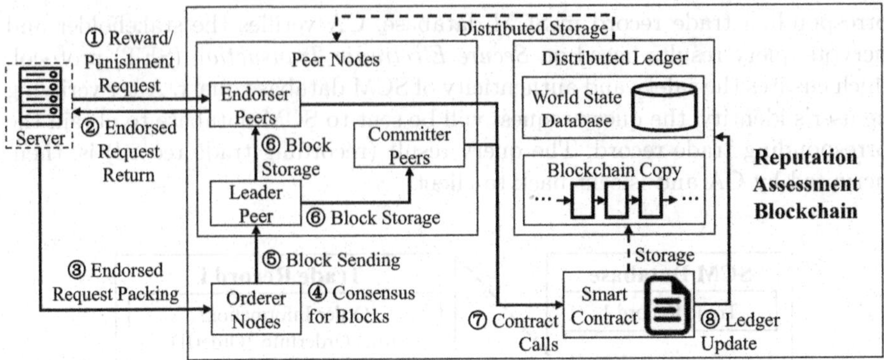

Fig. 3. The architecture of RAB in RTB-SCM.

peer and orderer nodes. Peer nodes store and maintain ledger copies and different roles are assigned to different peer nodes during the transaction process. Roles of peer nodes include endorser, leader and committer. Endorser peer nodes receive and validate transaction requests sent by clients, and then endorse validated transactions. After endorsement, valid transaction request is sent back to client and then sent to orderer. Only transaction requests with endorsements by all endorser peer nodes can be sent to orderer nodes. Leader peer node is connected to orderer nodes to receive and validate blocks. When a block is generated, leader peer node will receive this block from orderer nodes and then validate it. After validation, the block is broadcasted to all other peer nodes. Specially, there is only one leader peer node in blockchain network and a new leader peer node is elected when current leader peer node has failed. Committer peer nodes store and maintain a copy of distributed ledger. Committers can receive blocks forwarded by leader and update state information in its own ledger copy based on the information in the block. Specifically, every peer node is a committer, and each peer node can be more than one role at the same time.

The client sends the corresponding transaction request to endorser peer nodes for updating reputation status (transaction endorsement) once the quality status of a transaction (on SC) is confirmed. Endorser peer nodes validates and endorses the transaction before the endorsed transaction request is sent back to client. The transaction request is validated and is sent to orderer nodes for consensus when endorsed transaction requests are received from all endorser peer nodes.

Orderer nodes in RAB pack transaction requests into blocks and use consensus mechanism to create blocks. At the end of each block time, orderer nodes sort all received transaction records and pack sorted transaction records in a block. Then, orderer nodes ensure the global consistency of blocks by running consensus mechanism. Our approach implements a *Practical Byzantine Fault Tolerance* (PBFT) consensus mechanism [20] to achieve the consistency of transactions. In each round of PBFT consensus, one orderer node is delegated as primary node that is connected to clients to receive endorsed transaction requests and generates block at the end of each block time. PBFT consensus in RAB uses *Elliptic Curve Digital Signature Algorithm* (ECDSA) [13] and SHA-256 hash algorithm [10] to generate signatures. Briefly, SHA-256 algorithm compress a message into a 256-bit message digest, and ECDSA generates a key pair $Key = (d, Q)$ in which d is private key and Q is the public key. Message senders use d to encrypt transaction digest into digital signature; message receivers use Q to decrypt signature back to message digest and validate its authenticity.

The main phases of PBFT consensus mechanism are described as follows. Suppose there are $(3f+1)$ orderer nodes in the blockchain network. First, current primary node broadcasts a pre-prepare message to all other orderer nodes after packing a batch of transaction requests in a block. A pre-prepare message records the block message digest (d) and signature ($PrePre$). When receiving a pre-prepare message, an orderer node will validate its d and $PrePre$. Then when an orderer node has validated a correct pre-prepare message, it broadcasts a prepare message to all other orderer nodes consisting of d and signature Pre. Each orderer node validates d and Pre when receiving a prepare message. When an orderer has validated $2f$ correct prepare messages and has broadcast its own prepare message, it then broadcasts a commit message to all other orderer nodes. A commit message has the d, block (B), and signature ($Comm$). Only current primary node can accept a commit message and validate its message digest and signature. After validating $(2f + 1)$ correct commit messages with the same transaction sequence in block, current primary node sends a reply message to leader peer node. A reply message contains the block in which the sequence of transactions is global-consistent among orderers.

Next, the status information of each blockchain user is updated by all peer nodes after adding a block into blockchain. Specifically, current leader peer node validates a transaction when a reply message from current primary orderer node is received. The validated block is, then, broadcasted to all committer peer nodes. Each committer peer node executes smart contract (attached to the transaction record), which updates status information in its distributed ledger copy to realize ledger update. A peer node stores a copy of distributed ledgers that include a blockchain copy and a global state database. The blockchain copy records blocks (containing transaction records or smart contract codes). Global state database stores status information of blockchain user accounts. Each transaction record in blocks contains the function name and input variables of corresponding smart contract code. Peer nodes update account information in each copy of world state database by using transaction records to call smart contract.

As mentioned in prior sections, RAB uses tokens to assess users' reputations through running R&P-oriented smart contracts. We implement a token bank account to realize token R&P. Token bank account is kept by regulator or government, which holds enough tokens to circulate. The account of token bank is stored in each node of RAB; all authorized nodes can access token bank. Token bank grants a certain amount of tokens (e.g., 100 tokens) to a stakeholder when a user account is created in RTB-SCM. When a transaction occurs, the participant stakeholder will either rewarded or punished, depending on whether the

Algorithm 1. Quality Status Generation algorithm

Require: Initialization of parameters: *TRec*
Ensure: Output of parameters: *QInfo*

```
Cnt ← 0;
Standard ← corresponding quality standard of TRec.Type;
for each commodity record C_i in TRec.List[] do
    for each commodity attribute Attr_i in C_i do
        Compare Attr_i with corresponding attribute standard in Standard;
    end for
    if any attribute of C_i can not satisfy Standard then
        Cnt = Cnt + 1;
    end if
end for
if Cnt/TRec.Amount > Standard.InfRate then
    QInfo ← 0;
else
    QInfo ← 1;
end if
return QInfo;
```

quality issue is detected. The user account of the stakeholder is denoted by St. The reward action is operated by transferring tokens to St from token bank and the punishment action is operated by transferring tokens from St to token bank. We provide a sample setting of the R&P as follows. Token rewards will reward a stakeholder with 5% of its current token amount, and token punishments will punish a stakeholder with 20% of its current token amount. It is easy to observe that one token punishment needs five token rewards to make up its reputation.

By using token rewards as positive incentive and punishment as negative incentive, the incentive mechanism for reputation of stakeholders is realized. Smart contract of RAB in RTB-SCM is designed to realize token-based reputation R&P mechanism, which will be introduced in Sect. 3.

3 Algorithms

This section explains two major algorithms in RTB-SCM model, which are *Quality Status Generation* (QSG) algorithm and *Token-based Reputation Reward/Punishment* (TR2P) algorithm. QSG algorithm is designed for clients to generate the quality status information of a trade on SC. TR2P algorithm is designed for RAB to reward or punish tokens to realize reputation reward or punishment, which is embedded into smart contracts of RAB.

3.1 Quality Status Generation (QSG) Algorithm

QSG algorithm uses transaction records in SCM database queried by clients to generate corresponding quality status. For an input of transaction record

($TRec$) queried from SCM database, QSG compares each commodity information record in $TRec$ with built-in quality standard of corresponding commodity type ($Standard$) to determine the quality of each commodity. By comparing the proportion of inferior commodity in $TRec$ with inferior rate in $Standard$, the overall quality status of $TRec$ is generated, denoted by $QInfo$. The pseudo codes of QSG algorithm are given in Algorithm 1. The main phases of QSG algorithm are shown below:

1. Obtain commodity type $TRec.Type$ to find corresponding commodity quality standard $Standard$;
2. Entry an FOR loop to process all commodity records. The quality of each commodity record C_i in $TRec$ is obtained by entering an inner FOR loop to compare each commodity attribute $Attr_j$ of C_i with corresponding attribute standard in $Standard$. Any commodity record not satisfying $Standard$ will be counted by Cnt;
3. Once every C_i in $TRec$ has been processed, end the FOR loop. Calculate the inferior rate of $TRec$ by dividing Cnt with the amount of commodity $TRec.Amount$.
4. Generate the overall quality status information $QInfo$ based on inferior rate. $QInfo$ is assigned with 0 when the calculated inferior rate is larger than standard inferior rate $Standard.InfRate$, and $QInfo$ is assigned with 1 when the calculated inferior rate is within $Standard.InfRate$.

A brief analysis on the time complexity of QSG algorithm is given as follows. Assume that there are N_{rec} commodity records in a trade record, and assume that this type of commodity has N_{attr} attributes in commodity records. The generation of quality status information $QInfo$ requires processing all N_{rec} commodity records in its FOR loop. For each commodity record in this trade record, all of its N_{attr} attributes are compared with the quality standard in an inner FOR loop. Thus the time complexity of QSG algorithm is $O(N_{rec}N_{attr})$, in which N_{rec} is the amount of commodity records in the input trade record, and N_{attr} is the amount of commodity attributes in each commodity record.

When $QInfo$ is generated, a client will send the seller stakeholder ID $TRec.Seller$ along with $QInfo$ to blockchain network in a transaction request for updating reputation. $QInfo$ reflects overall quality status of corresponding transaction ($TRec$). $QInfo = 0$ means that quality issues occur in $TRec$. $QInfo$ is then used in smart contract to determine specific reputation reward or punishment. The algorithm running in smart contract is presented in Sect. 3.2.

3.2 Token-Based Reputation Reward/Punishment Algorithm

The *Token-based Reputation Reward/Punishment* (TR2P) algorithm uses parameters in transaction requests to implement token-based reputation reward or punishment. Parameters in TR2P transaction requests include the quality status $QInfo$ and the seller account ID $Seller$ of a certain trade. For a certain trade record $TRec$, $QInfo$ is generated by QSG algorithm illustrated above,

Algorithm 2. TR2P algorithm

```
Require: Initialization of parameters: QInfo, Seller
Ensure: Update of parameter: Seller.Balance
Balance ← Seller.Balance;
if QInfo = 0 then
    Punishment ← Balance × 0.2;
    Seller.Balance ← Seller.Balance − Punishment;
    TokenBank.Balance ← TokenBank.Balance + Punishment;
else
    Reward ← Balance × 0.05;
    Seller.Balance ← Seller.Balance + Reward;
    TokenBank.Balance ← TokenBank.Balance − Reward;
end if
```

and *Seller* is assigned with *TRec.Seller*, which is the seller ID in trade record *TRec*. Then TR2P algorithm updates the token balance of *Seller* and token bank account to realize token reward or punishment based on the value of $QInfo$. The pseudocode of TR2P algorithm is shown in Algorithm 2.

The main phases of TR2P algorithm is shown below:

1. Obtain the token balance of seller account *Balance* by assigning it with the amount of tokens of seller account *Seller.Balance*;
2. If $QInfo = 0$, then quality problems occurred in current trade. Calculate the amount of token for token punishment *Punishment*, and then transfer tokens with amount *Punishment* from user account to token bank as punishment.
3. If $QInfo! = 0$, then no quality problem occurred in current trade. Calculate the amount of token for token reward *Reward*, and then transfer tokens with amount *Reward* from token bank to user account as reward.

The analysis of time complexity of TR2P algorithm is given below. First, the scale of $QInfo$ is fixed with only two possible value 0 and 1. For any input seller account *Seller*, obtaining its token balance only requires 1 assignment operation. Each branch in the IF-ELSE structure in TR2P only requires 1 R&P token amount calculation and 2 token balance update operation. Thus each process in TR2P is finished in constant time, and the time complexity of TR2P algorithm is $O(1)$.

After receiving a new block, each peer node executes all transaction records in this block by calling smart contract with TR2P algorithm. TR2P algorithm is coded as a smart contract rule to update the token balance of stakeholders according to the input parameters in transaction records. After all transaction records are processed, the reputation of stakeholders is updated by token rewards and token punishments.

4 Experiment and the Results

In this section, we use a experiment to evaluate the feasibility of our RTB-SCM model. This experiment uses a specific use case of the reputation R&P of a simulated trade in SC. We developed a simulator of RTB-SCM model based on Hyperledger Fabric v1.4, which is a consortium blockchain platform based on Golang. The hardware used in our experimental evaluations included a PC with Ubuntu 16.04 64-bit OS, Intel Core i7-6700HQ CPU and 4G RAM.

The use case in our experiment included a trade record $TRec$ recording a trade of 4 pieces of rolled steel as commodity, and it also included the quality standard of rolled steel $Standard$. First, $TRec$ was sent to a client to generate its quality status information $QInfo$ by using QSG algorithm. Then, a reputation R&P transaction record with $QInfo$ was send to blockchain network to update the reputation of the seller stakeholder in $TRec$. The R&P process required the participation of token bank account, which was account $usr0000$ in our system.

Trade record $TRec$ records a trade with ID 0001 between stakeholder A and B. In trade $TRec$, A was the seller with ID $usr0001$ and B was the buyer with ID $usr0002$. Both A and B had 100 tokens in their accounts initially. The commodities in this trade was 4 rolled steel products. In this use case, the main quality index of rolled steel products included strength and toughness. Both strength and toughness could be obtained by sensors. Table 1 shows the detail of $TRec$, and Table 2 shows the detail of all 4 commodity information records in $TRec.List$. Besides, the quality standard of rolled steel product $Standard$ stored in clients is shown in Table 3.

Table 1. The detail of trade record $TRec$.

OrderID	Seller	Buyer	Type	Amount	List
0001	usr0001	usr0002	RolledSteel	4	$[C_1, C_2, C_3, C_4]$

Table 2. The detail of commodity information records $TRec.List$.

Commodity	[Strength (MPa), Toughness (J)]	Commodity	[Strength (MPa), Toughness (J)]
C_1	[238, 34]	C_2	[233, 35]
C_3	[239, 37]	C_4	[236, 32]

Table 3. The quality standard of rolled steel $Standard$.

Type	[Strength (MPa), Toughness (J)]	InfRate
RolledSteel	[235, 34]	0.20

```
func QSG(TRec1 TRec) (QInfo int) {
    var cnt int = 0
    var i, j int
    var InfRate float32
    for i = 0; i < TRec1.Amount; i ++ {
        for j = 0; j < len(TRec1.List[0]); j ++ {
            if TRec1.List[i][j] < Standard.List[j] {
                cnt ++
                break
            }
        }
    }
    InfRate = float32(cnt) / float32(TRec1.Amount)
    if InfRate > Standard.InfRate {
        QInfo = 0
    } else {
        QInfo = 1
    }
    return
}
```

Fig. 4. Partial code of QSG algorithm.

After buyer B sent a reputation update request with order number 0001 to a client C, C would query corresponding trade record $TRec$ from SCM database. The query result of $TRec$ was then used in QSG algorithm to generate quality status information $QInfo$. First, the type of commodity in $TRec$ was obtained by QSG algorithm, which was *RolledSteel*. Then QSG obtained the corresponding quality standard with commodity type *RolledSteel* using $TRec.Type$ = *RolledSteel*, which was *Standard*. For each commodity record $C_i(i = [1, 2, 3, 4])$ in $TRec.List$, QSG algorithm compared strength and toughness value in the record with corresponding quality index in *Standard*. According to Table 2, C_2 can not satisfy the strength request of *Standard*, and C_3 can not satisfy the toughness request of *Stanadrd*. The inferior rate of $TRec$ was then calculated, which was 0.5. Because the inferior rate of $TRec$ was higher than $Standard.InfRate$ which is 0.2, $QInfo$ was assigned by value 0 to represent that quality problem occurred in $TRec$.

Partial code of QSG algorithm in our experiment is shown in Fig. 4. These codes of QSG realized the quality status generation by comparing commodity attribute with quality standard. The outer FOR loop is to process each commodity record *TRec.List[i]* in trade record, and the inner FOR loop is to compare each attribute *TRec.List[i][j]* in a commodity record with corresponding attribute standard in *Standard*. Variable *cnt* counts the amount of unqualified commodities for inferior rate calculation. The value of $QInfo$ is assigned after the inferior rate is calculated and compared with $Standard.InfRate$.

Client then generated a transaction request $Request$ = {"*name*" : "*TR2P*", "*args*" : ["0", "*usr*0001"]} to RAB network. In transaction request *Request*, "*name*" indicates the function name of TR2P algorithm in smart contract, and "*args*" is used to pass the input arguments to smart contract. The input arguments are 0 and "*usr*0001", which corresponds to the value of $QInfo$ and *Seller* respectively. After endorsement and consensus process, *Request* was then

```
func (s *SmartContract) TR2P(stub shim.ChaincodeStubInterface, args []string) pb.Response {
    _QInfo,_ := strconv.Atoi(args[1])
    _Seller:= args[2]
    tokenAsBytes,err := stub.GetState(args[0])
    if err != nil {
        return shim.Error(err.Error())
    }
    token := Token{}
    json.Unmarshal(tokenAsBytes, &token)
    amount := token.balance(args[2])
    if _QInfo == 0 {
        PunishmentAmount := amount / 5
        token.transfer(_Seller, "usr0000", PunishmentAmount)
    } else {
        RewardAmount := amount / 20
        token.transfer("usr0000", _Seller, RewardAmount)
    }
    tokenAsBytes, err = json.Marshal(token)
    if err != nil {
        return shim.Error(err.Error())
    }
    err = stub.PutState(args[0], tokenAsBytes)
    if err != nil {
        return shim.Error(err.Error())
    }
    return shim.Success(tokenAsBytes)
}
```

Fig. 5. Partial code of smart contract rule of TR2P algorithm.

recorded into a block, which was sent to all peer nodes in RAB. Each peer node processed this newly received block, and *Request* triggered smart contract according to *Addr*. First, smart contract got the token balance of *Seller* in world state database according to its ID $usr0001$, which was 100. Then smart contract chose to punish *Seller* according to the value 0 of $QInfo$ and calculated the token amount of punishment, which was 20. A transfer from A's account $usr0001$ to token bank $usr0000$ was executed to finish the punishment.

Partial codes of smart contract rule of TR2P algorithm are shown in Fig. 5, which realized token R&P based on quality status information. Arguments $args$ in transaction request was obtained by smart contract. $QInfo$ was stored in *arg[1]* to determine the specific reward or punishment action, and seller account was stored in *arg[2]*. After a reward or punishment was finished, token object was compressed to a string by function $json.Marshal()$, and the string was then stored into world state database by function *stub.PutState*().

The result of reputation update for trade $TRec$ could be queried from RAB. The token balance of A was changed to 80. This indicates that seller stakeholder A had quality problem in trade $TRec$ and was punished with 20 token. Since tokens are linked to stakeholder reputation in SC, the reputation of A was decreased after reputation update due to its quality problem in $TRec$. We can observe that RTB-SCM model achieves reputation-based trustworthy SCM by updating the reputation of stakeholders based on quality status of trades.

5 Conclusions

This paper mainly focused on realizing reputation-based trustworthy SCM and proposed a smart contract-based reputation R&P approach. Our approach also used trade records to generate corresponding quality status information to determine specific reputation reward or punishment operation. Besides, our approach used consortium blockchain to record and maintain user reputation, in which tokens are used to realize reputation assessment. Smart contract was used to realize the reward and punishment of tokens, thus realizing reputation-based incentive management. The proposed algorithm realized quality status generation for trades and token R&P operation for stakeholders. Experiment with a specific use case of a trade was used to evaluate the feasibility of our proposed approach.

Acknowledgments. This work is partially supported by Beijing Natural Science Foundation (Grant No. 4204111), National Natural Science Foundation of China (Grant No. 61972034, 61902025), Natural Science Foundation of Beijing Municipality (Grant No. 4202068), Natural Science Foundation of Shandong Province (Grant No. ZR2019ZD10), Guangxi Key Laboratory of Cryptography and Information Security (No. GCIS201803), Henan Key Laboratory of Network Cryptography Technology (Grant No. LNCT2019-A08), Beijing Institute of Technology Research Fund Program for Young Scholars (Keke Gai, Peng Jiang).

References

1. Aich, S., Chakraborty, S., Sain, M., Lee, H., Kim, H.: A review on benefits of IoT integrated blockchain based supply chain management implementations across different sectors with case study. In: 2019 21st International Conference on Advanced Communication Technology, pp. 138–141, February 2019. https://doi.org/10.23919/ICACT.2019.8701910
2. Chen, S., Shi, R., Ren, Z., Yan, J., Shi, Y., Zhang, J.: A blockchain-based supply chain quality management framework. In: 2017 IEEE 14th International Conference on e-Business Engineering, pp. 172–176, November 2017. https://doi.org/10.1109/ICEBE.2017.34
3. Dahbi, A., Mouftah, H.T.: Supply chain efficient inventory management as a service offered by a cloud-based platform. In: 2016 IEEE International Conference on Communications, pp. 1–7 (2016)
4. Feng, T.: An agri-food supply chain traceability system for China based on RFID & blockchain technology. In: 2016 13th International Conference on Service Systems and Service Management, pp. 1–6 (2016)
5. Gai, K., Guo, J., Zhu, L., Yu, S.: Blockchain meets cloud computing: a survey. IEEE Commun. Surv. Tutor. **PP**(99), 1 (2020)
6. Gai, K., Wu, Y., Zhu, L., Qiu, M., Shen, M.: Privacy-preserving energy trading using consortium blockchain in smart grid. IEEE Trans. Ind. Inform. **15**(6), 3548–3558 (2019)
7. Gai, K., Wu, Y., Zhu, L., Xu, L., Zhang, Y.: Permissioned blockchain and edge computing empowered privacy-preserving smart grid networks. IEEE Internet Things J. **6**(5), 7992–8004 (2019)

8. Gai, K., Wu, Y., Zhu, L., Zhang, Z., Qiu, M.: Differential privacy-based blockchain for industrial internet-of-things. IEEE Trans. Ind. Inform. **16**(6), 4156–4165 (2019)
9. Gao, F., Zhu, L., Gai, K., Zhang, C., Liu, S.: Achieving a covert channel over an open blockchain network. IEEE Netw. **34**(2), 6–13 (2020)
10. Jung, M.Y., Jang, J.W.: Data management and searching system and method to provide increased security for IoT platform. In: 2017 International Conference on Information and Communication Technology Convergence (ICTC), pp. 873–878 (2017)
11. Li, H., Gai, K., Fang, Z., Zhu, L., Xu, L., Jiang, P.: Blockchain-enabled data provenance in cloud datacenter reengineering. In: The 2019 ACM International Symposium on Blockchain and Secure Critical Infrastructure, pp. 47–55, July 2019. https://doi.org/10.1145/3327960.3332382
12. Li, Z., Kang, J., Yu, R., Ye, D., Deng, Q., Zhang, Y.: Consortium blockchain for secure energy trading in industrial internet of things. IEEE Trans. Ind. Inform. **14**(8), 3690–3700 (2018)
13. Liu, Y., Liu, X., Tang, C., Wang, J., Zhang, L.: Unlinkable coin mixing scheme for transaction privacy enhancement of bitcoin. IEEE Access **6**, 23261–23270 (2018)
14. Malik, S., Dedeoglu, V., Kanhere, S.S., Jurdak, R.: Trustchain: trust management in blockchain and IoT supported supply chains. In: 2019 IEEE International Conference on Blockchain (Blockchain), pp. 184–193 (2019)
15. Qiu, M., Chen, Z., Niu, J., Zong, Z., et al.: Data allocation for hybrid memory with genetic algorithm. IEEE Trans. Emerg. Top. Comput. **3**(4), 544–555 (2015)
16. Qu, F., Haddad, H., Shahriar, H.: Smart contract-based secured business-to-consumer supply chain systems. In: 2019 IEEE International Conference on Blockchain, pp. 580–585, July 2019. https://doi.org/10.1109/Blockchain.2019.00084
17. Scully, P., Hbig, M.: Exploring the impact of blockchain on digitized supply chain flows: a literature review. In: 2019 Sixth International Conference on Software Defined Systems, pp. 278–283 (2019)
18. Shao, Z., Xue, C., Zhuge, Q., Qiu, M., Xiao, B., Sha, E.: Security protection and checking for embedded system integration against buffer overflow attacks via hardware/software. IEEE Trans. Comput. **55**(4), 443–453 (2006)
19. Tian, Z., Li, M., Qiu, M., Sun, Y., Su, S.: Block-DEF: a secure digital evidence framework using blockchain. Inf. Sci. **491**, 151–165 (2019)
20. Wang, Y., Cai, S., Lin, C., Chen, Z., Wang, T., Gao, Z., Zhou, C.: Study of blockchainss consensus mechanism based on credit. IEEE Access **7**, 10224–10231 (2019)
21. Wang, Y., Geng, X., Zhang, F., Ruan, J.: An immune genetic algorithm for multi-echelon inventory cost control of IoT based supply chains. IEEE Access **6**, 8547–8555 (2018)
22. Wu, H., Cao, J., Yang, Y., Tung, C.L., Jiang, S., Tang, B., Liu, Y., Wang, X., Deng, Y.: Data management in supply chain using blockchain: challenges and a case study. In: 2019 28th International Conference on Computer Communication and Networks, pp. 1–8, July 2019. https://doi.org/10.1109/ICCCN.2019.8846964
23. Zhu, L., Wu, Y., Gai, K., Choo, K.: Controllable and trustworthy blockchain-based cloud data management. Future Gener. Comput. Syst **91**, 527–535 (2019)

Content-Aware Anomaly Detection with Network Representation Learning

Zhong Li[1,2](✉), Xiaolong Jin[1,2], Chuanzhi Zhuang[1,2], and Zhi Sun[1,2]

[1] CAS Key Laboratory of Network Data Science and Technology, Institute of Computing Technology, Chinese Academy of Science, Beijing, China
{lizhong17b,jinxiaolong,zhuangcz,sunzhi17b}@ict.ac.cn

[2] School of Computer Science and Technology, University of Chinese Academy of Sciences, Beijing, China

Abstract. Given a network with node descriptions or labels, how to identify anomalous nodes and anomalous links in it? Existing methods (e.g., non-negative matrix factorization) mostly focuses on structural anomalies, without taking node descriptions or labels seriously into account. However, such information is obviously valuable for detecting anomalous nodes and links. On the other hand, network representation learning aims to represent the nodes and links in a network as low-dimensional, real-valued and dense vectors, so that the resulting vectors have representation and reasoning ability. It is straightforward that the reconstruction errors between normal node representations are small, while anomaly ones are not. Therefore, we propose a novel Content-Aware Anomaly Detection (CAAD) method based on network representation learning and encoder-decoder. The CAAD method learns structural and content representations with convolutional neural networks. By using the learned low-dimensional node representations, an encoder-decoder model is trained to perform anomaly detection in terms of reconstruction errors. Experiments on two synthetic datasets and one real-world dataset demonstrate that CAAD consistently outperforms the existing baseline methods. For more information and source codes of this study please visit https://github.com/lizhong2613/CAAD.

Keywords: Data mining · Network embedding · Anomaly detection · Link analysis

1 Introduction

In recent years, with the rapid development of Internet technology, more and more attention has been paid to network anomaly detection, which plays an increasingly vital role in such fields of fraud detection, intrusion detection, false-voting identification and zombie fan analysis, to name a few. Social networks are ubiquitous in our daily lives, e.g., the followee-follower network in Sina Weibo and the networks of accounts concerning the same topics in Zhihu. Given a network with node

M. Qiu (Ed.): ICA3PP 2020, LNCS 12454, pp. 50–64, 2020.
https://doi.org/10.1007/978-3-030-60248-2_4

descriptions or labels, how to identify anomalous nodes and links in it? Early studies on network anomaly detection are mainly based on connectivity by using the features of nodes (such as the number of triangles, the number of nodes in the ego-net, etc.) to characterize the pattern of normal links. If the features of a node differ very much from those of others in a given network, it will be regarded as being anomalous. Traditional anomaly detection often investigates network structures. But in real networks, most nodes contain text information or other meta-data, which has not been well applied in existing detection methods.

In real-world networks, it is straightforward that the more common meta-information there is between two nodes, the closer they are to each other. Thus, if two nodes do not have any identical meta-information and their structural distance is very far, a link between them will be anomalous. The corresponding nodes are the abnormal ones.

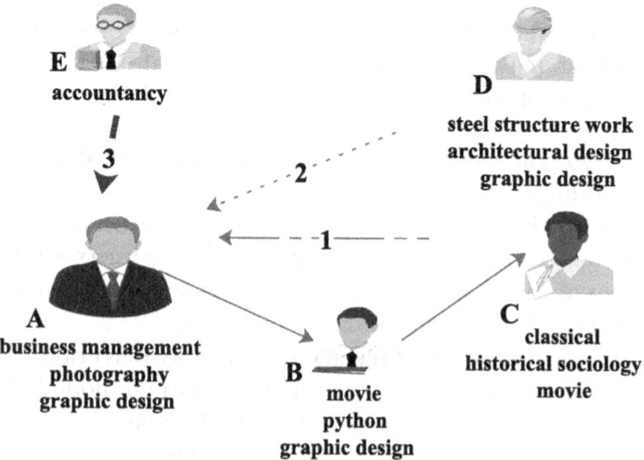

Fig. 1. A toy example of information network. The proposed CAAD method utilizes structural context and node content information.

A simple example of the model is given in Fig. 1. The solid lines in the diagram represent the real relationships, and the dashed ones represent the links that need to be verified for normality. As Nodes A and node C are connected, they are correlated in the low dimensional space or are closer to each other than to the others. Therefore, the link between Nodes A and C is normal. On the other hand, although Nodes A and D are not directly related and do not have the same neighbor nodes, the link between them is still normal because they have a common node label or similar node content information. Nevertheless, the link between Nodes E and A is anomalous, because the distance between them is very far, and there is no similarity in their node contents. If a node is connected to a number of anomalous links, this node will be an anomaly one, for example a zombie fan in social networks. It is expected that the consideration of two pieces of meta-information, network structure and node contents, is needed for network anomaly detection.

Network Structure. The anomaly detection methods as described in the early literature are mainly based on the information of network structures. The most representative ones include oddball [5] and Eigenspoke [10], which uses the probabilistic statistical distribution patterns of normal network structures to find out abnormal nodes. The decomposition of tensors is an effective method developed in recent years. CROSSSPOT [11], M-Zoom [12], D-cube [13] and NrMF [14] are some examples of tensor decomposition. The matrix decomposition method can be used for detection of network community and anomaly. Community detection based approaches attempt to overcome the limitations of statistical techniques by detecting outliers via identifying the bridging nodes or bridging links between different communities. However, all these methods are limited to network topology.

Node Content. The nodes with rich information in the networks are ubiquitous in our lives, e.g., labels and text contents. In Zhihu, for example, each account in the network, as a node, has some concerned text information, all of which is also important for network structure. Hence, this study comes up with an idea to learn normal links between nodes and spot abnormal links.

The contributions of this paper are as follows:

- It provides a general method for the network anomaly detection based on network embedding.
- While most network embedding methods investigate network structures for learning, this paper incorporates content of nodes and network topology into network representation learning. On this basis, it proposes a method that can be used for both link and node anomaly detection.
- It proposes an encoder-decoder scheme for anomaly detection. By learning a prediction model, this scheme can reconstruct 'normal' node links. Then the reconstruction errors are used to detect anomaly links. The anomaly metric of nodes is represented by the mean of the connected links' prediction errors. Experiments on two synthetic datasets and a real-world dataset show the effectiveness of the CAAD method.

The rest of the paper is organized as follows: The background and related works are introduced in Sect. 2. The problem formulation is given and an algorithm is proposed in Sect. 3. Then, the simulation experiments are presented and a summary is made.

2 Background and Related Works

2.1 Network Embedding

Network embedding represents nodes in a network with a low-dimensional dense vector (meaning that the dimension of vector space is far less than the total number of nodes), and preserves structural characteristics of the original network [4]. There is a great deal of literature exploring the graph embedding. To map every node in the original space to a low-dimensional space, mainly three models can be used for structure and property preserving: matrix factorization, random walk and deep neural networks. Inspired by Word2Vector, random walk generates nodes' local information by random paths over a network. The representative algorithms of random walk model include DeepWalk and Node2Vec, etc. Combined with the first- and second-order neighbor information of nodes, LINE [8] (Tang et al. 2015) takes minimizing KL divergence as the goal of neural network training, getting the final output result of low-dimensional vectors with first- and second-order proximity information.

2.2 Network Anomaly Detection

Aiming to find the nodes or the links which are much different from the others, network anomaly has definitions varying from application to application. There are many methods for network structure anomaly detection, e.g., power law models [1], spectral graph theory, MDL, and community detection. The common method is to build up a prediction model with the known normal data, and then use the prediction model to estimate new observation values. If the deviation is larger than a certain threshold, that data will be considered as abnormal data, e.g., a statistical power law that captures the distribution model, or several spikes are presumably caused by suspicious nodes. Spectral graph model and matrix decomposition are based on dimension reduction techniques. In real-world networks, the texts are attached to individual nodes to record nodes' attributes or other information. However, very few models are available to capture the nodes' content in the network, and even fewer are practical for large-scale applications, because the distances between content information are difficult to calculate.

2.3 Network Embedding Based Anomaly Detection

A broad variety of methods for anomaly detection based on network embedding are available in the literature [1–3]. For instance, NetWalk is used for anomaly detection in real time by learning the network representation, and it can be dynamically updated as the network evolves. In addition, NetWalk obtains the node representation by minimizing the distances between nodes in each walk derived from the node. Clique embedding technology and deep autoencoder reconstruction are also employed to dynamically detect network anomalies. Gao [3] computes the distances between different eigen decompositions as its anomalous score. However, most of the existing works on anomaly detection learn representations from network structures only.

3 Problem Formulation

Our goal is to detect suspicious nodes and links in a directed network with text information. Thus we give the basic notations and the definitions of the problem as:

Given a directed graph $G = (V, E, T)$, where V represents the nodes set, $E \in V \times V$ are links between the nodes, and T is the text content information set of the nodes. Link $e = (u, v) \in E$ denotes the link between the source node u and the target node v. The content S_v of the node $v \in V$ is represented by a word sequence $S_v = (w_1, w_2, ..., w_{nv})$, where $nv = |S_v|$. In this paper, network anomaly detection aims to find the link $e \in E$ and the node set $V_{suspect}$ which are suspicious and not likely to connect with each other, based on the network structure and node content.

Definition 1. Content-aware network embedding: Most of the existing researches on network embedding learn the representations from network structures only. Taking full use of both network structures and associated content information will facilitate learning the informative representations, which are a low-dimensional embedding $v \in R^d$ for each node in $v \in V$. $|V|$ is much smaller than the dimension of the embedding space.

Definition 2. Content-aware anomalous links detection: Given an network G, the task is to maximize the number of true anomalous links subset $L_{suspect}$. Every link in the subset has an 'irregular' features compared to the other links.

Definition 3. Content-aware anomalous links and node detection: Given an network G, the task is to maximize the number of true anomalous node subset $V_{suspect}$. Every node in the subset has an 'irregular' features compared to the other nodes. Intuitively, more abnormal nodes usually connect to more anomalous links.

4 The Proposed Content-Aware Anomaly Detection Method

The overall workflow of the content-aware network anomaly detection method will be introduced in this section. This method models the content and network topological structure, and the process of each step will be carefully illustrated.

4.1 Content-Aware Network Anomaly Detection Method

The proposed CAAD anomaly links detection method is shown in Fig. 2. In the first step, CAAD reads the structural information of each link of the input network and the text content information of the nodes, and then learns the representation V^t of node content through convolution processing. The V^s denotes the embedding of network structure, which can be learned by using the common approaches (e.g., DeepWalk [6], Node2Vec [7], LINE, etc.). The final representation acquired in this paper is a concatenation of two types of embedding.

As Fig. 1 example shows, the connected nodes (either directly connected or indirectly connected) or the nodes with similar content should be close in the embedded space. Given a normal link (u, v), $V_{random} = (V_1, V_2, ...V_n)$ is a

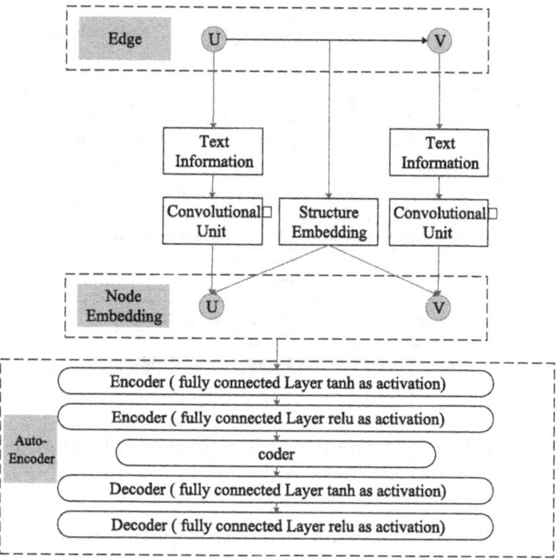

Fig. 2. An illustration of content-aware network anomaly detection of networks.

random node set. The predicted value of the link (u, v) should be greater than $(u, V_i), V_i \in V_{random}$. For each normal link (u, v) in V, the difference between (u, v) and (u, V_i) is used as the input of the Auto-Encoder model. Finally we gets a trained model which can distinguish whether the link is normal or not based on the threshold of the normal link.

An illustrative diagram of the content-aware node anomaly detection pipeline is shown in Fig. 3. This paper first obtains the latent network representation as shown in Fig. 2. We can formulate the every node's mean value of prediction error S_{nodei} by using the representation. We also extract from the initial network node features, including: degree centrality, in-degree, out-degree, average neighbor degree, in-degree density and out-degree density. The input vector V_f is obtained by concatenation of three vectors. $V_f = F_v \bigoplus S_{nodei} \bigoplus V$. By leveraging the content of the original node, common structure features and embedding features, this paper can find the anomaly node.

4.2 Network Embedding Objective

Content-aware network embedding can learn more informative representations than the traditional embedding methods. It learns two types of embedding: representation V^t of node content and the representation V^s of network structure. The final embedding is generated by naive concatenation of V^t and V^s. $V = V^s \bigoplus V^t$, $\bigoplus$ denoting the concatenation operation.

A lot of models can get the node embedding by taking node content into consideration, e.g., TADW, CENE, and CANE. After investigating different techniques for embedding, this paper employs CANE, which uses CNN for text modeling and performs better than others.

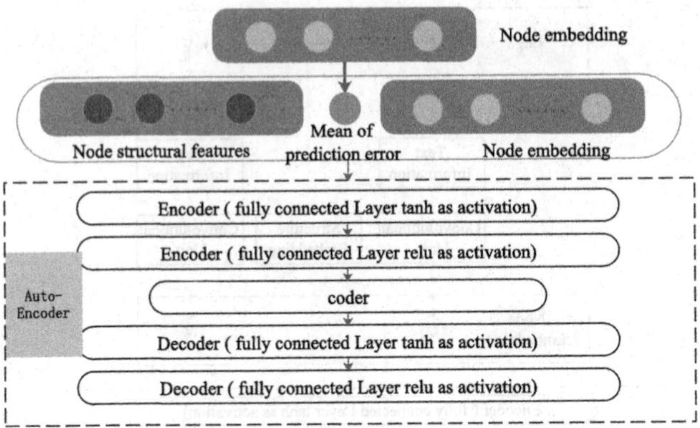

Fig. 3. An illustration of content-aware anomaly detection over nodes.

Following CANE (Tu et al., 2017 [15]), this paper defines the objective of content-aware network embedding as follows:

$$L = \sum_{e \in E} L(e) \tag{1}$$

where $L(e)$ is the objective of the link e, which consists of the structure-based objective $L_s(e)$ and the content-based objective $L_c(e)$ as shown below:

$$L(e) = \alpha L_s(e) + \beta L_c(e) \tag{2}$$

where α and β denote different weight of each part. For a node u, $N_S(u)$ denotes its neighborhood and $f(u)$ denotes its feature representation. Network embedding methods' objective is to maximize the log-probability:

$$L_s(e) = W_{u,v} \log p(N_s(u)|f(u)) \tag{3}$$

Depending on the structural network embedding method used, $p(N_s(u)|f(u))$ is defined differently. For example, the probability of v^s generated by u^s is defined as follows:

$$p(v^s|u^s) = \frac{exp(u^s \cdot v^s))}{\sum_{z \in V} exp(u^s \cdot z^s)} \tag{4}$$

where $L_c(e)$ is text-based objective, which is defined as follows:

$$L_c(e) = \gamma \cdot L_{cc}(e) + \delta \cdot L_{ct}(e) + \epsilon \cdot L_{tc}(e) \tag{5}$$

where γ, δ, ϵ is the weights of each part. $L_{cc}(e)$, $L_{ct}(e)$ and $L_{tc}(e)$ are defined as the same as in Eq. (3). For example $L_{cc}(e)$ can be defined as follows:

$$L_{cc}(e) = W_{u,v} \log p(v^c|u^c) \tag{6}$$

$L_{cc}(e)$ represents log-probability of a directed link when using the content-based embedding.

CNN model is employed to get the content embedding, and it can extract the local semantic dependency. The model contains three layers: looking-up, convolution and pooling. Details of the process can be obtained in CANE.

4.3 Auto-encoder Based Anomaly Detection

Auto-encoder is used for data dimension reduction or feature extraction. If the encoded data can be easily reconstructed to the original data through decoding, the encoded data will be deemed as normal data. Abnormal data cannot be reconstructed by using the features extracted from the normal data. The links with high reconstruction errors show that there are anomaly links. This model uses normal link instances to obtain the low-dimension model. On the basis of content-aware network embedding, given a normal link (u, v), $V_{random} = (V_1, V_2, ...V_n)$ is the random node set. The predicted value of the link (u, v) should be greater than $(u, V_i), V_i \in V_{random}$. For each normal link (u, v) in V, the difference between (u, v) and (u, V_i) is used as the input of the auto-encoder model. The model trained with normal data can reconstruct normal data very well. Finally, we get a trained model which can distinguish whether an link is normal or not according to the threshold of the normal link.

We define the input of the auto-encoder model as follows:

$$X_{u,v} = p(v^E|u^E) - p(v^E_{random}|u^E) \tag{7}$$

where $|X_{u,v}| = |V_{random}| = n$, u^E and v^E denotes the corresponding representation when using the embedding method E. The auto-encoder model is a neural network which is composed of an encoder and a decoder. A hidden layer h inside the auto-encoder generates code to represent the input. The auto-encoder learns to minimize the difference between the construction data and the original data. f_θ and g_ϕ represent the encoder and the decoder respectively.

The pseudo codes of the auto-encoder model are summarized in Algorithm 1.

As shown in Fig. 2, the architecture of both the encoder and the decoder have two hidden layers with different activation functions. The objective of the model is to get the mean squared error, which is defined as follows:

$$L = \frac{1}{n}\sum_{i=1}^{n} w_i \left(x_i - \hat{x}_i\right)^2 \tag{8}$$

For new observation data, we define the anomaly score A_i in Eq. 9, which is the same as the reconstruction error.

$$A_i = \sqrt{\sum_{j=1}^{D} \left(x_j(i) - \hat{x}_j(i)\right)^2} \tag{9}$$

We formulate a suspiciousness metric based on the reconstruction error as follows.

$$S_{nodei} = \frac{1}{K}\sum_{i=1}^{K} A_i \tag{10}$$

Algorithm 1. Auto-encoder based network anomaly detection

Require: a network embedding of a directed graph $G = (V, E, T)$ by using method E with training dataset X, test dataset $x^i, i = 1, ..., n$, and threshold φ

Ensure: anomaly link $x^{ai}, i = 1, ..k, k <= N$

$X_{u,v} = p(v^E|u^E) - p(v^E_{random}|u^E)$ as train input

$\theta, \phi \leftarrow$ train auto-encoder by using the train input

for $i = 0$ to N **do**
  calculate reconstruction error $error(i) = ||x_i - x_i^r||$
  if $error(i) > \varphi$ **then**
    x^i is an anomaly link
  else
    x^i is not a anomaly link
  end if
end for

where K denotes the number of links of corresponding $node_i$.

5 Experiments

In this section, experiments are conducted on two synthetic datasets and a real-world dataset, so as to verify the effectiveness of CAAD.

5.1 Benchmark Networks

The experiments involve three information networks, namely Cora, Zhihu, and Twitter, which are summarized in Table 1.

Table 1. The statistics of the networks used in experiments

Network information	Datasets		
	Cora	*Zhihu*	*Twitter*
# of Nodes	2,277	10,000	16,011,443
# of Links	5,214	43,894	5,384,160
# of Classes	7	–	2

Cora: As a paper citation network, it contains 2,277 machine learning papers and 5,214 citation links among these papers. All of these papers are divided into 7 categories. The network content is represented by a dictionary of 1,433 unique words.

Zhihu: Zhihu is an online Q&A community for users to carry out relevant discussions around a certain topic they are interested in. And users can also follow people with the same interests.

Twitter: This twitter graph is a real-world dataset, which contains information about the Twitter link structure. Every node in the dataset has a true/fake label obtained from Ben Gurion University in 2014 [16].

We use random generation to generate 100 abnormal links (ensuring that the newly generated links do not exist in the original data set). Since there is almost no relationship between the random nodes, a relationship between two unrelated nodes can be considered an abnormal link. In Twitter's data set, some nodes have many friends, but more users only have 0 users following or follow 0 users. Nodes with incomplete friend information are meaningless for network analysis. So we removed nodes that have 0 followers or followed 0 users. Finally, a network of 75624 nodes is obtained.

In order to obtain ground-truth anomaly accounts, relevant information of users is queried by calling the interface of Twitter. If a user is suspended, the account will be considered to be an anomaly one.

5.2 Baseline Methods

The CAAD algorithm proposed in this paper can be used for link anomaly detection and node anomaly detection.

For link anomaly detection, our approach is compared with the following baseline methods:

1. **Deepwalk + Logistics Regression:** Deepwalk learns the node representation by using the random walk method, which is a robust method. Logistic regression is a generalized linear regression analysis model, which is often used in data mining and for binary classification.
2. **Node2Vec + Logistics Regression:** Node2Vec, comprehensively considers DFS neighborhood and BFS exploration, and introduces an embedding method based on random walk.
3. **LINE + Logistics Regression:** LINE uses BFS to obtain context nodes: only the node with the maximum distance of two hops from a given node is considered as its adjacent node. In addition, negative sampling is used to optimize the skip-gram model.
4. **CNN Text:** CNN model is employed to encode the content information. The baseline uses a 300-dimensional text features as node representation.

Logistic regression and network representation learning are also used as the baseline. For example, **Deepwalk + Text + Logistics Regression, TADW + Logistics Regression, LINE + Logistics Regression,** and **NodeVec + Text + Logistics Regression** may be adopted.

For anomaly node detection, this paper's approach is compared with following baseline methods:

1. **Two-stage classifier** [17]. The first step of the model uses the classifier to calculate the nonexistence probabilities of links by based on the various features of the network. The second step uses these probabilities to calculate the abnormal metric of the nodes. The nodes with more abnormal links are the

abnormal nodes. In order to obtain better algorithm performance, we used different classifiers in the first stage, including logistic regression, random forest, and boosting classifiers, adaboost and gradient boosting.

2. **Isolation forest** [18]. Isolation forest (IForest) is a fast ensemble learning anomaly detection method, which can cut a data space with a random hyper plane, with each cut dividing the space into two parts. Intuitively, clusters with a high density can hardly be segmented before stop cutting, but those with abnormal data points in low-density spaces will be cut early.

5.3 Experiment Settings and Evaluation Metrics

The first step of our baseline is to obtain the representation by using the classical network embedding method. In our experiment, this paper use the OpenNE to get the embedding result, which is a standard NE/NRL (Network Representation Learning)training and testing framework. For DeepWalk, we set walk length to 80, and set number of walks to 10. For Node2Vec, the same parameter settings are used with $p = 0.25$ and $q = 0.25$. For LINE, this paper concatenates 1-step and 2-step information as the representation, and the negative sample number is 5. Default parameters are set for other embedding methods. The optimal configuration of weight values are $\alpha = 1, \gamma = 0.3, \delta = 0.3$, and $\epsilon = 0.3$, All of the embedding methods learns a 128-dimensional features and the content information will be represented as a 300-dimensional features. Since the ratio of the abnormal samples to the normal ones is very small, the positive and negative samples are not balanced in the data set. SMOTE [9] is an improved method based on the oversampling algorithm. As oversampling adopts the method of simply copying samples to increase the number of samples, the model may be easy to over-fit and lack generalization capabilities. The basic idea of the SMOTE [9] method is to analyze and use the samples to synthesize new samples. Mini-batch stochastic optimization and Adam Optimizer are used for training the model. For anomaly link detection, we adopt a standard evaluation metric AUC. In statistics and machine learning, AUC is a kind of evaluation to measure the quality of binary model, so as to indicate the probability that positive examples are ranked before negative ones. It is a kind of evaluation metrics to measure the quality of dichotomy model so as to identify the probability that the positive examples rank before the negative ones. The full name of AUC is the area under the receiver operating characteristic (ROC) curve. For anomaly node detection, Precision@k is used to evaluate the performance of label prediction and we take it as the evaluation metrics. The formula of Precision@k is:

$$Precision@k = \frac{|\{n_i | n_i \in N_o \cap N_p\}|}{|N_p|} \quad (11)$$

where N_p is the set of top k predicted anomalous nodes, N_o is the set of observed anomalous nodes. The results show the model precision at k value.

Table 2. Evaluation results on Zhihu

Method information	Splitting ratio					
	30%	*40%*	*50%*	*60%*	*70%*	*80%*
deepwalk + LR	0.71	0.73	0.73	0.74	0.75	0.76
node2vec + LR	0.82	0.814	0.823	0.818	0.83	0.83
LINE + LR	0.57	0.57	0.57	0.57	0.64	0.65
deepwalk + CNN Text + LR	0.754	0.784	0.804	0.786	0.797	0.820
node2vec + CNN Text + LR	**0.827**	0.828	0.862	0.848	0.855	0.841
LINE + CNN Text + LR	0.707	0.688	0.701	0.696	0.728	0.764
CAAD	0.792	**0.869**	**0.922**	**0.926**	**0.926**	**0.932**

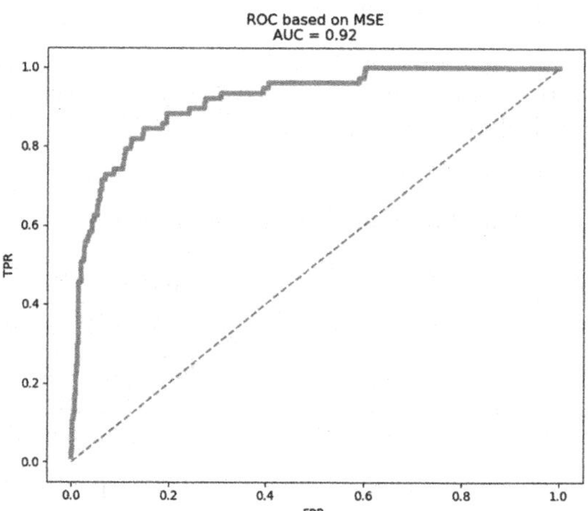

Fig. 4. ROC curve based on MSE AUC = 0.92

5.4 Experiment Results

The AUC values for anomaly detection with different methods on Zhihu are reported in Table 2. From Table 2, which shows: (1) on Zhihu dataset, CAAD significantly outperforms the baseline method including the combination of the traditional embedding method and the logistic regression method. (2) An embedded representation with text content is better than the approach that only uses network structure. (3) Different embedding methods with different splitting ratios will affect the final abnormal judgment result. However, by adding more training data, the result of abnormal detection will be improved a little, because the segmentation boundary between normal and abnormal links can be roughly predicted by embedding a small number of links. (4) Moreover, SMOTE technique can improve the performance of anomaly detection.

Figure 4 shows the receiver operating characteristic curve based on MSE. Figure 5 shows the scatter diagram of reconstruction mean square error (MSE)

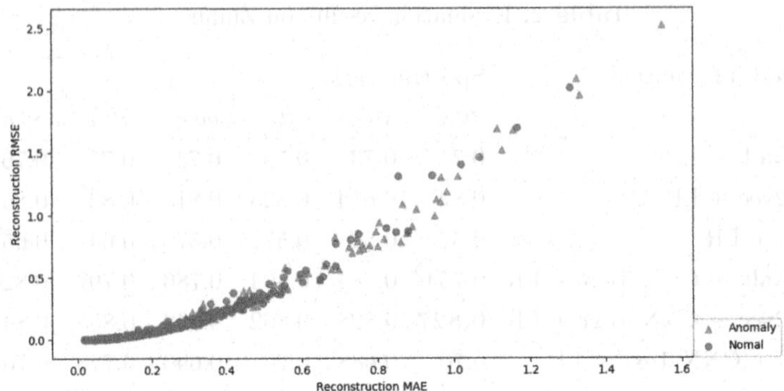

Fig. 5. Scatter diagram on reconstruction MAE and reconstruction RMSE

and reconstruction root mean square error (RMSE). As shown in Fig. 5, RMSE and MSE are obviously larger than normal.

Table 3. Evaluation results on Cora

Method information	Splitting ratio			
	80%	70%	60%	50%
Line 1st-order + LR	0.717	0.719	0.707	0.697
Line 1st-order + 2nd-order + LR	0.721	0.702	0.712	0.730
deepwalk + LR	0.762	0.702	0.812	0.726
node2vec + LR	0.769	0.761	0.721	0.643
Line 1st-order + CNN Text + LR	0.692	0.794	0.766	0.751
Line 1st-order + 2nd-order + CNN Text + LR	0.827	0.689	0.704	0.742
deepwalk + CNN Text + LR	0.845	**0.842**	0.807	**0.772**
node2vec + CNN Text + LR	0.835	0.813	0.819	0.583
CAAD	**0.862**	0.834	**0.832**	0.720

In Fig. 6 each line represents a Twitter precision at K of anomaly detection method: two-stage classifier methods and isolation forests. The precision at K of two-stage classifier with different first stage models is obtained by certain experiments, such as Random Forest, Logistic Regression, Adaboost and Bagging. Improvements in the precision at K for CAAD are presented in Fig. 6. The precision at 10, 50, 100, 200, 300, 400 and 500 is 0.1, 0.3, 0.22, 0.165, 0.13, 0.14 and 0.15, respectively. Therefore, experiments showed that CAAD can achieve improvements on synthetic and real data, compared with the other detection methods.

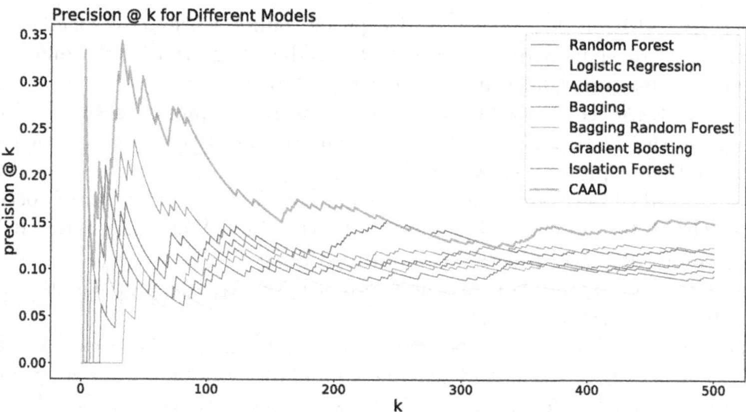

Fig. 6. precision@k for different models

6 Conclusion

In this paper, we proposed a new network anomaly detection method, called CAAD, based on network embedding and auto-encoder models. CAAD learns structural embedding and content embedding of nodes with CNN. On the basis of the learned low dimensional vectors of nodes, an auto-encoder model was proposed to perform anomaly detection based on reconstruction errors. Experiment results show that CAAD can achieve better improvement than the combination of the traditional embedding methods and classification models.

Acknowledgements. This work is supported by the National Key Research and Development Program of China, the National Natural Science Foundation of China under grants U1911401, 61772501, and U1836206, and the GF Innovative Research Program. We are grateful to the editor and anonymous reviewers for their constructive comments.

References

1. Yu, W., et al.: Netwalk: a flexible deep embedding approach for anomaly detection in dynamic networks. In: Proceedings of the 24th ACM SIGKDD International Conference on Knowledge Discovery & Data Mining (2018)
2. Hu, R., et al.: An embedding approach to anomaly detection. In: IEEE 32nd International Conference on Data Engineering (ICDE). IEEE (2016)
3. Gao, J., et al.: A spectral framework for detecting inconsistency across multi-source object relationships. In: 2011 IEEE 11th International Conference on Data Mining. IEEE (2011)
4. Fischer, A., et al.: Virtual network embedding: a survey. IEEE Commun. Surv. Tutorials **15**(4), 1888–1906 (2013)
5. Akoglu, L., McGlohon, M., Faloutsos, C.: Oddball: spotting anomalies in weighted graphs. In: Zaki, M.J., Yu, J.X., Ravindran, B., Pudi, V. (eds.) PAKDD 2010. LNCS (LNAI), vol. 6119, pp. 410–421. Springer, Heidelberg (2010). https://doi.org/10.1007/978-3-642-13672-6_40

6. Perozzi, B., Al-Rfou, R., Skiena, S.: Deepwalk: online learning of social representations. In: Proceedings of the 20th ACM SIGKDD International Conference on Knowledge Discovery and Data Mining. ACM (2014)
7. Grover, A., Leskovec, J.: node2vec: scalable feature learning for networks. In: Proceedings of the 22nd ACM SIGKDD International Conference on Knowledge Discovery and Data Mining. ACM (2016)
8. Tang, J., et al.: Line: large-scale information network embedding. In: Proceedings of the 24th International Conference on World Wide Web. International World Wide Web Conferences Steering Committee (2015)
9. Chawla, N.V., et al.: SMOTE: synthetic minority over-sampling technique. J. Artif. Intell. Res. **16**, 321–357 (2002)
10. Prakash, B.A., Sridharan, A., Seshadri, M., Machiraju, S., Faloutsos, C.: EigenSpokes: surprising patterns and scalable community chipping in large graphs. In: Zaki, M.J., Yu, J.X., Ravindran, B., Pudi, V. (eds.) PAKDD 2010. LNCS (LNAI), vol. 6119, pp. 435–448. Springer, Heidelberg (2010). https://doi.org/10.1007/978-3-642-13672-6_42
11. Jiang, M., et al.: A general suspiciousness metric for dense blocks in multimodal data. In: IEEE International Conference on Data Mining. IEEE (2015)
12. Shin, K., Hooi, B., Faloutsos, C.: M-Zoom: fast dense-block detection in tensors with quality guarantees. In: Frasconi, P., Landwehr, N., Manco, G., Vreeken, J. (eds.) ECML PKDD 2016. LNCS (LNAI), vol. 9851, pp. 264–280. Springer, Cham (2016). https://doi.org/10.1007/978-3-319-46128-1_17
13. Shin, K., et al.: D-cube: dense-block detection in terabyte-scale tensors. In: Proceedings of the Tenth ACM International Conference on Web Search and Data Mining. ACM (2017)
14. Tong, H., Lin, C.-Y.: Non-negative residual matrix factorization with application to graph anomaly detection. In: Proceedings of the 2011 SIAM International Conference on Data Mining. Society for Industrial and Applied Mathematics (2011)
15. Tu, C., et al.: Cane: context-aware network embedding for relation modeling. In: Proceedings of the 55th Annual Meeting of the Association for Computational Linguistics (Volume 1: Long Papers) (2017)
16. Kagan, D., Elovichi, Y., Fire, M.: Generic anomalous vertices detection utilizing a link prediction algorithm. Social Netw. Anal. Min. **8**(1), 27 (2018)
17. Kagan, D., Fire, M., Elovici, Y.: Unsupervised anomalous vertices detection utilizing link prediction algorithms. arXiv preprint arXiv:1610.07525 (2016)
18. Liu, F.T., Ting, K.M., Zhou, Z.-H.: Isolation forest. In: Eighth IEEE International Conference on Data Mining. IEEE (2008)

Blockchain Based Data Integrity Verification for Cloud Storage with T-Merkle Tree

Kai He[1,2], Jiaoli Shi[3(✉)], Chunxiao Huang[1], and Xinrong Hu[1,2]

[1] School of Math and Computer Science, Wuhan Textile University, Wuhan, China
[2] Hubei Clothing Information Engineering Technology Research Center, Wuhan, China
[3] School of Information Science and Technology, Jiujiang University, Jiujiang, China
shijiaoli@whu.edu.cn

Abstract. Blockchain based verification has recently been studied as a new emerging approach for cloud data integrity protection without a centralized third party auditor. Existing schemes are mostly limited to scalability barrier and search efficiency of blockchain, which restricts their capability to support large-scale application. To address the problem above, we propose a blockchain based integrity verification scheme for large-scale cloud data using T-Merkle Tree. In our design, data tags are generated by ZSS short signature and stored on blockchain, and a new verification method based on ZSS short signature is proposed. The integrity of cloud data can be efficiently and undeniably verified with the property of bilinear pairing by offloading the computation from verifier to prover. Furthermore, a new blockchain storage structure called T-Merkle tree and its search algorithm is designed to improves the storage utilization and supports binary search in a block. Finally, a prototype system based on Hyperledger Fabric is implemented to evaluate our scheme. Theoretic analysis and experimental results demonstrate the security and efficiency of our proposed scheme.

Keywords: Cloud storage · Blockchain · Data integrity verification · Blockchain query

1 Introduction

Cloud computing paradigm provides a elastic way to meet clients' dynamic demands through pay-on-demand. By remotely storing data in cloud, clients enjoy a number of advantages such as free from the burden of storage management, access at anytime and anywhere, avoidance of capital expenditure on hardware/software and staff maintenance. However, losing local control over their data may bring new and challenging security threats towards outsourced data. One big concern is how to verify the integrity of outsourced data, which is important given the fact that data stored in an untrusted cloud can be easily

M. Qiu (Ed.): ICA3PP 2020, LNCS 12454, pp. 65–80, 2020.
https://doi.org/10.1007/978-3-030-60248-2_5

lost or corrupted, due to hardware failures, human errors and malicious attacks. Data loss or server corruption with current cloud service providers was reported from time to time [1].

To check the integrity of cloud data, public verification is widely studied in recent years. In public verification scenario, each data shard is attached with a tag, and the integrity check relies on the correctness of these tags. The Third Party Auditor (TPA) is introduced to check the data integrity on behalf of client periodically. When verification is performed, the TPA issues a challenge with some random selected data shard indexes to cloud server, then the cloud server will compute the proofs based on the stored data and tags and return them to the TPA. Finally, the TPA can check the data integrity according to these proofs. For public verification in cloud storage, several benefits and essential requirements have been discussed in previous schemes [2]. However, it still suffers a range of drawbacks. First, the assumption of TPA being fully trusted by clients is unrealistic, since centralized TPA is vulnerable to both inside and outside security threats and may conspire with cloud server. Second, The TPA may delay previously the agreed tasks when receiving a large number of verification tasks in a short period, which beyond his capability. Therefore, the TPA may not be completely trusted [3].

Fortunately, the emergence of blockchain technology offers a new insight to address the above problems for its decentralization, transparency, tamper-proof and security [4]. Nevertheless, the design of building a decentralized verification scheme without TPA is still a big challenging. If storing client's data on blockchain directly, there is no need for data tags [5]. But it may limit its large-scale applications, since there is a significant obstacle of capacity and scalability for blockchain implemented based on Ethereum or Hyperledger Fabric. Besides, it's inconvenient for later utilization such as data access or share. Recently, Wang *et al.* proposed a blockchain-based private PDP scheme to check remote data integrity, they tried to store tags on blockchain while data is still stored in cloud [6]. However, their scheme has to query blockchain to get challenged tags during verification, which is inefficient and unpractical when the blockchain grows huge. Furthermore, data dynamic operations have not been investigated in their proposed scheme.

In this work, we proposed a blockchain based verification scheme for large-scale cloud data with T-Merkle Tree, aiming to ensure data integrity without TPA and support efficient query on blockchain. In our design, clients upload data to cloud server and store data tags on blockchain, and the data integrity can by checked by verifier efficiently and credibly. The main contributions made in our work can be summarized as follows:

Firstly, we propose a decentralized verification framework for cloud storage to get rid of a centralized TPA based on blockchain technology. It eliminates the trust requirements on TPA and increases the reliability of verification result. The data tags are generated by ZSS short signature [7], and a new verification method based on ZSS short signature is proposed. The efficiency of verification is improved by shifting computation from verifier to cloud server and blockchain with the property bilinear pairing.

Secondly, we design a new blockchain storage structure called T-Merkle tree and its searching algorithm, which improves the storage utilization and supports binary search on blockchain.

Thirdly, We have implemented a prototype system to evaluate the performance of the proposed scheme. Extensive security analysis shows that our scheme is secure and experiment results demonstrate our scheme is efficient.

In the rest of this paper, we discuss related work in Sect. 2, and introduce the system model in Sect. 3. We design an verification scheme in Sect. 4 and analyze its security properties in Sect. 5, and further evaluate the scheme through simulation studies in Sect. 6. Section 7 concludes the paper.

2 Related Work

2.1 Centralized Data Integrity Verification

Data integrity verification schemes are mainly divided into two types: Provable Data Possession (PDP) and Proofs of Retrievability (POR). Juels *et al.* [8] proposed a proof of retrievability (POR) model for protecting remote data integrity. Their scheme uses spot-checking and error correcting code to ensure both ownership and retrievability of files on profile service systems. Ateniese *et al.* [9] gave a definition of provable data possession (PDP) model for guaranteeing the possession of data files on untrusted servers. Wang et al. [2] introduced a third party auditor to check the integrity of cloud data on behalf of data owners. Based on these works, researchers have made a lot of effort on public auditing from several aspects such as privacy-preserve auditing, dynamic auditing and batch auditing. Wang *et al.* [10] implemented a PDP scheme that supports data dynamic operations by using the Merkle tree to ensure the correctness of data block. Yang *et al.* [11] protected the data privacy against TPA by encrypting the proofs and exploited index hash tables to support dynamic data during the public auditing process and further extended their work to support batch auditing.

Later on, many third party verification schemes have been proposed to focus on data share, availability, malicious or pretended third-party auditor and so on. Wang *et al.* [12] and Yuan *et al.* [13] worked on secure data verification of shared data storage with user revocation. Chen et al. [14] proposed schemes for regenerating code based cloud storage, which enables a client to easily check the integrity of random subsets of outsourced data against corruptions. Liu *et al.* [15] proposed an authorized public auditing method to dispel threats of unaccredited audit challenges from malicious or pretended auditors, and support fine-grained dynamic data update requests on variable sized file blocks.

2.2 Blockchain Based Data Integrity Verification

To avoid integrity verification relying on TPA, Liu *et al.* [16] proposed a blockchain based scheme for IoT data integrity service. However, they only implemented the fundamental function in a small scale. Liang *et al.* [17] designed an architecture to verify cloud data provenance, by adding the provenance data into blockchain transactions. Wang *et al.* [18] proposed a decentralized model to resolve the single point of trust problem and allow clients to trace the history of their data. Wang *et al.* [19] proposed a blockchain based data integrity scheme for large-scale IoT data to deal with the problems of large computational and communication overhead. The size of blockchain will increase fast since it needs several chains. Yu *et al.* [20] proposed a decentralized big data auditing scheme for smart city environments, they designed an blockchain instantiation called data auditing blockchain (DAB) that collects auditing proofs. In summary, most of existing blockchain based data integrity verification schemes focus on avoiding centralized TPA. In this paper, we focus on the issue of blockchain query and storage scalability.

3 Definitions and Preliminaries

3.1 System Model

We present a blockchain based verification system for cloud storage as described in Fig. 1. The system includes three different entities: Client, Cloud Server (CS) and Blockchain (BC). Client has a large volume of data to be stored in the cloud. They can be either individual consumer or corporation. Cloud Server is managed by cloud service provider and has huge storage space and computation resource to maintain Client's data. Blockchain stores the data signatures for Client, which is used to check data integrity.

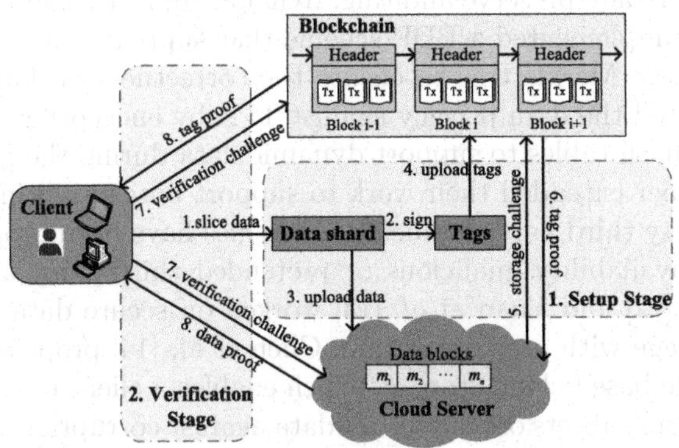

Fig. 1. Blockchain based verification model for cloud storage

The process of our scheme consists of two stages: setup stage and verification stage.

- *Setup stage:* (1) Client slices the data file into a set of data shards with equal size. (2) Client calculates a data tag for each data shard. (3) Client uploads the data shards to Cloud Server. (4) Client stores the data tags on Blockchain. (5) Cloud Server send a challenge to the Blockchain. (6) Blockchain response a tag proof, then Cloud Server verifies the integrity of the data which will be stored in cloud.
- *Verification stage:* (7) Client picks a challenge with a series of random values along with the data shard index set, and sends to Cloud Server and blockchain respectively. (8) Cloud Server computes the data proof and sends it to Client. Meanwhile, Blockchain also computes the tag proof and sends it to Client. If the proofs pass the verification, it means that the cloud data is intact; otherwise, the cloud data is thought to be not intact.

Due to the non-tamperability of blockchain, any Client or CS cannot modify the root or data tags stored on the blockchain, which makes integrity verification more believable. For simplicity, we take Client as the role of data verifier. In fact, the verification stage can be carried out by any entity, *i.e.*, client, cloud server providers, data consumer, or other third parties. Anyone who with the verification parameters can execute sampling verification periodically or when data is accessed.

3.2 Threat Model and Design Goals

In the adversary model, we consider threats from two different aspects: semi-trusted or untrusted servers and malicious client. The cloud server may not honestly store client's data and hide data loss or even discard the corrupted data to maintain its reputation. Client may intentionally upload incorrect data or tags to cheat the cloud server for compensation.

The proposed verification scheme aims to achieve the following goals: (1) *Decentralized verification.* Without relying on centralized TPAs, the data can be checked by any entity publicly. (2) *Low Blockchain storage.* The data tags stored on Blockchain with minimum storage cost. (3) *Efficient Blockchain Query.* Our scheme supports fast location of challenged tags through binary search rather than traverse the whole Blockchain. (4) *Efficiency.* The verification computation and communication overhead should be performed with minimum overhead.

3.3 Preliminaries

Blockchain realizes a decentralized, open, transparent, auditable, and tamper-proof record without the need of mutual trust, through data encryption, time stamping, and distributed consensus. Each operation in the blockchain is recorded by the way of a transaction. Multiple transactions constitutes a block, and lots of blocks are connected together to compose a blockchain. Blockchain consists of Blockchain header and Blockchain body. The header contains the hash value of the previous block, Timestamp (block creation time), Merkle root and Nonce (a random value). We add a tag index range field in our design to improve query efficiency. The Merkle hash tree comprises all data tags by transactions. In this way, all data tags are stored in blocks and the integrity of them can be ensured by the Merkle root. Since Merkle tree only stores hash value of data tags in leaf node, the data structure will be heavy and the efficiency of tag query will be low when the number of tags is large. To solve this problem, we construct the Blockchain body through T-Merkle tree.

4 The Proposed Blockchain Based Data Integrity Verification Scheme

In this section, we firstly describe the structure of T-Merkle tree. Secondly, we present the blockchain based data integrity verification scheme. Then, we give out the batch verification. Finally, we discuss our design from several aspects.

4.1 Structure of T-Merkle Tree

In order to reduce storage cost and increase the query efficiency, we improve the Merkle tree from three aspects: (1) Data tags are stored on each node rather than leaf node, that is, each node stores multiple successive tags. (2) Tag index is attached to each tag to support fast query. (3) A index range field which stands for the minimum and maximum index value of current node is inserted into the block header.

Definition 1. *A T-Merkle tree combines the characteristics of T-tree and Merkle hash tree. A node v_i of T-Merkle tree consists of minimum index value Min_i, maximum index value Max_i, a consecutive index and tag set $\{j, tag_j\}_{j\in[k]}$ and hash value $H(v_i)$. k is the number of data tags in each node. The node structure is described in Fig. 2.*

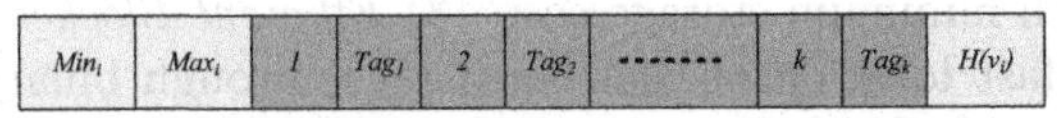

Fig. 2. Node structure of T-Merkle tree

The node hash value $H(v_i)$ is computed from the hash value of data node tags and the hash value of its children, that is,

$$H(v_i) = \begin{cases} h(h(v_i)||H(rchild)||H(lchild)) & \exists \text{ two children} \\ h(h(v_i)||H(rchild)) & \exists \text{ right child} \\ h(h(v_i)||H(lchild)) & \exists \text{ left child} \\ h(v_i) & \text{leaf node} \end{cases}$$

$h(v_i)$ is a hash value of current node, which equals to $h(h(tag_1)||h(tag_2)||...||h(tag_k))$, where h is a hash function and $||$ denotes concatenation operation.

When generating T-Merkle tree, we firstly construct a T-tree with the data tags, and then compute the hash value of each node. Figure 3 depicts an example of T-Merkle tree based on 14 data tags. For the sake of simplicity, we set the tag index from 1 to 14. Each node has two data tags means $k = 2$. The verifier with the authentic $H(root)$ requests for $\{Tag_5, Tag_6\}$ and requires the authentication of the received tags. The prover provides the verifier with the auxiliary authentication information $\Omega = <H(c), H(b), h(Tag_3), h(Tag_4), h(Tag_7), h(Tag_8)>$. The verifier can check $\{Tag_5, Tag_6\}$ by calculating $H(a) = h(h(Tag_5)||h(Tag_6)||H(c)||h(Tag_3|)|h(Tag_4))$, $H(root) = h(h(Tag_7)||h(Tag_8)||H(a)||H(b))$, and then checking if the calculated $H(root)$ is the same as the stored one. Thus, the authentication nature of Merkle hash tree is kept in our T-Merkle tree.

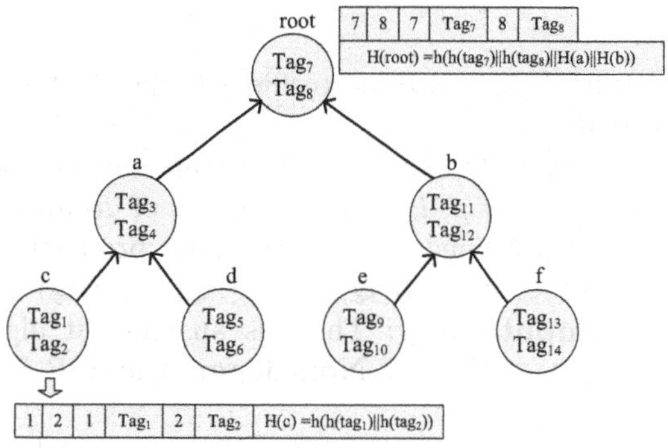

Fig. 3. Example of T-Merkle tree

To support efficient query, we embed a index range field of current block body into block header. When querying a tag with index *ind*, we traverse the blockchain from last block to old block by comparing index *ind* with index range field in block header. If index *ind* is in the some block, we find out the data tag through binary search in T-Merkle tree. The search algorithm of T-Merkle tree is shown in Algorithm 1. For example in Fig. 3, if we want to find the tag with index 5, we first compare 5 with minimum index value 7 and maximum index value 8 of root node. Since 5 is less than minimum index value 7, we access the left child of root node. As we known 5 is greater than the maximum index value 4 of node a, we then access its right child. Finally we get Tag_5 in node d.

4.2 Construction of the Proposed Scheme

In our scheme, data tags store on the blockchain, which makes the tags more safety due to the non- tamperability of blockchain. The tags are computed by ZSS short signature [7]. The ZSS scheme is based on the bilinear pairings, uses general cryptographic hash functions such as SHA-1 or MD5, and does not require special

Algorithm 1. Search algorithm of T-Merkle tree

Input: index *ind*
Output: data tag Tag_{ind}
foreach *block* **in** Blockchain
  if *ind* **in** index range of *block* **then**
    Acesss T-Merkle root of *block* , set $p = root$.
    if $p[Min] \leq ind \leq p[Max]$ **then**
      get Tag_{ind} in node p by comparing index value.
    else if $ind < p[Min]$ **then**
      set $p = p[leftchild]$.
    else // $ind > p[Max]$
      set $p = p[rightchild]$.
  else
    access next block.
end

hash functions unlike BLS. It is also more efficient than BLS signature, since it needs less pairing operations.

Let G_1 be a Gap Diffie-Hellman (GDH) group and G_2 be multiplicative cyclic groups with prime order p, P is the generator of group G_1. The mapping $e : G_1 \times G_1 \rightarrow G_2$ be a bilinear map with some properties. (1) **Bilinearity:**$e(aP, bQ) = e(P, Q)^{ab}$, $e(P + R, Q) = e(P, Q) \cdot e(R, Q)$ for all $P, Q, R \in G_1, a, b \in Z_p$. (2) **Computability:** There is an efficient algorithm to compute $e(P, Q)$ for all $P, Q \in G_1$. (3) **Non-degeneracy:** P is non-degenerate if $e(P, P) \neq 1$.

The proposed scheme includes two stages :setup stage and verification stage.

Setup Stage: Firstly, Client randomly chooses a value $sk \in Z_p$ as his private tag key and computes $pk = skP$ as public key. The private key can't calculate from the public key under the Inv-CDHP assumption.

Secondly, Client slices the data file into n shards with equal size as $F = \{m_1, m_2, ..., m_n\}$, and then generates a data tag for each shard m_i as:

$$Tag_i = \frac{1}{\mathcal{H}(m_i) + sk} P$$

Where $\mathcal{H}$ is a short signature security hash function such as SHA-1 or MD5. The tag set of data file F is $T = \{Tag_1, Tag_2, ..., Tag_n\}$.

Finally, Client uploads the data shard set F to cloud storage server and sends the tag set T to Blockchain. The tags are stored on the Blockchain through T-Merkle tree. Cloud server will verify the data shards before agreeing to store the data, the verification method is similar to as verification stage described below. The detail process of setup stage is shown in Fig. 4(a).

Verification Stage: Client (as verifier) randomly chooses a set of challenge index-coefficient $Chall = \{i, u_i\}_{i \in I}$, where $I = \{s_1, s_2, ..., s_c\}$, and c is a random block indexes subset of $[1, n]$ and u_i is a random value in Z_p. Client sends the *Chall* to the Cloud Server and Blockchain.

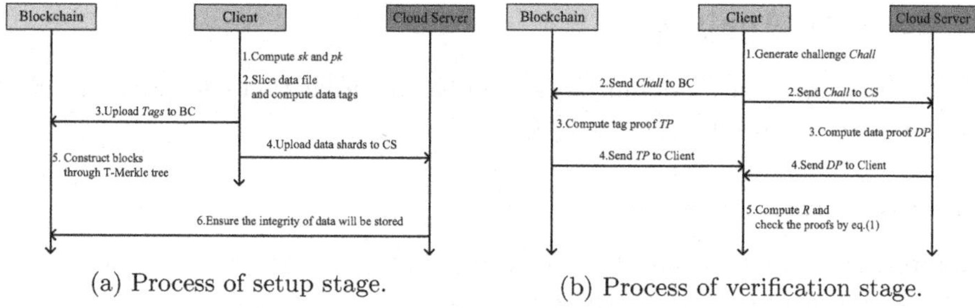

(a) Process of setup stage.

(b) Process of verification stage.

Fig. 4. Framework of proposed scheme.

After receiving the verification request $Chall$, Cloud Server generates the data proof DP and encrypts it with bilinear map as:

$$DP = e(\sum_{i \in I} u_i \mathcal{H}(m_i)P, P)$$

At the same time, Blockchain gets the challenged tags by search algorithm of T-Merkle tree, and then computes the tag proof as:

$$TP = e(\sum_{i \in I} \frac{u_i}{Tag_i} P^2, P)$$

Cloud Server and Blockchain returns the data proof DP and tag proof TP to Client separately.

Upon receiving the data proof DP and tag proof TP, Client firstly calculates a value $R = \sum_{i \in I} u_i pk$ with the public key pk and challenge $Chall$, and then checks the proofs as

$$TP = DP \cdot e(R, P) \tag{1}$$

If the Eq. (1) holds, it means the data file on the Cloud Servers is intact. The process of verification stage is illustrated in Fig. 4(b).

4.3 Discussion on Decentralized Verification

In centralized verification scheme, both data corruption and tag corruption can lead to verification failure. However, the data tags stored on Blockchain in our decentralized scheme is tamper-proof. When verification fails, we can know that the cloud data must be corrupted. By let Cloud Server and Blockchain encrypt the proofs with bilinear map, we move the verification computation from Client to Cloud Server and Blockchain. Moreover, Client can still check the proof without decryption. Since the public key and tags are unknown to Cloud Server, it can also help Client to keep anonymous from Cloud Server. Thus, the security and efficiency of the proposed scheme can be improved.

5 Security Analysis

We analyzed the security of our blockchain based verification scheme from three aspects: *correctness* and *unforgeability.*

5.1 Correctness

Theorem 1. *Given the data file F and its data tags T, the verifier can correctly check the integrity of the outsourced data.*

Prove the correctness of our scheme is equivalent to prove the correctness of verification Equation (1). Based on the properties of bilinear maps, the correctness of Equation (1) can be proved as follows

$$\begin{aligned}
&TP \\
&= e(\sum_{i\in I} \frac{u_i}{Tag_i} P^2, P) \\
&= e(\sum_{i\in I} u_i(\mathcal{H}(m_i) + sk)P, P) \\
&= e(\sum_{i\in I} u_i\mathcal{H}(m_i)P, P) \cdot e(\sum_{i\in I} v_i skP, P) \\
&= DP \cdot e(R, P)
\end{aligned}$$

5.2 Unforgeability

Theorem 2. *For the Cloud Server and Blockchain, it is computationally infeasible to forge a data and tag proof under our scheme.*

Following the adversary model and security game defined in [8,9], our scheme can be proved as follows.

Given the same challenge $Chall = \{i, u_i\}_{i\in I}$, the correct proof should be DP and TP. Suppose Cloud Server and Blockchain forge an incorrect proof DP' and TP'. If this incorrect proof can still pass the verification. Then, according to Equation (1), we have:

$TP' = DP' \cdot e(R, P)$.

Since (DP, TP) is a correct proof, we have:

$TP = DP \cdot e(R, P)$.

we can get:

$TP' \cdot TP^{-1} = DP' \cdot DP^{-1}$.

In our Scheme, it is impossible to forge a tag proof by Blockchain. We only consider that there are some attackers or malicious Cloud Server, tampering with the cloud data. this is, $DP' \neq DP$ and $TP' = TP$. Based on the properties of bilinear maps, we can learn that:

$$e(\sum_{i\in I} u_i\mathcal{H}(m_i')P, P) \cdot e(\sum_{i\in I} u_i\mathcal{H}(m_i)P, P)^{-1} = 1$$

By set $\Delta m = \sum_{i \in I} u_i \mathcal{H}(m_i') - \sum_{i \in I} u_i \mathcal{H}(m_i)$ to make:

$$e(\Delta m P, P) = e(P, P)^{\Delta m} = 1.$$

Since, $\Delta m \neq 0$, we have $e(P, P) = 1$, Which contradicts the assumption of P is non-degenerate if $e(P, P) \neq 1$.

We further assume that Blockchain is under attacked, the attacker or malicious Blockchain node forges a tag proof, namely, $TP' \neq TP$. Similar to above, without the private key sk, it is impossible to generate a forged m_i' and make equation $\frac{1}{\mathcal{H}(m_i')+sk} P = \frac{1}{\mathcal{H}(m_i)+sk} P$ holds.

From the analysis above, we can know that our verification scheme can resist malicious attacks.

6 Performance Evaluation

6.1 Performance Analysis

We now make a full performance comparison between our scheme and the state-of-the-art that includes both Yue's scheme [5] and Wang's scheme [19], from the aspects of storage structure of blockchain and verification efficiency. Yue's scheme [5] constructs a multi-branch Merkle tree based on data shards and uses the root of the hash tree stored on blockchain to verify data integrity. Wang's scheme [19] is more similarity to our scheme, which both store data tag on blockchain rather than the hash of data shard, but the structure of tree is different.

1) *Storage structure of blockchain.*

We compare the storage structure of blockchain from three aspects: generation complexity, storage overhead and query complexity. The storage structure of these three schemes is multi-branch Merkle tree, Merkle tree and T-Merkle tree respectively. The generation complexity refers to the number of hash operations during building tree. In our scheme, we will compute hash operations on each data shard and each node of T-Merkle tree. Similarity, the storage overhead is also related to the number of tree nodes. However, our scheme store data on each node, so the number of tree nodes is less than the other two schemes. Since we add some fields of index key in each node to support efficient search, the query complexity is also lower than others. Table 1 gives a detail comparison among three schemes.

2) *Verification efficiency.*

The verification efficiency is compared by computation and communication cost. The computation cost is estimated in terms of cryptographic operations. The communication cost is assessed according to the size of data proof and tag proof. From the Table 2, we can see that the communication cost of the proof in our scheme less than other two schemes, and the computation complexity of our scheme and Wang's Scheme [19] are both $O(c)$, while Yue's Scheme [5] is

Table 1. Comparison of storage structure.

Scheme	Generation complexity	Storage overhead	Query complexity
Yue's scheme [5]	$sum(Y)$	$(sum(Y)+n)\,\lvert p\rvert$	$O(log_m{}^n)$
Wang's scheme [19]	$sum(W)$	$(sum(W)+n)\,\lvert p\rvert$	$O(nlog_2{}^n)$
Our scheme	$n+\frac{n}{k}$	$(n+\frac{n}{k})\,\lvert p\rvert+(n+\frac{3n}{k})\,\lvert id\rvert$	$O(log_2{}^{\frac{n}{k}}+log_2{}^k)$

n is the total number of shards in the data file. m is the branch number of tree.k is the number of tags in a T-Merkle tree node.$|id|$ is the size of index. $|p|$ is group size. let $sum(Y)=m^0+m^1+...+m^{log_m{}^n}$ and $sum(W)=2^0+2^1+...+2^{log_2{}^n}$.

$O(clog_m{}^n)$. Furthermore, our scheme incurs less computation cost on the verifier side. It is because that we reduce the computing loads of the verifier by moving to the Cloud Server and Blockchain, which have stronger computation ability than the verifier.

Table 2. Comparison of verification efficiency.

Scheme	Communication cost		Computation cost		
	CS	BC	CS	BC	Verifier
Yue's scheme [5]	$c\,\lvert p\rvert$	$clog_m{}^n\,\lvert p\rvert$	$c\mathcal{H}$	$clog_m{}^n\mathcal{H}$	–
Wang's scheme [19]	$3\,\lvert p\rvert$	$c\,\lvert p\rvert$	$c\mathcal{A}+c\mathcal{H}$	$(c+1)\mathcal{A}+2\mathcal{M}$	$3\mathcal{P}$
Our scheme	$\lvert p\rvert$	$\lvert p\rvert$	$c\mathcal{A}+c\mathcal{H}+\mathcal{P}$	$c\mathcal{A}+2\mathcal{M}+\mathcal{P}$	$\mathcal{P}$

n is the total number of shards in the data file. m is the branch number of tree.c is the number of challenged shards. $\mathcal{A}$ denotes one addition in G, $\mathcal{M}$ denotes one multiplication in G $\mathcal{P}$ denotes one pairing operation, and $\mathcal{H}$ denotes one hashing operation.

6.2 Experimental Results

We evaluate the verification service and storage service on the 64 bits Ubuntu 18.04 system with an Intel Core 8 processor at 2.11 GHz and 16G of RAM. We build a blockchain network on Hyperledger Fabric 1.4.0 since it is the most widely used blockchain platform which supports smart contract. Algorithms are implemented based on the Pairing-Based Cryptography (PBC) library (pbc-0.5.14). The parameter of PBC we used is a.param, the group order is 160 bits and the base field order is 512 bits. All simulation results are the mean of 50 trials.

1) Evaluation of different structures of Merkle tree.

Firstly, we compare the tree generation efficiency versus the number of data shards. During experiments, we set the number of tag in a node of our scheme is $k=20$ and the number of branch in Yue's scheme [5] is $m=8$. As shown in their work, eight-branching tree gets the best performance. Assuming the total number of shards is from 2048 to 65536 (2048, 4096, 8192, 16384, 32768, 65536). From Fig. 5(a), it can be seen the linear relationship between the generation time and the number of shards. But our T-Merkle tree is better than Wang's Merkle tree and Yue's eight-branching Merkle tree, since our T-Merkle tree decreases the number of node and the depth of tree under the same total number of shards.

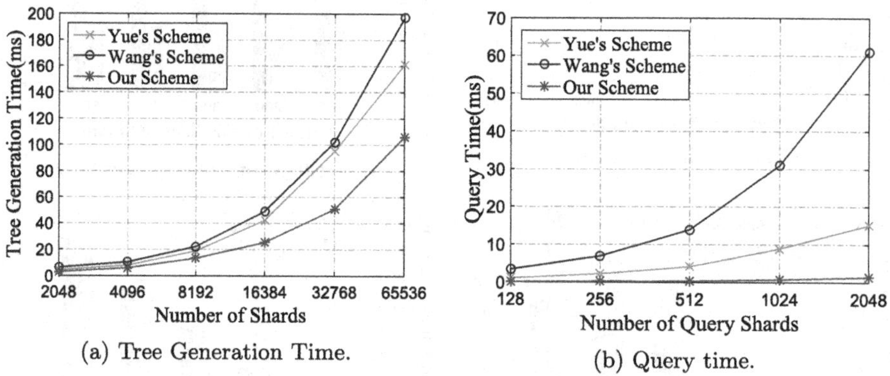

(a) Tree Generation Time.

(b) Query time.

Fig. 5. Comparison of different Merkle trees.

Secondly, we compare the query efficiency. As discussed before, the challenged shards in verification stage is very small, so we just set the number of test query shards from 128 to 2048 (128, 256, 512, 1024, 2048). From Fig. 5(b), we can see that the total query time of our scheme is low. When the number of query shard gets 1024, Our scheme only takes 1.4 ms, while Yue's scheme [5] needs 15.1ms and Wang's scheme [19] needs 54.7 ms. This is because our scheme supports binary scheme, while the other two schemes compare the hash value one by one when querying.

1) Evaluation of verification efficiency.

Firstly, we compare the verification time under different number of challenged shards. Yue's scheme I refers to the total number of shards is 65536 and Yue's scheme II refers to the total number of shards is 131072. Figure 6(a) shows the verification time increases when the number of challenged shards increases. The verification time of Yue's scheme is influenced by the total number of shards which is significantly inefficient when the data file is very large. When the challenged number is 500, the verification time in both our scheme and Wang' scheme are about 940 ms, no matter what the total number of shards is.

Secondly, we compare the verification time between our scheme and Wang's scheme from three sides: Cloud Server (CS), Blockchain (BC) and Verifier. From Fig. 6(b) to Fig. 6(d), we find out that the verification time on the Verifier side of our scheme is smaller than Wang's scheme, while the verification time on the BC side and CS side are larger. Since we move computation from Verifier side to BC and CS side, to improve the efficiency when the Verifier uses computation constrained device to check data integrity.

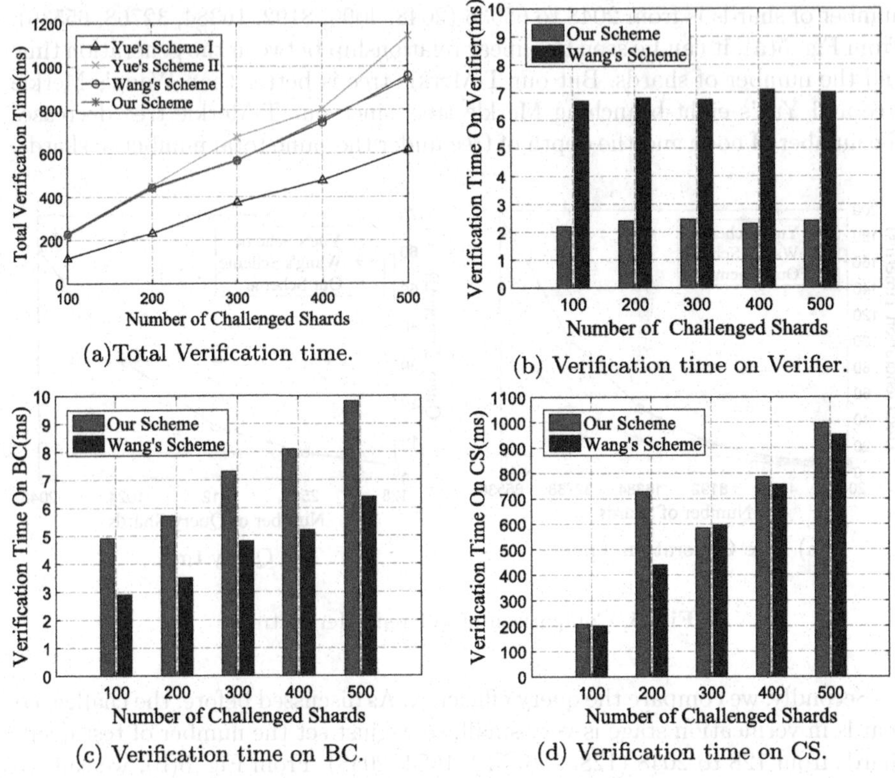

Fig. 6. Comparison of verification time under different challenged blocks.

7 Conclusion

We proposed a blockchain based verification scheme for large-scale cloud data. By utilizing the method of ZSS short signature and the technique of bilinear pairing, our scheme achieves efficiently data integrity verification by moving computation from verifier to cloud server and blockchain. We also designed a new T-Merkle tree to support binary search on Blockchain. The analysis and experimental results prove that our scheme is provably secure and efficient.

Acknowledgment. This work is supported by the Natural Science Foundation of Hubei Province (2018CBF109), Research Project of Hubei Provincial Department of Education (Q20191710), Project of Jiangxi Provincial Social Science Planning (17XW08), Science and Technology Research Project of Jiangxi Provincial of Education Department (GJJ180905). Team plan of scientific and technological innovation of outstanding youth in universities of Hubei province (T201807).

References

1. Amer, A., Long, D.D.E., Schwarz, S.J.T.: Reliability challenges for storing exabytes. In: International Conference on Networking and Communications (ICNC), vol. 2014, pp. 907–913 (2014)
2. Wang, C., Wang, Q., Ren, K., Lou, W.: Privacy-preserving public auditing for secure cloud storage. IEEE Trans. Comput. **62**(2), 362–375 (2013)
3. Zhang, Y., Xu, C., Lin, X., Shen, X.S.: Blockchain-based public integrity verification for cloud storage against procrastinating auditors. IEEE Trans. Cloud Comput. (2020)
4. Nakamoto, S.: Bitcoin: a peer-to-peer electronic cash system (2008). https://bitcoin.org/bitcoin.pdf
5. Yue, D., Li, R., Zhang, Y., Tian, W., Peng, C.: Blockchain based data integrity verification in P2P cloud storage. In: IEEE 24th International Conference on Parallel and Distributed Systems (ICPADS), pp. 561–568 (2018)
6. Wang, H., Wang, Q., He, D.: Blockchain-based private provable data possession. IEEE Trans. Dependable Secure Comput. (2020)
7. Zhang, F., Safavi-Naini, R., Susilo, W.: An efficient signature scheme from bilinear pairings and its applications. In: Bao, F., Deng, R., Zhou, J. (eds.) PKC 2004. LNCS, vol. 2947, pp. 277–290. Springer, Heidelberg (2004). https://doi.org/10.1007/978-3-540-24632-9_20
8. Juels, A., Kaliski, B.S.: PORs: proofs of retrievability for large files. In: Proceedings of ACM CCS, pp. 584–597 (2007)
9. Ateniese, G., et al.: Provable data possession at untrusted stores. In: Proceedings of ACM CCS 2007, pp. 598–610 (2007)
10. Wang, Q., Wang, C., Ren, K., Lou, W., Lin, J.: Enabling public auditability and data dynamics for storage security in cloud computing. IEEE Trans. Parallel Distrib. Syst. **22**(5), 847–859 (2011)
11. Yang, K., Jia, X.: An efficient and secure dynamic auditing protocol for data storage in cloud computing. IEEE Trans. Parallel Distrib. Syst. **24**(19), 1717–1726 (2013)
12. Wang, B., Li, B., Li, H.: Panda: public auditing for shared data with efficient user revocation in the cloud. IEEE Trans. Serv. Comput. **8**(1), 92–106 (2015)
13. Yuan, J., Yu, S.: Efficient public integrity checking for cloud data sharing with multi-user modification. In: Proceedings of IEEE INFOCOM, pp. 2121–2129 (2014)
14. Chen, C.H., Lee, P.C.: Enabling data integrity protection in regenerating-coding-based cloud storage: theory and implementation. IEEE Trans. Reliab. **64**(3), 840–851 (2015)
15. Liu, C., et al.: Authorized public auditing of dynamic big data storage on cloud with efficient verifiable fine-grained updates. IEEE Trans. Parallel Distrib. Syst. **25**(9), 2234–2244 (2014)
16. Liu, B., Yu, X.L., Chen, S., Xu, X., Zhu, L.: Blockchain based data integrity service framework for IoT data. In: Proceedings of ICWS, pp. 468–475, June 2017
17. Liang, X., Shetty, S., Tosh, D., Kamhoua, C., Kwiat, K., Njilla, L.: ProvChain: a blockchain-based data provenance architecture in cloud environment with enhanced privacy and availability. In: Proceedings of 17th IEEE/ACM International Symposium on Cluster, Cloud Grid Computing (CCGRID), Madrid, Spain, May 2017, pp. 468–477 (2017)
18. Wang, C., Chen, S., Feng, Z., Jiang, Y., Xue, X.: Block chain-based data audit and access control mechanism in service collaboration. In: Proceedings of ICWS, pp. 214–218 (2019)

19. Wang, H., Zhang, J.: Blockchain based data integrity verification for large-scale IoT data. IEEE Access **7**, 164996–165006 (2019)
20. Yu, H., Yang, Z., Sinnott, R.: Decentralized big data auditing for smart city environments leveraging blockchain technology. IEEE Access **7**, 6288–6296 (2019)

IM-ACS: An Access Control Scheme Supporting Informal Non-malleable Security in Mobile Media Sharing System

Anyuan Deng[1], Jiaoli Shi[1(✉)], Kai He[2], and Fang Xu[3]

[1] Jiujiang University, Jiujiang 332005, Jiangxi, China
shijiaoli@whu.edu.cn
[2] Wuhan Textile University, Wuhan 430073, Hubei, China
[3] Hubei Engineering University, Xiaogan 432000, Hubei, China

Abstract. Subscribers own their subscription credentials when they accesses mobile media. They cannot extend their subscription credentials without permission. We propose an Access Control Scheme supporting Informal non-Malleable security (IM-ACS) for mobile media. In our scheme, Attribute Based Encryption method is applied to bring de-couple and fine-grain features when publishers share their media data with subscribers. All-or-Nothing transfer method is used to ensure the efficiency when publishers encrypt their media data before upload it onto cloud server. The method of re-encryption on cloud is adopted to ensure that subscribers cannot obtain any acknowledge from the previous data accesses unless publisher allow them to. This scheme is analyzed to be secure.

Keywords: Mobile media · Access control · Attribute-Based Encryption · Non-malleable security

1 Introduction

In mobile media sharing system, publishers upload their mobile media online, and subscribers download any mobile media data by using their terminal such as a phone. The publish/subscribe model is widely accepted as a message passing method. The model removes coupling in space, time and synchronization between publishers and subscribers. It can be a basic technical support for media sharing system by providing data access anywhere at any time. But at the same time, there are some specific difficulties, such as its *huge data volume*, *various types* and *multiple replicas*. More than that, fine-grained access control must be considered when subscribers access media data. To be specific, all of unauthorized subscribers should not access the media data, and authorized subscribers just can access the part which is authorized for them.

ABE (Attribute-Based Encryption) method is suitable for media data fine-grained access control and media sharing system. As in Pub/Sub model, the ABE method also removes the tight coupling between data owners and users. CP-ABE (Ciphertext-Policy ABE, a kind of ABE) allows multiple subscribers access any piece of media data on-demand by using Data-Binding-Policy method. Meanwhile, media enterprises can be

M. Qiu (Ed.): ICA3PP 2020, LNCS 12454, pp. 81–92, 2020.
https://doi.org/10.1007/978-3-030-60248-2_6

saved from the construction of infrastructure, because cloud storage technology can provide data storage and access.

A subscribing credential is used to obtain authorized media data, and its malleable security must be considered. Using CP-ABE method, publishers encrypt media data to ciphertext. Then the ciphertext is uploaded onto cloud storage system, and accessed by subscribers with their credentials. These credentials should be malleable security according to the publishers' will. That is to say, some subscribers' credentials are malleable, but others are not. Some parts of a credential are malleable, but others are not. For example, a subscriber has paid a one-year fee for movies and got a credential during the year of 2019. He should not get the paid services of movies in the next year by modifying his old credential.

To address above challenge, this paper propose An Access Control Scheme Supporting Informal non-Malleable Security in Mobile Media Sharing System, which is abbreviated as IM-ACS. Section 2 presents the related works. Section 3 gives Background and Assumptions of this paper. Section 4 models the system of IM-ACS and its security requirements. Section 5 constructs the IM-ACS. Analysis and simulation of IM-ACS is given in Sect. 6 and Sect. 7 separately. Finally, Sect. 8 concludes this paper.

2 Related Works

In Ion [1], a scheme, based on ABE method, was implemented that supported *confidentiality for events and filters, filters that expressed very complex constraints on events,* and *did not require publishers and subscribers to share keys.* In Tariq [2], decouple feature was ensured based on IBE (Identity-Based Encryption). They completed users' identity authentication and content routing with data confidentiality. Teng [3] proposed a hierarchical attribute-based access control scheme with constant-size ciphertext. They also proved their scheme is of CCA2 security under the decisional q-Bilinear Diffie-Hellman Exponent assumption. Qiu [4] proposed a data fusion trust system for the multi-source and multi-formats of data exchange is presented by defining different trust levels. Their data fusion system can meet the latency requirements by using General Purpose Graphic Processing Unit (GPGPU) in their location-based PKI system. Tian [5] proposed a secure digital evidence framework using blockchain (Block-DEF). They stored the evidence information in the blockchain, and the evidence on a trusted storage platform. They also adopt the multi-signature technique to support the traceability and the privacy of evidence. However, all schemes above ignored the malleable security of subscribe credentials.

3 Background and Assumptions

3.1 NM-CCA2 and IND-CCA2

It is difficult to prove NM-CCA2, because that non-malleable attack is a computable problem, non-distinguish attack is a decision problem. Fortunately, NM-CCA2 is proved to be equal to IND-CCA2.

Furthermore, if there exists an encryption scheme that is IND-CPA-secure secure, then there exists another encryption scheme which remains m-bounded NM-CCA2-secure for any polynomial m [6].

3.2 ABE and Broadcast Encryption

ABE. It is constructed with four algorithms (*Setup, KeyGen, Encrypt* and *Decrypt*). The algorithm *Setup* generates public parameters PP and a Master Key MK. The algorithm *KeyGen* issues a Private Key SK for each subscriber according his attribute set *x*. The algorithm *Encrypt* produces a ciphertext *C by* associating the plaintext *M* with an access policy *P*. The algorithm *Decrypt* matches the policy *P* with user's private key.

$$\mathrm{p}(x) = \begin{cases} 1, & x \text{ matches } P \\ 0, & \text{others} \end{cases}$$

Authorized users will read the plaintext M only if $p(x) = 1$. ABE realizes data access control with the ensure of data confidentiality.

Broadcast Encryption. Multiple receivers can decrypt the same ciphertext which has been encrypted once. Publisher can definite a access policy *P*, and those subscribers, whose private key matches the policy *P*, can access the data. The number of authorized subscribers can be unlimited.

3.3 AONT(All-or-Nothing Transform)

AONT is constructed by a matrix-vector product.

$$\boldsymbol{G} * \boldsymbol{M} = \boldsymbol{Codeword}$$

Wherein, $\boldsymbol{G}$ means a generator or dispersal full rank matrix of k rows. $\boldsymbol{G}$ is invertible. $\boldsymbol{G}^{-1}$ means the reverse matrix of $\boldsymbol{G}$. That is, the plaintext $\boldsymbol{M}$ can be obtained from $\boldsymbol{Codeword}$ and $\boldsymbol{G}^{-1}$.

The matrix vector referred to above has the character of All-or-Nothing. If only someone gets any no less than $k - 1$ rows in $\boldsymbol{G}$, he can reconstruct $\boldsymbol{G}^{-1}$, and then obtain $\boldsymbol{M}$ from $\boldsymbol{Codeword}$.

4 Modeling

4.1 System Model of IM-ACS

Cloud storage architecture is considered as Broker in Pub/Sub system. IM-ACS model is designed as in Fig. 1 without considering the routing problem.

Our IM-ACS is constructed of four entities (publisher, subscriber, public cloud for data exchange, and attribute authority). Publisher issues events anytime and anywhere. Subscriber accesses events anytime and anywhere. Public cloud stores, matches and transfers events. Attribute Authority (abbreviated as AA) distributes authorities to event publish for publishers and issues access credentials for subscribers.

Wherein, AA is trusted, cloud is semi-trusted (that is, cloud server performs all request actions honestly, and is curious about the content of data).

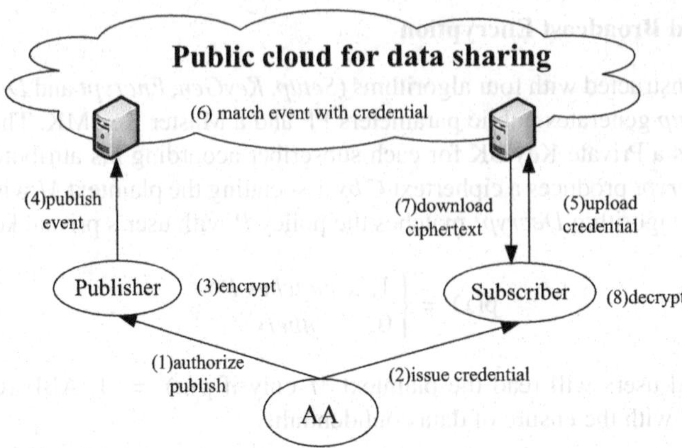

Fig. 1. System model of IM-ACS

4.2 Security Requirements

1) **Data Confidentiality.** Sensitive media data needs to be encrypted before being upload onto cloud, and unauthorized users, cloud servers or AA should not to access the data.
2) **Loose Coupling.** Publishers share media data with subscribers without prior key agreement. Publishers don't know who subscribes data event, and subscribers don't know who has published the data events.
3) **Collusion Resistance.** Either of authorized subscribers cannot obtain greater privileges by combining or modifying their private credentials.
4) **Informal Non-malleable Security.** Unauthorized subscribers cannot obtain more permissions by extending their credentials.

5 IM-ACS

5.1 Design of IM-ACS

(See Fig. 2)

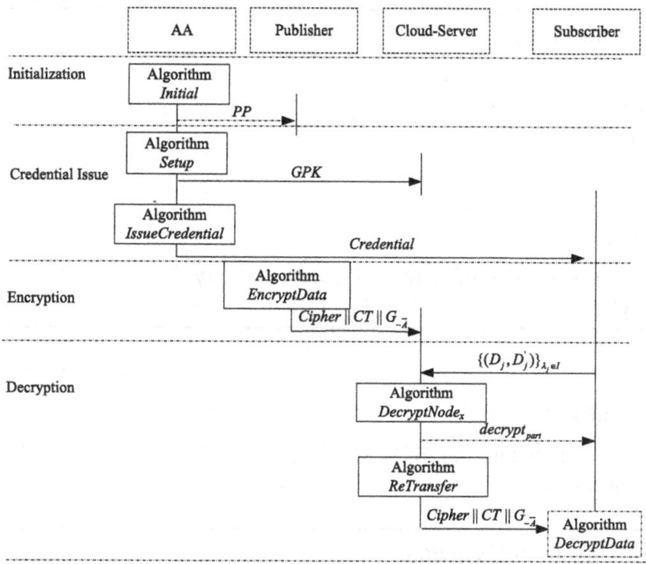

Fig. 2. Framework of IM-ACS

5.2 Construction of IM-ACS

Let G_m, G_n and G_x denote bilinear groups with the prime order *m*, *n* and *x* respectively, G_m and G_n are subgroups of G_x. Let *g* be a generator of G_x, φ be a generator of G_n. Let $e(\cdot,\cdot)$ be the bilinear map. $e(g,g) \neq 1$, $e(g,\varphi) = 1$. Let $H : \{0,1\}^* \rightarrow G_x$ be a hash function.

5.2.1 Initialization and Register

Algorithm *Initial*

Input: λ (security parameters)
Output: public parameters, the master key *MK*
1: AA chooses public parameters (G_m , G_n , G_x , $e(\cdot,\cdot)$ and H) as *PP*;
2: AA chooses random numbers $a,b \in Z_p^*$;
3: $MK_1 \leftarrow e(g,g)^a$;
4: $MK_2 \leftarrow g^{a-b}$;
5: $MK \leftarrow (MK_1, MK_2)$;
6: AA sends MK_1 and *PP* to publishers, and keeps MK_2 secretly;

Let S_U denote the set of subscribers, Let $|S_U|$ denote the number of subscribers. Algorithm ***Setup*** deals with subscribers' register.

Algorithm *Setup*

Input: $param$
Output: (GPK, GSK)
For t = 1 **to** $|S_U|$ **Do**
  $t, r_u, \varphi \in Z_p^*$; $v_t = \varphi^{\lambda^t}$;
  $GPK \leftarrow v_t^{r_u} \cdot g^b \cdot MK_2$;
  $GSK \leftarrow g^{a \cdot b + r_u} \cdot v_t^{r_u}$;
  GPK is sent to cloud, and GSK is kept secretly;
End For

5.2.2 Credential Issue

Let I denote the attributers set of a subscriber. Algorithm ***IssueCredential*** grants privilege to a subscriber.

Algorithm *IssueCredential*

Input: MK_2, PP
Output: *Credential*
AA chooses random numbers $r_1, r_2, \ldots, r_j, \ldots \in Z_p^*$ for each attribute of a subscriber;
$D \leftarrow g^b \cdot v_t^{-r_u}$;
For $\lambda_j \in I$ **Do**
  $D_j \leftarrow H(\lambda_j)^{r_j} \cdot g^{r_u}$;
  $D_j' \leftarrow g^{r_j}$;
End For
$Credential \leftarrow (D, \{(D_j, D_j')\}_{\lambda_j \in I})$;
AA issues *Credential* for the subscriber;

It is worthy to know that attribute combining may prompt the privilege of a subscriber. For example, a subscriber pays movies for the year of 2019 and education videos for the year of 2020. It is particularly unreasonable to define the subscriber's attributes set {2019, Movie, 2020, EducationVideo}. These sensitive attributes should be specified to be {2019Movie, 2020EducationVideo} to avoid effectively the attribute combining and then prompt their privileges.

ABE can provide fine-grained access control for media data, and thus AA produces subscribe credentials malleable. Even though it also supports fine-grained non-malleable control on subscribe credential by defining sensitive attribute set carefully.

5.2.3 Encryption

It is required to encode/decode media data quickly because of its stream features. Key part of media data is extracted, encrypted and considered as the object of access control.

$$\boldsymbol{M} = (m_0, m_1, \ldots, m_{k-1})^T$$

Let *Cipher* denote codeword after AONT transfer.

$$Cipher = (c_0, c_1, \ldots, c_{k-1})^T$$

S_T denotes leaf attribute set of access control tree *policy*, and R denotes the root attribute of the tree. Algorithm ***EncryptData*** transfers and encrypts media data before uploading it onto cloud.

Algorithm *EncryptData*

Input: MK_1, PP, M
Output: $Cipher \| CT \| G_{-\bar{A}}$

Publisher extracts the key media data as $M = (m_0, m_1, \ldots, m_{k-1})^T$;
Publisher chooses a generator matrix $G \in Z_q^{k \times k}$, and implements the AONT transform as $Cipher \leftarrow G \times M$;
For i=0 **to** $k-1$ **Do**
    Publisher chooses a random number $r_i \in Z_n$;
    Publisher chooses a random element from line i of G. $g_{i,r_i} \in G$;
    Publisher appends g_{i,r_i} to $\bar{A}$;
End For
Publisher chooses a random number $s \in Z_p^*$;
$\hat{C} \leftarrow \bar{A} \cdot e(g,g)^{as}$; $C \leftarrow g^s$;
For $\lambda_i \in S_T$ **Do**
    $C_i \leftarrow g^{q_i(0)}$;
    $C_i' \leftarrow H(\lambda_i)^{q_i(0)}$;
End For //The Lagrange's interpolation is used with $q_R(0) \leftarrow s$.
$CT \leftarrow (policy, \hat{C}, C, \{(C_i, C_i')\}_{\lambda_i \in S_T})$;
Publisher uploads $Cipher \| CT \| G_{-\bar{A}}$ onto public cloud; // $G_{-\bar{A}}$ denotes the incomplete matrix composed of these rest element after extracting $\bar{A}$ from G.

5.2.4 Decryption

Once a subscriber requests media data, the nearest cloud server runs the algorithm ***ReTransfer*** to retransfer *Cipher*. This retransferring can thus provide different *Cipher* to different subscribers. Moreover, the same subscriber will get different *Cipher* when he/she applies data at different time points.

Algorithm *ReTransfer*

Input: $Cipher \| CT \| G_{-\bar{A}}$
Output: $Cipher \| CT \| G_{-\bar{A}}$

Cloud server chooses random numbers $s', y, k \in Z_p^*$;
$Cipher \leftarrow k \cdot Cipher$;
$\hat{C} \leftarrow k \cdot \hat{C} \cdot e(C, GPK)^{s'} \cdot e(v_i^y, GPK)$;
$C \leftarrow C^{s'}$;
$C' \leftarrow C \cdot v_i^y$;
$C_i \leftarrow C_i^{s'}$;
$C_i' \leftarrow C_i'^{s'}$;
$CT \leftarrow (policy, \hat{C}, C, C', \{(C_i, C_i')\}_{\lambda_i \in S_T})$;
$G_{-\bar{A}} \leftarrow k \cdot G_{-\bar{A}}$;
Cloud server sends $Cipher \| CT \| G_{-\bar{A}}$ to the subscriber;

Let S denote an attribute set of a subscribe credential, $decrypt_{part}$ denotes the result of the algorithm $DecryptNode_x$ running on the set I_x, and I_x denotes the set of z(children of node x). Algorithm $DecryptNode_x$ runs from the root node R of the access tree *policy* in a recursive way.

Algorithm *DecryptNode$_x$*

Input: CT, $\{(D_j, D_j')\}_{\lambda_j \in I}$
Output: $decrypt_{part}$

If x is a leaf node **Then**
  λ_i is set to be an attribute of the node x;
  If $\lambda_i \in S$ **Then** $DecryptNode_x \leftarrow \dfrac{e(D_i, C_i)}{e(D_i', C_i')}$;
  If $\lambda_i \notin S$ **Then** $DecryptNode_x \leftarrow \perp$;
End If
If x is a non-leaf node **Then**
  $i \leftarrow index(z)$;
  $s_x \leftarrow \{index(z), z \in I_x\}$;
  $F_x \leftarrow \prod_{z \in I_x} F_z^{\Delta_{i,s_x}(0)}$;
End If
If x is the root node **Then** $decrypt_{part} \leftarrow F_x$;

A subscriber runs the algorithm *DecryptData* to obtain the key media data M after he/she retrievals $\widetilde{Cipher}||\widetilde{CT}||\widetilde{G_{-\bar{A}}}$ from the cloud server.

Algorithm *DecryptData*

Input: SK, $Cipher \| CT \| G_{-\bar{A}}$

Output: M

1: The subscriber sends $\{(D_j, D_j')\}_{\lambda_j \in I}$ to a cloud server, and calls the algorithm $DecryptNode_x$ to obtain $decrypt_{part}$;

2: The subscriber calculates:

$$D \leftarrow D \cdot GSK;$$

3: The subscriber calculates:

$$k \cdot \bar{A} \leftarrow \frac{\hat{C} \cdot decrypt_{part}}{e(C, D) e(C', GPK)};$$

4: The subscriber reconstructs $k \cdot G$ by $k \cdot \bar{A}$ and $k \cdot G_{-\bar{A}}$;

5: The subscriber calculates the reverse matrix of $k \cdot G$,that is $\frac{1}{k} \cdot G^{-1}$;

6: The subscriber calculates $\frac{1}{k} \cdot G^{-1} \cdot Cipher$, and then obtains the key media data M;

---

6 Analysis of IM-ACS

1) Security Analysis

Data Confidentiality. The key data of the mobile media is transferred and encrypted by AONT and CP-ABE separately before being uploaded onto cloud. Authorized subscribers can decrypt their authorized media data using their subscribing credentials. Unauthorized ones cannot access the media data.

Loose Coupling. Publishers and subscribers need not to negotiate the key beforehand. All media data ciphertext is uploaded and stored on cloud, multiple subscribers can request and decrypt ciphertext using their credentials.

Collusion Resistance. Either of authorized subscribers hold different g^{r_u}. Caused by security assumption, they cannot combine their credentials to get more privileges.

Non-malleable Security. When the subscribe credential is issued, sensitive attribute should be defined exactly. When different subscribers request the same key media data, they retrieval different ciphertexts, decrypt these different ciphertexts, and then obtain the same key media data. For a subscriber, different time points will lead to different ciphertexts. In a word, a subscriber cannot piece together a new ciphertext. With all the above methods made, our scheme achieves informal non-malleable access control over media data.

2) Performance Analysis

Table 1 shows the comparison of computational cost on each entities between our scheme and three existing schemes (Li's scheme [7], Hur's scheme [8] and Teng's scheme [3]).

Our scheme spends less computation cost than three other schemes on decryption on the subscriber. Most notably, our scheme can realize informal non-malleable security by retransferring all of ciphertext, which spends necessary computational cost on cloud. It assumed to omit the top-level domain authority DA in Teng's scheme for ease of comparison between the other methods.

Table 1. Computational analysis

Algorithms	Li's [7]	Hur's [8]	Teng's [3]	Our IM-ACS
IssueCredential (run on AA)	$(2 + 3\|S\|)$ *Exp*	$(1 + 3\|S\|)$ *Exp*	$(2\|S\|)$ *Exp*	$(2 + 2\|S\|)$ *Exp*
EncryptData (run on Publisher)	$(2 + 2\|S_T\|)$ *Exp*	$(2 + 2\|S_T\|)$ *Exp*	6 *Exp*	$(2 + 2\|S_T\|)$ *Exp* + *Pair*
DecryptData (run on Subscriber)	$(\|S\| + \log\|S_T\|)$ *Exp* + $(3\|S\| + 1)$ *Pair*	$(\|S\| + \log\|S_T\|)$ *Exp* + $(2\|S\| + 1)$ *Pair*	$2\|S_T\|$ *Exp* + 6 *Pair*	2 *Pair*
DecryptNode$_x$ (run on cloud)	–	–	–	$(\|S\| + \log\|S_T\|)$ *Exp* + $2\|S\|$ *Pair*
ReTransfer (run on cloud)	–	–	–	$(3 + 2\|S_T\|)$ *Exp* +2 *Pair*

$|S|$ denotes the number of attributes in a credential, and $|S_T|$ denotes the number of attributes in a ciphertext. *Exp* denotes an exponent calculation, and *Pair* denotes a pairing calculation.

7 Simulation

The simulation has been done on Ubuntu system with the Pairing-Based Cryptography library. The elliptic curve is chosen as α, order of all groups as 160 bit, and field size as 512 bit. Times are the mean of 10 trials to avoid the results of accidents.

In Fig. 3, both $|S|$ and $|S_T|$ are set to be 5 by default. In Fig. 3(a), the computational cost, spent to encrypting on a publisher, is linear with $|S|$ in all schemes except for the one in Teng's scheme with 6 exponents. Figure 3(b) shows that the computational cost, spent to decrypting on a subscriber, is linear with $|S|$ obviously in all schemes except for our IM-ACS. Figure 3(c) shows that we might not be conscious of the $|S_T|$ in all schemes, which influences the computational cost slightly. Figure 3(d) shows that cloud server has borne the most of the computational cost of decryption in our IM-ACS.

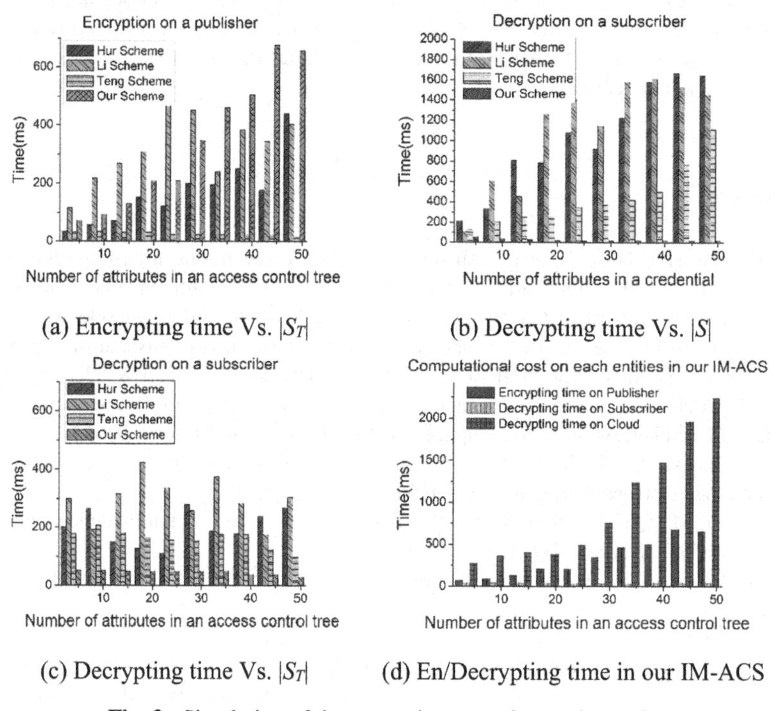

(a) Encrypting time Vs. $|S_T|$

(b) Decrypting time Vs. $|S|$

(c) Decrypting time Vs. $|S_T|$

(d) En/Decrypting time in our IM-ACS

Fig. 3. Simulation of time spent in encryption or decryption

8 Conclusion

For security requirements in a mobile media data sharing system, we propose an Access Control Scheme supporting Informal non-Malleable security (IM-ACS) for mobile medias by ABE, AONT and re-transform on cloud. On the one hand, our scheme realizes one-to-many encryption and fine-grained access control based on ABE. On the other hand, our scheme is informal non-malleable because that subscribers cannot obtain any acknowledge from the previous data accesses unless publisher allow them to. According our performance analysis, IM-ACS spends less computation cost than three other schemes on decryption on the subscriber.

Acknowledgements. This work is supported by the Project of Jiangxi Provincial Social Science Planning (No. 17XW08), the Science and Technology Research Project of Jiangxi Provincial of Education Department (No. GJJ180905), the Project of Hubei Provincial Science Foundation (No. 2018CFB109), the Research Project of Hubei Provincial Department of Education (No. Q20191710), and the MOE (Ministry of Education in China) Project of Humanities and Social Sciences (No. 20YJAZH112).

References

1. Ion, M., Russello, G., Crispo, B.: An implementation of event and filter confidentiality in pub/sub systems and its application to e-Health. In: 17th ACM Conference on Computer and Communications Security, CCS2010, Chicago, IL, United states, 2010, pp. 696–698
2. Tariq, M.A., Koldehofe, B., Rothermel, K.: Securing broker-less publish/subscribe systems using identity-based encryption. IEEE Trans. Parallel Distrib. Syst. **25**, 518–528 (2014)
3. Teng, W., Yang, G., Xiang, Y., Zhang, T., Wang, D.: Attribute-based access control with constant-size ciphertext in cloud computing. IEEE Trans. Cloud Comput. **5**, 617–627 (2017)
4. Qiu, H., Qiu, M., Lu, Z., Memmi, G.: An efficient key distribution system for data fusion in V2X heterogeneous networks. Inf. Fusion **50**, 212–220 (2019)
5. Tian, Z., Li, M., Qiu, M., Sun, Y., Su, S.: Block-DEF: a secure digital evidence framework using blockchain. Inf. Sci. **491**, 151–165 (2019)
6. Pass, R., Shelat, A., Vaikuntanathan, V.: Bounded CCA2-Secure Non-Malleable Encryption. MIT (2006)
7. Jiguo, L., Wei, Y., Jinguang, H., Yichen, Z., Jian, S.: User collusion avoidance CP-ABE with efficient attribute revocation for cloud storage. IEEE Syst. J. **12**, 1767–1777 (2018)
8. Hur, J.: Improving security and efficiency in attribute-based data sharing. IEEE Trans. Knowl. Data Eng. **25**, 2271–2282 (2013)

Blockchain-Based Secure and Privacy-Preserving Clinical Data Sharing and Integration

Hao Jin[1], Chen Xu[1], Yan Luo[1](✉), and Peilong Li[2]

[1] Department of Electrical and Computer Engineering, UMASS Lowell, Lowell, MA, USA
{Hao_Jin,Yan_Luo}@uml.edu, chen_xu@student.uml.edu

[2] Department of Computer Science, Elizabethtown College, Elizabethtown, PA, USA
lip@etown.edu

Abstract. This paper is an exploration to securely integrate geo-scattered medical data silos and to provide privacy-preserving data sharing via blockchain and cryptography. We leverage broadcast encryption, key regression, data privacy classification, blockchain and smart contract holistically to address such a challenge. In our design, a patient's medical records are divided into multiple parts with different sensitivities according to personal privacy requirements. Each part is encrypted with a symmetric encryption key and it is broadcast encrypted to a predefined user set. Data encryption keys are generated by a key regression scheme, where keys used to encrypt low-sensitivity data parts can be derived from keys used to encrypt high-sensitivity data parts. With such a mechanism, we can enforce strict access and privacy control on medical data. Furthermore, we use Ethereum blockchain to connect various institutions to provide efficient data sharing, and we design smart contracts to implement the business logic of medical data sharing.

Keywords: Blockchain · Security · Privacy · Smart contracts

1 Introduction

The era of cloud computing and big data demand medical data to be shared among various third-party medical institutes to conduct analytics for better healthcare service and new treatment. However, privacy protection regulations, such as the Health Insurance Portability and Accountability Act (HIPAA) [2] in the United States and the General Data Protection Regulation (GDPR) [1] in Europe, necessitate secure storage and sharing of medical data and may inflict severe penalties for data breach events.

Electronic medical records (EMR) are usually stored in private hospital databases due to privacy and security reasons. Those medical silos mostly reside in separated network domains, which leads to the lack of interoperability in medical data integration and sharing, and it further puts a barrier to big data

M. Qiu (Ed.): ICA3PP 2020, LNCS 12454, pp. 93–109, 2020.
https://doi.org/10.1007/978-3-030-60248-2_7

analytics across multiple medical organizations, especially when a patient's medical records are scattered across multiple hospitals.

Hence, how to efficiently integrate these insular medical databases for comprehensive medical care without violating privacy regulations has become a difficult problem. Recently, some blockchain-based approaches [4,9,15,20,21] are proposed to address this problem. Unfortunately, these schemes made a strong assumption that hospitals storing medical information rarely undergo cyber attacks or compromises, and consequently they store plaintext medical information. Nevertheless, the status quo is that over 50% of security breaches occur in the medical industry, and almost 90% of healthcare institutions had their data exposed or stolen, according to a 2014 study [3]. In this context, we believe that data encryption and secure key management are critical to enhance the security and privacy of medical information.

Our motivation comes from two challenging problems existing in the state-of-the-art schemes on medical data sharing:

- how to securely and efficiently integrate scattered healthcare data silos to provide statewide or national data sharing? Conventional solutions usually build a private network to connect all hospitals to segregate it from public Internet, which means countless efforts and money need to be invested on the network infrastructure.
- how to securely manage encryption keys if we choose to store medical information encrypted? If we let patients issue keys to thousands of users and institutes, it will be a huge burden for them. A feasible way is to use some advanced cryptographic primitive to secure the storage and distribution of encryption keys.

To address these problems, we propose a new approach to provide secure and privacy-preserving clinical data sharing in a statewide scale, where broadcast encryption, key regression, data privacy classification, blockchain, and smart contract are leveraged holistically to achieve this goal with controllable granularity on how and whom the information is to be shared. By using blockchain to connect scattered hospitals and relying on the automatic executions of smart contracts to implement the business logic of medical data management, we can reduce costs on infrastructure construction and make data sharing secure and auditable without modifying existing healthcare infrastructure significantly. This is a worthy attempt to fuse scattered medical records throughout various hospitals for holistic and comprehensive data analysis.

Our contribution mainly lies in the following aspects. First, medical data are classified into multiple parts with different privacy levels through a customized policy, and each data part is encrypted with a key corresponding to its data privacy level. Encryption keys are generated by using a key regression scheme [11] where a high-level key can derive all lower-level keys. Hence, all keys compose a key sequence where different privacy levels are implicitly associated with keys in the sequence.

Second, encryption keys are further broadcast encrypted [5] with a predefined user set so that only users in this set can decrypt the broadcast cipher message to recover the key. These broadcast messages are securely stored on the blockchain to let authorized users conveniently obtain them. As a result, a data owner (patient) can enforce access and privacy control by providing a certain set of users an appropriate key with which they decrypt a corresponding data part.

Third, we adopt Ethereum [6] blockchain to connect various hospitals to securely manage medical data. We design smart contracts to implement the business logic of clinical data sharing and integration. Only a small amount of metadata (i.e., broadcast encryption messages, Ethereum addresses) are stored on chain, which imposes limited burden to the blockchain network, as our evaluation shows.

2 Background

2.1 Preliminaries

Broadcast Encryption. Broadcast encryption (BE) was firstly introduced in [10] and then improved in [5], which let an owner encrypt a message to a subset of users. Only users in the subset can decrypt the broadcast message to recover the original data. A BE scheme consists of three algorithms.

$Setup(1^{\lambda}, n)$. This algorithm takes as input a security parameter 1^{λ} and the number of receivers n, outputs a public key pk and n private keys $d_1, d_2, ..., d_n$.

$Encrypt(pk, S)$. This algorithm is run by the broadcaster to encrypt a message to a user subset S. It taks as input a public key pk and a user subset $S \subseteq 1, 2, ..., n$, outputs (Hdr, K). Hdr is the BE header and K is a message encryption key embedded in Hdr. Let M be a message to be broadcast to S and C_M be the encryption result of M under key K. The broadcast message consists of (S, Hdr, C_M).

$Decrypt(pk, S, i, d_i, Hdr)$. This algorithm is run by an authorized user to decrypt the broadcast message. It taks as input a public key pk, a user subset S, a user id $i \in 1, 2, ..., n$, the private key d_i for user i, and the BE header Hdr, outputs the message encryption key K, which can be used to decrypt the broadcast body C_M and get message M.

Key Regression. Key regression (KR) [11] allows a sequence of keys to be generated from an initial key and a secret master key. Only the possessor of the secret key can rotate the key forward to a new version in the sequence. A user who knows the current key can produce all earlier-version keys in the sequence. Specifically, a key regression scheme has two properties, namely, given the i-th key K_i, it is: (1) easy to compute the keys K_j for all previous time periods $j < i$; (2) computationally infeasible to compute the keys K_l for future time periods $l > i$.

Key regression is usually adopted in cryptographic storage to enforce lazy revocation policy to avoid data re-encryption caused by key updates, which can reduce the number of keys to be retained to provide scalable key management. We leverage the forward secrecy property of key regression to enable multiple-level privacy classification of medical information.

Blockchain and Smart Contract. Blockchain can be categorized into permissionless and permissioned. Permissionless or public blockchains allow every user to participate in the network by creating and verifying transactions and adding blocks to the ledger. Bitcoin [14] is the most famous example, which applies a Proof of Work (PoW) consensus that requires participating nodes to add blocks by solving a computational puzzle. While Ethereum, another mainstream representative, uses a combination of Proof of Work and Proof of Stake [8]. In contrast, permissioned or consortium blockchains maintains an access control layer in their network to allow certain actions to be performed only by certain kinds of nodes. Hyperledger [7] is an increasingly popular, collaborative permissioned blockchain that adopts BFT-SMART algorithm [19] as its consensus protocol.

Smart contract is a small program placed on the blockchain to execute certain kinds of computation. Considering the limitation of the script language in Bitcoin, we turn to Ethereum [6] since its script language (Solidity) for smart contracts is Turing-complete.

2.2 Related Work

MedRec [4] firstly proposed a decentralized EMR management system based on blockchain and provided a functional prototype implementation. K. Peterson et al. [15] designed a proof-of-interoperability consensus to ensure transaction data be in conformance with fast healthcare interoperability resources (FHIR) in cross-institutional health data sharing. However, they did not mention how the medical data are organized, stored, and accessed.

Q. Xia et al. proposed MedShare [20], which is a blockchain-based framework with data provenance and auditing among healthcare providers. K. Fan et al. proposed MedBlock [9], a hybrid blockchain-based architecture to secure EMR data management, where nodes are divided into endorsers, orderers and committers as in Hyperledger Fabric [7]. Their proposal of using asymmatric encryption to encrypt medical information may bring huge overhead due to involved expensive asymmatric en/decryption. A. Zhang et al. [23] proposed BBPS, a hybrid blockchain-based personal health information (PHI) system, where a private blockchain is used to store PHI for each hospital and a consortium blockchain is used to keep secure indexes of the PHI data. But how these two blockchains interact with each other is not clearly mentioned.

Besides, some studies propose to combine blockchain with advanced cryptographic primitives to provide fine-grained access control and privacy protection. Guo et al. [12] proposed to combine blockchain with a multi-authority attribute-based signature scheme to secure the storage and access of electronic health records. However, their scheme encapsulates and stores health records in on-chain blocks, which makes its scalability a problem because the amount of medical data can be very large. Han et al. [17] proposed to use selective encryption, data fragmentation, and information dispersion to secure the sharing of health data in medical cyber-physical systems. They further extended their idea by introducing invertible Discrete Wavelet Transform (DWT) into the design to

Table 1. Notations and keys

Notation	Description
K_E	Content encryption key
$E_K(m)$	Symmetric encryption of m with key K
$E_B(m, S)$	Broadcast encryption of m to a user set S
d_i	Private key for i-th user in BE scheme
(pk_i, sk_i)	Public-private key pair of the i-th user
PL_i	Privacy level of the i-th data field in records
APL_j	Allowed privacy level of user u_j

strengthen the data fragmentation and dispersion method and gave solid performance evaluation in [16].

3 Design

3.1 System and Trust Model

There are three roles in our framework. 1) **Healthcare provider**, who stores and manages patients' encrypted medical data; 2) **Patient**, who is the owner of his medical information and relies on care providers to store and manage his data. A patient is responsible for enforcing access control of his encrypted data. 3) **Third-party user**, who may be a professional medical researcher and need to analyze a large amount of clinical data. He should get patients' approval before accessing their medical records.

We assume healthcare providers are semi-trusted, which means that they behave appropriately most of the time, but they also have motives to violate regulations by sending a third-party user more data than he is allowed to access. Therefore, providing patient-centric access control over his medical data is of critical importance. Blockchain can take the role of a public ledger to store involved transactions and smart contracts for the purpose of recording, auditing, and tracking of medical data usage, and the underlying p2p network connects all participants (hospitals, patients, and medical researchers). We assume the cryptographic primitives (hash, signature, encryption) used in blockchain are secure enough to resist potential attacks. Furthermore, Ethereum generates a private-public key pair according to the Elliptic Curve Digital Signature Algorithm (ECDSA) for each participant.

3.2 Broadcast Encryption-Based Access Control

Current research work on medical data sharing [4,9] implies a strong trust assumption on hospitals, i.e., hospitals will not leak patients information and their networks are safe. However, this is not always true when we further consider the storage and maintenance problem of medical data. As usual, a patient's data is stored in a hospital's private database managed by the IT department. The problem is to what extent can a patient believe that the hospital will never leak his data or have it stolen, either intentionally or unintentionally? According to [3], data leakage events in medical industry happen frequently due to administrative faults, software loopholes, and network intrusions, etc. Hence, storing plaintext medical records in hospital databases is a risky policy, especially for long-term preserved data.

Furthermore, state-of-the-art blockchain-based approaches [4,9,15] still face the same security threats as in conventional network-based storage since medical data are stored off-chain. Thus it should be made clear that blockchain itself cannot ensure the security of off-chain data. Due to this reason, we propose to encrypt medical records before storing them in hospital's databases. To securely manage encryption keys and enforce strict access control, our framework adopts broadcast encryption and key regression. Specifically,

1. We use broadcast encryption (BE) [5] to securely store encryption keys used to encrypt medical information, where adding or revoking a user is efficient.
2. We devise a data privacy classification policy by associating an indepedent key with a privacy level, where a key regression mechanism is adopted to enforce privacy control and facilitate key management.
3. We provide a privacy policy maker to help a patient to customize his personal privacy policy or use a default one.

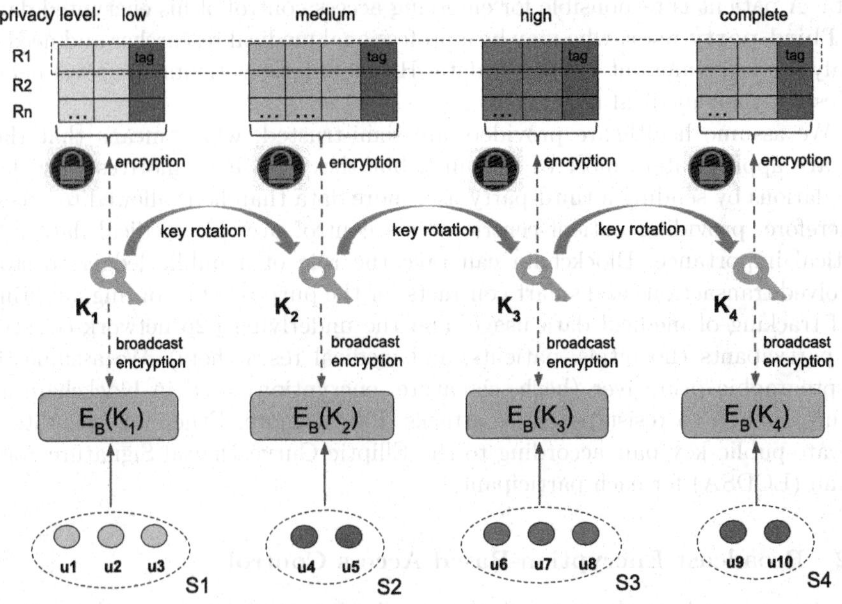

Fig. 1. Broadcast encryption of keys and key regression based privacy classification. Low-sensitivity keys can be derived from a high-sensitivity key, and each key in the sequence is associated with a data privacy level.

We adopt a two-level encryption policy for medical information, as depicted in Fig. 1. Firstly, a patient's medical information is encrypted with a symmetric key K_E, which is chosen by the patient; Then, the encryption key is broadcast encrypted to a pre-defined user set S to get the broadcast message $E_B(K_E, S)$. Only users in set S can decrypt the broadcast message to recover the content encryption key K_E. Notations and keys are illustrated in Table 1.

A broadcast encryption system consists of three algorithms, namely, *Setup*(), *Encrypt*() and *Decrypt*(). The *Setup* phase generates a private key d_i for each user in a pre-defined user set S. On decryption, an authorized user can use his private key d_i to decrypt the broadcast message to recover the content key. A patient needs to execute a broadcast encryption to generate a BE header according to the following equation.

$$Hdr = (C_0, C_1), C_0 = g^t, C_1 = v \cdot \prod_{j \in S} g_{n+1-j}^t \tag{1}$$

where t is a random element in $\mathbb{Z}_p$ and g is a generator of group $\mathbb{G}$ of prime order p. The key will be set as $K = e(g_{n+1}, g_1)^t$ by the *Encrypt*() algorithm. Due to the space limitation, we do not introduce all algorithm details of a broadcast scheme here, which can be found in [5].

The bottom part of Fig. 1 shows broadcast encryption of four different keys K_1, K_2, K_3, K_4, where each key is associated with a corresponding user set $S_i (1 \leq i \leq 4)$. Moreover, each key is used to encrypt a medical data part which is tagged with a specific privacy level, as will be discussed in the following section.

User Addition and Revocation. The reason we choose broadcast encryption scheme instead of attribute based encryption (ABE) [18] is based on a fact, namely, BE scheme [5] provides more efficient method of adding a new user or revoking an existing user.

Recall the broadcast cipher message is $Hdr = (C_0, C_1)$. To add a new user $u \in 1, ..., n$, simply set C_1 as $C_1 \cdot g_{n+1-u}^t$; to revoke a user u, set C_1 as C_1 / g_{n+1-u}^t. Both operations are actually a multiplication or division on group elements. However, in current ABE schemes [13,22], user revocation is still a challenging task for its significant computation overhead.

3.3 Key Regression-Based Privacy Classification

Encrypting data with one single symmetric key essentially provides an all-or-nothing access policy. For a specific user, if he has the key, he can decrypt the data; otherwise, he can not. However, in many situations, we need to classify data into multiple parts and each part requires different privilege to access. A feasible way to achieve this is to encrypt each part with a different key. But this would leave the management of these separated keys a tedious problem, especially when the data classification is fine-grained, which implies that a large number of keys need to be securely managed.

To address this issue, we proposed to use key regression technique to keep a relation between keys. Specifically, we adopt the KR-SHA construction in [11] for key regression. From an initial state stm_{max}, a sequence of states $(stm_1, ..., stm_{max})$ can be obtained by hashing the previous state, where we have

$$stm_i = SHA256(stm_{i+1}) \tag{2}$$

Correspondingly, we can get a sequence of keys $K_1, K_2, ..., K_{max}$ used for data encryption, where each key

$$K_i = SHA256(0^8||stm_i) \tag{3}$$

The number of rotations in a state sequence depends on the number of privacy levels in classified medical information. Let's assume a patient's data are divided and tagged with privacy levels $\{PL_1, PL_2, ..., PL_n\}$, where n denote the number of privacy levels. Then the key sequence can be represented as $K_1, K_2, ..., K_n$, where each key is derived as Eq. 3. Given a state value stm_i corresponding to a specific key K_i, it is easy to derive all previous keys $K_1, K_2, ..., K_{i-1}$. Hence, a key K_i possesses a higher privilege than its predecessors $K_j (j < i)$.

Using this property, we can build a privacy control policy by encrypting different parts of a patient's medical information with different keys in the sequence. Data fields labelled with a high privacy level require high access privilege, thereby they should be encrypted with a key having a large subscript in the sequence. For example, in the sequence $K_1, K_2, ..., K_n$, K_0 has the lowest privilege, while K_n has the highest privilege.

In Fig. 1, the dashed box at the top represents a complete medical record $R1$ that is divided into four parts, each part is attached with a tag denoting its privacy level(i.e., low, medium, high and complete). Keys K_1, K_2, K_3, K_4 are used to encrypt these four data parts, and each key is broadcast encrypted. Among them, K_4 can be used to derive other three keys. Hence, a user who has key K_4 can recover the broadcast message $E_B(K_4, S_4)$, and further can recover all plaintext record fields. After he recovers K_4, he simply derives previous keys K_3, K_2, K_1 by backward key rotations (according to Eq. 2 and 3), and uses these keys to decrypt corresponding parts of the patient's medical records. For instance, users in set $S4$ can access all parts of the medical records while users in set $S1$ can only access the first part labelled with a "low" tag.

Conclusively, the key sequence $K_1, K_2, ..., K_n$ for a patient's medical data encryption corresponds to a series of privacy levels $\{PL_1, PL_2, ..., PL_n\}$ used to describe the sensitivity of n data parts. Accordingly, the broadcast encryption result of these keys $E_B(K_1, S_1), E_B(K_2, S_2), ..., E_B(K_n, S_n)$ are stored in patient-hospital smart contract, as will be introduced in the following section (Fig. 2).

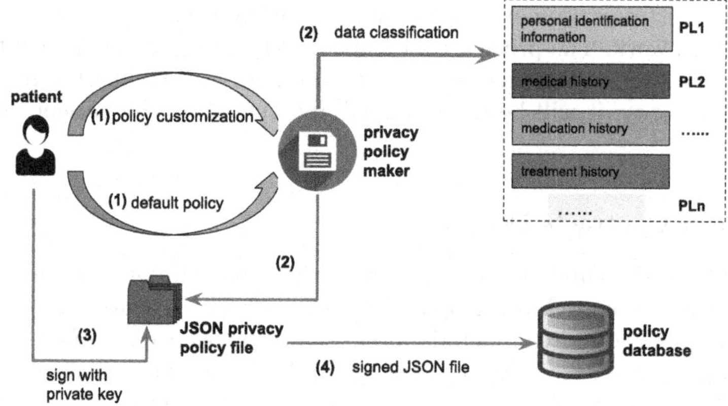

Fig. 2. Process of privacy policy maker.

Privacy Policy Maker. Fine-grained access control implies two levels: 1) the finely differentiated roles a user may take on; and 2) the finely differentiated data parts according to their sensitivity levels. In medical data sharing, the first level needs the adoption of cryptographic primitives with rich semantics support to secure the storage of content encryption keys. We address this issue by adopting broadcast encryption. The second level needs to classify a patient's medical information into multiple parts that are labelled with different sensitivities through a privacy policy.

We devise a privacy policy maker to automate the data classification task. When a patient visits a hospital and generates his/her medical data, the privacy maker will devide the patient's medical information into personal idetification information, medical history, medication history, and treatment history, etc. Such a devision can be very finely grained to satisfy a patient's privacy requirements. The following JSON (JavaScript Object Notation) policy file depicts an example of a default policy with four privacy levels, i.e., low, medium, high, and confidential.

```
var PrivacyClassification = {
  "policy-type":  "default",
  "privacy-levels": 4,
  "personal-identification-information":
  {
    name: "medium"
    address: "high"
    social-security-no: "confidential"
  },
  "personal-medical-history": "low",
  "family-medical-history": "high",
  "personal-medication-history": "low",
  "personal-treatment-history": "medium",
};
```

During data classification, each data part will be attached with a privacy level (i.e., an integer value) through a default privacy policy or a customized policy. It should be noted that even a customized policy can be quickly generated based on previous customization experiences and default policies without involving too much manual work. The privacy policy maker generates a JSON policy file and stores it in a health domain's policy database and related smart contract. To ensure its security, it will be timestamped and signed by the patient using his private key.

3.4 Smart Contracts Design

To efficiently and securely track and share medical data throughout various medical institutions, we use blockchain and smart contracts to achieve this goal. We store small pieces of critical metadata on chain, such as access control information, broadcast message of encryption keys. While patients' medical data are encrypted and stored off chain in private databases of healthcare providers.

In our framework, blockchain network acts as a bridge to connect various isolated medical databases where hospitals and medical institutes participate the network as nodes. Smart contracts brings the integration of scattered clinical data through contract creations and executions. Critical metadata such as data ownership, access permissions, broadcast encryption messages, privacy classification policies, and Ethereum addresses of involved entities, are stored in relevant contracts. In Ethereum, a contracts is compiled to byte code and the resulting bytes are included in a transaction to be persisted onto the blockchain [6] storage. Afterwards, one can interact with the smart contract with other transactions.

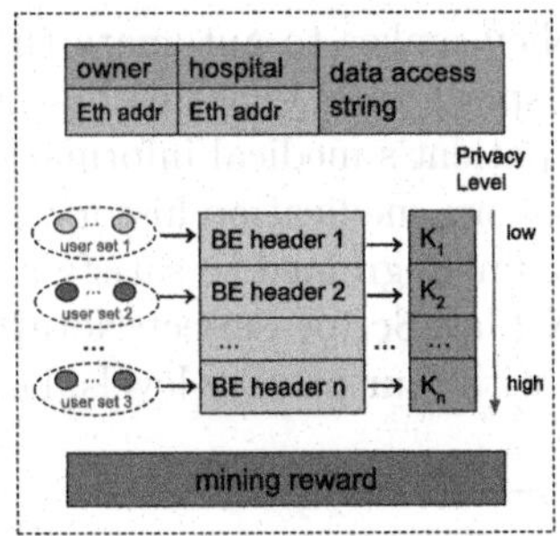

Fig. 3. Patient hospital contract

We define three kinds of contracts in our design to implement the business logic of medical data management. It is worth noting that the contract execution that enforces metadata modification will be carried out by legitimate transactions after necessary authentication.

Address mapping contract (AMC), keeps identity string and Ethereum address mappings for all involved patients, healthcare providers and third-party users. During system running, AMC is responsible for converting an entity's name string into its Ethereum address, which is unique in the blockchain network. When a new user or a new provider is added to the system, its name or

Algorithm 1. Patient Hospital Contract

```
Require: H: a hospital, the creator of this contract
Require: P: a patient
Struct BEheader{
    string cipher;
    byte[] user_list;
};
mapping(uint ⇒ BEheader) BE_list;
function PHC_CREATE(address H, address P)
    Require msg.sender is H and its address is registered
    Require P is registered by H
    P defines his privacy levels PL_1, PL_2, ..., PL_n
    P classifies his data into D_1, D_2, ..., D_n
    P chooses a initial state for key regression
    P derives the key sequence K_1, K_2, ..., K_n
    P encrypts data and gets E_{K_1}(D_1), ..., E_{K_n}(D_n)
    H computes BE headers E_B(K_1, S_1), ..., E_B(K_n, S_n)
    H provides the location of medical data for P

function PHC_ADDUSER(address U(user), address P, int j)
    Require msg.sender is H and its address is registered
    Require P is registered by H
    P defines the new user's privacy level PL_j
    P add the user to a corresponding set S_j
    H re-computes the BE header for the updated user set E_B(K_j, S_j)
    H updates BEheader list
function PHC_REVOKEUSER(address U(user), address P, int j)
    Require msg.sender is H and its address is registered
    Require P is registered by H
    Require U is in the user set S_j
    P revoke the user from some set S_j = S_j − U
    H re-computes the BE header for the updated user set E_B(K_j, S_j)
    H updates BEheader list
```

identity string and corresponding Ethereum address will be registered by the AMC contract.

Patient-hospital contract (PHC), defines the relationship between a patient and a hospital, where the patient has the ownership of his medical records and the hospital has the stewardship. Figure 3 depicts such a process, in which n privacy levels are classified. Correspondingly, n user sets are defined and n data encryption keys are chosen by the owner, which are broadcast encrypted. A BE header denotes the ciphertext message of an encryption key K_i that is encrypted to a user set S_j. The top-to-bottom arrow indicates that the privacy level is increasing from K_1 to K_n, hence a key with a higher subscript has higher privilege and can access more information.

An authorized user can decrypt the corresponding broadcast header to recover a content encryption key. For each third-party user that is allowed by the patient to access his medical information, a corresponding allowed privacy level (APL) is defined to restrict his access privilege. This process is actually carried out by putting the user in a user set, as is illustrated in Fig. 3. Algorithm 1 depicts the process of PHC contract creation.

User-hospital contract (UHC), records the user's access behavior to a hospital. In the *setup* phase of boradcast encryption, a BE private key is generated for each user in the set. With his private key, a user can decrypt the broadcast message to recover the content. When a user requests to access a patient's information, the hospital finds the user's privacy level by checking which set he belongs to. Then the hospital sends the corresponding BE private key to the user, along with a data access string.

3.5 Execution Logic of Smart Contracts

We show the business logic of medical data management by explaining the process of a registered user visiting a patient's medical data.

Firstly, the user u_i needs to send an access request to a hospital with the Ethereum addresses of the hospital and the patient being included. On receiving the request, the hospital searches for the corresponding PHC contract, finds the corresponding user set S to which u_i belongs and the broadcast message $E_B(S, K_j)$. If it is the first time for the user u_i to access medical data, the hospital should send the BE private key d_i to the user(see Sect. 2.1), which is used to decrypt the broadcast message to recover the data encryption key K_j. According to our privacy classification policy, user u_i can only decrypt a specific broadcast encryption message to recover an encryption key K_j matching his allowed privacy level APL_i. With this key, the user u_i can derive all lower-level keys $K_m(m \leq j)$, so that he can access record fields with matching and lower privacy levels.

Then the hospital searches his private database and returns data fields whose privacy level are less or equal than APL_u. In the meantime, the hospital generates a new UHC contract which records the user's access behavior, and posts this new contract on the blockchain.

On receiving the data fields from the hospital, what the user needs to do is to search for the related PHC contract to get the broadcast encryption message $E_B(S, K_j)$ and to perform a broadcast decryption to recover the key K_j.

From this use case, we can see that a third-party user's data access does not need the patient to be involved, which simplifies the data access process without sacrificing security. From a real-world perspective, this is of some significance since patients are usually equipped with mobile devices with limited computation and storage capability.

Adding a new institute to the blockchain or revoking an existing institute from the network is similar to the aforementioned process except that the AMC contract needs to be updated. We omit the details here.

4 Security Analysis

In this section, we analyze the security and privacy fulfillment of our design goals, including confidentiality, integrity and privacy.

Confidentiality. We adopt a two-level encryption policy to guarantee the confidentiality of medical data, where content keys are used to encrypt medical data and a broadcast encryption scheme is used to encrypt content keys. In this sense, to decrypt cipher-text medical information, an attacker has to break a broadcast encryption system to recover the content key. Hence, the confidentiality guarantee is dependent on the security of the adopted BE scheme, whose security has been proved in [5]. On the other hand, an attacker may deduce some information by logging access patterns and frequency of some data fields. How to hide these information is beyond the scope of this paper.

Integrity. In our design, each medical record is attached with a signature signed with the owner's private key. Any other party can verify its integrity with the owner's public key if they have a complete access of the entire record. Due to the security of a secure signature scheme, an unauthorized user cannot forge a valid signature of a medical record without the private signing key.

Privacy. Our data privacy classification policy can let a patient decide which part of his medical information can be accessed by what kind of users. Hence it can provide flexible privacy control and protection. The security of our key regression based privacy control policy comes from two sides: 1) each privacy level PL_i corresponds to a key which is used to encrypt a data part labelled with PL_i. The key is broadcast encrypted. Hence, security of the key storage depends on the security of our adopted broadcast encryption scheme. 2) all encryption keys constitute a key sequence. Due to the forward secrecy property of a key regression scheme, it is infeasible for a user with a lower-level key in the sequence to derive a higher-level key, which is proved under the random oracle model by [11]. Hence, the security of a key regression scheme guarantees that our privacy classification policy will leak no more information than what a user is allowed to access.

5 Implementation and Evaluation

5.1 Cryptographic Overhead

Table 2. Test environment

Item	Specifications
Host CPU	2 x Intel E5-2643 v3 @3.40 GHz
CPU cores	6 core
Host memory	32 GB ECC Memory
Disk	1024 GB at 7200 RPM
Host OS	Ubuntu 16.04.4

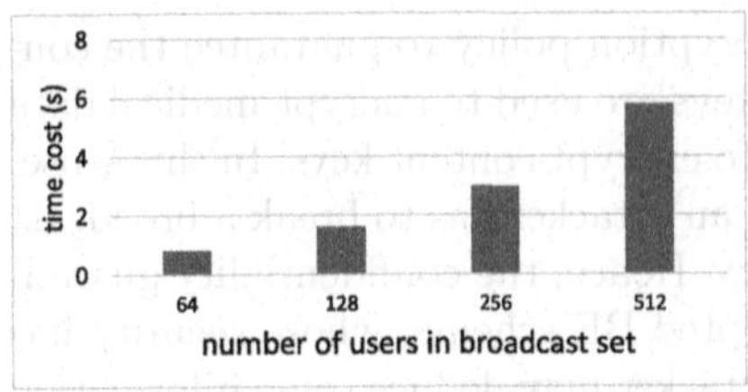

Fig. 4. Setup time cost of broadcast encryption

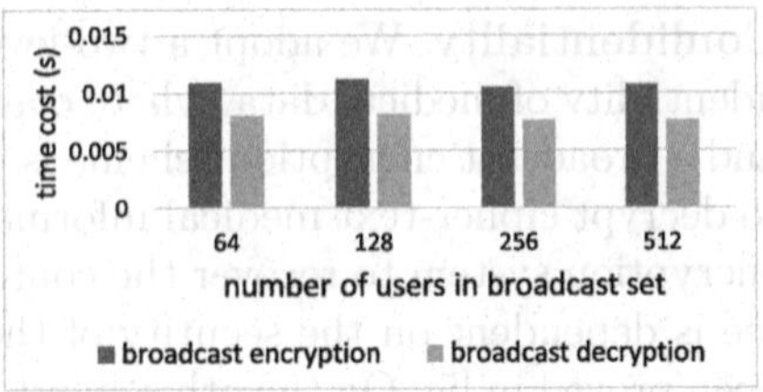

Fig. 5. Time cost of broadcast encryption and decryption

To evaluate the overhead brought by broadcast encryption, we have developed a prototype with C language and the pbc_bce library (version 0.0.1), which is a software implementation of Boneh-Gentry-Waters broadcast encryption scheme [5]. The test environment, including the hardware and operating system, is depicted in Table 2.

Figure 4 depicts the setup overhead of broadcast encryption, where the BE system needs to generate a private key for each user. The involved operation is mainly an exponentiation on group elements. Hence its overhead is nearly linear to the number of users involved, e.g., when the user set contains 64 or 128 users, its time cost is 0.8 or 1.61 s. Fortunately, this setup phase will be executed only once before data encryption.

For broadcast encryption and decryption, their overhead almost keep constant when the number of users varies from 64 to 512. This is because the dominating factor is the group paring operation and exponentiation, whose execution times is fixed. The involved group multiplication is linear to the number of users, however, whose cost is less expensive. In Fig. 5, broadcast encryption costs about 11 ms while decryption costs about 8 ms. This demonstrates that even the broadcast encryption operation is carried out by patients, it will not put a heavy burden on them.

Finally, according to Sect. 3.3, the key regression overhead caused by key derivation is actually two hash operations for each key rotation operation, which is really fast on modern devices so this cost can almost be omitted.

5.2 Smart Contract Overhead

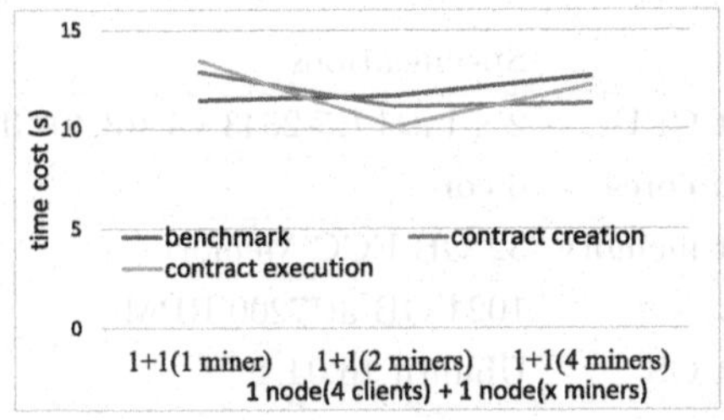

Fig. 6. Time cost of smart contract creation and execution (Color figure online)

We divide the smart contract test into two steps. Firstly, we run an Ethereum client node in a server to mine blocks without including any transactions, whose mining time can be used as a benchmark, as indicated by the blue curve in Fig. 6. Secondly, we test the time cost of smart contract creation and execution under Ethereum test platform. Specifically, we deployed two servers in our lab network, whose hardware and software environment is depicted in Table 2. To simulate the business logic of clinical data management, one server is deployed with three Ethereum Geth clients to denote a provider, a patient, and a user respectively; the other server are equipped with mining clients with their number varying from 1 to 4.

Figure 6 illustrates the creation overhead of smart contracts (orange curve) and execution overhead of contract functions (gray curve). We can see that both overheads are between 10 and 14 s, which falls in the typical time range (10 to 20 s) in Ethereum. It is worth noting that both the creation and execution of a smart contract include the time period from creating or executing a smart contract to the point when it is included in a successfully mined block. We found the execution time of smart contracts is mainly decided by the block mining operation, which is further decided by the PoW consensus mechanism adopted in current Ethereum.

Finally, our test result includes 90 trials of smart contract execution and block mining. The average time of smart contract creation and execution is about 11.9 s, which is 2.49 s less than current Ethereum block mining time (14.39 s, November 11, 2019). This is reasonable, because our test environment is a LAN, where the propagation time for a transaction to be broadcast to the majority of peers can almost be omitted.

Discussion. We choose Ethereum and its smart contract language solidity to implement the business logic of medical data management is because of its code maturity. However, there are also other options such as Hyperledger. Actually, the gas consumption necessary for smart contract execution in Ethereum is a big burden for its applications, especially considering the fact that medical data access events happen frequently. In the future, we will consider our implementation in Hyperledger Fabric to increase its throughput to fit the high-frequent daily events of medical data management.

6 Conclusion and Future Work

This work is an attempt to provide secure medical data integration and sharing among healthcare providers, patients, and third-party researchers without privacy violation. Based on previous studies, we find that many state-of-the-art schemes choose to store patients' medical data in plaintext format in hospital databases. We regard it as a serious weakness which potentially may bring data leakage events. To address this problem, we adopt broadcast encryption to encrypt data encryption keys, and combine it with key regression and data classification policy to enforce fine-grained access and privacy control. Three kinds of smart contracts are designed to integrate underlying components with

Ethereum blockchain to implement the business logic of clinical data management. Our performance evaluation shows the cryptographic overhead brought by broadcast encryption and key regression is reasonable, and the execution time of smart contracts falls in an expected time range.

Acknowledgment. The authors would like to thank the anonymous referees for their reviews and insightful suggestions to improve this paper. This work is partially supported by the National Science Foundation of USA (Award No. 1547428, No. 1738965).

References

1. General data protection regulation (2016). https://eugdpr.org/the-regulation/
2. Summary of the HIPAA security rule (2017). https://www.hhs.gov/hipaa/for-professionals/security/laws-regulations/
3. https://www.healthcareitnews.com/projects/biggest-healthcare-data-breaches-2018-so-far (2018)
4. Azaria, A., Ekblaw, A., Vieira, T., Lippman, A.: MedRec: using blockchain for medical data access and permission management. In: International Conference on Open and Big Data (OBD), pp. 25–30. IEEE (2016)
5. Boneh, D., Gentry, C., Waters, B.: Collusion resistant broadcast encryption with short ciphertexts and private keys. In: Shoup, V. (ed.) CRYPTO 2005. LNCS, vol. 3621, pp. 258–275. Springer, Heidelberg (2005). https://doi.org/10.1007/11535218_16
6. Buterin, V.: A next-generation smart contract and decentralized application platform. White Paper (2014)
7. Cachin, C.: Architecture of the hyperledger blockchain fabric. In: Workshop on Distributed Cryptocurrencies and Consensus Ledgers, vol. 310 (2016)
8. Ethereum: Proof of stake FAQ (2014). https://github.com/ethereum/wiki/wiki/Proof-of-Stake-FAQ
9. Fan, K., Wang, S., Ren, Y., Li, H., Yang, Y.: MedBlock: efficient and secure medical data sharing via blockchain. J. Med. Syst. **42**(8), 136 (2018)
10. Fiat, A., Naor, M.: Broadcast encryption. In: Stinson, D.R. (ed.) CRYPTO 1993. LNCS, vol. 773, pp. 480–491. Springer, Heidelberg (1994). https://doi.org/10.1007/3-540-48329-2_40
11. Fu, K., Kamara, S., Kohno, T.: Key regression: enabling efficient key distribution for secure distributed storage (2006)
12. Guo, R., Shi, H., Zhao, Q., Zheng, D.: Secure attribute-based signature scheme with multiple authorities for blockchain in electronic health records systems. IEEE Access **776**(99), 1–12 (2018)
13. Hur, J., Noh, D.K.: Attribute-based access control with efficient revocation in data outsourcing systems. IEEE Trans. Parallel Distrib. Syst. **22**(7), 1214–1221 (2011)
14. Nakamoto, S.: Bitcoin: a peer-to-peer electronic cash system (2009). http://bitcoin.org/bitcoin.pdf
15. Peterson, K., Deeduvanu, R., Kanjamala, P., Boles, K.: A blockchain-based approach to health information exchange networks. In: Proceedings of the NIST Workshop Blockchain Healthcare, vol. 1, pp. 1–10 (2016)
16. Qiu, H., Noura, H., Qiu, M., Ming, Z., Memmi, G.: A user-centric data protection method for cloud storage based on invertible DWT. IEEE Trans. Cloud Comput. (2019)

17. Qiu, H., Qiu, M., Memmi, G., Liu, M.: Secure health data sharing for medical cyber-physical systems for the healthcare 4.0. IEEE J. Biomed. Health Inform. **24**, 2499–2505 (2020)
18. Sahai, A., Waters, B.: Fuzzy identity-based encryption. In: Cramer, R. (ed.) EUROCRYPT 2005. LNCS, vol. 3494, pp. 457–473. Springer, Heidelberg (2005). https://doi.org/10.1007/11426639_27
19. Sousa, J., Bessani, A., Vukolic, M.: A byzantine fault-tolerant ordering service for the hyperledger fabric blockchain platform. In: 48th Annual IEEE/IFIP International Conference on Dependable Systems and Networks (DSN), pp. 51–58 (2018)
20. Xia, Q., Sifah, E.B., Asamoah, K.O., Gao, J., Du, X., Guizani, M.: MeDShare: trust-less medical data sharing among cloud service providers via blockchain. IEEE Access **5**, 14757–14767 (2017)
21. Yang, H., Yang, B.: A blockchain-based approach to the secure sharing of healthcare data. In: Proceedings of the Norwegian Information Security Conference (2017)
22. Yu, S., Wang, C., Ren, K., Lou, W.: Attribute based data sharing with attribute revocation. In: Proceedings of the 5th ACM Symposium on Information, Computer and Communications Security, pp. 261–270. ACM (2010)
23. Zhang, A., Lin, X.: Towards secure and privacy-preserving data sharing in e-health systems via consortium blockchain. J. Med. Syst. **42**(8), 140 (2018)

Blockchain Meets DAG: A BlockDAG Consensus Mechanism

Keke Gai[1(✉)], Ziyue Hu[1], Liehuang Zhu[1(✉)], Ruili Wang[2], and Zijian Zhang[3]

[1] School of Cyberspace Security, Beijing Institute of Technology, Beijing 100081, China
{gaikeke,3120191003,liehuangz}@bit.edu.cn

[2] School of Natural and Computational Sciences, Massey University, Palmerston North 4442, New Zealand
Ruili.wang@massey.ac.nz

[3] School of Computer Science, University of Auckland, Auckland, New Zealand
zhang.alex@auckland.ac.nz

Abstract. With the advent of the blockchain technology, low throughput and scalability have gradually become technical bottlenecks. A DAG (Directed Acyclic Graph)-based blockchain system is deemed to be a potential solution to addressing both issues. However, constructing consensus protocol to meet the requirement of the consistency in a networked environment is an unsolved challenge. In this paper, we propose a novel DAG-oriented consensus mechanism. Specifically, our approach sorts and merges original blocks from a DAG structure and re-construct a single-chain-based blockchain system; hence, consensus in DAG can be achieved on new formed blocks through running the proposed global ordering scheme and block mergence operations. Blockchain-related functions can be retrieved from splitting merged blocks.

Keywords: Blockchain · Directed Acyclic Graph · BlockDAG · Consensus mechanism · Graph theory

1 Introduction

Blockchain technology has been introduced to the public recently due to its distinct characteristics, such as tamper-resistant and transparent consensus. From Bitcoin to Ethereum, blockchain has experienced the technology alternate jumping from single purpose of cryptocurrency to smart contract-based multi-purpose applications [1,6,9]. With the increasing extent of adoption, blockchain has attracted remarkable attentions from both academia and the industry. However, a notable restriction in blockchain implementations is that transactions/service throughput remains a low level, which has not reached an equative capability yet to the existing systems, e.g., Visa and AliPay [3,7,13].

In most current deployments of blockchain, a newly added block uses a hash pointer to attach to its father block. All blocks follow the mechanism so that it

M. Qiu (Ed.): ICA3PP 2020, LNCS 12454, pp. 110–125, 2020.
https://doi.org/10.1007/978-3-030-60248-2_8

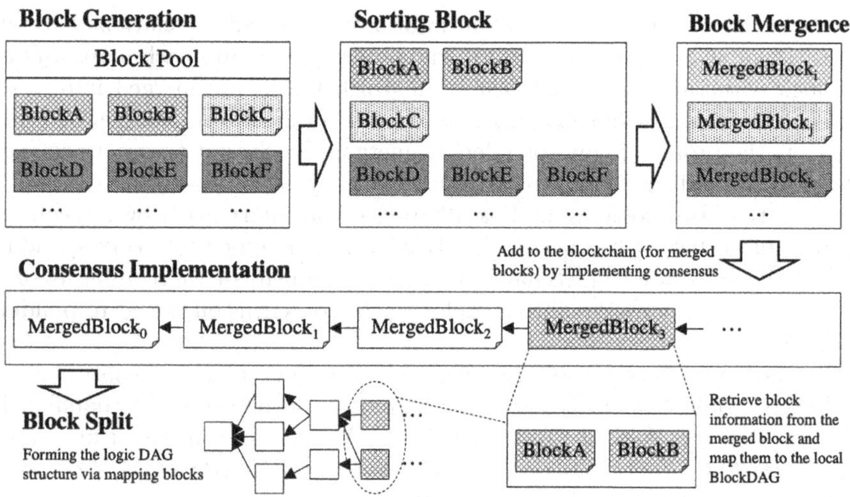

Fig. 1. Technical architecture of the proposed BlockDAG.

formulates the single-chain creation method [8,14,16]. The single-chain structure introduces the limitation, like limiting throughputs capability and the scalability.

Emerging blockchain technology with *Directed Acyclic Graph* (DAG), also known as BlockDAG, is deemed an alternative technical idea to solve the bottleneck issue above [2,4]. Since Lernet *et al.* [10] proposed the concept of DAG-chain in a while paper, using the DAG to replace the single-chain structure presents more benefits than improving the topology of blockchain systems. The improvement made by BlockDAG technology underlies the amelioration of the block structure as well as embracing DAG networking structure [11,12]. The involvement of network features and graph theory-related optimizations has lighted up a potential access to solving the low throughput issue of blockchain. Specifically, as a network structure, DAG inherent has an advantage in high-throughputs and scalability due to the setting of block forks [5,15].

BlockDAG theoretically possesses merits in efficiency enhancement; however, constructing a BlockDAG system still encounters a few technical obstacles, such as establishing a consensus over the network structure. Due to the network topology of BlockDAG, broadly existing block forks further amplify the impact of the network latency on consensus. Even though BlockDAG can create more blocks in a certain period than that of a (single-chain) blockchain due to the advantage of parallel computing, the challenge remains on ensuring each transaction is validated and is recorded only once.

Addressing all challenges above, in this work we focus on constructing a consensus mechanism that is adoptable in a DAG-enabled blcockahin system. The basic idea of the proposed consensus is to transfer a network-oriented problem to a single-chain problem, which relies on a proposed rule for merging blocks. Differ from blocks in the typical blockchain, the concept of *Block Mergence* (BM) records blocks containing transaction information as well as relations between blocks rather than pure transactions. The proposed BM approach maps a DAG to a logic structure on a single chain, in which blocks are merged into a few blocks on one chain by implementing a pre-defined rule. We emphasize the "mergence" as the proposed concept refers to merging blocks not transactions, e.g., different from Fabric consensus. A blockchain system is the actual carrier for recording the DAG logic as well as all transaction information over a consensus. It implies that each state of BlockDAG will be mapped to a corresponding state of the BM-based blockchain. An accomplishment of the consensus on a BM-based blockchain is equal to completing the consensus on the corresponding BlockDAG.

The primary objective of this work is to achieve a global consensus for BlockDAG, while considering high throughput performance and scalability. In order to reach this goal, the proposed BlockDAG system consists of five major phases, which are *Block Generation* (BG), *Sorting Block* (BS), BM, *Consensus Implementation* (CI), and *Block Splits* (BS).

Figure 1 presents the high architecture of the proposed approach. At the first phase, BG is mainly responsible for creating original blocks (before merging blocks). When a transaction is generated, it will be added to the block pool of the nearest blockchain node. The node validates all transactions in the block pool and packs them up into a few blocks, which will be added to the system block pool awaiting for block validations. The second phase, SB, sorts all unvalidated blocks in the block pool. A sorting algorithm is proposed in this work, which makes a sequence for unvalidated blocks in the DAG structure. Both double spending and consensus conflict problems have been addressed by the proposed approach. Next, the fourth phase BM refers to the mergence manipulation over the unvalidated blocks during the sorting operation, which is getting ready to implement a global consensus. In addition, our proposed consensus algorithm will be ran in the phase of CI, in which a consensus for DAG will be completed by implementing a blockchain-oriented consensus. Finally, BS phase is a process that splits those merged blocks into original states as well as locate blocks to on-premise BlockDAG structure.

Main contributions of this work are summarized in the following:

1. We propose a new BlockDAG model with a global block ranking algorithm for constructing consensus, which conducts two types of blocks to formulate the sequential order, namely, parent and child blocks. Our approach determines the block sequence by estimating the relations between parent blocks, by which forged blocks and double-spend attacks can be eliminated. We propose a layered sorting algorithm for ordering blocks in DAG, in order to defend adversary blocks and double spending attacks.
2. This work proposes a consensus mechanism for BlockDAG, which simplifies the consensus problem in DAG by transferring the network structure to a single-chain structure. We conduct a method that merges a few blocks into one block while considering the preceding-succeed relation of merged blocks.

Mathematical model has been established in this work that proved model had an advantage in offering high-throughputs with a good scalability.

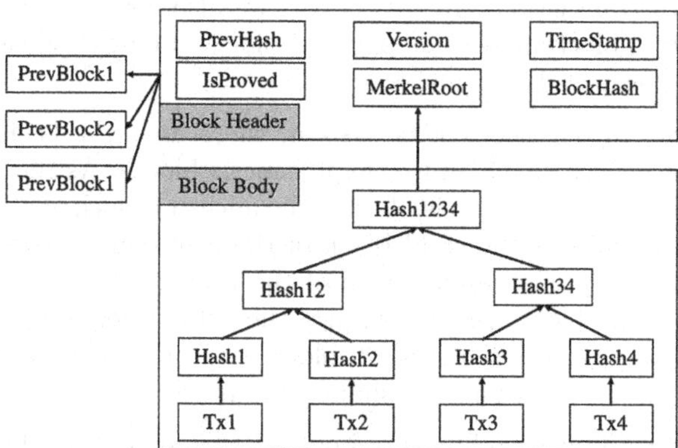

Fig. 2. Illustration of block structure in our BlockDAG.

The rest of the paper is organized by the following order. Section 2 introduces the detailed model design and the working procedure of our BlockDAG system. In addition, core algorithms are described in Sect. 3. Moreover, Sect. 4 shows the implementation and evaluation of our BlockDAG system. Finally, Sect. 5 draws a conclusion of our work.

2 Concepts and the Proposed Model

2.1 Main Concepts and Definitions

In our model, we define a few key concepts, including Leaf-block, Merged-Blockchain, MergedBlock, and BlockPool. A **BlockDAG** allows blocks to be connected with each other by implementing a DAG network structure. We give a definition a G = {V, E} as DAG, where $V = \{v_1, v_2, v_3, \ldots, V_n\}$ is the set of blocks, and $E \subseteq V \times V$ is the edge set which describe the precedence relationship among the Blocks in V.

A **block** is responsible for recording transaction data, denoted by $Block = \langle Header, Body \rangle \in V$, where $Header = \langle PrevHashs, Version, TimeStamp, IsProved, MerkelRoot, BlockHash \rangle$, and $Body = \{T_x, Count\}$. Specifically, we have $PrevHashs = \{PrevHash_1, PrevHash_2, PrevHash_3, \ldots, PrevHash_n\}$ be a set of hash values for its parent block, which defines connection relationship between blocks. $Header$ identifies the block. $Version$ refers to the system's version number; $TimeStamp$ refers to the block time; $Isproved$ is a binary value ($True$ or $False$) that describes whether block is validated; $MerkelRoot = Hash(Tx_1, Tx_2, \ldots, Tx_n)$ is a hash value deriving from all transactions in a block, which can be considered the presentation of *Body*'s content in the *Header*; $BlockHash = Hash(Header, Body)$ is a hash value from $\langle Header, Body\rangle$ as a label for a block. Specifically, *Body* records all transaction information in the block. $T_x = \{Tx_1, Tx_2, Tx_3, \ldots, Tx_n\}$ denotes a set of transactions and *Count* is the number of transactions. In this paper, we also consider all blocks physically recorded in a blockchain to be physical blocks. Figure 2 illustrates a block structure of our proposed system.

Genesis Block: In our BlockDAG system, there are $(g + 1)$ genesis blocks, including 1 root genesis block and g shape genesis blocks. As the starting node of the entire DAG, the root genesis block is the ancestor block of all other blocks; the shape genesis block is the child block of the root genesis block. The shape genesis block has only one parent block of the root genesis block; the number of the shape genesis block, m, determines the bifurcation speed of the DAG structure. For example, when g takes 3, block 0 becomes the root genesis block and blocks 1, 2, and 3 turn into the shape genesis blocks.

Logical Blocks: All physical blocks of LogicalBlock are classified by our defined rules, which are divided into multiple logical "equivalent" logical blocks. That is to say, logical blocks are a type of physical blocks meeting a certain characteristics set. "Equivalence" here refers to the equivalence in the parent-child relationship of the block. Logical blocks are nodes in BlockDAG, while physical blocks are not. In other words, BlockDAG consists of the genesis block and the logic block.

Growth Cycle: BlockDAG jumps into a growth stage once genesis block(s) is/are created. New blocks are continuously generated and are added to BlockDAG. A growth cycle is a period between two adjacent waves of new blocks.

Leaf-Blocks: Leaf-blocks denote blocks whose in-degree is equal to 0, which is a collection of all logical blocks without sub-blocks in BlockDAG when a new growth cycle starts. The number of leaf blocks is an m. The initial leaf block is the system's shape genesis block; the number of leaf blocks g is equal to the number of shape genesis blocks m.

MergedBlock: Different from the traditional block concepts, a *MergedBlock* records original blocks rather than transactions' information. A $Merged\ Block = \langle MergedHeader, MergedBody\rangle \in MergedV$, where $MergedHeader = \langle \text{PrevHash}, Version, TimeStamp, MerkelRoot\rangle$ and $MergedBody = \langle Block, Count\rangle$. MergedHeader represents the identity of the block. Different from the *Header* in *Block*, *PrevHash* in *MergedHeader* stores its parent block's hash value that is a value rather than a set. *MergedBody* records all blocks' information in *MergedBlock*, where $Block = \{Block_1, Block_2, Block_3, \ldots, Block_n\}$ is a set of *Blocks* and *Count* is the number of *Blocks*.

2.2 BlockDAG System Installation

Our BlockDAG system fully keeps the concept of the block, which differs from methods in IOTA and ByteBall. Transaction information is recorded in original blocks but the consensus is constructed over the merged blocks. In this section, we mainly introduce two crucial aspects in the system installation, which include network installation over blocks and the expansion of the system.

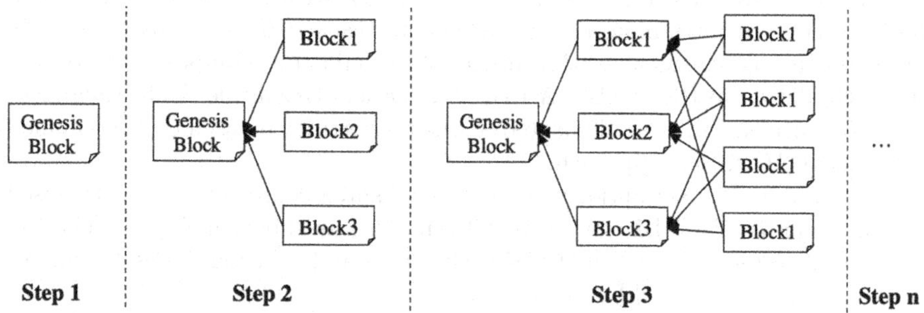

Fig. 3. Implementation process of the BlockDAG's installation and expansion.

Network Installation over Blocks: The chain-connection mechanism in our approach is similar to classic blockchain systems, which relies on hash pointer-based methods. A major difference is that the previous block's field stored in the header is a hash set. A block can have several previous blocks.

Assume that there exists a block with $BlockB \in V$. If its previous blocks are A_1, A_2, A_3, ..., A_n ($A_i \in V$, i = 1, 2, 3, ..., n), there will be a $B.Header.PrevHashs = \{A_1.Header.BlockHash, A_2.Header.BlockHash, A_3.Header.BlockHash, \ldots, A_n.Header.BlockHash\}$ and $\{\langle B, A_1\rangle, \langle B, A_2\rangle, \langle B, A_3\rangle, \ldots, \langle B, A_n\rangle\} \in E$.

BlockDAG Installation and Expansion: In BlockDAG, when a new block is created, a hash pointer will be used to refer it to n blocks from the Leaf-Blocks in the local nodes of BlockDAG. Suppose there exists a new Block C, the node will select n Leaf-Blocks as the parent block of Block C. This is a basic mechanism for all nodes. When Block C is validated after verification, there will be $C \in V$, and $e = \langle C, D\rangle | D \in Leaf - Blocks \in E$ after the consensus. The system eventually expands into a DAG structure that is composed of blocks by running this mechanism. For instance, Fig. 3 illustrates a schematic diagram of the process. The number of the shape genesis blocks (m) is 3; the number of the parent blocks of the first layer of physical blocks (n) is 2.

2.3 System Design

As mentioned in previous sections, a DAG structure is implemented as the topological structure between nodes, since the system is designed for higher-level throughput and scalability. A key issue in a DAG-based blockchain is to solve the global consensus problem.

The design objective of the proposed approach is described in the following. Assume that there exists a BlockDAG, denoted by $G = \{V, E\}$. We merge a number of blocks (*Block* $\in V$) into a group of new blocks (*MergedBlock* $\in$ *MergedV*) in order to a BlockDAG can be logically presented in a manner of chain structure ($MergedBlockChain$). In a line with the prior presentation, the BlockDAG is corresponding with a certain $MergedBlockChain$. Therefore, we utilize the consensus algorithm in the blockchain to accomplish the consensus manipulation in BlockDAG. A PBFT (Practical Byzantine Fault Tolerance) consensus mechanism is implemented as the consortium blockchain is combined with BlockDAG in our approach.

There are mainly six major phases accomplishing consensus in our proposed system, including BG, BL, SB, BM, CI, and BS, as shown in Fig. 1. The followings will present functionality of each phase and relations between phases through a sample of BlockDAG, $G = \{V, E\}$.

(I) *BG (Block Generation) Phase.* This phase is responsible for creating blocks in BlockDAG. When a new transaction is triggered, the transaction will be collected by the nearby node and be added to its block pool. When blocks in the pool match a certain rule of the mergence, the node will pack blocks into a new block. For a BlockDAG (G), the original state $V = \{\emptyset\}$. The system firstly will generate a genesis block as the first block of the system, $GenesisBlock = \{Header, Body\} \in V$. After the genesis block is created, each time when a new block is created, $Leaf - Blocks$ will be generated by the node of this block. All $Leaf - Blocks$ created by the node will be quoted when the $Block$ is validated, that is $Block.Header.PrevHashs = \{Leaf - Block_i.BlockHash — i = 1, 2, \ldots, len(Leaf - Blocks)\}$. After the entire consensus is completed, the $Block$ is validated so that it turns into a new $Leaf - Block$ and all its quoted blocks are no longer $Leaf - Blocks$. A block will be added to the system's *BlockPool* when it is successfully generated by a node, which will be used for further sorting operations and validations in the next phase.

(II) *BL (Block Layering) Phase.* BL is a core that classifies physical blocks generated by the nodes according to the parent blocks of the physical blocks. The physical blocks of the same type are merged into logical blocks, and the same block is generated within a time All logical blocks of are blocks of the same layer. A logical block is a block that actually exists in the layered BlockDAG. Among them, the $0th$ layer of the BlockDAG is the root genesis block; the first layer is the shape genesis block; the third layer is the logical block after the physical blocks that directly take the shape genesis block as the parent block.

(III) *SB (Sorting Block) Phase.* The SB phase is one of the core components in our system. The Sorting Block module is one of the core parts in this system, and the Block Sorting module is responsible for sorting the Blocks in the BlockPool and verifying the transactions therein. After Block Layering, a layered BlockDAG structure is formed. Block Sorting is sorted according to the layered structure of BlockDAG. The hierarchical structure has a natural sorting advantage. The logical blocks with a small number of layers are ranked first, and the blocks with a large number of layers are ranked backward.

For ordering transactions, the first step is that transactions are sorted in the order to the block index, following a First-in-First-Out strategy over indices. Next, for transactions in one block, the sorting depends on the transaction time. After steps above, we can obtain a global sorting for all transactions.

A examination for detecting the conflict of the transaction is processed after completing the global sorting operation. Once the conflict is detected, the system will retrain the previous order and mark the later one an invalid, such that double-spend attacks can be avoided. The SB phase is implemented by the main nodes of the consensus protocol.

(IV) *BM (Block Mergence) Phase.* This phase merges the ordered *Blocks* that are obtained from the SM phase into a group of *MergedBlocks*. In essence, *Blocks* in *BlockPool* are in order, $BlockPool = \{Block_1, Block_2, \ldots, Block_n\}$. Based on this state, the system merges all *Blocks* into one/ a few *Merged Block(s)*, denoted by $MergedBlock.MergedBody = \{Block_1, Block_2, \ldots, Block_x, Count\}$. Those nodes that are responsible for sorting operations are called *Sorting Nodes.* Once sorting nodes receive *MergedBlock*, they duplicate blocks to other nodes following the consensus protocol, in order to achieve global consensus over DAG.

This phase represents the core idea of our approach, which relies on a method of merging original blocks so that BlockDAG structure can be mapped to a chain. It means that BlockDAG is in a line with the *MergedBlockChain*, which transfers a networked consensus problem to a blockchain consensus problem. The basic unit of our BlockDAG system is a *Block*, in which the block body is formed by a transaction Merkel tree and records transaction information. In the *MergedBlockChain*, the basic unit is a *MergedBlock* and its block body is formed by a block Merkel tree and records *Block*'s information.

(V) *CI (Consensus Implementation) Phase.* The CI phase is a process for implementing *MergedBlockChain* protocol that is responsible for achieving a consensus over *MergedBlockChain*. We refer to the PBFT consensus protocol (one of common protocols in consortium blockchain) in our system. Assume that a *MergedBlockA* is obtained from the BM phase, the main node within PBFT protocol will duplicate the PBFT protocol used by *MergedBlockA* to other nodes. Other nodes add the *MergedBlockA* to the local *MergedBlockChain* once they receive *MergedBlockA*, denoted by $A \rightarrow MergedV$. A consensus is completed once $A \rightarrow MergedV$ is done.

In our system, operations at other phases are for BlockDAG. The CI phase's operations are designed for achieving consensus in blockchain. A blockchain consensus is used to solve the BlockDAG consensus problem. After the system completes the consensus over the merged blocks, other nodes can retrieve original blocks from splitting the *MergedBlocks*. A global consensus can be achieved.

(VI) *BS (Block Split) Phase.* The last phase of our BlockDAG system is the BS phase, which updates states of local nodes by using validated *MergedBlocks* and eventually completes the consensus operation. For example, nodes in Merged-BlockChain consensus phase.

3 Algorithms

In the proposed BlockDAG system, three critical components include the block sorting, block merging, and block splitting implementations. The challenging issue, as mentioned in prior sections, is to construct a sorting algorithm fitting in the BlockDAG structure, while considering the block mergence/ splits for matching the requirement of the consensus.

3.1 BlockDAG Sorting (BDS) Algorithm

The function of the BlockDAG sorting algorithm is to sort all logical blocks in the entire BlockDAG, solve the transaction conflict problem, and merge logical blocks to facilitate the merger blockchain consensus (according to the global sorting results).

Algorithm 1. BlockDAG Sorting (BDS) Algorithm (blocks, globalBlockOrder)

```
Input: blocks []block, globalBlockOrder *block    //sorting results of existing blocks
Output: blockOrder *block    //block sorting results after adding new blocks
for block ← from block_1 to block_n do
    e, d ← call GetTheAverageAndVariance(block)
    E[block] ← e    //average of parent block order
    D[block] ← d    //variance of parent block order
end for
call Sort(E, D)    //sort by variance
for block_1, block_2, ...in E do
    call AddBlockToOrder(block, blockOrder)
        //add the block to the sorting result
    call AddBlockToOrder(block, globalBlockOrder)
        //add the block to the local block sorting sequence
end for
return blockOrder
```

To address BlockDAG's partial ordering issue, we propose a solution as follows. First, we layer the DAG structure and utilize the advantage of the layered structure to realize a global sorting in BlockDAG. Ordering logical blocks depends on the block's layer. For those blocks with the same layer, the order relies on the layer of each block's parent block(s). The hierarchical structure of sorting is the key in our approach to solving global sorting issue in DAG.

The input of BDS algorithm include the set *LogicalBlock* deriving from the same layer generated within a block time and a global sorting sequence of BlockDAG blocks (right before the execution of BDS algorithm). The output is the sorting result of the set *LogicalBlock*.

The algorithm is mainly composed of following parts:

GetTheVariance. The function *GetTheVariance* is to calculate the variance of the position of each block's parent blocks in the global sorting order. We configure that the block with a smaller variance has a higher-level priority as a preference setting.

AddBlockToOrder. The function *AddBlockToOrder* enables the **Orderer** to add block(s) to the current sorting. When a block's credibility reaches the threshold, the block will be added to the **Orderer**'s global block sorting queue (deemed higher-preference blocks); otherwise, blocks are added to the waiting queue (deemed lower-preference blocks). After all higher-preference blocks are added to the **Orderer**'s global block sorting queue, lower-preference blocks are added to queue one-by-one. This scheme evaluates the variable "credibility" so that blocks are ordered by a configured preference. Double spend attacks and malicious nodes attacks can be prevented.

Algorithm 2. Block Mergence Algorithm (blocks,*leader_DAG,*leader_mbc)

```
Input: blocks []block, leader_DAG *BlockDAG, leader_mbc *MergedBlockChain
Output: mergedblock MergedBlock
for block1, block2, block3, ...in blocks do
    if !Verified(block, VerifiedBlocks) then
        continue;        //discard invalid blocks
    end if
    call  Add(VerifiedBlocks, block)
end for
bmt = call  BlockMerkelTreeGen(VerifiedBlocks)
    //create new mergedblock
mergedblock → body → treebody = bmt
mergedblock → body → block_num = call len(VerifiedBlocks)
mergedblock → head → preMergedBlockHash = leader_mbc → tail
mergedblock → head → timestamp = call GetTimeStamp(system.Now())
mergedblock → head → merkelRoot = bmt → root
call leader_DAG.AddBlocks(VerifiedBlocks)        //add physical block
call leader_mbc.AddMergedBlocks(mergedblock)     //add merged block
return mergedblock
```

Pseudo codes of the algorithm are shown in Algorithm 1. In Algorithm 1, *blockOrder* is an ordered sequence of blocks. It is the result of the sorting algorithm; E and D are data structures which record the blocks' indexes and their parent blocks' average block order number and variance; *globalBlockOrder* is a ordered sequence of existing blocks. First, the Algorithm 1 call the function GetTheAverageAndVariance to calculate every block's e and d, and record them in E and D. Then, the Algorithm 1 sort the blocks by blocks' variance and get the result. Finally, executor of the Algorithm 1 add the blocks to *blockOrder* and its *globalBlockOrder*.

3.2 Block Mergence Algorithm (BMA)

BMA algorithm is to verify and pack up non-consensus blocks in the *UnprovedBlock* pool into *Merged Blocks*, in preparation for the next step of consensus. The leader node in PBFT will run this algorithm during the phase of merging blocks. Inputs of the algorithm include the ordered block set in the *UnprovedBlock* pool (obtained from the block sorting), local blockDAG structure (*leader_DAG*), and the merged blockchain structure (*leader_mbc*). The output is a merged block that contains all verified blocks. The pseudo code of the algorithm is described as Algorithm 2. It should be noted that the right arrows in Algorithm 2 all indicate pointers.

In Algorithm 2, *VerifiedBlocks* is an array of verified blocks; *bmt* is the merkel tree of blocks in the block body; *mergedblock* is the final merged block. First, the algorithm traverse the *UnprovedBlock* pool by validating blocks and adding verified blocks to *VerifiedBlocks*. The blocks that don't pass the validation will be discarded. After that, the function *BlockMerkelTreeGen* use the *VerifiedBlocks* array to generate the block merkel tree (*bmt*). Then, the algorithm fills the new mergedblock with the variables that we've got, like *bmt*. Finally, we get a new *mergedblock* and the leader node who execute this algorithm update its BlockDAG with this new *mergedblock*.

4 Experiment and the Results

We have evaluated the BlockDAG sorting algorithm proposed in this paper and compared the test results with the *Phantom* sorting algorithm. The specific details of the experiment are as follows.

The equipment used in this simulation experiment was a personal laptop with Windows 10 pro OS. The hardware configuration included a computer (DELL Inspiron 14R 5420) with an Intel(R) Core(TM) i5-3210M CPU @ 2.50 GHz 2.50 GHz, a 4 GB RAM and a hard driver WDC wd7500bpvt-75a1yt0.

The performance of the sorting algorithm was tested in this experiment. When testing BlockDAG sorting algorithm, we choosed to compare our scheme with Phantom algorithm, and compared the sorting time consuming of the two algorithms for the same BlockDAG structure. We used Golang to implement our sorting algorithm and *Phantom* sorting algorithm respectively, and sorted the same BlockDAG structure on the same computer. Golang version is go1.13.1, and the Golang IDE we used is Goland whose version is 2019.2. *BlockDAG.A* is a simple BlockDAG structure which is composed of 11 blocks, while *BlockDAG.B* expands the A, its blocks have more complex relationship. Specific experiments and results are described below.

4.1 BlockDAG Use Cases

In this section, we used two algorithms to sort two different BlockDAGs, respectively. *BlockDAG.A* was a simple DAG structure that was composed of 11

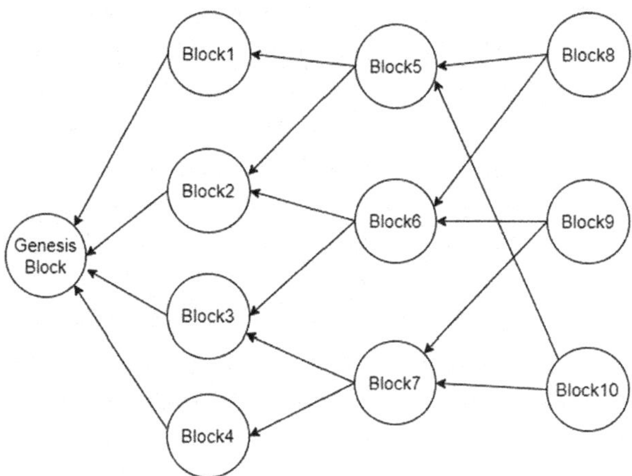

Fig. 4. DAG structure of BlockDAG (A).

Table 1. Results collected from Experiment (A)

Indicator	Our algorithm		Phantom algorithm
D	Step 1	0, 0, 0, 0	
	Step 2	0.25, 0.25, 0	
	Step 3	1, 4.22, 4.00	
Time-consuming	Step 1	7.0065 ms	39.041 ms
	Step 2	6.9281 ms	
	Step 3	6.0024 ms	
	Total	19.937 ms	39.041 ms
Sorting result	1, 2, 3, 4, 7, 5, 6, 8, 10, 9		1, 2, 3, 5, 4, 6, 9, 8, 7, 10

blocks, while *BlockDAG.B* expanded the A and its blocks had more complex relationship. Experiment details and results are shown as follows.

The *BlockDAG.A* consisted of one genesis block and ten Blocks. The results of the two sorting algorithms for the BlockDAG structure were shown in Table 1. Table 1 depicted the result of Experiment (A). We could see that our algorithm continued to order the BlockDAG as it expanded, which meant the global block order did not change once it is determined and the block order gradually expanded as the BlockDAG expanded. Thus, the sorting process for the BlockDAG A was divided into three steps, each of which was the sorting of the newly created block. However, the Phantom algorithm sorted the whole BlockDAG structure everytime, and as the BlockDAG structure growed, the global block order produced by the Phantom algorithm may change. Finally, the sorting sequence of blocks obtained by our algorithm was 1, 2, 3, 4, 7, 5, 6, 8, 10, 9, the time consuming was 19.937 ms; the Phantom algorithm's sequence was

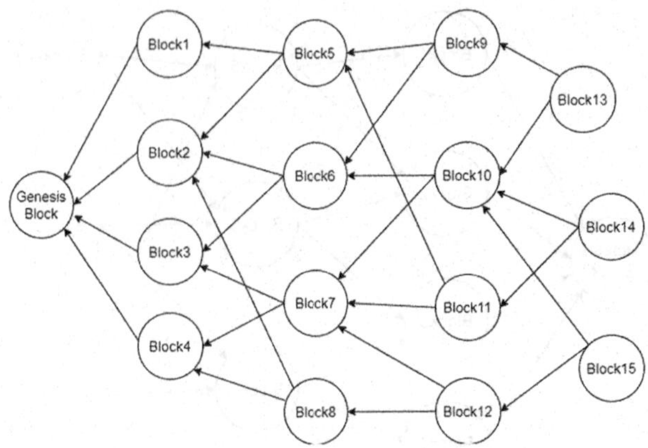

Fig. 5. DAG structure of BlockDAG (B).

Table 2. Results collected from Experiment (B)

Indicator	Our approach		Phantom
D	Step 1	0, 0, 0, 0	
	Step 2	0.25, 0.667, 0	
	Step 3	1, 6.889, 2.25	
	Step 4	4	
Time-consuming	Step 1	5.0027 ms	45.331 ms
	Step 2	4.0027 ms	
	Step 3	4.5015 ms	
	Step 4	1.0012 ms	
	Total	14.5081 ms	45.331 ms
Sorting result	1, 2, 3, 4, 7, 5, 6, 8, 10, 9, 11		2, 3, 4, 1, 7, 6, 5, 8, 9, 11, 10

1, 2, 3, 5, 4, 6, 9, 8, 7, 10, time consuming was 39.041 ms. It showed that our algorithm had an advantage in efficiency (Fig. 4).

Compared to *BlockDAG.A*, *BlockDAG.B*'s blocks had a more complicate relationship with each other. *BlockDAG.B* could be considered as an extension of *BlockDAG.A*. BlockDAG (B)'s structure is shown as Fig. 5. The results of the two sorting algorithms for the BlockDAG (A) are shown in Table 2.

BlockDAG.B could be considered an extension of *BlockDAG.A.* it is highly scalable, so the result of sorting *BlockDAG.A* and *BlockDAG.B* is quite similar, Table 2 showed the result of Experiment (B). Because the *BlockDAG.B* could be considered as an extension of *BlockDAG.A*, and our algorithm was highly scalable, so the result of sorting *BlockDAG.A* and *BlockDAG.B* was quite similar. However, there was an observable change on the block global sorting result made by *Phantom*. It implied that our scheme was superior to *Phantom* in both stability and efficiency (low latency); therefore, our approach would perform better than Phantom in the complex network.

4.2 BlockDAG Random

In the experiment in this section, we set two random parameters m and n to randomly generate a set of BlockDAG structures, where the parameter m determines the number of blocks at each layer of BlockDAG and each block has n parent blocks. We generated 10 BlockDAG structures when the total number of BlockDAG layers were 5 and 7, respectively. The purpose of setting different layers was to observe the impact made by the number of layers on the performance. Comparisons were executed in 20 rounds to compare our scheme with Phantom in performance and results were given in Table 3. Specifically, the first 10 rounds of experiments's subject were BlockDAGs with 5 layers and the other 10 rounds were examined by the 7-layer setting.

Table 3. Results collected from experiment (random)

Round	1	2	3	4	5	6	7	8	9	10
Number of layers	5									
Number of blocks	13	14	16	18	20	26	31	42	49	63
Our time-consuming (ms)	17.462	16.998	19.652	22.576	31.478	39.082	52.838	69.926	89.254	112.323
Phantom time-consuming (ms)	39.054	41.208	44.652	57.970	68.823	89.542	99.358	136.875	177.640	228.312
Round	11	12	13	14	15	16	17	18	19	20
Number of layers	7									
Number of blocks	19	25	32	45	50	63	84	97	116	125
Our Time-consuming (ms)	38.054	50.580	61.830	79.954	88.559	120.663	195.325	218.875	267.778	293.925
Phantom time-consuming (ms)	58.249	92.533	108.544	145.268	180.002	232.980	319.778	396.540	511.121	559.829

Figures 6 and 7 depicted the relationship between the time cost and the number of blocks under different layer settings. Considering the fixed setting of the number of layers, when the number of blocks increased, time costs of the two schemes were both growing, but our scheme's time cost grew slower than Phantom. In addition, when the number of blocks was roughly the same, the growth of the number of layers had an impact on both schemes. Our scheme had a better performance than Phantom, as our scheme retained the previous step's results.

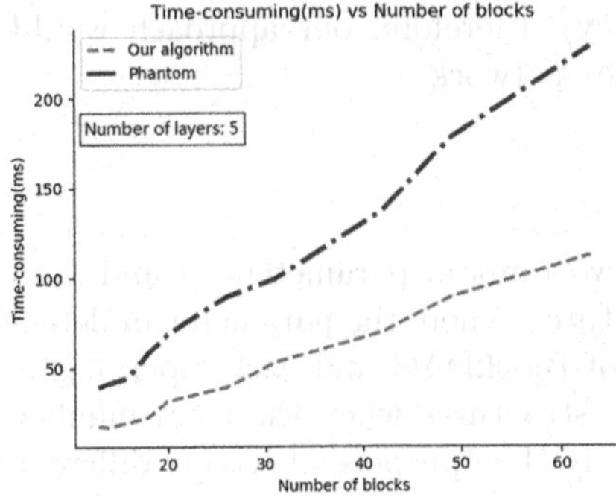

Fig. 6. Time-consuming vs number of blocks when number of layer is 5.

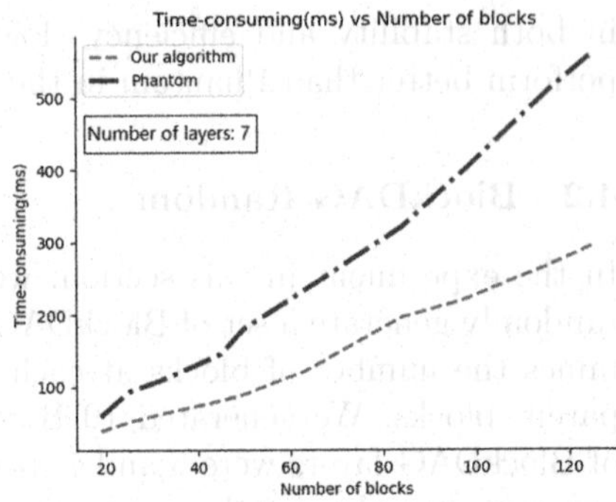

Fig. 7. Time-consuming vs number of blocks when number of layer is 7.

In summary, we observed that the complexity of DAG structure had a positive relationship with the time consumption. Our approach has an advantage in time saving, comparing with *Phantom*. The time consumption on sorting operations was efficiently stable and throughput capability had been dramatically increased.

5 Conclusions

In this paper, we propose a BlockDAG model and a consensus mechanism designed for it, and give the core algorithm BlockDAG sorting algorithm and block merge algorithm. Experiments show that the consensus mechanism can meet the requirements of BlockDAG for consistency, and the sorting algorithm is efficient and stable. Finally, we solved the consensus needs of BlockDAG through this consensus mechanism, and used BlockDAG to improve the transaction throughput of the blockchain system. The results show that through our consensus mechanism, the entire system has excellent characteristics of high scalability, high robustness, and high transaction throughput.

Acknowledgement. This work is partially supported by Natural Science Foundation of Beijing Municipality (Grant No. 4202068), National Natural Science Foundation of China (Grant No. 61972034), Natural Science Foundation of Shandong Province (Grant No. ZR2019ZD10), Guangxi Key Laboratory of Cryptography and Information Security (No. GCIS201803), Henan Key Laboratory of Network Cryptography Technology (Grant No. LNCT2019-A08), Beijing Institute of Technology Research Fund Program for Young Scholars (Dr. Keke Gai).

References

1. Bartoletti, M., Zunino, R.: BitML: a calculus for bitcoin smart contracts. In: ACM SIGSAC Conference on CCS, Toronto, Canada, pp. 83–100 (2018)

2. Benčić, F., Žarko, I.: Distributed ledger technology: blockchain compared to directed acyclic graph. In: 38th ICDCS, Vienna, Austria, pp. 1569–1570. IEEE (2018)
3. Dinh, T., Liu, R., Zhang, M., et al.: Untangling blockchain: a data processing view of blockchain systems. IEEE TKDE **30**(7), 1366–1385 (2018)
4. Gai, K., Choo, K., Qiu, M., Zhu, L.: Privacy-preserving content-oriented wireless communication in internet-of-things. IEEE IoT J. **5**(4), 3059–3067 (2018)
5. Gai, K., Guo, J., Zhu, L., Yu, S.: Blockchain meets cloud computing: a survey. IEEE Commun. Surv. Tutorials **PP**(99), 1 (2020)
6. Gai, K., Wu, Y., Zhu, L., Qiu, M., Shen, M.: Privacy-preserving energy trading using consortium blockchain in smart grid. IEEE TII **15**(6), 3548–3558 (2019)
7. Gai, K., et al.: Permissioned blockchain and edge computing empowered privacy-preserving smart grid networks. IEEE IoT J. **6**(5), 7992–8004 (2019)
8. Guo, J., Gai, K., Zhu, L., Zhang, Z.: An approach of secure two-way-pegged multi-sidechain. In: Wen, S., Zomaya, A., Yang, L.T. (eds.) ICA3PP 2019. LNCS, vol. 11945, pp. 551–564. Springer, Cham (2020). https://doi.org/10.1007/978-3-030-38961-1_47
9. Kosba, A., et al.: Hawk: the blockchain model of cryptography and privacy-preserving smart contracts. In: S & P, San Jose, CA, USA, pp. 839–858. IEEE (2016)
10. Lerner, S.: DagCoin: a cryptocurrency without blocks. White paper (2015)
11. Lewenberg, Y., Sompolinsky, Y., Zohar, A.: Inclusive block chain protocols. In: Böhme, R., Okamoto, T. (eds.) FC 2015. LNCS, vol. 8975, pp. 528–547. Springer, Heidelberg (2015). https://doi.org/10.1007/978-3-662-47854-7_33
12. Wan, Z., Cai, M., Lin, X., Yang, J.: Blockchain federation for complex distributed applications. In: Joshi, J., Nepal, S., Zhang, Q., Zhang, L.-J. (eds.) ICBC 2019. LNCS, vol. 11521, pp. 112–125. Springer, Cham (2019). https://doi.org/10.1007/978-3-030-23404-1_8
13. Xu, C., Wang, K., et al.: Making big data open in edges: a resource-efficient blockchain-based approach. IEEE TPDS **30**(4), 870–882 (2018)
14. Zamani, M., Movahedi, M., Raykova, M.: RapidChain: scaling blockchain via full sharding. In: ACM SIGSAC Conference on CCS, Toronto, Canada, pp. 931–948 (2018)
15. Zander, M., Waite, T., Harz, D.: DAGsim: simulation of DAG-based distributed ledger protocols. ACM SIGMETRICS PER **46**(3), 118–121 (2019)
16. Zhu, L., Wu, Y., Gai, K., Choo, K.: Controllable and trustworthy blockchain-based cloud data management. Future Gener. Comput. Syst. **91**, 527–535 (2019)

Dynamic Co-located VM Detection and Membership Update for Residency Aware Inter-VM Communication in Virtualized Clouds

Zhe Wang, Yi Ren(✉), Jianbo Guan, Ziqi You, Saqing Yang, and Yusong Tan

College of Computer, National University of Defense Technology, Changsha 410073, China
wangzhe_nudt@163.com, {renyi,guanjb}@nudt.edu.cn

Abstract. Inter-VM traffics are witnessed as one of the dominating costs for massive parallel computation jobs in virtualized clouds nowadays. One way to improve the efficiency is to exploit shared memory facilities provided by the hypervisor or the host domain to bypass traditional network path for co-located VMs. Thus it is important to be capable of determining whether two VMs are co-located on the same physical node or not. Existing approaches employ either static or dynamic method to maintain co-located VM membership. Since static methods are not adaptive to support runtime detection of VM existence changes, polling based dynamic methods are adopted, with which the response time to membership change is hold up by pre-configured polling cycle. Moreover, the overhead for the membership change and the scalability over the number of co-located VMs are not considered. In this paper, we propose CoKeeper, a dynamic event driven co-located VM detection and membership update approach for residency aware inter-VM communication. CoKeeper responses faster than polling based methods since the membership updates are immediately visible after the events of VM existence changes. Experimental results show that the response time of CoKeeper is more than 10 times lower than that of polling based method and more stable. It achieves lower CPU and network overhead as well as co-located VM scalability. With real network intensive application workloads, we validate that CoKeeper can ensure user level and Linux kernel transparency even in cases of dynamic VM existence changes.

Keywords: Inter-VM communication · Residency aware · Virtualized cloud · Co-located VM detection · Co-located VM membership update · Event driven

1 Introduction

Virtualization has been widely adopted in constructing the infrastructure of cloud computing platforms for its features such as flexibility, elasticity and ease of management. It provides the advantages of enabling multiple virtual machine (VM) to run on a single physical server to provide cost-effective consolidation, while preserving logical isolation of these co-located VMs. However, for network intensive applications, such as HPC MPI

Z. Wang and Y. Ren—These authors contributed equally to this work.

M. Qiu (Ed.): ICA3PP 2020, LNCS 12454, pp. 126–143, 2020.
https://doi.org/10.1007/978-3-030-60248-2_9

applications, Hadoop MapReduce jobs, and Memcached processing, high density network traffic usually generates heavy workload stress for I/O data moving and I/O event processing across the VM boundaries [1–3]. Inter-VM communication inefficiency is witnessed as a known problem for virtualized cloud environments, even the communication overhead between co-located VMs can be as high as that between VMs residing on different hosts [4–7].

Recent research advances in network I/O virtualization has focused on improving inter-VM network communication performance through SDN and NFV [8, 9]. Representative technologies include SR-IOV, Virtio and DPDK [10–14]. They improve the communication performance either by reducing overhead between the VMs and the physical device or by supporting direct access to the packets on NIC without going through OS kernel. They cannot bypass the network stack and are not designed to mitigate the overhead of inter-VM communication between co-located VMs.

With current trend towards more cores and increasing computing capabilities per physical server, the opportunity for more co-located VMs deployed on same physical host is growing. Researchers have proposed approaches to reduce the latency and enhance the bandwidth of communication between co-located VMs. Among these approaches, residency aware inter-VM communication optimizations have been made to enable co-located VMs to transfer data via pre-established fast shared memory channels, bypassing traditional long network I/O path for VMs residing on separate physical hosts [2, 14–23]. If the VMs are co-located, they will transfer data via shared memory based channel (in local mode), otherwise via traditional network path (in remote mode).

However, co-located VM membership are evolving due to dynamic addition of new VMs or the removal of existing VMs caused by events such as VM creation, VM migration in/out and VM shutdown on current host machine. Moreover, with the requirements of live migration enabled load balance, fault tolerance or higher resources utilization based VM replacement and deployment, the frequency of such events is even increasing. Furthermore, VMs are by design not aware of the existence of one another directly due to the abstraction and isolation support by virtualization, which makes it difficult to determine whether two VMs are co-located.

There are two categories of methods to maintain co-located VM membership in a deterministic way: static and dynamic. The static method collects or registers the membership of co-located VMs during system configuration time prior to runtime [18, 19]. In contrast to static method, dynamic method automatically detects the changes of co-located VM membership and updates the membership transparently and dynamically [2, 15–17]. For Xen, dynamic polling method emerged, which is asynchronous and needs centralized management by the privileged domain (Dom0), represented by XenLoop as a typical implementation [7]. XenLoop periodically gathers co-located VM membership from VMs residing on the same host. The co-location information is not be sent to VMs locating on the same host until next polling cycle after the membership changes, which introduces delayed updates and may lead to some level of inconsistency. Moreover, whether the membership changes or not, the overhead for the polling is inevitable. And the overhead for periodical status collection rises along with the increasing number of co-located VMs.

In this paper, we focus on Xen based approaches. We propose CoKeeper, a dynamic event driven method for co-located VM detection and membership update. We design CoKeeper with an attempt to meet following criteria that we believe are required for residency aware inter-VM communication optimizations:

- **Fast response**: the detection and update of the co-location should be synchronous and the co-located VM membership refreshing is immediately visible upon the occurrence of VM dynamic addition or removal with acceptable response time, to avoid errors that may arise due to delayed update.
- **Low overhead**: the detection of VM addition or removal is supposed to consume less computing and network resources than existing methods. Thus the protocol for co-located VM membership update should be designed and implemented to work in a fashion of low CPU and network overhead.
- **Transparency**: neither intervention from the cloud provider nor application level programming efforts from the cloud tenant are required. Neither the applications nor OS kernel need to be modified to incorporate the proposed mechanism.
- **VM scalability**: due to strengthened hardware and enhanced software stack, the number of co-located VMs deployed on the same physical server is increasing. The proposed method should keep stable in overhead while the number of VMs deployed on the same host rises.

To demonstrate the ability of CoKeeper, we implement a prototype system of it to satisfy above desirable criteria. For CoKeeper, no user level intervention is required. It is user level and OS kernel transparent. And it does not require centralized involvement of Dom0 and provides fresher and more consistent VM co-location information. In contrast to polling approach with fixed polling cycles, CoKeeper is less CPU consuming, and its communication overhead is low and stays stable while the number of co-located VMs increases. We implement CoKeeper on Xen as an integrated part of our XenVMC, a residency aware inter-VM communication accelerator for virtualized clouds.

The rest of the paper is organized as follows. Section 2 analyzes existing work. Section 3 discusses several key issues for designing and implementing a dynamic co-located VM detection and membership update mechanism. Section 4 presents our implementation. Section 5 demonstrates the experimental evaluations and the results. Section 6 concludes the paper.

2 Related Work

2.1 Detecting VM Co-residency with Heuristics for Cross-VM Side Channel Scenario

VMs in virtualized clouds offers large scale and flexible computing ability. However, it also introduces a range of new vulnerabilities. Malicious entities in an attack VM can extract sensitive information from other VMs via cross-VM side channel attacks, which breaks the isolation between VMs. The primary goal of security oriented heuristic methods is to ascertain if a VM is co-located via carefully designed measurements imposed on network traffic, CPU cache, etc., over monitored cross-VM side channel by extracting characteristics such as workload or time consumed [24–27].

HomeAlone is an initial representative work of using side channels to monitor for co-located attack VM [24]. It exploits side channels via the L2 memory cache to detect undesired co-residency as a defensive mechanism. The cloud tenant coordinates its collaborating VMs so that they keep selected portions of cache silent for a period of time and measure whether it has been accessed by other VMs during the resulting quiescent period, which suggests the presence of a potential attack VM which is sharing the same host. From the perspective of a malicious VM, achieving co-residency with a victim VM in the cloud allows it to launch various side-channel attacks that target information leakage. [27] undertakes extensive experiments on EC2 to explore the ways and the effectiveness with which an attacker can achieve co-residency with a victim VM. It discovers timing based side channels that can be used to ascertain co-residency. [25, 26] propose co-located watermarking, a traffic analysis attack that allows a malicious co-located VM to inject a watermark signature into the network flow of a target VM. It leverages active traffic analysis by observing data such as the throughput of traffic received to determine VM co-residency.

For heuristics methods, the intervention of cloud provider or application level programming efforts from cloud tenant is required. Instead of maintaining the membership of all the co-located VMs, the detection is conducted only between the attacker and the victim. And it usually takes more than a few seconds to get the analysis results. For high accuracy detection, even more time is required. High overhead and probabilistic inaccuracy prevent this category of approaches to be used into residency aware inter-VM communication optimizations.

2.2 Co-located VM Detection and Membership Update for Shared Memory Based Inter-VM Communication

For existing static co-located VM detection methods, co-location information is prefigured [17, 18]. And no membership update is allowed during runtime, unless the static file with membership information is modified and the co-located VM list is updated manually via extended API. XWAY refers to a static file that lists all co-located VMs. IVC initially registers the co-location information in the backend driver of Dom0 in a static manner. Since static method is not user level transparent and is not capable of automatically detecting runtime co-located VM membership update, dynamic method emerged [7]. For Xen, the update of co-located VM changes can be detected by examining isolated domain information provided by XenStore [28]. XenStore is a configruration and status information storage space hosted by Dom0 and shared by all do-mains. The information in XenStore is organized hierarchically as a directory in a file system. Each domain gets its own path in the store, while Dom0 can access the entire path. The domains can be notified of the status changes or updates in XenStore by setting watchers on items of interest in the hierarchy.

In previous work, only XenLoop and MMNet support dynamic method. XenLoop provides a polling approach by which a domain discovery module in Dom0 periodically scans all guest entries (items) in XenStore, where each entry represents its state. Then the module advertises the updated information to all the co-located VMs covered by the existed entries and the VMs update their local co-located VM lists. MMNet does not require a coordinator in Dom0. Each VM on the physical node writes to the XenStore directory to advertise its presence and watches for membership updates. When VM addition, removal or migration occur, the IP routing tables are updated accordingly. Technical details of MMNet are not available.

Different from heuristic methods for security scenario, existing work for shared memory based Inter-VM communication achieve lower overhead and can check co-location with deterministic results. And they usually provide the protocols for co-located VM detection and membership update for all the VMs on a physical node instead of only for two or a subset of the co-located VMs.

3 Design Issues

3.1 Design Goals

VM live migration support is an important capability provided by Xen platform. When VM migration occurs, if co-located VM membership is not updated accordingly with fast response, race conditions may occur. For instance, if VM_1 on host A is migrated to host B and its communicating peer VM_2 on node A is not notified about the change, VM_1 will still try to communicates with VM_2 on host A via the previously established shared memory channel, which will lead to connection failure and shared memory release error since VM_1 is no longer present on host A. Similarly, no matter for guest VM addition or for running guest VM removal, it is also necessary to ensure that the co-location information is made visible to other VMs on the same physical node immediately after the event occurs and is kept up to date.

The primary goal of our design is to achieve fast response upon events such as VM addition, removal or migration, to avoid possible inconsistency caused by race conditions due to untimely perception and handling of co-located VM membership changes. All the operations for status collecting and update are supposed to finish in milliseconds. Low overhead is another goal. To reduce resource consumption and to avoid race conditions, we propose to employ the idea of actions on events in our design. Thus the co-located VM membership changes will be caught once they occur by the watchers set on XenStore items. Transparency is an important feature from the viewpoint of system administrator and end users. We believe that by encapsulating expected functionalities into loadable OS kernel modules, instead of putting them at user level, less overhead of data exchanging between user applications and OS kernel will be introduced. Another design goal is to ensure VM scalability. With polling method, it is Dom0 who gathers co-located VM membership from all VMs one by one, which introduces nearly linear overhead proportional to the number of co-located VMs. We need to keep the mechanism work stably even when the number of VMs deployed on the same physical node changes from small to large.

3.2 Analysis of VM Membership Modifications

xl is a representative tool provided by Xen for the administrator to manage VMs. It provides a set of console instructions with parameters that affect the existence status of operated VMs. We analyze *xl* management commands over the VMs and find that the commands which could change a guest VM's existence status can be semantically converted into three basic operations or their combination: VM addition, VM removal and VM migration. We summarize the substitution in Table 1.

Table 1. *xl* commands on guest VMs and their semantic substitutes.

xl command	Usage	Semantic substitution
create	Create a new VM	Add a new VM
reboot	Reboot a running VM	Remove a running VM, add a new VM
shutdown	Shutdown a running VM	Remove a running VM
migrate	Migrate a running VM	Migrate a running VM
destroy	Destroy a running VM	Remove a running VM
restore	Restore a suspended VM	Add a new VM
(VM crash)	(A running VM crashes)	Remove a running VM

3.3 Event and Message Based Light Weight Protocol

Upon the occurrence of VM existence status changes, actions are supposed to be invoked directly or indirectly in Dom0 and DomU to react to and propagate such changes accordingly by exchanging information and triggering related handlers in a collaborative way. However, there is no available protocol in Xen for such synchronous processing. Therefore, we propose CoKeeper, an event and message based lightweight protocol, which serves as a fundamental facility to exchange residency status for dynamic co-located VM detection and the membership update.

CoKeeper enables fast reaction and information propagation both inside and across the boundaries of VMs. Events and messages are different in that events are only visible to the VM where it is created while messages are transmitted between VMs. When a VM existence status change is captured, an event will be created directly inside a guest domain or the privileged domain, which will be captured by the VM where it is locally created, leading to operations defined in the handlers and therefore invoking message exchange across the boundaries of VMs and accordingly generates corresponding events in related VMs indirectly. Different from *Xen event channel*, CoKeeper is light-weight and does not lead to context switches across VMs and virtual machine monitor (VMM).

3.4 Co-located VM Detection and Membership Update: Overview

As an indispensable part of XenVMC kernel modules, CoKeeper enables XenVMC to be aware of dynamic co-located VM existence changes and to update the membership adaptively to the changes. The overview of CoKeeper and its blue-colored components is shown in Fig. 1. CoKeeper's interactive control flows spanning in the host domain and guest domains are also illustrated. The data delivery path and other control flow from a sender VM to a receiver VM by XenVMC are also given, with both its backend and frontend kernel modules concerned. With watchers registered on specific items in XenStore, CoKeeper can detect the membership changes. Each domain maintains a local co-located VM list of the membership. With the self-defined event and message protocol, CoKeeper refreshes the list for each VM dynamically.

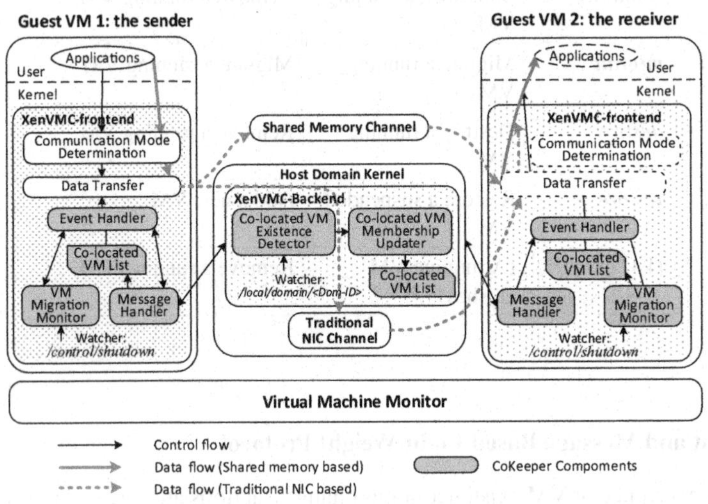

Fig. 1. CoKeeper in XenVMC: overview.

Co-located VM Detection. One of the key issues for co-located VM detection is to catch co-located VM existence status changes. Therefore, we register watchers for related items of XenStore in both Dom0 and DomU. As shown in Fig. 1, to discover VM addition and removal, in the co-located VM existence detector, one watcher is registered on item "*/local/domain/<Dom-ID>*" by Dom0. This item is only accessible by Dom0 and serves as a monitor of the guest VM's existence. For VM live migration support, one of the key issues is to perceive the migration so that it is possible to prepare for the migration beforehand. Thus, in VM migration monitor of every DomU (shown in Fig. 1), we register a watcher on the item of "*/control/shutdown*" in XenStore since its value changes from null to "*suspend*" to indicate that the VM is going to migrate. To pause the original migration process for possible communication mode switches, pending data handling, etc., we unregister its original watcher and reregister it into the FIFO watcher list, so that it is invoked after the pre-handling.

Co-located VM Membership Update. Since VMs are not directly aware of the existence of one another, so we use the proposed event and message protocol to update the co-located VM lists of VMs on the same physical node. The basic idea of the event driven approach is that: we define a series of events and messages, and implement their handlers; on the detection of VM co-location change events, the VM that discovers the change refreshes its local co-located VM list first, then it propagates invoked messages upon the events, the VMs that receive the messages update their local co-located VM list synchronously. If a VM is communicating with another guest VM, and it is going to migrate in or out of current node, then before the migration: i) pending data remained from former local/remote connections need to be handled properly to avoid data loss, ii) communication mode switches from local to remote, or vice versa, iii) release the shared memory if previous communication mode is local.

4 Implementation

To ensure the feature of high transparency, we encapsulate the functional components of CoKeeper into XenVMC's frontend kernel modules in DomUs and its backend kernel module in Dom0. Since Linux kernel modules can be dynamically loaded into or unloaded from the kernel, CoKeeper can be seamlessly incorporated into the kernel on demand transparently as an integral part of XenVMC modules. We implement CoKeeper on Xen-4.6 with Linux kernel 3.16 LTS.

4.1 Events and Messages: Definition and Implementation

Dom0 and DomU play different roles in the processes of co-location VM status change detection and the membership update. We define the self-defined events and messages Dom0 and DomU separately as illustrated in Table 2 and Table 3.

Dom0 is responsible to handle three types of events and receives two types of messages as far as co-located VM membership is concerned. And DomU handles five types of events and receives four types of messages. Dom0 and DomU coordinate to propagate the status changes and update the co-located VM membership.

We implement event handlers and message processing mechanism in both Dom0 and DomU. For event handling, Dom0 and DomU exclusively maintain an event link list, which organized as a FIFO queue. The handlers are woken up by the actions of inserting new events into the event link. In addition, a kernel thread runs in either Dom0 or DomU when CoKeeper is enabled, aiming to handle events in the event link list. We implement the message mechanism in the XenVMC kernel modules to ensure user level transparency and fast message processing. We define *ETH_P_VMC*, a new Network Layer packet type in Linux kernel, through which messages are transmitted from Dom0 to DomU or conversely.

4.2 Event and Message Based Protocol: A Global Perspective

VM Addition. For a guest VM addition, the initialization function of CoKeeper creates a *VM Self-Joining* event. Then the event handling thread in this DomU is woken up and

Table 2. *xl* commands on guest VMs and the events created accordingly.

xl command	Event created	Handled by
create	*VM Self-Joining*	DomU
	DomU Registration	Dom0
reboot	*DomU Deletion*	Dom0
	VM Self-Joining	DomU
	DomU Registration	Dom0
shutdown	*DomU Deletion*	Dom0
migrate	*VM Self-Preparing to Migrate*	DomU (before migration)
	VM Self-Joining	DomU (after migration)
destroy	*DomU Deletion*	Dom0
restore	*VM Self-Joining*	DomU
	DomU Registration	Dom0

Table 3. Messages invoked and exchanged across VM boundaries.

Message	Sender	Receiver	Event created indirectly
DomU Registration	DomU	Dom0	*DomU Registration*
DomU Migrating	DomU	Dom0	*DomU Migrating*
VM Registration Response	Dom0	Newly added DomU	*Other VM Addition*
Other VM Addition	Dom0	All DomUs	*Other VM Addition*
Other VM Deletion	Dom0	All DomUs	*Other VM Deletion*
Other VM Migrating	Dom0	All DomUs	*Other VM Migrating*

generates a *DomU registration* message and sends the message to Dom0, which creates a *DomU Registration* event accordingly, which is dealed with by the handling thread in Dom0 by adding the new DomU into its local co-located VM list. Afterwards, Dom0 sends a *VM Registration Response* message to the new DomU, which contains the info of all the other co-located VMs and triggers the update of co-located VM list of the new DomU. Then Dom0 sends an *Other VM addition* message to all the other DomUs, with the information of the newly added VM. After receiving the message, these DomUs creates *Other VM addition* events and update their co-located VM list accordingly. The process is shown in Fig. 2.

VM Removal. Once a DomU is going to be removed, Dom0 will detect the status change of the item "*/local/domain/<Dom-ID>*" in XenStore by the watcher registered on it. Then Dom0 creates a *DomU Deletion* event. And the event handler removes the DomU from the co-located VM list and creates an *Other VM Deletion* message, which is sent to all the other co-located DomUs and the DomU to be removed is deleted from their

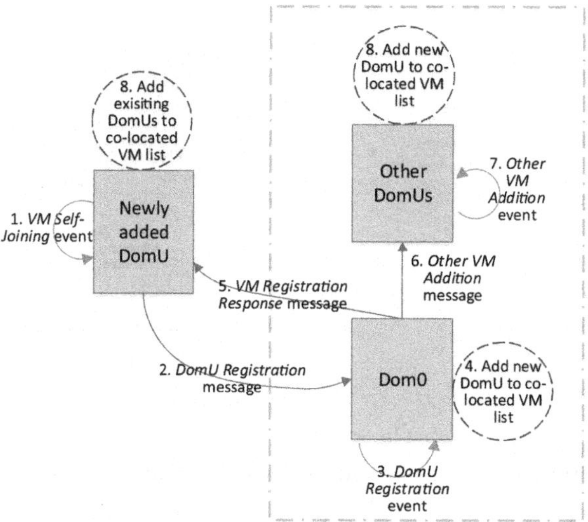

Fig. 2. The process of co-located VM membership update: VM addition.

co-located VM lists. Then all the other DomUs should deal with remained operations related to the DomU to be removed, such as transferring the remained data (if any) in shared memory channel and releasing shared memory buffer. The process is illustrated in Fig. 3.

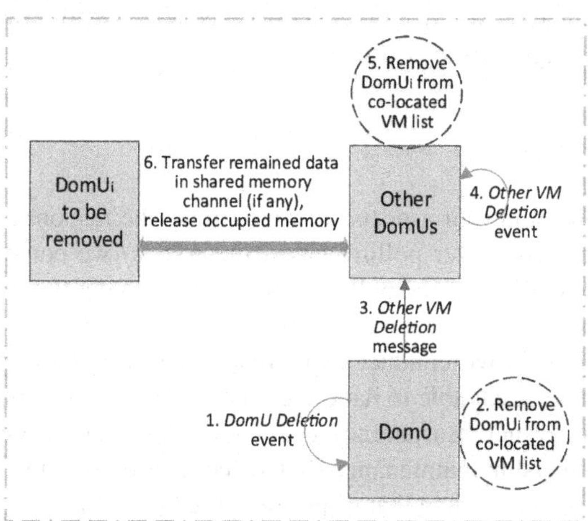

Fig. 3. The process of co-located VM membership update: VM removal.

VM Migration. The process for VM migration out is shown in Fig. 4. When a DomU is ready to migrate, the item "*/control/shutdown*" in XenStore for this DomU turns from null to "*suspend*".

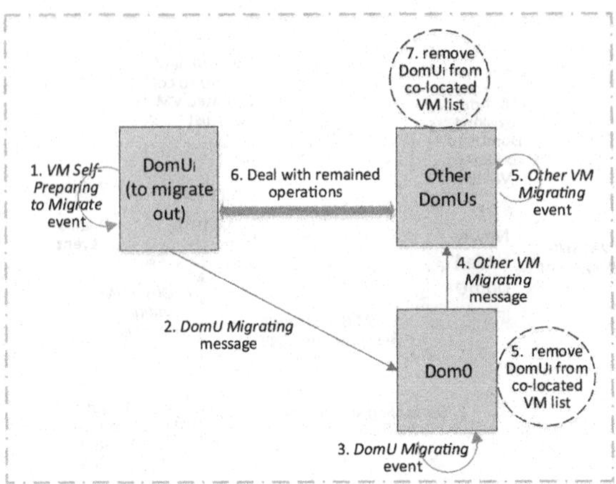

Fig. 4. The process of co-located VM membership update: VM migration.

The change will be caught by the watcher, which leads to the creation of a *VM Self-Preparing to Migrate* event in this DomU. Then the handler is woken up and creates a *DomU Migrating* message and send it to Dom0. Dom0 receives the message and creates a *DomU Migrating* event. Dom0 handles this event by broadcasting an *Other VM Migrating* message to all the other DomUs on this physical node and removing the DomU preparing to migrate from its co-located VM list. While other DomUs receive the message, they create an *Other VM Migrating* event, which invokes the event handling thread to deal with remained operations related to the DomU to migrate. And the VM prepared to migrate is removed from local co-located VM lists of all the other DomUs. The process of VM migration in is similar.

5 Evaluation

CoKeeper are evaluated in four aspects: 1) we evaluate the response time to verify the effectiveness of CoKeeper over polling based method, 2) we compare the CPU and network overhead of CoKeeper and polling based method under two circumstances, i.e., no co-located VM membership change, and membership changes due to VM addition, deletion, or migration, 3) we conduct experiments with real network workloads and confirm that CoKeeper is capable to handle co-located VM membership changes, with user level and Linux kernel transparency guaranteed, 4) we show that the overhead of co-located VM membership maintenance for CoKeeper is more stable when the number of co-located VMs increases than that of polling based method.

Among Xen based existing work, only XenLoop and MMNet support dynamic co-located VM membership maintenance. Since MMNet is not open source, XenLoop is used as a representative implementation of the polling based method for comparison. Note that we do not compare CoKeeper with heuristics based approaches for cross-VM side channel scenario. The reasons are that it often takes a few seconds to detect co-located VMs and the results come out in a probabilistic way, which make this kind of approaches not feasible for fast inter-VM communication.

The experiments are carried out in our two-server test bed. The larger server has 8 Xeon E5 2.4 GHz CPU and 64 GB main memory. The other one has an Intel i5-4460 3.2 GHz CPU and 4 GB main memory. All the data reported was run on Xen-4.6 and Linux kernel 3.16 LTS. To enable necessary comparison, we transplant XenLoop to the target Linux kernel and Xen versions without modifying their functionalities.

5.1 Evaluation of Response Time

We measure the response time between the event of co-located VM existence change happens and the point of time when the co-located VM lists of all the VMs on the same physical node are updated. For polling based method, the response time is a composition of three parts: i) the time from the event occurs to the beginning of next polling point, ii) the time for the detection of membership change, iii) the time to update the membership for all the co-located VMs. For CoKeeper, only the detection time and the update time are needed. The comparison is carried out mainly on the larger server, and the other server serves as a destination machine for VMs on the larger one to migrate out.

The first experiment is conducted under the scenarios that the number of co-located VMs varies from *1* to *10*. Co-located VM membership change events are arranged in the sequence of VM_1 addition,..., VM_{10} addition, VM_{10} removal,..., VM_1 removal, VM_1' addition,..., VM_{10}' addition, VM_{10}' migration out,..., and VM_1' migration out. Since the process of VM migration in is similar to that of VM addition, we do not put VM migration in into the tests. We evaluate the response time by running the tests with above sequence for three times for both polling based method represented by XenLoop and for CoKeeper. During the tests, all the clock of VMs are synchronized with that of Dom0 and the synchronization accuracy is controlled below 100 ns. And the polling cycle of XenLoop is set to 1000 ms. Figure 5 shows the experimental results.

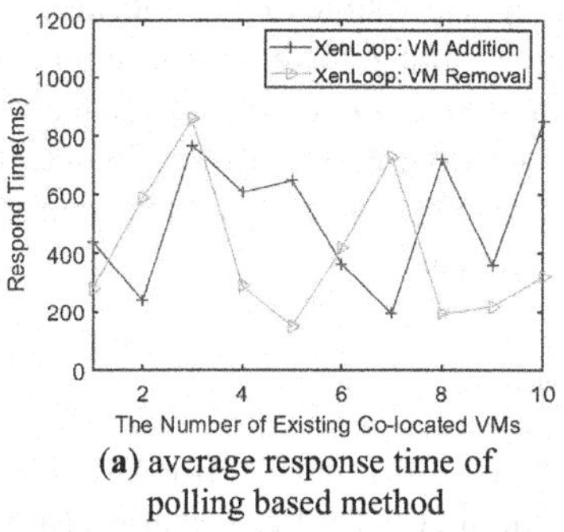

(**a**) average response time of polling based method

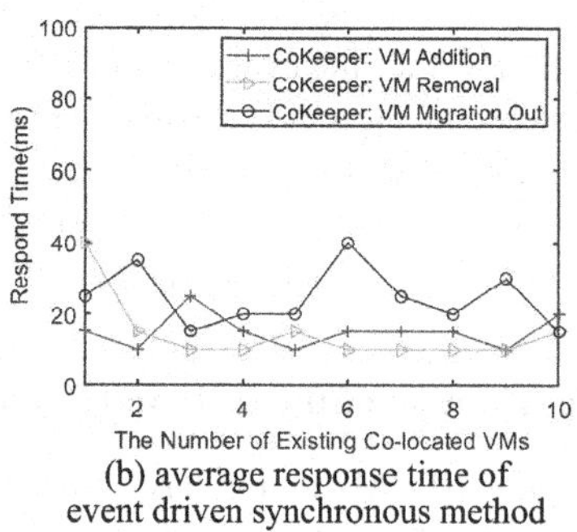

(b) average response time of event driven synchronous method

Fig. 5. Average response time of polling based method and event driven method.

Figure 5(a) shows that the response time of polling based method fluctuates from 100 ms to 900 ms when the number of co-located VMs ranges from *1* to *10*. As presented in Fig. 5(b), CoKeeper's response time is between 10 ms to 40 ms, which is more than 10 times lower than that of polling based method and more stable. The reason is that for

CoKeeper, instead of collecting the membership changes periodically, the changes are immediately visible to all the other VMs with the synchronous notification approach.

We also evaluate the response time of polling based method with different polling cycle configurations. We set the cycle to 500 ms, 1000 ms, and 2000 ms respectively. Under each configuration, we randomly issue *xl* commands which lead to membership changes. Figure 6 shows the average response time of XenLoop and CoKeeper under the scenarios with *10* and *20* deployed co-located VMs. Each bar in Fig. 6 represents an average response time of *10* events under the same configuration.

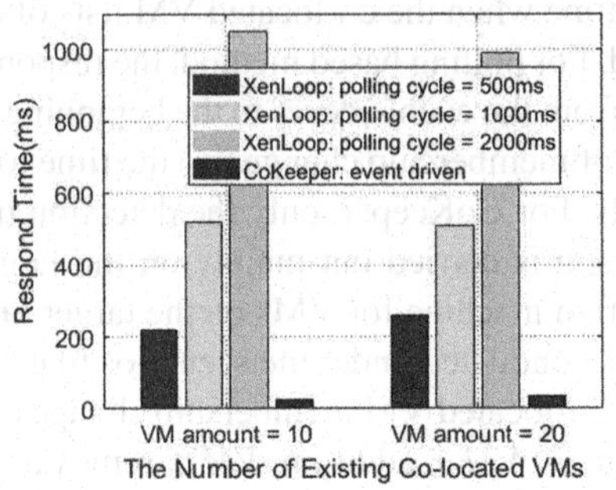

Fig. 6. Average response time comparison.

From the experimental results, the average response time for polling based method is affected by the polling cycle since it must wait for next polling point to detect the events and thereafter to handle them. The average response time is about a half of the cycle, which can be improved if we set the polling cycle to a smaller value (it will bring other problems, such as higher overhead). The results show that the response time of CoKeeper is between 10–40 ms, which makes the update visible to its co-located VMs immediately after the event occurs. CoKeeper achieves much lower response time than polling base method. It is capable of avoid race condition as far as de facto frequency of co-located VM membership changes concerns. In addition, we find that event detection and message exchange contribute to most of the consumed time, while the time spent on event and message handling is only a few milliseconds.

5.2 Evaluation of the CPU and Network Overhead

We deploy *10* VMs on the larger server, each with 1 VCPU and 1 GB RAM. The we arrange two experiments, each lasts for 30 min. The first is for the scenario of no co-located VM membership changes, the results of which show that the CPU overhead of polling based method is higher than that of CoKeeper, and both are stable. And for network overhead, we count how many network packets have been sent during the tests. The results show that network overhead of polling based method is higher. The reason is that polling based method needs to check the status in the beginning of every cycle. However, event driven method only needs to initiate the co-located VM list when a VM is deployed and its shared memory based acceleration is enabled, no additional maintenance is needed when VM existence status does not change. The second experiment is for the scenario when co-located VM membership changes. For the *10* deployed VMs, the membership changes every 3 min. We randomly pick a group of changes in the sequence as illustrated in Table 4. We measure the CPU overhead and network packets for membership maintenance every 2 min.

Table 4. Co-located VM membership changes.

Point of time (min)	3	6	9	12	15	18	21	24	27	30
Membership changes	VM removal	VM migration	VM addition	VM removal	VM addition	VM addition	VM removal	VM migration	VM removal	VM addition

The results are shown in Fig. 7.

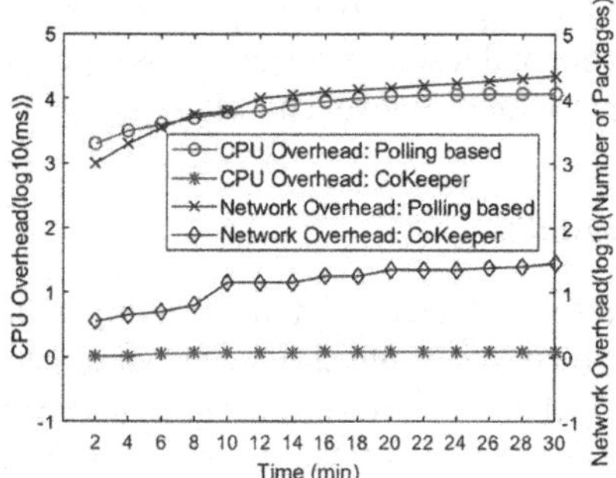

Fig. 7. CPU and network overhead: polling based method vs. CoKeeper.

The overhead of polling based method increases rapidly over time. From the *2*nd min to the *30*th min, the CPU overhead rises from 10^3 ms to 10^4 ms, and the network overhead goes up from *10*^{3} packets to over *10*^{4} packets. When co-located VM membership changes, the overhead of event driven method increases. The CPU and network overhead values are quite low, but not zero. Generally, for CoKeeper, the CPU overhead is a few milliseconds, and the network overhead is about tens of packets. This is because that the event driven method will not create events and messages until co-located VM existence status changes.

5.3 User Level and OS Kernel Transparency

We conduct a series of tests with VM addition, VM migration, and VM removal to validate if CoKeeper ensures user level and OS kernel transparency with four widely used real world standard network applications in Linux systems: *ssh, scp, samba* and *httpd.* For each group of tests, to enable the shared memory based inter-VM communication optimization, we load XenVMC's frontend and backend modules into the kernels of guest or host domains of our two-server test bed.

First, VM_1 and VM_2 are deployed separately on the smaller server and larger server. The tests are as following: 1) login to VM_2 from VM_1 and operate with *ssh*, 2) use *scp* to copy files from VM_2 to VM_1, 3) startup *samba* server in VM_2 and mount VM_1's shared directory in VM_2, 4) *httpd*. Startup *httpd* server in VM_2 and browse its html pages in VM_1 with a Web browser. From the system logs, we see that the two VMs communicates in remote mode.

Then VM_3, a newly added VM, are deployed on the larger server. And VM_3 communicates with VM_2 with the same sequence of operations, where *ssh, scp, samba* and *httpd* are used as application level workloads. And system logs indicate that they communicate automatically with local mode. During the communication, VM_3 is live migrated from the larger server to the other server. The migration is detected and handled correctly without data loss. And the communication between VM_3 and VM_2 transparently switch from local mode to remote mode. After all the communication completes, we use *scp* to copy files from VM_3 to VM_1. The logs show that VM_3 and VM_1 communicate in local mode. Then VM_1 is shutdown. From the logs, we find that the co-located VM lists of both the host domain and VM_3 are updated accordingly.

In above tests, we use the original network applications and Linux kernels without code modifications. Experimental results show that CoKeeper is capable of maintain the co-located VM membership with user level and Linux kernel transparency.

5.4 Co-located VM Scalability

We measure the time spent for co-located VM membership maintenance under different numbers of VMs deployed on the same physical node to evaluate the correlation between the number of co-located VM and the overhead. The experiments are conducted on the larger server. The number of co-located VMs increases one by one in each experiment, with a random VM existence status change event for each experiment. If it is the event of VM removal or migration out that occurs, an additional VM will be deployed after this experiment to ensure the number of co-located VMs increases from *1* to *20* for the set of experiments as expected.

The results are illustrated in Fig. 8. For polling based method, from the detection of the event to the co-located VM lists of all the domains are updated, the time spent is almost linearly in proportion to the number of co-located VMs. The main reason is that for polling based method, the guest VMs cannot get the status of the host VM, and the messages generated upon each membership change must be sent sequentially. Thus, when the number of co-located VMs turns large, the guest VMs will spend a considerable percentage of time on collecting co-located VM membership information. The overhead can be mitigated by setting the polling cycle with a bigger value. However, this brings other problems such as larger response time.

Comparing with polling based method, the time for event driven co-located VM membership maintenance is much lower and more stable when the number of VMs deployed on the same physical node increases. The reason lies in that event driven method only reacts when co-located VM membership changes.

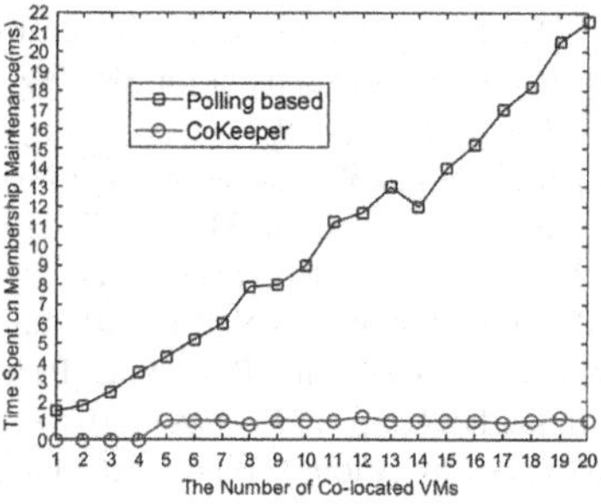

Fig. 8. The comparison of co-located VM scalability.

6 Conclusion and Future Work

We have present CoKeeper, a dynamic co-located VM detection and membership update approach for residency aware inter-VM communication in virtualized clouds. CoKeeper detects co-located VM membership changes synchronously and refreshes the membership immediately upon the occurrence of VM dynamic addition, removal, or migration with faster response time, compared with polling based dynamic method. Experimental results show that CoKeeper achieves much lower CPU and network overhead. Encapsulated as integral parts of XenVMC modules, CoKeeper can be seamlessly incorporated into the kernels transparently without modifying Linux kernel. And no extension and modification to current applications are needed. So legacy application can benefit from CoKeeper without code modification and recompilation. CoKeeper is more stable and introduces less overhead than polling based method when the number of co-located VMs scales. For the future work, we plan to apply our approach to a wider range of real-life software, and we are working on how to extend the approach to real situations such as environments with network failures.

Acknowledgment. The authors would like to thank Renshi Liu for his previous partial work on XenVMC. The work in this paper is supported by Natural Science Foundation of China (Under Grant NO.: 61872444 and U19A2060). Zhe Wang and Yi Ren contributed equally to this work.

References

1. Zhang, J., Lu, X., Panda, D.K.: Designing locality and NUMA aware MPI runtime for nested virtualization based HPC cloud with SR-IOV enabled InfiniBand. In: Proceedings of 13th ACM SIGPLAN/SIGOPS International Conference on Virtual Execution Environments (VEE 2017), Xi'an, China, 7–8 April 2017, pp. 187–200 (2017)
2. Zhang, Q., Liu, L.: Workload adaptive shared memory management for high performance network I/O in virtualized cloud. IEEE Trans. Comput. **65**(11), 3480–3494 (2016)
3. Ahmad, F., Chakradhar, S.T., Raghunathan, A., Vijaykumar, T.N.: Shufflewatcher: shuffle-aware scheduling in multi-tenant MapReduce clusters. In: Proceedings of the 2014 USENIX Conference on USENIX Annual Technical Conference, Philadelphia, PA, USA, 19–20 June 2014, pp. 1–12. USENIX Association (2014)

4. Li, J., Xue, S., Zhang, W., Ma, R., Qi, Z., Guan, H.: When I/O interrupt becomes system bottleneck: efficiency and scalability enhancement for SR-IOV network virtualization. IEEE Trans. Cloud Comput. **7**(4), 1183–1196 (2019)
5. Zhou, F., Ma, R., Li, J., Chen, L., Qiu, W., Guan, H.: Optimizations for high performance network virtualization. J Comput. Sci. Technol. **31**(1), 107–116 (2016)
6. Zhang, J., Lu, X., Panda, D.K.: Performance characterization of hypervisor and container based virtualization for HPC on SR-IOV enabled InfiniBand clusters. In: Proceedings of IEEE International Symposium on Parallel and Distributed Processing, Workshops and PHD Forum (IPDPSW), Chicago, IL, USA, pp. 1777–1784 (2016)
7. Ren, Y., et al.: Shared-memory optimizations for inter virtual machine communication. ACM Comput. Surv. **48**(4), 49:1–49:42 (2016)
8. Martini, B., Gharbaoui, M., Adami, D., Castoldi, P., Giordano, S.: Experimenting SDN and cloud orchestration in virtualized testing facilities: performance results and comparison. IEEE Trans. Netw. Service Manage. **16**(3), 965–979 (2019)
9. Linguaglossa, L.: Survey of performance acceleration techniques for network function virtualization. Proc. IEEE **107**(4), 746–764 (2019)
10. Overview of single root I/O virtualization (SR-IOV). http://msdn.microsoft.com/en-us/library/windows/hardware/hh440148%28v=vs.85%29.aspx. Accessed 28 Mar 2020
11. Xu, X., Davda, B.: A hypervisor approach to enable live migration with passthrough SR-IOV network devices. ACM SIGOPS Operating Syst. Rev. Spec. Top. **51**(1), 15–23 (2017)
12. Russell, R.: Virtio: towards a de-facto standard for virtual I/O devices. ACM SIGOPS Operating Syst. Rev. **42**(5), 95–103 (2008)
13. Intel Corporation. Intel data plane development kit: getting started guide for Linux. http://doc.dpdk.org/guides/linux_gsg/index.html. Accessed 15 Jul 2019
14. Hwang, J., Ramakrishnan, K.K., Wood, T.: NetVM: high performance and flexible networking using virtualization on commodity platforms. In: Proceedings of the 11th USENIX Symposium on Networked Systems Design and Implementation (NSDI), Seattle, WA, USA, 2–4 April 2014, pp. 445–458 (2014)
15. Wang, J., Wright, K.L., Gopalan, K.: XenLoop: a transparent high performance inter-VM network loopback. In: Proceedings of 17th International Symposium on High Performance Distributed Computing (HPDC), pp. 109–118. ACM, New York (2008)
16. Ren, Y., et al.: An efficient and transparent approach for adaptive intra- and inter-node virtual machine communication in virtualized clouds. In: Proceedings of 25th IEEE International Conference on Parallel and Distributed Systems (ICPADS 2019), Tianjin, China, 4–6 December 2019, pp. 35-44 (2019)
17. Radhakrishnan, P., Srinivasan, K.: MMNet: an efficient inter-VM communication mechanism. In: Proceedings of Xen Summit, Boston, USA, 23–24 June 2008
18. Huang, W., Koop, M., Gao, Q., Panda, D.K.: Virtual machine aware communication libraries for high performance computing. In: Proceedings of the 2007 ACM/IEEE Conference on Supercomputing (SC). ACM, New York. Article No. 9 (2007)
19. Kim, K., Kim, C., Jung, S., Shin, H., Kim, J.: Inter-domain socket communications supporting high performance and full binary compatibility on Xen. In: Proceedings of the 4th ACM SIGPLAN/SIGOPS International Conference on Virtual Execution Environments (VEE), Seattle, WA, USA, 5–7 March, pp. 1–10 (2008)
20. Zhang, X., McIntosh, S., Rohatgi, P., Griffin, J.L.: XenSocket: a high-throughput interdomain transport for virtual machines. In: Cerqueira, R., Campbell, R.H. (eds.) Middleware 2007. LNCS, vol. 4834, pp. 184–203. Springer, Heidelberg (2007). https://doi.org/10.1007/978-3-540-76778-7_10
21. Diakhaté, F., Perache, M., Namyst, R., Jourdren, H.: Efficient shared memory message passing for Inter-VM communications. In: César, E., et al. (eds.) Euro-Par 2008. LNCS, vol. 5415, pp. 53–62. Springer, Heidelberg (2009). https://doi.org/10.1007/978-3-642-00955-6_7

22. Macdonell, A.C.: Shared-memory optimizations for virtual machines. PhD thesis, Department of Computing Science, University of Alberta, Canada (2011)
23. Zhang, J., Lu, X., Jose, J., Li, M., Shi, R., Panda, D.K.: High performance MPI library over SR-IOV enabled InfiniBand clusters. In: Proceedings of 21st Annual IEEE International Conference on High Performance Computing (HiPC 2014), Goa, India, 17–20 December 2014
24. Zhang, Y., Juels, A., Oprea, A., Reiter, M.K.: HomeAlone: co-residency detection in the cloud via side-channel analysis. In: Proceedings of the 2011 IEEE Symposium on Security and Privacy (SP2011), Berkeley, CA, USA, 22–25 May 2011, pp. 313–328 (2011)
25. Bates, A., Mood, B., Pletcher, J., Pruse, H., Valafar, M., Butler, K.: Detecting co-residency with active traffic analysis techniques. In: Proceedings of the 2012 ACM Workshop on Cloud Computing Security Workshop (CCSW2012), Raleigh, North Carolina, USA, 19 October 2012, pp. 1–12 (2012)
26. Bates, A., Mood, B., Pletcher, J., Pruse, H., Valafar, M., Butler, K.: On detecting co-resident cloud instances using network flow watermarking techniques. Int. J. Inf. Secur. **13**(2), 171–189 (2013). https://doi.org/10.1007/s10207-013-0210-0
27. Atya, A.O.F., Qian, Z., Krishnamurthy, S.V., Porta, T.L., McDaniel, P., Marvel, L.: Malicious co-residency on the cloud: attacks and defense. In: 36th IEEE Conference on Computer Communications (INFOCOM 2017), Atlanta, GA, USA, 1–4 May 2017, pp. 1–9 (2017)
28. Xenstore protocol specification. http://xenbits.xen.org/docs/4.6-testing/misc/xenstore.txt. Accessed 21 Mar 2020

An Attack-Immune Trusted Architecture for Supervisory Intelligent Terminal

Dongxu Cheng(✉), Jianwei Liu, Zhenyu Guan, and Jiale Hu

Beihang University, Beijing 100191, China
xu94xu@163.com

Abstract. With the wide application of the mobile Internet, many aspects such as mobile payment, mobile office, private-data protection, security control of terminals and so on will face huge secure pressure. For that reason, this paper presents a supervisory control architecture based on secure SOC system with attack-immune and trustworthiness for intelligent terminals, which possesses ability for dynamic integrity measurement (DIM) without interference and trusted escrow application, meanwhile, this trusted architecture is fully verified based on FPGA prototype system. Compared with other schemes, this trusted architecture has higher security and dynamic integrity measurement efficiency, which can thoroughly supervise running of mobile OS and applications.

Keywords: Dynamic integrity measurement · SOC · Supervisory control · Intelligent terminals · Trusted architecture

1 Inroduction

At present, malicious attacks against intelligent terminals emerge in endlessly, which makes the security application based on mobile internet face great risk. Accordingly, all kinds of security architecture are presented to solve this problem. The Mobility Phone Work Group (MPWG) affiliated to the Trusted Computing Group (TCG) has presented Mobile Trusted Module(MTM) Specification as early as 2007, however, there are some security and applicability problems in MTM for terminal equipment to be solved [1]. In MTM Specification, the realization of trusted smart mobile terminals depends largely on the implementation of these terminals' providers to Trusted Execution Environment (TEE). Early in system boot, TEE will be running before Trusted Platform Module (TPM), hence TPM first must believe that TEE is secure and reliable and then builds the subsequent trusted chain and trusted environment.

As ARM CPU begins to provide TrustZone hardware security features [2], terminal security and trusted architecture based on these features is built by many people. Hang Jiang, Rui Chang et al. [3] propose a secure boot scheme based on ARM TrustZone technology, and construct the independent secure operation environment to resist all kinds of attacks faced by terminal systems and hardware devices in their start-up phase. Jian Xiao, Zhi Yang et al. [4] propose a secure system for mobile terminals, improving security performance of the running environment through TrustZone and fingerprint

M. Qiu (Ed.): ICA3PP 2020, LNCS 12454, pp. 144–154, 2020.
https://doi.org/10.1007/978-3-030-60248-2_10

identification technology. Based on the TrustZone assistant environment, Sandro Pintoet et al. [5] develop a set of virtualization technology for embedded system to support the safe and fast switching of multiple client operation systems (OS). In reference [6], a two-stage boot verification scheme based on the TrustZone is proposed to verify and record the integrity of the program experienced at startup, detect the startup trace at runtime and check whether the software meets the security requirements.

Although ARM's TrustZone technology provides a secure running environment for user's sensitive programs, the running of programs in the TrustZone security zone will interrupt the operation of other programs in the OS, and the real-time monitoring of system resources or process measurement will lead to large system overhead. In addition, because the ARM CPU core is shared with other kernel programs and user processes of the OS, it is difficult for the super monitor programs and DIM programs running in the TrustZone area to be escrowed by the third-party, that is, the security application design is separated from the common application development.

In order to solve the problem of TOC-TOU [7], which is difficult to avoid in static integrity measurement, people put forward DIM technology to enhance the security and trustworthiness of the system. Chen Mu, Li Rui et al. [8] present a measurement architecture of secure OS for mobile terminals, which integrates the functions of system security configuration checking, environment vulnerability scanning, application static and dynamic security testing. In reference [9], a program integrity measurement information base based on predefined creation is put forward, the process is divided into three stages: preparation, running and blocking suspension, which is conducted by real-time DIM, this DIM has better performance. Kai-Oliver Detken, Marcel Jahnke et al. [10] present a scheme of DIM based on Loadable Kernel Modules (LKM), this scheme implements a set of dynamically adjustable benchmarks as Reference Value Generator (RVG), then DKM calls RVG and TPM interface to complete integrity measurement of the running programs. In paper [11], Liu Zi-wen and Feng Deng-guo present a DIM architecture of OS which depends on trusted computing. This architecture can carry out real-time DIM and monitoring of the active processes or modules in the system according to the demand situation, and basically solve the problem of TOC-TOU.

The DIM mechanism proposed in the above literature all depends on the protection at the OS level. Although it can measure the process dynamically, the system overhead is large. When the kernel module of OS is compromised by malicious viruses and Trojans, the security of the DIM mechanism is difficult to be guaranteed.

To sum up, there are still many deficiencies in various intelligent terminal security solutions, for example, the MTM of TCG needs TEE support, ARM TrustZone technology relies on the main CPU and has high CPU computing overhead. The traditional DIM mechanism is limited by the host OS, so it is difficult to resist the attack of kernel level malware. Therefore, with the support of the National Nature Science Foundation of China through project 61972018, the trusted startup mechanism, DIM security factors and TEE security mechanism of intelligent terminal are fully studied. A new trusted architecture of intelligent terminal is proposed. The principle is verified on FPGA prototype system, and the security performance of the new architecture is analyzed. The architecture can effectively deal with various security threats faced by intelligent terminals.

2 The New Trusted Architecture of Intelligent Terminal

At present, the Main Processor System (MPS) of mainstream mobile phones, tablet computers and other intelligent terminals has evolved into a multi-core RISC CPU architecture, and the on-chip SRAM memory has also reached the MB level, which has appropriate material conditions, so it is convenient to improve the SOC system architecture, enhance its security and credibility, and cope with increasingly security risks.

The SOC hardware architecture of a typical high-performance multi-core intelligent terminal is shown in Fig. 1.

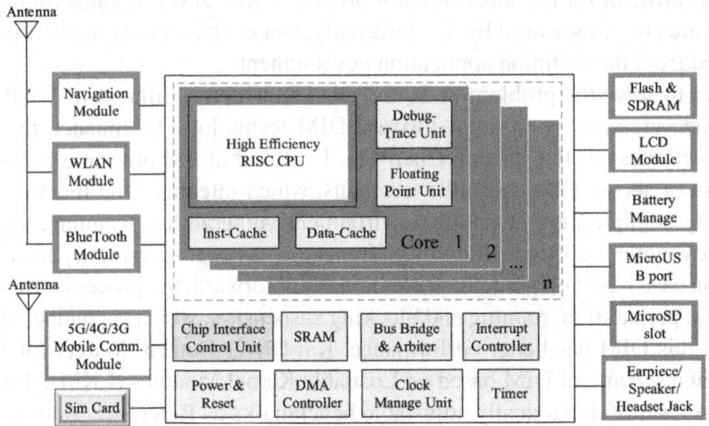

Fig. 1. Main processor structure of mainstream intelligent terminal

Based on the SOC system of the main processor in Fig. 1, by introducing a lightweight embedded CPU core, a small amount of SRAM storage space, eFlash (or off-chip flash) as non-volatile storage space and appropriate hardware control design based on on-chip bus, we have constructed a new trusted architecture as shown in Fig. 2. The dashed box in Fig. 2 is a new small SOC system (Note: Trusted Supervision System (TSS)). The operating frequency of lightweight RISC CPU is below 200 MHz, and cache is not required. In ASIC circuit implementation, the expected chip area of small SOC system should be less than 1/50 of the whole chip area. The capacity of on-chip SRAM2 is usually between 128 KB and 256 KB, which not only meets the needs of trusted software stack program running, but also does not occupy too much chip area. The TSS can fully access the slave devices on the SOC system of the main processor through Bus Bridge1. Furthermore, by setting Bus Arbiter, the read-write operation of its on-chip bus slave devices in the main processor SOC system can be controlled, conversely, it will not work.

The TSS integrates the physical noise generator and cryptography acceleration engine. The trusted software stack can call them to realize the required cryptographic service function. The key sensitive data such as trusted measurement root, PCR and

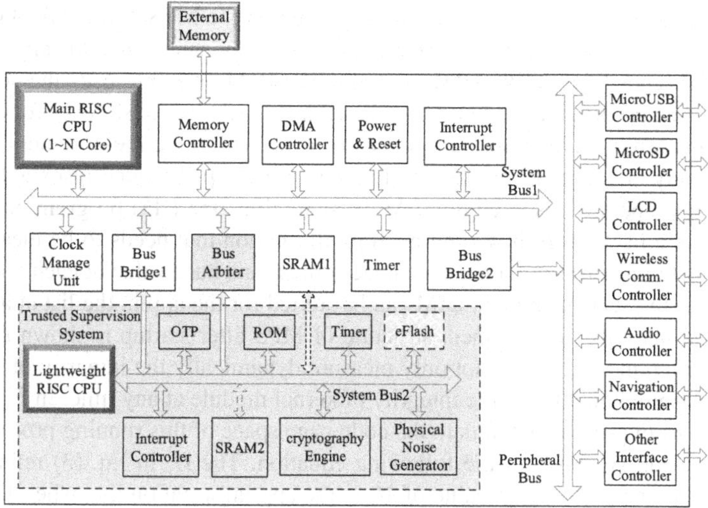

Fig. 2. Trusted enhanced architecture of MPS of mainstream intelligent terminal

master key can be directly stored in the on-chip eFlash, because the MPS cannot access the components of the TSS, which ensures the data security.

3 Trusted Startup and DIM

3.1 Static Trusted Measurement Process in System Startup

(a) In hardware design, make sure that the TSS starts before the main processor of intelligent terminal (in reset state). Then run the static measurement program directly from eFlash.
(b) The static measurement program measures the integrity of the boot program of the main processor SOC in the External Memory through Bus Bridge1. If ok, it will start the main processor of the intelligent terminal and run the boot program, and copy the OS image to SRAM1 by setting DMA controller.
(c) The static measurement program reads the OS image and measures its integrity through the path shown by the dotted arrow pointing to SRAM1 in Fig. 2. SRAM1 is a dual port Memory or other optimized structure.
(d) The MPS starts to run the embedded OS, and the Lightweight Measurement Agent (LMA) is loaded as the kernel module of the embedded OS.
e) In the TSS, the DIM program interacts with the LMA to verify that the program integrity and the program loading base address are legal, and the static trusted measurement process is ended.

3.2 DIM Based on Trusted Enhanced Architecture of Intelligent Terminal

When the embedded OS runs on the MPS, the DIM program of the TSS can read the OS image and LMA in real time through the SRAM1 monitoring channel, and conduct real-time measurement. For other key apps running on the MPS, the LMA obtains the code page memory address, size information and program name when the app is loaded and run, and sends them to the DIM program. The DIM program reads the code page content of the app program specified by the address and size in SRAM1 for dynamic real-time measurement, and no main processor is needed in the measurement process. The LMA only obtains relevant information when the app is loaded or exited, and its impact on app operation is negligible. At the same time, the DIM program of the TSS will maintain a list of currently running app information that needs to be measured in real time, including process name, code page address and hash value of each code page. According to the security policy, DIM can be started for the app in the list at any time. The detailed integrity measurement structure of MPS after startup is shown in Fig. 3. The DIM program in TSS can not only measure dynamically the integrity of key app, but also measure dynamically the integrity of kernel module at any time. In Fig. 3, The $M_i(1 \leq i \leq n)$ in process A refers to the code page space of this running process. DIM operation of app is shown in the following equation. The H_i in Eq. (3) refers to the pre-calculated and stored hash value of M_i in the app information list, when V_i, H_i are all equal in $1 \leq i \leq n$ range, *MeasureIsOK* is True, otherwise it is False; Eq. (4) extends all the measures of M_{prog} to *PCR*, and forms the trusted chain data that can be called and verified by security policy.

$$M_{prog} = M_1||M_2||\cdots||M_n, (M_i \cap M_j = \varphi;\ \ i \neq j) \tag{1}$$

$$V_i = Hash(M_i)(1 \leq i \leq n) \tag{2}$$

$$MeasureIsOK = (V_1 == H_1)\&(V_2 == H_2)\&\cdots\&(V_n == H_n) \tag{3}$$

$$PCR = Hash(PCR||V_1||V_2||\cdots||V_n) \tag{4}$$

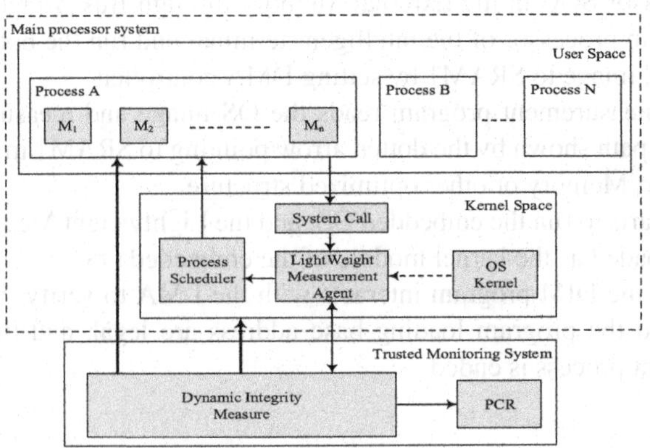

Fig. 3. DIM structure based on TSS

4 Security Properties of Trusted Enhanced Architecture

The trusted enhanced architecture of intelligent terminal proposed in this paper has the following security properties.

4.1 Attack Immunity

Assuming that the attack of intelligent terminals is mainly at the hacker level, the attacker mainly uses various vulnerabilities to implant Trojans, viruses and other malicious programs and carries out the attack, but cannot tamper with and crack the main processor chip through physical attack means. For the side channel attack of cryptographic algorithm engine circuit, there are corresponding chip physical protection measures and design patterns for security protection, which is another level of attack and defense problem. It is not directly related to the security attributes of the trusted enhancement architecture described in this paper, nor discussed in this paper.

The security programs such as integrity measurement, trusted computing and cryptographic service running in the TSS are physically isolated from the SOC system of the main processor. The embedded operating system and its applications running on the MPS cannot access the memory space of the TSS. Ensure that the security program of the TSS can run safely and credibly, and be completely immune to the attack of malicious program.

4.2 DIM Without Interference

At the system level, according to the system configuration information provided by the intelligent terminal providers and the characteristics of the specific main CPU and embedded OS, the DIM program can locate the starting memory address space and size of the OS kernel, key modules and the LMA, and measure dynamically the integrity in real time.

At the level of DIM of key app, based on the information of code page memory address and size obtained and written into the specific memory space by the LMA when the key app is loaded, the DIM program reads the relevant information of the key app, and measures dynamically its integrity accordingly. Once the measurement process starts, the main CPU system will not be occupied, this measure operation does not cause any interference and performance loss to the running embedded OS and key app program.

At the security level, because DIM can not only measure the kernel modules and programs of embedded OS, but also the key app programs of users, it can measure the kernel and user programs comprehensively at any time, which has high security and credibility characteristics.

4.3 Trusted Escrow Based on Third-Party

Because the program development, loading operation, upgrading and maintenance of the TSS are independent of the MPS, when the intelligent terminal provider completes the development and verification of the MPS, it can completely open the programming model of the TSS to the professional trusted third party, and the third party completes the design and development of the trusted program, which is conducive to the formation of industry standard and accepting supervision of relevant national departments t,so as to reduce security loopholes.

4.4 Super Monitoring

When the intelligent terminal works normally, the main RISC CPU, lightweight CPU, DMA controller and other master devices read and write the interface of each slave device through the bus arbitration logic and decoding logic circuit shown in Fig. 4 based on the system bus. If the TSS detects system abnormality based on integrity measurement program, it needs to control intelligent terminal functions such as touch-screen input and display output, network communication, navigation, photographing, recording, etc. The specific implementation process is: write control word directly to Bus Arbiter through System Bus2 of TSS in Fig. 2, forbid Bus Arbiter to provide decoding service for main RISC CPU or DMA controller, and only output meaningless protocol data or response timeout according to peripheral type, so as to ensure that the MPS program does not crash due to peripheral denial of service.

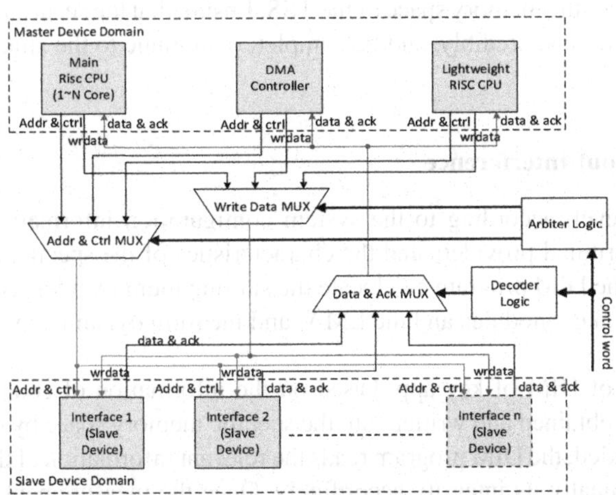

Fig. 4. Control structure chart of peripheral interface based on bus arbiter

Further, the TSS can be given more related security functions, such as monopolizing the mobile communication connection channel, resetting the main CPU, monitoring the current program pointer, executing identity authentication protocol at startup, and upgrading the MPS image file safely.

5 Principle Verification of New Trusted Architecture

5.1 Prototype Design

Based on a single FPGA chip of Intel's cyclone V series (model: 5CSEMA5F31C6N, ARM CPU hard core is integrated in the chip), we build a principle verification system of a new trusted architecture as shown in Fig. 5. Since the SRAM module in FPGA can pre store data at the beginning of power on, the OTP and ROM modules in Fig. 2 can be simulated by SRAM in FPGA, while the on-chip eFlash as non-volatile memory can be simulated by configuring the flash chip of FPGA.

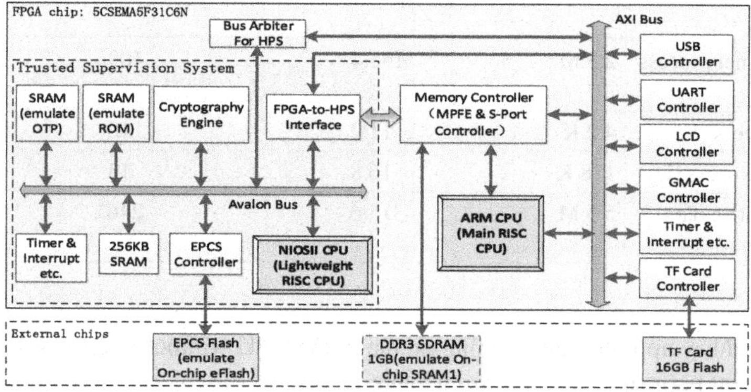

Fig. 5. Prototype verification framework based on FPGA for new trusted architecture

By using SOPC technology, a TSS with 32-bit Nios II CPU as the RISC core is built in FPGA chip. The MPS with 32-bit ARM processor as the main CPU core is monitored. The ARM CPU system runs embedded Linux OS, and the OS image is stored on the TF card. Nios II CPU directly accesses Memory Controller module through FPGA-to-HPS interface module in the way of master device, reads and writes DDR3 SDRAM program memory, and accesses and controls all peripheral modules of ARM CPU system through Bus Arbiter For HPS module.

5.2 Trusted Startup

In Fig. 5, the FPGA-to-HPS interface module is connected with the AXI bus, which enables the Nios II CPU to access the TF card and read the Linux kernel image stored on the TF card. The TSS runs the trusted software protocol stack and the static integrity measurement program to realize the integrity measurement, secure storage and integrity reporting functions. Firstly, reset ARM CPU, measure the static integrity of bootloader program code and Linux image data, then start bootloader to load Linux image, and then start ARM CPU to allow Linux system to run. Linux system is responsible for calling LMA. The TSS enters the DIM stage.

5.3 Dim

As the main memory of ARM system, the 1 GB DDR3 SDRAM in Fig. 5 can be accessed efficiently by Nios II CPU and ARM CPU, can run OS and user programs, and TSS can access it through physical channel and conduct DIM of MPS. According to the principle of Sect. 4.2, we run the DIM program to successfully measure the integrity of the kernel

Table 1. Measurement performance of processes and data in accessible memory

Category (K: in kernel address Space; U: in user address Space)	Actual measured memory size (Byte)	Actual measured time (ms)	Expected measured time (ms)
Partial kernel process (K)	2.3 M	155.4	109
LMA process (K)	4.2 K	11.2	10
Key app process (U)	128 K	16.8	15
User sensitive data loaded (U)	5.5 M	315.6	246

module and key app program running on the ARM CPU without interference, and the measurement performance is shown in Table 1.

The actual measurement time of process and data in Table 1 is mainly composed of:

$$t_{total} = t_{agent\_comm} + t_{sdram} + t_{hash\_calc} + t_{linux\_access} + t_{dynamic\_meas} \quad (5)$$

Where t_{agent_comm} represents the time taken by the DIM program running in TSS to interact with LMA in the Linux main system. t_{sdram} represents the time taken to read data from the address to be measured in the off-chip SDRAM. t_{hash_calc} represents the time taken for the data to be sent to the hash hardware computing engine and return the result. t_{linux_access} indicates that programs running in the Linux main system need to load programs and data from the off-chip SDRAM in order to refresh the instruction of data cache of ARM CPU, and compete with the DIM program in TSS to access SD RAM, resulting in the increase time of dynamic measurement. $t_{dynamic\_meas}$ represents the time taken by DIM program to run SDRAM interface driver, call hash engine interface and generate log report. After analysis, t_{sdram} and t_{hash_calc} change linearly with the size of the measurement memory. $t_{dynamic\_meas}$ is basically the same as the software and hardware environment of the TSS is fixed. When the secure and trusted SOC system is ASIC implemented, the program can be stored by integrating the dual-port SRAM in the chip to avoid competing access, and t_{linux_access} can be optimized to 0. t_{agent_comm} is basically fixed, because the LMA will maintain a list of addresses of kernel processes, key app processes and sensitive data running in memory, Through the test, the running time is in the order of milliseconds. The DIM program can quickly obtain the memory addresses of processes and data to be measured from the agent, so t_{agent_comm} is very short and negligible. The expected measurement time is composed of:

$$t_{total} = t_{sdram} + t_{hash\_calc} + t_{dynamic\_meas} \quad (6)$$

The impact of t_{agent_comm} and t_{linux_access} will not be considered. The actual is calculated based on SDRAM interface rate, hash hardware engine processing rate and the running performance of DIM on the NIOS II CPU.

In the aspect of attack detection for process integrity, by implanting malicious programs in the Linux system kernel to tamper with the specified kernel module, LMA process or key app process, the DIM program in the TSS can give an alarm prompt within the actual measured time in Table 1.

5.4 Third-Party Trusted Escrow and Super Monitoring

The architecture shown in Fig. 5 realizes the hardware system simulation of the core function module of the architecture in Fig. 2, and fully realizes the third-party escrow and super monitoring functions described in Sect. 4.

5.5 Experimental Analysis

Based on the principle verification system of FPGA, the new architecture is emulated. The trusted startup, DIM and trusted escrow experiments will truly reflect the security related performance of intelligent terminals.

The DIM in Fig. 5 only has the overhead of t_{agent_comm}. LMA program does not measure it, only gets the memory address of the process to be measured. The running time consumed by DIM on the Main CPU is usually in the order of microseconds, which can be ignored. However, the traditional real-time DIM costs a lot in the host system, at least in milliseconds. For example, in reference [12], the measurement time of 25 processes reaches 93.224 ms. In reference [13], the DIM time of conventional linux commands is in the order of seconds. Therefore, the DIM advantage of the new architecture is obvious.

6 Conclusion

By improving the current mainstream intelligent terminal SoC system, this paper proposes an attack-immune super monitoring trusted architecture for intelligent terminal, discusses the principle of trusted start and DIM under this architecture in detail. This architecture supports the non-interference DIM of the OS kernel and user program, which is superior to the DIM architecture based on the current OS kernel integrated with the process integrity measurement program, and supports the third-party trusted escrow and super monitoring of the intelligent terminal MPS. By building a FPGA prototype verification system, we verify the security attributes of the new trusted enhanced architecture to ensure that the theory and practice are consistent. In the future, we will further optimize the internal bus interconnection structure of the trusted architecture, and give it more secure and trusted functions by expanding the LMA.

Acknowledgement. This work is supported by the National Nature Science Foundation of China through project 61972018.

References

1. Grossschadl, J., Vejda, T., Page, D.: Reassessing the TCG specifications for trusted computing in mobile and embedded systems. In: IEEE International Workshop on Hardware-oriented Security & Trust. IEEE (2008)
2. Alves, T., Felton, D., Alves, T., et al.: Trustzone: integrated hardware and software security. ARM white paper. White Paper (2004)
3. Jiang, H., Chang, R., Ren, L., et al.: Implementing a ARM-based secure boot scheme for the isolated execution environment. In: 2017 13th International Conference on Computational Intelligence and Security (CIS). IEEE (2017)
4. Xiao, J., Yang, Z., Hu, X., Liu, Y., Li, D.: TrustZone-based mobile terminal security system. In: 2018 Chinese Automation Congress (CAC), Xi'an, China, pp. 3981–3985 (2018)
5. Pinto, S., Pereira, J., Gomes, T., et al.: Towards a TrustZone-assisted hypervisor for real-time embedded systems. IEEE Comput. Arch. Lett. PP(99) (2016)
6. González, J., Hölzl, M., Riedl, P., Bonnet, P., Mayrhofer, R.: A practical hardware-assisted approach to customize trusted boot for mobile devices. In: Chow, Sherman S.M., Camenisch, J., Hui, Lucas C.K., Yiu, S.M. (eds.) ISC 2014. LNCS, vol. 8783, pp. 542–554. Springer, Cham (2014). https://doi.org/10.1007/978-3-319-13257-0_35
7. Bratus, S., D'Cunha, N., Sparks, E., et al.: TOCTOU, traps, and trusted computing. In: International Conference on Trusted Computing-challenges & Applications. DBLP (2008)
8. Chen, M., Li, R., Li, N., et al.: A mobile terminal operating environment security measurement framework. In: International Conference on Intelligent Computation Technology & Automation. IEEE Computer Society (2017)
9. Hu, M., Yang, Y., Lv, G., Liu, X.: Research on dynamic integrity measurement strategy based on measurement point distribution. In: 2018 Chinese Control And Decision Conference (CCDC), Shenyang, pp. 220–223 (2018)
10. Detken, K., Jahnke, M., Rix, T., Rein, A.: Software-design for internal security checks with dynamic integrity measurement (DIM). In: 2017 9th IEEE International Conference on Intelligent Data Acquisition and Advanced Computing Systems: Technology and Applications (IDAACS), Bucharest, pp. 367–373 (2017)
11. Liu, Z.W., Feng, D.G.: TPM-based dynamic integrity measurement architecture. Dianzi Yu Xinxi Xuebao/J. Electron. Inf. Technol. **32**(4), 875–879 (2010)
12. Du, R., Pan, W., Tian, J., et al.: Dynamic integrity measurement model based on vTPM. Chin. Commun. **15**(2), 88–99 (2018)
13. Wei, C., Song, S., Hua, W., et al.: Operating systems support for process dynamic integrity measurement. In: IEEE Youth Conference on Information, Computing & Telecommunication. IEEE (2009)

A Simulation Study on Block Generation Algorithm Based on TPS Model

Shubin Cai(✉), Huaifeng Zhou, NingSheng Yang, and Zhong Ming

Shenzhen University, Shenzhen, China
shubin@szu.edu.cn

Abstract. Blockchain is a emerging decentralized infrastructure and distributed computing paradigm. However, the low TPS performance of blockchain technology can not meet the performance requirements of large-scale and high concurrency in application reality. A polling discrete event simulation platform is designed to investigate the performance of PoW based block generation algorithm. The operation of block generation algorithm is simulated from three aspects: network topology level, message queue of communication and protocol of PoW algorithm. The result shows that when the block size is 1 MB, the average relative error between the experimental results and the fixed TPS is 13.00%, and when the block size is 4 MB, the average relative error between the experimental results and the fixed TPS is 15.25%. Experiment result shows that the simulation platform can be use to investigate the transaction performance effectively.

Keywords: Blockchain · Bitcoin · Simulation model · TPS

1 Introduction

Blockchain is an emerging decentralized architecture and distributed computing paradigm underlying Bitcoin and other cryptocurrencies, and has recently attracted intensive attention from governments, financial institutions, high-tech enterprises, and the capital markets. In 2008, Nakamoto took the lead in proposing bitcoin a virtual currency based on blockchain technology without the endorsement of third-party trust institutions [1]. As the underlying technology of bitcoin, blockchain technology first appeared in Nakamoto's bitcoin white paper. Blockchain technology not only solves the cost problem of traditional transaction which needs third party endorsement, but also establishes a secure network using encryption keys to achieve secure communication [2]. However, due to the low throughput of blockchain applications such as bitcoin, the number of business requests that can be processed per unit time is very limited, so the existing blockchain technology is difficult to be applied in large-scale and high concurrency business situations. How to optimize the throughput performance of blockchain has become the key research point of blockchain technology to large-scale commercial application.

M. Qiu (Ed.): ICA3PP 2020, LNCS 12454, pp. 155–169, 2020.
https://doi.org/10.1007/978-3-030-60248-2_11

1.1 Block Structure

The block is mainly consists of two parts (Fig. 1): body and header. The header mainly includes a random binary value *nonce,* the degree of difficulty, the timestamp of the generated block, the root of MHT (Merkle Hash Tree), a hash pointer pointing to the last executed block and a random binary value. For Pow algorithm, every node would change the random value to get a different hash value [6]. The body contains an ordered set of client requests to be executed.

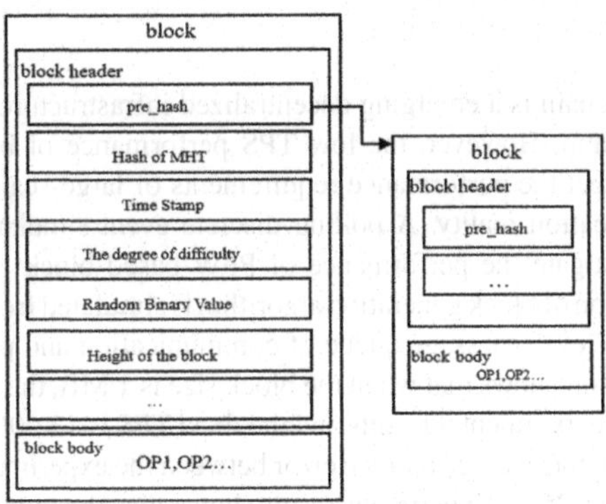

Fig. 1. Block structure.

In the block header, the block height points out the execution order of blocks. Each blockchains have a unique block which height is 0 (called "The Foundation Block"). When a node joins the network of the blockchain, the foundation block is written to the system initialization system state. The hash pointing to the previous block is a cryptographic credential to avoid someone tampering with data maliciously. The timestamp is used to reord the time when the block was generated.

The block body summary is a summary of the collection of client requests in the whole block body. Block chain technology introduces MHT as a means to verify a large amount of data. By constructing MHT, the block generation algorithm can check the integrity of multiple client request data in the scenario of only transmitting limited length of summary data. First, the node selects an executable ordered subset from the cached unexecuted client request according to its local node status. The node then generates a summary of client requests in the block by constructing MHT (Fig. 2).

Where *Data bolck* $i(1 \leq i \leq n)$ is the request from the client and these requests are sorted according to the execution order. Through a hash function, any length of binary string information will be converted into a fixed length of binary string digest. Then, the abstracts of these client requests are placed in the leaves of MHT and MHT is constructed from the leaves to the tree roots. Among them, the binary value in the non leaf node is the string splicing operation of the binary value in the child node, and then input the hash generation function.

When a node finds an ordered set of client requests that can be executed, it traverses

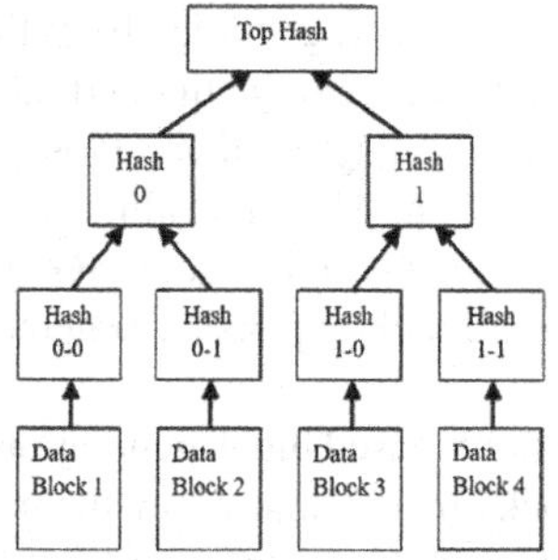

Fig. 2. Construction process of MHT

the range of *nonce* and tries to find a value such that

$$f(header) <= 2^{D-L} \tag{1}$$

In this formula, L is the length of the output of the hash function. In bitcoin, the digest function adopts SHA-256 standard one-way hash algorithm, so the length is 256 bits. If the difficulty coefficient D is 10, it means that the 256 bit binary number output by the block header through the hash function must meet the requirement that the first 10 bits of binary number are 0. Then the probability that the service node generates a block randomly can satisfy the formula is 2-246. If the block which generated by the service node satisfies the Eq. (1), it means that this node successfully generates a block, and when a node finds a block which satisfies Eq. (1), it will writes the block to the local database and updates the state immediately.

In the bitcoin, each node can generate blocks independently, so many blocks of the same height (height greater than 0) but different block contents maybe appear in the system, which are called parallel blocks. Besides, the different chains that are bifurcated from parallel blocks are called branches and the branch chain containing the largest number of blocks is called the main chain (Fig. 3). According to the research of ittay Eyal and Emin gun sirer, if more than 75% of the nodes in bitcoin have the same state, then the state of the system is consistent [5].

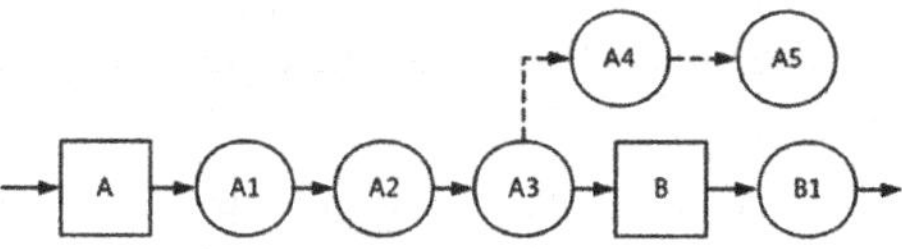

Fig. 3. Parallel blocks

1.2 Related Research

From the perspective of network topology, the research on optimizing the throughput of block generation algorithm can be divided into two types: master-slave structure and packet structure. If we classify the existing studies from the point of view of changing the parameters that affect throughput, it can be divided into pseudo-random election to reduce delay and increase block size. Algorand and Gosig are the algorithms to reduce the delay of pseudo-random election. The algorithms that belong to the direction of increasing block volume include segwit and bitcoin-NG, ghost and elastico.

Pieter wiulle first proposed a block expansion scheme called segwit in December 2015 [7].

Bitcoin-NG algorithm and POW based block generation algorithm need to obtain the qualification of generating blocks by calculating mathematical problems, but nodes in bitcoin-NG algorithm can continuously generate multiple blocks that can be generated without computing mathematical problems after finding feasible solutions [8].

Elastico algorithm adopts grouping method to increase the number of blocks generated per unit time [9]. Elastico algorithm divides nodes into several fixed capacity packets, and each group uses practical Byzantine fault tolerance (PBFT) to generate blocks independently [11].

For PBFT algorithm, malkhi, Nayak and Ren et al. [12] proposed flexible BFT, which realized state machine replication in synchronous networks. Flexible BFT only requires the network to be synchronized in the commitment step, that is, it only uses the network delay parameter in the commitment step, and does not need network synchronization in other steps, so as to improve the overall efficiency.

Ghost algorithm rewrites the chain structure of the blockchain into a tree structure, but the system still traverses the tree hierarchically and executes the blocks in each node of the tree in sequence. Moreover, ghost algorithm allows the system to increase the number of blocks generated by the system in unit time [10].

2 Modeling

In this part, we model TPS Based on PoW-based block generation algorithm and analyze the influence of block size on TPS in different network scales.

2.1 The Derivation of TPS Model Derivation Based on PoW Algorithm

First, we assuming that the system works from time 0 until time t. During that time the system generates $N_e(t)$ blocks with V size. We suppose M(v) denotes the number of client requests in a block. Then, the TPS of the block-generating method based on PoW in time $[0, t]$ can be expressed as:

$$T = \frac{N_e(t) \cdot m(V)}{t} = \frac{N_e(t) \cdot (V - V_0)}{tL} \tag{2}$$

In the PoW algorithm, as mentioned above, the blocks generated by the system can be divided into two categories: one is the blocks on the main chain, which is the longer one and the other is the blocks on the non-main chain. Over time, only blocks of the main chain will eventually be written to the local databases. Therefore, the TPS of PoW algorithm in time period $[0, t]$ can be expressed as:

$$T = \frac{(N_b(t) - N_w(t)) \cdot (V - V_0)}{tL} \tag{3}$$

Deduce the Total Number of Generated Blocks

When the number of nodes, which is n, in the network is large enough, the probability p_i of generating blocks by a single node is relatively small. At that time, the number of blocks generated in the network will subject to Poisson distribution $P(\lambda)$. Then the λ can be expressed as:

$$\lambda = \sum_{i=1}^{n} p_i \tag{4}$$

The probability of generating α blocks by PoW algorithm during $[0, t]$ time can be expressed as

$$Pr(N(t) = \alpha) = \frac{\left(\sum_{i=1}^{n} p_i t\right)^{\alpha} e^{-\sum_{i=1}^{n} p_i t}}{\alpha!} \tag{5}$$

The expected value of the total number of blocks which is generated by the system during $[0, t]$ should be expressed as:

$$E(N(t)) = \sum_{\alpha=0}^{\infty} \alpha \cdot \frac{\left(\sum_{i=1}^{n} p_i t\right)^{\alpha} e^{-\sum_{i=1}^{n} p_i t}}{\alpha!} = t \sum_{i=1}^{n} p_i \tag{6}$$

Derivation of the Number of Horizontal Blocks Generated by the System

In this part, we assuming that all nodes have the same blockchain data (each node has the same state), and node i generates the next block B on the main chain after time t. Within the time block B is propagated to all nodes, the probability that the system generates block a parallel to block B can be expressed as:

$$Pr(N_w = \alpha) = \sum_{j=1}^{n} Pr\left(N\left(t_{i,j}\right) = \alpha | N_e(t) = 1\right) \tag{7}$$

Because the behavior of nodes generating blocks does not affect each other, nodes in the same state can still try to generate parallel blocks with the same height as the received blocks before receiving the newly generated blocks and updating the local state. Therefore, the number of parallel blocks generated in the process of synchronizing blocks can be expressed as the expected value of Eq. (6):

$$E(N_w) = \sum_{\alpha=0}^{\infty} \alpha \cdot Pr\left(N\left(t_{i,j}\right) = \alpha\right) \tag{8}$$

Then the expected value of the number of effective blocks generated by the PoW algorithm during $[0, t]$ should be expressed as:

$$E(N_e(t)) = E(N(t)) - E(N_w(t)) \tag{9}$$

The Derivation of Delay of Synchronized Block

The delay of synchronized blocks includes three parts: propagation delay, transmission delay and waiting delay on communication links. The sending delay is the time cost for the sender to send out the block, which is related to the block size and node bandwidth. When the link bandwidth is constant, the transmission delay increases linearly with the increase of block size. Propagation delay is the time cost of data transmission on the link, which is related to the distance between nodes. The longer the transmission path length is, the greater the delay is. Waiting delay is the time cost of data waiting to be forwarded in the router and other relay equipment. In the hypothesis of this paper, we ignore the interference factor of waiting delay.

Round-Trip Time (RTT) is an index to measure the delay between nodes, which indicates the total delay that occurs from the sending end to the receiving end after it receives the acknowledgement. During the transmission of each packet, the sender node will send the next packet only after the receiving node returns the confirmation information. Therefore, the transmission delay can be expressed as the product of round-trip delay $RTT_{i,j}$ and the number of packets between two nodes. In addition, we assume that each communication node will send as much data as possible, then the message delay $d_{i,j}$ generated when establishing synchronization block between any two nodes can be expressed as:

$$d_{i,j} = \frac{V}{B} + RTT_{i,j} \cdot \left\lceil \frac{V}{MSS} \right\rceil \tag{10}$$

In this formula, V denotes the size of the block, B denotes the message sending rate of the node, and MSS denotes the maximum transmission unit.

In the PoW-based block generation algorithm, gossip algorithm is used as the broadcast algorithm of synchronous blocks [3]. When a node generates a block or receives a block higher than its own local block completely, it will randomly select k nodes from the local routing list to establish communication and synchronize blocks with these nodes. According to Pierre Fraigniaud and proof of Gossip algorithm, it shows that at least 75% of the nodes in the system update a block's communication rounds with a probability of not less than $1 - n^{-k+O(1)}$, the expected value of $lg\, n + klglgn$. Then the average delay of the block propagating to 75% of the nodes can be expressed as:

$$(lg\, n + klglgn) \cdot \overline{d_{i,j}} \tag{11}$$

In this equation, $\overline{d_{i,j}}$ is the average delay, which is D, of any two nodes i and j in the network, which should be expressed as:

$$\overline{d_{i,j}} = \frac{1}{N^2}\sum_{i=1}^{N}\sum_{j=1}^{N}\left(\frac{V}{B} + RTT_{i,j} \cdot \left\lceil \frac{V}{MSS} \right\rceil\right) = \frac{V}{B} + \overline{RTT} \cdot \left\lceil \frac{V}{MSS} \right\rceil \tag{12}$$

Where, $\overline{RTT}$ is the average round trip delay between all nodes in the network of the system, which should be expressed as:

$$\overline{RTT} = \frac{1}{N^2}\sum_{i=1}^{N}\sum_{j=1}^{N} RTT_{i,j} \tag{13}$$

The Integration of the Model

In the previous paper, we have expressed the expression of the total number of blocks generated in the system, the expression of horizontal blocks generated in the system and the expression of block synchronization delay. Next, we will integrate these expressions. Because the height of parallel blocks is the same, the nodes that generate parallel blocks do not have nodes that synchronize blocks. With the decrease of the number of nodes in the system that are not synchronized, the probability of generating parallel blocks in the system per unit time will decrease. Therefore, we will deduce the number of nodes that are not synchronized during the time when the system synchronizes blocks. We note that $f(t)$ represents the proportion of nodes that generate a new block with a higher height after the beginning of the time period t. Under the probability of not less than $1 - n^{-k+O(1)}$, the probability that more than 75% of nodes can generate parallel blocks before synchronization is completed can be expressed as:

$$Pr(N_w = \alpha) = \int_0^{(lgn+klglgn)\overline{t_{i,j}}} \frac{\left(f(t)t\sum_{i=1}^{n} p_i\right)^{\alpha} e^{-f(t)t\sum_{i=1}^{n} p_i}}{\alpha!} dt \tag{14}$$

According to Demers Alan's research [4], we Suppose $i(t)$ is the proportion of the number of nodes that have completed the synchronization block in the total number of system points, then we can get:

$$\begin{cases} \frac{df}{dt} = -fi \\ \frac{di}{dt} = fi - \frac{1}{k}(1-f)i \\ f + i = 1 \end{cases} \tag{15}$$

Besides, when the first higher block was generated in the system, $f(0) = \frac{n-1}{n}$, $i(0) = \frac{1}{n}$. Then the proportion $f(t)$ should be expressed as

$$f(t)\left(1 + \frac{kt}{Nk - Nktf(t) + Nt(1-f(t))}\right) = \frac{N-1}{N} \tag{16}$$

If (15), (12), and (6), are brought into (3), the expectation of TPS can be expressed as the following formula.

$$E(T) = \left(\sum_{i=1}^{n} p_i - \frac{\int_0^{(lgn+klglgn)\overline{t_{i,j}}} f(\beta)\beta d\beta \cdot \sum_{i=1}^{n} p_i}{t}\right) \cdot \frac{V - V_0}{L} \tag{17}$$

At this point, we have expressed the expected value of TPS of PoW algorithm.

3 Simulation Experiment

3.1 Structure of Simulation Experiment

Bitcoin is a well-known representative application of block generation algorithm based on pow. So, we use the standard configuration parameters in bitcoin configuration file to build the simulation platform, and observe the influence of block size on TPS in different network sizes.

We use Java as the development language and build a polling discrete event simulation platform with a virtual timer. All events in the platform are triggered by this virtual time.

Our simulation platform has the following four structure: 1) service node: equivalent to the physical server with bitcoin running protocol deployed in reality. 2) Link: in our experiment, any two nodes are connected through P2P link. 3) Packet: our simulation platform encapsulates application layer data to network layer. 4) Block: block is the smallest unit of application layer data encapsulation in block generation algorithm.

We mainly simulate the operation of PoW algorithm from three layers: network topology layer (network layer), Message queue for communication (protocol layer) and the protocol of workload proof algorithm (application layer). Among them, the simulation node and the PoW algorithm are combined to realize. Considering that the network scale in the simulation experiment may be very large, if a thread is set for each simulation node by using multithreading technology, the time cost of thread switching will be very large. Therefore, instead of using multithreading technology to implement each simulation node and physical link, we use state machine model to construct the behavior and state of simulation node and simulation node in workload proof algorithm. The advantage of this method is that it can reduce the performance burden caused by a large number of thread context switches.

3.2 Simulation of Network Layer

There are three parts of network layer simulation: RTT, network topology and IP message delivery mechanism. When the total number of nodes is less than or equal to 1000, we use fully connected rule diagrams to represent the network topology; when the number of nodes is more than 1000, we use the complex network model of ER random diagrams to represent the network topology of Bitcoin. When constructing Er random graph model, we need to set the probability of edge connection. In the experiments of different network scales, we set the probability of any node connecting other nodes as a, and set a counter for each node to record the number of routing table entries locally recorded by each node. When the counter value reaches 1000, the current node stops discovering other nodes.

In the simulation experiment, we use the 100 Mbps Ethernet to connect virtual nodes in the simulation experiment. Because RTT is dynamic, each data message in the simulation experiment generates RTT of current message transmission when it is sent. In addition, we will let the nodes randomly select the nodes in their routing table and complete the missing entries from the routing tables of these nodes until the counter value reaches 1000.

Because we have to simulate the randomness of the message sequence generated by the nodes competing for the channel, we use a queue in the link simulation, which is used to store the messages sent by each simulation node to the link (Fig. 4). When the timer is increased by 1, the link will process the messages added in the last unit time before all other simulation nodes work. The link has three main jobs: first is sending messages already placed in the Execute Queue (EQ) to the target simulation node. Then, adding messages already placed in the Cache Queue (CQ) to EQ. At last, adding messages in the message queue sent by the simulation node to CQ.

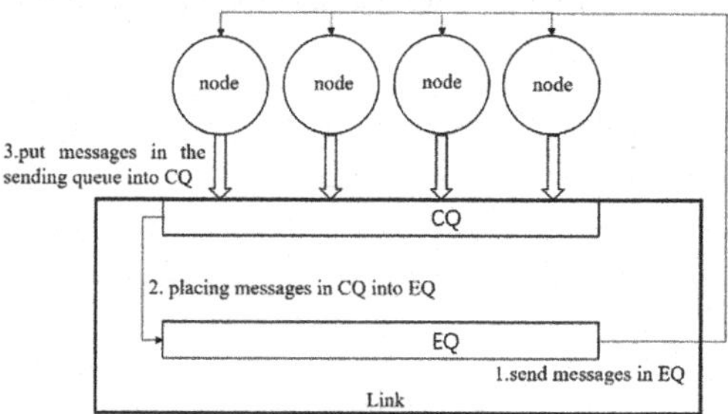

Fig. 4. Link working diagram

Firstly, the link will process the messages in the EQ queue. For IP packets, we can use a five tuple $\langle src, dst, L, T, TCP\rangle$ to describe them. Among them, and represents the sender node ID and receiver node ID of the IP message, represents the length of the IP message, represents the time when the *IP* message arrives at the receiver node, and is the content of the upper *TCP* message. Then, the link places the messages in the message queue of the simulation node into CQ. At the same time, the link calculates the current IP message length of each sending queue and the time when the message is sent to the target node. The last step is to put the messages in CQ into EQ. When processing the messages of virtual nodes, the system always traverses each virtual node from small to large according to the node number. Therefore, the messages put into CQ message queue by nodes at the same time have partial order relationship based on node number. In order to eliminate the deviation caused by the partial order relation, we shuffle the messages in CQ queue and put them into EQ according to the order after shuffling. In our simulation experiment, we use Fisher Yates algorithm to detect IP packets from CQ queue to EQ queue.

3.3 Simulation of Transport Layer

There are two steps in transport layer simulation, one is message queue of virtual node, and the other is that port is responsible for TCP protocol. When the TCP connection between the two virtual nodes is established, both nodes will assign a message queue and port number locally to transmit block data.

In this simulation experiment, the key of transport layer simulation simulates the process of data transmission by TCP protocol. And the most important is to achieve two functions: timeout retransmission and window sliding mechanism. Therefore, we use two pointers, *Begin* and *End,* to simulate window sliding in TCP protocol. The range between this two pointers is the window size. A priority queue Q is used to record all the sent and unacknowledged TCP messages in the message queue without TCP connections. Each TCP message in queue Q can be represented by four tuples <RTO, SerialNumber, Begin, End>, where RTO is the message timeout time, SerialNumber is the serial number of the TCP message, Begin is the starting offset position of the TCP message, and End is the end offset position of the TCP message. The timeout timer records the time when the message queue requests the latest receipt of the return message. In the simulation experiment, we use and update algorithm of the timeout timer defined in RFC2988 (Algorithm 1).

```
Algorithm 1: RTO Update Algorithm
// All numerical units are seconds
RTO := 3
//Receive the message and calculate the first RTT
SRTT := RTT
RTTVAR := 0.5 * RTT
RTO := SRTT + max{0.5, 4 * RTTVAR }
// Next receive the other TCP messages to get the latest RTT'
//and update RTO
RTTVAR := (1 - b) * RTTVAR + b * |SRTT - RTT'|
SRTT := (1 - a) * SRTT + a * RTT'
RTO := SRTT + max{0.5, 4 * RTTVAR }
```

3.4 Simulation of Application Layer

There are two steps in transport layer simulation: implements the block-generating algorithm and the state machine model of nodes.

In the simulation experiment, we assume that the physical performance parameters of all simulation nodes are identical, and each node has a unique network ID. Besides, each node can establish multiple TCP connections in parallel during network communication. We stipulate that the maximum number of TCP connections that each node can establish is 125, which is set according to the parameters in the bitcoin standard configuration file. We analyze the behavior and state changes of the nodes when they execute the PoW algorithm, and then construct the state machine model of the simulation nodes (Fig. 5).

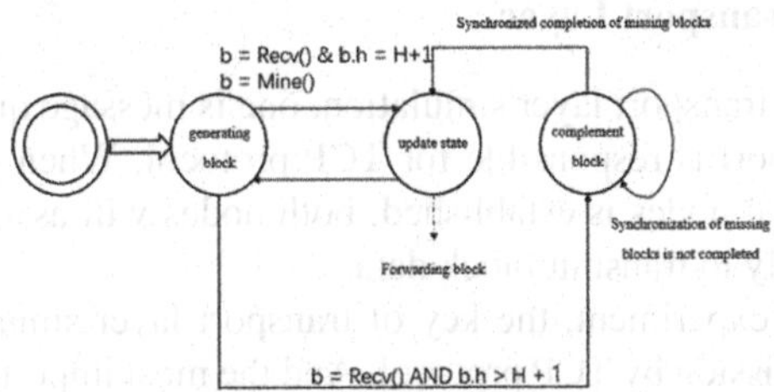

Fig. 5. Node state machine

After the simulation node starts, it directly enters the process of exhausting random numbers and sets the status to "generated block". We use the random number generated by pseudo-random number generator as the basis for simulation nodes to find blocks in unit time. When a node successfully generates a block or receives a block with a height equal to H + 1, it will enter the "update state". The simulation node adds its own block chain height by 1 and writes the block into the local block array. When the simulation node receives a block whose height is greater than, it indicates that the local blockchain of the simulation node has block missing phenomenon. Therefore, the node stops generating blocks and switches the system state to "completion block"(Fig. 6).

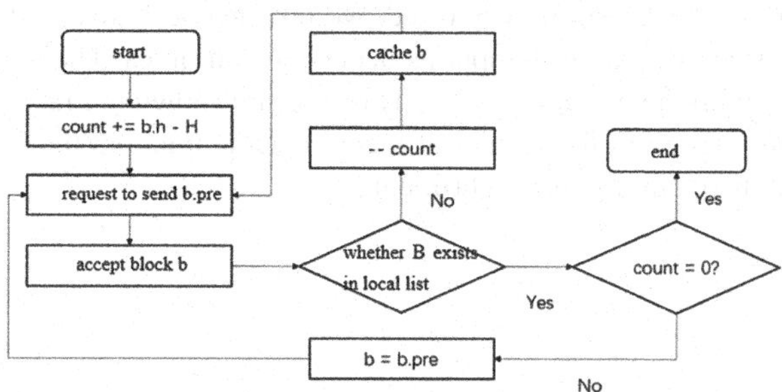

Fig. 6. Completion block

3.5 Design of Simulation Experiment Parameters

The purpose of this part is to study the influence of block size on TPS in different network sizes. We use the control variable method to observe the TPS performance of pow based block generation algorithm in parameter scenarios with different network sizes and block sizes. The parameters of our simulation experiment range from 0 to 10000 nodes (step size is 500 nodes), and the block volume is 1 MB to 20 MB (step size is 1 MB).

The block formatted in the simulation experiment uses the format specified in the Bitcoin protocol. The data length of each client request is about 627B, while the sum of the length of the standard block header plus the length of other description fields of the block in bitcoin is 80 bytes, so the maximum number of 1672 client requests can be saved in a 1 MB block. Due to the performance limitation of the machine, we can't make the workload proof algorithm execute too long, so we adopt another method to measure the message delay. We let the cluster finish its work when it generates 1000 blocks in a row. Then we count how many blocks are saved by more than half of the simulation nodes in the blocks with the height of 1 to 1000. After reading, they are stored in a two-dimensional matrix and written into the external experimental result file. Therefore, we need to design a structure in each simulation node to record the local block and the corresponding write time. We assign each simulation node an array with the size of 1000 to simulate the local persistent storage. When the node receives a higher block or generates a new block, the newly generated block is written to the local persistence device.

Two parameters, network size n and block size V, are needed to start the simulation experiment. Because many experiments with different block sizes will be executed in the same network scale, we will separate the code for generating network topology and write the generated network topology data to external storage device for reuse.

4 The Result and Analysis of Simulation Experiment

In the experiment, we simulate the implementation of pow algorithm in a static P2P network with a specified scale, and record the time when the block is generated and when each node writes the block to the local state. Then we calculate the time interval between the birth of each block and its acceptance by more than 75% of the nodes in the network.

So, according to the questions we asked, we take network size and block volume as independent variables and throughput as dependent variables. Then we compare the simulation results with the model results. The results are shown in Fig. 7. Among them, the red data points represent the results of model calculation, and the blue data points represent the results of simulation experiments.

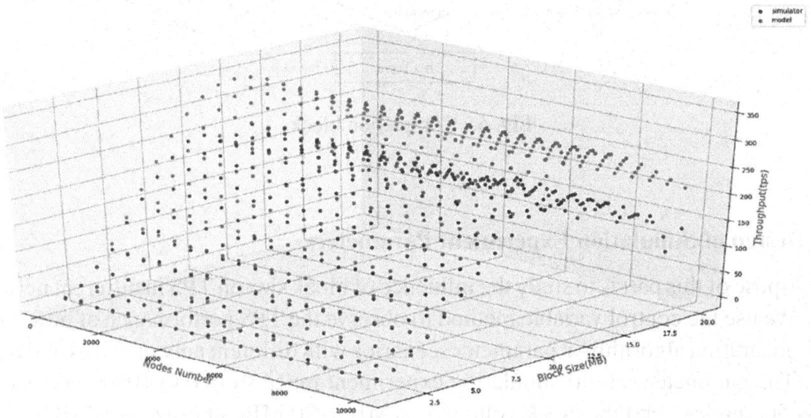

Fig. 7. Simulation results of block volume change (Color figure online)

From the Fig. 8, it can be seen that the results of simulation experiments are always less than the results of model calculation. This is because the model that proposed by us assumes that both the sender and the receiver work at full load when the nodes communicate with each other, but in the simulation experiment, they are not necessarily full work.

With the increase of block size, the simulation results and model calculation data show a trend of first rising and then decreasing. The proposed model does not consider the network resource contention, so the model has poor ability to describe scenarios with large block size. When the block volume ranges from 1 MB to 15 MB, the average difference between the model TPS and the experimental TPS is 12.76 tps and the standard deviation is 7.08 tps. The average relative error between model TPS and experimental TPS is 16.01%.

In order to verify that the proposed model can effectively predict the volume of selected blocks under the constraint of maximizing throughput, we further collate the data. We collate the comparison of the optimal block size calculated by the model and the optimal block size obtained from the experimental results in different scale networks

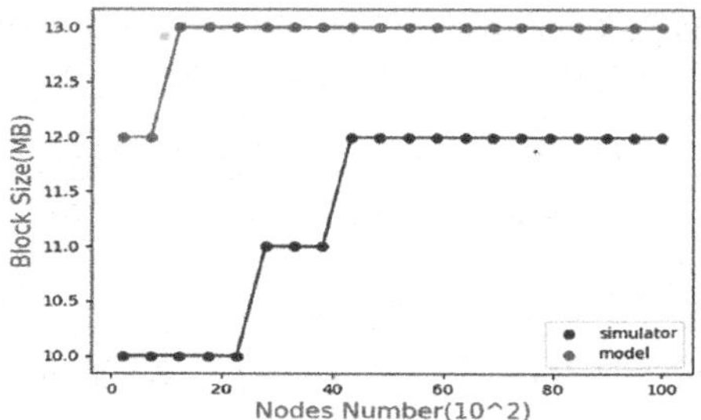

Fig. 8. Block size in different network sizes when TPS maximizes

when TPS is maximized (Fig. 8). It can be seen that the average absolute error between the optimal block size calculated by the model and the experimental block size is 1.55 MB, and the standard deviation is 0.74 MB. The average relative error is 14.25%.

In order to prove the validity of our conclusions, we collect the date when the number of nodes falls within the range from coindance, and grab the block data generated by bitcoin in the corresponding date from bitcoin browser, and compare the actual running data of bitcoin with the data obtained from simulation and model. The comparison results are shown in Table 1 and Table 2.

Table 1. Comparison of TPS at block size of 1 MB

Number of nodes	TPS			
	Actual value	Fixed value	Simulated value	Model value
4000	3.7	10.9	12.0	15.9
4500	4.1	12.1	11.9	15.9
5000	5.8	9.3	11.5	15.9
5500	4.6	13.1	16.8	15.8
6000	5.1	11.4	13.1	15.8

From the Tables 1 and 2, it can be seen that in most cases, the results of simulation experiments and numerical calculations are much higher than the actual TPS of bitcoin. There are two reasons for this, one is In bitcoin, the length of client requests is not strictly constrained. The other is that there is a certain deviation between the expected value and the observed value of the time interval of the generated block Considering these two factors, the actual TPS of bitcoin deviates greatly from the simulation results and model values.

Because actual data cannot reach the peak of TPS without a lot of transactions happening, it is lower than the results of experiments and calculations. Therefore we calculated a set of fixed data. First, we count the set of dates (*DS*) that meet the required block size and all within the range of nodes. Then, we count the number of blocks that bitcoin generates and writes to the main chain every day. Based on the above two data,

Table 2. Comparison of TPS at Block Size of 4 MB

Number of nodes	TPS			
	Actual value	Fixed value	Simulated value	Model value
6000	11.7	32.7	45.5	46.2
6500	16.9	45.1	41.3	46.1
7000	20.3	43.1	45.0	46.0
7500	13.2	35.4	44.8	45.9
8000	21.8	43.1	44.6	45.8
8500	15.7	46.9	44.5	45.7
9000	12.6	43.5	48.3	45.6
9500	22.7	34.8	44.1	45.5
10000	13.1	40.2	44.0	45.4

we can get the correction value of the corresponding node number as follows

$$\frac{m(V) \cdot \sum_{i=1}^{|DS|} DS_i}{|DS| \cdot 86400} \tag{18}$$

From the comparison results, it can see that when the block size is 1 MB, the average difference between the model calculation results and the modified value is 4.50 tps, the variance is 1.30 tps, and the average relative error is 16.90%. The average difference between the experimental results and the modified TPS is 1.78 TPS, the variance is 1.17 TPS, and the average relative error is 13.00%. When the block size is 4 MB, the average difference between the model calculation results and the modified value is 5.53 tps, the variance is 4.48 tps, and the average relative error is 15.25%. The average difference between the experimental results and the modified TPS is 5.52 tps, the variance is 3.77 tps, and the average relative error is 14.93%.

5 Conclusion

In this paper, we first build a mathematical model of TPS Based on PoW block generation algorithm. Then, we use the standard parameters of the configuration file in bitcoin and the statistical data of previous studies to construct the simulation experiment. Compared with the simulation results and the calculated values of the model, we analyze how increasing the block size affects TPS in different scale networks. Through the result, It can be found that when the block size is 1 MB, the average relative error between the experimental results and the fixed TPS is 13.00%, and when the block size is 4 MB, the average relative error between the experimental results and the fixed TPS is 15.25%. Experiment result shows that the simulation platform can be use to investigate the transaction performance effectively.

References

1. Nakamoto, S.: Bitcoin: a peer-to-peer electronic cash system [OL] (2008). http://bitcoin.org/bitcoin.pdf
2. Gai, K., Qiu, M., Xiong, M., Liu, M.: Privacy-preserving multi-channel communication in edge-of-things. Future Gener. Comput. Syst. **85**, 190–200 (2018)
3. Fraigniaud, P., Giakkoupis, G.: On the bit communication complexity of randomized rumor spreading. In: Proceedings of the twenty-second annual ACM symposium on Parallelism in algorithms and architectures, pp. 134–143. ACM (2010)
4. Allavena, A., Demers, A., Hopcroft, J.E.: Correctness of a gossip based membership protocol. In: Proceedings of the Twenty-Fourth Annual ACM Symposium on Principles of Distributed Computing, pp. 292–301. ACM (2005)
5. Eyal, I., Sirer, E.G.: Majority is not enough: bitcoin mining is vulnerable. Commun. ACM **61**(7), 95–102 (2018)
6. Zheng, Z., Xie, S., Dai, H., et al.: An overview of blockchain technology: architecture, consensus, and future trends. In: 6th IEEE International Congress on Big Data. IEEE (2017)
7. Gavin, A.: Bitcoin improvement proposal 101 [OL] (2015). https://github.com/bitcoin/bips/blob/master/bip-0101.mediawiki
8. Eyal, I., Gencer, A.E., Sirer, E.G., et al.: Bitcoin-NG: a scalable blockchain protocol. In: 13th USENIX Symposium on Networked Systems Design and Implementation, pp. 45–59 (2016)
9. Luu, L., Narayanan, V., Zheng, C., et al.: A secure sharding protocol for open blockchains. In: Proceedings of the 2016 ACM SIGSAC Conference on Computer and Communications Security, pp. 17–30. ACM (2016)
10. Sompolinsky, Y., Zohar, A.: Accelerating bitcoin's transaction processing. Fast Money Grows on Trees, Not Chains. IACR Cryptology ePrint Archive, 2013(881) (2013)
11. Sousa, J., Bessani, A., Vukolic, M.: A byzantine fault-tolerant ordering service for the hyperledger fabric blockchain platform. In: 2018 48th Annual IEEE/IFIP International Conference on Dependable Systems and Networks, pp. 51–58. IEEE (2018)
12. Malkhi, D., Nayak, K., Ren, L.: Flexible byzantine fault tolerance [EB/OL] (2019). https://arxiv.org/pdf/1904.10067.pdf

Improving iForest for Hydrological Time Series Anomaly Detection

Pengpeng Shao[1], Feng Ye[1](✉), Zihao Liu[2], Xiwen Wang[1], Ming Lu[3], and Yupeng Mao[4]

[1] College of Computer and Information, Hohai University, Nanjing, China
yefeng1022@hhu.edu.cn
[2] Nanjing Branch of China Telecom, Nanjing, China
[3] Jiangsu Water Resources Department, Nanjing, China
[4] NanJing Yiting Internet of Things Technology Co., Ltd., Nanjing, China

Abstract. With the increasing number of installed hydrological sensors, the data from these sensors usually contain a variety of abnormal values due to network congestion, equipment failure, or environmental influence. To deal with the anomaly on a larger scale of hydrological sensor data, a series of algorithms have been proposed. However, they are usually based on the ideas of distance or classification, which usually bring pretty high time complexity. To solve this problem, a detection algorithm called AR-iForest is proposed. It is an algorithm for hydrological time series anomaly detection based on the isolation forest. Firstly, the features of hydrological data are extracted and mapped it to a high-dimensional space. Before using the isolation forest in high-dimensional space for anomaly detection, the Auto-Regressive model is used first to predict the current data and calculate the confidence interval. Only the data not in the confidence interval needs to be detected. Secondly, a measure of the effectiveness of trees in the isolation forest is proposed. This method selects the tree with the best classification effect through continuous iteration. Finally, the proposed algorithm is integrated into the window of the big data platform Flink to give a performance evaluation. The experimental results show that the proposed algorithm increases the AUC value from 90.60% to 96.72%, and the detection time is reduced by 52.23%.

Keywords: Anomaly detection · Hydrological time series · Window-based detection · Apache flink · Isolation forest

1 Introduction

Time series analysis is an important aspect of data mining technology in practical applications. The main characteristics of hydrological time series are uncertainty, nonstationarity, multiple time scale changes, and multiple branches [1]. Hydrological sensor time series usually contain a variety of abnormal values due to network congestion, equipment failure, or environmental influence. To efficiently and accurately process hydrological sensor data and detect outliers timely, a variety of anomaly detection methods [2–5] have been proposed. These currently used algorithms are usually based

M. Qiu (Ed.): ICA3PP 2020, LNCS 12454, pp. 170–183, 2020.
https://doi.org/10.1007/978-3-030-60248-2_12

on the ideas of distance or clustering, which often have higher time complexity in the big data processing. The isolation Forest (iForest) [6] algorithm is based on isolation, which avoids a large number of numerical calculations, has optimized time complexity, and focuses on "outlier" data with excellent accuracy performance. Therefore, it gradually attracted extensive attention from academia and industry. However, the existing research on isolation forest and their optimization algorithms still have the following shortcomings:

1. In general, abnormal data is always a small amount, but it takes a lot of time to detect all data.
2. The algorithm itself has some randomness. The uneven quality of the tree leads to a relatively large deviation in the detection results.
3. There are few studies on its application to time series anomaly detection, especially in the field of hydrology. A large number of sensors produce a large amount of time-series data, and the application of anomaly detection algorithms has not considered combining with big data platforms, which is contrary to the real environment.

In response to the above problems, combined with the current research on the iForest and hydrological data anomaly detection algorithm, we have made the following improvements. 1) After cleaning the original sensor data and extracting data features, we first predict the data based on the AR autoregressive model and calculate the confidence interval. For all instances in the feature space that is within the confidence interval, the program automatically skips the detection of these instances. 2) We introduced an indicator for determining the effectiveness of iTree. An algorithm for iteratively selecting the optimal forest is proposed to improve the detection quality and the stability of the results. 3) We integrate iForest with the current popular big data processing platform Apache Flink [7] and use its window mechanism to update the AR-iForest model in real-time to respond to changes in data features. At the same time, this integration provides the ability to handle infinite streams.

This paper will be organized as follows. Section 2 introduces the research status and the related work of hydrological anomaly detection. Section 3 details the proposed method for anomaly detection. In Sect. 4, experiments were conducted using real hydrological sensor data from 70 hydrological stations. Section 5 is the results and discussion, which gives the effectiveness of the proposed algorithm based on the experimental results. Finally, the conclusion and research prospect are given.

2 Related Work

At present, for anomaly detection of streaming data, Toliopoulos [8] combined with Apache Flink, realized continuous outlier mining of streaming data. But its distance-based calculation method cannot avoid high time complexity and the calculation process still needs to consume a lot of time. In [9], Bandaragoda proposed the nearest neighbor method which is based on the iForest algorithm, combined with density or distance based anomaly detection can effectively detect local anomalies. It is more efficient than pure distance or density-based methods, but due to the calculation of hyperspheres, its time complexity is higher than iForest. Xu [10] proposed an algorithm based on simulated annealing algorithm to find the best forest and solve the stability of abnormal scores. However, the establishment of its model requires cross-validation of all data, which increases the complexity to a certain extent. Ding [11] proposed the concept of conditional anomaly detection, creating different combinations of attribute selection and its value for anomaly detection and using it in the field of fraud detection. Aryal [12] optimized iForest based on relative mass theory and improved the problem that the native iForest is insensitive to local abnormal points. In [13], Zou proposed an online anomaly detection system based on an isolation forest, which implements anomaly detection of containers in a cloud computing environment and improves detection accuracy and efficiency by assigning different weights to different resources. Ma [14] proposed a hybrid model based on iForest-LSTM to predict short-term load. The role of iForest is to filter out abnormal data, but the filtering quality is not evaluated.

For anomaly detection of hydrological time series, the most common method is based on distance or density. Although its precision has reached a high level [2], it is not suitable for the current big data environment. On the one hand, excessive time complexity results in poor anomaly analysis performance. Sun [3] proposed a hydrological time series anomaly detection algorithm based on time series prediction and SVM classification. This type of algorithm works well on small-scale data sets, but it cannot handle large-scale data. The method based on distance detection is to set a certain distance function to calculate the distance of the data points. When one point is too far away from the other points, it is regarded as an abnormal point. Vy [4] proposed a variable-length anomaly detection algorithm in the time series. The first step is to segment time series, and then calculate the anomaly factors for each mode to get the distance between them. Finally, the distance is judged based on the anomaly factor distance. It is convenient for users to use and the time complexity is relatively low, but it is not sensitive to local abnormal points. Yu [5] proposed the use of sliding window prediction based hydrological time series anomaly detection. However, it is aimed at the daily average water level. The data volume is too small and time complexity is relatively high.

Apache Flink is an open-source distributed streaming processing framework. It provides Exactly-once semantics with the support of message queues that can provide data consistency guarantees such as Kafka [15], which can support the dumping of large amounts of sensor data (especially when the network is unstable) to ensure that data is not lost or duplicated. Karimov [16] showed that Flink usually has performance advantages in current big data stream processing platforms under the same experimental environment.

To sum up, it is obvious that there are few studies on time series detecting based on iForest. There is also no targeted optimization for the time spent in the evaluation phase of the data. Therefore, we proposed a detection method called AR-iForest based on isolation forest for hydrological sensor time series anomalies. Through the combination with the current popular big data platform, AR-iForest has realized the processing of the never-ending data stream from the sensor in the real application environment. Our experiments are based on real-time series streams from sensors of multiple local hydrological monitoring stations.

3 Methodology

This section will introduce the proposed method in detail. The algorithm used will be explained first, followed by a detailed implementation process for hydrological data anomaly detection.

3.1 Building Isolation Forest

Isolation forest [6] is an efficient anomaly data detection algorithm based on the isolation idea. The main work of this paper is to optimize iForest to make it suitable for hydrological time series and to better detect anomalies in the series. For the convenience of description, we briefly give the important concepts and design ideas of iForest algorithm. The "anomaly" object is closer to the root node of the tree, that is, the path length is shorter, it reflecting that only requires very few conditions and split times to distinguish the "anomaly" object from other objects. Given a bunch of data sets *D* where all attributes of *D* are numeric variables, the process of constructing iTree is described as Algorithm 1.

Algorithm 1: $iTree(x, c, h)$

```
Inputs: X - training samples input, c - current
height of the new iTree, h - height limit
Output: an iTree
1:  if c >= h OR |X|<=1
2:    return exNode{Size ← |X|}
3:  else
4:    Randomly select an attribute α
5:    Randomly select a split point v between the
maximum and minimum values of attribute α in X
6:  Xl ← filter(X, α < v),Xr ← filter(X, α >= v)
7:  return inNode{ Left←iTree(Xl, c+1, h)
                   Right←iTree(Xr, c+1, h)
                   SplitAttr ← α, SplitValue ← v}
```

By building a specified number of iTrees to form an isolation forest, then the sequence data can be detected through the isolation forest. The detection process is to execute the record on the iTree to find out which leaf node the test record falls on. After traversing the entire forest, the program calculates the expectation of path length h(x). The authors mentioned that log(n) and n−1 normalization does not guarantee bounded and inconvenient comparisons [4]. It uses a slightly more complicated normalization formula as follows:

$$S(\mathrm{x}, \mathrm{n}) = 2^{\frac{-\mathrm{E(h(x))}}{\mathrm{c(n)}}}$$

$$c(\mathrm{n}) = 2(\ln(\mathrm{n}-1) + \xi) - 2^{\frac{n-1}{n}},\ \xi \text{ is Euler constant.} \quad (1)$$

$S(x, n)$ is the comprehensive anomaly score of x in the iTree composed of n samples of training data. Its value range is (0, 1]. Normal data score is close to 0, abnormal data score is close to 1. If most of the training samples have $S(x, n)$ close to 0.5, it means that there are no obvious outliers in the entire data set.

3.2 ITreesort Algorithm

The construction process of iTree has certain randomness, and the performance of distinguishing data is also very different. To reduce this instability, an algorithm for measuring the effectiveness of iTree has been proposed. We added an attribute to iTree, namely effectiveness. It is an integer, indicating which iTree can quickly identify these samples in all the given abnormal samples (normal data only occupies a small part). Our work was inspired by SA-iForest [10], but the judgment of the validity of the tree in it requires a traversal of all nodes of the tree and cross-validation. The abnormal data is always quickly isolated, that is, it appears at a not too deep level. When the depth of iTree is found to be 4 through experiments, the abnormal points can be distinguished. The algorithm is described as Algorithm 2.

```
Algorithm 2: iTreeSort(exs,sL)
Inputs: exs - samples to be detected, sL - Limited
search length
for (ex in exs)
    for (iTree in iForest)
        // find ex falls in each iTree
        pathLength = h (ex)
        // if ex falls on the first few layers
        if (pathLength <= sL)
              iTree.efficitiveness++
        end if
sortIForest according to iTree.effectiveness DESC
```

We use an ordered list to store iForest in the program, and based on the effectiveness of each tree, maintain this ordered list according to ***iTreeSort***, such as updating and sorting. Last step in it is to sort effectiveness in reverse order.

3.3 Data Feature Extraction

We first give the definition of the hydrological time series:

Definition 1: Hydrological time series stream. It is an ordered set of elements consisting of record value and record time, denoted as ($Z = <(v_1, t_1), (v_2, t_2), \ldots>$), where $z_i = (v_i, t_i)$ represents the observation value corresponding to the time, and the recording time is strictly increased.

In order to describe the changes of time series data and detect abnormal values in time, some data features need to be extracted from them. The water level is often an important factor. In addition, we also define two other data features, including the first-order difference and the mean of adjacent water levels. The former is a feature related to water level changes, and the latter describes the relationship between the current water level and the nearby average. Therefore, the feature spaces defined by the three can be described as {Level, Diff, Mean}. According to definition 1, they can be described as:

$$\text{Level} = <(v_1), (v_2), (v_3), \ldots>$$

$$\text{Diff} = <(v_1 - v_0), (v_2 - v_1), (v_3 - v_2), \ldots>$$

$$\text{Mean} = <(v_0 + v_1 + v_2)/3, (v_1 + v_2 + v_3)/3, \ldots> \quad (2)$$

According to the above formula, the sequence is mapped to the three-dimensional space as $\mathbb{Z}$:

$$\mathbb{Z} = <((Level_i, Diff_i, Mean_i))>, \text{ where } i >= 1 \quad (3)$$

3.4 Auto Regressive-IForest Detection

Auto Regressive (AR) model is a statistical method to deal with time series. It uses the previous periods to predict value in this period.

The AR model requires the sequence to be stable, so we differentiate the sequence before the data prediction to obtain a stationary sequence. In this paper, AR model is used to make single-value predictions. We first based on the m water level records in the first window to initialize the AR model for data prediction. As data flows into a window, the AR model generates predictions based on the most recent m data, and the predictions will form a sequence:

$$P_i = <(level_i, diff_i, mean_i)>, \text{ where } i > m \quad (4)$$

Then we calculate the confidence interval based on the value of the prediction sequence.

$$\text{Confidence interval} = P_i.level\left[\text{lower, upper}\right] \quad (5)$$

Lower and upper in (5) represent the lower limit and upper limit of the confidence interval of the predicted water level.

To determine whether $\mathbb{Z}_i$ in (3) is inside the confidence interval, initially the confidence level was set very low. If $\mathbb{Z}_i$ is within the confidence interval of the predicted value, we have enough confidence that the water level is within the normal fluctuation range inside. In this way, the program skips the detection of $\mathbb{Z}_i$. In fact, most of the water level data is normal. Doing so can prevent these normal data from entering iForest for anomaly judgment, thereby reducing the evaluation time of a large number of normal data and improving program running efficiency. In order to make the prediction results more accurate, as the data continues to flow into the window, the AR model is continuously updated. On the other hand, in order to reduce the complexity of the algorithm, the historical data used when updating is always the most recent m data.

3.5 Program Execution

Based on the above content in this section, here we explain the detailed process of applying the above content to the abnormal detection of hydrological sensor time series data. The entire anomaly detection process is divided into two parts, namely the data preprocessing and the window-based AR-iForest algorithm for instance detection.

Preprocessing
After extracting the data features, the time sequence is mapped to three-dimensional space, and the first m instances in the sequence are used to initialize the AR model and iForest. We use these m instances to create more trees than program need for iForest initialization. Then sort their effectiveness based on Algorithm 2, and retain the first part of trees with higher effectiveness scores to construct iForest.

Window-Based Detection
Next, as the data flows into Apache Flink's window, anomaly detection is performed inside the window. Algorithm 4 explains its detailed steps. Note: For the purpose of making the classification results more stable, after calculating the anomaly score, we set the maximum and minimum values of the anomaly score to two landmarks. For an instance's anomaly score, if it is closer to the maximum, it will be marked as anomaly. The closer it is to the minimum, it will be marked as normal. As shown in Algorithm 3 below.

Algorithm 3: ***sequenceDetect(fl)***

```
Inputs:  wl - instances in the window, c - confidence
Outputs:  nl - normalList, al - anomalyList
1: Initialize anomalyList
2: Initialize normalList
3: scoresList = iForest.predict (fl)
4: for (score in scoresList)
       if score is more closer to scoreList.max
           add to anomalyList.
       if score is more closer to scoreList.min
           add to normalList.
   return anomalyList and normalList
```

Step 1: Initialize a FuzzyList in the window to store suspicious data that is not in the confidence interval. Initialize an AnomalyList to store the detected abnormal instances.

Step 2: With the continuous inflow of data in the window, the AR model is updated and a prediction sequence is generated.

Step 3: Calculate the confidence interval and store instances that are not in it.

Algorithm 4: $AR\_iForest(wl,c)$

```
Inputs: wl - instances in the window, c - confidence
Outputs: nl - normalList, al - anomalyList
1:  initialize fuzzyList.
    initialize anomalyList.
    for (instance in wl)
2:    ARmodel.predict()
      update ARmodel use instance
3:    calculate confidence interval(c) = [lower,upper]
      if (instance.level ∈ [lower,upper])
        add instance to normalList, continue
      else
        fuzzyList.add(data)
4:  anomalyList = sequenceDetect(fuzzyList)
5:  if(anomalyList.size/window.size > threshold)
6:    wl as traindata, reconstruct 50 iTrees, then
    merging them into iForest
7:    add sequenceDetect(anomalyList) to anomalyList,
    normalList, respectively
8:    sortITree(anomalyList, lengthLimit)
9:    remove iTree in iForest where iTree.index >= 100
10: return anomalyList, normalList;
```

Step 4: After all the elements in the window are calculated, the program will detect instances in FuzzyList based on Algorithm 3.

Step 5: If the ratio of abnormal instances in this window is greater than a certain threshold, program will perform steps 6–9. Otherwise it will skip to step 10.

Step 6: We use the instances in the current window, retrain to generate 50 iTrees and merge them into iForest.

Step 7: Re-detect anomalyList.

Step 8: Call ***iTreeSort*** to sort the effectiveness of the tree.

Step 9: Remove iTrees with poor effectiveness from iTreeList, ensuring that the number of trees is 100.

Step 10: The calculation of this window is completed and anomaly is output.

4 Experimental Setup

4.1 Preprocessing

Figure 1a shows the data extracted from the local hydrological monitoring station. Then, we extracted the data features from the hydrological monitoring station. In order to facilitate the comparison, it is normalized. A certain piece of data is mapped to 3D space and shown in Fig. 1b. It can be seen that normal data appears in the form of clusters, while anomaly is few and scattered.

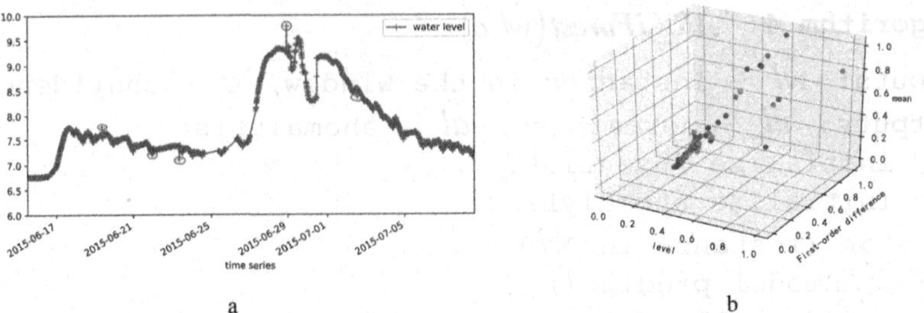

Fig. 1. 1a shows the water level data of the 60403100 hydrological monitoring stations, the red circle indicates the data abrupt position. 1b shows the extracted three-dimensional data feature. (Color figure online)

4.2 Framework

In order to judge the effectiveness of the proposed method, we combine Kafka to simulate data anomaly detection in a real environment. Our experiment is based on the following architecture (Fig. 2):

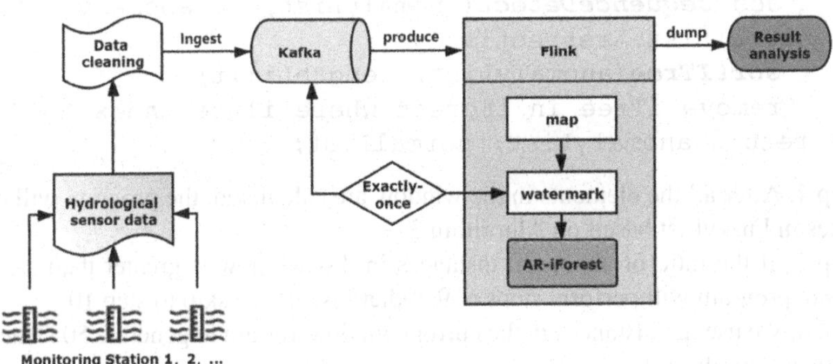

Fig. 2. Overall architecture diagram of anomaly detection algorithm integrated with Flink and using real hydrological monitoring station data for experiments.

The cleaned data is read into the specified topic of the Kafka pipeline, Flink consumes the data and first uses the map operator to perform object parsing on the received data. Then, based on the watermark feature provided by Flink and Kafka's redistribution mechanism, exactly-once semantics can be implemented in the data production process to ensure data consistency and avoid data errors caused by equipment failures or network congestion. After the above steps, we apply the AR-iForest to the sliding window provided by Flink. Finally, the calculation results of each window are aggregated and dumped into a file for result analysis.

5 Results and Discussion

It should be noted that, the iForest algorithm uses the parameters set to the recommended values if not otherwise specified in the following experiments. The sub-sampling size is 256, and the number of iTree is 100. We first studied the effect of confidence on the prediction performance of the AR-iForest algorithm. The results are shown in Table 1.

Table 1. Experimental results with different confidence

Confidence	Time (ms)	TPrate	FPrate	Precision	Accuracy
30%	4161	97.38%	0.03%	85.43%	99.96%
50%	3719	96.64%	0.02%	87.95%	99.97%
70%	3177	96.83%	0.02%	87.81%	99.97%
80%	2733	97.20%	0.02%	90.14%	99.97%
90%	**2225**	**97.76%**	**0.02%**	**91.93%**	**99.98%**
95%	1769	87.13%	0.01%	94.15%	99.97%
99%	1088	73.13%	0.00%	95.84%	99.94%

It can be seen from Table 1 that as the confidence increases, the time consumed by the program shows a downward trend. This is because the increase of confidence expands the upper and lower limits of the confidence interval, so that more instances appear in the confidence interval and skip detection. True Positive Rate (TPrate) reached its highest value at 90% confidence level. However, with the further increase of confidence, TPrate showed a sharp downward trend—the continued expansion of the limit may cause it to cover up some outliers. When the confidence reaches 99%, most of the instances will appear inside the confidence interval. The AR-iForest algorithm will skip the detection of these instances, resulting in a very small number of False Positive, which will lead to a high Precision (TP/(TP + FP)). Table 1 also confirms this conclusion.

Secondly, we compared the time cost and metrics performance of different algorithms in processing hydrological sensor data. Tested algorithms include K-Means, Local Outlier Factor (LOF), iForest, AR-iForest, one-class support vector machine (1-SVM). For the first two algorithms, their different K values were tested, from 5 to 100. Then the optimal K is selected as the parameter. For 1-SVM, we used the Radial Basis Function as the kernel function. The experimental results are shown in Table 2.

As can be seen from Table 2, because the iForest algorithm avoids a large number of numerical calculations based on distance or density, it has a lower time complexity and its AUC value is relatively high. This shows that after extracting data features from the hydrological data, the iForest algorithm can be used to detect outliers more effectively (relative to distance or density-based methods). AR-iForest performs better, and its AUC value is higher than the original iForest algorithm. Since the streaming data may trigger the update of the AR-iForest model, the time spent in model training is relatively long. However, in the evaluation stage, the test time is greatly shortened due to the avoidance of the evaluation of a large amount of normal data. The execution time of the program has been shortened by more than 52.23%.

Besides, we give a comparison of the space complexity and time complexity of the proposed algorithm and the native iForest algorithm.

Table 2. Performance comparison of different algorithms

	AUC	Time (ms)
K-Means	0.73	61032
LOF	0.88	101885
1-SVM	0.84	77460
iForest	0.91	4658 Train: 82 Eval: 4576
AR-iForest	0.97	2225 Train: 1281 Eval: 944

Where e is the number of instances that do not appear in the confidence interval in the data set. After analyzing the above time complexity, combined with the working principle of AR-iForest, we can conclude that AR-iForest can greatly reduce the detection time of time series data, because most of the normal data does not participate in iForest detection, this part of the data has been filtered out by the AR model, that is, $e \ll n$ in Table 3.

Table 3. The complexity of iForest and AR-iForest

	iForest	AR-iForest
Time complexity	$O(nt\psi log\psi)$	$O(et\psi \log \psi)$
Space complexity	$O(t\psi)$	$O(t\psi)$

Next, we tested the effect of different sample sizes on the AR-iForest algorithm. The results are shown in Fig. 3.

For the experiments involved in this paper, since the training data used when iForest is updated is all the data in the window, the subsampling size is consistent with the window size. The proposed AR-iForest performs relatively stable under different subsampling sizes. Through our experiment, the optimal sampling size is 256. In general, the proposed method is much better than the iForest algorithm.

Finally, to illustrate the effectiveness of the selection mechanism for iTree, the anomaly score in a certain window is calculated according to the current model during the experiment. From Fig. 4, we can see that most data points of normal water levels are filtered by the AR model, and only extremely low anomaly scores with little fluctuation are obtained. The proposed selection mechanism of iTree removes those trees with poor performance. The anomaly has a pretty short path length, thus obtaining a very high anomaly score, and the difference between the scores is enough to distinguish normal data from abnormal data. While iForest has a certain degree of instability, and it is reflected in some records that are difficult to distinguish. As shown in the data marked in the red circle in Fig. 4, they are more likely to be classified as "outliers" after prediction, resulting in errors happened.

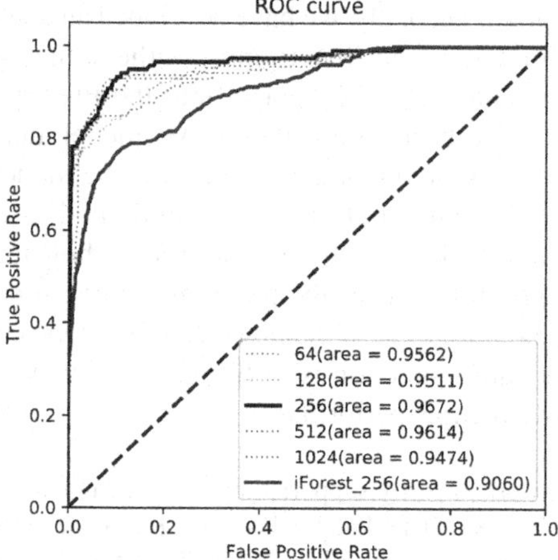

Fig. 3. The effect of different sub-sampling sizes on the algorithm, the curve based on the optimal value is displayed in bold.

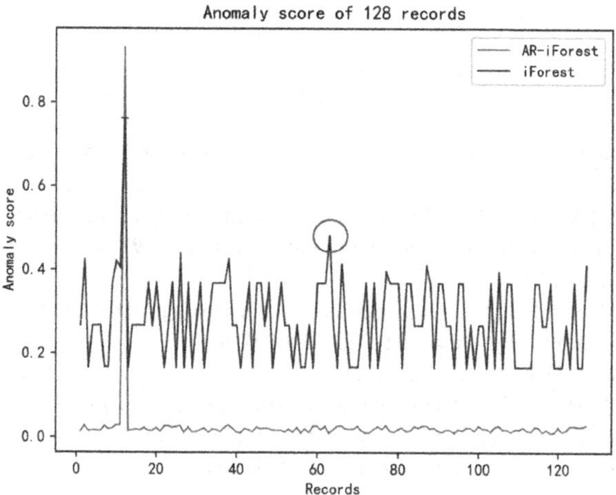

Fig. 4. For the data in a certain window, the figure shows the difference between the anomaly scores predicted by iForest and AR-iForest. (Color figure online)

6 Conclusion

The main contribution of this paper is to propose a detection method called AR-iForest. In order to make iForest suitable for anomaly detection of time series data in the hydrological industry, we propose a joint model based on Auto-Regressive (AR) model and iForest. First, we extract the data features of the sequence data, then predict the data features based on the AR model, and calculate the confidence interval based on the predicted value. We only perform anomaly detection on the data in the sequence that does not appear in the confidence interval, thereby skipping part of the data and improving the efficiency of the program. Secondly, we have developed an algorithm for ranking the effectiveness of trees. If a tree is sensitive to most outliers, then they will have higher effectiveness. Finally, we integrated AR-iForest into the Flink platform and used the window mechanism provided by it to update the AR model and iForest in real-time. Through this integration with the big data platform and the model update mechanism, the proposed AR-iForest has the ability to handle unlimited data streams, and it can also efficiently and accurately detect outliers. The results show that compared with the iForest, the proposed algorithm reduces the program running time by 52.23%, and the AUC increases from 90.60% to 96.72%. It shows that the proposed algorithm has great advantages. The next research direction is to further improve the algorithm to reduce running time and improve metrics performance.

Acknowledgments. This work is partly supported by the Fundamental Research Funds for the Central Universities B200202185, 2018 Jiangsu Province Key Research and Development Program (Modern Agriculture) Project under Grant No. BE2018301, 2017 Jiangsu Province Postdoctoral Research Funding Project under Grant No. 1701020C, 2017 Six Talent Peaks Endorsement Project of Jiangsu under Grant No. XYDXX-078, Research on the Analysis System of Hydrological Big Data under Grant No. 818116816.

References

1. Wu, D.: Research and Application of Hydrological Time Series Similarity Pattern. HoHai University, pp. 1–2 (2007)
2. Talagala, P.D., Hyndman, R.J., Miles, K.S., Kandanaarachchi, S., Muñoz, M.A.: Anomaly detection in streaming nonstationary temporal data. JCGS **29**(1), 13–27. https://doi.org/10.1080/10618600.2019.1617160
3. Sun, J.S., Lou, Y.S., Chen, Y.J.: Outlier detection of hydrological time series based on ARIMA-SVR Model. Comput. Digit. Eng. **02**, 225–230 (2018)
4. Vy, N.D.K., Anh, D.T.: Detecting variable length anomaly patterns in time series data. In: Proceedings of DMBD, Bali Island, Indonesia, June 2016, pp. 279–287 (2016)
5. Yu, Y.F., Zhu, Y.L., Wan, D.S.: Time series outlier detection based on sliding window prediction. J. Comput. Appl. **34**(8), 2217–2220 (2014)
6. Liu, F.T., Ting, K.M., Zhou, Z.H.: Isolation forest. In: Proceedings of ICDM, Pisa, Italy, December 2008, pp. 413–422. https://doi.org/10.1109/icdm.2008.17
7. Carbone, P., Katsifodimos, A., Ewen, S., Markl, V., Haridi, S., Tzoumas, K.: Apache flinkTM: stream and batch processing in a single engine. In: Proceedings of ICDE, Seoul, South Korea, vol. 38, no. 4, pp. 28–38 (2015)
8. Toliopoulos, T., Gounaris, A., Tsichlas, K., Papadopoulos, A., Sampaio, S.: Continuous outlier mining of streaming data in flink (2019). arXiv:1902.07901
9. Bandaragoda, T.R., Ting, K.M., Albrecht, D., Liu, F.T., Wells, J.R.: Efficient anomaly detection by isolation using nearest neighbour ensemble. In: Proceedings of ICDMW, Shenzhen, China, pp. 698–705 (2014). https://doi.org/10.1109/icdmw.2014.70
10. Xu, D., Wang, Y., Meng, Y., Zhang, Z.: An improved data anomaly detection method based on isolation forest. In: Proceedings of ISCID, HangZhou, China, December 2017 (2017)
11. Ding, Z.G., Fei, M.R.: An anomaly detection approach based on isolation forest algorithm for streaming data using sliding window. In: Proceedings of ICONS, ChengDu, China, September 2013, pp. 12–17 (2013)

12. Aryal, S., Ting, K.M., Wells, J.R., Washio, T.: Improving iForest with relative mass. In: Proc. PAKDD, TaiWan, China, May 2014, pp. 510–521 (2014)
13. Zou, Z., Xie, Y., Huang, K., Xu, G., Feng, D., Long, D.: A docker container anomaly monitoring system based on optimized isolation forest. In: IEEE TCC, to be published. https://doi.org/10.1109/tcc.2019.2935724
14. Ma, Y., Zhang, Q., Ding, J., Wang, Q., Ma, J.: Short term load forecasting based on iForest-LSTM. In: Proceedings of ICIEA, Xi'an, China, pp. 2278–2282 (2019)
15. Apache Kafka. https://kafka.apache.org/
16. Karimov, J., Rabl, T., Katsifodimos, A., Samarev, R., Heiskanen, H., Markl, V.: Benchmarking distributed stream data processing systems. In: Proceedings of ICDE, Paris, pp. 1507–1518 (2018)

Machine Learning-Based Attack Detection Method in Hadoop

Ningwei Li, Hang Gao, Liang Liu(✉), and Jianfei Peng

Nanjing University of Aeronautics and Astronautics, Nanjing, China
{lnw,gaohang,liangliu,pengjf}@nuaa.edu.cn

Abstract. With the development of big data and cloud computing, big data clusters have become the main targets of attackers. However, The security of big data clusters is not guaranteed well. Based on our previous work, this paper proposes a big data cluster attack detection method based on Machine Learning. This method extracts 46 types of features related to the execution of cluster tasks, which solves the problem that the existing detection scheme has few features and detects a single attack type. Considering the impact of too many features on the detection results, this paper uses filter to extract key features, which can improve the accuracy of detection; Finally, this paper uses different machine learning algorithms to train different attack data extracted on different tasks. Experiments show that the attack detection scheme in this paper has an accuracy of over 88% for different attack.

Keywords: Hadoop · Big data · Security · Machine learning · Attack detection

1 Introduction

With the development of big data, its security issues become more prominent [1,2]. The scale of big data platforms has evolved from the traditional three major components to a huge ecosystem composed of more than 60 related components. If one problem occurs with one of these components, it will have a significant impact on the security of the cluster [3].

The research is mainly targeted at Hadoop in this paper. The current research is mainly focused on improving the performance of Hadoop, but lacking attack detection. So it is important to analyze security vulnerability and detect attacks in cluster [4]. In addition, machine learning is also widely used in many scenarios, but in many studies, machine learning and big data are combined to determine

This work is supported by the Aeronautical Science Foundation of China under Grant 20165515001, the National Natural Science Foundation of China under Grant No.61402225, State Key Laboratory for smart grid protection and operation control Foundation, and the Science and Technology Funds from National State Grid Ltd.(The Research on Key Technologies of Distributed Parallel Database Storage and Processing based on Big Data).

M. Qiu (Ed.): ICA3PP 2020, LNCS 12454, pp. 184–196, 2020.
https://doi.org/10.1007/978-3-030-60248-2_13

the security vulnerabilities of other platforms or software. Machine learning is rarely applied to attack detection on big data platforms. Although Hadoop has some security mechanisms, it is still not enough to ensure the security of big data clusters. At present, the following problems still exist:

- Data nodes have no access control mechanism. The compromised internal nodes may gain access to intermediate data through the shuffle protocol of the task tracker;
- The broken internal node can monitor the tasks assigned to the node and change the execution status of the task at will;
- The Hadoop security mechanism lacks monitoring of the network, the speed of network will affect the data transmission efficiency during the shuffle phase between nodes;
- Nodes interact through the heartbeat protocol, and the internal nodes that are compromised can adjust the heartbeat time of the nodes at will (without exceeding the system timeout time);

Although there are some schemes that use machine learning to detect attacks, these schemes extract fewer attack features and detect a single type of attack [5]. In our previous work, four attack models have been proposed based on the above security vulnerabilities, namely MR Schedule Attack, Data Attack, Heartbeat Attack, and Shuffle Attack [6]. This paper mainly uses these machine learning algorithms to detect these security vulnerabilities, the main contributions are as follows:

- We propose a big data cluster attack detection scheme based on machine learning;
- We extracted 46 characteristics related to task execution. These characteristics not only include the characteristics of the node itself, such as CPU, disk, memory, network, etc., but also include the characteristics of Hadoop cluster, such as dfs, yarn, and other components, which solved the problem of original machine learning algorithm has few features and detects a single attack type;
- Considering that 46 types of data features will generate a lot of redundant information if all of them are used, it will have a great impact on the accuracy of different attack detection results. We use filter to extract key features for different attacks, which improve the accuracy of detection;
- The detection scheme proposed in this paper is implemented in an actual environment, using Support Vector Machine Algorithm (SVM), Logistic Regression Algorithm, Random Forest Algorithm, K-Nearest Neighbor Algorithm (KNN) to detect different attack types under different tasks, which verified the effectiveness of the attack detection method in this paper, and provided a new idea for the establishment of a complete security protection mechanism in big data clusters.

The rest of this paper is organized as follows: We introduce the background and related work int Sect. 2. Then we introduce the machine learning detection framework of this paper in Sect. 3. In Sect. 4, we verify the effectiveness of this detection framework by experiments. Finally, we conclude the paper in Sect. 5.

2 Related Work

In this section, we first give a brief introduction to our previous work, then details the existing methods of Hadoop to detect attacks.

2.1 Previous Work

In our previous work, four attack schemes were proposed for Hadoop security vulnerabilities. This paper mainly detects these four attack schemes. We will briefly introduce our previous work.

Our first attack scheme is MR Schedule Attack, which mainly uses Hadoop's speculative scheduling mechanism to extend the execution time of tasks. In Hadoop, if a task is detected to be slow, a speculative task will be launched to replace it. We can estimate the completion time of each task. Specifically, hadoop estimates the completion time of the task based on their execution time and execution progress, as shown in Eq. (1).

$$TimeComplete = T_{now} + (1 - P) * T/P \tag{1}$$

Equation (1) represents the calculation method used by Hadoop to estimate task completion time. Where $TimeComplete$ represents the completion time of the task; P represents the progress of the task and can be approximately equal to the percentage of input data that has been read to the entire input data; T represents the time that the task has been executed; T_{now} represents current time. The completion time of the speculative task is shown in Eq. (2).

$$STimeComplete = T_{now} + T_{mean} \tag{2}$$

$STimeComplete$ represents the completion time of speculative task; T_{mean} represents the average execution time of each task of the same type. The execution time of a speculative task is defined as the average execution time of all tasks that have been executed. If $TimeComplete$ is more than $STimeComplete$, the speculator service will create a new task to replace the task.

Suppose a task pauses execution during process p, and the time consumed is T. When $T \div P > T + T_{mean}$, the speculative task will be executed. Therefore the longest delay time is shown in Eq. (3).

$$DelayTime = \frac{T_{mean}}{\frac{1}{p} - 1} \tag{3}$$

Suppose we pause the execution of the task at the progress p. According to Eq. (3), the maximum pause time is:

$$PauseTime = DelayTime + p * T_{mean} \tag{4}$$

The second type is called Heartbeat Attack. The communication way between namenode and datanode is heartbeat. The heartbeat time is common 3 s. When the datanode has not sent heartbeat for a long time, the namenode determines that the connection to the datanode has been interrupted, and the datanode can not continue to work. Namenode defines it as "dead node". This time is timeout. We can extend the heartbeat response time until timeout, which can extend the execution time of jobs.

The third is Data Attack. We simulate an attacked node, monitor the task manager of the node, and obtain task information assigned to the node, so that the intermediate data generated by the task can be attacked. During the execution of the Map task, Attackers in the malicious node inject a new attack module in the node manager. The module is responsible for monitoring tasks assigned to this node. Then the attackers will get the jobId and taskId. By using jobId and taskId, the attackers can modify the data of this task, delete some data or add some irrelevant data. Then in the group stage will get the wrong key-value pair, which will cause the execution result of the entire job to be wrong.

The fourth is Shuffle Attack, which mainly uses Hadoop's shuffle mechanism to slow down the task execution speed by extending the shuffle time [7]. By limiting the bandwidth in the cluster, the execution time of this process can be extended.

2.2 Related Work

In the paper [8], the authors proposed a new system architecture that can detect internal attacks by utilizing the characteristics of data replication on each node in the system. This solution requires specific modifications to the hardware equipment and the Hadoop ecosystem, which is more complicated to implement and is not suitable for general situations.

The paper [9] is aimed at Spark's abnormal task detection for big data processing platform, but it can also be applied to Hadoop platform. This article proposes an innovative tool called LADRA, which is used for log-based abnormal task detection and root cause analysis using Spark logs. LADRA detects abnormal tasks by examining the features extracted from the logs, and analyzes them through a neural network model to find the root cause. The shortcoming of this paper is that the number of features extracted from the log is relatively small, which is suitable for the attack detection proposed by the author, and it is not suitable for general situations.

In the paper [10], the authors proposed a runtime performance analysis method to predict the memory usage of data nodes. This method mainly relies on the replication characteristics of big data clusters, models memory mapping as two-dimensional time series data, and applies LSTM recurrent neural networks to predict memory usage. Run-time analysis will increase task execution time and reduce task performance.

In the paper [11], the author mainly uses machine learning algorithms to detect network attacks inside big data clusters. The data set in this paper is not the real Hadoop task execution data and is not representative.

3 Attack Detected Based on Machine Learning

In this section, we discuss the process of using machine learning for attack detection in big data clusters.

3.1 System Structure

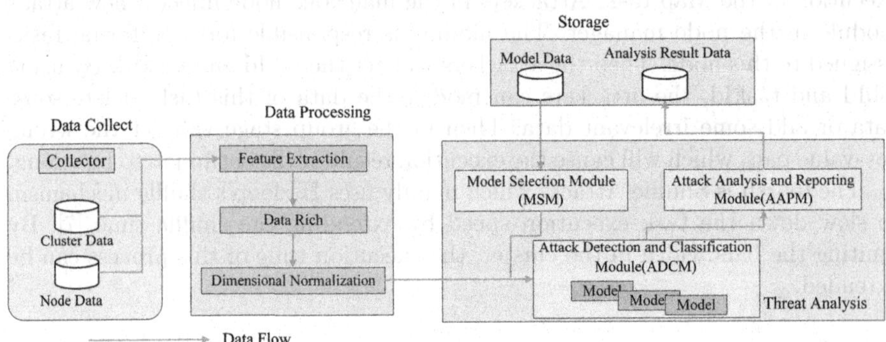

Fig. 1. The system architecture that includes four components to detect attacks.

Figure 1 shows the system architecture of the detection method in this paper. The required data is collected in the Data Collect stage. After the Data Collect stage, the collected data will be turned into trainable vector set in the Data Processing stage. The train model is selected by the Model Selection Module. The Attack Detection and Classification Module uses this model to detect and classify the data which has been processed, and the Attack Analysis and Reporting Module saves the results.

The main parts of the system architecture are Data Collect and Data Processing.

Data Collect. We use the open source tool Ganglia to collect data. Ganglia contains two important components: gmond and gmetad. gmond is responsible for collecting information of each node, and gmetad is responsible for collecting all the information which collected by gmond.

The data collected in this paper mainly includes two types: Cluster Data and Node Data.

- Cluster Data: the data of the big data cluster when performing tasks, including dfs data, yarn data, etc.
- Node Data: the data of the node itself, such as memory size, disk size, network, etc.

Threat Analysis. Threat Analysis contains four important modules: Data Processing, Model Selection Module (MSM), Attack Detection and Classification Module (ADCM) and Attack Analysis and Reporting Module (AAPM).

Data Processing: includes three functions:

- Feature Extraction: select and extract the data as the input vector of the Attack Detection Model;
- Data Rich: If the Attack Detection Model requires some additional data as input, this part can add these additional data to the input vector;
- Dimension Normalization: Unify the dimensions of the data in the input vector to prevent results' deviations caused by the inconsistencies in the dimensions.

MSM: The function of MSM is to select a suitable training model. We convert the collected data into the input vector of each detection and classification model through data preprocessing. Then the user can select the appropriate model from the model database for attack detection.

ADCM: After the user selects a suitable detection model, the model can be used to perform attack detection on the collected data that has been prepared. The function of this module is to match the input vector of the collected data with the selected detection model, and then to perform attack detection and classification on the input vector.

AAPM: This module mainly analyzes the results of attack detection and reports the detection results to the master node, which determines the subsequent processing. In addition, the module will also store the results in the database.

Result Storage. Result Storage contains two types of data: Model Data and Analysis Result Data.

- Model Data: This data is the training model data. Users can choose to use a certain model during the attack detection process, including KNN, RF, LR, SVM etc.
- Analysis Result Data: This data is to save the attack detection results, so that the user can analyze the node status.

3.2 Attack Detection Process Based on Machine Learning

Figure 2 shows the process of Attack Detection and Classification Model establishment based on machine learning. The model establishment could be divided into four steps: select feature, generating data set, naming data set, evaluating Attack Detection and Classification Model. We repeatedly train the model by adjusting the model parameters to obtain the model with the best accuracy and performance. The data set will be divided into train set and test set:

Select Features: As shown in Table 1, we provide 46 types of features to choose. In order to prevent overfitting, we need to choose appropriate features. This

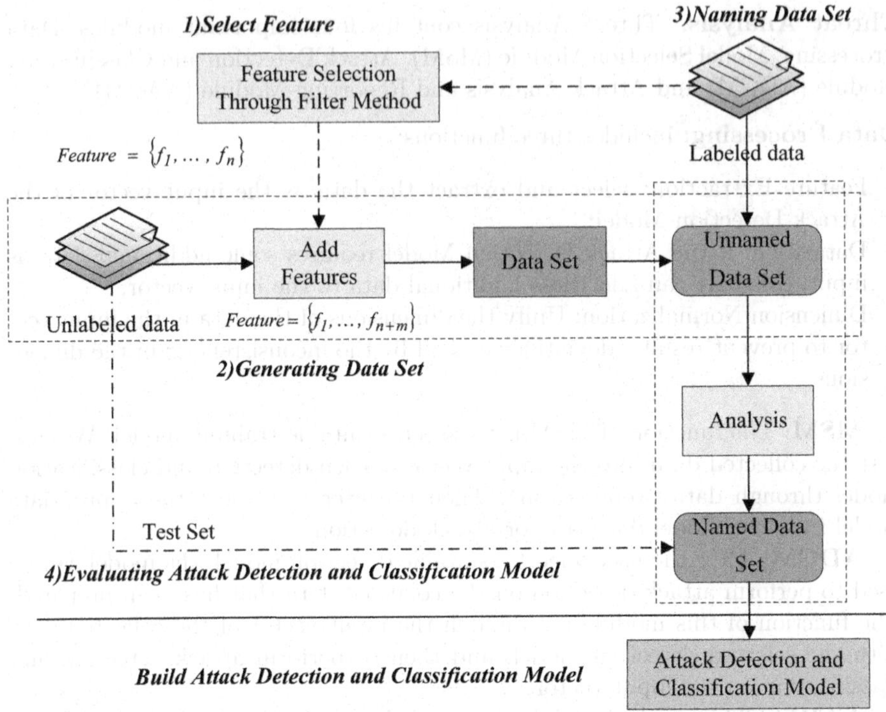

Fig. 2. Attack detection and classification process with four key steps. In the Select feature stage, the importance is used for feature selection.

paper selects the features according to their importance. Assuming that there are $n(n > 0)$ data sets, and the data corresponding to the feature f is $(x_1, x_2, ..., x_n)$, the importance of the feature f is shown in Eq. 5:

$$I_f = \frac{S}{M} \tag{5}$$

where I_f represents the importance of the feature f; S represents the standard deviation of $(x_1, x_2, ..., x_n)$ and M represents the average value of $(x_1, x_2, ..., x_n)$. The calculation of S and M is shown in Eqs. 6, 7:

$$M = \frac{\sum_{i=1}^{n}(x_i)}{n} \tag{6}$$

$$S = \sqrt{\frac{\sum_{i=1}^{n}(x_i - M)^2}{n}} \tag{7}$$

The method is to calculate the importance of each feature, then sort according to the magnitude of the importance, finally select features according to a specific threshold or the number of features. Because the magnitude of the feature importance actually represents the amount of information contained in the feature, the smaller importance value may be because the value of the feature is

Table 1. 46 optional detection features

Category	No	Name	Description
Network	1	byte-in	Network receiving speed
	2	byte-out	Network sending speed
	3	pkts-in	Incoming packets per second
	4	pkts-out	Packets sent out per second
CPU	5	cpu-idle	CPU idle percentage
	6	cpu-system	Kernel space occupied cpu percentage
	7	cpu-user	User space occupied cpu percentage
dfs	8	dfs.datanode.BlocksRead	Number of blocks read from disk
	9	dfs.datanode.BlocksWritten	Number of blocks write to disk
	10	dfs.datanode.BytesRead	Read total bytes
	11	dfs.datanode.BytesWritten	Write total bytes
	12	dfs.datanode.HeartbeatsAvgTime	Average heartbeat time
	13	dfs.datanode.HeartbeatsNumOps	Total number of heartbeats
	14	dfs.datanode.IncrementalBlock ReportsAvgTime	Average time to report block to namenode
	15	dfs.datanode.IncrementalBlock ReportsNumOps	Number of block reports to namenode
	16	dfs.datanode.ReadBlock OpAvgTime	Average time to read data blocks
	17	dfs.datanode.ReadBlock OpNumOps	Number of data blocks read
	18	dfs.datanode.ReadsFrom LocalClient	Number of local reads
	19	dfs.datanode.ReadsFrom RemoteClient	Number of remote reads
	20	dfs.datanode.SendDataPacket BlockedOnNetworkNanosAvgTime	Average block sending time on the network
	21	dfs.datanode.SendDataPacket BlockedOnNetworkNanosNumOps	Number of blocks sent on the network
	22	dfs.datanode.SendDataPacket TransferNanosAvgTime	Average time of sending packets on the network
	23	dfs.datanode.SendDataPacket ·TransferNanosNumOps	Number of packets sent on the network
	24	dfs.datanode.WriteBlock OpNumOps	Number of data blocks written
	25	dfs.datanode.WritesFromLocal Client	Number of local writes
	26	dfs.datanode.WritesFromRemote Client	Number of remote writes
Disk	27	disk-free	Free disk capacity
	28	disk-total	Total disk capacity
JVM	29	jvm.metrics.gcCount	Total GC times
	30	jvm.metrics.logError	JVM error times
	31	jvm.metrics.logWarn	JVM warn times
	32	jvm.metrics.memHeapUsedM	JVM uses heap memory size
mapred	33	mapred.MRAppMetrics. MapsCompleted	Number of completed map tasks
	34	mapred.MRAppMetrics.Maps Running	Number of running map tasks
	35	mapred.MRAppMetrics. ReducesCompleted	Number of completed reduce tasks
	36	mapred.MRAppMetrics. ReducesRunning	Number of running reduce tasks
	37	mapred.ShuffleMetrics. ShuffleConnections	Number of shuffle connections
	38	mapred.ShuffleMetrics. ShuffleOutputBytes	number of shuffle output bytes
	39	mapred.ShuffleMetrics. ShuffleOutputsFailed	Number of shuffle failures
	40	mapred.ShuffleMetrics. ShuffleOutputsOK	Number of successful shuffles

(continued)

Table 1. (*continued*)

Memory	41	mem_shared	Shared memory size
	42	mem-cache	Memory cache size
	43	mem-free	Free memory capacity
	44	mem-total	Total memory capacity
rpc	45	rpcdetailed.rpcdetailed. HeartbeatNumOps	Number of heartbeat operations detected by rpc
	46	rpcdetailed.rpcdetailed. SendHeartbeatNumOps	Number of sent heartbeats detected by rpc

relatively fixed. It is not useful for our detection and classification, so we can eliminate the smaller importance.

Generating Data Set: In this step, we use the open source tool Ganglia to extract cluster data and node data, and then analyze the original data set to extract the feature data to generate unlabeled input data. It should be noted that the feature data is not the 46 kinds of feature data shown in Table 1, but the data corresponding to the selected feature.

Naming Data Set: We use supervised machine learning algorithms for training and detection, so we need to label the unlabeled data. There are four types of tags: Normal, Heartbeat Attack, Data Attack, and Shuffle Attack. We label the data generated by different attacks and use different machine learning algorithms to generate the training models.

Evaluating Attack Detection and Classification Model: After generating the training models, we compare the performance and accuracy of machine learning algorithm models by using test sets. Then we compared the detection efficiency and accuracy of different machine algorithms for different attacks. Users can choose the appropriate training model for attack detection.

4 Implementation and Evaluation

This section introduces the implementation process of our attack detection scheme, which mainly includes the following parts: (1) the settings of big data clusters; (2) Through four machine learning algorithmsK-NearestNeighbor (KNN), Random Forest (RF), Logistic Regression (LR), Decision Tree (DT) and Support Vector Machine (SVM) to validate our attack detection model, and analyze the experimental results.

4.1 Cluster Settings and Workloads

We set up a small Hadoop cluster. The cluster contains three nodes, one of which is namenode and the rest are datanodes. We used a standard computer to implement the machine learning detection architecture of this paper. This architecture runs on the Keras & Tensorflow framework. The configuration of the Hadoop cluster and computer is shown in Table 2. Besides, the workload we use is the standard Hibench benchmark. Hibench benchmark is a test framework designed by Intel to evaluate the performance of big data clusters. We selected four benchmarks from it: RandomTextWriter, WordCount, TeraSort, and Wordmean.

Table 2. Equipment configuration in experiment

Attribute	Value	
Purpose	Hadoop cluster	Attack detection computer
Processor	Intel Core(TM) i5-8300H	Intel Core(TM) i7-2600
Compute units	1	8
Storage	20 GB	256 GB
Operating system	Centos7	Windows10
Software	Hadoop-2.9.1 & Ganglia	Keras& Tensorflow

4.2 Performance Evaluation

In this section, we will use KNN, RF, LR, DT and SVM algorithms to detect Heartbeat Attack, Data Attack, and Shuffle Attack.

The experimental steps are:

- Use RandomTextWriter to randomly generate 5G files;
- Use the generated files to execute WordCount, TeraSort, and WordMean tasks under normal circumstances to generate normal type data sets;
- Use the generated files to execute the WordCount, TeraSort, and WordMean tasks in the Heartbeat Attack, Data Attack, and Shuffle Attack environments to generate attacked data sets;
- The KNN, RF, LR, DT and SVM algorithms are brought into the system structure for attack detection. During the detection process, the proportion of the train set and test set will be continuously adjusted to evaluate machine learning algorithms' efficiency and accuracy.

Table 3. Average training time of different algorithms

	Times
KNN	3066.646
DT	248.653
RF	485.228
SVM	842.313
LR	5.648

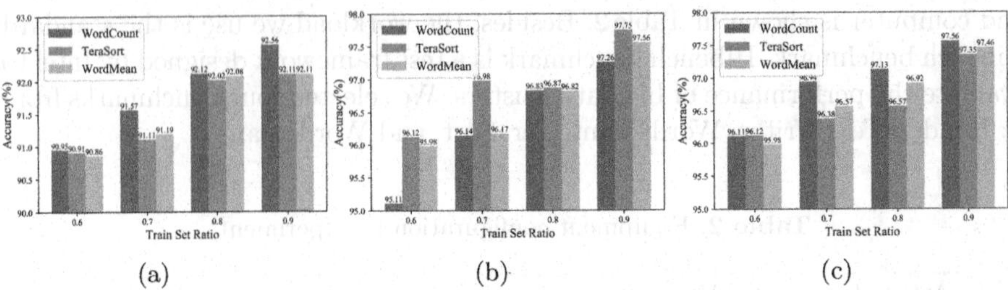

Fig. 3. KNN attack detection results under different jobs, different attacks and different training set ratios. (a) Data attack (b) Shuffle attack (c) HeratBeat attack

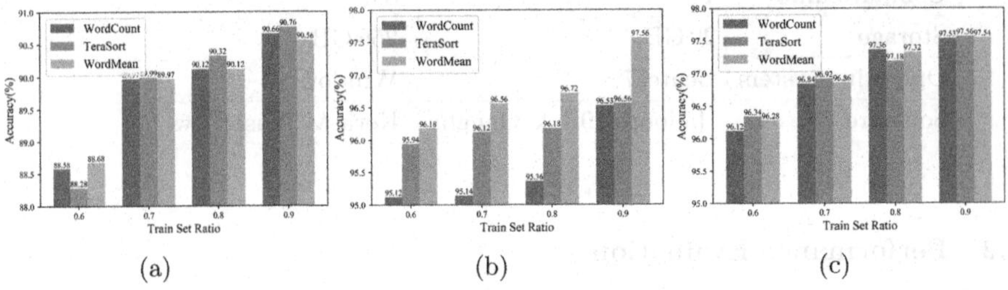

Fig. 4. RF attack detection results under different jobs, different attacks and different training set ratios. (a) Data attack (b) Shuffle attack (c) HeratBeat attack

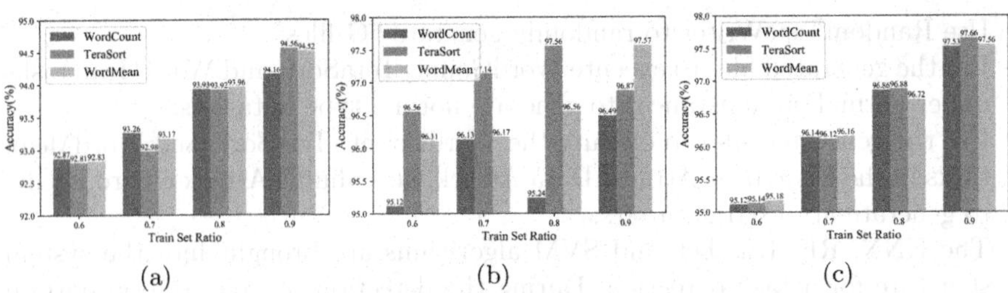

Fig. 5. SVM attack detection results under different jobs, different attacks and different training set ratios. (a) Data attack (b) Shuffle attack (c) HeratBeat attack

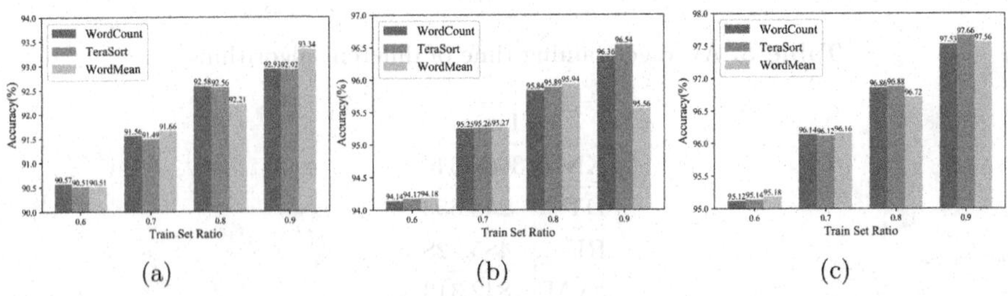

Fig. 6. LR attack detection results under different jobs, different attacks and different training set ratios. (a) Data attack (b) Shuffle attack (c)HeratBeat attack

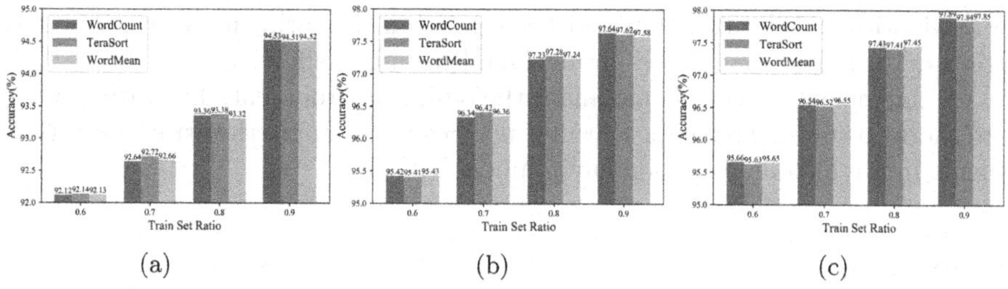

Fig. 7. DT attack detection results under different jobs, different attacks and different training set ratios. (a) Data attack (b) Shuffle attack (c) HeratBeat attack

The experimental results are shown in Fig. 3, Fig. 4, Fig. 5, Fig. 6, Fig. 7. Besides, the average training time is shown in Table 3. It can be seen from the results that the smaller the proportion of the training set, the worse the accuracy obtained, but the accuracy will be greater than 85%. In addition, regardless of the detection algorithm, the detection results of data attack are worse than shuffle attack and heartbeat attack. This is because the data attack belongs to the internal attack. It is concealed, and feature extraction is difficult. The detection of the shuffle attack mainly extracts the network state features, and the detection of the heartbeat attack mainly monitors the heartbeat frequency features. These features have a large difference from the normal state, so the detection results are more accurate.

For KNN algorithm, its disadvantage is that the training time is too long, and it is not suitable for real-time detection of attacks.

For RF algorithm, we can be seen from Fig. 4 that the accuracy are basically the same as KNN algorithm, but the time consumed is about one-sixth that of the KNN algorithm.

For SVM algorithm, it can be seen from Fig. 5 that the overall performance is better than KNN and RF algorithm. However, the training time of the SVM algorithm is about twice the RF algorithm. If we want a higher accuracy without considering the time factor, we can choose to use the SVM algorithm.

For LR algorithm, the overall performance is worse than other algorithms. But the training time of LR is much less than the other three algorithms. If we consider online detection, we can use LR algorithm.

For the DT algorithm, its training time is half of the SVM algorithm, but the accuracy is only slightly lower than the SVM algorithm. It is the most cost-effective algorithm for offline detection.

5 Conclusion

This paper designs a framework for machine learning to detect attacks based on our previous work, and proposes 46 types of task features of Hadoop systems. We implemented the attack detection scheme of this paper in Hadoop system, and evaluated the advantages and disadvantages of KNN algorithm, RF algorithm,

SVM algorithm, DT algorithm and LR algorithm. Experiments prove that the attack detection scheme based on machine learning is effective.

We implemented our scheme under the offline environment. The future work is to choose a suitable machine learning algorithm and apply the attack detection scheme in a real-time detection environment.

References

1. Win, T.Y., Tianfield, H., Mair, Q.: Big data based security analytics for protecting virtualized infrastructures in cloud computing. IEEE Trans. Big Data 1 (2017). https://doi.org/10.1109/TBDATA.2017.2715335
2. Gao, Y., Fu, X., Luo, B., et al.: Haddle: a framework for investigating data leakage attacks in hadoop. In: IEEE Global Communications Conference (2015). https://doi.org/10.1109/GLOCOM.2015.7417387
3. Fu, X., Gao, Y., Luo, B., et al.: Security threats to hadoop: data leakage attacks and investigation. IEEE Netw. **31**(2), 67–71 (2017). https://doi.org/10.1109/MNET.2017.1500095NM
4. Wang, J., Wang, T., Yang, Z., et al.: SEINA: a stealthy and effective internal attack in hadoop systems. In: International Conference on Computing. IEEE (2017). https://doi.org/10.1109/ICCNC.2017.7876183
5. Kim, H., Kim, J., Kim, Y., et al.: Design of network threat detection and classification based on machine learning on cloud computing. Cluster Comput. **22**(Suppl 1), 1–10 (2019). https://doi.org/10.1007/s10586-018-1841-8
6. Li, N., Gao, H., Liu, L., et al.: Attack models for big data platform hadoop. In:2019 IEEE 5th International Conference on Big Data Security on Cloud (BigDataSecurity), pp. 154–159. https://doi.org/10.1109/BigDataSecurity-HPSC-IDS.2019.00037
7. Guo, Y., Rao, J., Cheng, D., et al.: iShuffle: Improving Hadoop Performance with Shuffle-on-Write[J]. IEEE Transactions on Parallel and Distributed Systems **28**(6), 1649–1662 (2017). https://doi.org/10.1109/TPDS.2016.2587645
8. Aditham, S., Ranganathan, N.: A system architecture for the detection of insider attacks in big data systems. IEEE Trans. Dependable Secure Comput. **15**(6), 974–987 (2018). https://doi.org/10.1109/TDSC.2017.2768533
9. Aditham, S., Ranganathan, N.: A system architecture for the detection of insider attacks in big data systems. IEEE Trans. Dependable Secure Comput. **15**(6), 974–987 (2018). https://doi.org/10.1109/TDSC.2017.2768533
10. Aditham, S., Ranganathan, N., Katkoori, S.: LSTM-based memory profiling for predicting data attacks in distributed big data systems. In: 2017 IEEE International Parallel and Distributed Processing Symposium Workshops (IPDPSW). IEEE (2017). https://doi.org/10.1109/IPDPSW.2017.76
11. Kurt, E.M., Becerikli, Y.: Network intrusion detection on apache spark with machine learning algorithms. In: Pimenidis, E., Jayne, C. (eds.) EANN 2018. CCIS, vol. 893, pp. 130–141. Springer, Cham (2018). https://doi.org/10.1007/978-3-319-98204-5_11

Profiling-Based Big Data Workflow Optimization in a Cross-layer Coupled Design Framework

Qianwen Ye[1], Chase Q. Wu[1(✉)], Wuji Liu[1], Aiqin Hou[2], and Wei Shen[3]

[1] Department of Computer Science, New Jersey Institute of Technology, Newark, NJ 07102, USA
{qy57,chase.wu,wl87}@njit.edu

[2] School of Information Science and Technology, Northwest University, Xi'an 710127, Shaanxi, China
houaiqin@nwu.edu.cn

[3] School of Informatics Science and Technology, Zhejiang Sci-Tech University, Hangzhou 310018, Zhejiang, China
shenwei@zstu.edu.cn

Abstract. Big data processing and analysis increasingly rely on workflow technologies for knowledge discovery and scientific innovation. The execution of big data workflows is now commonly supported on reliable and scalable data storage and computing platforms such as Hadoop. There are a variety of factors affecting workflow performance across multiple layers of big data systems, including the inherent properties (such as scale and topology) of the workflow, the parallel computing engine it runs on, the resource manager that orchestrates distributed resources, the file system that stores data, as well as the parameter setting of each layer. Optimizing workflow performance is challenging because the compound effects of the aforementioned layers are complex and opaque to end users. Generally, tuning their parameters requires an in-depth understanding of big data systems, and the default settings do not always yield optimal performance. We propose a profiling-based cross-layer coupled design framework to determine the best parameter setting for each layer in the entire technology stack to optimize workflow performance. To tackle the large parameter space, we reduce the number of experiments needed for profiling with two approaches: i) identify a subset of critical parameters with the most significant influence through feature selection; and ii) minimize the search process within the value range of each critical parameter using stochastic approximation. Experimental results show that the proposed optimization framework provides the most suitable parameter settings for a given workflow to achieve the best performance. This profiling-based method could be used by end users and service providers to configure and execute large-scale workflows in complex big data systems.

Keywords: Big data workflows · performance optimization · workflow profiling · stochastic approximation · coupled design

M. Qiu (Ed.): ICA3PP 2020, LNCS 12454, pp. 197–217, 2020.
https://doi.org/10.1007/978-3-030-60248-2_14

1 Introduction

Many large-scale applications in science, industry, and business domains are generating colossal amounts of data on a daily basis, now commonly termed as "big data". The processing and analysis of such data increasingly rely on workflow technologies for knowledge discovery and scientific innovation.

Big data workflows typically consist of several computing modules[1] with intricate dependencies, each of which could be as simple as a serial program as in most traditional applications or as complex as a parallel job in big data systems. The execution of such workflows goes far beyond the capability and capacity of single computers and parallel computing is widely recognized as a viable solution to support data- and network-intensive workflows.

In recent years, significant progress has been made in almost every aspect of the hardware and software for big data computing, including the hardware upgrade of computers, the bandwidth improvement of network infrastructure, and the emergence of reliable and scalable data storage and analysis platforms such as Hadoop MapReduce/Spark that can support parallel processing of colossal amounts of data. For example, in China, the most popular search engine of Baidu, the e-commerce transaction service of Alibaba, and the social network platform of Tencent, all depend on Spark-based solutions at scale. Particularly, Tencent has eight hundred million active users generating over 1 PB of data per day, processed on a cluster consisting of more than 8,000 computing nodes [1].

However, even with these cutting-edge technologies, we have not seen the corresponding performance improvement, because workflow performance largely depends on how big data systems are configured and used. For example, workflow makespan or end-to-end delay, which is the most commonly concerned performance metric, is affected by multiple layers of big data systems, including the inherent properties (scale and topology) of the workflow, the parallel computing engine it runs on, the resource manager that orchestrates distributed resources, the file system that stores data, as well as the parameter setting of each layer.

Optimizing workflow performance is challenging because the compound effects of the aforementioned layers are complex and opaque to end users. Generally, tuning their parameters requires an in-depth understanding of big data systems, and the default settings do not always yield optimal performance. Due to the dynamics of workflow execution in distributed environments and the lack of accurate performance models for parallel computing, it is generally very difficult to analytically derive the best system configuration for a given workflow. Oftentimes, even production systems may run with misconfigured parameters and present performance bottleneck, which significantly limits the productivity of end users and the utilization of expensive computing resources. To end users, who are primarily domain experts, it is a daunting task to troubleshoot the performance issue and identify possible system misconfiguration.

---

[1] A module is a processing unit, executed in serial or parallel, in a workflow, and is also referred to as a job or subtask in some context.

In this paper, we propose a profiling-based cross-layer coupled design framework to determine the most suitable configuration with a recommended parameter settings for each layer, including workflow structure, computing engine, and cluster infrastructure to optimize workflow performance. Such a profiling-based framework requires the exploration of a large parameter space across the entire technology stack of big data systems by running a large set of workflow experiments with different parameter settings. However, sweeping through the entire parameter space is practically infeasible, due to the large number of tunable parameters and their corresponding seemingly infinite value ranges. Thus, exhaustive profiling is generally unacceptable, especially when the overhead of profiling is much higher than the time needed for actual workflow execution.

Therefore, we design and employ two approaches to reduce the number of experiments needed in our framework to accelerate the exploration of the parameter space for optimal performance: i) identify a subset of critical parameters with the most significant influence on workflow performance through feature selection; and ii) minimize the search process within the value range of each critical parameter using stochastic approximation. Experimental results show that the proposed optimization framework provides the most suitable parameter settings for a given workflow to achieve the best performance. This profiling-based method could be used by end users and service providers to configure and execute large-scale workflows in complex big data systems.

In sum, our work makes the following contributions to the field of big data computing with a focus on workflow execution:

- We design a profiling-based workflow optimization framework, which considers multiple layers in the entire technology stack of Hadoop MapReduce/Spark and provides end users with configuration recommendations to optimize workflow performance in terms of makespan.
- We employ feature selection to determine critical parameters with the most significant influence on workflow performance and apply stochastic approximation to minimize the time needed to explore the parameter space.
- We implement and test the proposed method with extensive experiments, which show that it yields satisfactory performance while significantly reducing profiling overhead compared with traditional methods.
- The proposed workflow optimization framework and the profiling minimization approaches are generic and hence applicable to other big data systems to systematically determine the most suitable system configuration.

The rest of this paper is organized as follows. In Sect. 2, we survey related work. In Sect. 3, we present a coupled design framework for cross-lay optimization of big data workflows. In Sect. 4, we investigate the effects of different parameters and propose a profiling-based approach to parameter setting. In Sect. 5, we discuss the rationale of using the Simultaneous Perturbation Stochastic Approximation algorithm and provide its convergence in our problem. In Sect. 6, we present the overall system configuration recommendation framework. We conduct experiments for performance evaluation in Sect. 7.

2 Related Work

Application-layer configuration for performance improvement of scientific workflows has been investigated in various contexts. Grid-based workflow management systems such as Taverna [2], Kepler [3] and Pegasus [4] seek to minimize workflow makespan by adjusting workflow-level parameters to group and map workflow components. Kuma *et al.* proposed an integrated framework, which is capable of supporting performance optimization along multiple dimensions of the parameter space [5]. In [6], in order to support domain-specific parameter-based optimization, end users are required to provide performance and quality models of expected application behaviors to the system. This approach is not readily applicable to many situations, where the relationships between parameters and performances cannot be directly modeled with an analytical form. In [7], Holl *et al.* proposed an optimization phase between the planning phase and the execution phase of the common scientific workflow life cycle to assist end users in determining the optimal setup of a specific workflow. This phase indeed helps end users configure workflows properly, however, in a try-and-error manner, which could be very time-consuming. In the aforementioned work, only parameters at the workflow layer are considered, while neglecting the influence of the configuration of other layers in the execution system.

There also exist a number of efforts on parameter setting in the layer of computing engine for parallel processing. In [8], Gounaris *et al.* investigated how parameters in Spark affects the performance of Spark jobs and proposed a systematic method for parameter tuning.In addition, many machine learning-based approaches have been proposed to tune Spark and Hadoop parameters. In [9], a binary classification and multi-classification machine learning model is proposed to accelerate the tuning of Spark parameters. In [10], Liao *et al.* proposed a machine learning model to automatically tune Hadoop MapReduce configuration parameters. In [11], Wu *et al.* proposed a profiling and performance analysis-based framework for Hadoop configuration, which considers parameters not only in the MapReduce layer but also in the Hadoop Distributed File System (HDFS) layer. In [12], Li *et al.* proposed an online performance tuning system that monitors a job's execution, tunes its associated performance-related parameters based on historical statistical data, and provides fine-grained control over parameter configuration.

Different from the existing work, we consider a generic end-to-end delay minimization problem for big data workflows consisting of multiple MapReduce/Spark jobs, investigate the cross-layer execution stack, analyze the compound effects of parameters in all layers, and propose a coupled design framework that determines and recommends the most suitable parameter setting.

3 A Coupled Design Framework for Cross-layer Optimization of Big Data Workflows

Large-scale workflows executed in big data systems such as Hadoop are complex processes that involve distributed storage and parallel computing.

Figure 1 illustrates a technology stack of Hadoop with multiple layers, each of which has a non-negligible impact on workflow performance, as detailed below:

- The top layer defines abstract big data workflows comprised of data-intensive computing jobs with execution dependencies. Such workflows are typically modeled as a Directed Acyclic Graph (DAG), where each vertex represents either a MapReduce or a Spark job, and each directed edge represents the execution dependency and data flow between two adjacent jobs.
- The second layer is the parallel computing engine, which orchestrates data processing by marshaling distributed nodes, running multiple tasks in parallel, managing all communications and data transfers between various parts of the system, and meanwhile providing redundancy and fault tolerance. Here, we consider MapReduce and Spark as they are the two most widely used parallel computing engines in Hadoop system.
- The third layer is the resource manager of the cluster, which is responsible for keeping track of the resource use on each data node, allocating distributed system resources to various jobs or applications submitted by different users, and scheduling tasks for execution on different nodes. We focus on the Yet Another Resource Negotiator (YARN), which was first introduced in Hadoop 2 to improve MapReduce implementation, and has been generalized to support different computing engines including Spark and Storm.
- The fourth layer is the distributed file system, which lays down the foundation for big data processing, and is responsible for managing data storage across multiple nodes and making it tolerant to node failures without suffering data losses. In this paper, we consider Hadoop Distributed File System (HDFS) as it is commonly used for distributed storage of big data in Hadoop.

Correspondingly, we list in Table 1 all performance-related parameters in these technology layers, namely, MapReduce, Spark, YARN, and HDFS. Note that any of these parameters could create a bottleneck and hence limit workflow performance. All of these parameters can be controlled in the system configuration, such as the degree of parallelism, the amount of computing resources allocated to each task, and the job scheduling policy.

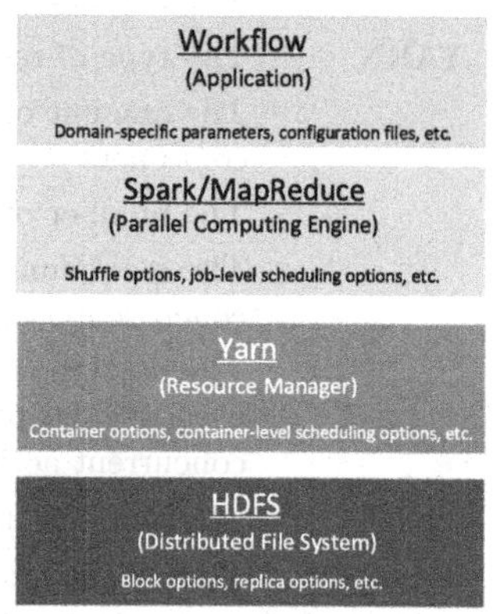

Fig. 1. A coupled design framework for big data workflow optimization.

In general, parameter tuning helps improve the performance of workflow execution in big data systems, but it usually requires considerable familiarity with the system and in-depth domain knowledge in parallel and distributed computing, which, unfortunately, many end users lack. Researchers have achieved remarkable results through parameter tuning in various layers, including job scheduling scheme [13–16], Spark and MapReduce parameter tuning [17–22], and resource manager configuration [21,23]. However, the parameter setting in one layer may have an impact on the parameter setting in other layers, which has not been largely explored. The proposed coupled design framework is to account for the coupled effects of different parameters across multiple layers in the technology stack to achieve optimal workflow performance.

Table 1. Performance-related parameters in MapReduce/Spark, YARN, and HDFS layers.

Layers	Parameters
MapReduce	The number of reduce tasks per job
	The fraction of the number of maps in the job, which should be completed before reduces are scheduled for the job
	Should the outputs of the maps be compressed before being sent across the network?
Spark	The number of executors for one Spark application
	The amount of memory to use per executor process
	The number of cores to use on each executor
	The number of partitions in RDDs returned by transformations such as join and reduceByKey, and parallelize when not set by user
	The amount of storage memory immune to eviction
	Whether to compress map output files?
	How long to wait to launch a data-local task before giving up and launching it on a less-local node?
	The scheduling mode between jobs submitted to the same SparkContext, which can be set to FAIR or FIFO
	The number of cores to allocate for each task
YARN	The type of resource scheduler
	The amount of physical memory, in MB, that can be allocated for containers
	The number of vcores that can be allocated for containers
	The maximum number of applications in the system, which can be concurrently active both running and pending
	The maximum percentage of resources in the cluster, which can be used to run application masters (it controls the number of concurrent active applications)
HDFS	The number of block replications
	The Sblock size for new files, in bytes

4 Parameter Effects and Workflow Profiling

4.1 An Empirical Study of Parameter Effects

In the workflow layer, the inherent properties of the workflow such as input data size, workflow topology, and implementation language would affect workflow performance. In addition, the performance of workflow execution is also affected by the parameter settings in various layers.

To examine such parameter effects, we conduct a set of exploratory experiments. For illustration, we collect and present the experimental results of workflow execution on a local cluster consisting of eight virtual machine instances to show how different early shuffle rates affect workflow performance. We plot in Fig. 2(a) the workflow end-to-end delay measurements with different early shuffle rates using four MapReduce workflows, including 1) library checkout data analysis from New York City Library (labeled as library checkout), 2) flight data analysis (labeled as flight), 3) parking violation analysis in New York City (labeled as parking violation), and 4) 311 service request analysis in New York City (labeled as service request), all of which are based on online public data repositories [24–27]. If the shuffling process starts earlier, at each time point, there is less data on the fly, and data transfer is less likely to become the bottleneck of workflow execution. However, if reduce tasks are scheduled earlier, then the computing resources allocated to reduce tasks are reserved and may lead to resource waste and consequently undermine the workflow execution performance. From these performance measurements, we observe that this parameter affects the workflow performance of flight data analysis the least and exhibits different performance patterns for parking violation analysis, service request analysis, and library checkout analysis.

To illustrate the effects of different parameters in Spark on job execution performance, we compare the execution time of these four different workflows with three different executor memory sizes, as shown in Fig. 2(b). Theoretically, a larger memory size allocated to each executor would lead to a shorter execution time for a given Spark job. We observe in Fig. 2(b) that: i) for library checkout, flight data analysis, and parking violation analysis, it is clear that increasing memory improves workflow performance; ii) for 311 service request analysis, which has many independent modules, there is an intensive resource competition, which leads to performance degradation as the memory size increases.

We compare the end-to-end delay of these four workflows in Fig. 2(c) with different node locality delays, which define the number of missed scheduling opportunities, after which the scheduler attempts to schedule rack-local containers. On a busy cluster, when an application requests a particular node where its input data are stored, there is a good chance that other containers are running on it at the time of the request. The obvious measure is to immediately loosen the locality requirement and allocate a container on the same rack. However, waiting a short time before scheduling to another node may significantly increase the chance of getting a container on the requested node, and therefore increase the efficiency of the cluster. On the other hand, waiting too long may exceed the time saved by avoiding data transfer. In Fig. 2(c), we observe that these workflows exhibit different performance patterns in response to this parameter.

Similar to native file systems, HDFS processes data in blocks, but with a larger size (e.g., 128 MB in most systems). By default, each block is replicated three times to increase reliability. Therefore, when a file is written into HDFS, it is firstly split into multiple blocks according to the configured block size and each

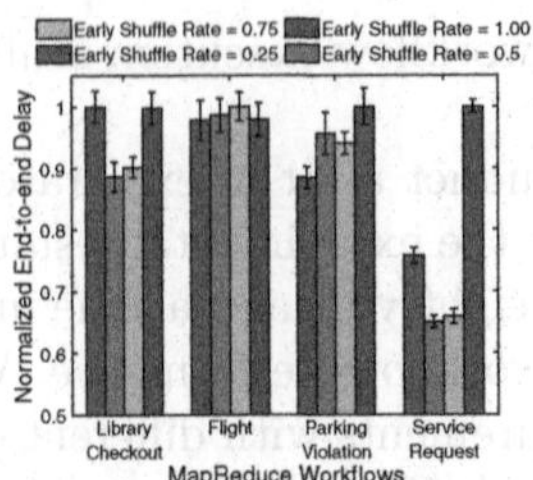

(a) MapReduce layer: comparison of end-to-end delay with different early shuffle rates

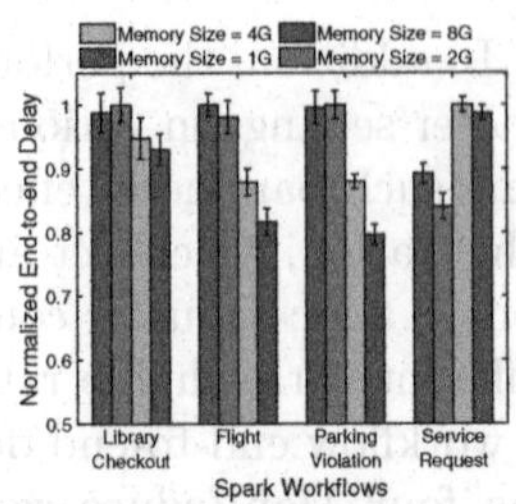

(b) Spark layer: comparison of execution time with different executor memory sizes

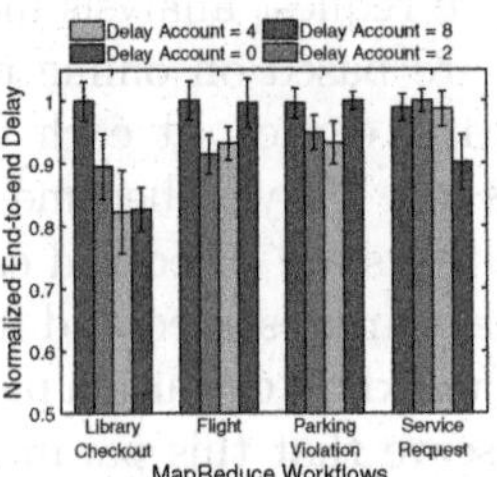

(c) YARN layer: comparison of end-to-end delay with different node locality delays

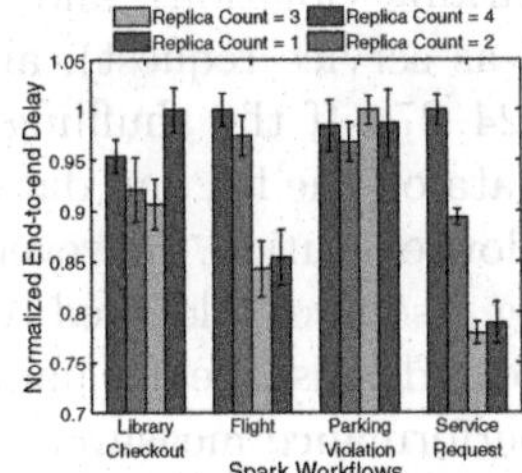

(d) HDFS layer: comparison of end-to-end delay with different replica numbers

Fig. 2. Illustration of parameter effects in different layers on the performance of four workflows.

block would be replicated and distributed across the entire cluster. Figure 2(d) shows the impact of block replication number on the end-to-end delay of the aforementioned four workflows. Theoretically, if there are more replicas stored on the cluster, there would be a higher chance that each node has a local data block to process. In this case, With sufficient computing resources, the node does not have to retrieve data over the network to process, thereby reducing the end-to-end delay. On the other hand, with a larger replication number, the time for writing a file into HDFS would increase, hence negatively affecting workflow performance. Also, since the output of MapReduce and Spark jobs is written into HDFS automatically, the time for writing the output of each module in the workflow would increase, hence negatively affecting workflow performance. We observe that the impact of this parameter on the performance of these workflows is different. Particularly, the parameter effect for library checkout data analysis is not monotonous: the end-to-end delay decreases in the beginning and then increases as the replication number increases.

4.2 Workflow Performance Profiling

A workflow profile $WP_w(G(\theta), C, \theta)$ is a control-response plot illustrating how a set θ of system parameters affect the performance $G(\theta)$ of workflow w running on cluster C. The workflow profile can be obtained by varying θ to exhaust the combination of parameter values on cluster C and measuring the corresponding average end-to-end delay T.

Effectiveness of Workflow Profiling. To illustrate the effectiveness of workflow profiling in improving workflow execution performance, we run the Spark and MapReduce workflows for 311 service request analysis for 10 times, each with the default setting and a customized setting, respectively, and plot the performance measurements in Fig. 3. These measurements show that the customized parameter setting significantly reduces the workflow end-to-end delay in comparison with the default setting. We observe in Fig. 3 that the customized setting makes more improvement over the default setting on MapReduce workflows. This is because, for a MapReduce job, the degree of parallelism in the reduce phase is 1 by default, which hurts the performance. On the other hand, Spark distributes the job workload to all executors, which results in a higher degree of parallelism.

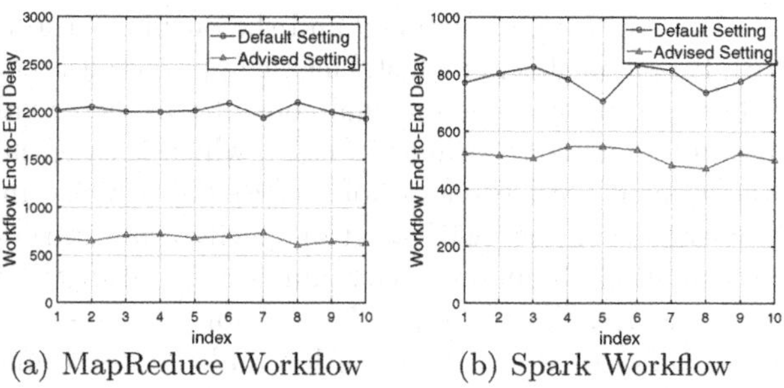

Fig. 3. Performance comparison of 311 service request analysis workflow with the default setting and a customized setting.

Overhead of Exhaustive Search-Based Workflow Profiling. The goal of workflow performance profiling is to determine the control parameter values θ^*, at which the workflow end-to-end delay T reaches the global minimum. An exhaustive search-based approach for workflow profiling is prohibitively time-consuming. Take a MapReduce workflow as an example, where we consider only one control parameter in each layer, including the number of reduce tasks ($m \in \{1, 2, ..., M\}$) in MapReduce, memory size constraint ($n \in \{1, 2, ..., N\}$) in Gigabytes for containers in YARN, and block size ($b \in \{32, 64, ..., 2^k\}$ MBytes) in HDFS. It would take a total time of $O(k \cdot N \cdot M \cdot \Delta T)$, where ΔT is the time taken by a single workflow profiling test, and is typically on the order of minutes or even hours. In the experiment of flight data analysis based on MapReduce, even if we only vary the reduce task count from 1 to 5, the memory size constraint from 1 GB to 8 GB, and the block size from 32 MB to 512 MB, a single profiling test takes around 9 min and the exhaustive search takes 1,125 min (around 19 h) in total to complete. Such a time overhead may be comparable with or even longer than the total time required by big data workflow execution, which is typically on the order of hours, days, or weeks. As the number of control parameters increases, the time to obtain a complete workflow profile rapidly increases, making the exhaustive search approach practically infeasible.

5 SPSA-Based Profiling Optimization

To obviate the need for conducting exhaustive profiling, we propose a Simultaneous Perturbation Stochastic Approximation (SPSA) [28] based profiling method to quickly determine the most suitable parameter setting in each layer.

5.1 Rationale on the Use of SPSA

As shown in Table 1, many components and factors are involved in the process of workflow execution and may affect workflow performance observed by the end user. Since our work is focused on workflow profiling with respect to end-to-end delay, the execution dynamics could be treated as a "black box" system, where the input is the set of control parameters θ and the output is the workflow end-to-end delay $G(\theta)$. Based on this model, stochastic approximation algorithms are suitable for quickly determining the optimal control parameter values, because they i) do not require an explicit formula of $G(\theta)$, which is essentially unknown, but only its measurement with random noise $G(\theta) + \xi$, which can be obtained by running a "one-time profiling" with a set of specific values; ii) do not require any additional information about system dynamics or input distribution; and iii) are of low time complexity. These are highly desirable features as they not only account for the dynamics of workflow execution and the randomness in performance measurements, but also significantly reduce the computational overhead compared with exhaustive profiling.

SPSA algorithms are similar to gradient descent methods to find a global minimum. In comparison with other methods, it has distinctive desirable features: i) the gradient approximation of SPSA requires only two measurements of the objective function, regardless of the dimension of the control parameters; ii) SPSA produces instant results without requiring a large amount of historical data; iii) SPSA has a proven convergence property, and thus its performance is theoretically guaranteed.

5.2 Stochastic Approximation (SA) Methods

In our profiling problem, the average workflow end-to-end delay is denoted as a function of $G(\theta)$ of control parameters θ. The goal is then to find the most suitable control parameters value θ^* that minimize $G(\theta)$ within the feasible space Θ. Following the standard Kiefer-Wolfowitz Stochastic Algorithm (KWSA) [29], the standard stochastic approximation form is given by,

$$\hat{\theta}_{k+1} = \hat{\theta}_k - a_k \cdot \hat{g}_k(\hat{\theta}_k), \tag{1}$$

where $\hat{\theta}_k$ is the set of control parameter values in the k-th iteration, $\hat{g}_k(\hat{\theta}_k)$ is the simultaneous perturbation estimate of $G(\theta)$'s gradient $g(\theta) = \frac{\partial G}{\partial \theta}$ at the iteration $\hat{\theta}_k$ based on the measurements of $G(\theta)$, and a_k is a nonnegative scalar gain coefficient.

The measurement of workflow end-to-end delay with noise, denoted as $y(\theta)$, is available at any value $\theta \in \Theta$, and given by $y(\theta) = G(\theta) + \xi$, where ξ is the noise incurred by system dynamics during workflow execution. In fact, $y(\theta)$ is the observed average workflow end-to-end delay of a single profiling test with a specific θ.

The essential part of Eq. 1 is the gradient approximation $\hat{g}_k(\hat{\theta}_k)$. With simultaneous perturbation, all elements of $\hat{\theta}_k$ are randomly perturbed at the same time to obtain two loss measurements of $y(\cdot)$. For the two-sided SP gradient approximation, this leads to

$$\hat{g}_k(\hat{\theta}_k) = \frac{y(\hat{\theta}_k + c_k \Delta_k) - y(\hat{\theta}_k - c_k)}{2c_k \Delta_k} \begin{bmatrix} \Delta_{k1}^{-1} \\ \Delta_{k2}^{-1} \\ \vdots \\ \Delta_{kp}^{-1} \end{bmatrix}, \tag{2}$$

where p is the number of control parameters under consideration, the mean-zero p-dimensional random perturbation vector, $\Delta_k = [\Delta_{k1}^{-1}, \Delta_{k2}^{-1}, \cdots, \Delta_{kp}^{-1}]^T$, is independent and symmetrically distributed around 0 with finite inverse $E\left|\Delta_{ki}^{-1}\right|, (i = 1, 2, \cdots, p)$, and c_k is a positive coefficient. Because the numerator is the same in all p components of $\hat{g}_k(\hat{\theta}_k)$, the number of loss measurements needed to estimate the gradient in SPSA is two, regardless of the dimension of p.

5.3 Convergence of SPSA-Based Workflow Profiling

The convergence of the proposed SPSA-based workflow profiling method is important as it affects both the quality of workflow profiling results and the efficiency of parameter setting recommendation. To explore the applicability of SPSA in the profiling optimization problem and investigate its convergence property, we validate the conditions that lead to the convergence in the context of workflow profiling.

As pointed out by Spall in [30] (pp. 161), the conditions for convergence can hardly be all checked and verified in practice due to the lack of knowledge on $G(\theta)$. We provide the following arguments to justify the appropriateness of SPSA in solving the profiling optimization problem.

According to Theorem 7.1 in [30] (pp. 186), if Conditions $B.1''$ – $B.6''$ hold and θ^* is a unique minimum of $G(\theta)$, then for SPSA, θ_k almost surely converges to θ^* as $k \to \infty$.

Conditions $B.1''$, $B.4''$, and $B.6''$ are the most relevant since we govern the gains sequence a_k and c_k and the random perturbation Δ_k. The coefficient sequences a_k and c_k we choose (See Sect. 6.2) and the symmetric Bernoulli ± 1

distribution we follow to generate the simultaneous perturbations $\{\Delta_{ki}\}$ easily validate Conditions $B.1''$. We generate the simultaneous perturbations $\{\Delta_{ki}\}$ to ensure that Δ_{ki} is a mutually independent sequence, which is independent of $\hat{\theta}_1, \hat{\theta}_2, \cdots, \hat{\theta}_k$. The noise in workflow execution is mainly caused by the dynamics and randomness of computer systems and network devices. Since the loss measurement $y(\cdot)$ captures both positive noise and negative noise, the long-term conditional expectation of the observed noise is considered to be zero, i.e., for all k, $E[\varepsilon_k^+ - \varepsilon_k^- | \{\hat{\theta}_1, \hat{\theta}_2, \cdots, \hat{\theta}_k\}, \Delta_k] = 0$. Since $\{\Delta_{ki}\}$ is generated following the symmetric Bernoulli ± 1 distribution with a probability of 0.5 for each outcome of either $+1$ or -1, $E[\Delta_{ki}^{-1}]$ is uniformly bounded. In addition, the loss measurements $y(\hat{\theta}_k \pm c_k \Delta_k)$ are bounded by the full computing/networking capacity of a given cluster, so the ratio of measurement to perturbation $E[y(\hat{\theta}_k \pm c_k \Delta_k)\Delta_{ki}]$ is uniformly bounded over i and k. Hence, Condition $B.4''$ holds.

Condition $B.2''$ and $B.3''$ impose the requirement that $\hat{\theta}_k$ (including the initial condition) is close enough to θ^* so that there is a natural tendency for an analogous deterministic algorithm to converge to θ^*. These two conditions are intuitively valid in the profiling scenario because: i) the main requirement for these conditions, i.e., $sup_{k \geq 0} \left\| \hat{\theta}_k \right\| < \infty$, can be satisfied since the control parameter values in all layers are both finite positive integer numbers; ii) since the feasible regions of the control parameters in workflow execution is finite and mapped to a limited range of the iterative variables, $\hat{\theta}_k$ (including the starting point) is sufficiently close to θ^*; iii) in workflow execution environments, due to the system dynamics and randomness, different runs with identical parameter settings may yield different end-to-end delays, which makes θ^* be not a single point but an "acceptable area" in order to tackle this profiling problem.

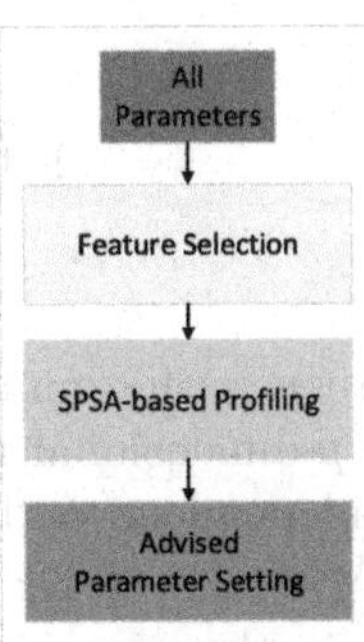

Fig. 4. Architecture of the proposed workflow configuration advisor.

$B.5''$ asks $G(\theta)$ to be three-times continuously differentiable and bounded by R^p. The end-to-end delay is obviously bounded by the cluster computing power and network capacity among all dimensions of the control parameter θ. However, the smoothness and differentiability of $G(\theta)$ is very difficult to verify since $G(\theta)$ is practically unknown and spans over a parameter space of discrete values. For this profiling problem, we assume $G(\theta)$ meets this condition.

In addition, it is worth pointing out that although the end-to-end delay should have a unique theoretical peak over the feasible control parameter space in a given distributed system, it has been observed that in our environments, due to the system dynamics and randomness, identical parameter values may yield different end-to-end delays in different runs, which makes the uniqueness of θ^* practically unverifiable. However, the loss measurement $y = G(\theta) + \xi$ indeed shows a trough property over the feasible control parameter space in the experiments.

6 A Workflow Configuration Advising Framework

We design a workflow configuration advising framework as shown in Fig. 4, which conducts workflow profiling across the entire technology stack of big data systems and provides end users with recommended configurations to deploy big data workflows for actual execution. This framework integrates two strategies to avoid the need of conducting exhaustive profiling and minimize the number of profiling runs while still ensuring the recommendation of a satisfactory parameter setting: i) it identifies a subset of critical parameters to be investigated; ii) it reduces the search space within the value range of each critical parameter.

6.1 Information-Theoretic Feature Selection

Not all parameters play an equally important role in workflow performance, and we, therefore, should focus on the parameters with the most significant influence. Towards this goal, we employ an information-theoretic feature selection method, the nearest neighbor method [31], to filter out a dominating subset of control parameters using historical dataset.

For the control parameter X, we take random samples x_i and repeat workflow execution for multiple times, collect the end-to-end delay measurements y_i and construct a historical dataset $D = \{x_i, y_i\}_{i=1}^{N}$, where N denotes the total number of data points collected in D. For a given control-response data point i, we first identify the k-th nearest neighbor of this point among data points with the same configuration x_i in terms of Euclidean distance, and denote the distance as d. Note that k is a customizable parameter. Then count the number m_i of nearest neighbors in D that lie within distance d to data point i (including the k-th nearest neighbor with the same configuration). The amount of information I_i data point i carries is calculated as

$$I_i = \phi(N) - \phi(N_{x_i}) + \phi(k) - \phi(m_i), \tag{3}$$

where N_{x_i} is the number of data points who has configuration x_i, ϕ is the digamma function [32]. The mutual information of X and Y is calculated by averaging I_i over all N data points in D:

$$I(X,Y) = \phi(N) - \langle\phi(N_x)\rangle + \phi(k) - \langle\phi(m)\rangle, \tag{4}$$

where $\langle\rangle$ is an averaging operator. Based on the estimated mutual information between X and Y, we decide to select X as a candidate critical parameter by thresholding the score of the mutual information.

6.2 SPSA-Based Profiling Process

The SPSA-based workflow profiling process is described as follows.

Step 1: Set counter index $k = 0$. Choose an either fixed or random starting point within the feasible space of the control parameters. Also, choose nonnegative coefficients a, c, A, α, and γ in the SPSA gain sequences $a_k = \frac{a}{(K+1+A)^\alpha}$ and $c_k = \frac{c}{(k+1)^\gamma}$. Practically effective (and theoretically valid) values for α and γ are 0.602 and 0.101, respectively [33]. For the choice of a, A and c, please refer to Sect. 6.2.

Step 2: Generate by Monte Carlo a p-dimensional random perturbation vector Δ_k, where each of p components of Δ_k is independently generated from a zero-mean symmetric Bernoulli ± 1 distribution with probability of $\frac{1}{2}$ for each ± 1 outcome.

Step 3: Conduct workflow profiling to collect two performance observations $y_k^+ = y(\hat{\theta}_k + c_k \Delta_k)$ and $y_k^- = y(\hat{\theta}_k c_k \Delta_k)$ with c_k and Δ_k from Step 1 and Step 2.

Step 4: Compute the simultaneous perturbation approximation to the unknown gradient $g(\hat{\theta}_k)$ using Eq. 2.

Step 5: Use the standard stochastic approximation form (see Eq. 1) to update $\hat{\theta}_k$ to a new value $\hat{\theta}_{k+1}$.

Step 6: Return to Step 2 with $k+1$ replacing k, and repeat the above process until the termination condition is met (please refer to Sect. 6.2).

Termination Condition. Many efforts have been made to establish the termination condition for SA-based methods since the Kiefer-Wolfowitz SA algorithm was first proposed [29]. Our algorithm employs the following two practical rules to guarantee its termination with satisfactory performance.

- **Upper bound - the $\mathcal{M}$ rule:** our algorithm terminates the profiling process when the total number of profiling iterations exceeds a threshold $\mathcal{M}$, which is typically set to be much less than the maximum number of profiling tests of the exhaustive profiling approach, i.e., $\mathcal{M} \ll \prod_{i=1}^{k} NV_i$, where k is the number of control parameters and NV_i is the number of possible values of the k-th control parameter. Note that when $\mathcal{M} = \prod_{i=1}^{k} NV_i$, our algorithm rolls back to the exhaustive search.
- **Impeded progress - the $\mathcal{R}$ rule:** if the number of consecutive profiling iterations that do not produce any performance improvement compared with the best one observed so far exceeds an upper bound $\mathcal{R}$, our algorithm terminates the profiling process for the given workflow.

Restrictions on Parameters. Restrictions have to be placed on some parameters to prevent unnecessary profiling runs and reaching the upper limit of the cluster capacity. For the parameters in the discussion, we confine the CPU count and memory size for a YARN container to be smaller than the capacity of computing nodes. Similarly, we confine the block size of HDFS to be smaller than the storage capacity of computing nodes and to be larger than 32 MB as an excessively small block size would jeopardize workflow performance [34,35].

Choice of Gain Sequences. The coefficients a_k and c_k of the SPSA-based method need to satisfy the following conditions to guarantee the convergence:

$$a_k > 0, c_k > 0; a_k \to 0, c_k \to 0; \sum_{k=0}^{\infty} a_k = \infty, \sum_{k=0}^{\infty} \frac{a_k^2}{c_k^2} < \infty. \quad (5)$$

The choice of the gain sequence is critical to the performance. With the Bernoulli ± 1 distribution of Δ_k, α and γ are specified in Step 1 with values commonly adopted in practice. Step sizes a and c are empirically determined based on the size of the search space.

Control Parameter Calculation. The control parameters used in our SPSA-based profiling algorithm are denoted as $\theta = \left[s_1^{'}\ s_2^{'}\ ...\ s_p^{'}\right]^T$, and set to be positive values within a reasonably selected range to ensure a comparable magnitude for each parameter dimension. These parameters are scaled and mapped to decide the actual parameter values in each iteration. In calculating the actual values of the parameters, we perform rounding operations in the case where the intermediate results are fractional.

The profiling unit of each critical parameter is denoted by $\mu_i, i \in 1, 2, \cdots, p$. For the profiling unit, we calculate its corresponding actual value for performance observation (i.e., y_k^+ and y_k^-) as $s_i = round(\lambda_i(s_i^{'}) \cdot \mu_i)$, where $\lambda(\cdot)$ are a scaling function that scales all control parameters into the same magnitude. For example, with $\lambda_i(s_i^{'}) = 2^{s_i^{'}}$, $s_i^{'}$ increases exponentially as the original value of $s_i^{'}$ increases.

We apply different scaling functions to different control parameters as their actual values exhibit different patterns. For instance, we consider i) linear type, where the actual values of such parameters grow linearly, such as early shuffle rate in MapReduce, and replication count in HDFS; ii) exponential type, where the actual values of such parameters grow exponentially, such as block size in HDFS. For linear type, we apply a min-max normalization approach to implement the scaling function; while for exponential type, we apply both an exponential scaling function and a min-max normalization approach to perform scaling.

7 Performance Evaluation

In this section, we evaluate the performance of the proposed workflow optimization framework. We first conduct experiments to test our profiling algorithm, and then demonstrate the advising procedure and illustrate the performance benefits of our algorithm compared with random walk [36], tabu search [37] and linear regression [38] plus dual-simplex algorithm [39].

7.1 Experiment Setting and Workflow Description

In our experiment, we establish a homogeneous cluster of three racks, each of which has two computer nodes installed. On the cluster, we install Apache

Hadoop 2.9.5 [40], Apache Spark 2.4.4 [41] and Apache Oozie 4.3 [42], which is a workflow engine that automatically dispatches each component MapReduce/Spark job in a workflow once all of its preceding jobs finish execution.

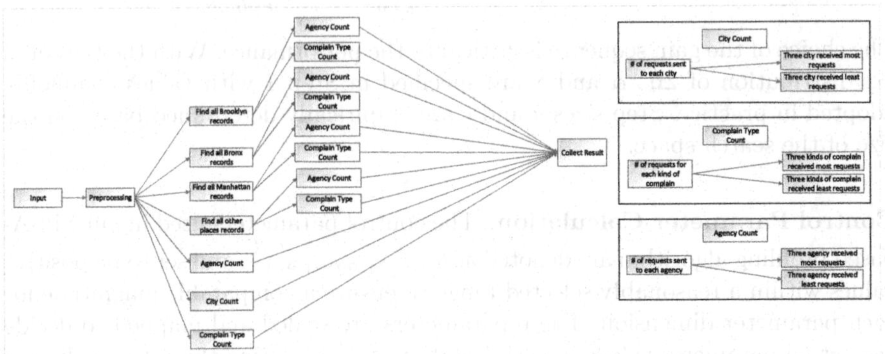

Fig. 5. The topology of the workflow for 311 service request data analysis.

We download four public datasets online: i) airline on-time performance dataset [24], ii) New York public library checkout records [25], iii) New York 311 service request record [27], and iv) New York parking violation record [26]. We implement MapReduce and Spark workflows with complex topologies for each dataset. For illustration, Fig. 5 shows a representative topology of the workflow for 311 service request data analysis.

Table 2. Critical parameters through feature selection.

Layers	Parameters
MapReduce	mapreduce.job.reduce.slowstart.completedmaps
	number of reducers
Spark	num-executors
	spark.executor.memory
	spark.executor.cores
Yarn	yarn.scheduler.capacity.node-locality-delay
	maximum-am-resource-percent
HDFS	dfs.replication
	dfs.blocksize

7.2 Experimental Results

We consider all parameters tabulated in Table 1, which do not play an equally important role in workflow performance. In fact, the combined setting of some critical parameters across different layers dominates the workflow execution time. Since exhaustive profiling by sweeping through the entire value range of all parameters is extremely time-consuming, we first use information-theoretic feature selection to identify a subset of critical parameters as shown in Table 2, all of which have a non-negligible influence on workflow performance.

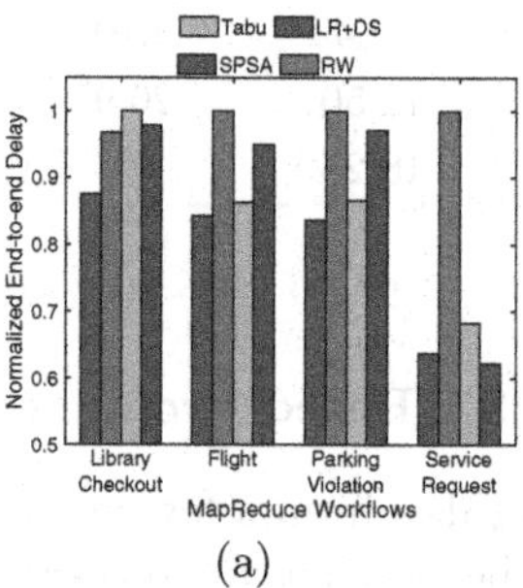

(a)

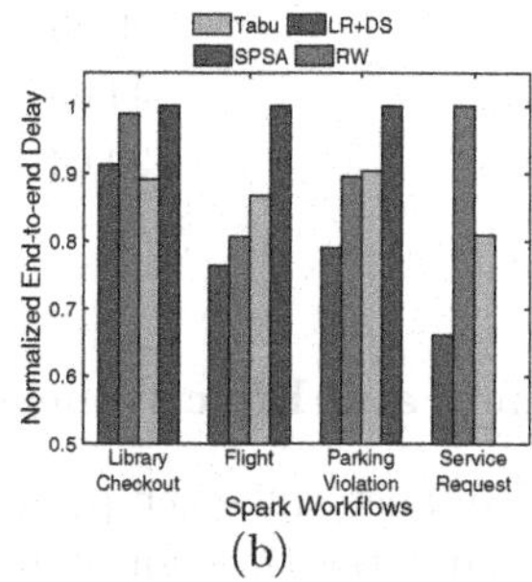

(b)

Fig. 6. Comparison of end-to-end delay between different settings produced by SPSA-based profiling (SPSA), random walk (RM), Tabu search (Tabu) and linear regression plus dual-simplex optimization algorithm (LR+DS).

We compare our SPSA-based profiling with Random Walk, Tabu Search, Linear Regression plus dual-simplex and default Hadoop/Spark in terms of workflow execution time (ED). The ED performance improvement of the SPSA-based profiling algorithm over other algorithms in comparison is defined as:

$$Imp(Other) = \frac{ED_{Other} - ED_{SPSA}}{ED_{Other}} \times 100\%, \tag{6}$$

where ED_{other} is the ED achieved by other algorithms, and ED_{SPSA} is the ED achieved by the SPSA-based profiling algorithm. The improvement result is tabulated in Table 3, where the upper side of the table is for MapReduce workflow (referred to as MR WFs in the table) and the lower side is for Spark workflow. Note that there is no result listed for linear regression plus dual-simplex for Spark service request workflow, because the proposed setting leads to the depletion of all cluster resources and the workflow execution is therefore suspended. Figure 6 shows the comparison of normalized end-to-end delay between these algorithms. These measurements show that our algorithm consistently outperforms the other algorithms in comparison.

Table 3. The percentage of ED improvement made by SPSA-based profiling over Default Setting (D), Random Walk (RW), Tabu Search (Tabu) and Linear Regression plus Dual-Simplex (LRDS).

MR/Spark Workflows	imp(D)	imp(RW)	imp(Tabu)	imp(LRDS)
Library	38.95	15.81	2.43	11.24
Flight	32.71	9.49	12.54	10.62
Parking	24.17	16.45	3.42	13.9
Service	66.62	36.29	6.65	−2.28
Library	48.84	5.36	11.96	23.63
Flight	29.83	7.55	−2.44	8.70
Parking	44.63	11.71	12.50	20.97
Service	32.04	33.84	18.25	–

7.3 Efficiency and Effectiveness of SPSA-Based Profiling

As mentioned in Sect. 4.2, each profiling run takes from a few minutes to hours. Therefore, to limit the time for identifying the best parameter setting, we consider the following constraints: i) for all algorithms, we set the maximum number of iterations to be 20; ii) for SPSA-based algorithm, we set $\mathcal{R} = 3$, which is the number of consecutive profiling iterations that do not produce any performance improvement over the best one observed so far; iii) additionally, for Tabu search, the tabu list size is set to be 3. We tabulate the average number of profiling runs in Table 4 for comparison.

Table 4. The average number of profiling runs for SPSA-based profiling (SPSA), Random Walk (RW), Tabu Search (Tabu), Linear Regression plus Dual-Simplex (LRDS), and Exhaustive Profiling (EP).

	SPSA	Random	Tabu	LRDS	EP
Number of runs	20	20	200	21	77,725

We further investigate if the parameter setting computed by the optimization framework from small-scale workflows can be applied to large-scale workflows. We sample a small amount of data from the original dataset and use the proposed profiling method to provide advice on the most suitable parameter settings. We apply these recommended settings to workflows processing the original dataset and plot the corresponding performance measurements in Fig. 7. These results show that the parameter setting recommended by the profiling approach from a small-scale workflow can significantly improve the performance of large-scale workflows in comparison with the default settings.

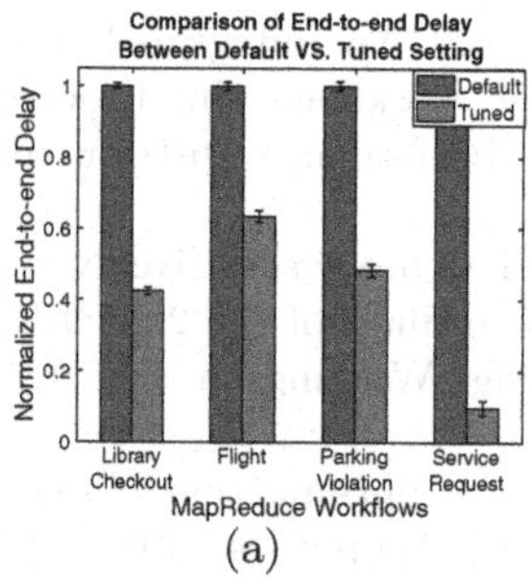

(a)

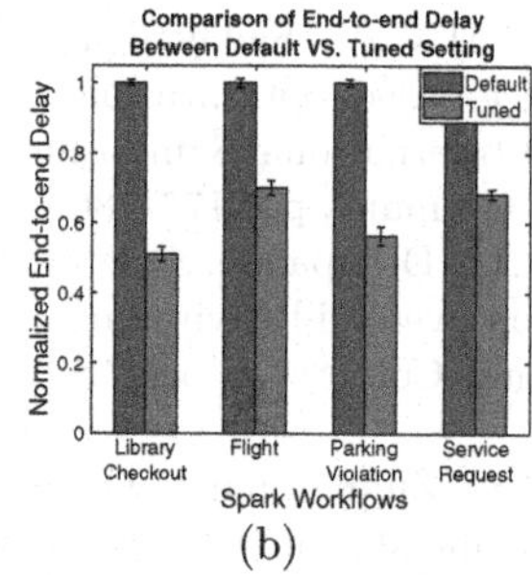

(b)

Fig. 7. Comparison of end-to-end delay between the default and recommended parameter settings across MapReduce/Spark, YARN and HDFS on large-scale workflows.

8 Conclusion and Future Work

We proposed a profiling-based coupled design framework for big data workflows, which provides end users with the most suitable system configuration to improve workflow performance in big data systems. Our framework employs feature selection and SPSA algorithm to accelerate the search process by obviating the need for exhaustive profiling.

It is of our future interest to provide configuration advice in finer granularity in the following aspects: i) recommend a suitable parallel computing engine, since for the same workflow, its performance varies considerably in Spark and MapReduce; ii) recommend parameter settings on a per-job basis, since not all jobs in the same workflow achieve its best performance with the same setting; iii) explore the most effective way to map parameter settings from small workflows to large ones; and iv) improve the performance of the SPSA-based profiling method, including gradient approximation averaging, step size adaptation, local optima prevention, convergence speed acceleration, and intelligent termination conditions, for faster profiling and more accurate recommendation.

Acknowledgments. This research is sponsored by U.S. National Science Foundation under Grant No. CNS-1828123 with New Jersey Institute of Technology.

References

1. Zaharia, P.W.M., Xin, R.S., et al.: Apache spark: a unified engine for big data processing. Commun. ACM **59**(11), 56–65 (2016)
2. Oinn, J.F.T., Addis, M., et al.: Taverna: a tool for the composition and enactment of bioinformatics workflows. Bioinformatics **20**, 3045–3054 (2004)
3. Ludascher, I.A.C.B.B., et al.: Scientific workflow management and the Kepler system. Spec. Issue Workflow Grid Syst. **18**, 1039–1065 (2005)
4. Deelman, E., Blythe, J., et al.: Pegasus: mapping scientific workflows onto the grid. In: Dikaiakos, M.D. (ed.) AxGrids 2004. LNCS, vol. 3165, pp. 11–20. Springer, Heidelberg (2004). https://doi.org/10.1007/978-3-540-28642-4_2

5. Kumar, G.M.V.S., Sadayappan, P., et al.: An integrated framework for performance-based optimization of scientific workflows. In: Proceedings of the 18th ACM International Symposium on High Performance Distributed Computing, Garching, Germany, pp. 177–186 (2009)
6. Chiu, G.A.D., Deshpande, S., et al.: Cost and accuracy sensitive dynamic workflow composition over grid environments. In: Proceedings of the 2008 9th IEEE/ACM International Conference on Grid Computing, Washington, DC, USA, pp. 9–16 (2008)
7. Holl, M.P.S., Zimmermann, O., et al.: A new optimization phase for scientific workflow management systems. Future Gener. Comput. Sci. **36**, 352–362 (2014)
8. Counaris, A., Torres, J.: A methodology for spark parameter tuning. Big Data Res. **11**, 22–32 (2018)
9. Wang, B.H.G., Xu, J., et al.: A novel method for tuning configuration parameters of spark based on machine learning. In: IEEE 18th International Conference on High Performance Computing and Communications, Sydney, NSW, Austrilia (2016)
10. Liao, G., Datta, K., Willke, T.L.: Gunther: search-based auto-tuning of mapreduce. In: Wolf, F., Mohr, B., an Mey, D. (eds.) Euro-Par 2013. LNCS, vol. 8097, pp. 406–419. Springer, Heidelberg (2013). https://doi.org/10.1007/978-3-642-40047-6_42
11. Wu, A.G.D., et al.: A self-tuning system based on application profiling and performance analysis for optimizing hadoop mapreduce cluster configuration. In: 20th Annual International Conference on High Performance Computing, Bangalore, India (2014)
12. Li, S.M., Zeng, L., et al.: MRONLINE: MapReduce online performance tuning. In: Proceedings of the 23rd International Symposium on High-Performance Parallel and Distributed Computing, New York, NY, USA, pp. 165–176 (2014)
13. Shu, T., Wu, C.: Performance optimization of *H*adoop workflows in public clouds through adaptive task partitioning. In: Proceedings of the IEEE INFOCOM, Atlanta, GA, USA, 1–4 May 2017
14. Wu, C., Lin, X., Yu, D., Xu, W., Li, L.: End-to-end delay minimization for scientific workflows in clouds under budget constraint. IEEE Trans. Cloud Comp. **3**(2), 169–181 (2015)
15. Yun, D., Wu, C., Gu, Y.: An integrated approach to workflow mapping and task scheduling for delay minimization in distributed environments. JPDC **84**, 51–64 (2015)
16. Ye, Q., Wu, C.Q., Cao, H., et al.: Storage-aware task scheduling for performance optimization of big data workflows. In: The 8th IEEE International Conference on Big Data and Cloud Computing, Melbourne, Australia, 11–13 December 2018
17. Wang, B.H.E.G. , Xu, J.: A novel method for tuning configuration parameters of spark based on machine learning. In: 2016 IEEE 18th International Conference on HPC and Communications, Sydney, NSW, Australia, 12–14 December 2016
18. Petridis, P., Gounaris, A., Torres, J.: Spark parameter tuning via trial-and-error. In: Angelov, P., Manolopoulos, Y., Iliadis, L., Roy, A., Vellasco, M. (eds.) INNS 2016. AISC, vol. 529, pp. 226–237. Springer, Cham (2017). https://doi.org/10.1007/978-3-319-47898-2_24
19. Gounaris, A., Torres, J.: A methodology for spark parameter tuning. Big Data Res. **11**, 22–32 (2018)
20. Jia, G.C.E.Z., Xue, C.: Auto-tuning spark big data workloads on POWER8: prediction-based dynamic SMT threading. In: 2016 International Conference on Parallel Architecture and Compilation Techniques (PACT), Haifa, Israel, 11–15 September 2016

21. Holmes, A.: Hadoop in Practice. Manning Publications Co., Greenwich (2012)
22. Li, S.M.E.M., Zeng, L.: MRONLINE: MapReduce online performance tuning. In: Proceedings of the 23rd International Symposium on High-Performance Parallel and Distributed Computing, Vancouver, BC, Canada, 23–27 June 2014
23. Ding, D.Q.E.X., Liu, Y.: Jellyfish: online performance tuning with adaptive configuration and elastic container in hadoop yarn. In: 2015 IEEE 21st International Conference on Parallel and Distributed Systems (ICPADS), Melbourne, Australia, 14–17 December 2015
24. Flight Data. http://stat-computing.org/dataexpo/2009/the-data.html
25. Library Checkout Data. https://data.seattle.gov/Community/Checkouts-by-Title/tmmm-ytt6
26. Parking Violation Data. https://data.cityofnewyork.us/City-Government/Open-Parking-and-Camera-Violations/nc67-uf89
27. Service Request Data. https://data.cityofnewyork.us/Social-Services/311-Service-Requests-from-2010-to-Present/erm2-nwe9
28. Spall, J.C.: Multivariate stochastic approximation using a simultaneous perturbation gradient approximation. IEEE Trans. Autom. Control **37**, 332–341 (1992)
29. Kiefer, J.W.J.: Stochastic estimation of the maximum of a regression function. Ann. Math. Stat. **23**(3), 462–466 (1952)
30. Spall, J.C.: Introduction to Stochastic Search and Optimization: Estimation, Simulation, and Control. Wiley, Hoboken (2005)
31. Ross, B.: Mutual information between discrete and continuous data sets. PLOS ONE **9**(2), 1–5 (2014)
32. Abramowitz, M., Stegun, I.: Handbook of Mathematical Functions with Formulas, Graphs, and Mathematical Tables. Dover Publishing Inc., New York (1972)
33. Spall, J.C.: Implementation of the simultaneous perturbation algorithm for stochastic optimization. IEEE Trans. Aerosp. Electron. Syst. **34**, 817–823 (1998)
34. Heger, D.: Hadoop performance tuning-a pragmatic & iterative approach. CMG J. **4**, 97–113 (2013)
35. White, T.: Hadoop: The Definitive Guide. O'Reilly Media Inc., Sebastopol (2012)
36. Lawler, G., Limic, V.: Random Walk: A Modern Introduction. Cambridge University Press, Cambridge (2010)
37. Glover, F.: Tabu search: a tutorial. Informs J. Appl. Anal. **20**(4), 1–185 (1990)
38. Montgomery, E.A.P.D.C., Vining, G.: Introduction To Linear Regression Analysis, vol. 821. Wiley, Hoboken (2012)
39. Nocedal, J., Wright, S.: Numerical Optimization. Springer, Heidelberg (2006)
40. Apache, Hadoop (2016). http://hadoop.apache.org
41. Spark (2016). http://spark.apache.org
42. Oozie (2016). https://oozie.apache.org

Detection of Loose Tracking Behavior over Trajectory Data

Jiawei Li[1], Hua Dai[1,2(✉)], Yiyang Liu[1], Jianqiu Xu[3], Jie Sun[4], and Geng Yang[1,2]

[1] Nanjing University of Posts and Telecommunications, Nanjing 210023, China
a2460336398@163.com, lyyadj15@163.com, {daihua,yangg}@njupt.edu.cn
[2] Jiangsu Security and Intelligent Processing Lab of Big Data, Nanjing 210023, China
[3] Nanjing University of Aeronautics and Astronautics, Nanjing 211016, China
jianqiu@nuaa.edu.cn
[4] Jiangsu Sealevel Data Technology Co., Ltd., Nanjing 210016, China
sunjie_nuaa@126.com

Abstract. The improvements in location-acquisition technologies make massive trajectory data available. Discovering objects that move together over trajectory data is beneficial in many applications. In this paper, a novel concept called loose tracking behavior is proposed to investigate the problem of detecting objects that travel together with a target. We develop two algorithms to solve the problem. The first one is a straightforward approach and the second one is an improved algorithm where we develop a prefix tree index structure for the trajectories encoded by Geohash to enhance the detection efficiency. Finally, the effectiveness of the proposed concepts and the efficiency of the approaches are validated by extensive experiments based on real trajectory datasets.

Keywords: Trajectory pattern mining · Loose tracking behavior · Spatio-temporal data mining · Trajectory data mining · Moving object

1 Introduction

With the development of location-acquisition and mobile computing technologies, trajectory data from people, vehicles and animals is growing exponentially. Mining such data enables us to discover usable knowledge about movement behavior which can be applied in a wide range of applications and services, including animal behavior study [1], urban planning [2], social recommendations [3], and location-aware advertising [4], to name but a few.

An essential data analysis task in trajectory data is discovering moving together patterns [5], which aims to find a group of objects moving together

Supported by the National Natural Science Foundation of China under the grant Nos. 61872197, 61972209 and 61872193; the Postdoctoral Science Foundation of China under the Grand No. 2019M651919; the Natural Research Foundation of NJUPT under the grand No. NY217119.

M. Qiu (Ed.): ICA3PP 2020, LNCS 12454, pp. 218–230, 2020.
https://doi.org/10.1007/978-3-030-60248-2_15

for a certain time period. These patterns all need a group of moving objects to be geometrically close to each other for at least some minimum time duration. Mining moving together patterns has been extensively studied in the literature such as *flock* [6,7], *convoy* [8,9], *swarm* [10], *traveling companion* [11,12], and *gathering* [13,14].

However, none of these works focus on detecting whether there are objects moving together with a target. Xu et al. first proposed their tracking behavior in [15]. That is, given a query trajectory o_q and a distance threshold d, they aim to find whether there is any object keeping some distance to o_q at each time point of o_q. But their definition of tracking behavior might not be practical in real life because there are often cases that trackers hardly keep close to the target all the time. In other words, it may result in the loss of interesting objects. For example, a tracker may deliberately keep away from the target at a certain time to avoid being easily exposed. Consider four trajectories o_1, o_2, o_3, o_4 during $[t_1, t_4]$ in Fig. 1, no tracker can be found by their definition although o_2 intuitively travels together with o_q and only leaves o_q at t_3.

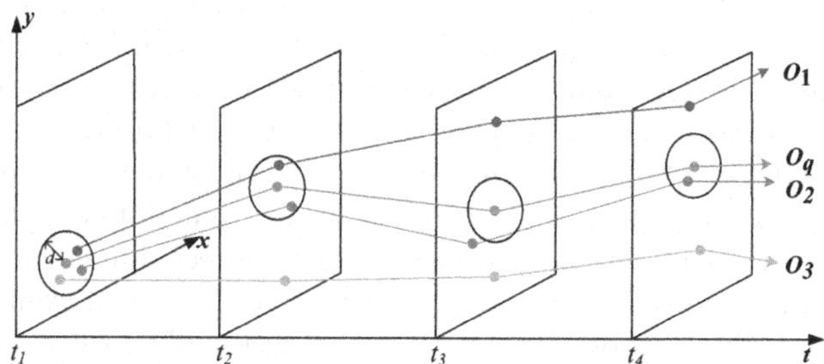

Fig. 1. Detect loose tracking behavior

In this paper, we propose a more general type of tracking behavior called loose tracking behavior. It allows trackers to leave the target at some timestamps, which is helpful to find interesting objects that cannot be discovered by Xu's definition of tracking behavior. Accordingly, we propose two algorithms to detect the loose tracking behavior. The first algorithm, Loose Tracking Behavior Detection algorithm (LTBD), is a straightforward method which directly follows the problem definition. The second algorithm is an improved algorithm named LTBD+, where the distance query is executed at each time point of the query trajectory. Each trajectory is encoded by Geohash and prefix trees are built before the query to improve the query efficiency.

The contributions of this paper are summarized as follows: (1) A novel concept, loose tracking behavior, is proposed to detect objects that track a target. (2) We introduce the LTBD and LTBD+ algorithm to solve the problem. In LTBD+, prefix tree index for encoded trajectories are built to enhance the detection efficiency. (3) Experiments on real datasets demonstrate the effectiveness and efficiency of our methods.

The remainder of this paper is organized as follows. The related work is reviewed in Sect. 2. Then we give the necessary preliminaries and formulate the focal problem of this paper in Sect. 3. Efficient solutions for detecting the loose tracking behavior over trajectory data are presented in Sect. 4. We test effectiveness and efficiency by extensive experiments in Sect. 5. Finally, Sect. 6 concludes this paper.

2 Related Work

Moving together patterns mining is an important data analysis task for moving object trajectories and has been variously formulated in the literature. *Flock* [6,7] is defined as a group of objects that travel together within a disc of a fixed size for k consecutive timestamps. However, the preset circular disc may not be able to well delineate the shape of a group in reality. *Convoy* [8,9], an extension of flock, is proposed to avoid rigid restrictions on the sizes and shapes of a moving group by employing the density-based spatial clustering. While both flock and convoy put strict requirement on consecutive time period, Li et al. proposed *swarm* in [10], which is a more general type of moving together patterns. This pattern has more relaxed temporal constraint. The *traveling companion* proposed in [11] can be considered as an online (and incremental) discovery pattern of convoy and swarm by using a data structure called traveling buddy. Naserian et al. [16,17] proposed *loose travelling companion* pattern, which is a loose pattern as objects that belong to one group are coming from different clusters. The *gathering* pattern [13,14] was proposed in order to detect some incidents such as celebrations and parades, in which objects join in and leave an event frequently. It allows the membership of a group to evolve gradually rather than requires the objects in the same group always stay the same. There are also some works allow objects to temporally join or leave a group including *moving clusters* [18] and *evolving convoy* [19].

Another line of trajectory pattern mining is trajectory clustering, which aims to find representative paths or common trends shared by different moving objects [20–22]. Different from moving together patterns, it emphasizes the geometric or spatial closeness of object trajectories and does not consider their temporal aspects.

However, the above-mentioned patterns all show the behavior for a group of objects. In contrast, we focus on finding objects that travel together with a target, which is an *individual behavior*. The tracking behavior proposed by Xu et al. [15] is like our work, but it has strict requirement that the tracker must always keep a certain distance to the target, which may miss interesting objects in real life.

3 Preliminaries and Problem Definition

3.1 Notations and Preliminaries

For the sake of clarity, we firstly introduce the main notations used in this paper:

- o : the original trajectory of an object
- o' : the encoded trajectory of o
- t : time point
- $o(t)$: the location of o at time point t
- $o'(t)$: the Geohash code of o at time point t
- O_{DB} : moving object database
- d : the distance threshold
- θ : the variation threshold in the definition of loose tracking behavior
- $T(o)$: the lifetime of o
- $Tra(o)$: the tracker set of o
- $D(o_u(t), o_v(t))$: the distance between two objects at time point t
- $T(o_u, o_v)$: the total time when distance between two objects is less than d

Geohash. Geohash is a public domain geocode system which encodes a geographic location into a short string of letters and digits. [23] Each string of Geohash code represents a grid region, where the points share this string. Geohash has the following features: (1) Each cell grid has a globally unique Geohash code corresponding to it. (2) It can encode the two-dimensional latitude and longitude coordinates into a one-dimensional string. (3) The length of a Geohash string determines its accuracy. The accuracy decreases as the length of Geohash string becomes shorter. Due to these features, places close to each other tend to have similar prefixes. The longer a shared prefix is, the closer the two places are.

Prefix Tree. The prefix tree is a tree data structure which is used to store a set of strings. Each node corresponds to a prefix of string shared by its descendants and the root node represents the empty string. It takes $O(L)$ time to look up a string, where L is the length of the search string. In this paper, we build the prefix tree index to quickly find objects that share a common Geohash code prefix.

3.2 Problem Definition

Let $O_{DB} = \{o_1, o_2, ..., o_n\}$ be the set of all moving object trajectories in the database.

Definition 1. Trajectory. The trajectory of a moving object in O_{DB} is represented as a sequence of points $\{(x_1, y_1, t_1), (x_2, y_2, t_2), ..., (x_m, y_m, t_m)\}(t_1 < t_2 < ... < t_m)$ where (x_i, y_i) are the coordinates at time t_i. t_1 and t_m are the start time and end time of the trajectory respectively.

In our paper, we consider a practical trajectory database model, where each trajectory may have a different length from others and each location of a trajectory may be sampled irregularly (i.e., points are sampled with different time intervals). Therefore, we create the virtual points by applying linear interpolation for the trajectories that have no sampled location at a given time point t_i.

Definition 2. Loose Tracking Behavior. Given a query trajectory o_q, a distance d and a parameter $\theta \in [0, 1]$, let $T(o_u, o_v)$ return the total time when the distance between o_u and o_v is less than d and $T(o_q)$ return the lifetime of o_q. Then the trackers of o_q are the object set $Tra(o_q) = \{o \in O_{DB} \mid \frac{T(o, o_q)}{T(o_q)} \geq \theta\}$.

Consider the example in Fig. 1, we can find $T(o_q) = 4$, $T(o_1, o_q) = 2$, $T(o_2, o_q) = 3$ and $T(o_3, o_q) = 1$. If we set $\theta = 1$, it is the tracking behavior proposed in [15] that the tracker of o_q must always be within the distance d to o_q. As a result, no tracker can be found under this condition. But if we set $\theta = 0.7$, there is only $\frac{T(o_2,o_q)}{T(o_q)} = 0.75 \geq 0.7$ so we can find that o_2 is o_q's tracker. It should be noted that the value of θ cannot be too small, which will lead to the object close to o_q in a little time being counted as a tracker. In this example, o_3 is only within the distance d of o_q at t_1 and intuitively travels in a totally different direction from o_q. Therefore, it can not be considered as the loose tracking behavior. Probably this is treated as *passby* or *overtaken*.

4 Detecting Loose Tracking Behavior

In this section, we present two algorithms for detecting loose tracking behavior over trajectory data. The first algorithm is a straightforward implementation of the problem definition. The second one, an improved algorithm, enhances the detection efficiency by developing a prefix tree index structure for the Geohash-encoded trajectories.

4.1 Basic Loose Tracking Behavior Detection Algorithm

We first propose the basic loose tracking behavior detection algorithm (LTBD). It takes the object trajectories set O_{DB}, the query trajectory o_q and parameters d and θ as input and outputs all o_q's trackers. Detailed procedures are shown in Algorithm 1.

Algorithm 1. LTBD(O_{DB}, o_q, d, θ)

Input: the object set O_{DB}, the query trajectory o_q, the distance threshold d and the threshold θ

Output: the tracker set $Tra(o_q)$

$Tra(o_q) = \emptyset$;
for each $o \in O_{DB}$ **do**
  $T(o, o_q) = 0$;
  for each time t of o_q **do**
    if $D(o(t), o_q(t)) < d$ **then**
      $T(o, o_q) = T(o, o_q) + 1$;
    end if
  end for
  if $\frac{T(o,o_q)}{T(o_q)} \geq \theta$ **then**
    Add o into $Tra(o_q)$;
  end if
end for
return $Tra(o_q)$;

In Algorithm 1, we first compute the distance between o_q and each trajectory in O_{DB} at each time point of o_q. Then the $T(o, o_q)$ will increase by one if the distance between o and o_q is less than d. Finally the trajectory satisfying $\frac{T(o,o_q)}{T(o_q)} \geq \theta$ is inserted into the result set $Tra(o_q)$. Obviously the LTBD algorithm has $O(|O_{DB}| \times T(o_q))$ computational cost, where $|O_{DB}|$ is the number of objects in O_{DB} and $T(o_q)$ is the lifetime of o_q.

4.2 Prefix Tree Index

In the LTBD algorithm, each trajectory in O_{DB} must be taken into account even though it is actually far away from the query trajectory o_q and can't contribute to the result. For the purpose of enhancing detection efficiency, we build an index structure based on the prefix tree for the trajectories encoded by Geohash algorithm to prune the search space. The construction algorithm of prefix tree index is given in Algorithm 2.

Algorithm 2. BuildPrefixTree(O_{DB}, L)

```
Input: the object set O_DB and the length parameter L
Output: the prefix tree set P
P = ∅;
for each time t do
   Construct a new root node of a prefix tree p(t);
   cur = root;
   for each o' ∈ O_DB do
      for each char c in the L-length prefix of o'(t) do
         if cur.childNodes.containsKey(c) = false then
            cur.childNodes.put(c, new Node);
         end if
         cur = cur.childNodes.get(c);
      end for
   end for
   Store the id of o into cur;
   Add p(t) into P;
end for
return P;
```

We insert the L-length prefix of each trajectory's Geohash code at the same time point into a prefix tree p. The children of a node are stored in a map structure where the key is a character and the value is a node. All leaf nodes are at the same level and store the set of trajectory id, which means trajectories in the same leaf node share a common Geohash prefix.

Given a query trajectory o_q, we only need to search the L-length prefix of $o'_q(t_i)$ in the prefix tree $p(t_i)$ to get nearby objects at time point t_i. The search procedures is presented in Algorithm 3.

Algorithm 3. Search(s, p)

Input: the search string s and the prefix tree p
Output: the object id set D

```
D = ∅;
cur = root;
for each char c in the L-length prefix of s do
    if cur.childNodes.containsKey(c) = false then
        return ∅;
    end if
    cur = cur.childNodes.get(c);
end for
for each object id in cur do
    Add id into D;
end for
return D;
```

According to Algorithm 3, it takes $O(L)$ time to find objects that share the L-length prefix with $o'_q(t_i)$. Figure 2 shows an example of the prefix tree index when $L = 4$ at time t_i. We can quickly find o_5 and o_6 share the prefix 'wx5h' with o_q if $o'_q(t_i) =$ 'wx5h20bc', but if $o'_q(t_i) =$ 'ts3gxvdp', we can find no object that shares the prefix 'ts3g' with o_q.

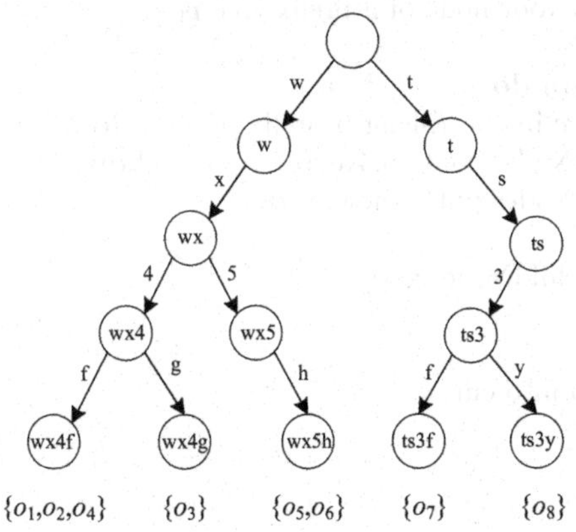

Fig. 2. Prefix tree index example

4.3 Improved Loose Tracking Behavior Detection Algorithm

Based on the prefix tree index, we next present an improved loose tracking behavior detection algorithm (LTBD+) where distance query is used. The prefix tree index can reduce the query time and thus enhance the the detection efficiency. The pseudocode of LTBD+ is shown in Algorithm 4.

Algorithm 4. LTBD+

Input: the object set O_{DB}, the query trajectory o_q, the prefix tree set P, the distance threshold d and the threshold θ

Output: the tracker set $Tra(o_q)$

$Tra(o_q) = \emptyset$;
for each time t of o_q **do**
  $D = \text{Search}(o'_q(t), p(t))$;
  for each $i \in D$ **do**
    if $D(o_i(t), o_q(t)) < d$ and $o_i \notin Tra(o_q)$ **then**
      $T(o_i, o_q) = T(o_i, o_q) + 1$;
      if $\frac{T(o_i,o_q)}{T(o_q)} \geq \theta$ **then**
        Add o_i into $Tra(o_q)$;
      end if
    end if
  end for
end for
return $Tra(o_q)$;

At each time point of o_q, we first take advantage of the index to get its nearby objects id set D by searching the Geohash prefix in a prefix tree. For each id i in D, $T(o_i, o_q)$ will increase only if o_i is not in the result set $Tra(o_q)$ and $D(o_i(t), o_q(t))$ is less than d after executing the distance query. Finally, the object satisfying $\frac{T(o_i,o_q)}{T(o_q)} \geq \theta$ is reported as the tracker of o_q and added into $Tra(o_q)$. Let $|D|$ be the number of trajectory id in D and we know a single Search(s, p) requires $O(L)$ time, then LTBD+ has $O((|D| + L) \times T(o_q))$ computational cost.

We proceed to illustrate algorithm LTBD+ by using the example shown in Fig. 3 with the parameter $\theta = 0.7$. There are 4 objects $\{o_1, o_2, o_3, o_q\}$ and 4 timestamps $\{t_1, t_2, t_3, t_4\}$. Each sub-figure is a snapshot of objects at each timestamp.

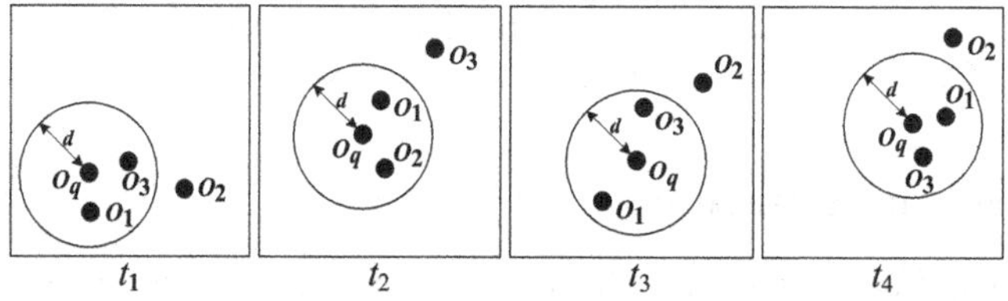

Fig. 3. Snapshots of objects at t_1 to t_4

The execution steps of the LTBD+ is shown in Table 1. The lifetime of o_q is 4 and we get o_1 and o_3 are within a given distance d of o_q at time t_1. Then, we have $\frac{T(o_1,o_q)}{T(o_q)} = \frac{T(o_3,o_q)}{T(o_q)} = 0.25 < \theta$ so no tracker can be found, and so on. Eventually, we get the final determined $Tra(o_q)$ is $\{o_1, o_3\}$.

Table 1. Execution Steps of LTBD+

Timestamp	$\frac{T(o_1,o_q)}{T(o_q)}$	$\frac{T(o_2,o_q)}{T(o_q)}$	$\frac{T(o_3,o_q)}{T(o_q)}$	$Tra(o_q)$
t_1	0.25	0	0.25	$\emptyset$
t_2	0.5	0.25	0.25	$\emptyset$
t_3	0.75	0.25	0.5	$\{o_1\}$
t_4	1	0.25	0.75	$\{o_1, o_3\}$

5 Experiments

In this section, we conduct a series of experiments to evaluate the effectiveness and efficiency of our proposed concepts and algorithms based on two real datasets. All the algorithms are implemented in Java language and the experiments are performed using Intel CPU 2.30 GHz system with 8 GB memory.

5.1 Dataset and Parameter Setting

We use two real datasets for the purpose of studying the performance of our methods in a real-world setting. The details of each dataset are described as follows:

(1) The taxi dataset contains the GPS logs of Beijing taxis recorded during one week [24,25]. We extract 5,000 trajectories of taxis on the same day, and there are 998,245 GPS records in total.
(2) The truck dataset consists of 276 trajectories of 50 trucks delivering concrete to several construction places around Athens metropolitan area in Greece for 33 days [26]. In the experiment, we regard each trajectory as a different truck's trajectory to increase the number of objects in the dataset.

In our experiments, we randomly select a trajectory from O_{DB} as the query trajectory o_q and search its trackers. The default parameters are: d = 1000 m, $\theta = 0.8$, $|O_{DB-taxi}| = 5000$ and $|O_{DB-truck}| = 250$.

5.2 Performance Evaluation

Effectiveness. Figure 4 shows the number of trackers found by LTBD and LTBD+. We vary d when $\theta = 0.6$, $\theta = 0.8$ and $\theta = 1$, respectively. We can see that more trackers are found with increasing d. This is because the larger d means more objects are included. In addition, as θ increases, the stricter requirement on consecutive time period often leads to us finding fewer trackers. When $\theta = 1$, which is the tracking behavior proposed in [15], we find no tracker in the taxi dataset and only one tracker in the truck dataset although d is large enough. The results show that our definition of loose tracking behavior is more flexible and practical.

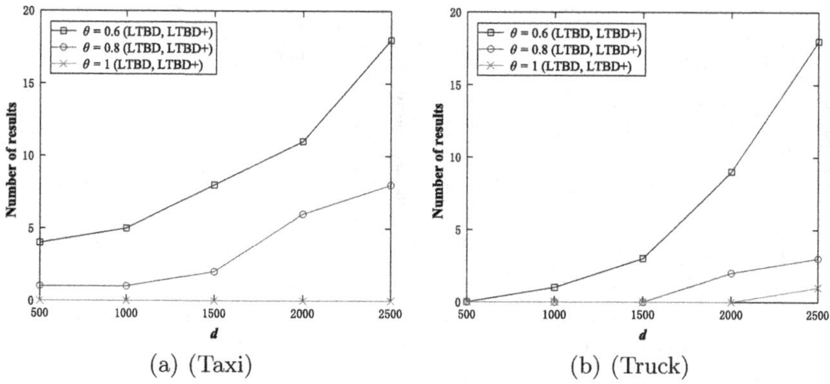

(a) (Taxi) (b) (Truck)

Fig. 4. Number of trackers w.r.t. d and θ

Efficiency w.r.t. $|O_{DB}|$. Fig. 5 depicts the effect of $|O_{DB}|$ by comparing LTBD and LTBD+. It is easy to see that LTBD+ outperforms LTBD by at least one order of magnitude. We also note that LTBD+ is relatively insensitive to the change of $|O_{DB}|$. The reason is that in LTBD+, we only focus on trajectories that share a common Geohash prefix. Therefore, the number of trajectories that can contribute to the result increases slowly while we raise the size of $|O_{DB}|$.

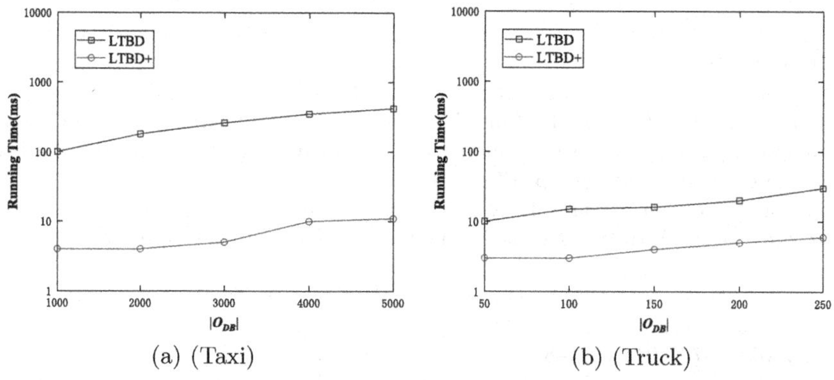

(a) (Taxi) (b) (Truck)

Fig. 5. Running time w.r.t. $|O_{DB}|$

Efficiency w.r.t. d. The running time of two algorithms w.r.t. d is shown in Fig. 6. In LTBD+, we set the length of common prefix $L = 5$ when $500\text{ m} \leq d \leq 1000\text{ m}$ and $L = 4$ when $1000\text{ m} < d \leq 2500\text{ m}$. It is obvious that the parameter d only affects the running time of LTBD+ because the longer L can prune more trajectories which are far away from the query trajectory.

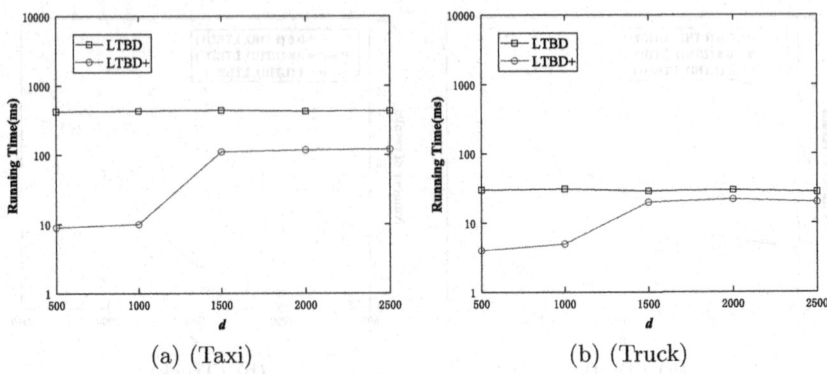

(a) (Taxi) (b) (Truck)

Fig. 6. Running time w.r.t. d

6 Conclusion

In this paper, we study the problem of detecting loose tracking behavior over trajectory data. Different from the previous work, we aim to find objects that track a target with a relaxed temporal constraint. We propose the LTBD and LTBD+ algorithm to find trackers and build prefix tree index in LTBD+ to improve the detection performance. Experiments based on real trajectory datasets demonstrate the effectiveness and the efficiency of our methods.

References

1. Wang, Y., Luo, Z., Qin, G., Zhou, Y., Guo, D., Yan, B.: Mining common spatial-temporal periodic patterns of animal movement. In: 2013 IEEE 9th International Conference on e-Science, pp. 17–26. IEEE (2013)
2. Zheng, Y., Liu, Y., Yuan, J., Xie, X.: Urban computing with taxicabs. In: Proceedings of the 13th International Conference on Ubiquitous Computing, pp. 89–98 (2011)
3. Bao, J., Zheng, Yu., Wilkie, D., Mokbel, M.: Recommendations in location-based social networks: a survey. GeoInformatica **19**(3), 525–565 (2015). https://doi.org/10.1007/s10707-014-0220-8
4. Guo, L., Zhang, D., Cong, G., Wei, W., Tan, K.-L.: Influence maximization in trajectory databases. IEEE Trans. Knowl. Data Eng. **29**(3), 627–641 (2016)
5. Zheng, Yu.: Trajectory data mining: an overview. ACM Trans. Intell. Syst. Technol. (TIST) **6**(3), 1–41 (2015)
6. Gudmundsson, J., van Kreveld, M.: Computing longest duration flocks in trajectory data. In: Proceedings of the 14th Annual ACM International Symposium on Advances in Geographic Information Systems, pp. 35–42 (2006)
7. Gudmundsson, J., van Kreveld, M., Speckmann, B.: Efficient detection of motion patterns in spatio-temporal data sets. In: Proceedings of the 12th Annual ACM International Workshop on Geographic Information Systems, pp. 250–257 (2004)
8. Jeung, H., Yiu, M.L., Zhou, X., Jensen, C.S., Shen, H.T.: Discovery of convoys in trajectory databases. Proc. VLDB Endowment **1**(1), 1068–1080 (2008)

9. Jeung, H., Shen, H.T., Zhou, X.: Convoy queries in spatio-temporal databases. In: 2008 IEEE 24th International Conference on Data Engineering, pp. 1457–1459. IEEE (2008)
10. Li, Z., Ding, B., Han, J., Kays, R.: Swarm: mining relaxed temporal moving object clusters. Proc. VLDB Endowment **3**(1–2), 723–734 (2010)
11. Tang, L.-A., et al.: On discovery of traveling companions from streaming trajectories. In: 2012 IEEE 28th International Conference on Data Engineering, pp. 186–197. IEEE (2012)
12. Lu-An Tang, Yu., Zheng, J.Y., Han, J., Leung, A., Peng, W.-C., La Porta, T.: A framework of traveling companion discovery on trajectory data streams. ACM Trans. Intell. Syst. Technol. (TIST) **5**(1), 1–34 (2014)
13. Zheng, K., Zheng, Y., Yuan, N.J., Shang, S.: On discovery of gathering patterns from trajectories. In: 2013 IEEE 29th International Conference on Data Engineering (ICDE), pp. 242–253. IEEE (2013)
14. Zheng, K., Zheng, Y., Yuan, N.J., Shang, S., Zhou, X.: Online discovery of gathering patterns over trajectories. IEEE Trans. Knowl. Data Eng. **26**(8), 1974–1988 (2014)
15. Xu, J., Zhou, J.: Detect tracking behavior among trajectory data. In: Cong, G., Peng, W.-C., Zhang, W.E., Li, C., Sun, A. (eds.) ADMA 2017. LNCS (LNAI), vol. 10604, pp. 872–878. Springer, Cham (2017). https://doi.org/10.1007/978-3-319-69179-4_64
16. Naserian, E., Wang, X., Xu, X., Dong, Y.: Discovery of loose travelling companion patterns from human trajectories. In: 2016 IEEE 18th International Conference on High Performance Computing and Communications; IEEE 14th International Conference on Smart City; IEEE 2nd International Conference on Data Science and Systems (HPCC/SmartCity/DSS), pp. 1238–1245. IEEE (2016)
17. Naserian, E., Wang, X., Xiaolong, X., Dong, Y.: A framework of loose travelling companion discovery from human trajectories. IEEE Trans. Mob. Comput. **17**(11), 2497–2511 (2018)
18. Kalnis, P., Mamoulis, N., Bakiras, S.: On discovering moving clusters in spatio-temporal data. In: Bauzer Medeiros, C., Egenhofer, M.J., Bertino, E. (eds.) SSTD 2005. LNCS, vol. 3633, pp. 364–381. Springer, Heidelberg (2005). https://doi.org/10.1007/11535331_21
19. Aung, H.H., Tan, K.-L.: Discovery of evolving convoys. In: Gertz, M., Ludäscher, B. (eds.) SSDBM 2010. LNCS, vol. 6187, pp. 196–213. Springer, Heidelberg (2010). https://doi.org/10.1007/978-3-642-13818-8_16
20. Lee, J.-G., Han, J., Whang, K.-Y.: Trajectory clustering: a partition-and-group framework. In: Proceedings of the 2007 ACM SIGMOD International Conference on Management of Data, pp. 593–604 (2007)
21. Li, Z., Lee, J.-G., Li, X., Han, J.: Incremental clustering for trajectories. In: Kitagawa, H., Ishikawa, Y., Li, Q., Watanabe, C. (eds.) DASFAA 2010. LNCS, vol. 5982, pp. 32–46. Springer, Heidelberg (2010). https://doi.org/10.1007/978-3-642-12098-5_3
22. Gaffney, S.J., Robertson, A.W., Smyth, P., Camargo, S.J., Ghil, M.: Probabilistic clustering of extratropical cyclones using regression mixture models. Clim. Dyn. **29**(4), 423–440 (2007)
23. Wikipedia: Geohash. https://en.wikipedia.org/wiki/Geohash (2020)
24. Yuan, J., Zheng, Y., Xie, X., Sun, G.: Driving with knowledge from the physical world. In Proceedings of the 17th ACM SIGKDD International Conference on Knowledge Discovery and Data Mining, pp. 316–324 (2011)

25. Yuan, J., et al.: T-drive: driving directions based on taxi trajectories. In Proceedings of the 18th SIGSPATIAL International Conference on Advances in Geographic Information Systems, pp. 99–108 (2010)
26. Chorochronos: Datasets & Algorithms. http://www.chorochronos.org/

Behavioral Fault Modelling and Analysis with BIP: A Wheel Brake System Case Study

Xudong Tang[1], Qiang Wang[2], and Weikai Miao[1,3](✉)

[1] Shanghai Key Lab for Trustworthy Computing, East China Normal University, Shanghai, China
wkmiao@sei.ecnu.edu.cn

[2] SUSTech Academy for Advanced Interdisciplinary Studies, Southern University of Science and Technology, Shenzhen, China

[3] Shanghai Key Laboratory of Computer Software Testing and Evaluating, Shanghai, China

Abstract. BIP (Behavior-Interaction-Priority) is a component-based framework supporting rigorous design of complex systems. Systems are modelled by a set of components and connectors. Behavioral fault modelling and analysis refers to an integration of model based system design and safety analysis. In this paper, we integrate fault tree based safety analysis into BIP model and apply statistical model checking to verify system specification and calculate probability of fault issues. We also trace the simulation result to confirm the extended system model without fault keeps consistence of the nominal system model. We illustrate an airplane wheel brake system meeting the industry standards as case study to show its advantage in analyzing faulty behavior of safety-critical systems in aerospace practice.

Keywords: Model based design · Model based safety analysis · Fault tree analysis · Statistical model checking · BIP

1 Introduction

There is a growing trend to advocate the methodology of model based system design (MBSD) and model based safety analysis (MBSA), as a response to the strong demand for minimizing errors during the development of safety critical systems such as aircraft and aerospace systems [13,31]. Generally speaking, MBSD refers to the type of approach that uses an abstract system model to describe the nominal behavior and the system architecture involving both software and hardware components. When the system model is built with a formal modelling language, various system level analysis and verification can then be performed on this formal model to ensure the correctness of the design. Moreover, low level implementations can also be automatically derived from the design models through code generation.

M. Qiu (Ed.): ICA3PP 2020, LNCS 12454, pp. 231–250, 2020.
https://doi.org/10.1007/978-3-030-60248-2_16

MBSA moves one step further to take into account the faulty behavior of software and hardware components, and then augments the nominal system model with fault models. This extended system model can then be used to study the safety and reliability properties when in presence of one or more faults. In practice, MBSD and MBSA has been highly recommended in the SAE (Society of Automotive Engineers) guidance for the development of complex aerospace systems (i.e., Aerospace Recommended Practice ARP4754A [4] and ARP4761 [3]). Particularly, SAE provides an aircraft wheel brake system (WBS) as an example of applying ARP4754A and ARP4761 into practice in Aerospace Information Report AIR6110 [5].

BIP (Behavior-Interaction-Priority) [9,10] is a component based framework that advocates the methodology of rigorous system design. Rigorous system design can be understood as the process that derives correct-by-construction implementations from high level system models automatically. BIP provides a formal modelling language that is expressive enough for describing both the functional behavior and the architecture of complex systems. It also provides a set of tools that automates the rigorous system design process. SBIP [28] is a statistical extension of BIP that supports formal modelling and statistical analysis of systems exhibiting stochastic behaviors. Currently, both BIP and SBIP are widely used in the modelling and analysis of nominal system models (i.e., system models without faulty behaviors).

In this work, we propose to apply the BIP framework for fault modelling and safety analysis and perform a case study on the aircraft wheel brake system provided in SAE AIR6110 [5]. In a related work [15], WBS has been also investigated as a case study to promote their approach to model based safety analysis. However, in their approach fault models are injected to the formal system model in a monolithic manner such that the integration process can obscure the nominal behavior of the system and create additional overheads for safety analysis. Our fault modelling is modular and features a clear separation between nominal behavior and fault behavior that makes the WBS model extendable and understandable. We also adopt the statistical approach to model checking the extended system model that uses less time and memory. To this end, the main contributions of this work can be summarized as followings.

1. For each reference system architecture of the WBS, we have built a formal model in BIP that is amenable to various formal safety analysis and verification. We also propose a method to derive fault models from fault trees, and integrate nominal system model with fault models. The extended system model is a stochastic system where faulty behaviors and failure rate are both manifested in the model.
2. We propose to use the Statistical Model Checking [22,32,35] approach to simulate the failure injected stochastic WBS model, and automatically verify whether the BIP system model satisfies the AIR6110 requirements using the statistical model checker SBIP for BIP [28]. We report interesting findings and confirm the necessity of the iteration in AIR6110 in the experiments of comparing several WBS architectures. We also observe the performance of

SBIP during probability estimation and give some suggestions for using BIP and SBIP tool chains.

The remainder of the paper is organized as follows. Section 2 presents the related work on model based safety analysis. Section 3 introduces the BIP framework and statistical model checking for BIP. Section 4 presents an overview of the SAE AIR6110 Wheel Brake System. In Sect. 5, we present the integration process of nominal system model and faulty behavior. Section 6 provides the verification methods and experimental results. Section 7 concludes this paper and discuss some future works.

2 Related Work

Model Based Safety Analysis (MBSA) [23] aims to generate fault injected models for safety analysis. In [26], MBSA has been distinguished into two categories according to how the faulty models are constructed. The first category performs MBSA on the extension of the design models used in development process with failure mode models. In this line of work, the ESACS and ISAAC project develops Extended System Model and Failure Injection methods [7,16]. This approach maximizes the consistency between system development and safety analysis. However, a complete extended system may be intractable for analysis tools. The second category creates dedicated models that are specifically for the goal of safety assessment Engineers can avoid unnecessary complexity by adjusting the level of detail. Modelling tools such as HiP-HOPS and AltaRica are used to build models for failure analysis specifically following this approach.

A number of tools have been extended to support both kinds of MBSA approaches, e.g. Architecture Analysis and Design Language (AADL) [6,20] and the integration of HiP-HOPS with Matlab Simulink [30]. In [19] a dialect of AADL has been proposed to validate failure injected model using a satellite system as case study. With the development of the revised Error Model (EMV2) Annex [18] for AADL, engineers have applied AADL on different systems (e.g., Flight Control [36], Fire Alarm [34] and other cyber-physical systems [33]) to capture key features of failure and error propagations from system models. In [17] a translator from AADL to AltaRica has been proposed and applied to a flight control system. In [29], a transformation from AADL model to HiP-HOPS model has been studied. However, no tool has demonstrated absolute advantages over other tools in performing MBSA.

3 Preliminaries

In this section, we briefly introduce the modelling framework of BIP and the statistical model checking for BIP.

The BIP framework [8,14,24] advocates a component-based design methodology for building systems from atomic components. In this work, we rely on the fragment of BIP with multiparty synchronization and data transfer. Formally, an atomic BIP component is defined by an automaton extended with data and port.

Definition 1 (BIP component). *Given a finite set of variables $\mathbb{V}$, an atomic BIP component is defined as a tuple $B = \langle \mathbb{V}, \mathbb{L}, \mathbb{P}, \mathbb{E}, \ell \rangle$, where 1. $\mathbb{L}$ is a finite set of control locations; 2. $\mathbb{P}$ is a finite set of communication ports; 3. $\mathbb{E} \subseteq \mathbb{L} \times \mathbb{P} \times \mathcal{F}_{\mathbb{V}} \times \mathcal{E}_{\mathbb{V}} \times \mathbb{L}$ is a finite set of transition edges extended with guards in $\mathcal{F}_{\mathbb{V}}$ and operations in $\mathcal{E}_{\mathbb{V}}$; 4. $\ell \in \mathbb{L}$ is an initial control location.*

Transition edges in a component are labeled by ports, which form the interface of the component. We assume that, from each control location, every pair of outgoing transitions have different ports, and the ports of different components are disjoint. In other words, transitions with the same ports in the component are not enabled simultaneously. Given a component violating such assumptions, one can easily transform it into the required form by renaming the ports, while retaining the BIP expressiveness power. To ease the presentation, we denote in the sequel by $id(p)$ the unique component where port p is defined.

We denote by $\mathcal{B} = \{B_i \mid i \in [1, n]\}$ a set of components. The coordination of components is specified by using interactions.

Definition 2 (Interaction). *An interaction for $\mathcal{B}$ is a tuple $\gamma = \langle g, \mathcal{P}, f \rangle$, where $g \in \mathcal{F}_{\mathbb{V}}$, $f \in \mathcal{E}_{\mathbb{V}}$ and $\mathcal{P} \subseteq \bigcup_{i=1}^{n} \mathbb{P}_i$, $\mathcal{P} \neq \emptyset$, and for all $i \in [1, n]$, $|\mathcal{P} \cap \mathbb{P}_i| \leq 1$.*

Intuitively, an interaction defines a guarded multiparty synchronization with data transfer: when the guard g of an interaction $\mathcal{P}$ is enabled, then the data transfer specified by f can be executed, and after that the transitions labelled by the ports in γ can be taken simultaneously. We denote by Γ a finite set of interactions.

A BIP model is constructed by composing a number of components with interactions.

Definition 3 (BIP Model). *A* BIP *model $\mathcal{M}_{\text{BIP}}$ is a tuple $\langle \mathcal{B}, \Gamma \rangle$, where $\mathcal{B}$ is finite set of components, and Γ is a finite set of interactions for $\mathcal{B}$.*

A state of a BIP model is denoted by a tuple $c = \langle \langle l_1, \mathbf{V}_1 \rangle, \ldots, \langle l_n, \mathbf{V}_n \rangle \rangle$, where for all $i \in [1, n]$, $l_i \in \mathbb{L}_i$ and $\mathbf{V}_i$ is a valuation of $\mathbb{V}_i$. A state c_0 is initial if for all $i \in [1, n]$, $l_i = \ell_i$ and $\mathbf{V}_i$ is the initial valuation of $\mathbb{V}_i$. A state c is an error if for some $i \in [1, n]$, l_i is an error location. We say an interaction $\gamma \in \Gamma$ is enabled on a state c if for every component $B_i \in \mathcal{B}$, such that $\gamma \cap \mathbb{P}_i \neq \emptyset$, there is an edge $\langle l_i, \gamma \cap \mathbb{P}_i, g_i, f_i, l_i' \rangle \in \mathbb{E}_i$ and $\mathbf{V}_i \models g_i$. Then we define the semantics of a BIP model by a labeled transition system.

Definition 4 (Labeled transition system of BIP). *Given a* BIP *model $\mathcal{M}_{\text{BIP}} = \langle \mathcal{B}, \Gamma \rangle$, the labeled transition system it defines is a tuple $\mathcal{T}_{\text{BIP}} = \langle \mathcal{C}, \Sigma, \mathcal{R}, \mathcal{C}_0 \rangle$, where*

1. *$\mathcal{C}$ is the set of states,*
2. *$\Sigma = \Gamma$,*
3. *$\mathcal{R}$ is the set of transitions, and we say that there is a transition from a state c to another state c', if there is an interaction γ such that,*
   (a) *γ is enabled in c;*
   (b) *for all $B_i \in \mathbb{B}$ such that $\gamma \cap \mathbb{P}_i \neq \emptyset$, there is an edge $\langle l_i, \gamma \cap \mathbb{P}_i, g_i, f_i, l_i' \rangle \in \mathbb{E}_i$, then $\mathbf{V}_i' = \mathbf{V}_i[\mathbb{V}/f_i(\mathbb{V})]$;*
   (c) *for all $B_i \in \mathbb{B}$ such that $\gamma \cap \mathbb{P}_i = \emptyset$, $l_i' = l_i$ and $\mathbf{V}_i' = \mathbf{V}_i$.*
4. *$\mathcal{C}_0$ is the set of initial states.*

In the stochastic extension of BIP, we can add stochastic behavior to atomic components by declaring probabilistic variables. A probabilistic variable x^P is attached to given distributions μ_x^P and can be updated on transitions using the attached distribution. The semantics is thus fully stochastic. We refer to [28] for details.

Given a stochastic system model $\mathcal{M}_{\text{BIP}}$ and a property specification ϕ, statistical model checking refers to a series of simulation-based techniques that can be used to answer two questions: (1) Qualitative: Is the probability for $\mathcal{M}_{\text{BIP}}$ to satisfy ϕ greater or equal to a certain threshold θ ? and (2) Quantitative: What is the probability for $\mathcal{M}_{\text{BIP}}$ to satisfy ϕ? Let B_i be a discrete random variable with a Bernoulli distribution of parameter p. Such a variable can only take 2 values 0 and 1 with $Pr[B_i = 1] = p$ and $Pr[B_i = 0] = 1 - p$. In our context, each variable B_i is associated with one simulation of the system. The outcome for B_i, denoted b_i, is 1 if the i-th simulation satisfies ϕ and 0 otherwise. The approaches and algorithms to solve the above questions are out of the scope of this paper. We refer to [25] for more information.

4 The AIR6110 Wheel Brake System

The Wheel Brake System (WBS) description is introduced in Aerospace Information Report 6110 (AIR6110) as a contiguous aircraft system development process example. According to AIR6110, WBS is a detailed function of an aircraft designated model S18. The hypothetical S18 aircraft is a two engine passenger aircraft designed to carry 300 to 350 passengers up to 5000 nautical miles at 0.84 mach, and has an average flight duration of 5 h.

The WBS provides braking on the main gear wheels used to provide safe retardation of the aircraft during taxiing and landing phases, and in the event of a rejected take-off. The wheel brakes also prevent unintended aircraft motion when parked, and may be used to provide differential braking for aircraft directional control. A secondary function of the WBS is to stop main gear wheel rotation upon gear retraction. Braking on the ground is commanded manually, via brake pedals, or automatically (autobrake) without the need for pedal application.

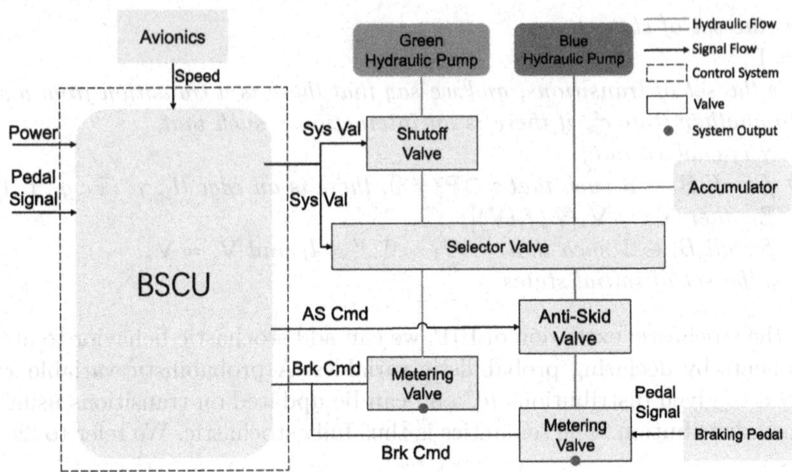

Fig. 1. The WBS BIP model with the nominal behavior

4.1 The WBS Architecture and Nominal Behavior

Figure 1 shows the WBS BIP model with the nominal behavior. The WBS is composed of an electronic control system and a physical system. The majority of the electronic control system is Braking System Control Unit (BSCU). The WBS receives several signals including the brake pedal position from upper level avionics system and electrically forwards them to the BSCU. The BSCU also receives two power inputs from two independent power supply resources. As the result of computation, the BSCU in turn produces the system validity command, anti-skid command and braking command to the physical system. The physical system includes two hydraulic pressure lines which are supplied by the green/blue hydraulic pump respectively.

Operation Mode. There are three operation modes for physical system. In *normal mode*, the wheel brake is supported by the main hydraulic circuit, refers to the green hydraulic circuit. In *alternate mode*, the wheel brake is supported by a second hydraulic circuit. This mode is standby and is selected automatically when the normal system fails. An accumulator supplies the *emergency mode* when blue hydraulic supply is lost and the normal mode is not available.

Braking System Control Unit (BSCU). According to the AIR6110, the BSCU is composed of two independent channels, each channel has its own power supply and avionics system inputs. Each channel has a command subsystem and a monitor subsystem. The monitor system generates system validity command and the command system calculates anti-skid command and braking command. The BSCU will make an ultimate judgement call between each command output by the two channels respectively.

Hydraulic Pump. In nominal system behavior, both the green and blue hydraulic pumps provide enough hydraulic pressure for their green/blue hydraulic circuit respectively. An accumulator is also a hydraulic pump to provide an emergency reserve of hydraulic pressure for blue hydraulic circuit in emergency mode.

Shutoff Valve. The shutoff valve responds the system validity command from BSCU to decide whether to apply the hydraulic pressure to the selector valve in green hydraulic circuit or not. The system validity command is modelled in BIP as a boolean value.

Selector Valve. The selector valve controls the switch between green and blue hydraulic circuits mechanically. It outputs appropriate pressure from green hydraulic pump, and switches to blue hydraulic circuit as soon as it detects a lack of pressure in the green hydraulic circuit. In BIP model, the component selector valve only outputs pressure from either the green hydraulic circuit input or blue hydraulic circuit input at a time.

Anti-Skid Valve. The anti-skid valve follows anti-skid command to control hydraulic pressure to the metering valve. It is used to restrict the hydraulic pressure to the wheel brake in order to prevent locking of the wheel. Wheel skid happens when the wheel is locked but the vehicle keeps a relative slid speed to the ground. We consider a loss of anti-skid function as a fault and will integrate it into nominal BIP model.

Metering Valve. Metering valve, or metering servo valve controls pressure to the demanded level and provides regulation for the anti-skid function.

4.2 The WBS Architecture Mode in BIP

According to the AIR6110, the system architecture evolves throughout the development life cycle and is tightly coupled with the requirements development (especially interface requirements) and is not finished until the requirements associated with the architecture have been validated.

We follow the standard to advance our BIP model. As a result, our WBS BIP model has four versions corresponding to the four architectures in the AIR6110. They are numbered from ARCH1 to ARCH4. Each architecture is obtained after design choices of different types.

ARCH1. The architecture one is regarded as a high level wheel brake system architecture to be analyzed against the system level functions operational and safety requirements, and any design constraints that have been identified early in the standard.

ARCH2. Modified braking system architecture ARCH2 implements the function of ARCH1 and meets various of derived requirements listed in the Wheel Brake System Preliminary Safety Assessment (PSSA). As a feedback of PSSA, things to consider include but are not limited to:

- The modified architecture shall have at least two independent hydraulic pressure sources.
- The modified architecture shall have dual channel BSCU and multimode brake operations to provide the required redundancy.

ARCH3. Following the result of the WBS trade study, the development of architecture three is designed with one BSCU housing two independent systems, each BSCU subsystem has independent command and monitor channels.

ARCH4. Since architecture three has been simulated and the results of the modelling for each system component are that the schematic of the braking system architecture does not work and there are some mistakes in the schematic. Architecture four is established to avoid the risk of hydraulic supply to wheel brake being possible in normal mode by accumulator. An input is added in ARCH4 to the selector valve corresponding to the validity of the control system. Pedal position signal is input in front of the anti-skid valve in blue hydraulic pressure line and the accumulator is moved to the front of the selector valve.

5 Integrating Fault Trees into WBS BIP Model

5.1 Behavioral Fault Modelling

In this section, we describe how to integrate a component's faulty behavior from system fault tree [21] into the BIP model. First, we decompose the system fault tree. Then we deduce faulty behaviors from each leaf node of the system fault tree. Next, we generate a fault component for each faulty behavior according to its nominal behavior BIP model. Afterwards, we put the behavioral fault component and the nominal behavior component together with a manager component deciding and monitoring the activation of both fault and nominal behavior components. Finally, we modify the connection between each component to ensure the input and output ports are the same as the original BIP component. The result of the integration is a fault-based BIP component.

Example 1. We take a commonly used valve component as an example. In general, a valve is used to control the passage of hydraulic pressure. When the valve is open, the output hydraulic pressure is equal to the input hydraulic pressure, indicates that the valve currently allows hydraulic pressure to pass. When the valve is close, the output hydraulic pressure is zero, indicates a rejection of hydraulic passage. We model the nominal behavior of valve in BIP, as shown in Fig. 2.

The description of leaf nodes depends on the chosen of system tree. We consider a fault tree that takes the loss of valve as basic fault event. We subjectively judge that a loss of valve occurs when the valve is stuck at open or it is stuck at close.

For stuck-at-open fault, the output hydraulic pressure is always the same as the input, while for stuck-at-close fault, the output hydraulic pressure is always zero.

The integration steps. We conclude our methodology of integrating the leaf node of fault tree into the valve component shown in Fig. 2 by the following steps:

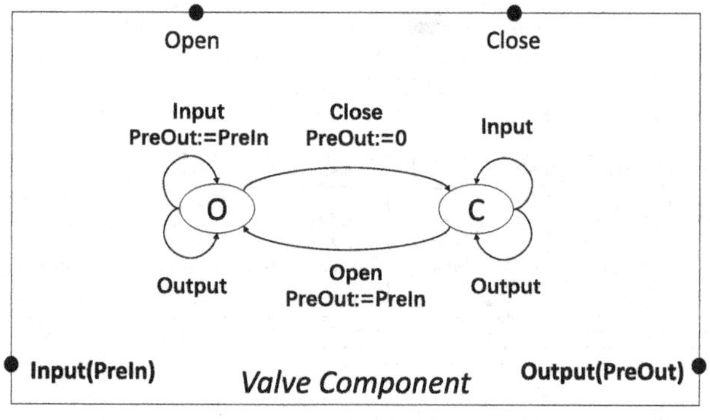

Fig. 2. The nominal behavior of valve in BIP

Step 1: Decompose the fault tree. We pay attention to those basic events which may cause the top level fault. In this example, we consider the loss of valve as leaf node decomposed from the fault tree.

Step 2: Deduce the faulty behaviors from the leaf node. We deduce faulty behaviors from the basic event *loss of valve*. We take two faulty behaviors into consideration, they are introduced as follows:

- For stuck-at-open fault, the output hydraulic pressure is always the same as the input.
- For stuck-at-close fault, the output hydraulic pressure is always zero.

Step 3: Generate BIP component with faulty behavior. Comparing with the nominal behavior valve component, the valves with stuck-at-open/close fault show an eternally unblocked/blocked for both *Open* and *Close* status respectively. According to step 3, the faulty behavior valve components are generated as follows:

- For both faulty behavior components, the transition among *Open* and *Close* status will no longer change the value of *PreOut*.
- For stuck-at-open component, the action of transition from status *Close* to *Close* under event *Input* is *PreOut:=PreIn*, just the same as transition from status *Open* to *Open* under event *Input*.
- For stuck-at-close component, the action of transition from status *Open* to *Open* under event *Input* is *PreOut:=0*, just the same as transition from status *Close* to *Close* under event *Input*.

Step 4: Design a manager component for deciding and monitoring the activation of both fault and nominal behavior components. In this example, we design a manager component with three states, each of them represents one kind of valve status. Once the state executes a self circulation, an event will be sent to the corresponding behavioral component to activate it. The three states switch through internal port *Trigger* which is controlled by BIP engine to support a stochastic process of faulty behavior occurred during simulation.

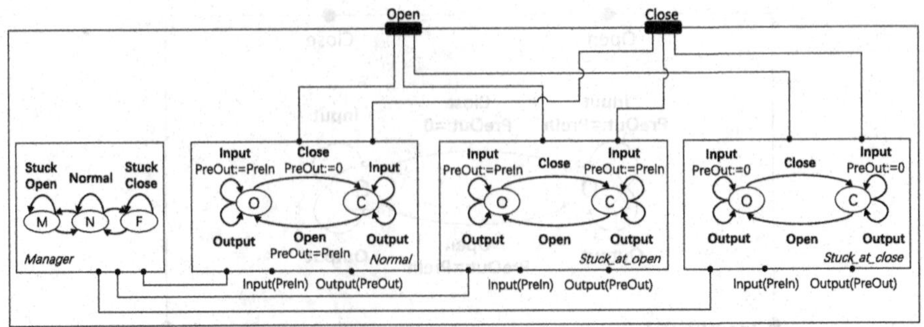

Fig. 3. An integration example of common valve component

Step 5: Since the original component has been integrated with faulty behavior, it keeps the same external port for connection as before but has been extended a lower level to contain various of components designed in the previous steps. Modifying the upward connectors from components to the compound and extending each component with an external port receiving activation event. Confirming the extended system model without fault keeps consistence of the nominal system model through observing their transition of status, internal and external ports and connectors.

As a result of integration, Fig. 3 shows the valve compound containing a valve component with nominal behavior, two valve components with faulty behavior and a manager component. Using this BIP model, we integrate failure into the software behavior instead of separating the hardware failure from software system. As hierarchies of complex components, the integrated compound helps us observe the system behavior through graphical representation of BIP model. Also, further simulation and validation of tracing faulty behavior using stochastic BIP engine becomes possible.

5.2 Fault Trees for the WBS

Figure 4 shows a fault tree for "Loss of wheel braking" event, which is based on the AIR6110 description. Loss of wheel braking is caused either through the Loss of operation of physical system or due to the loss of BSCU. We focus on the leaf nodes which represent faulty behavior for their component respectively. Notice that the expansion of the fault tree nodes "Loss of BSCU channel 1/2" are carried out but not included in Fig. 4 for the sake of brevity.

We deduce the faulty behaviors for every component from the fault tree shown in Fig. 4. Also, we consider the compound *BSCU Channel 1/2* as the lowest level of component for the sake of brevity. A statistics of deduced faulty behaviors is given in Table 1. The selector valve's faulty behavior is loss of its selector function, while each BSCU channel has three faulty behaviors including "loss of power supply", "loss of monitor system" and "command system does not operate".

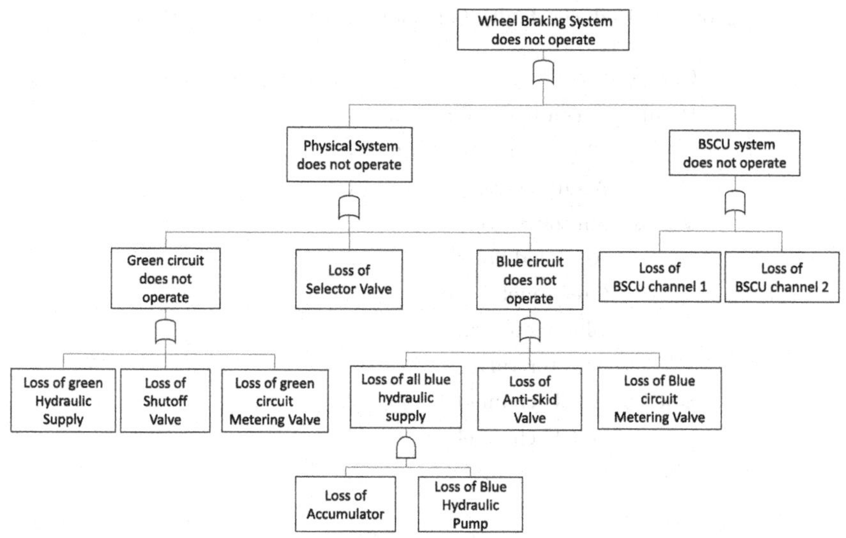

Fig. 4. Fault tree for a loss of WBS occurrence under ARCH4

5.3 Extended WBS Model with Faults

In this section, we give a whole view of modified BIP model which is integrated with fault tree introduced in Sect. 5.1 using the methodology provided in Sect. 5.2.

Table 2 shows metrics for the different architectures. We focus on observing the diversities between the adjacent architectures and the differences between the nominal system architecture and extended system architecture. The amount of max depth of each architecture provides an overview of the usage of hierarchies of complex components. The count of internal ports indicating the number of inner transitions with guard in every component while the count of external ports indicating the number of outer interface of each component.

We can conclude that the later the design version of the architecture, the greater the scale of the BIP model. ARCH1 is established as high level architecture without regard to redundancy. Thus, the design of blue hydraulic line and selector valve are not included in ARCH1. ARCH2, ARCH3 and ARCH4 show large delta comparing with ARCH1. There are differences in the design of BSCU between ARCH2 and ARCH3. For ARCH2, the standard only requires a redundency of BSCU which results in a WBS model including two identical BSCU systems. For ARCH3, motivated by trade studies on ARCH2, the structure is modified from two BSCUs to a BSCU with dual channels. Though the decision of dual channels improve the fault tolerance, it increases the number of ports and variables. ARCH3 and ARCH4 show the closest metrics for the only little change is the input of selector valve and the location of accumulator. From another perspective, as the result of integration, the extended BIP model with faulty behavior presents a larger scale than the nominal BIP model.

Table 1. Deduced faulty behavior statistics under ARCH4

Component		Number of fault(s)
Pumps	Green hydraulic pump	1
	Blue hydraulic pump	1
	Accumulator	1
Valves	Shutoff valve	2
	Anti-skid valve	2
	Green meter valve	2
	Blue meter valve	2
	Selector valve	1
BSCU	BSCU channel 1	3
	BSCU channel 2	3

The graphical representation of the BIP model can separate the correct behavior from the faulty behavior to a great extent. This helps us tracing how a single point of failure impacts the system and whether it will cause a top level failure or not.

The nominal WBS model and the extended model described in this section can be found at https://github.com/Rosaugo/WBSFaultModelingBIP.

6 Verification Method and Experiments

In this section, we present our experiments on the wheel brake system BIP model conducted. We introduce how we apply Statistic BIP (SBIP) tool chains on verifying stochastic model. We are interested in verifying the safety and other requirements introduced in Sect. 6.1. Through the experiment results, we demonstrate that our verification method can help system engineers improve the role of specification and promote the iteration of the system architecture.

To carry out the experiments, a critical problem is that, since we have integrated various of faulty behaviors into the model, its transitions become fully stochastic. Engineers who use classical model checking algorithms will face with problems caused by stochastic complex components (e.g., state explosion). On the other hand, the classical model checking always gives the answer whether a model meets a given specification or not. But it can not be used to answer whether our faulty behavior model meets any specification for it is purposely modelled to contain faulty behaviors triggered stochastically. As a better perspective of observation, we'd like to carry out the experiments to answer such kind of questions, e.g. *For every time the system runs, what is the probability of failure,* etc. Hence, our verification aim changes from assessing the overall correctness of the entire system to reasoning on system requirements in a quantitative manner.

Table 2. BIP Nominal/Fault model statistics

		ARCH1	ARCH2	ARCH3	ARCH4
Component types	Nominal	12	12	15	15
	Extended	24	24	25	25
Compound types	Nominal	1	1	2	2
	Extended	8	8	9	9
Max depth	Nominal	2	2	3	3
	Extended	4	4	4	4
Variables	Nominal	38	45	90	91
	Extended	102	115	134	135
Internal ports	Nominal	18	22	30	30
	Extended	104	114	128	128
External ports	Nominal	33	39	64	65
	Extended	96	109	126	127

Our idea is close to those introduced in [11,12]. Setting a probability distribution on the internal port *Trigger* which controls the component to behave as a nominal component or component with faulty behaviors. Using SMC engine (SBIP) to give a randomly sampled set of simulations with the selected distribution. Then we estimate the probability that system meets the given specification.

Here comes another problem about the choice and realization of probability distribution, to use statistical model checking techniques to estimate the probability of properties, an authentic probability distribution should be selected. To model the distribution of failure occurrence in our model, SBIP provides two methods. One method is using specific annotations to tag a component port which supports several built-in density functions (normal, gamma, chi-square). Another method is through the use of standard library's function such as *rand()*, or importing external probability distribution function library.

The remainder of this section is organized as follows. First, we introduce several categories of standard requirements of WBS in quantitative manner. Then we discuss our choice of probability distribution and estimation methods. Finally, the experimental results for each requirement are presented.

6.1 WBS Requirements

The AIR6110 document contains several requirements for the WBS. These can be grouped into two main categories: Requirements corresponding to safety, e.g. *the loss of all wheel braking shall be extremely remote*, and others, e.g. *the WBS shall have at least two hydraulic pressure sources.*

Safety requirements are divided into the WBS requirements, e.g. *Undetected inadvertent wheel braking on one wheel w/o locking during takeoff shall be extremely improbable* , and subsystem requirements derived from WBS requirements, e.g. *Loss of a signal BSCU shall be less than* 5.75×10^{-3} *per flight.*

Table 3. WBS requirements specification

Requirement ID	Type	Description	Trace from
S18-WBS-R-0321	Safety	The probability of "Loss of all wheel braking" shall be extremely remote	N/A
S18-WBS-R-6108	Safety	The probability of "Loss of Normal Brake System Hydraulic Components" shall be less than 3.3×10^{-5} per flight	N/A
S18-BSCU-R-0002	Safety	The probability of "BSCU Fault Causes Loss of Braking Commands" shall be less than 3.3×10^{-5} per flight	S18-WBS-R-6104
S18-BSCU-R-0003	Safety	The probability of "Loss of a single BSCU" shall be less than 5.75×10^{-3} per flight	S18-WBS-R-6105
S18-WBS-2975	Design Decision	The accumulator shall be attached to the blue hydraulic line before the selector valve	N/A

Table 3 gives an overview of several requirements we focus on. These five requirements described in AIR6110 document are designed to be handled through different stages of system development. In some safety requirements such as S18-WBS-R-0321, the precise lower bound of probability of failure (e.g., "extremely remote") is defined in [2].

6.2 Experiments and Results

The AIR6110 document provides a sample value for every node in fault tree indicating its budgeted probability of occurrence. We take this sample as reference and use a constant to represent the failure rate instead of using the common failure rate (e.g., Bathtub curve, Poisson distribution). There is a general agreement in Military and Aerospace standards that replacing curve distribution with constant. This practice is valid and useful for complex units, systems and electronics.

As introduced in Sect. 6.1, there are two methods to model the distribution of failure occurrence: built-in functions for normal distribution, gamma distribution and chi-square distribution, or library functions imported supporting other kinds of distribution. We choose the random() function from standard library to set probability distribution on the internal port *Trigger*. For example, the budgeted probability of failure "*Loss of Hydraulic Components*" is asserted to 3.3×10^{-5} per flight. This reflects in the manager component of the compound *Hydraulic Pump* setting its switch probability to 3.3×10^{-5} for each trace during system simulation. Table 4 shows the budgeted probability of failure occurrence for every component containing faulty behavior.

Table 4. Budgeted probability of failure

Failure component		Budgeted Prob. of failure occurrence
Power	Power supply	1.00×10^{-5}
Pumps	Green hydraulic pump	3.30×10^{-5}
	Blue hydraulic pump	
	Accumulator	
Valves	Shutoff valve	1.30×10^{-5}
	Anti-skid valve	
	Green meter valve	
	Blue meter valve	
	Selector valve	1.30×10^{-5}
BSCU	BSCU command subsystem	5.75×10^{-3}
	BSCU monitor subsystem	5.75×10^{-3}

Probability Estimation. We are mainly interested in estimating the probability that the top level failure (e.g., "*Loss of wheel braking*") is smaller than the given bound (e.g., "3.3×10^{-5}"). The SMC engine provides statistical testing algorithm called Probability Estimation (PE) [22] for stochastic systems verification. It is helpful for engineers to learn the estimated probability of top level failure on every step of complex system development.

We apply our methodology to every modified WBS architecture and here we consider modified WBS ARCH4 which is the most advanced among all the experimented architectures as a sample scenario. In ARCH4, accumulator is set before the selector to prevent unexpected hydraulic pressure supply. Loss of WBS is caused either through loss of BSCU or due to the physical system does not operate. The top level properties are derived from the requirements and stated as Linear Temporal Logic (LTL) formulas. We use standard library's function *rand()* to realize the uniform distribution [1] replacing on the internal port *Trigger*. We use the PE to estimate probabilities for LTL formulas.

For requirement S18-WBS-R-0321, the SBIP generates 647500 traces in 6 h, observes no trace with the occurrence of failure "Loss of all wheel braking", thus we provide the result as "*not able to conclude*". Comparing with the *confidence* set for PE and the demanding probability for the concept "extremely remote" defined in [2], we basically conclude whether ARCH4 satisfies the requirement S18-WBS-R-0321 can not be defined currently.

For a subsystem requirement S18-BSCU-R-6108, the failure is caused either through loss of BSCU Command System or due to the BSCU Monitor System does not operate. ARCH4 provides a single BSCU with dual command/monitor subsystems inherited from ARCH3. The SBIP generates 235440 traces in 1 h and 40 min, observes 7 traces with the occurrence of failure, giving the rate of 2.97×10^{-5}. Comparing with the *confidence*, we conclude that ARCH4 satisfies the requirement S18-BSCU-R-6108.

We provide the whole experiment result with architecture comparison in the next section. Our experiments are performed on a 64-bit Ubuntu 14.04 LTS machine with 2.0 GHz Inter CORE i7-8550U CPU and 2133 MHz 16 GB memory, the SBIP version is 2.2.3. Notice that for the limitation of time and hardware, we do not conclude any estimation of probability less than 10^{-7} in order of magnitude.

Architectures Comparison. Following the methods above, we generate a comparison of all the four architectures on the results of PE as shown in Table 5. The overall results are based on the following evaluation criteria:

Table 5. Overall experiment result

	Time	Traces	Nb. failure	Probability	Pass/Fail
ARCH1					
S18-WBS-R-0321	6:00:00	1079100	1	9.27×10^{-7}	Fail
S18-WBS-R-6108	1:15:12	359700	11	3.06×10^{-5}	Pass
S18-BSCU-R-0002	0:29:31	120000	694	5.78×10^{-3}	Fail
S18-BSCU-R-0003	0:29:31	120000	694	5.78×10^{-3}	Pass
S18-WBS-2975	–	–	–	–	Fail
ARCH2					
S18-WBS-R-0321	6:00:00	925000	0	0	Not able to conclude
S18-WBS-R-6108	1:27:30	294300	9	3.05×10^{-5}	Pass
S18-BSCU-R-0002	1:21:00	240000	8	3.33×10^{-5}	Pass
S18-BSCU-R-0003	0:42:18	120000	696	5.80×10^{-3}	Pass
S18-WBS-2975	–	–	–	–	Fail
ARCH3					
S18-WBS-R-0321	6:00:00	647500	0	0	Not able to conclude
S18-WBS-R-6108	1:40:00	235440	8	3.40×10^{-5}	Pass
S18-BSCU-R-0002	6:00:00	647500	0	0	Pass
S18-BSCU-R-0003	6:00:00	647500	0	0	Pass
S18-WBS-2975	–	–	–	–	Fail
ARCH4					
S18-WBS-R-0321	6:00:00	647500	0	0	Not able to conclude
S18-WBS-R-6108	1:40:00	235440	7	2.97×10^{-5}	Pass
S18-WBS-2975	–	–	–	–	Pass

- Time limit is set to 6 h.
- Each trace contains at least a system execution cycle.
- The *precision* of PE is 0.005
- The *confidence* of PE is 0.05

- The design decision requirement is judged through graphical representation of BIP model or BIP code.

We subjectively omit the estimation of BSCU subsystems on ARCH4 due to the reuse of BSCU design between ARCH3 and ARCH4.

Here we give some architecture comparisons. The ARCH1 is a basic architecture, but is also the weakest. It is confirmed with a higher failure rate of the top level requirement "S18-WBS-R-0321" than the other three architectures. It also fails to pass the subsystem requirement (e.g., S18-BSCU-R-0002) for it is proposed for assessing architectures in later versions, not ARCH1. It performs well in verifying requirement S18-WBS-R-6108 for this requirement is a check with low level component which is not effected by the iteration of architecture.

We take ARCH2 and ARCH3 together for comparison. After the redundency of BSCU is added to ARCH2 and ARCH3, the two architectures reach the lower limit of probabilities of subsystem requirements considering the precisions. We also find the change from two independent BSCUs to a single BSCU with dual channels does not matter the top level failure rate too much.

The comparison of ARCH3 with ARCH4 is through an extra requirement S18-WBS-R-0323 which aims to detect an inadvertent wheel braking occurrence. In the limited estimation process, ARCH3 reports a higher rate of failure occurrence than ARCH4 for this is exactly what the design of ARCH3 lack of. Our experiments confirm the necessity of the iteration from ARCH3 to ARCH4. This iteration is also motivated by a design decision requirement S18-WBS-2975, which is only satisfied by the advanced ARCH4.

Due to the limitation of time and hardware, and the characteristics of PE algorithm, we can not conclude any estimation of probability less than 10^{-7} in order of magnitude. Thus we can not give an assertion for the requirements with some terminology (e.g., extremely remote) representing the lower bound of its rate.

7 Discussion and Future Work

It is hard to generate a formal WBS model by using the AIR6110 document without any background knowledge or help from engineers. Also, functional modelling and fault-tree-based safety analysis are often conducted separately. The construction of fault trees is an engineering practice which requires a deep understanding with the complex real-world system behavior. But the experienced engineers are always not able to learn the state-of-the-art formal notations or proof knowledge. The engineers need models with clear graphical representation and semantics. It is better for them to use a modelling method which supports the reuse of components because the architectures are always similar with each other during the system iteration.

The scalability of our approach depends on the fault trees. In the case study of this paper, we reuse the leaf nodes between different fault trees. With the increasing scale of system, the reuse of behavioral fault components will become more urgent.

The probability distribution of failure rate also matters the runtime verification. The bathtub curve is generally regarded as a particular form of the failure rate function. But the failure rate is generally assumed constant for complex units, systems and electronics, especially in aerospace standards. The exponential distribution is suitable for situations where the failure rate of occurrence is constant. But it is continuous and not a built-in function of SBIP. So we would like to use the discrete uniform distribution for runtime verification.

We want to use model checking as a formal method for verifying the correctness of modified BIP models. As a part of this work, the estimation of S18-WBS-R-6108 has proved the integration of green hydraulic pump with faulty behavior is correct. Nevertheless, we hope to reuse other model checking algorithms to do this verification, this may need an encoding from BIP to other languages (e.g., NuXmv).

Another direction is improving the simulation process to estimate the probability of rare event, i.e. failure rate of S18-WBS-R-0321. The requirements of most complex safety-critical systems can also be considered as the negation of safety properties which are rare events. A predictable method is using some state-of-the-art methods [27,37] to reduce the scale of sample traces.

8 Conclusion

This paper proposes a model of the wheel brake system based on behavioral fault modelling approach as well as a verification technique based on statistical model checking. Our design tightly follows the AIR6110 with a result of four architectures. As far as we know, this is the first BIP model for wheel brake system. Taking the advantages of nominal BIP model and stochastic systems, our extended BIP model is (1) more easier to extend/modify, (2) capable to retrieve stochastic fault information on the wheel brake system, (3) able to conduct MBSA in SBIP tool chains using SMC approach.

Acknowledgment. This work is supported by the NSFCs of China (No.61872144 and No. 61872146). The authors wish to thank: (1) Braham Lotfi Mediouni from RSD Grenoble Research Center for helpful discussion of SBIP, (2) Sam Procter from Software Engineering Institute of Carnegie Mellon University for helpful discussion of the AIR6110 wheel brake system case study.

References

1. Uniform Distribution (Continuous). https://www.mathworks.com/help/stats/uniform-distribution-continuous.html
2. Advisory Circulars (ACs) 25.1309-1A - System Design and Analysis (1988)
3. ARP4761 - Guidelines and methods for conducting the safety assessment process on civil airborne systems and equipment (1996)
4. ARP4754A - Guidelines for Development of Civil Aircraft and Systems (2010)
5. AIR6110 - Contiguous Aircraft/System Development Process Example (2011)

6. Aerospace Structures: Society of Automotive Engineers, Architecture Analysis and Design Language (AADL), Annex E: Error Model Annex (Annex Volume 1), April 2011
7. Akerlund, O., et al.: ISAAC, a framework for integrated safety analysis of functional, geometrical and human aspects, January 2007
8. Basu, A., et al.: Rigorous component-based system design using the BIP framework. IEEE Softw. **28**, 41–48 (2011)
9. Basu, A., Bozga, M., Sifakis, J.: Modeling heterogeneous real-time components in BIP. In: Fourth IEEE International Conference on Software Engineering and Formal Methods (SEFM 2006), pp. 3–12 (2006)
10. Basu, A., Bensalem, S., Bozga, M., Bourgos, P., Sifakis, J.: Rigorous system design: the BIP approach. In: Kotásek, Z., Bouda, J., Černá, I., Sekanina, L., Vojnar, T., Antoš, D. (eds.) MEMICS 2011. LNCS, vol. 7119, pp. 1–19. Springer, Heidelberg (2012). https://doi.org/10.1007/978-3-642-25929-6_1
11. Basu, A., Bensalem, S., Bozga, M., Caillaud, B., Delahaye, B., Legay, A.: Statistical abstraction and model-checking of large heterogeneous systems. In: Hatcliff, J., Zucca, E. (eds.) FMOODS/FORTE -2010. LNCS, vol. 6117, pp. 32–46. Springer, Heidelberg (2010). https://doi.org/10.1007/978-3-642-13464-7_4
12. Basu, A., Bensalem, S., Bozga, M., Delahaye, B., Legay, A., Sifakis, E.: Verification of an AFDX infrastructure using simulations and probabilities. In: Barringer, H., et al. (eds.) RV 2010. LNCS, vol. 6418, pp. 330–344. Springer, Heidelberg (2010). https://doi.org/10.1007/978-3-642-16612-9_25
13. Biehl, M., Chen, D.J., Törngren, M.: Integrating safety analysis into the model-based development toolchain of automotive embedded systems. ACM SIGPLAN Not. **45**, 125–132 (2010)
14. Bliudze, S., et al.: Formal verification of infinite-state BIP models. In: Finkbeiner, B., Pu, G., Zhang, L. (eds.) ATVA 2015. LNCS, vol. 9364, pp. 326–343. Springer, Cham (2015). https://doi.org/10.1007/978-3-319-24953-7_25
15. Bozzano, M., et al.: Formal design and safety analysis of AIR6110 wheel brake system. In: Kroening, D., Păsăreanu, C.S. (eds.) CAV 2015. LNCS, vol. 9206, pp. 518–535. Springer, Cham (2015). https://doi.org/10.1007/978-3-319-21690-4_36
16. Bozzano, M., et al.: ESACS: an integrated methodology for design and safety analysis of complex systems, June 2003
17. Brunel, J., et al.: Performing safety analyses with AADL and AltaRica. In: Bozzano, M., Papadopoulos, Y. (eds.) IMBSA 2017. LNCS, vol. 10437, pp. 67–81. Springer, Cham (2017). https://doi.org/10.1007/978-3-319-64119-5_5
18. Delange, J., Feiler, P.: Architecture fault modeling with the AADL error-model annex. In: 2014 40th EUROMICRO Conference on Software Engineering and Advanced Applications, pp. 361–368 (2014)
19. Ern, B., Nguyen, V.Y., Noll, T.: Characterization of failure effects on AADL models. In: Bitsch, F., Guiochet, J., Kaâniche, M. (eds.) SAFECOMP 2013. LNCS, vol. 8153, pp. 241–252. Springer, Heidelberg (2013). https://doi.org/10.1007/978-3-642-40793-2_22
20. Feiler, P., Rugina, A.: Dependability modeling with the architecture analysis and design language (AADL), January 2007
21. Haasl, D.F., Roberts, N.H., Vesely, W.E., Goldberg, F.F.: Fault Tree Handbook
22. Hérault, T., Lassaigne, R., Magniette, F., Peyronnet, S.: Approximate probabilistic model checking. In: Steffen, B., Levi, G. (eds.) VMCAI 2004. LNCS, vol. 2937, pp. 73–84. Springer, Heidelberg (2004). https://doi.org/10.1007/978-3-540-24622-0_8
23. Joshi, A., Miller, S.P., Whalen, M., Heimdahl, M.P.E.: A proposal for model-based safety analysis. In: 24th Digital Avionics Systems Conference, vol. 2, pp. 13 (2005)

24. Konnov, I., Kotek, T., Wang, Q., Veith, H., Bliudze, S., Sifakis, J.: Parameterized systems in BIP: design and model checking. In: Proceedings of the 27th International Conference on Concurrency Theory (CONCUR 2016), pp. 30–31. Schloss Dagstuhl-Leibniz-Zentrum fuer Informatik (2016)
25. Larsen, K.G., Legay, A.: Statistical model checking past, present, and future. In: Margaria, T., Steffen, B. (eds.) ISoLA 2014. LNCS, vol. 8803, pp. 135–142. Springer, Heidelberg (2014). https://doi.org/10.1007/978-3-662-45231-8_10
26. Lisagor, O., Kelly, T., Niu, R.: Model-based safety assessment: review of the discipline and its challenges. In: The Proceedings of 2011 9th International Conference on Reliability, Maintainability and Safety, pp. 625–632 (2011)
27. L'Ecuyer, P., Mandjes, M., Tuffin, B.: Rare Event Simulation using Monte Carlo Methods, pp. 17–38. Wiley, Hoboken (2009)
28. Mediouni, B.L., Nouri, A., Bozga, M., Dellabani, M., Legay, A., Bensalem, S.: $\mathcal{S}$BIP 2.0: statistical model checking stochastic real-time systems. In: Lahiri, S.K., Wang, C. (eds.) ATVA 2018. LNCS, vol. 11138, pp. 536–542. Springer, Cham (2018). https://doi.org/10.1007/978-3-030-01090-4_33
29. Mian, Z., Bottaci, L., Papadopoulos, Y., Mahmud, N.: Model transformation for analyzing dependability of AADL model by using HiP-HOPS. J. Syst. Softw. **151**, 258–282 (2019)
30. Papadopoulos, Y., Maruhn, M.: Model-based synthesis of fault trees from Matlab-Simulink models, pp. 77–82, August 2001
31. Paulitsch, M., Reiger, R., Strigini, L., Bloomfield, R.: Evidence-based security in aerospace: from safety to security and back again, pp. 21–22, November 2012
32. Sen, K., Viswanathan, M., Agha, G.: Statistical model checking of black-box probabilistic systems. In: Alur, R., Peled, D.A. (eds.) CAV 2004. LNCS, vol. 3114, pp. 202–215. Springer, Heidelberg (2004). https://doi.org/10.1007/978-3-540-27813-9_16
33. Wei, X., Dong, Y., Sun, P., Xiao, M.: Safety analysis of AADL models for grid cyber-physical systems via model checking of stochastic games. Electronics **8**(2), 212 (2019)
34. Wei, X., Dong, Y., Yang, M., Hu, N., Ye, H.: Hazard analysis for AADL model, pp. 1–10, August 2014
35. Younes, H.: Planning and verification for stochastic processes with asynchronous events, pp. 1001–1002, January 2004
36. Zhang, T., Jiang, Y., Ye, J., Jing, C., Qu, H.: An AADL model-based safety analysis method for flight control software. In: 2014 International Conference on Computational Intelligence and Communication Networks, pp. 1148–1152 (2014)
37. Zuliani, P., Baier, C., Clarke, E.: Rare-event verification for stochastic hybrid systems. In: HSCC 2012 - Proceedings of the 15th ACM International Conference on Hybrid Systems: Computation and Control, April 2012

H2P: A Novel Model to Study the Propagation of Modern Hybrid Worm in Hierarchical Networks

Tianbo Wang[1,2(✉)] and Chunhe Xia[2(✉)]

[1] School of Cyber Science and Technology, Beihang University, Beijing, China
wangtb@buaa.edu.cn

[2] Beijing Key Laboratory of Network Technology, Beihang University, Beijing, China
xch@buaa.edu.cn

Abstract. Network worms have become a critical security threat to the Internet in recent years. Due to highly non-uniform distribution of vulnerable hosts and the wide application of NAT technique, they restrain the spread of traditional scan-based worm to some extent. With the rapid growth of social software, worms can spread from physical networks to social networks. Though social worms fix scanning worm deficiencies, it cannot hit some hosts not in social networks because of the dependence on the topological structure. Modern hybrid worm combines the advantages of two types of worms. On the one hand, it makes use of *important scanning* which results from Important Sample in statistics that scan IP-address space. On the other hand, it exploits a *reinfection-trigger* mechanism in social networks. Because of the potential damages, it is important for us to gain a deep understanding of the propagation dynamics of modern hybrid worm. In this paper, we present an analytical **H**ierarchical-**Hy**brid **P**ropagation (H2P) model for characterizing worm propagation. We perform comprehensive theoretical analyses and experimental evaluation to validate our simulation model. The results show that our model presented in this paper achieves a greater accuracy in characterizing the propagation of modern hybrid worms.

Keywords: Hybrid worms · Hierarchical networks · Propagation dynamics · H2P model

1 Introduction

Hybrid worms and their variants have been a persistent network security threat in recent years, causing Internet services to become temporarily inaccessible, and leading to financial loss and privacy disclosure. For example, the Nimda worm in 2001 infected windows 95-XP hosts (workstations) through Email and direct

This work was supported by the National Natural Science Foundation of China (Grant No. U1636208, No. 61862008, No. 61902013.) and the Beihang Youth Top Talent Support Program (Grant No. YWF-20-BJ-J-1038.).

M. Qiu (Ed.): ICA3PP 2020, LNCS 12454, pp. 251–269, 2020.
https://doi.org/10.1007/978-3-030-60248-2_17

scanning methods. The Stuxnet worm in 2010 and the Flame worm in 2012 attacked a large number of computers by using USB device between acquaintances and exploiting some flaws to spread in local networks. According to Symantec Corporation's [1] report on official Internet security threats, the frequency and virulence of their propagation outbreaks have increased dramatically in recent years. For social networking layer, Email remains the medium of choice for cybercriminals, as criminals seek to leverage the trust people have in their social relationships to spread malicious codes. For physical networking layer, attackers focus on how to exploit any vulnerability to compromise hosts through efficient scanning strategies.

1.1 Motivation

Due to the combination epidemiology and social engineering, the propagation of modern hybrid worm has the characteristics of **interdisciplinarity** and **multidimension** [2]. In order to help researchers understand and predict potential damages of a worm, some models have been presented to study worm propagation dynamics. Valid propagation models can be used as testbeds to: 1) estimate the infection trend before a worm occurs in reality; 2) evaluate some countermeasure for restraining worm propagation. Recently, there exist some scan-based models [3–6] and topology-based models [7–10] to characterize and predict the infection trend of worms. However, they have considered only modeling propagation of scan-based worms and topology-based worms, respectively. Thus, previous models cannot characterize the propagation dynamics of hybrid worms in the hierarchical networks.

In this paper, in order to address the above-mentioned problems of characterizing hybrid worm propagation in hierarchical networks, we propose a **H**ierarchical-**H**ybrid **P**ropagation model for characterizing modern hybrid worm.

1.2 Contributions

The major contributions of this research are summarized as follows:

- We carry out extensive analysis on multi-behaviors of modern worms, *scanning propagation* and *social network propagation*, which affect the accuracy of current analytical models.
- We propose an analytical H2P model on the propagation of modern hybrid worm, which helps us to better understand modern hybrid worm propagation.
- We conduct a series of experiments to evaluate the accuracy of the proposed model. The results show that the H2P model is more accurate than the state-of-art models.

The rest of this paper is organized as follows: Related work are presented in Sect. 2. Section 3 states the problems in modeling the propagation dynamics of hybrid worms. In Sect. 4, we a novel H2P model for characterizing modern hybrid worm propagation. We implement a series of experiments to evaluate the correctness of the H2P model in Sect. 5. We conclude this paper in Sect. 6.

2 Related Work

In the last decade, there have been many research achievements and findings that have focused on two different types of worms: the scan-based worms and the topology-based worms. The former relies on infected hosts scanning vulnerable hosts with different scanning strategies, such as random, localized and selective scanning ones in two surveys [3, 11]. In recent years, researchers have found several new classes of scan-based worms. The presented model [12] characterizes important-scanning worm propagation taking advantage of a vulnerable-host distribution. The paper [4] investigates a camouflaging worm to intelligently manipulate its scan traffic volume over time. The paper [5] introduces a game-theoretical formulation to the spread of a self-disciplinary worm. The paper [6] proposes an accurate propagation model to comprehensively study the permutation-scanning worm.

The topology-based worms rely on a topological structure to infect their neighbors. With the development of social networks and engineering, scholars have new findings on social worm propagation. Chen et al. [7] present the Markov model incorporating both detailed information of topological structure and statistical dependence characteristic. An email worm model is discussed in [8] to account for the checking email behavior. In order to solve the two problems in previous models: temporal dynamics and spatial dependence, Wen et al. [9] propose the SII model. Meanwhile, [13] presents the reinfection process of modern email malware and overcomes the underestimation in previous works. Wang et al. [10, 14] present the SADI model to characterize the spread of social worms in hierarchical networks, and quantify the spreading ability.

However, scan-based models mainly focus on the differential continuous-time and the difference discrete-time equations. The former cannot characterize discrete propagation events, and the latter cannot describe random user behaviors. Although topology-based models adopt the stochastic discrete-time method can solve the problems above, these models pay attention to social worms and have an independent assumption. Finally, some researchers characterize some hybrid worms [15, 16]. However, they cannot analyze propagation dynamics and user behaviors from an analytical perspective.

3 Problem Statement

The distribution of vulnerable-hosts in the Internet is highly non-uniform over the IPv4 address space [17]. Meanwhile, traffic analysis data of Code Red II show 60% of infected hosts have private or dynamic IP addresses through NAT technique [18]. Importance-sampling estimates the vulnerable-host distribution by collecting infection information [12]. Due to the relative weakness of intranet defense, local-preference-based scanning can easily infect hosts in the same subnet. Thus, the combination of two methods (i.e. *enhanced combination-scanning*) is considered as a more effective scanning strategy by hackers, as shown in Sect. 4.1. However, due to the existence of NAT, infected hosts cannot scan

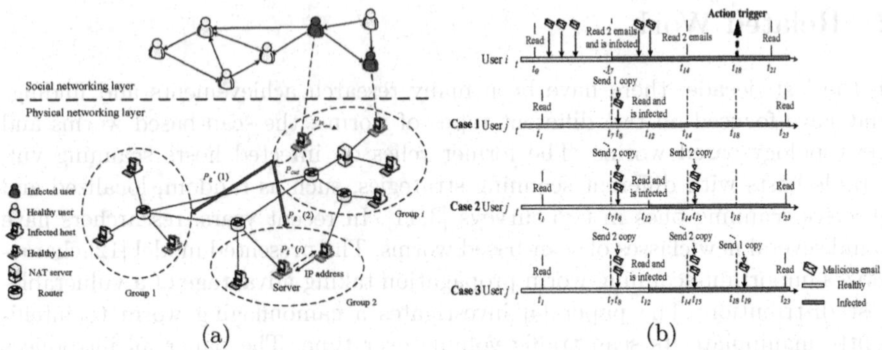

Fig. 1. The propagation scenario of modern hybrid worm. The left subfigure (a) illustrates the hierarchical network and the scanning strategy in the physical networking layer. The right subfigure (b) illustrates the topological propagation mechanism in the social networking layer. User i reads two malicious emails and another two at t_7 and at t_{14}, respectively. And then triggers (i.e. open the mailbox) t_{18}. Case 1: nonreinfection; Case 2: reinfection; Case 3: reinfection-trigger.

vulnerable hosts in different NAT subnets efficiently. Email is the most common spreading carrier for topology-based worms. And email-based method with one new mechanism, namely *reinfection-trigger*, is far more aggressive in spreading throughout social network [19]. However, they rely on the topological structure, and cannot infect other vulnerable hosts not in the social topology. Thus, the hybrid worm combines with the advantages of two kinds of worms to get more aggressive. A technical study will be introduced later. Meanwhile, these observations become the motivation of our work.

The propagation scenario of modern hybrid worm is shown in Fig. 1. Suppose that the network is composed of g groups. In the group i, the global routing subnet prefix is $/n_i$, the size of IP addresses is $\Omega_i = 2^{32-/n_i}$. Secondly, there are h_i NAT subnets, which have different subnet prefixes $/n_{j|i}$. And the size of IP addresses is $\Omega_{j|i} = 2^{32-/n_{j|i}}$, but these addresses are only routable in the NAT subnets. For the convenience of readers, we list major notations of this paper in Table 1. The two types of propagation behavior for hybrid worms are relatively independent by analyzing the propagation load of the mixed worm. Thus, we analyze propagation mechanisms of hybrid worms from different networking layers as follows:

Scan-based behavior in the physical networking, as shown in Fig. 1(a): (1) *learning stage*. Each infected host (client) performs uniform scanning. Once a susceptible host is infected, it reports its IP address to the worm sever. The worm server records these IP addresses until reaching required sample size, it estimates the group distribution $P_g(i)$ based on collected data, and sends the corresponding group scanning distribution $P_g^*(i)$ to all infected hosts on the list. (2) *scanning stage*. Upon receiving $P_g^*(i)$, the scanning strategy of a worm client is converted to enhanced combination-scanning. And then the infected host will

Table 1. Major notations used in this paper.

Symbol	Explanation
N	The number of susceptible hosts in the heterogeneous network
g	The number of groups in the Internet
N_i	The number of susceptible hosts in the group i
$N_{j\|i}$	The number of susceptible hosts in the NAT subnet j of group i
$N^t_{j\|i}$	The number of susceptible hosts in the NAT subnet j of group i at time t
Ω	The number of address space which the worm needs to scan in its propagation
Ω_i	In the group i, the size of address space which are reached by global routing
$\Omega_{j\|i}$	In the group i, the size of address space of the NAT subnet j
h_i	The number of the NAT subnets in the group i
s	The scanning rate of worm
p_{in}	The probability of scanning the intra-area network
p_{out}	The probability of scanning the inter-area network
$P_g(i)$	Group distribution: the actual percentage of vulnerable hosts in group i, i.e. $\frac{N_i}{N}$
$\hat{P_g}(i)$	The estimated percentage of vulnerable hosts in group i, i.e. $\frac{\hat{N_i}}{N}$
$P^*_g(i)$	The probability of the worm scan hitting group i
C_i	The number of worm clients' IP addresses from group i
C	The number of clients' IP addresses collected on the worm server
I_t	The number of infected hosts in the heterogeneous network at time t
I^t_i	In the group i, the number of infected hosts reached by global routing at time t
$I^t_{j\|i}$	The number of infected hosts in the NAT subnet j of group i at time t
p_{ij}	The propagation probability from user i to node j
$r(t)$	The recovery function of hosts, which provides the probability for any host to be immunized at time t
$open^t_i$	The event of user i checking the mailbox at time t
t'	The arbitrary time between user i last checking messages and the current time t (excluding t), which records the time when a new message comes to a social account
q_n	The probability of user i checking message notification at time t
NB_i	The neighbor set of user i in the social networking layer
$P_g(i, \Omega_i)$	The actual percentage of vulnerable hosts in Ω_i addresses of group i. i.e. $\frac{N_i}{\Omega_i}$
$P_g(\hat{i}, \Omega_i)$	The estimated percentage of vulnerable hosts in Ω_i addresses of group i. i.e. $\frac{\hat{N_i}}{\Omega_i}$

scan the intra-area (i.e. local) network with the probability of p_{in}, and select a susceptible host in the same $/n_{j|i}$ NAT subnet or $/n_i$ network randomly, and scan the inter-area (i.e. global) network with the probability of $p_{out} = 1 - p_{in}$ and select a susceptible host in the $/n_i$ network with the probability of $P^*_g(i)$.

Topology-based behavior in the social networking, as shown in Fig. 1(b): Reinfection indicates that a previous infected user can get infected again whenever the user reads a malicious message, while sending out a malicious copy. Action trigger indicates a user can send malicious emails to members of the contact list whenever the user opens a mailbox.

Suppose that a user i gets infected and sends malicious emails to user j. In case 1 of the non-reinfection, although user i reads two malicious emails at t_7, the infected user will send only one malicious copy to user j at t_7. The malicious copy arrives at user j at t_8. When user j checks the mailbox at t_{12} and reads the malicious email from user i, user j gets infected. User j will not receive any malicious emails from user i after t_8. Nevertheless, in case 2 of the reinfection, user j will receive two malicious emails from user i at t_8. Furthermore, after user j gets infected at t_{12}, when user i reads two malicious emails at t_{14}, user j will receive two malicious emails from user i at t_{15}. In case 3 of the action trigger (i.e. open the mailbox due to message checking period or message notification), user i has been infected at t_7. When the user opens the mailbox at t_{18}, a malicious copy will be sent to user j. Thus this maneuver can enhance the spreading ability of email worms.

4 Modeling Propagation Dynamics of the Hybrid Worm

In this section, **on the one hand**, we model the scanning strategy for the scan-based behavior in the Sect. 4.1 and the events of social user for the topology-based behavior in the Sect. 4.2, respectively. **On the other hand**, we propose a H2P propagation model for charaterizing modern hybrid worm propagation in the Sect. 4.3.

4.1 Calculating the Group Scanning Distribution for the Scan-Based Behavior

When a worm scans group i, Ω_i hosts in this group are targeted by that scan with the same likelihood. That is, a susceptible computer in group i has a probability of $\frac{1}{\Omega_i}$ to be hit by a worm scanning this group. Thus, a susceptible host h in the group i is hit with the probability

$$P_{hit}(h) = P_g^*(i) \cdot \frac{1}{\Omega_i} \tag{1}$$

The events of any susceptible host being hit are assumed to be independent, the number of scans required until the first scan hits an appointed susceptible computer in the group i, denoted by X_i, follows a geometric distribution

$$P(X_i = j) = P_{hit}(h) \cdot [1 - P_{hit}]^{j-1} \tag{2}$$

Then the expected number of scans required until this susceptible computer is hit is

$$E[X_i] = P_{hit}(h)^{-1} = \frac{\Omega_i}{P_g^*(i)} \tag{3}$$

Therefore, we randomly choose a susceptible computer in the Internet, the average number of scans required until the first scan hits this computer, denoted by Avr_h, is

$$Avr_h = \sum_{i=1}^{g} P_g(i) \cdot E[X_i] = \sum_{i=1}^{g} P_g(i) \cdot \frac{\Omega_i}{P_g^*(i)} \tag{4}$$

In order to find a good scanning strategy, we need to minimize the Avr_h

$$\begin{cases} \text{minimize } \sum_{i=1}^{g} P_g(i) \cdot \frac{\Omega_i}{P_g^*(i)} \\ \text{subject to } \sum_{i=1}^{g} P_g^*(i) = 1 \quad \text{and} \quad \sum_{i=1}^{g} P_g(i) = 1 \end{cases} \tag{5}$$

We obtain the *optimal group scanning distribution* $\hat{P_g^*}(i)$ with Lagrange Multipliers

$$\hat{P_g^*}(i) = \frac{\sqrt{\Omega_i \cdot P_g(i)}}{\sum_{j=1}^{g} \sqrt{\Omega_j \cdot P_g(j)}} \tag{6}$$

In the real world, N and N_i are unknown in advance. Thus, the vulnerable-host distribution $P_g(i)$ is unknown. In order to calculate $\hat{P_g^*}(i)$, we obtain an unbiased estimator [20] is

$$\hat{P_g}(i) = \frac{C_i}{C} \tag{7}$$

Thus, the enhanced combination-scanning strategy is to select target IP addresses as follows:

- The probability p_{in} to scan an address within its own "/n" prefix network;
- The probability $p_{out} \cdot \hat{P_g^*}(i)$ to scan an address within the group i of the inter-area network.

4.2 Modeling Events of Social User for the Topology-Based Behavior

We introduce an M by M square matrix to describe a social networking layer as follows:

$$\begin{pmatrix} p_{11} & \cdots & p_{1M} \\ \vdots & p_{ij} & \vdots \\ p_{M1} & \cdots & p_{MM} \end{pmatrix} p_{ij} \in [0, 1] \tag{8}$$

where p_{ij} represents the propagation probability from user i to user j. $p_{ij} = 0$ means that there is no connection between them.

Let random variable T^i_{check} denote the period of checking emails, and we define a random variable $open^t_i$. We have $open^t_i = 1$ if user i is checking emails at time t. Otherwise, $open^t_i = 0$.

$$P(open^t_i = 1) = \begin{cases} 0, & \text{otherwise} \\ 1, & t \bmod T^i_{check} = 0 \\ q_n, & t = t' \end{cases} \tag{9}$$

where $t = t'$ denotes once an email comes to a social account, the social application will pop up a notification for user i at time t.

On the other hand, users check new messages periodically. Thus, it is necessary to introduce variable t' to obtain the number of unread emails until current time t in Fig. 2.

$$\begin{cases} t - T^i_{check} \le t' < t, & \text{if } open^t_i = 1 \\ t - (t \bmod T^i_{check}) \le t' < t, & \text{otherwise} \end{cases} \tag{10}$$

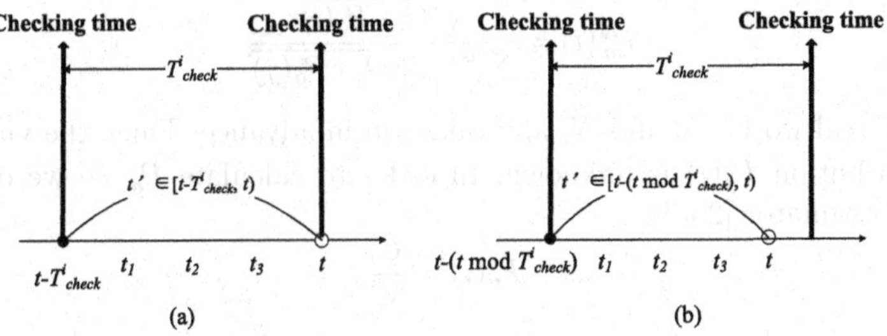

Fig. 2. Two cases of variable t': (a) User checks new emails at current time t; (b) User dose not check new emails at current time t.

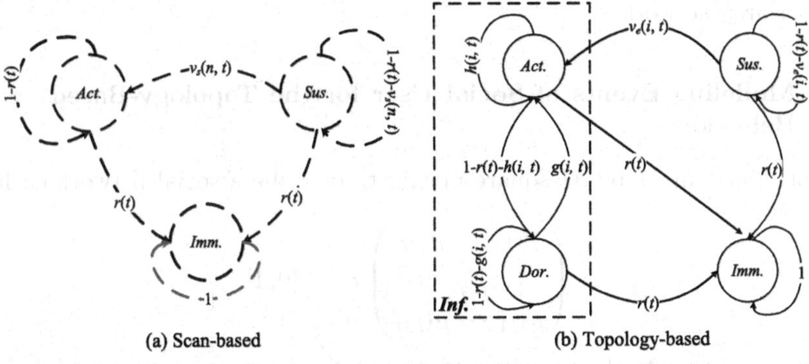

Fig. 3. State transition of a node in hierarchical networks. The left subfigure (a) illustrates state transition of a node for the scan-based behavior. The right subfigure (b) illustrates state transition of a node for the topology-based behavior.

4.3 H2P Propagation Model

We formalize the states of nodes and the hierarchical network information. Let random variable X^t_n denote the state of a host n at time t.

$$X^t_n = \begin{cases} \textbf{\textit{Hea.}}, \ \ healthy \begin{cases} \textbf{\textit{Sus.}}, & susceptible \\ \textbf{\textit{Imm.}}, & immune \end{cases} \\ \textbf{\textit{Inf.}}, \ \ infected \begin{cases} \textbf{\textit{Act.}}, & active \\ \textbf{\textit{Dor.}}, & dormant \end{cases} \end{cases} \tag{11}$$

As shown in Fig. 3, the dotted line and the solid line describe transition processes of the scan-based and the topology-based behaviors, respectively.

Calculating $n(t)$. Let the values 0 and 1 denote the healthy state and the infected state, respectively. Given a topology of the hierarchical network with N nodes, the expected number of infected nodes at time t, $n(t)$ can be calculated as in

$$\begin{aligned} n(t) &= E\left[\sum_{n=1}^{N} X_n^t\right] = \sum_{n=1}^{N} E\left[X_n^t\right] = \sum_{n=1}^{N} P\left(X_n^t = 1\right) \\ &= \sum_{n=1}^{N} P\left(X_n^t = Inf.\right) \end{aligned} \tag{12}$$

According to Fig. 3, we derive the computation of $P(X_n^t = Inf.)$ by the following equations:

$$P(X_n^t = Inf.) = v(n,t) \cdot P(X_n^{t-1} = Sus.) + (1 - r(t)) \cdot P(X_n^{t-1} = Inf.) \tag{13}$$

$$P(X_n^t = Sus.) = 1 - P(X_n^t = Inf.) - P(X_n^t = Imm.) \tag{14}$$

$$P(X_n^t = Imm.) = P(X_n^{t-1} = Imm.) + r(t) \cdot [1 - P(X_n^{t-1} = Imm.)] \tag{15}$$

Once we obtain the values of $v(n,t)$ and $r(t)$, the value of $P(X_n^t = Inf.)$ can be computed by the above Eqs. (13), (14) and (15).

Calculating $v(n,t)$. In the social networking layer, there are three preconditions for any user being by malicious emails: (1) the host has not been immunized; (2) the user is checking the mailbox; (3) the user reads those malicious emails, which are represented by $s(i,t)$.

$$\begin{aligned} \underbrace{v(n,t)}_{hybrid} &= 1 - (1 - \underbrace{v_e(i,t)}_{topology-based})(1 - \underbrace{v_s(n,t)}_{scan-based}) \\ v_e(i,t) &= s(i,t) \cdot P(open_i^t = 1) \cdot [1 - r(t)] \\ v_s(n,t) &= \{(1 - NAT_{in})[1 - (1 - \frac{1}{\Omega_i})^{(p_{out} \cdot \hat{P_g^*}(i) \cdot s \cdot I_{t-1} + p_{in} \cdot s \cdot I_i^{t-1})}] \\ &+ NAT_{in} \cdot [1 - (1 - \frac{1}{\Omega_i})^{p_{out} \cdot \hat{P_g^*}(j) \cdot s \cdot I_{j|i}^{t-1}} \cdot (1 - \frac{1}{\Omega_{j|i}})^{p_{in} \cdot s \cdot I_{j|i}^{t-1}}]\} \cdot [1 - r(t)] \end{aligned} \tag{16}$$

where NAT_{in} denotes whether the host is in a certain NAT subnet. $NAT_{in} = 1$ means that the host is in a certain NAT subnet. Otherwise, $NAT_{in} = 0$.

Then, we can compute the probability $s(i,t)$ of user i reading emails from an infected neighbor j, as in

$$s(i,t) = 1 - \prod_{j \in NB_i} [1 - p_{ji} \cdot P(X_j^{t'} = Act.|X_i^{t-1} = Sus.)] \tag{17}$$

Note: For the convenience of the next discussion, we use $\Gamma^{t_j|t_i}_{j=a|i=b}$ instead of $P(X^{t_j}_j = a|X^{t_i}_i = b)$, where $a, b \in \{Act., Sus., Dor, Inf.\}$

We disassemble the Eq. (17) by excluding $t-1$ from the range of value t', and obtain the iteration equation $s(i,t)$ as follows:

$$s(i,t) = 1 - (1 - s(i,t-1) \cdot [1 - P(open^{t-1}_i = 1)]) \cdot \prod_{j \in NB_i} (1 - p_{ji} \cdot \Gamma^{t-1|t-1}_{j=Act.|i=Sus.}) \tag{18}$$

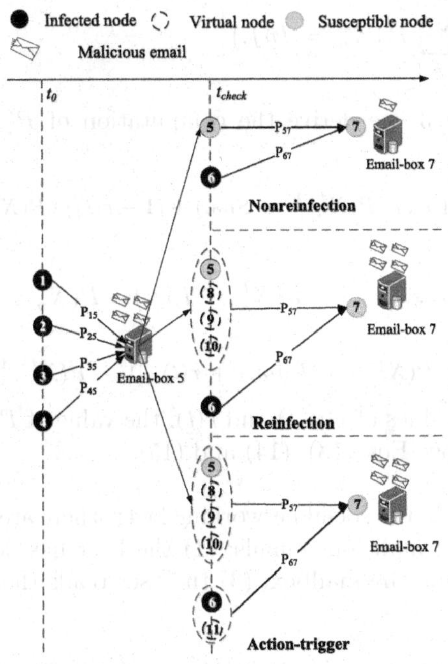

Fig. 4. The illustration of virtual nodes in the reinfection-trigger case.

Extending NB_i. In order to characterize the **reinfection-trigger** mechanism, we use virtual nodes to represent the kth infection event caused by reading kth malicious email or open the mailbox, as shown in Fig. 4. Nodes 1, 2, 3, and 4 send malicious emails to node 5. When node 5 reads those emails, the user gets infected. If node 5 reads three other emails, he will send triple email copies to node 7. The spreading process of extra email copies is *equivalent* to three virtual nodes (8, 9 and 10) sending emails to node 7. The virtual node 8 denotes the possible spreading if node 5 reads the second email. Virtual nodes 9 and 10 denote the similar spreading processes. Moreover, when infected node 6 opens his mailbox or some specific events are triggered, the node will send out one email copy to node 7. It is *equivalent* to a virtual node 11 sending one email to node 7.

Thus, we need to extend NB_i into a new neighbor set of node i, E_i, which includes three subsets: NB_i^N, NB_i^{VR} and NB_i^{VA}. And we revise the Eq. (18) as in

$$\begin{aligned} s(i,t) = 1 - (1 - s(i,t-1) \cdot [1 - P(open_i(t-1) = 1)]) \cdot \\ \prod_{j \in NB_i^N} (1 - p_{ji} \cdot \Gamma^{t-1|t-1}_{j=Act.|i=Sus.}) \cdot \\ \prod_{j \in NB_i^{VR}} (1 - p_{ji} \cdot \Gamma^{t-1|t-1}_{j=Act.|i=Sus.}) \cdot \\ \prod_{j \in NB_i^{VA}} (1 - p_{ji} \cdot \Gamma^{t-1|t-1}_{j=Act.|i=Sus.}) \end{aligned} \tag{19}$$

First, the subset NB_i^{VR} includes the virtual nodes which represent the extra infection event caused by users reading more than one email (i.e. NB_7^{VR} = nodes 8, 9 and 10). We introduce $\eta_{ji}(t)$ to denote the probability of user i being infected by user j at time t.

$$\begin{cases} \eta_{ji}(t) = \Gamma^{t'|t-1}_{j=Act.|i=Sus.} \cdot P(open_i(t-1) = 1) \cdot p_{ji} \cdot [1 - r(t)] \\ \overline{\eta_{ji}(t)} = 1 - \eta_{ji}(t) \end{cases} \tag{20}$$

Given that the set R_w^m consists of w elements which are all possibilities that we randomly select m elements from the set E_i. The value k represents the order of emails that each user may read, and the value k_m represents the maximal number of emails that each user may read. Thus, we can convert the reinfection process into a permutation problem. Then, we calculate the value of $P(X_j^{t-1} = Act.|X_i^{t-1} = Sus.)$ as follows:

$$\begin{cases} \Gamma^{t-1|t-1}_{j\_1=Act.|i=Sus.} = \Gamma^{t-1|t-1}_{j=Act.|i=Sus.}, k = 1 \\ \Gamma^{t-1|t-1}_{j\_k=Act.|i=Sus.} = \sum_k^{k_m} [\sum_{w=1}^{C^k_{\|E_i\|}} (\prod_{a \in R_w^k} \eta_{ai} \cdot \prod_{b \in E_i - R_w^k} \overline{\eta_{bi}})] \\ = \Big\{ \Gamma^{t-1|t-1}_{j\_(k-1)=Act.|i=Sus.} - [\prod_{j=1}^{k-1} \eta_{ji}(t) \prod_{j=k}^{\|E_i\|} \overline{\eta_{ji}(t)} + \cdots \\ + \prod_{j=1}^{\|E_i\|-k+1} \overline{\eta_{ji}(t)} \prod_{j=\|E_i\|-k+2}^{\|E_i\|} \eta_{ji}(t)] \Big\}, k \geq 2 \end{cases} \tag{21}$$

where $j \in NB_i^N, j\_k \in NB_i^{VR}$. If the user reads the kth email, The number of computing $\eta_{ki}(t)$ multiplication combinations is $\sum_k^{k_m} C^k_{\|E_i\|}$. And then if the user checks at most k_m emails, the computation for each node is totally $k_m \sum_k^{k_m} C^k_{\|E_i\|}$. Noticeably, it is computationally too expensive to obtain the result. It results in using Bernoulli approximation to calculate the average value $\xi_i(t) = \frac{1}{\|E_i\|} \cdot \sum_{k \in \|E_i\|} \eta_{ki}(t)$ for $\eta_{ki}(t), k \in \| E_i \|$, and we revise the Eq. (21) as follows:

$$\begin{cases} \Gamma^{t-1|t-1}_{j\_1=Act.|i=Sus.} = \Gamma^{t-1|t-1}_{j=Act.|i=Sus.}, k = 1 \\ \Gamma^{t-1|t-1}_{j\_k=Act.|i=Sus.} = \Big\{ \Gamma^{t-1|t-1}_{j\_(k-1)=Act.|i=Sus.} \\ -C^{k-1}_{\|E_i\|} [\xi_i(t)]^{k-1} \cdot [1 - \xi_i(t)]^{\|E_i\|-k+1} \Big\}, k \geq 2 \end{cases} \tag{22}$$

Second, the subset NB_i^N includes the real neighbors of user i (i.e. NB_7^N=nodes 5 and 6). Suppose that $\tau(t, C_{i\to j})$ as the beginning time of the propagation path. $\Psi_m(t)$ the probability for node j being infected by node i through the mth path $C_{i\to j}$, and $\delta(C_{i\to j})$ denotes the dependence probability of the propagation path. There are totally $NUM(l)$ l-hop $(1 \le l \le L)$ propagation path from node i to node j, and L is the maximum length of the propagation path.

$$\begin{aligned}
&\Gamma^{t-1|t-1}_{j=Act.|i=Sus.}\\
&= \left[1 - \frac{1 - v(j,t-1)}{\prod_{l=1}^{L}\prod_{m=1}^{NUM(l)}(1-\Psi_m(t-1))}\right] \cdot P(X_j^{t-2} = Sus.)\\
&= \left[1 - \frac{1 - v(j,t-1)}{\prod_{l=1}^{L}\prod_{m=1}^{NUM(l)}[1-\delta(C_{i\to j})P(X_i^{\tau(t,C_{i\to j})} = Act.]}\right] \cdot\\
&\quad P(X_j^{t-2} = Sus.)
\end{aligned} \tag{23}$$

Third, the subset NB_i^{VA} includes the virtual nodes which represent the extra infection event caused by infected nodes triggering certain actions (i.e. NB_7^{VA} = node 11). The random variable *action*(t) denotes whether the event is triggered at time t. If the user performs a certain action and the event are triggered, *action*$(t) = 1$. Otherwise, *action*$(t) = 0$. Assuming the event is triggered through the user opening the mailbox periodically, as in

$$P(action_j(t) = 1) = \begin{cases} 0, & \text{otherwise} \\ 1, & open_j(t) = 1 \end{cases} \tag{24}$$

And then we obtain

$$\begin{aligned}
\Gamma^{t-1|t-1}_{j=Act.|i=Sus.} &= P(action_j(t-1) = 1) \cdot \Gamma^{t-1|t-1}_{j=Inf.|i=Sus.}\\
&= P(action_j(t-1) = 1) \cdot [\Gamma^{t-1|t-1}_{j=Act.|i=Sus.} + \Gamma^{t-1|t-1}_{j=Dor.|i=Sus.}]
\end{aligned} \tag{25}$$

$$\begin{aligned}
&\Gamma^{t-n|t-n}_{j=Dor.|i=Sus.}\\
&= \Big\{[1 - r(t-n) - h(i,t-n)] \cdot \Gamma^{t-n-1|t-n-1}_{j=Act.|i=Sus.} +\\
&[1 - r(t-n) - g(i,t-n)] \cdot \Gamma^{t-n-1|t-n-1}_{j=Dor.|i=Sus.}\Big\} \cdot [1 - r(t-n)]
\end{aligned} \tag{26}$$

where $n = 1, 2, \cdots, t-1$, and $\Gamma^{0|0}_{j=Dor.|i=Sus.} = P(X_j^0 = Dor.)$

We consider $g(j,t)$ and $h(j,t)$ to be the probabilities of node j being infected at the dormant state and the active state, respectively. There are four preconditions: (1) the host is dormant or active; (2) the user is checking new messages; (3) the host is not recovered and immunized; (4) there are some neighbors in the active state at time t' , or there are some unread messages in the user j's mailbox at time $t-1$.

$$g(j,t) = \Gamma^{t|t-1}_{j=Act.|i=Dor.} = \Big\{[1 - \prod_{z\in E_j}(1 - P(X_z^{t'} = Act.)] + \sum_{k_m<m\leq\|E_i\|} P(X_{j\_m}^t = Act.)\Big\}\cdot P(X_j^{t-1} = Dor.)\cdot P(open_j(t) = 1)\cdot[1 - r(t)] \tag{27}$$

$$h(j,t) = \Gamma^{t|t-1}_{j=Act.|i=Act.} = \Big\{[1 - \prod_{z\in E_j}(1 - P(X_z^{t'} = Act.)] + \sum_{k_m<m\leq\|E_i\|} P(X_{j\_m}^t = Act.)\Big\}\cdot P(X_j^{t-1} = Act.)\cdot P(open_j(t) = 1)\cdot[1 - r(t)] \tag{28}$$

5 Model Validation

In order to validate the rationality and the correctness of our proposed model, according to existing simulation models [3,4,6,13,21], and [12], we developed a compatible propagation simulator using C++ and Matlab R2014a. We produce random numbers by C++ TR1 library extensions in experiments. The topologies of the hierarchical network adopted for evaluation include the social networking layer and the physical networking layer. The social network is the real-world topology (i.e. Facebook), which has 45,814 nodes. In the IPv4 physical network, we suppose that the vulnerable population $N = 50,000$ and the scan rate $\eta = 5$ [3]. There are 23 NAT subnets, which have 78,000 hosts. The final results are averaged by 100 runs of experimental results, and each run of the spread begins with two random infected nodes.

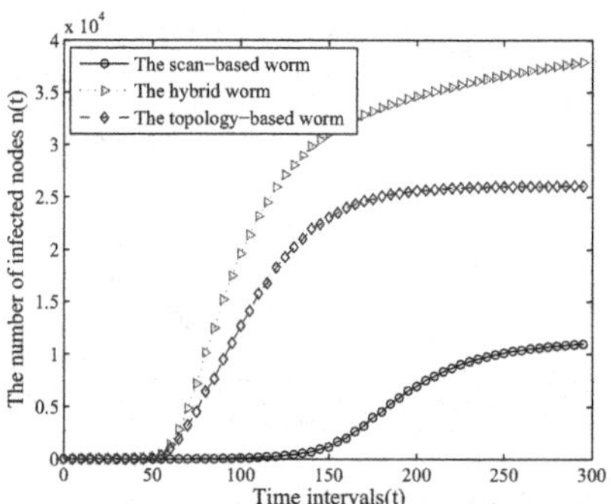

Fig. 5. The infection scale of different spreading behaviors. Checking time: $T^i_{check} \sim N(40, 20^2)$; The scan rate$\eta = 5$, $p_{in} = 0.7$, $p_{out} = 0.3$.

In the experimental procedure, we model the propagation of modern hybrid worm in discrete time. **Firstly**, we construct the hierarchical network consisting of physical networking layer and social networking layer. The former topology has global routing accessibility and NAT function, and the latter topology has the semi-directed and assortative properties based on the real-world topology. Each node in the social networking layer corresponds to one node in the physical networking layer. **Secondly**, we deploy related events of social behaviors for user simulation nodes, and deploy scanning strategies for host simulation nodes, respectively. A host stays in one of four states at any time: susceptible, active, dormant and immune. And then the state of any host is decided by state transition diagram. Each copy of the worm on an infectious host sends out a sequence of infection attempts during its lifetime. At each infection attempt, the worm infects some other hosts through scanning strategies and user checking malicious message. **Thirdly**, we take the infection scale data obtained in the simulation as the benchmark, and then evaluate the accuracy of our proposed model against the benchmark.

5.1 Variability in the Spreading Ability of the Hybrid Worm

To examine the spreading ability of the hybrid worm, we characterize the scan-based behavior, the topology-based behavior and the hybrid behavior, respectively. Figure 5 shows that the hybrid worm is more aggressive to spread in networks than the other two worms. There are two main reasons as follows: 1) Due to the NAT technique, the scan-based worm cannot hit other vulnerable hosts not in the same NAT subnet. 2) By social networks, the topology-based worm can infect vulnerable hosts in different NAt subnets. However, the limitation of the topological structure results in not infecting vulnerable hosts in the social network. Thus, modern hybrid worm combines with their advantages, and has a greater potential threat.

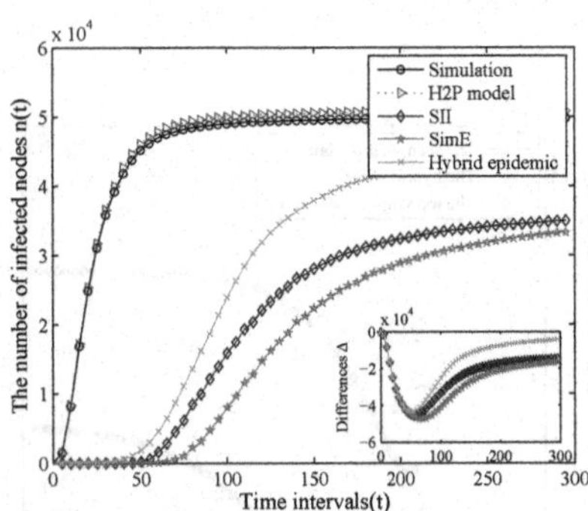

Fig. 6. The infection scale of different spreading behaviors. Checking time: $T^i_{check} \sim N(40, 20^2)$; The scan rate$\eta = 5$, $p_{in} = 0.7$, $p_{out} = 0.3$.

5.2 Comparison with Previous Models

To evaluate the accuracy of our H2P model, we conduct a series of experiments. Firstly, in order to exclude the impact of the recovery process on worm propagation, the experiments are carried out without the recovery process ($r(t) = 0$)

in this section. We let the infection probability p_{ij} and the checking time T^i_{check} follow Gaussian distributions $N(0.5, 0.2^2)$ and $N(40, 20^2)$, respectively. These parameter settings come from previous works of [3,8,13] and [15].

The inset of Fig. 6 shows that other models except the H2P model result in underestimating the infection scale. The main reason is that none of them can characterize the hybrid behaviors based on scanning strategy and social networks. Figure 6 shows that our H2P model can characterize the propagation dynamics of the hybrid worm more accurately than previous models [8,13,15].

5.3 Evaluating the Performance of Our H2P Model

We also evaluate the impact of various parameters on the performance of the modeling.

Impact of Different Propagation Mechanisms in the Modeling. According to the above statements in Sect. 3, we mainly analyze how different propagation mechanisms influence the worm spread. The event of opening message-box is associated with the period T^i_{check} of user i checking messages, and then the event of message notification is associated with the probability q_n of checking message notification. As shown in Fig. 7, we can see that message notification can spread faster than others, because a user will check the message as long as new messages come into a message-box. Compared with message notification, the reason of spreading slower of opening message-box mechanism is that this event is only triggered at the time of checking messages even if there is a new message in the message-box. However, it spreads faster than reinfection, because no matter whether there are malicious messages, the spreading process can be triggered whenever a user checks a message-box in the infected host. Compared with nonreinfection, the reinfection can spread faster. The reason is that when a user reads more than one malicious message, it will not send only one malicious copy like nonreinfection.

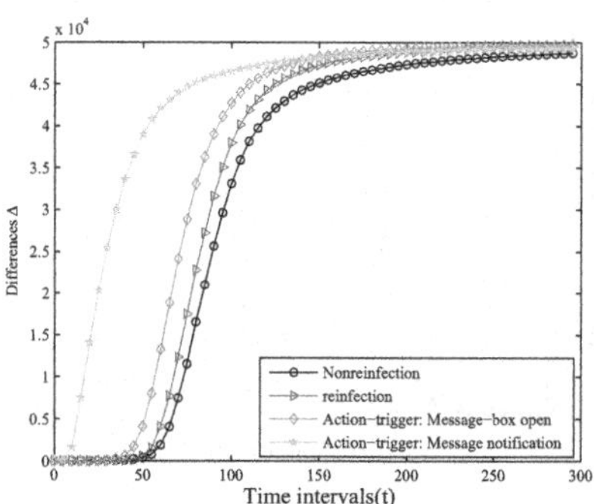

Fig. 7. Impact of different propagation mechanisms. Checking time: $T^i_{check} \sim N(40, 20^2)$; $q_n \sim N(0.5, 0.2^2)$; The scan rate$\eta = 5$, $p_{in} = 0.7$, $p_{out} = 0.3$.

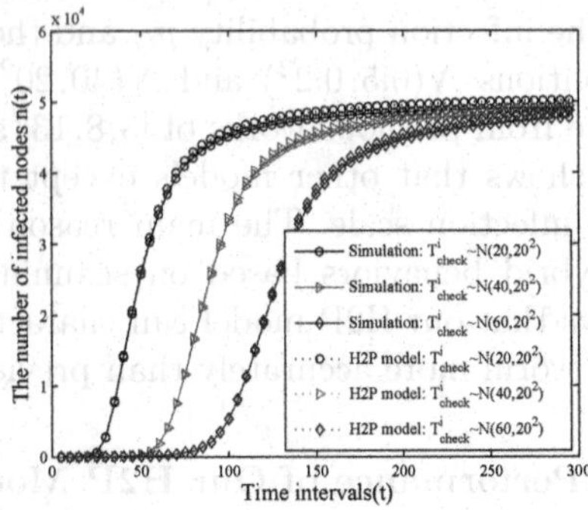

Fig. 8. The accuracy with different values of T^i_{check}.

Impact of Checking Time. T^i_{check} We evaluate the effects of the parameter T^i_{check}. Figure 8 shows that the worm spreads faster with the decrease of T^i_{check}. The less T^i_{check} indicates that users check social accounts more frequently, and then it results in the worm infecting their hosts with higher probability.

Impact of Users' Vigilance. Modern malware infects users' hosts when users check their social accounts and read their malicious messages. Users' vigilance determines how many malicious messages are read by the users. Higher user's vigilance means less malware emails are read. Thus, the vigilance of users determines the number of virtual nodes for each user in the modeling. We introduce the variable k_m to denote the maximal number of each user reading malicious messages, and then analyze the effect of users' vigilance quantitatively. Figure 9 shows that the differences (Δ) are large with varied values of k_m. the difference reaches about 15,000 at maximum. The results show that lower users' vigilance will speed up the outbreak of worms in the real world.

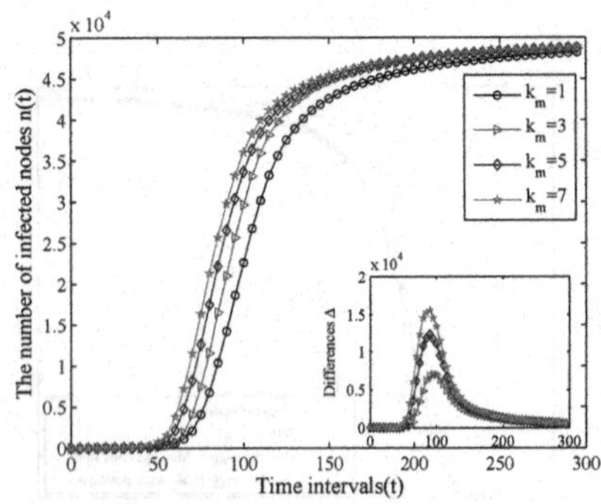

Fig. 9. The effect with different values of k_m.

5.4 Real-World Samples of Network

In the real world, the spread of most worms is typically impossible to be tracked directly by researchers. Meanwhile, hybrid worms refer to users' privacy. Thus, it should be pointed out that there is no real data of hybrid worms available for the evaluation of our proposed model. However, in order to justify rationality and correctness of our model, we use some similar realistic data to verify our proposed model.

Microsoft IIS is one of the most widely deployed Web servers throughout the world. Symantec has documented several high-severity vulnerabilities affecting it. Their characteristics render these vulnerabilities attractive targets for future blended threats. These applications with Microsoft IIS vulnerabilities become the target of a highly destructive malicious code "Nimda".

The date of Nimda worm is obtained by CAIDA [22]. On Sept. 18th incidents.org and its partner organization DShield.org received a huge number of reports of increased HTTP probing and network slowdowns. Further, the activity jumped dramatically at approximately 13:00 GMT and then proceeded to taper off in the following hours. The histogram in Fig. 10 shows the hourly breakdown for number of unique sources generating the activity on the same day. This worm mainly has Email and local preference scanning methods of transferring itself to different computers and networks. Thus, we revise the Eq. (16), as in

$$v(n,t) = 1-(1-v_e(i,t))(1-v_s(n,t)) = 1-1-s(i,t)\cdot P(open_i^t=1)\cdot[1-r(t)]\} \cdot \{1-\{[1-(1-\frac{1}{\Omega})^{p_{out}\cdot s\cdot I_{t-1}}\cdot(1-\frac{1}{\Omega_i})^{p_{in}\cdot s\cdot I_i^{t-1}}] \tag{29}$$

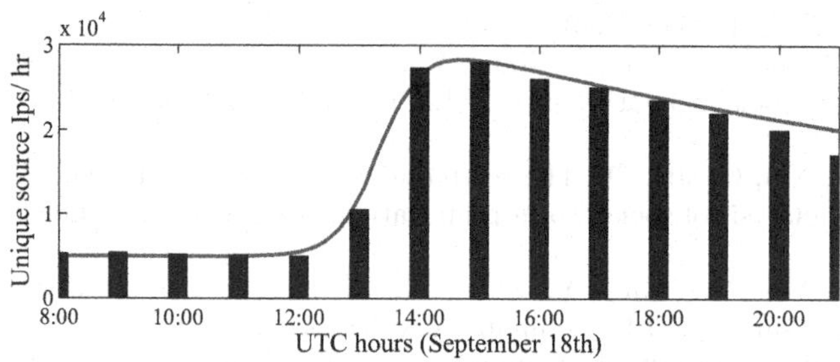

Fig. 10. Hourly distinct source IPs probing port 80 on Sept. 18th.

Figure 10 shows that our model can also characterize multi-behavior propagation of the hybrid worm very well, and verify the rationality of our H2P model.

6 Conclusion

In this paper, based on the problem statement and analysis, a novel H2P model is proposed for the propagation of modern hybrid worm. This model can address two critical processes: the important scanning and reinfection-trigger. The experimental results show that our model can more comprehensively and accurately characterize the hybrid propagation behavior. The work in this paper is not only a guide to understand propagation mechanism of hybrid worms, but also an important step towards the more practical and optimal strategies of restraining the spread of malicious codes in hierarchical networks.

References

1. Haley, K., Johnson, N., Fulton, J.: Symantec internet security threat report 2015. Symantec Corporation, Technical report, April 2016
2. Egele, M., Scholte, T., Kirda, E., Kruegel, C.: A survey on automated dynamic malware-analysis techniques and tools. ACM Comput. Surv. (CSUR) **44**(2), 6–54 (2012)
3. Zou, C.C., Towsley, D., Gong, W.: On the performance of internet worm scanning strategies. Perform. Eval. **63**(7), 700–723 (2006)
4. Yu, W., Wang, X., Calyam, P., Xuan, D., Zhao, W.: Modeling and detection of camouflaging worm. IEEE Trans. Dependable Secure Comput. **8**(3), 377–390 (2011)
5. Yu, W., Zhang, N., Fu, X., Zhao, W.: Self-disciplinary worms and countermeasures: modeling and analysis. IEEE Trans. Parallel Distrib. Syst. **21**(10), 1501–1514 (2010)
6. Manna, P.K., Chen, S., Ranka, S.: Inside the permutation-scanning worms: propagation modeling and analysis. IEEE/ACM Trans. Networking **18**(3), 858–870 (2010)
7. Chen, Z., Ji, C.: Spatial-temporal modeling of malware propagation in networks. IEEE Trans. Neural Networks **16**(5), 1291–1303 (2005)
8. Zou, C.C., Towsley, D., Gong, W.: Modeling and simulation study of the propagation and defense of internet e-mail worms. IEEE Trans. Dependable Secure Comput. **4**(2), 105–118 (2007)
9. Wen, S., Zhou, W., Zhang, J., Xiang, Y., Zhou, W., Jia, W.: Modeling propagation dynamics of social network worms. IEEE Trans. Parallel Distrib. Syst. **24**(8), 1633–1643 (2013)
10. Wang, T., Xia, C., Jia, Q.: The temporal characteristic of human mobility: modeling and analysis of social worm propagation. IEEE Commun. Lett. **19**(7), 1169–1172 (2015)
11. Wang, Y., Wen, S., Xiang, Y., Zhou, W.: Modeling the propagation of worms in networks: a survey. IEEE Commun. Surv. Tutorials **16**(2), 942–960 (2014)
12. Chen, Z., Ji, C.: Importance-scanning worm using vulnerable-host distribution. In: GLOBECOM 2005, IEEE Global Telecommunications Conference, vol. 3, pp. 6. IEEE (2005)
13. Wen, S., et al.: Modeling and analysis on the propagation dynamics of modern email malware. IEEE Trans. Dependable Secure Comput. **11**(4), 361–374 (2014)
14. Wang, T., Xia, C., Wen, S., Xue, H., Xiang, Y., Tu, S.: SADI: a novel model to study the propagation of social worms in hierarchical networks. IEEE Trans. Dependable Secure Comput. **16**, 142–155 (2015)
15. Zhang, C., Zhou, S., Chain, B.M.: Hybrid epidemics a case study on computer worm conficker. PloS One **10**(5), e0127478 (2015)
16. Liu, Q., Jing, J., Wang, Y.: An improved method of hybrid worm simulation. In: The Ninth International Conference on Web-Age Information Management, WAIM 2008, pp. 612–618 . IEEE (2008)

17. Moore, D., Paxson, V., Savage, S., Shannon, C., Staniford, S., Weaver, N.: Inside the slammer worm. IEEE Secur. Priv. **1**(4), 33–39 (2003)
18. Abu Rajab, M., Monrose, F., Terzis, A.: On the impact of dynamic addressing on malware propagation. In: Proceedings of the 4th ACM Workshop on Recurring Malcode, pp. 51–56. ACM (2006)
19. Serazzi, G., Zanero, S.: Computer virus propagation models. In: Calzarossa, M.C., Gelenbe, E. (eds.) MASCOTS 2003. LNCS, vol. 2965, pp. 26–50. Springer, Heidelberg (2004). https://doi.org/10.1007/978-3-540-24663-3_2
20. Xing, S., Paris, B.-P.: Measuring the size of the internet via importance sampling. IEEE J. Sel. Areas Commun. **21**(6), 922–933 (2003)
21. Yan, G., Chen, G., Eidenbenz, S., Li, N.: Malware propagation in online social networks: nature, dynamics, and defense implications. In: Proceedings of the 6th ACM Symposium on Information, Computer and Communications Security, pp. 196–206. ACM (2011)
22. Mooree, D., Shannon, C.: The nimda worm (2001). http://www.caida.org/data/

Design of Six-Rotor Drone Based on Target Detection for Intelligent Agriculture

Chenyang Liao(✉), Jiahao Huang, Fangkai Zhou, and Yang Lin

Jingdezhen Ceramic Institute, Jingdezhen, China
117060100218@stu.jci.edu.cn

Abstract. At present, most of the agricultural drones on market coverage spray use completed garment medicine technology, simply for reducing the manual labor, which is a big waste of pesticide, and do harm to the environment deeply. So, to improve the efficiency of the usage of the precision of pesticide spraying on the crops, and improve the efficiency of plant protection, a kind of agricultural drone is purposed based on artificial intelligence. A raspberry with a trained deep learning model could identify the area of plants with diseases through an intelligent camera, and control the drone hovering and spraying pesticide. Experimental results show that the proposed drone have good grades in identifying the area of plants with diseases and automatic spraying.

Keywords: Agricultural drone · Smart farming · Target detection · Fault-tolerant mechanism · Flight strategy optimization · Precision agriculture · Six-rotor control

1 Introduction

Since the concept of agriculture 4.0 was put forward, our country gradually put forward the requirements of agricultural precision. The traditional agricultural plant protection method can not meet the requirements of modern agriculture. The traditional low-efficiency pesticide spraying method not only wastes human resources and financial resources, but also has a huge impact on the environment. The development of new agricultural plant protection has become an inevitable trend.

Although Chinese agricultural drone technology started late compared with the United States, Canada, Australia and other developed agricultural countries, it developed rapidly. A group of drone manufacturers represented by DJI, Jifei and Yihang win their own place in the global drone market, mastering the core technology of independent intellectual property rights. In this paper, a new type of six-rotor intelligent agricultural drone is proposed, which uses the most popular artificial intelligence technology at present, and adopts the deep learning models for the identification of crops with diseases that developed and trained by ourselves. The whole machine was designed to be disassembled quickly, and the six-rotor model was selected for the balance of efficiency and power, the power system was driven by electricity, and the apm2.8 open source flight control board was used for secondary development.

M. Qiu (Ed.): ICA3PP 2020, LNCS 12454, pp. 270–281, 2020.
https://doi.org/10.1007/978-3-030-60248-2_18

Our research is aimed at promoting the development of modern precision agriculture, providing a precision solution led by UAV for modern agriculture, effectively reducing the cost of agricultural plant protection, reducing the environmental pollution caused by agricultural plant protection, and improving the efficiency of agricultural plant protection.

The agricultural drones researched in this project have the advantages of adapting to multiple terrains and targeting a variety of crops. They can be applied not only to plateaus and mountainous areas, but also to plains and hills, which can contribute to made in China 2025 and agricultural 4.0 eras.

The next jobs, we will discuss the rationality and advancement of the agricultural drone researched in this topic from flight control principles, overall design, experimental results and three important related works of APM flight control, target detection, flight control.

2 Related Work

2.1 APM Development

The flight control platform is mainly realized by the APM module. APM is a flight controller with signal processor. The core processor is a 8 bit ATMEGA2560 microprocessor with powerful computing capabilities and enhanced advanced RISC architecture and can be operated in full static. By executing powerful instructions in a single clock cycle, the ATMEGA2560 can achieve a throughput of nearly 1MIPS/Hz, with three embedded interface types of SPI, USART, I2C, and its CPU processing speed is 16 MHz, which is able to optimize energy consumption for processing speed.

Through a Micro USB cable or a set of wireless data transmission connection, you can set and download the program to the APM with a mouse click, without programming knowledge and other hardware devices such as download cable. APM could realize automatic take-off, automatic landing, waypoint flight, automatic return and various autopilot performances.

2.2 Target Detection

The drone's Hawkeye "firefly" 6s–4k high-definition motion camera can adapt to various climates and environments, capture crop images in real time and transmit them to the raspberry Pi (Fig. 1).

Raspberry Pi receives image data, and carries out accurate recognition and analysis through the trained target detection model. The target detection model uses the migration learning algorithm based on convolutional neural network, and the pre training model is Inception V3. The data we collected are enhanced to be suitable for convolutional neural network algorithm, and GPU is used to accelerate the training.

Through the target detection model, the drone can accurately identify diseased crops, guide the flight control system to the target point, and direct the drone spray system or the ground pipeline system to process (Fig. 2).

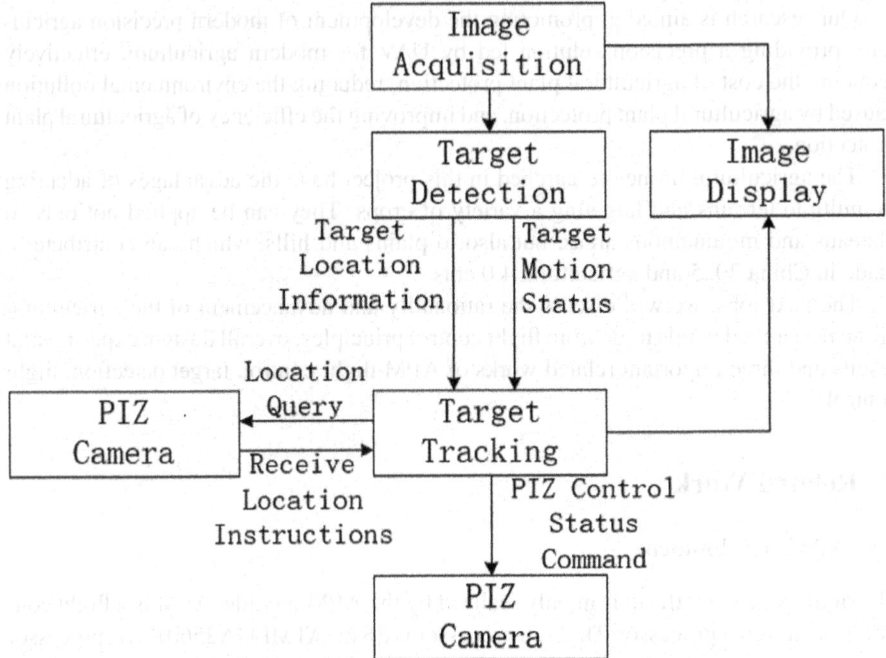

Fig. 1. Design of target detection

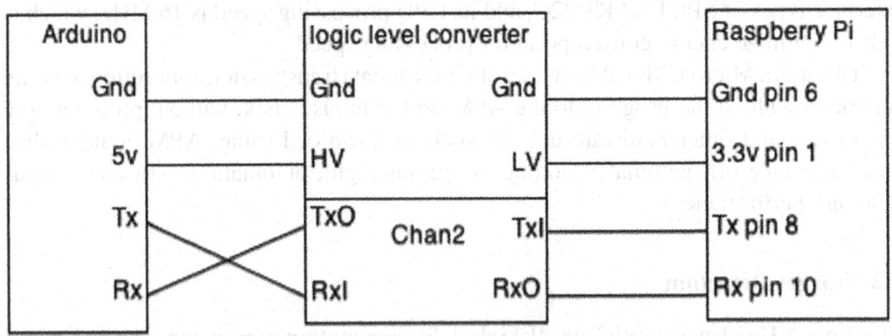

Fig. 2. Communication between Arduino and Raspberry Pi

2.3 Flight Control

The drone is susceptible to various external factors during its flight. The control system obtains the attitude information of the drone through serious of sensors, and the PID control algorithm parses the rotation speed of the motor, and sends it to the Electronic Speed Control through the I2C port to control the flight attitude. The cascade PID control adopts two-level PID of angle and angular velocity, with strong stability, accuracy and rapidity. The motor is controlled by a double loop, the main measuring unit and the main controller form the closed loop of the outer ring, and the auxiliary measuring unit and the sub controller form the closed loop of the inner ring. Cascade control uses a measuring unit and two feedback loops to form a closed loop to acquire and overcome system interference, correct errors to maintain a steady state.

When the disturbance attitude angle of the drone does not change, the main controller cannot predict the angle error of the system, but the measurement unit of the sub-controller can sense the change in angular velocity and can feed back to the sub-controller, the secondary controller performs PID control on the angular velocity error to obtain an output, which is handed over to the actuator to enable the drone to quickly eliminate the error and maintain a steady state. The main controller obtains a desired posture angle. The posture expectation minus the measured value of the main measurement unit is the input error of the main controller. The main control controller obtains the expected angular velocity through PID control.

3 Method Principle

3.1 Design of Fault Tolerant Flight Control Algorithm

The kinematic model of a six-rotor drone can be obtained from the literature:

$$\begin{cases} S = \vec{v} \\ \\ \vec{v} = gz_y - R\frac{N_s Z_b}{m} \end{cases} \tag{1}$$

S represents the position vector. *v* represents the speed vector. *Ns* represents the total lift. *R* represents the transformation matrix between the body coordinate system $o_j x_j y_j z_j$ and the ground inertial coordinate system $o_d x_d y_d z_d$, *m* represents the mass, *g* represents the gravitational acceleration. z_j, z_d represents the unit vector of $o_j x_j y_j z_j$ and $o_d x_d y_d z_d$.

$$\begin{cases} \xi = K\Omega \\ J\Omega = -\Omega \times J\Omega + LU \end{cases} \tag{2}$$

ξ represent the Euler angle of the drone, Ω represents the angular velocity vector around the center of mass, *J* represents the inertia matrix, *K* represents the transformation matrix between angular rate around the body axis and Euler angular rate. *L* represents equivalent force arm of corresponding rolling, pitching and yaw channels, according to the structure of different models, the value of L is different. U_c represents the virtual control variables, which show the equal effect of rolling, pitching, and yaw channel

$$\begin{cases} U_c = NF \\ F = MU_c \end{cases} \tag{3}$$

F represents the actual input of the drone, which show the lift produced by the six brushless motor. $N \in R^{3\times 6}$ represents the transformation matrix of virtual control quantity synthesized by actual input quantity, which is decided by the mechanical structure of the drone. $M \in R^{6\times 3}$ represents the transformation matrix of virtual control quantity allocated to actual input quantity, which is the generalized inverse matrix of *N*, and satisfy $NM = I_3$.

3.2 Design of Control Law

Using sliding mode control algorithm to design proper sliding surface, get the switching function s(x) and control the input U_c, and define the error vector.

$$\xi_e = \xi_d - \xi \tag{4}$$

The design of sliding mode:

$$h = \xi_e + k\xi_e \tag{5}$$

ξ_d represents the Euler angle to be reached. TO make the system be: $h \to 0$, $\xi_e \to 0$, we could define d as a positive constant matrix, selecting law of approach.

$$h = -\varepsilon\,\mathrm{sat}(h), (h > 0) \tag{6}$$

The sat(h) is saturation forcing number

$$\mathrm{sat}(h) = \begin{cases} 1 & h > \sigma \\ \frac{h}{\sigma} & -\sigma < h < \sigma \\ -1 & h < -\sigma \end{cases} \tag{7}$$

The value of σ decides the thickness of boundary layer, and 1/σ decide how fast the saturated fault number changes in the boundary layer. According to (4) (5) (6):

$$\begin{aligned} h =& \xi_e + k\xi_e = (\xi_d - \xi) + k(\xi_d - \xi) \\ =& -\varepsilon\,\mathrm{sat}(h) \end{aligned} \tag{8}$$

Then we could calculate U_c;

$$U_c = L^{-1}\left[JW^{-1}(\xi_d + k(\xi_d - \xi)) + JW^{-1}(\varepsilon\,\mathrm{sat}(h) + ch) + (\Omega \times J\Omega)\right] \tag{9}$$

In order to ensure that the control law can achieve the same asymptotical stability of the flight control system, Lyapunov function is selected;

$$V = \frac{1}{2}s^T J\mathrm{sat}(h) \tag{10}$$

Because of ε and J are positive constant matrix, we could know that: $\dot{V} \le 0$.

According to Lyapunov stability theorem, when $h \to 0$, the state of the system reaches the sliding surface finally, and slides on the sliding surface to achieve global asymptotic stability, $\xi_e \to 0$, ξ keep tracking ξ_d.

Among all the failures of six rotor drone, the failure of actuator (brushless, motor, rotor) is common. Especially in the process of cruise with electricity, once the brushless motor is jammed, the common flight control program can't detect the jammed fault of the motor in time, which will output the wrong control command, resulting in the output confusion of the drone, which will lead to the drone rollover, crash and other accidents. What's more, once a failed drone collides with a transmission line or transmission tower, it will often lead to secondary disasters in the power grid, resulting in serious power production safety accidents.

By redesigning the electronic governor, we can continuously monitor the operation state of the brushless motor and obtain the speed information of the brushless motor Once the brushless motor is detected to be stuck, it will change from the normal control mode to the fault control mode. Without changing the original control law, by adjusting the control distribution matrix, the fault-tolerant flight of the drone under the fault can be realized.

Take No. 1 motor stuck as an example, after the fault occurred:

$$\begin{cases} U_c = T^f F^f \\ F^f = M^f U_c \end{cases} \tag{11}$$

Because No. 1 motor can no longer provide enough lift, we could make $T_f = TQ$, T is the transformation matrix of the actual input synthesized virtual control quantity under the normal condition of all the brushless motors, and Q is the weight coefficient of the motor, $Q = \left(a_y\right)_{6\times 6}$, and

$$a_y = \begin{cases} 1\; i = j \; and \; i \neq 1 \\ 0\; i \neq j \; or \; i = 1 \end{cases} \tag{12}$$

According the value of T^f, we could find M^f, the generalized inverse matrix of T^f, and make $T^f M^f = I_3$. The physical significance of M^f is to eliminate the influence of the fault on the original control law by not taking the failed motor 1 into account the virtual control quantity. Flight control program get the result of M^f according to calculation, decomposing virtual control quantity U_c to get actual input quantity T^f.

4 Experiments and Evaluation

4.1 Design of the Framework

The main components of six-rotor drone are: drone framework, flight control system, navigation and positioning system, image processing system, automatic obstacle avoidance system. The drone system could be separated to airborne part and ground station, airborne part is in charge of drone attitude control, ground station is in charge of mission planning and video processing (Fig. 3).

The drone has six rotors, and it's of foldable structure, which is able to be put into a case and easy to carry out. Its spraying system is in the center of the framework, with two hoses extending from the center. The drone also has shields around the rotors to protect it from being destroyed. The drone's flight control system is APM2.8, which use ATMEGA2560 main processor. The central processor is the core module of the control system of the drone, which is often referred to as MCU. The attitude angle of the drone in the pitch channel is the control object of the cascade PID controller, and its actuator is composed of six brushless motors. In the application of cascade PID, the attitude angle of drone is the input information of outer loop angle control system, while the attitude angle speed is the input information of inner loop angle speed control system. It is mainly used

Fig. 3. Drone framework design

to achieve the collection of sensor information, real-time calculation of body attitude angle, transmission of flight data, control of motor speed. The main control core captures the PWM signal on the remote controller, decodes it, and then obtains the required target attitude. The flight control system has three sensors: three axis gyroscope sensor, triaxial accelerometer sensor, triaxial magnetoresistance sensor. The three sensors are connected with DSP. The drone uses 40A Electronic Speed Control and 2212 brushless motor to ensure the power of flight, they connect with the main processor through PWM.

There are some other sensors being used to check the condition of the drone: optical flow sensor, air pressure sensor, ultrasonic sensor and GPS. The NEO-M8N GPS module is used to obtain the latitude and longitude data. The NEO-M8N GPS module includes the HMC5883L digital compass, which keeps low power consumption, has high sensitivity and excellent receiving ability. The height of the drone is measured by the MS5611 air pressure sensor, MS5611 air pressure sensor is a new generation high-resolution sensor with SPI and I2C bus interface. 3DR wireless receiver is a data transmission module of APM autopilot. The data transmission module receives a frequency of 433 MHz, and the effective distance is about 500 m. Plug in the computer with a USB cable and connect to the ground station after debugging. The wireless receiver can transmit flight data to the ground station in real time, which is beneficial to monitor and debug the drone. The image processing system is formed of five parts: PTZ controller, camera, image transmission system transmitter, wireless receiver and communication module. The PTZ controller and wireless receiver connect with main processor through PWM, and the communication module connect with main processor through UART (Fig. 4).

The ground station is formed of five parts: control handle, remote control, communication module, image transmission system receiver and computer. The control handle or the remote control could control the drone through the 2.4 G wireless receiver, the drone could also be controlled by mobile phone or computer. Mission planning, air parameter adjustment, video display, voice synthesis, and viewing flight records can all be realized in the ground station (Fig. 5).

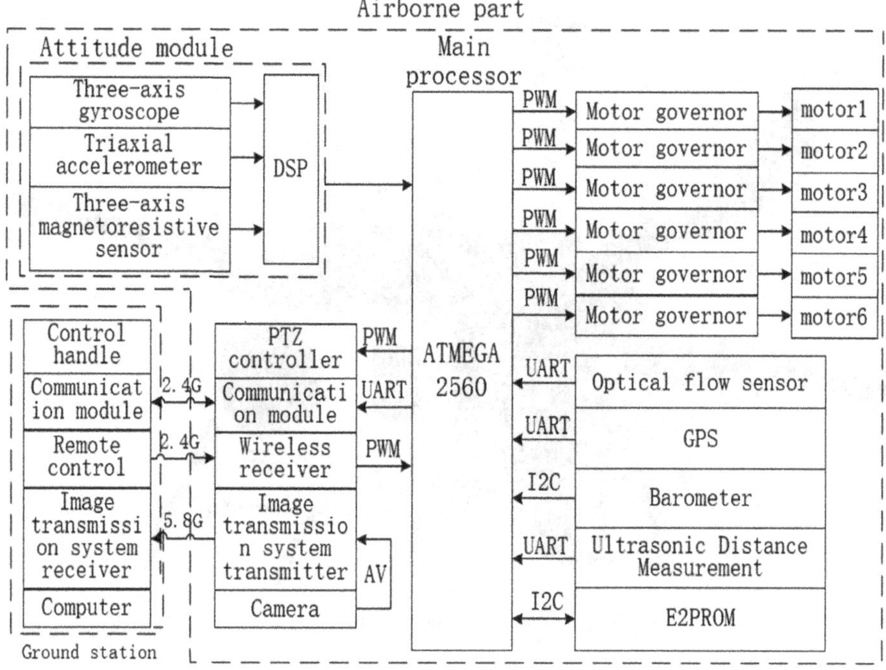

Fig. 4. Components of six-rotor drone

The drone camera transmits images to the raspberry pi, then the raspberry pi detects diseased crops with the trained models and control the drone to hover and spray pesticides intelligently. At the same time, the images will be transmitted to the image transmission system receiver, a phone or a computer. When a tutor occurs fault, the drone could check out the fault and then adjust flight control in time with the help of fault tolerant flight control algorithm.

4.2 Experimental Results and Analysis

The experiment was carried out using a self-built drone flight platform, the flight control board used APM2.8, which made the drone have better stability when flying, and the embedded platform used a Raspberry Pi. The training process is mainly divided into two steps: the first step is to match the priori frame in the picture with the true value; the second step is to determine the loss function. The matching principle of the SSD's a priori frame and the real value has two main points: firstly, for each real value in the picture, find the a priori frame with the largest overlap with it, and then match the a priori frame with it; secondly, guarantee Positive and negative samples are balanced to ensure that the ratio of positive and negative samples is close to 1:3.

After determining the training samples, then determine the loss function. The loss function is obtained by the weighted sum of position error and confidence error, and its

Fig. 5. Drone camera

formula is:

$$L(x, c, l, g) = \frac{1}{N}\left(L_{conf}(x, c) + \alpha L_{loc}(x, l, g)\right) \tag{13}$$

In the formula, N represents the number of positive samples in the prior frame. Here $x^{p_{ij}} \in \{1, 0\}$ represents an indicator parameter, when $x^{p_{ij}} = 1$ indicates that the i-th prior box matches the j-th true value. For the position error, it uses Smooth L1Loss, which is defined as follows:

$$L_{loc}(x, l, g) = \sum_{i \in Pos}^{N} * \sum_{m \in \{cx, cy, wk\}} x_{ij}^{k} \mathrm{smooth}_{L1}\left(l_i^m - \widehat{g_i^m}\right) \tag{14}$$

For the confidence error, it uses SoftmaxLoss, which is defined as follows:

$$L_{conf}(\mathrm{x}, \mathrm{c}) = -\sum_{i \in Pos}^{N} x_{ij}^{k} \log\left(\widehat{c_i^0}\right) \tag{15}$$

Here:

$$\widehat{c_i^p} = \frac{\exp(c_i^p)}{\sum_p \exp(c_i^p)} \tag{16}$$

This experiment uses crops as the target, so this paper first trains the model to identify crops. Here the training model uses GPU to accelerate the training process. Firstly, install the development environment required by the Tensorflow Object Detection API and test it on the official Demo. Then use the data for training and testing to make your own model. For preparation, downloading 1,000 pictures of crops from the Internet, dividing them into training set and test set. And set the label "crop", then start training. After the model training, we use the trained model to check the effect. We randomly selected several photos of crops to test the accuracy of the model for crops identification. The recognition rate of the model in the test set is about 95% (Fig. 6).

Fig. 6. Soybean plant diseases and pests identification

5 Conclusion

Based on the above research results, the intelligent agricultural six-rotor drone researched in this project can be more suitable for the farmland terrain of most precision agricultural solutions with improved algorithm, and improve the safety of the drone. Intelligent agricultural six-rotor drone can accurately identify diseased crops through the trained deep learning model, which has a good recognition rate. The overall experimental effect of intelligent agriculture six-rotor drone appears good. The precision agriculture could reduce the cost and improve the efficiency under the condition of polluted environment. I believe it could contribute to the development of Intelligent Agriculture in the era of industry 4.0, and help with the development of more efficient and environmentally friendly precision agriculture.

Acknowledgment. This work is supported by Jingdezhen Ceramic Institute College Students Innovation and Entrepreneurship Training Program under Grant No. 201810408016.

References

1. Alonzo, M., et al.: Mapping tall shrub biomass in Alaska at landscape scale using structure-from-motion photogrammetry and lidar. Remote Sens. Environ. **245**, 111841 (2020)
2. Zhu, B., Zhu, J., Chen, Q.: A bio-inspired flight control strategy for a tail-sitter unmanned aerial vehicle. Sci. China (Inform. Sci.). **63**, 1–10 (2020). http://kns.cnki.net/kcms/detail/11.5847.TP.20200529.1556.003.html
3. Librán-Embid, F., Klaus, F., Tscharntke, T., Grass, I.: Unmanned aerial vehicles for biodiversity-friendly agricultural landscapes - a systematic review. Sci. Total Environ. **732**, 139204 (2020)
4. Wu, J.: Extracting apple tree crown information from remote imagery using deep learning. Comput. Electron. Agric. **174**, 105504 (2020)
5. Johansen, K.: Mapping the condition of macadamia tree crops using multi-spectral UAV and WorldView-3 imagery. ISPRS J. Photogrammetry Remote Sens. **165**, 28–40 (2020)
6. Kerkech, M., Hafiane, A., Canals, R.: Vine disease detection in UAV multispectral images using optimized image registration and deep learning segmentation approach. Comput. Electron. Agric. **174**, 105446 (2020)
7. Wu, D., Johansen, K., Phinn, S., Robson, A., Tu, Y.-H.: Inter-comparison of remote sensing platforms for height estimation of mango and avocado tree crowns. Int. J. Appl. Earth Observations Geoinf. **89**, 102091 (2020)
8. Freitas, H., Faiçal, B.S., e Silva, A.V.C., Ueyama, J.: Use of UAVs for an efficient capsule distribution and smart path planning for biological pest control. Comput. Electron. Agric. **173**, 105387 (2020)
9. Hu, G., Yin, C., Wan, M., Zhang, Y., Fang, Y.: Recognition of diseased Pinus trees in UAV images using deep learning and AdaBoost classifier. Biosyst. Eng. **194**, 138–151 (2020)
10. Yang, Q., Shi, L., Han, J., Yu, J., Huang, K.: A near real-time deep learning approach for detecting rice phenology based on UAV images. Agric. Forest Meteorol. **287**, 107938 (2020)
11. Gašparović, M., Zrinjski, M., Barković, Đ., Radočaj, D.: An automatic method for weed mapping in oat fields based on UAV imagery. Comput. Electron. Agric. **173**, 105385 (2020)
12. Science – Agronomy: Study Findings from University of Seville Broaden Understanding of Agronomy (Deep Learning Techniques for Estimation of the Yield and Size of Citrus Fruits Using a Uav)
13. Remote Sensing: Researchers' Work from Universita degli Studi Mediterranea di Reggio Calabria Focuses on Remote Sensing (Applications of UAV Thermal Imagery in Precision Agriculture: State of the Art and Future Research Outlook). Agric. Week (2020)
14. Machine Learning - Artificial Intelligence: Research from king Abdullah university of science and technology (KAUST) broadens understanding of artificial intelligence (predicting biomass and yield in a tomato phenotyping experiment using UAV imagery and random forest). Rob. Mach. Learn. (2020)
15. Machine Learning: Polytechnic university Torino researchers publish new data on machine learning (UAV and machine learning based refinement of a satellite-driven vegetation index for precision agriculture). Agric. Week (2020)
16. Networks - Computer Networks: Researchers from university of Western Macedonia report new studies and findings in the area of computer networks (a compilation of UAV applications for precision agriculture). Agric. Week (2020)

17. Engineering: Investigators from Beihang university target engineering (path planning for UAV ground target tracking via deep reinforcement learning). Technol. News Focus (2020)
18. Patent Application: patent application titled software process for tending crops using a UAV published online (USPTO 20200113166). Defense Aerosp. Week (2020)
19. Huang, H., Lan, Y., Yang, A., Zhang, Y., Wen, S., Deng, J.: Deep learning versus object-based image analysis (OBIA) in weed mapping of UAV imagery. Int. J. Remote Sens. **41**(9), 3446–3479 (2020)
20. Zheng, L., Ji, R., Sun, H., Yang, Z., Li, M.: Plant protection UAV operation recommendation using storm framework. IFAC PapersOnLine **52**(30), 213–218 (2019)
21. Fang, P., Lu, J., Tian, Y., Miao, Z.: An improved object tracking method in UAV videos. Procedia Eng. **15**, 634–638 (2011)
22. Wang, X., Sun, H., Long, Y., Zheng, L., Liu, H., Li, M.: Development of visualization system for agricultural UAV crop growth information collection. IFAC PapersOnLine **51**(17), 631–636 (2018)
23. Torres-Sánchez, J., Peña, J.M., de Castro, A.I., López-Granados, F.: Multi-temporal mapping of the vegetation fraction in early-season wheat fields using images from UAV. Comput. Electron. Agric. **103**, 104–113 (2014)
24. Michaelsen, E., Meidow, J.: Stochastic reasoning for structural pattern recognition: an example from image-based UAV navigation. Pattern Recogn. **47**(8), 2732–2744 (2014)
25. Quintero, S.A.P., Hespanha, J.P.: Vision-based target tracking with a small UAV: optimization-based control strategies. Control Eng. Pract. **32**, 28–42 (2014)
26. Lecun, Y., Bottou, L., Bengio, Y., et al.: Gradient-based learning applied to document recognition. Proc. IEEE **86**(11), 2278–2324 (1998)
27. LNCS Homepage. http://www.springer.com/lncs. Accessed 21 Nov 2016

Consensus in Lens of Consortium Blockchain: An Empirical Study

Hao Yin[1], Yihang Wei[1], Yuwen Li[1], Liehuang Zhu[2(✉)], Jiakang Shi[3], and Keke Gai[2(✉)]

[1] School of Computer Science and Technology, Beijing Institute of Technology, Haidian District, Beijing 100081, China
yinhao@bit.edu.cn, weiyongjian200708@126.com, yw.li@hotmail.com
[2] School of Cyberspace Security, Beijing Institute of Technology, Beijing 100081, China
{liehuangz,gaikeke}@bit.edu.cn
[3] China Mobile Research Institute, Xichen District, Beijing 100032, China
shijiakang@chinamobile.com

Abstract. Blockchain emerges as a public decentralized ledger system in recent years. Compared to the traditional distributed database, the blockchain realizes trustless property over the distributed network but consumes more computing resources and processing time. In blockchain, the consensus algorithm is the key component that guarantees such a significant property. To reach better performance, people adjust the network assumptions and classifies the blockchain into three types of the public, consortium, and private. Since consortium blockchain is prevalent studied and more practical, in this paper we investigate various representative consensus in the lens of consortium blockchain. By comparative analyzing those consensus algorithms, we find that deterministic consensus can speed up the transaction process in consortium blockchain. The related experiments are also conducted to understand what mainly causes the consensus delay. The results show that communication complexity seriously influences the algorithm performance. From this empirical study, we suggest that the message transmission path can be an optimized method to make research in future work.

Keywords: Consortium blockchain · Consensus algorithm · Byzantine fault tolerance · Performance · Empirical study

1 Introduction

Blockchain [21] establishes a decentralized database over a peer-to-peer (P2P) network, in which users do not need to trust each other nor a centralized organization. To build such a system, users run the open-source program code on the local computer to become a node. The blockchain can help users to spread the transactions and make an agreement via a consensus protocol over the whole

M. Qiu (Ed.): ICA3PP 2020, LNCS 12454, pp. 282–296, 2020.
https://doi.org/10.1007/978-3-030-60248-2_19

network. Since every transaction is witnessed by a majority of nodes in the network, blockchain builds a *trustless* [26] system from scratch. This property can cut down lots of costs in production activities, but blockchain's performance is not ideal for most practical applications.

Technically, we can optimize the consensus protocol to improve blockchain performance in creating blocks and processing transactions [7,32,33]. In general, we classify the type of blockchain according to the permission to make different assumptions for the nodes in the consensus protocol. Blockchain [6,34] is roughly divided into public, consortium, and private, where the nodes have trends to be stable and trustworthy. The weaker trust assumption leads to a simple consensus protocol thus getting better performance. It is a trade-off between performance and trustless. From a practical view, consortium blockchain takes both of them into account and very adopts to a real production environment.

However, existing consensus protocols often consider the scalability of public blockchains, or just apply the traditional distributed consistent algorithm into the consortium blockchain. It is rarely based on the features of consortium blockchain to reach optimal performance. In this paper, we revisit various representative consensus protocols and comparatively analyze them in the lens of consortium blockchain. From a highly abstract perspective, we can understand the key factors that affect the performance of the consensus protocol. Under this inspiration, we give a piece of advice in consensus optimization for future work. The contributions in our paper are as follows.

1. We synthesize the representative consensus algorithms and classify them into three types. In this way, we try to revisit these consensus algorithms without too many details, helping understand the essential design better.
2. We investigate those consensus algorithms in the lens of consortium blockchain, elaborating the process of reaching consensus under a high-level framework. Some features can be used to optimize performance.
3. We demonstrate the main reason for the performance bottleneck and take experimental verification. Such an empirical study inspires that communication path optimizing may improve the performance of consortium blockchain.

The rest of the paper is organized as follows. We give a revisit of the representative consensus in Sect. 3. Then we make a comparison and discussion in Sect. 4 from the perspective of consortium blockchain. Section 5 shows the experiments we conduct to verify the performance bottleneck. Finally, we conclude this paper in Sect. 6.

2 Related Work

The consensus is the main factor of performance bottleneck in transaction processing. There are a lot of works attempting to design different blockchain consensus against some specific scenarios [9,20], which can improve the performance for practical usage. Some survey works collect those schemes and discuss various views, making it easy to know about the study progress.

Bano et al. [1] conducted a systematic and comprehensive study of blockchain consensus protocols, where the consensus is classified into three types: proof-of-work (PoW), proof-of-X (PoX), and hybrid protocols. Xiao et al. [30] introduced a framework to analyze fundamental differences of various blockchain consensus protocols. Five core components are identified, including block proposal, block validation, information propagation, block finalization, and incentive mechanism, to make evaluations. Garay et al. [10] presented a roadmap for studying the consensus problem. They perform a landscape of consensus research in the Byzantine failure model, aiming to present a new class of blockchain consensus protocols.

Sankar et al. [28] discussed these proposed well-optimized Byzantine fault tolerant consensus protocols and analyzed their feasibility and efficiency in meeting the characteristics. Nguyen et al. [24] presented a review of the Blockchain consensus algorithms by categorizing them into two main groups, which are proof-based consensus and voting-based consensus. Salimitar et al. [27] took focus on a typical internet of things (IoT) network, which consists of several devices with limited computational and communications capabilities. Gudgeon et al. [12] structured the rich research on layer-two transactions, categorizing the research into payment and state channels as well as commit-chains. While the blockchain is used only as a recourse for disputes. Wang et al. [29] investigated sharding protocols used in blockchain, providing a systematic and comprehensive review of blockchain sharding techniques.

In our work, we make a classification of blockchain consensus protocols according to the ways of reaching an agreement. For blockchain, the longest chain principle is a type of probabilistic consensus while the message interaction is a type of deterministic consensus. We review these existing consensus protocols from the perspective of consortium blockchain for practical applications.

3 Consensus Mechanism

The core of blockchain consensus protocols is to select a leader for bookkeeping. The input of the consensus protocol is a sequence of transactions generated by nodes, and the outputs are the encapsulated data block appending to the blockchain. According to the ways of outputting block, we can classify blockchain consensus protocols into three categories: probabilistic consensus algorithms, deterministic consensus algorithms, and hybrid consensus algorithms.

3.1 Probabilistic Consensus

This type of consensus usually relies on the chain structure reaching a final distributed agreement. All peers in the blockchain network adopt the same strategy to determine which block should be appended to the current blockchain. Although temporary forks sometimes happen, they eventually converge to a particular chain to reach a consensus.

Proof of Work. Each node in a bitcoin [23] system solves a complicated but easy-to-verify puzzle. Once the majority of the whole network approves the solution, the node will obtain the bookkeeping rights for the block appending this block to the end of the blockchain. Such a difficult puzzle can be described as searching a random number to make the hash value of block less than or equal to the target value. The bitcoin system regulates the average block generation time to about 10 min by adjusting the search difficulty. If the average block generation time is less than 10 min observed from the previous 2016 blocks, the target value will be reduced to increase the difficulty.

Proof of Stake. Peercoin [15] first proposed an alternative blockchain consensus protocol called proof-of-stake (PoS) to address the problem of computing resource waste in PoW. The nodes with higher equity obtain the bookkeeping rights with higher probability. The PoS protocol does not require external physical input, which is more environmentally friendly than PoW. Furthermore, the 51% attacks in consensus protocols refer to the stake all miners hold instead of computing power. The advantage of PoS is that it can shorten the time needed to reach a consensus. It is also highly efficient and saves energy because it does not need a lot of computing power to solve problems.

Proof of Capacity. Like the PoW algorithm, the miners in Burstcoin [19] also need to solve complex problems to compete for mining rights. But the difference is that the most complex calculation can be cached through the design of proof-of-capacity (POC) consensus. It can consume storage space to replace with computing time. In POC, the miner needs to store results on the hard disk, and then the miner finds the required data in the previously generated cache data only to generate a new block. The more cached data the miner stores, the greater the chance of getting the bookkeeping right, so this consensus can encourage that the miner chooses larger storage space rather than greater computing power.

Proof of Elapse Time. This consensus method [5] was proposed by Intel in 2016 and is regarded as the core consensus algorithm of Hyperledger Sawtooth. The implementation of this consensus method needs the support of a trusted execution environment (TEE). Specifically, the proof-of-elapse-time (PoET) consensus algorithm used in Sawtooth is based on Intel's SGX (software guard extension) technology to ensure that trusted code is executed reliably. By running this consensus protocol, a new node downloads the trusted algorithm code and loads it into the environment of SGX. To simulate computing PoW solution, the node waits for the appropriate random waiting time, and then packs a block and broadcasts the block and the proof to network.

Proof of Burn. The allocation of mining rights by proof-of-burn [13] (PoB) is based on the user's initiative to "burn" the token. During this process, tokens that are intentionally destroyed are used as a way to "invest" in the blockchain to prove investment in the network. The more tokens a user destroys in the system, the more likely they are to be selected as the next leader. PoB consensus ensures similar security to PoW without consuming computing resources. In the algorithm, a token is burned by sending it to a verifiable public "devourer" address without a corresponding private key. Therefore, the token sent to this address will not be used by anyone and cannot be circulated anymore.

3.2 Deterministic Consensus

This type of consensus applies a traditional distributed agreement protocol to the blockchain area. Each block is treated as a batch of requests that is ready to decide on the distributed network. The difference from traditional consistent algorithms is that there is a chain to link the sequential blocks and no fixed participants in the network.

Practical Byzantine Fault-Tolerance. Miguel Castro and Barbara Liskov [4] in 1999 proposed a practical Byzantine fault-tolerant algorithm (PBFT) to solve the efficiency in byzantine general problem. This algorithm reduces the message complexity from exponential level to polynomial level and makes it feasible in a practical system. The PBFT algorithm usually assumes that the network has $n = 3f + 1$ nodes with tolerate fault f. At any time, there are only a master node and other slave nodes. The master node drives the protocol of message exchange to reach an agreement on at least $f + 1$ honest nodes. Every node runs under the same configuration from the protocol, which is called a *view*. The view change is triggered when the master node failed to make a consensus.

Paxos. The Paxos [18] algorithm is a distributed consistency algorithm based on message passing. The algorithm solves the problem of distributed consistency under non-byzantine assumptions. Nodes may encounter network delay or even shut down without any response. The algorithm runs in an asynchronous network, which can tolerate message loss, delay, disorder, and repetition. There is also a master node that writes a consistent result to other slave nodes using two-stage message multicast. The algorithm uses a majority approach to ensure a f fault tolerance under the network scale of $2f + 1$.

Raft. The Raft [25] algorithm adopts a more simple requirement than the Paxos, which only considers a single proposer to write results. The algorithm must ensure that all nodes execute the same sequence of instructions and reach a consistent state after every round of the protocol. The nodes are divided into three states: leader, candidate, and follower. The state of nodes transition among these three states, which determined by an election procedure. After the leader node's election is completed, other nodes will transition to follower state and set the election timer. The leader node sends heartbeat packets to other nodes, resetting the follower nodes' timer. In this state, the raft algorithm will copy logs from the leader to followers.

Kafka. The consensus in a Kafka [17] is the leader-follower model where the heartbeat detection is implemented by Zookeeper. Followers will follow the leader to copy and remain consistent with the leader. If the leader fails, a new leader will be selected from the followers. The Kafka algorithm adopts the publish-subscribe model to sync messages. The producer of the message generates the message and submits it to the Kafka cluster. After sorting and consensus by the Kafka cluster, it is obtained by the subscribers. Kafka divides multiple topics according to the message type, while each topic can contain multiple partitions for redundancy. Kafka cluster saves messages for a period of time or until the number of messages exceeds a threshold. Consumers are required to actively poll for new messages.

3.3 Hybrid Consensus

This type of consensus combines the structure of blockchain and traditional distributed agreement protocols. Generally, it uses the probabilistic consensus to elect potential blocks and applies deterministic consensus to decide which block should be the final result.

Algorand. The Algorand [11] consensus is mainly divided into two steps. First, a generation group is randomly selected to generate new blocks, and then each new block is signed and broadcast. Second, a verification group is randomly selected to verify the new block broadcast, and in the verification group through an improved Byzantine protocol for consensus. In this process, the key thing is to ensure that all participants in each group are randomly and fairly selected. To solve the problem, the Algorand uses a verifiable random function (VRF) to designate the valid members. It is a hash function with asymmetric key technology, where the output is pseudorandom and can be publicly verified.

Elastico. Elastico [22] is a method based on the idea of fragmentation that divides the nodes in the network into several smaller committees. And then the consensus uses a byzantine fault tolerance protocol to reach consensus in each committee and deal with disjointed transaction sets. The node establishes its own identity through computing a PoW solution and joins the committee. By mixing the PoW, sharding, and BFT technologies, Elastico achieves a linear expansion of throughput.

Ouroboros. The Ouroboros [14] designed the Verifiable Secret Sharing protocol to reliably generate multi-party true random numbers. This method solves the problem that some nodes in the multi-party random number protocol fail to generate random numbers due to malicious or dropped lines. This algorithm forms a Merkle tree with all the rights and interests, where the leaf node of the tree is the equity value of an equity owner. The weight of the non-leaf node is the sum of the rights and interests of the left and right subtrees. According to the random sequence, we start from the root of the Merkle tree, select the left and right subtrees, and finally reach the leaf node to select the block node.

Snow White. In the Snow White [2], there is a hash function that uses a random number seed to determine whether the members of the committee were leaders at each step. At the beginning of each era, there is redistribution for committees. First, the node finds the latest block with a timestamp of 2ω, where the prefix of this block will be used to confirm the members of the next era committee. Then, the node finds the latest block in the local blockchain with a timestamp before ω. Through the prefix of this block, we can get the random number seed of the hash function in the next era.

4 Comparison and Discussion

In essence, blockchain is a distributed database system that aims to reach an agreement on transaction log over the distributed network. The chain structure of blockchain brings new benefits and also challenges to the traditional distributed consistent algorithm. We comparatively analyze these types of consensus algorithms mentioned above.

4.1 Deterministic v.s. Probabilistic

Deterministic consensus and probabilistic consensus are the two types of consensus with the biggest difference. The former ensures that each block output by the algorithm is valid, while the latter usually relies on the longest chain principle to make a new block valid with high probability.

To reach an agreement over the distributed network, there must exist a message (i.e. a new block in blockchain) proposed by a leader. In deterministic consensus, there is only one leader at each unit time. The leader broadcasts a new block and executes several rounds of message interaction to guarantee that most nodes indeed record the same block. We assume the network scale is $n = 3f + 1$, and the leader collects $n - f$ confirmation messages from all other nodes. We need al least $(n - f) - f > f$ confirmations to make a decision. In probabilistic consensus, there are many leaders at the same time. This type of consensus pre-define a puzzle to limit the number of leaders. The consensus relies on the longest chain principle to get confirmation as many as possible. The longest chain principle $(n > 2f)$ implies that the proposed blocks in the longest chain have reached an agreement with a high probability as long as a majority of nodes, i.e. $n - f > f$, adopt the same strategy.

As shown in Fig. 1(a), since only one leader in each unit time, the deterministic consensus has to change the views from the network to choose a new leader when the current leader encounter faults. Such a leader election method only supports the network with a quorum. In terms of message interaction, consensus protocols use two-phase (e.g. Raft) or three-phase (e.g. PBFT) commit to decide on the proposed single block. As shown in Fig. 1(b), the probabilistic consensus

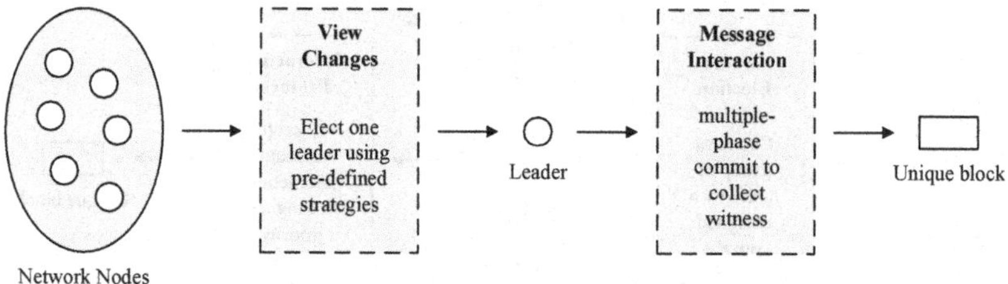

(a) The process in determinisitc consensus

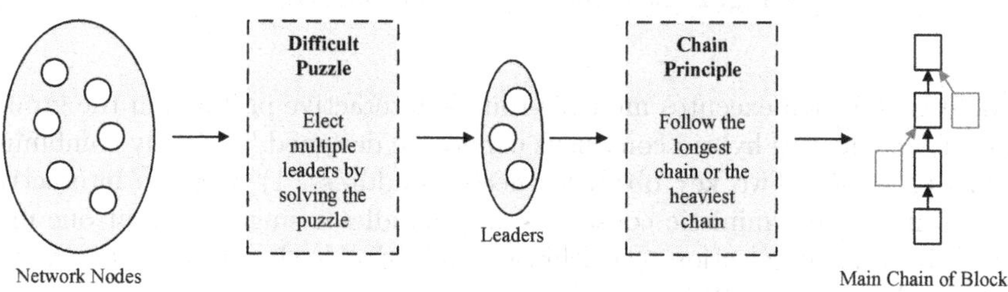

(b) The process in probabilistic consensus

Fig. 1. The difference between deterministic consensus and probabilistic consensus.

lets the nodes solve a difficult puzzle to prove that they are valid leaders. In this process, consensus protocols consume different resources to design the puzzle. For example, PoW uses computing power, PoS uses virtual tokens, PoC uses local storage. The elected leaders simply broadcast those proposed blocks, relying on the chain principle (e.g. the longest chain or the heaviest chain strategies) to choose a block to append into the blockchain.

From the above comparison, we can find that using chain principle rather then message interaction can also design a consensus protocol although it is probabilistic. The chain structure provides an option that we can move the communication cost in the second process to the computing cost in the first process. It allows us to get stronger fault tolerance in a more dynamic network. Especially, when the scale of the network is big enough, such a probabilistic consensus can show better performance because of no message interaction.

4.2 Probabilistic v.s. Hybrid

Probabilistic consensus has great potential in a large-scale network because it gives up the complex message interaction. Instead, the actual agreement process is solved by a chain principle, which is fast in outputting a new block. But incurring forks in the chain structure is an unstable factor of consensus.

To let public blockchain more practical, the hybrid consensus replaces the chain principle with the original message interaction. It first selects a group of

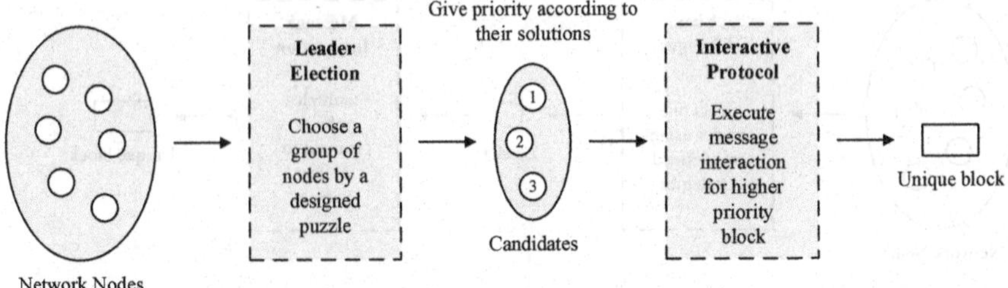

Fig. 2. The process in hybrid consensus.

candidates and then executes multiple-rounds interactive protocol in the group. However, an effective hybrid consensus cannot be designed by simply combining the two processes. Two key obstacles need to address. 1) Message interactive protocol in the deterministic consensus only handles a single block at one unit time. Thus, a leader election algorithm needs to decide the prepared block over the selected candidates. 2) Leader election algorithm in the probabilistic consensus substitutes static view changes strategy, which forces a message interactive protocol to allow the competition of several proposed blocks.

For better understanding, we use a similar framework to depict hybrid consensus in Fig. 2. Different from the probabilistic consensus that relies on chain principle to finally commit a block, the hybrid consensus can decide which block to commit after the first process completes. Due to the second process actually does the consensus work, we re-introduce the types of node faults into the hybrid consensus, resulting in that the ability to fault tolerance is consistent with the used message interactive protocol. To maintain this fault tolerance, the algorithm needs to securely narrow down a consensus group by randomly sampling the candidates over the distributed network. For example, Algorand adopts VRF to select a consensus group, where the size of the group can be adjusted by some parameters. Elastico leverages PoW to divides the network into several shardings. Likewise, the size of the sharding can be adjusted.

Compared to probabilistic consensus, hybrid consensus indeed has a better performance in reaching an agreement. There are neither forks in the blockchain nor too many message interactions in the network. For the public blockchain, the hybrid consensus is a good choice because it keeps the feature of a large-scale dynamic network. However, consortium blockchain can clearly specify a group of consensus nodes, skipping the leader election process.

4.3 Deterministic Consensus

The byzantine assumption lets each pair of nodes keep a point-to-point communication channel against the attacks from malicious nodes. To learn the difference between from non-byzantine fault assumption, we use an ideal model in secure multi-party computation (SMPC) to build a broadcast communication channel, assuming an ideal center that delivers messages to all nodes.

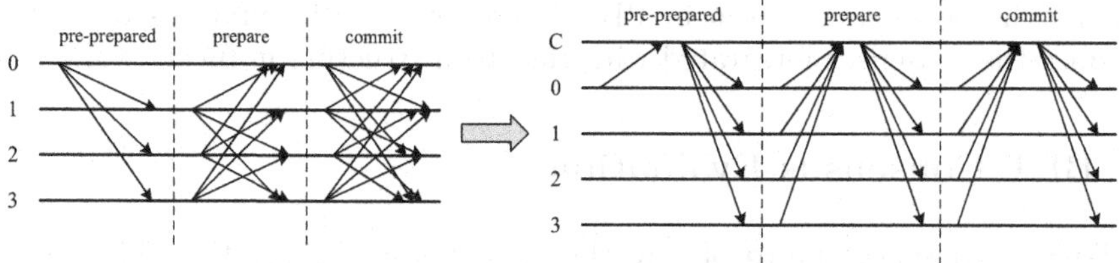

Fig. 3. The process of PBFT in the ideal model.

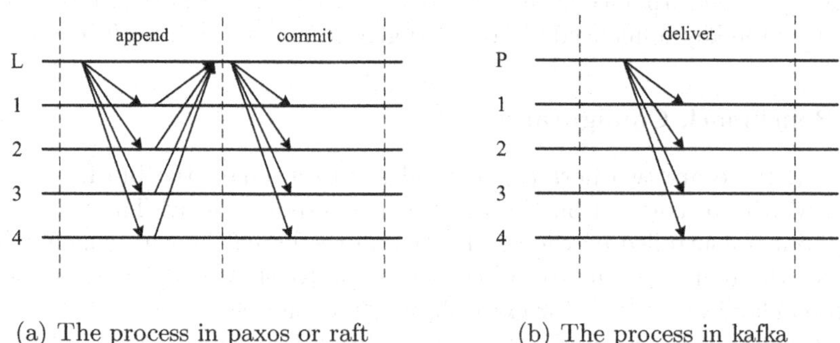

(a) The process in paxos or raft

(b) The process in kafka

Fig. 4. The process of non-byzantine fault consensus.

As shown in Fig. 3, we omit the phases of request from users and the reply for users, simulating the process of PBFT consensus protocol in the ideal model. The ideal center becomes a hub to handles the votes from different nodes, sending the next phase's messages to each node. The only difference is that the complexity of the message is reduced because of the ideal center. The first two phases convince a node that a majority of nodes witness the same message, and the last phase notifies that a majority of nodes keep the same state for this message. Thus, we can safely reach an agreement over most of the honest nodes.

In contrast, it is easier for the consensus protocol with non-byzantine fault. We also illustrate the process of two-phase consensus protocol (e.g. Paxos and Raft) in Fig. 4(a). Since messages in the non-byzantine fault assumption cannot be tampered with, nodes only check that if a majority of nodes witness the same message. Moreover, Kafka uses zookeeper service to inspect whether the nodes are online or not, thus discarding the check phase. As shown in Fig. 4(b), Kafka more likes a backup system that syncs messages to replica nodes.

Although consortium blockchain organizes a relatively fixed group of nodes using a permission mechanism, only less trust to be needed between different organizations. A blockchain system is ought to establish trust based on the weaker trust assumption, which is why we are more interested in the blockchain instead of cloud computing [8]. As we discussed before, chain structure can benefit consensus in a probabilistic way. To improve performance, we need to validate the bottleneck reason in PBFT and get some inspirations from it to optimize the consensus integrated with the chain structure in future work.

5 BFT Consensus Evaluation

BFT-based consensus protocols are the best choice of consortium blockchain, meeting both performance and application requirements. We investigate open-source code implementations and select the BFT-SMART [3] for deployment, evaluating the performance at different request sizes and network scales. Finally, we analyze the key influenced factors and give some advice in performance.

5.1 Benchmark Configuration

Generally, there are two performance indexes to evaluate the blockchain consensus, which are transaction throughput and response delay. The transaction throughput is controlled by the batch size and the batch timeout, which is hard to reflect the real performance on consensus protocol. We prefer to adopt the response delay to evaluate how each phase delays the system.

BFT-SMART is a high-performance Byzantine fault-tolerant algorithm library developed in Java language (with support for version 1.8 and above). This library executes the consensus under the network with $n = 3f + 1$ servers, following the three phases PROPOSE, WRITE, and ACCEPT to multicast messages. Only the PROPOSE phase contains the request from clients. The latter two phases only send a summary of the request, namely the hash value. We adjust some software configuration to measure the specific delay as follows. 1) The batch timeout is set to a negative value, which means that a new round of consensus starts right now after the last round completes. 2) The batch size is set to 1, which means that each block has only one request message, where servers do not cache requests. 3) the number of reply threads is set to 0, which means that servers use the main thread to reply consensus results to clients. Based on these configurations, we let a client constantly feeds one request to the consensus network, collecting the overall delay on average.

Table 1. Host configuration

Host operating system	Windows 10 64 bit
Host processor	Intel(R) Core(TM) i5-5250U CPU @1.60 GHz
Docker CPU	2 CPUs
Docker memory	2 GB
Docker container	library/java:latest

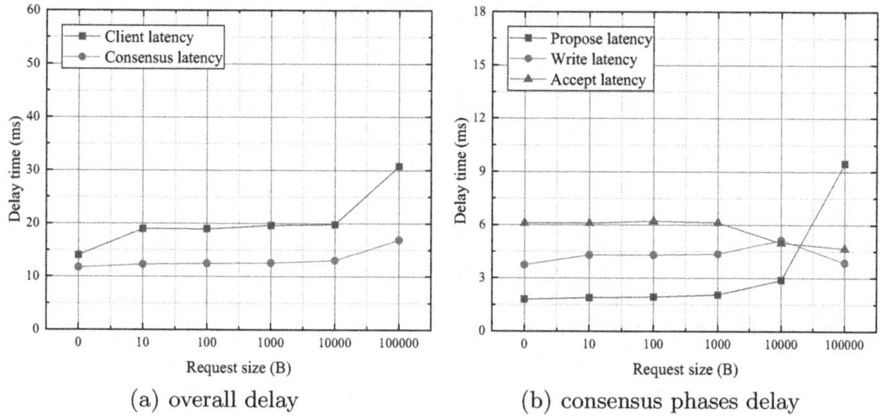

(a) overall delay (b) consensus phases delay

Fig. 5. The latency when increasing request size.

5.2 Latency Experiments

Setup. Table 1 shows the main configuration applied. We use a docker container to build the consensus network because it is easy to deploy with good compatibility. The BFT-SMART library is added to the running container to execute the benchmark programs and output results. We conduct our experiments on a laptop that installed the latest docker engine.

In our experiments, the severs are connected consisting of a network while a client just connects to one of them to send requests every an interval of time. Here we let the client send 1000 requests in total and the interval time is 1 second. Each server collects 100 samples to compute the average latency of the consensus process. In this way, the client can output the response latency and the servers can output the consensus and each phase latency. We learn the performance bottleneck at the two aspects of overall delay and consensus phases delay.

First, we conduct the latency experiments when increasing the request size from the client. As shown in Fig. 5(a), the overall delay is not so large when the request size is small enough. But if the request message is too large, it will cost too much time in transmitting the data. As a result, the latency raises a lot when the request size over 100 KB. Especially for client latency, it needs to collect at least $f + 1$ replies, causing the latency grows faster than consensus latency. Figure 5(b) shows the latency of each phase under the same situation. These three phases together consist of the whole consensus latency, where request size has mainly effect on the Propose phase.

Second, we also conduct the latency experiments when adjusting the network scale, where the request size is set to 0 in all experiments. We let the number of byzantine fault nodes be 1 to 6, thus having the network scale $n = 3f + 1$ from 4 to 19. As shown in Fig. 6(a), the overall delay significantly increases when the network scale becomes large. The message interaction in consensus causes large communication complexity. Especially when the network scale over 13 in

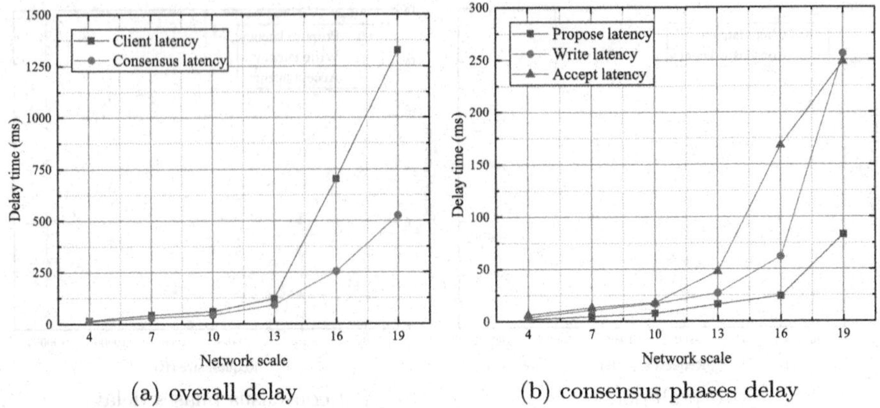

(a) overall delay

(b) consensus phases delay

Fig. 6. The latency when increasing network scale.

our experiment, each server needs to collect enough votes from others, which leads to a network burden. Figure 6(b) shows the latency of each phase under the same situation. In the propose phase, servers only receive the message that ready to agree. In the last two phases, the pairwise servers need to exchange messages and thus causing the main latency of consensus.

The experimental results verify that the latency of each phase and overall delay present the same effect even though we run the servers in parallel. The massive protocol messages in the network cause a large latency. Thus, reducing message complexity is indeed a way to optimize consensus performance. In future work, we will study potential methods and consider some features of consortium blockchain to dynamically optimize the consensus in different cases. First, we can adopt an aggregate signature, such as Byzcoin [16], to cut down on extra communication, where a well-designed optimal routing path can be applied. Second, the chain structure can help accelerate the rate of block generation, like Hotstuff [31]. The phases in consensus can be parallelized. Third, we can supervise the consensus nodes in consortium blockchain, using the technique of proof of authority (POA) to reduce the risk of byzantine faults.

6 Conclusion

Consensus is an essential component in the blockchain, which can build a trustless system over the distributed network. This paper does an empirical study on the blockchain consensus algorithms. We revisit the representative consensus algorithms via classifying them into three types. Then, we carefully compare and discuss the difference between the types of consensus algorithms. Compared to probabilistic consensus, consortium blockchain is more adapted to deterministic consensus for better performance. To find out the main reason for the performance bottleneck, we choose BFT-SMART library to make an evaluation for BFT consensus algorithm. The experimental results verify that message complexity leads to the delay in consensus. Finally, we conclude three aspects of improvement thoughts, which are an aggregate signature, chain structure, and proof of authority, to optimize the consensus for future work.

Acknowledgements. This work is supported by Ministry of Education - China Mobile Research Fund Project (Grant No. MCM20180401). Natural Science Foundation of Beijing Municipality (Grant No. 4202068), National Natural Science Foundation of China (Grant No. 61972034, 61902025), Natural Science Foundation of Shandong Province (Grant No. ZR2019ZD10), Guangxi Key Laboratory of Cryptography and Information Security (No. GCIS201803), Henan Key Laboratory of Network Cryptography Technology (Grant No. LNCT2019-A08), Beijing Institute of Technology Research Fund Program for Young Scholars (Dr. Keke Gai).

References

1. Bano, S., et al.: Consensus in the age of blockchains. arXiv preprint arXiv:1711.03936 (2017)
2. Bentov, I., Pass, R., Shi, E.: Snow white: provably secure proofs of stake. IACR Cryptology ePrint Archive, p. 919 (2016)
3. Bessani, A., Sousa, J., Alchieri, E.E.P.: State machine replication for the masses with BFT-SMART. In: IEEE/IFIP International Conference on Dependable Systems & Networks (2014)
4. Castro, M., Liskov, B., et al.: Practical Byzantine fault tolerance. In: OSDI, pp. 173–186 (1999)
5. Chen, L., Xu, L., Shah, N., Gao, Z., Lu, Y., Shi, W.: On security analysis of proof-of-elapsed-time (PoET). In: Spirakis, P., Tsigas, P. (eds.) SSS 2017. LNCS, vol. 10616, pp. 282–297. Springer, Cham (2017). https://doi.org/10.1007/978-3-319-69084-1_19
6. Gai, K., Wu, Y., Zhu, L., Qiu, M., Shen, M.: Privacy-preserving energy trading using consortium blockchain in smart grid. IEEE Trans. Industr. Inf. **15**(6), 3548–3558 (2019)
7. Gai, K., Wu, Y., Zhu, L., Xu, L., Zhang, Y.: Permissioned blockchain and edge computing empowered privacy-preserving smart grid networks. IEEE Internet Things J. **6**(5), 7992–8004 (2019)
8. Gai, K., Guo, J., Zhu, L., Yu, S.: Blockchain meets cloud computing: a survey. IEEE Commun. Surv. Tutorials **PP**(99), 1 (2020)
9. Gai, K., Wu, Y., Zhu, L., Zhang, Z., Qiu, M.: Differential privacy-based blockchain for industrial internet-of-things. IEEE Trans. Industr. Inf. **16**(6), 4156–4165 (2019)
10. Garay, J., Kiayias, A.: SoK: a consensus taxonomy in the blockchain era. In: Jarecki, S. (ed.) CT-RSA 2020. LNCS, vol. 12006, pp. 284–318. Springer, Cham (2020). https://doi.org/10.1007/978-3-030-40186-3_13
11. Gilad, Y., Hemo, R., Micali, S., Vlachos, G., Zeldovich, N.: Algorand: scaling Byzantine agreements for cryptocurrencies. In: Proceedings of the 26th Symposium on Operating Systems Principles, pp. 51–68 (2017)
12. Gudgeon, L., Moreno-Sanchez, P., Roos, S., McCorry, P., Gervais, A.: SoK: off the chain transactions. IACR Cryptol ePrint Arch, p. 360 (2019)

13. Karantias, K., Kiayias, A., Zindros, D.: Proof-of-burn. In: Bonneau, J., Heninger, N. (eds.) FC 2020. LNCS, vol. 12059, pp. 523–540. Springer, Cham (2020). https://doi.org/10.1007/978-3-030-51280-4_28
14. Kiayias, A., Russell, A., David, B., Oliynykov, R.: Ouroboros: a provably secure proof-of-stake blockchain protocol. In: Katz, J., Shacham, H. (eds.) CRYPTO 2017. LNCS, vol. 10401, pp. 357–388. Springer, Cham (2017). https://doi.org/10.1007/978-3-319-63688-7_12
15. King, S., Nadal, S.: PPCoin: peer-to-peer crypto-currency with proof-of-stake. Self-published paper, 19 August 2012
16. Kokoris-Kogias, E., Jovanovic, P., Gailly, N., Khoffi, I., Gasser, L., Ford, B.: Enhancing bitcoin security and performance with strong consistency via collective signing. Appl. Math. Model. **37**(8), 5723–5742 (2016)
17. Kreps, J., Narkhede, N., Rao, J., et al.: Kafka: a distributed messaging system for log processing. In: Proceedings of the NetDB, vol. 11, pp. 1–7 (2011)
18. Lamport, L., et al.: Paxos made simple. ACM SIGACT News **32**(4), 18–25 (2001)
19. Larsson, T., Thorsén, R.: Cryptocurrency performance analysis of Burstcoin mining (2018)
20. Li, H., Gai, K., Fang, Z., Zhu, L., Xu, L., Jiang, P.: Blockchain-enabled data provenance in cloud datacenter reengineering. In: ACM International Symposium on Blockchain and Secure Critical Infrastructure, pp. 47–55 (2019)
21. Li, X., Jiang, P., Chen, T., Luo, X., Wen, Q.: A survey on the security of blockchain systems. Future Gener. Comput. Syst. **107**, 841–853 (2020)
22. Luu, L., Narayanan, V., Zheng, C., Baweja, K., Saxena, P.: A secure sharding protocol for open blockchains. In: The 2016 ACM SIGSAC Conference (2016)
23. Nakamoto, S., Bitcoin, A.: A peer-to-peer electronic cash system. Bitcoin (2008). https://bitcoin.org/bitcoin.pdf
24. Nguyen, G.T., Kim, K.: A survey about consensus algorithms used in blockchain. J. Inf. Process. Syst. **14**(1), 101–128 (2018)
25. Ongaro, D., Ousterhout, J.: In search of an understandable consensus algorithm. In: {USENIX} Annual Technical Conference, pp. 305–319 (2014)
26. Pilkington, M.: Blockchain technology: principles and applications. In: Research Handbook on Digital Transformations. Edward Elgar Publishing (2016)
27. Salimitari, M., Chatterjee, M.: A survey on consensus protocols in blockchain for IoT networks. arXiv preprint arXiv:1809.05613 (2018)
28. Sankar, L.S., Sindhu, M., Sethumadhavan, M.: Survey of consensus protocols on blockchain applications. In: 2017 4th International Conference on Advanced Computing and Communication Systems (ICACCS), pp. 1–5. IEEE (2017)
29. Wang, G., Shi, Z.J., Nixon, M., Han, S.: SoK: sharding on blockchain. In: 1st ACM CAFT, pp. 41–61 (2019)
30. Xiao, Y., Zhang, N., Lou, W., Hou, Y.T.: A survey of distributed consensus protocols for blockchain networks. IEEE COMST **22**(2), 1432–1465 (2020)
31. Yin, M., Malkhi, D., Reiter, M.K., Gueta, G.G., Abraham, I.: HotStuff: BFT consensus in the lens of blockchain. arXiv preprint arXiv:1803.05069 (2018)
32. Zheng, Z., Xie, S., Dai, H., Chen, X., Wang, H.: An overview of blockchain technology: Architecture, consensus, and future trends. In: 6th IEEE International Congress on Big Data (2017)
33. Zhu, L., Wu, Y., Gai, K., Choo, K.: Controllable and trustworthy blockchain-based cloud data management. Future Gener. Comput. Syst. **91**, 527–535 (2019)
34. Zhu, L., Gai, K., Li, M.: Blockchain Technology in Internet of Things. Springer, Cham (2019). https://doi.org/10.1007/978-3-030-21766-2

Collaborative Design Service System Based on Ceramic Cloud Service Platform

Yu Nie[1(✉)], Yu Liu[1], Chao Li[1], Hua Huang[1(✉)], Fubao He[1], and Meikang Qiu[2]

[1] Jingdezhen Ceramic Institute, Jingdezhen 333403, Jiangxi, China
{nieyu,lichao1005,huanghua,hefubao}@jci.edu.cn,
1820043003@stu.jci.edu.cn

[2] Department of Computer Science, Texas A&M University-Commerce, Commerce, TX 75428, USA
meikang.Qiu@tamuc.edu

Abstract. Currently enterprise information service system based on the Internet is becoming more and more important for operating activities of enterprises, such as the production and manufacture of enterprises. To improve the efficiency of ceramic products' design and production, the paper designed and implemented a collaborative design service system based on ceramic cloud service platform. In the proposed system, the principle and technology of workflow are adopted to control and manage the product collaborative design process. In this paper, a collaborative design business process model is presented, and the system structure and engine of collaborative design based on ceramic cloud service platform are also proposed and realized. The application effect of the proposed system shows that it can be used for cross-regional personnel or teams to participate in the design and development of the same products, to improve the efficiency of product design.

Keywords: Collaborative services · Collaborative design process · Ceramic enterprise information · Ceramic cloud service platform

1 Introduction

With the development of economic globalization, industrial specialization and social division of labor, some industries in the world have undergone big changes, and the development of industries has shown a trend of horizontal integration [1]. More and more enterprises attach importance to and adopt the enterprise information service system based on the Internet, and the production and operation activities of enterprises are increasingly affected by enterprise informatization [2], including ceramic industry.

Collaborative design service system based on ceramic cloud service platform, as the important component of the design service system of the ceramic cloud service platform, using the designing ideas of workflow engine, deploying and running with the architecture of sub-platform middleware, can fulfil the requirements of collaborative service task release and management. It is more friendly and intuitive to show the basic situation of the main task and sub-task, dividing a main task into several sub-tasks according to

M. Qiu (Ed.): ICA3PP 2020, LNCS 12454, pp. 297–308, 2020.
https://doi.org/10.1007/978-3-030-60248-2_20

the need, selection, acceptance and payment for the sub-task. At the same time, it can support graphical operation interface. The system runs in the form of middleware or sub-platform, which is easy to be compatible with all kinds of heterogeneous systems, with beautiful interface, easy expandability, stable and reliable system.

The goal of collaborative design service system based on ceramic cloud service platform is to solve the cross-regional system-based collaborative design management of ceramic products, i.e., to build collaborative design process control based on workflow, to integrate the whole process of collaborative design about ceramic products based on system into system management, and to provide document sharing, information or task process control for all design service participants of collaborative design service tasks, so as to realize the monitoring and management of completion process for the expected design tasks.

In this system, the principle and technology of workflow are applied to the collaborative design process control and management of ceramic products based on cloud platform, to achieve the cross-regional designers or teams supported by based on the collaborative design platform to be involved in the same product design, and to achieve the whole process management of ceramic product design.

Based on the principle and mode of the workflow, the collaborative design process management function is realized.

- Users with design requirements register at the platform to complete the registration of collaborative design task in the system, and then the collaborative design service is divided into several collaborative sub-tasks to release into the system;
- All collaborative design tasks need to be approved by the system collaborative design task before they are released to the ceramic cloud service platform;
- The design service units, designers or design teams registered in the platform can receive or access tabular data of collaborative design in time in the cloud platform, and can participate in real-time bidding collaborative design tasks for product collaborative design;
- After the designer of the winning bid has completed the collaborative design tasks, he can submit the final design plan in the system and the collaborative design process will be completed after the collaborative design publisher audit. The designer makes further improvement for the design according to the revised suggestions made by the publisher and approved by the other party;
- Paying for the design cost through the system.

The main functions of the system include: collaborative design task management, user management, financial management task, model management and other related functions.

2 Design of the Collaborative Design Service System

2.1 Workflow-Driven Collaborative Design Module

Workflow is one part of computer-supported collaborative work. The concept originated in the field of production organization and office automation. It is a concept for the daily work with fixed program activities. The purpose of it is to break down the work into well-defined tasks or roles, according to certain rules and procedures to carry out these tasks and monitor them so as to achieve the purpose of improving work efficiency, better control of the process, enhance customer service, effective management of business processes [3]. When the user places the demand of product integrated design, the service demand of product integrated design is decomposed under the support of the platform [4–6], and the product collaborative design process is entered. After the demand timing bidding, the winning designer (platform dynamic service resource) is confirmed, and the dynamic design service resource (professional designer) designated in the platform is activated to complete the corresponding function design in time. As soon as all the functions represented by the process nodes of the workflow have been designed, the final product design service is completed. With the support of the work engine, the platform can put the scattered dynamic design service resources into a complete application combination, and accomplish the complex design service collectively [7, 8].

1) Business process model of Collaborative design service

Order $\Omega = \{\omega_1, \varpi, \ldots, \varpi_n\}, \Psi = \{< i, j > |(0 < i \le n \wedge 0 < j \le n) \wedge i \ne j\}$
Platform user zone $\Gamma = \{\gamma_1, \gamma_2, \ldots, \gamma_l\}$, Area of WF $\Delta = \{\delta_1, \delta_2, \delta_3, \ldots\}$
Action zone of WF $\Sigma = \{\sigma_1, \sigma_2, \ldots, \sigma_m\}$
Node action (trigger) sequence zone $\mathrm{O} = \{\ (\sigma_i, t_j)|1 \le i \le m, j \ge 1\}$
So platform collaborative design business process model WF can be expressed as:

$$
\begin{aligned}
W_f &= (N, F) \\
f_k &=< l, m > |1 \le \{l, m\} \le n, f_k \in F, F \subseteq \Psi \\
n_i &= (\delta_i, O_i), \delta_i \in \Delta, n_i \in N, N \subseteq \Omega, O_i \subseteq \mathrm{O}
\end{aligned} \tag{1}
$$

Platform collaborative design business process model WF can be shown in Fig. 1.

2) Collaborative design process instances

When the users and designers of the platform participate in collaborative design business based on WF, the equivalent mapping between the user and the role in the process WF will be established by the users and the winning designers according to its different uses in the collaborative business to realize the collaborative design business under the control of platform users participating in WF.

Business process instance I is Eq. (2):

$$
\begin{aligned}
I &= (P, W_f, S) \\
P &\subseteq \Gamma \\
S &= \{E(P_i, \delta_j)|p_i \in P \wedge \delta_j \in \Delta\}
\end{aligned} \tag{2}
$$

Among them, p, as the platform designers participating in this collaborative design business, is a subset $P \subseteq \Gamma$ of the platform user space, including the collaborative design sponsors. S maps the user p of the platform to the user (operator) δ in the workflow engine by means of the user, role mapping (matching) in the business process, such as (p, δ),

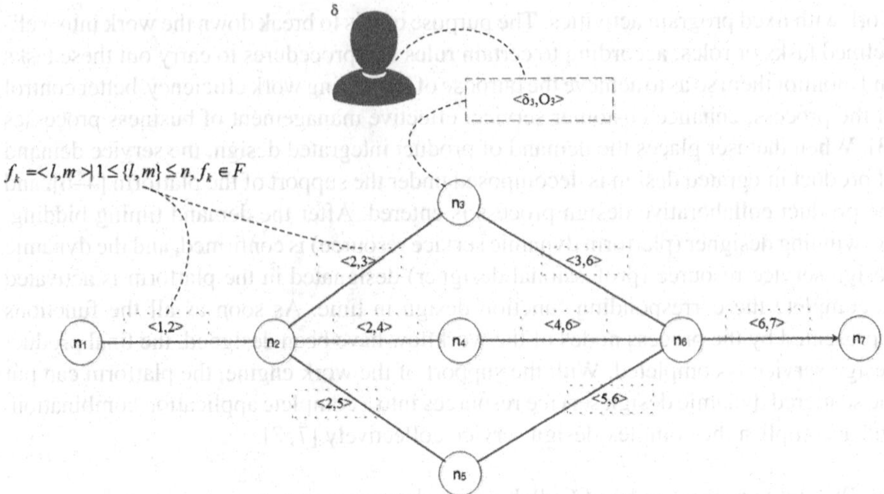

Fig. 1. Collaborate design business processes

so that the designers and the sponsors can participate in the collaborative design under the control of the related business process through the user δ of the workflow engine (Fig. 2).

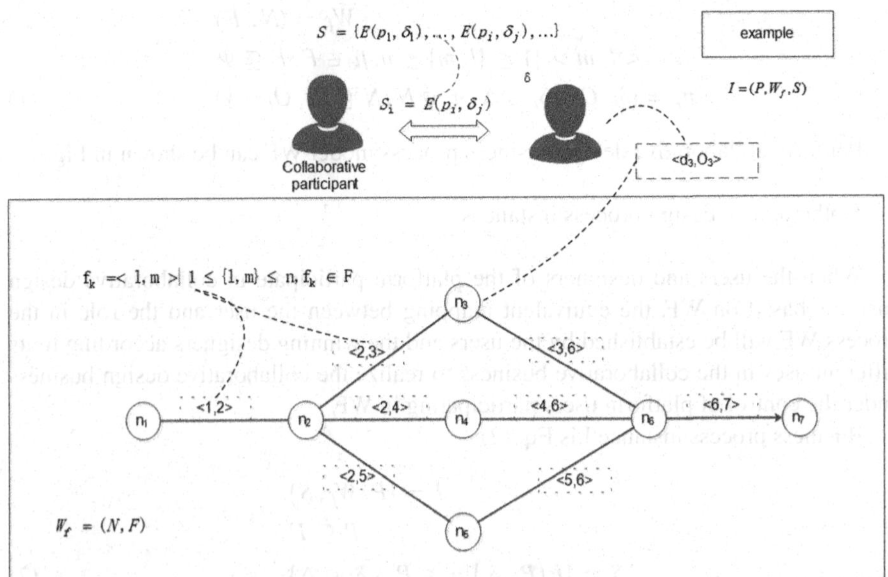

Fig. 2. The instance of collaborative design process

3) $p - \delta$ mappings

$p - \delta$ mapping refers to the specific collaborative design instance I. Platform users p participate in collaborative design in a way that plays a specific role δ in the workflow WF.

When designing a collaboration, one must design specific collaborative business processes for the workflow engine w_f, and specify the business bearers(δ) in the w_f ($w_f \in W_f$) for the business process nodes.

Platform collaborative design completing is supported by collaborative design instance $i(i \in I)$. The problem is how the collaborative design platform user p establishes equivalent mapping relationship with business bearers in workflow engine.

The collaborative design platform user p undertakes the equivalent mapping with the business bearer δ in the workflow engine, which involves the one timely to respond to the data information responded by workflow to the platform user p, and the other is to send the trigger event information of the node trigger σ in the engine to the user p and give the user response information to the event as a feedback to the engine to activate and ensure the correct operation of the business process w_f.

The equivalent mapping E of Collaborative design user p and workflow engine role δ are made up of three parts is Eq. 3. As shown in Fig. 3.

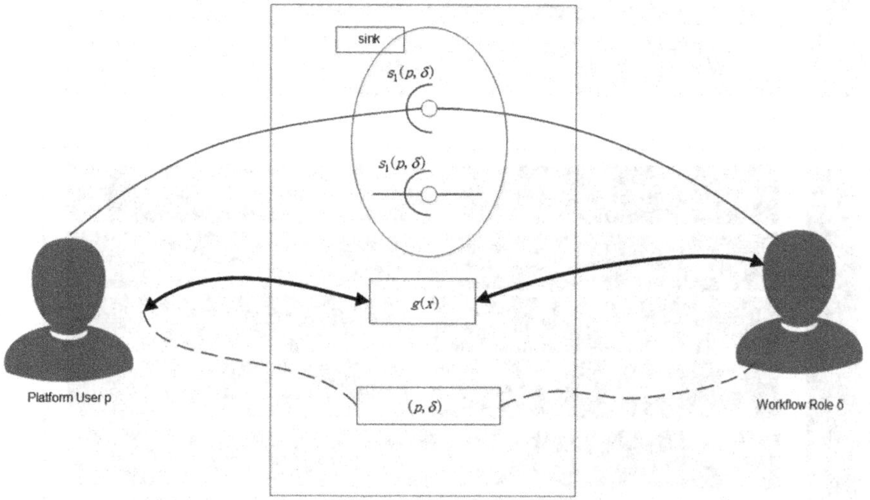

Fig. 3. The equivalent mapping relationship between the user and the workflow engine role in collaborative design service system

$$E(p, \delta) = ((p, \delta), g(x), S) \tag{3}$$

(p, δ) -means matching between users and roles. $g(x)$ -is the data transformation channel for user p and process, to realize the data transfer between user and role δ. S-transfers the task and triggered response of the role δ in the workflow engine to p end of the platform user, and makes the user have all the functions of the role δ in the business process, to realize the platform user p participate in the collaborative design based on the workflow engine.

2.2 Architecture of the Collaborative Design Service System

The collaborative design service system based on ceramic cloud service platform is mainly composed of server and client, which are the subsystems of ceramic cloud service platform. The development and implementation of the system require a strong scalability and flexibility of the system to adapt to various complex situations. The system of system design has clear levels, clear function modules and independent modules. Based on the above requirements, the collaborative design system of ceramic products based on Internet is implemented by three-tier architecture, which mainly consists of client-side presentation layer, server-side business logic layer and data storage persistence layer.

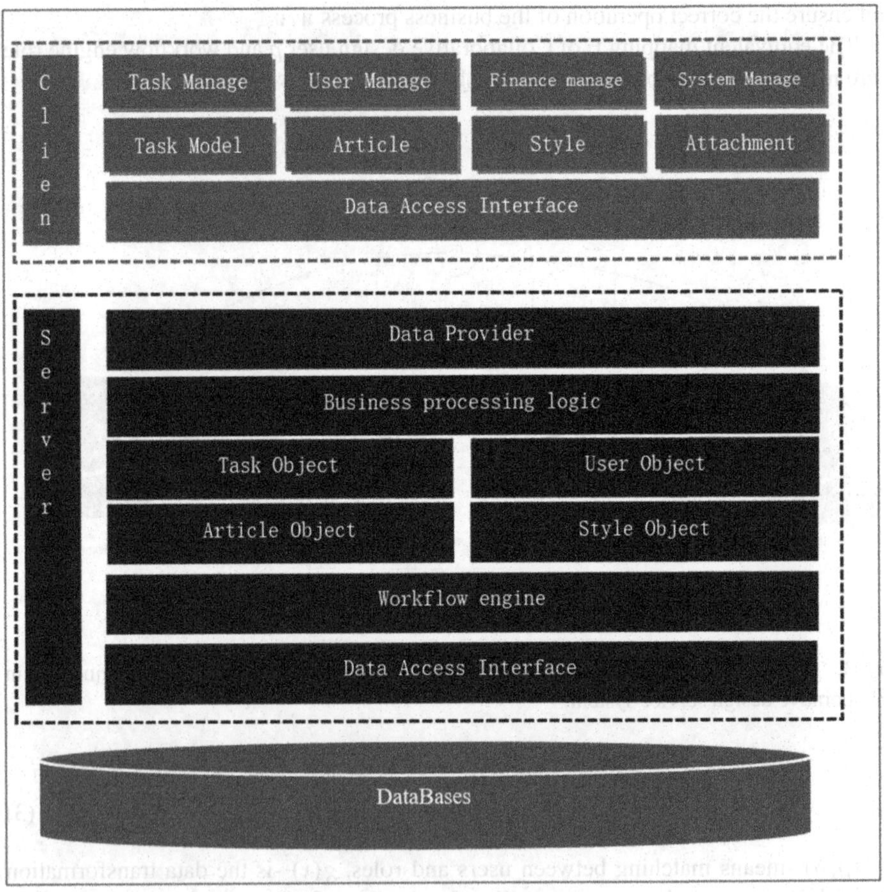

Fig. 4. System structure diagram

The system communicates with the data through the data access interface and server. The data is analyzed and displayed on the client side through the task object management module, so system module mainly includes task object management module, user object module, information object module and style object module. The server-side business logic layer includes: data supply service layer, business logic processing, task object, user object, consulting object, workflow engine, data access interface, etc. As shown in Fig. 4.

2.3 Collaborative Design Engine

The function of the collaborative design process management engine is to automatically match the collaborative design process with the same applications and tasks according to the type of collaborative design task, and to manage the progress of each collaborative design task in the system [9, 11]. First, a collaborative requirement task is issued by the user. Then, the sub-task is selected and completed by the design unit, and finally the whole task is completed.

In the collaborative design process, the collaborative design engine consists of one or more workflow executors. Workflow executors are actually task schedulers for collaborative design processes and allotters for designing resources. The main functions of the workflow executor are to explain the process definition, to be responsible for debugging the operation of the process, to create and manage the operation of the process instances, to schedule the operation of the activity and to create the work items to be processed, and to maintain the workflow control data and related data and the work list of the users [10, 12]. As shown in Fig. 5.

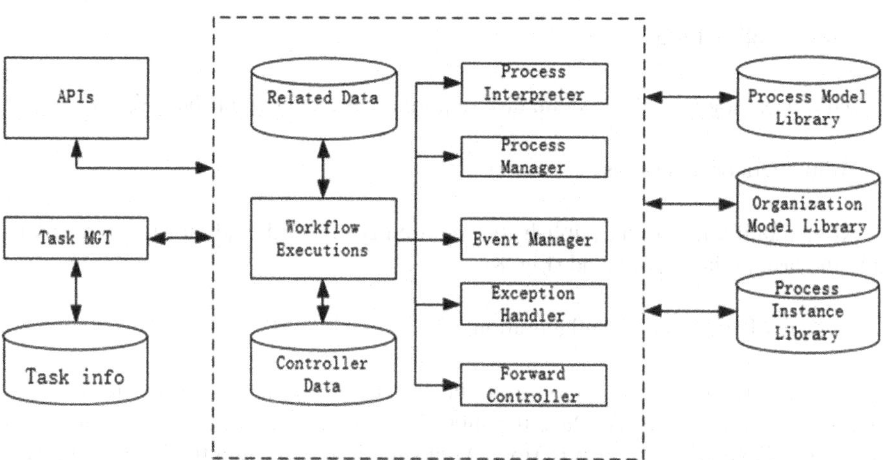

Fig. 5. Architecture of collaborative design engine

3 Implementation and Application of Collaborative Design Service System

3.1 Collaboration Instances

Cases of the case analysis come from the ceramic product cloud design service platform [13, 14], which is developed and operated by Jiangxi Ceramic Enterprise Information Engineering Technology Research Center [15], and is a comprehensive cloud platform developed for collaborative design services between upstream and downstream enterprises in the ceramic industry chain. The business scope of the platform covers the business processes of material design, product design, design and processing among enterprises, mainly realizing the design business interaction and collaborative service management among enterprises. At present, there are more than 400 enterprise users of ceramic cloud service platforms, which are divided into four categories: ceramic generation and manufacturing enterprise, art design service enterprise, art design studio and individual distributor.

Collaborative design service instance of ceramic products under the control of collaborative design engine based on ceramic cloud service platform can be shown in Fig. 6.

3.2 Realization of Collaborative Business Process

1) Users Publish Tasks.

After users log in to the system, task requirements can be published (Fig. 7).

2) Administrator review tasks.

After requirement users publish the requirement task, the administrator needs to audit the task in the background (Fig. 8).

3) Designer Participate in Collaboration.

First, the designer needs to log in to the system. Then the task instructions can be seen in the task hall. Finally, Once the subtasks have been designed, one just needs to click "I want to hand in the manuscript". After submission, the user that requires release will review and propose further amendments (Fig. 9).

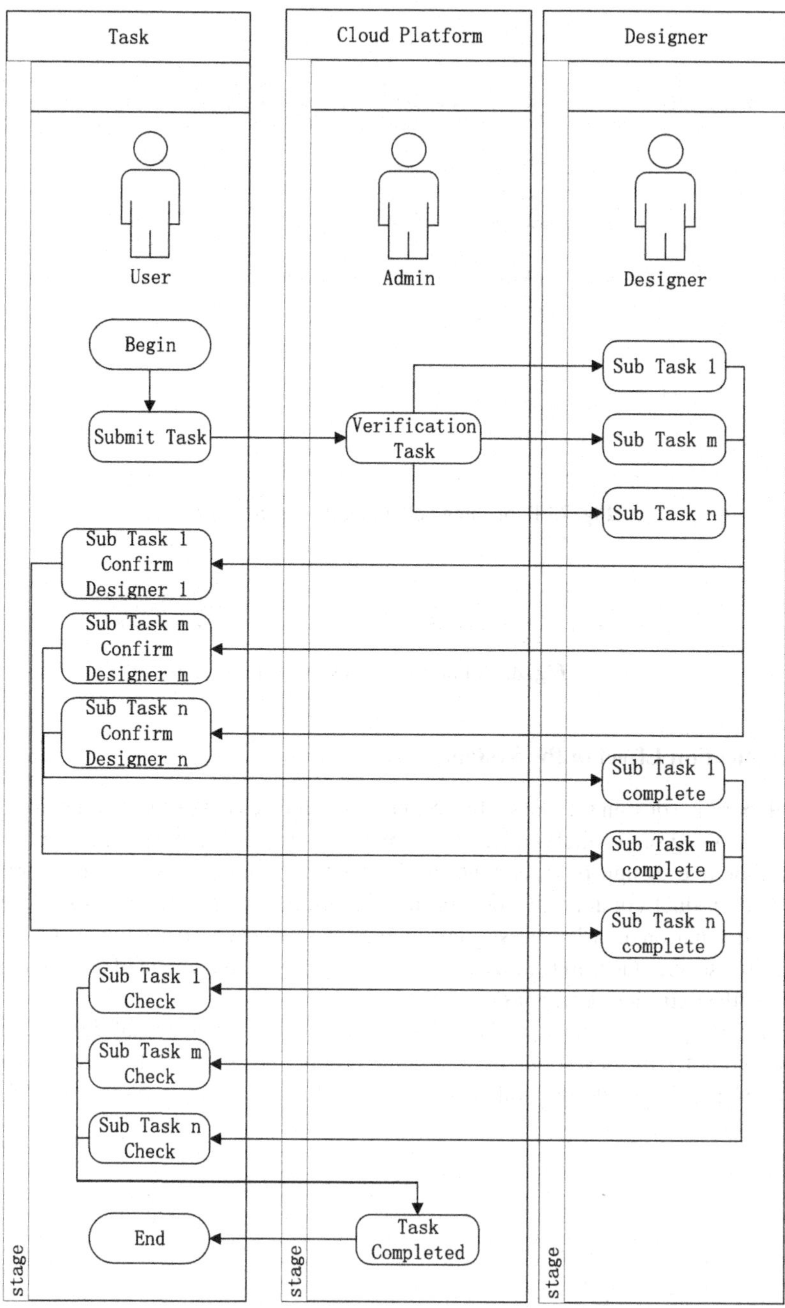

Fig. 6. An example of cooperative design service for ceramic products

Amount: ¥ 30000 CNY

Please describe the subtasks in the form of a word document, upload the task description document in the second step

Subtask 1: Allow 2 designers to participate, each will get 10000 CNY.

Subtask 2: Allow 2 designers to participate, each will get 10000 CNY.

Subtask 3: Allow 2 designers to participate, each will get 10000 CNY.

+ Add — Remove

Deadline: 2020-05-05 yyyy-mm-dd

Next

Fig. 7. Publishing collaborative design tasks

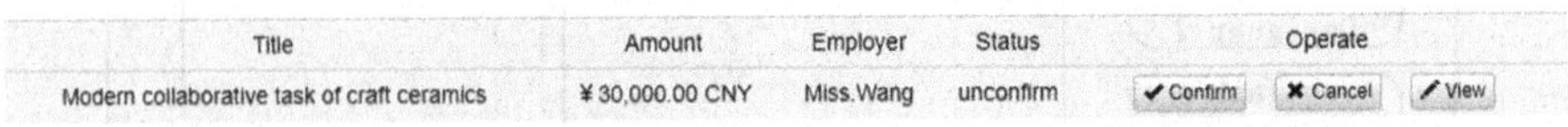

Title	Amount	Employer	Status	Operate		
Modern collaborative task of craft ceramics	¥ 30,000.00 CNY	Miss.Wang	unconfirm	✔ Confirm	✖ Cancel	View

Fig. 8. Administrator review tasks

3.3 Application Effect of the System

The collaborative design service system based on ceramic cloud service platform tackles with the cross-regional collaborative design management of ceramic products. Based on the collaborative design process control of workflow, it integrates the whole process of ceramic product collaborative design into system management, provides document sharing, information or task process control for all design service participants of collaborative design service task, and realizes the monitoring and management of the completion process of the expected design task.

It not only greatly improves the collaborative design efficiency of ceramic products and promotes the development of ceramic product design field, but also solves the problem of insufficient design ability of many small and medium ceramic enterprises.

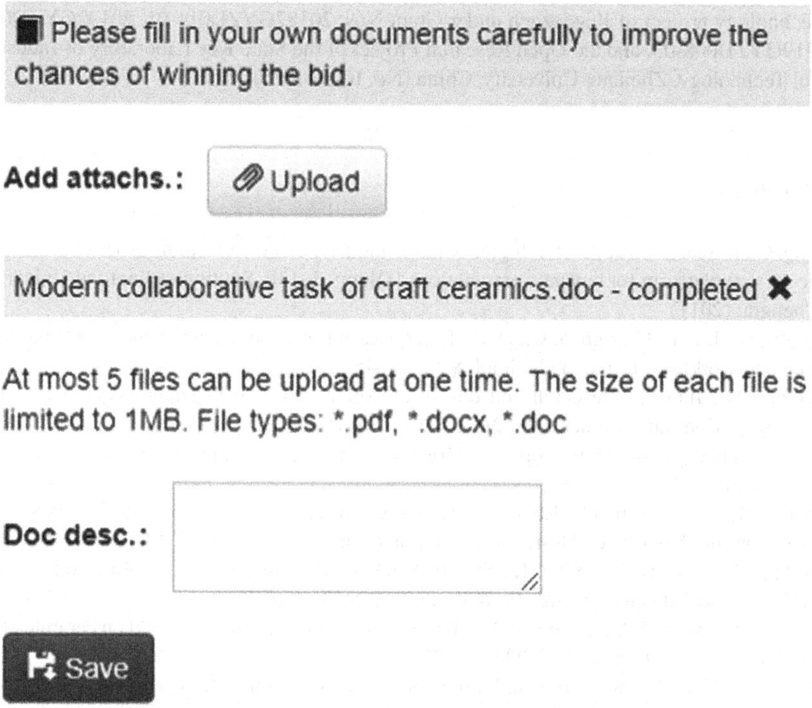

Please fill in your own documents carefully to improve the chances of winning the bid.

Add attachs.: Upload

Modern collaborative task of craft ceramics.doc - completed ✖

At most 5 files can be upload at one time. The size of each file is limited to 1MB. File types: *.pdf, *.docx, *.doc

Doc desc.:

Save

Fig. 9. Designers participate in collaborative design services

4 Conclusion

Based on the analysis of workflow-driven collaborative design and the research of the design service system, this paper presents a collaborative design service scheme combining ceramic cloud service platform. The scheme adopts the design idea of workflow engine and runs with the architecture of sub-platform middleware. It can meet the requirements of collaborative service task release and management. It is more friendly and intuitive to show the basic situation of the main tasks and sub-tasks. It can also divide a main task into several sub-tasks according to need, selection, acceptance and payment for the sub-tasks, and can support graphical operation interface. This paper is based on the actual project research. The research process is aimed to solve the problem of collaboration between modern design services with the help of open source workflow flexibility, scalability and other powerful functions. The focus of this stage is to design the collaborative design system into a third-party service platform and the next stage will be the expansion of the system providing a standard interface for new information systems.

Acknowledgment. This work is supported by the Key Research and Development Plan of Jiang Xi province under Grant No. 20171ACE50022 and the Natural Science Foundation of Jiang Xi province under Grant No. 20171BAB202011, the science and technology research project of Jiang Xi Education Department under Grant Nos. GJJ180730, GJJ180727, GJJ181520, and the Science and Technology project of Jingdezhen under Grant Nos. 20182GYZD011-01, 20192GYZD008-01, 2019GYZD008-03, and the Open Research Project of the State Key Laboratory of Industrial Control Technology, Zhejiang University, China (No. ICT 20025).

References

1. Jing, C.: Research on key technologies of multi-industry chain collaboration network and public service platform for business associations [D], pp. 8–130. Southwest Jiaotong University, Chengdu (2011)
2. Qijun, D., Jin, Y., Changhua, C., et al.: Enterprise automation management system based on web and workflow. Exp. Sci. Technol. **8**(5), 44–45 (2010)
3. Yixiao, W., Jian, Z.: Research and design of collaborative management system based on workflow. Comput. Technol. Dev. **24**(7), 233–236 (2014)
4. Saul Greenberg: Computer-supported cooperative work and groupware. Semant. Sch. **34**(7), 68–79 (2015)
5. Jung, M., Woongsup, C., Ravi, K., Hong, G., Jung, S., Kim, H.: Service Model for Collaborating Distributed Design and Manufacturing. Ulrike Sattler, CEUR-WS (2014)
6. Hongri, F., Yusheng, L., Ying, L.: SysML-based model integration for online collaborative design of mechatronic systems. In: ICED, pp. 237–246 (2013)
7. Nie, Y., Ziping, J., Yilai, Z., et al.: Collaborative design service model based on ceramic cloud platform. Chin. J. Ceram. **1**(2), 234–238 (2018)
8. Yilai, Z., Hua, H.: Research and implementation of ceramic cloud technology service integration platform. Chin.Ceram. **1**(02), 41–47 (2015)
9. Qingxuan, G., Shuping, Y., Dehai, Y., Yu, Z.: Research on process reengineering of product collaborative development in enterprise information environment. J. Beijing Inst. Technol. **30**(4), 496–500 (2010)
10. Yongqing, J.: Web application of enterprise workflow management system. Wuhan Iron Steel Technol. **43**(4), 35–39 (2005)
11. Cao, Y., Liu, Y.S., Fan, H.R., Fan, B.: SysML-based uniform behavior modeling and automated mapping of design and simulation model for complex mechatronics. Comput. Aided Des. (2012). https://doi.org/10.1016/j.cad.2012,05.001
12. Fan, H.R., Liu, Y.S.: Integration of system-level design and detailed design models of mechatronic systems based on SysML and step ap 203 standard', Technical Presentation. In:Proceedings of the ASME 2012 IDETC/CIE 2012, Chicago, IL, USA (2012)
13. Chao, L., Yilai, Z.: Full-text retrieval optimization of ceramic cloud based on ant colony algorithm. Fujian Comput. **1**(1), 1–3 (2014)
14. Zhonghua, W., Yongkang, P.: Research and development of ceramic design service resource management system. Fujian Comput **1**(7), 29–30 (2015)
15. Wang, (ed.): Jiangxi ceramic enterprise information engineering technology research center. China Ceram. Ind. **1**(4), 80–80 (2019)

Towards NoC Protection of HT-Greyhole Attack

Soultana Ellinidou[1(✉)], Gaurav Sharma[1], Olivier Markowitch[1], Jean-Michel Dricot[1], and Guy Gogniat[2]

[1] Cybersecurity Research Center, Université Libre de Bruxelles, Brussels, Belgium
{soultana.ellinidou,gsharma,olivier.markowitch,jdricot}@ulb.ac.be
[2] Lab-STICC, Université de Bretagne Sud, Lorient, France
guy.gogniat@univ-ubs.fr

Abstract. As the number of processing cores is increasing dramatically, the communication among them is of high importance. Network-on-Chip (NoC) has direct access to all resources and information within a System-on-Chip (SoC) by rendering it appealing to attackers. In this paper a novel Hardware Trojan (HT) assisted Denial of Service (DoS) attack, called Greyhole attack is introduced. The HT-Greyhole attack targets the routers within NoC by causing performance decrease and packet loss increase. However, during an HT-Greyhole attack certain packets, which are arriving towards the router, are dropped which makes it hard to detect. In this paper, we design a detection and defense method, against HT-Greyhole attack, which is based on Software Defined Network-on-Chip (SDNoC) architecture. The results demonstrate that by using the proposed defense method the packet loss decreases by 76% under Transpose traffic, 77% under BitReverse and 50% under Uniform traffic.

Keywords: Hardware Trojans · Network-on-Chip · Greyhole attack · HT DoS attacks · SDNoC

1 Introduction

Malicious hardware modifications at different stages of its life cycle create major security concerns in the field of electronics. The Hardware Trojan (HT) attacks emerged as a major security threat for Intellectual Properties (IPs) blocks, Integrated Circuits (ICs), Printed Circuit Boards (PCBs), and System-on-Chip (SoC). Specifically, these attacks introduce a malicious modification of a circuit during the design or fabrication process in an untrusted design house or foundry, in which untrusted people, design tools, or components are involved [1]. Such modifications can lead to abnormal functional behavior of a system, degrade performance and provide covert channels or backdoors by which an attacker can leak sensitive information.

Since the number of processors and cores on a single chip is increasing, the interconnection between them becomes significant. A key challenge is to provide secure and reliable communication in the SoC, even in the case of an untrusted

M. Qiu (Ed.): ICA3PP 2020, LNCS 12454, pp. 309–323, 2020.
https://doi.org/10.1007/978-3-030-60248-2_21

Network-on-Chip (NoC) IP inserted into it. Since the NoC has direct access to all communication resources and information flow within the SoC, attackers have a strong motivation to exploit its possible vulnerabilities. In recent literature, a vast number of HT attacks, which are mainly focusing on NoC, have been introduced [2–4]. As far as the hardware methods for the detection and defense of the HT-attacks targeting NoC, it is observed that most of them are employed in the Network Interfaces (NIs) [2], which connect the IP cores and routers, some of them on the links between routers [5] and very few on the routers [3].

Following the literature, the common assumption is that a NoC is supplied to a SoC integrator and there is a possibility that it is already compromised with a HT [6]. In order to activate the HT, a malicious circuit has already been inserted during the design time of the IP block and a malicious program can activate it later at runtime. The possible attacks due to infected NoC IP block are:

- **Snooping**: In this case, illegal monitoring is performed by an untrusted router within the path, which tracks the number of packets between source and destination IP cores.
- **Corruption of the data**: A malicious router can modify the content of the incoming flits and the route of the packets.
- **Spoofing**: A malicious router copies and replays packets, which may lead to the malfunction or eviction of sensitive data.
- **Denial of service (DoS)**: The denial or distributed denial of service can make the resources unavailable to legitimate IPs.

In this paper, we consider a specific HT assisted DoS attack, called the Greyhole attack, which targets the routers of a NoC within a SoC. The Greyhole attack is a well known attack from Wireless Sensor Networks (WSN) [7,8]. In case of a Greyhole attack, a malicious router blocks certain packets from its neighboring routers instead of forwarding them. Hence, critical packets, that are forwarded to a Greyhole router, are captured and can not arrive to their destinations. In order to detect and mitigate a malicious router within a network, some of the security mechanisms encountered in the literature are: data partitioning, key management, key generation, localization and trust management [8].

However, despite the large amount of research contributions in WSN about the Greyhole attack, this attack has not been introduced in the field of electronics and more specifically in NoC context. Hence our main contributions are summarized as follows:

- the description and activation of an HT-Greyhole attack in NoC context,
- the exploration of Software Define Network on Chip (SDNoC) as a potential solution for NoC protection,
- a security management mechanism relying on SDNoC, as key proposal in order to identify malicious routers,
- depending on the position of affected routers, a route exclusion approach is presented in order to mitigate the impact of the attack.

SDNoC provides secure paths in presence of untrusted routers and assures that the packets will be successfully delivered to their destination.

The rest of the paper is organized as follows: In Sect. 2 we discuss the related work. Thereafter, in Sect. 3 the SDNoC concept is presented. In Sect. 4 the launching of HT-Greyhole attack is introduced, followed by the detection strategy and the defense approach, described in Sect. 5 and Sect. 6 respectively. The evaluation results are discussed in Sect. 7, followed by conclusion and future work in Sect. 8.

2 Related Work

There is no existing literature on HT-Greyhole attacks, however since Greyhole attacks are variants of Blackhole attacks, the related literature of HT-Blackhole attacks is presented. In 2018 the HT-Blackhole attack targeting the NoCs was introduced [9], the authors investigate not only the Blackhole but also the Sinkhole attack in the context of NoC. Specifically, they focus on the effects of the attack by measuring the packet loss rate, considering the number of HTs and their distribution in NoC. They provide a theoretical detection method, where a global manager injects detection request packets to randomly selected routers in order to find the suspicious one. Though the main disadvantage of the detection method is that it can only detect HTs which are always on trigger mode. A defense method is also presented, where each router keeps a record of neighbors, which is updated by the global manager and needs to be checked by the router itself before taking routing decisions.

Afterwards, in [10], an analysis of the HT-Blackhole attack considering the area and power overhead of the malicious router was presented. Precisely, the authors presented the influence of the number of HT-Blackhole routers along with their distribution in the NoC. Another contribution by the same authors is presented in [11], where they proposed a secure protocol with runtime detection and protection of HT-Blackhole attack. The proposed secure protocol protects the system from HT-Blackhole attacks, but it increases dramatically the overhead due to the large number of acknowledgment (ACK) packets that need to be exchanged between the routers for each data packet transmission from a source to a destination router.

3 SDNoC

The launching part can individually be performed on different NoC architectures. However, for our defense and detection approach we target a NoC alternative, called SDNoC [12–14]. The SDNoC concept is inspired by the well known Software Defined Networking (SDN) technology [15]. The main idea of SDNoC is to minimize router complexity by exporting the routing logic to a centralized controller which has a general view of the network and can take routing decisions efficiently. According to the authors of [12,14], SDNoC could possibly be adaptable for SoCs thanks to its advantages: 1) it reduces the hardware complexity, 2) it has high re-usability and 3) it has flexible management of communication policies. However, there are also some challenges that should be taken into account, in particular the high overhead for path selection in software against hardware based approaches and the centralized controller can be a single point of failure.

The only difference between SDNoC and NoC is that the SDNoC manages routing in an adaptive manner with the help of a centralized controller. In Fig. 1, an SDNoC architecture is depicted, which consists of 16 routers and a centralized controller. The routers are interconnected with the neighboring routers through links and with a Processing Element (PE) through the NI. The centralized controller is interconnected with the routers through direct links, as depicted in Fig. 1. Furthermore, it sends configuration packets and manages the routing of the packets in an efficient manner.

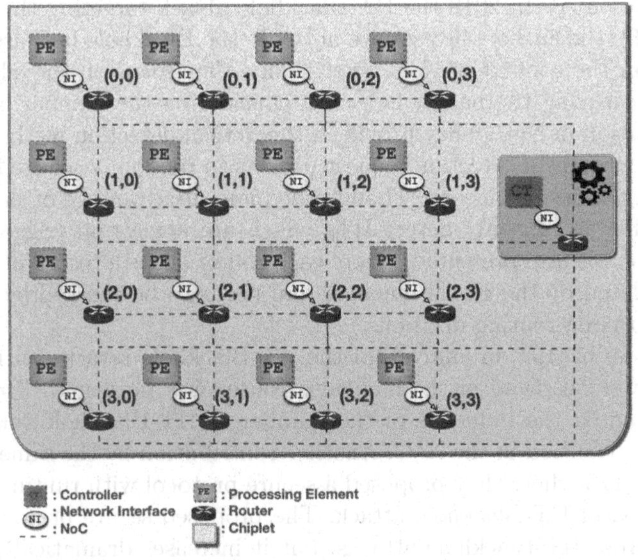

Fig. 1. Software Defined Network-on-Chip (SDNoC) architecture

4 Launching of HT-Greyhole Attack

HTs can be inserted into the pipeline of a Virtual Channel (*VC*) router according to [16] and at each input port of a router. The main HT is placed on the *VC Allocator* and the other HTs are synchronized with it through a control signal [9]. A HT structure consists of three modules: the Trigger, the Configuration and the Greyhole function module. The placement of the HT-Greyhole in a NoC router is shown in Fig. 2, where specifically a malicious HT-Greyhole router architecture is illustrated. The router consists of 5 input/output ports, 5 *VC*s, 5 buffers with a counter (C), a *VC allocator*, a *Crossbar switch*, a *Switch allocator*, a *TrustTable* and *Flow Tables.* The five ports correspond to the four cardinal directions and the local direction which connects the router with the PE through the NI. The router employs a pipelined design with speculative path selection to improve performance. The router consists of a two-stage, pipelined architecture. The first stage is responsible for routing and the second stage is responsible for crossbar traversal. In this work, the functionality of the router is described with respect to a 2D mesh interconnect. A HT is placed in each input port: South, North, East, West, Local, which are synchronized with the main HT-Greyhole which is placed on the *VC allocator.* More details for the HT insertion in one port in a single cycle VC-based router can be found in [17].

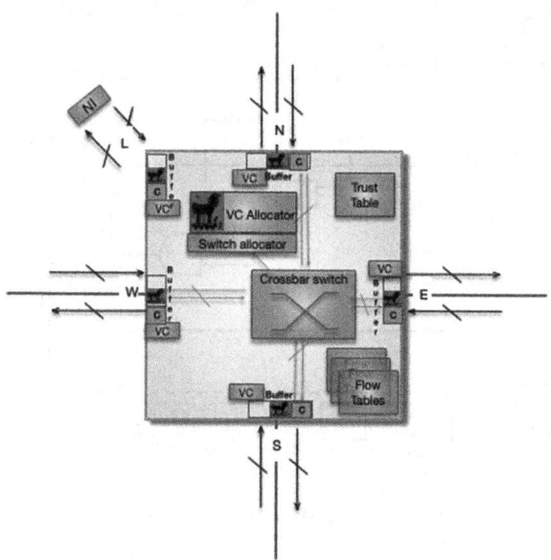

Fig. 2. HT-Greyhole router.

Before an attack is launched, a configuration packet should be sent to the target router by an attacker through a malicious program running on the given IP core connected to the router. The configuration packet, which is depicted in Fig. 3, consists of the following fields:

- **Config cmd**: is the field of a packet that consists of a specific bit pattern (e.g. 00110101), which states as a HT configuration packet.
- **Trigger**: has 2 modes: Always Activated (AA) and Destination Based (DB). An AA trigger HT is always active, while a DB trigger is activated only when the destination ID of an incoming packet is identical with the target ID of the configuration packet.
- **Packet Type**: declares the type of packet which is either signal or data packet.

- **Activation Signal:** could be on or off depending on the activation of the HT.
- **Target ID**: specifies the target address for the DB trigger.
- **Interceptor ID**: in case that an HT is launched, every data packet Destination ID will be replaced with the Interceptor ID.

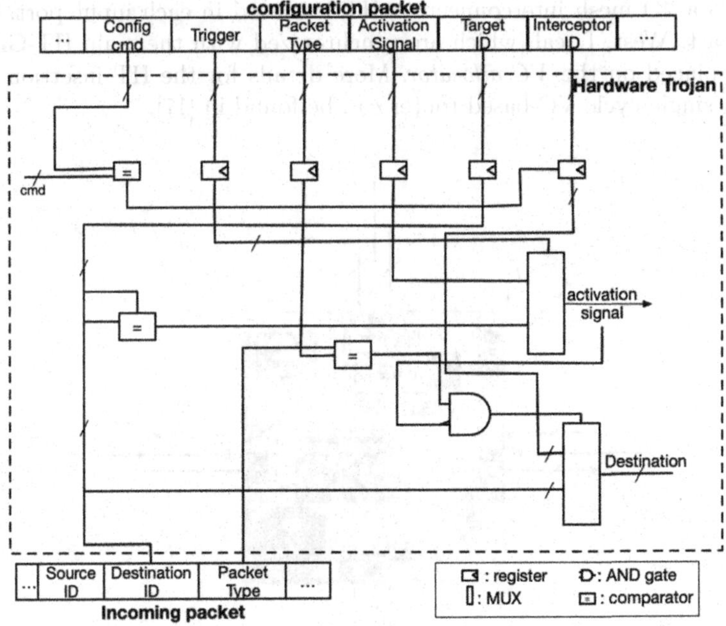

Fig. 3. HT design on circuit level.

After the configuration packet has been delivered to the target router, the HT configuration information will be saved in a set of registers (Fig. 3). When a HT has been configured, it can be activated by the trigger module. More precisely the steps for launching an attack are the following:

- **Step 1**: An attacker sends a configuration packet through a malicious program to the target router.
- **Step 2**: The HT, placed in the target router, receives the configuration packet and updates its configuration information.
- **Step 3**: The trigger module chooses its mode based on the trigger field data stored in the registers.
- **Step 4**: An activation signal is generated by trigger, by taking into account the trigger mode. As for AA mode an activation signal is generated all the time, while for the DB mode the activation signal is set to on when the destination ID of the incoming packet matches with the Target ID of the register.

- **Step 5**: The attacker specifies in the configuration packet the type of the packet that needs to be dropped. If the type of the packet matches with the type of the incoming packet then we move to the next step (in our scenario the signal packets are normally processed and the data packets are dropped).
- **Step 6**: Launch the attack according to the signal and the packet type.
- **Step 7**: If the Packet Type is *data* then the Destination ID of the incoming packet will be replaced with the Interceptor ID. If the Packet Type is *signal*, the Destination ID will not be modified.

5 Detection

HT assisted DoS attacks are hard to be detected due to their low silicon footprint, small power and area consumption but also due to their conditional activation during the run time. Specifically in the HT attack presented in [9], the area and the power consumption are 0.07% and 0.02% of a NoC router and in [10] the malicious router area and power increase are 1.98% and 0.74%, respectively.

Our detection strategy has been designed in order to specifically detect a HT-Greyhole attack in the context of SDNoC. Based on our architecture, each router has a counter in each port (Fig. 2), which is incremented every time that a new packet is imported and it is decremented every time that a packet is exported. The results are saved in the *TrustTable*, which includes all the values for each port. The routers are responsible to periodically send the *TrustTable* along with their *RouterID*, to the controller. The controller calculates and chooses the routes for each individual source and destination by storing them in the table *Routes*. The value k indicates the 4 different directions north, east, south, west.

As soon as the controller receives a *TrustTable*, it uses the Algorithm 1 in order to find out which routers are considered as suspects. In the algorithm, the controller checks if any input of the *TrustTable* is less than a threshold value (tv), which value can be chosen depending on traffic pattern or buffer holding capacity. More details about tv value can be found in the Sect. 7.

Since a malicious router can modify its *TrustTable* and pretend that it is non-malicious, the only option is to be detected through their neighbors. Hence the algorithm searches the previous hop (*neighbor*) of the given *RouterID* and afterwards it clarifies if the direction of the *neighbor* matches with the direction of the port value of *RouterID* within the *TrustTable*. If so then the neighbor is considered as suspect.

Since the controller calculates the table *Suspect* of the given *RouterID*, it will also check the tables *Suspect* of the other *RouterID*'s. If a suspect appears at least in two different *Suspect* tables, because each router could have at least two neighbor routers, the suspect router will be considered as malicious.

The detection method is less costly in terms of overhead and complexity since the control links between routers and controller are utilized and the only router side operation is to calculate a *TrustTable*, which includes the values of the 4 counters (4-bit each), and to send it through the control links to the controller.

Algorithm 1 Detection Algorithm

```
Data: Routes[ ][ ], TrustTable[ ][ ], RouterID, a=0
for k=1:4 do
    if TrustTable[k][2] < tv then
        for j=1:Routes.rows() do
            for t=1:Routes.column() do
                if Routes[j][t] == RouterID then
                    neighbor == Routes[j-1] [t];
                    if TrustTable[k][1] == direction.neighbor() then
                        a=a+1;
                        Suspect[a]=neighbor;
                    end
                end
            end
        end
    end
end
```

6 Defense

As the proposed detection strategy has already identified the malicious routers and their positions, a route exclusion approach is presented in order to mitigate the attack. The controller executes the defense approach which consists of following three phases:

- **Route Exploration Phase:** Given a source and a destination the controller computes a set of admissible routes based on Odd-Even (OE) routing algorithm [18] and it stores them in a table. OE is a turn model routing algorithm, lightweight and deadlock-free. Among the existing turn model routing algorithms OE tends to provide better performance and higher adaptiveness than the others.
- **Untrusted Paths Phase:** From the detection algorithm, the controller already has a list with the malicious routers. Hence, in this phase, it has as input the set of admissible routes from the previous phase, which are checked if they include any malicious router. The routes that include a malicious router are marked as untrusted and the rest of the routes as trusted.
- **Selection Phase:** The inputs in this phase are all the trusted routes from a given source to a destination. In the classic OE routing algorithm, a random route is chosen among the admissible ones. However, in our case the controller chooses the least congested route among the admissible ones by calculating the link load (l_i) of the routes. The l_i corresponds to the number of flits per second that are passing through the link. The designed formula in order to avoid the highly-loaded links and routers within the route is depicted in (1).

$$S = \sum_{i=0}^{L_t} l_i. \tag{1}$$

Where S is the computed score for each admissible trusted route and L_f is the number of links along the route.
Among the scores of the different routes, the route with smallest score value is selected and indeed it represents the least congested route. Note that the controller's knowledge concerning the data network state (via the link load) is gathered by immediate inputs from different routers and their *TrustTable* computations. Nevertheless for the initial route computation there is no available score, hence a random route is chosen among the admissible ones, offered by OE.

7 Evaluation

In order to perform an attack but also to evaluate our detection and defense strategy, simulations were performed with Garnet2.0 [19], which is a NoC model implementation within the gem5 simulator. The traffic generated by the processing cores is based on the traffic injection rate (tir), which is expressed as the average number of packets injected by the cores into the network per clock cycle ($0 < tir \leq 1$). Each processing core will generate a packet every $1/tir$ clock cycles on average, but the actual time at which the packets are transmitted is random. It should be noted that for each scenario we perform 40 iterations, of which the average value of throughput and packet loss is calculated.

An 8×8 topology is simulated, by taking into account that it contains 1, 3 and 6 HT-Greyhole routers. Furthermore, three different traffic scenarios have been evaluated: Transpose, BitReverse and Uniform.

As far as the evaluation of the detection strategy is concerned, binary classification is used. Binary or binomial classification is the task of classifying the elements of a given set into two groups (predicting which group each one belongs to) on the basis of a classification rule [20]. Considering a two-class prediction problem, where the outcomes are labeled either as Positive (P) or Negative (N). In this case there are four possible outcomes from a binary classifier. If the outcome from a prediction is P and the actual value is also P, then it is called a True Positive (TP); however if the actual value is N then it is said to be a False Positive (FP). Conversely, a True Negative (TN) has occurred when both the prediction outcome and the actual value are N and False Negative (FN) is when the prediction outcome is N, while the actual value is P. These four counts constitute a confusion matrix shown in Table 1.

Table 1. Confusion matrix

		True condition	
Predicted condition	Total population	Positive (P)	Negative (N)
	Positive (P)	True Positive	False Positive
	Negative (N)	False Negative	True Negative

Table 2 presents the measurements, that we took into account, for binary classification based on the values of confusion matrix. In the context of confusion matrix, Receiver Operating Characteristic (ROC) is a graphical representation plot that illustrates the diagnostic ability of a binary classifier system as its discrimination threshold is varied. ROC graphs are two-dimensional graphs in which True Positive Rate (TPR) is plotted on the y axis and False Positive Rate (FPR) is plotted on the x axis. When using normalized units, the Area Under the Curve (AUC) is equal to the probability that a classifier will rank a randomly chosen P instance higher than a randomly chosen N one [21]. The Area Under Curve (AUC) measure gives a better view about the algorithms capability of distinguishing between classes. The higher AUC, the better model is at predicting the positive and negative values.

Table 2. Measurements for binary classification

Measurement	Abbr	Formula	Explanation
Sensitivity, Recall or True Positive Rate	TPR	$\frac{TP}{TP+FN}$	Effectiveness of a classifier to identify positive labels
Miss rate or False Negative Rate	FNR	$\frac{FN}{FN+TP}$	Probability of identifying positive labels as negative
Accuracy	ACC	$\frac{TP+TN}{TP+TN+FP+FN}$	Overall effectiveness of a classifier
Area Under the Curve	AUC	$\frac{1}{2}(\frac{TP}{TP+FN} + \frac{TN}{TN+FP})$	Classifier's ability to avoid false classification

By using binary classification, we identify 27 different scenarios of an 8×8 topology, where we took into account, different traffics (Transpose, BitReverse, Uniform), different number of HT (1, 3, 6) and different value of tv (0, -10, -100) of the detection algorithm. For our scenario a malicious node (HT-Greyhole router), which considered as negative and non-malicious node, which considered as positive, in that setting:

- **TP**: Non-malicious node correctly identified as non-malicious.
- **FP**: Malicious node incorrectly identified as non-malicious.
- **TN**: Malicious node correctly identified as malicious.
- **FN**: Non-Malicious node incorrectly identified as malicious.

Figure 4 represents the ROC curves of the different scenarios by taking into account the number of HT-Greyhole routers within the network, under different traffic scenarios and different tv values. From the graphs, it is obvious that the ROC curves for the $tv = -100$ tend to be ideal for all scenarios, hence our algorithm is able to better distinguish between positive and negative values for this threshold value. As fas as the Accuracy (ACC) of the algorithm is concerned, for $tv = 0$ the ACC is between 73.4% and 92.2%, for $tv = -10$ the ACC was between 81.2% and 98.4% and for $tv = -100$ the ACC is between 95.2% and 100%. However it worths to be mentioned that for some test cases for $tv = 0$ and $tv = -10$, we noticed FN values, hence the FPR will be higher. As far as the AUC value is concerned for the $tv = -100$, it is between 0.965 and 1, which means that in some test cases it is perfect (AUC=1) and in other cases tend to be perfect $(0.95 < AUC < 1)$.

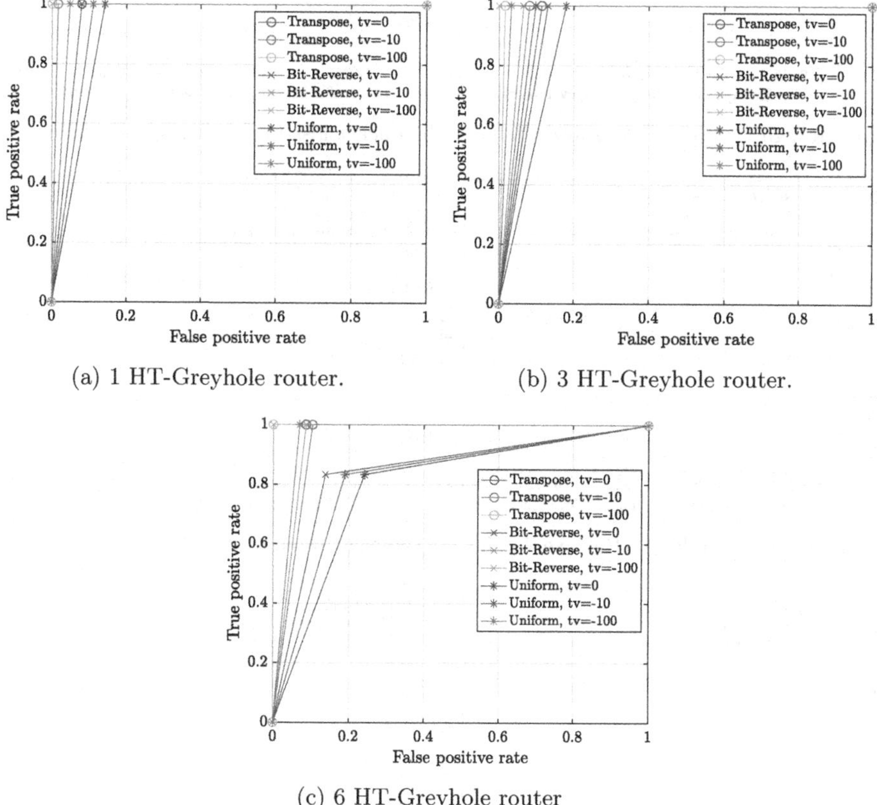

(a) 1 HT-Greyhole router.

(b) 3 HT-Greyhole router.

(c) 6 HT-Greyhole router

Fig. 4. Roc curve diagrams for 1, 3, 6 HT-Greyhole routers with tv = 0, −10, −100 and for Transpose, BitReverse, Uniform traffic.

Figure 5 depicts a scenario of 1 HT-Greyhole router. More precisely, on Fig. 5(a), (c), (e) the average throughput under different injection rates (0.015–0.024) is presented for Transpose, BitReverse and Uniform traffic respectively. In Fig. 5b, d, f the packet loss rate is shown under different injection rates. From the figures it is obvious that there is an increase of the packet loss rate and a

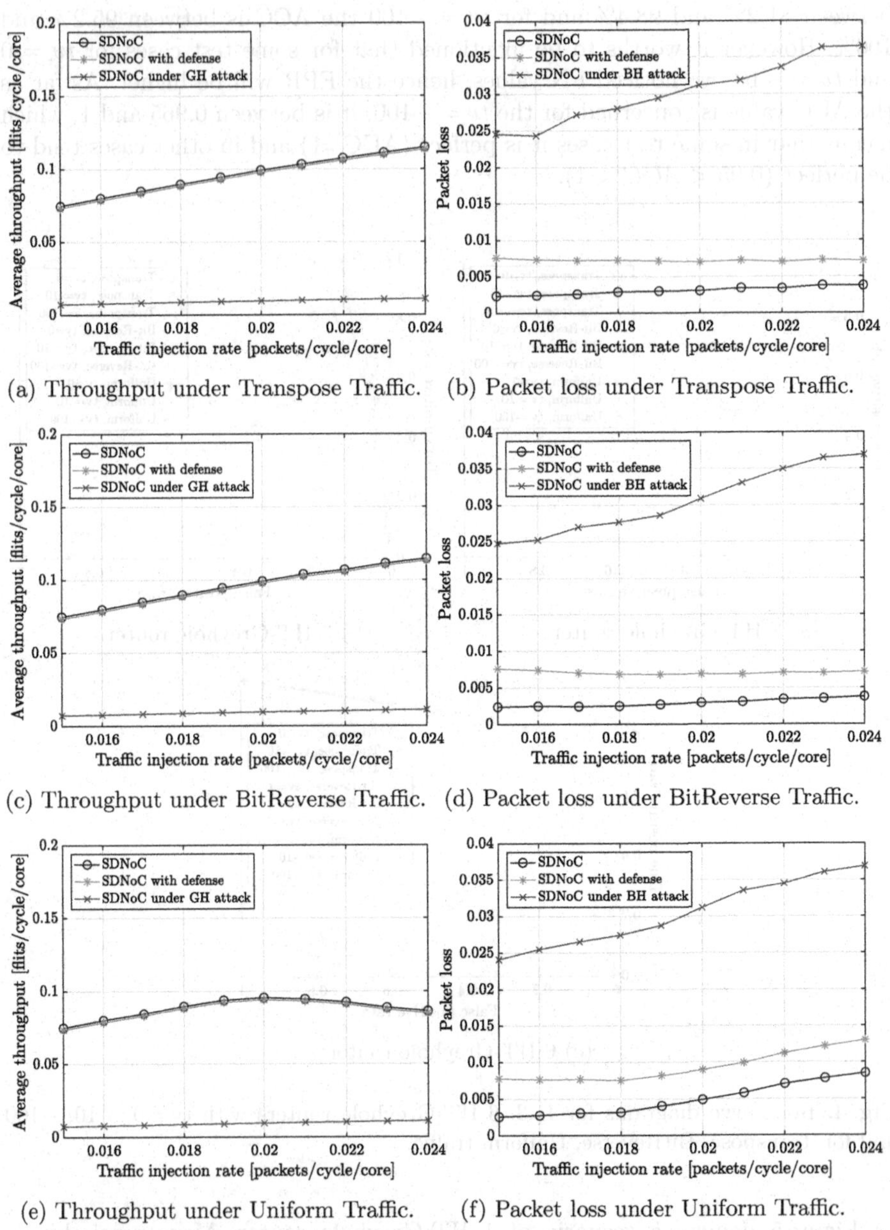

(a) Throughput under Transpose Traffic. (b) Packet loss under Transpose Traffic.

(c) Throughput under BitReverse Traffic. (d) Packet loss under BitReverse Traffic.

(e) Throughput under Uniform Traffic. (f) Packet loss under Uniform Traffic.

Fig. 5. Throughput and packet loss graphs for 1 HT-Greyhole router under Uniform, Transpose and BitReverse traffic scenarios.

decrease of the throughput of the SDNoC when the network is under attack compared to when the network works normally. Furthermore, when our defense part is employed on SDNoC, it is noticeable that the throughput values of SDNoC with defense and the throughput values of normal SDNoC tend to be identical. Precisely, under the higher injection rate we have an increase of 3% under Uniform, Transpose and BitReverse traffics of the overall packet loss rate between SDNoC and SDNoC under GH attack. As far as the average throughput, it is decreased by 8% under Uniform traffic, 10% under Transpose and BitReverse traffic. Thus the detection of this attack is a very difficult process.

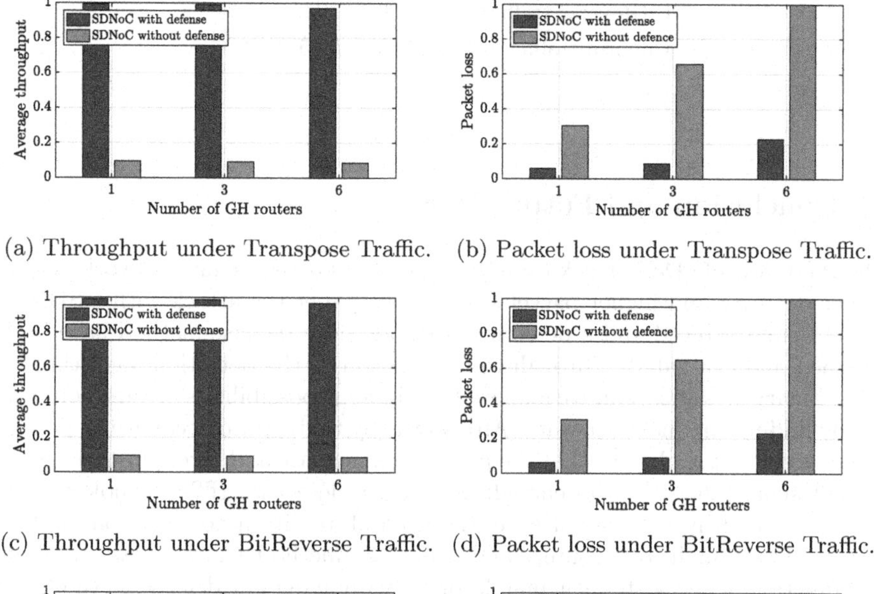

(a) Throughput under Transpose Traffic. (b) Packet loss under Transpose Traffic.

(c) Throughput under BitReverse Traffic. (d) Packet loss under BitReverse Traffic.

(e) Throughput under Uniform Traffic. (f) Packet loss under Uniform Traffic.

Fig. 6. Throughput and packet loss graphs for 1, 3, 6 HT-Greyhole routers.

In Fig. 6, three different scenarios are presented in each graph. In the first scenario, only one HT-Greyhole router is considered, second scenario considers three HT-Greyhole routers and in third instance, there are six HT-Greyhole routers. Figure 6(a), (c), (e) depict the *normalized* average throughput under

Transpose BitReverse and Uniform traffic respectively. Figure 6(b), (f), (d) show the *normalized* packet loss rate. By taking into account these results, the packet loss improvement is shown in Table 3. As far as the throughput is concerned, by applying our defense method it improved between 63–66% for Uniform traffic and 88–89% for Transpose and BitReverse traffics.

Table 3. Packet loss improvement with defense method.

# HT-Greyhole router	1	3	6
Transpose traffic	27.3%	56.8%	76%
BitReverse traffic	27.6%	56.2%	72%
Uniform traffic	23.6%	50.5%	66%

8 Conclusion and Future Work

The HT-Greyhole DoS attack targeting NoC can possibly cause network performance decrease and higher packet loss. In this paper the attack within SDNoC context is introduced and a detection method and a defense method have been designed and evaluated. Through the evaluation of the detection algorithm by using binary classification, we explore the different possibilities of threshold values by finding the most accurate. Afterwards by taking into account the performance results, it is obvious that the packet loss increase and throughput decrease are not significant (3%–0%) enough in order to detect an HT-Greyhole router, due to its stealthy behavior. Hence, the need of an alternate detection method able to detect malicious routers and a defense method which allows the normal function of the systems is mandatory. By applying our defense method, the interconnection system continues to function normally by improving the overall packet loss 23.6%–7% and the average throughput 63%–9%. As a future work, more measurements in the context of power and area consumption of this attack could be considered together with the time of HT-Greyhole router detection and its affect on the system.

References

1. Bhunia, S., Tehranipoor, M.: The Hardware Trojan War. Springer, Cham (2018). https://doi.org/10.1007/978-3-319-68511-3
2. Ancajas, D.M., Chakraborty, K., Roy, S.: Fort-NoCs: mitigating the threat of a compromised NoC. In: Proceedings of the 51st Annual Design Automation Conference, pp. 1–6. ACM (2014)
3. Frey, J., Yu, Q.: Exploiting state obfuscation to detect hardware trojans in NoC network interfaces. In: 2015 IEEE 58th International Midwest Symposium on Circuits and Systems (MWSCAS), pp. 1–4. IEEE (2015)

4. Hussain, M., Guo, H.: Packet leak detection on hardware-trojan infected NoCs for MPSoC systems. In: Proceedings of the 2017 International Conference on Cryptography, Security and Privacy, pp. 85–90. ACM (2017)
5. Boraten, T., Kodi, A.K.: Mitigation of denial of service attack with hardware trojans in NoC architectures. In: 2016 IEEE International Parallel and Distributed Processing Symposium (IPDPS), pp. 1091–1100. IEEE (2016)
6. Rajesh, J.S., Chakraborty, K., Roy, S.: Hardware trojan attacks in SoC and NoC. In: Bhunia, S., Tehranipoor, M.M. (eds.) The Hardware Trojan War, pp. 55–74. Springer, Cham (2018). https://doi.org/10.1007/978-3-319-68511-3_3
7. Tripathi, M., Gaur, M.S., Laxmi, V.: Comparing the impact of black hole and gray hole attack on LEACH in WSN. Procedia Comput. Sci. **19**, 1101–1107 (2013)
8. Martins, D., Guyennet, H.: Wireless sensor network attacks and security mechanisms: a short survey. In: 2010 13th International Conference on Network-Based Information Systems, pp. 313–320. IEEE (2010)
9. Zhang, L., Wang, X., Jiang, Y., Yang, M., Mak, T., Singh, A.K.: Effectiveness of HT-assisted sinkhole and blackhole denial of service attacks targeting mesh networks-on-chip. J. Syst. Archit. **89**, 84–94 (2018)
10. Daoud, L., Rafla, N.: Analysis of black hole router attack in network-on-chip. In: 2019 IEEE 62nd International Midwest Symposium on Circuits and Systems (MWSCAS), pp. 69–72. IEEE (2019)
11. Daoud, L., Rafla, N.: Detection and prevention protocol for black hole attack in network-on-chip. In: Proceedings of the 13th IEEE/ACM International Symposium on Networks-on-Chip, pp. 22. ACM (2019)
12. Ellinidou, S., Sharma, G., Kontogiannis, S., Markowitch, O., Dricot, J.-M., Gogniat, G.: MicroLET: a new SDNoC-based communication protocol for chipLET-based systems. In 2019 22nd Euromicro Conference on Digital System Design (DSD), pp. 61–68. IEEE (2019)
13. Berestizshevsky, K., Even, G., Fais, Y., Ostrometzky, J.: SDNoC: software defined network on a chip. Microprocess. Microsyst. **50**, 138–153 (2017)
14. Cong, L., Wen, W., Zhiying, W.: A configurable, programmable and software-defined network on chip. In: 2014 IEEE Workshop on Advanced Research and Technology in Industry Applications (WARTIA), pp. 813–816. IEEE (2014)
15. McKeown, N.: Software-defined networking. INFOCOM Keynote Talk **17**(2), 30–32 (2009)
16. Jerger, N.E., Krishna, T., Peh, L.-S.: On-chip networks. Synth. Lect. Comput. Archit. **12**(3), 1–210 (2017)
17. Dimitrakopoulos, G., Psarras, A., Seitanidis, I.: Microarchitecture of Network-on-Chip Routers. Springer, New York (2015). https://doi.org/10.1007/978-1-4614-4301-8
18. Chiu, G.-M.: The odd-even turn model for adaptive routing. IEEE Trans. Parallel Distrib. Syst. **11**(7), 729–738 (2000)
19. Agarwal, N., Krishna, T., Peh, L.-S., Jha, N.K.: GARNET: a detailed on-chip network model inside a full-system simulator. In: 2009 IEEE International Symposium on Performance Analysis of Systems and Software, pp. 33–42. IEEE (2009)
20. Sokolova, M., Lapalme, G.: A systematic analysis of performance measures for classification tasks. Inf. Process. Manage. **45**(4), 427–437 (2009)
21. Fawcett, T.: An introduction to ROC analysis. Pattern Recogn. Lett. **27**(8), 861–874 (2006)

Cross-shard Transaction Processing in Sharding Blockchains

Yizhong Liu[1,3], Jianwei Liu[2(✉)], Jiayuan Yin[2], Geng Li[2], Hui Yu[1], and Qianhong Wu[2]

[1] School of Electronic and Information Engineering, Beihang University, Beijing, China
{liuyizhong,yhsteven}@buaa.edu.cn

[2] School of Cyber Science and Technology, Beihang University, Beijing, China
{liujianwei,yinjianyuan,ligeng,qianhong.wu}@buaa.edu.cn

[3] Department of Computer Science, University of Copenhagen, Copenhagen, Denmark

Abstract. Sharding blockchains could improve the transaction throughput and achieve scalibility, making the application fields of the blockchain technology more extensive. Cross-shard transactions account for a large fraction of transactions in a sharding blockchain, so the processing method of cross-shard transactions is of vital importance to the system efficiency. In this paper, we focus on the study of cross-shard transaction processing methods. Firstly, a summary of cross-shard transaction processing methods for sharding blockchains is given. Secondly, we propose RSTBP, which is built on the basis of a two phase commit protocol. In RSTBP, an input shard runs an intra-shard consensus algorithm, i.e., a Byzantine fault tolerance (BFT) algorithm, to process multiple inputs of different transactions simultaneously. For each input, a corresponding proof of availability is generated and sent to the relevant shards. Compared with previous schemes, the number of BFT calls is reduced by hundreds of times when processing the same number of transactions. Thirdly, RSTSBP is designed by making some modifications to RSTBP. The proofs of availability are constructed according to different shards. The Merkel tree structure is different from that of RSTBP to cut down message complexity of the proofs. Both of the two schemes are proved to satisfy the consistency, liveness and responsiveness properties, and improve the cross-shard transaction processing efficiency.

Keywords: Sharding blockchains · Byzantine fault tolerance · Cross-shard transactions · Scalability · Responsiveness

1 Introduction

Since proposed by Bitcoin [1], the blockchain technology is developing rapidly with applications to many different fields. A blockchain could create a decentralized public ledger that everyone is able to access [2]. The data on the ledger, usually in the form of transactions, is temper-resistant.

M. Qiu (Ed.): ICA3PP 2020, LNCS 12454, pp. 324–339, 2020.
https://doi.org/10.1007/978-3-030-60248-2_22

However, in order for the blockchain technology to be applied more widely, some issues and limitations need to be addressed [3]. One of the most important problems is the low transaction throughput. Many financial scenarios require a throughput of thousands or even tens of thousands transactions per second, which is currently not achieved by most blockchain systems [4].

To improve transaction throughput, sharding blockchains are proposed, which combine the idea of sharding with the blockchain technology [5]. In a sharding blockchain, nodes participating in the consensus are divided into different shards, and in each shard, shard members are responsible for generating and maintaining a specific blockchain [6]. Therefore, the whole system is composed of multiple shards, corresponding to multiple parallel blockchains. For nodes in a shard, most of the communication is done inside the shard. Transactions are assigned to shards according to certain rules. For a transaction whose inputs and outputs are managed by different shards, the commitment of the transaction needs to be done jointly by these corresponding shards. This type of transactions, namely cross-shard transactions, accounts for the vast majority of transactions in a sharding blockchain. The greater the number of shards, the greater the proportion of cross-shard transactions. For example, when the number of shards is 3 and 16, the fraction of cross-shard transactions is 96.3% and 99.98%, respectively [7]. Therefore, the processing method of cross-shard transactions has a great impact on the efficiency and throughput of sharding blockchains.

At present, most of the existing cross-shard transaction processing solutions are based on the two phase commit (2PC) protocol [8] which contains a prepare phase and a commit phase. In the prepare phase, all input shards are required to generate an availability certificate to prove that an input is available or not. An input is available means that it has not been spent by any transaction, nor is it locked. To produce such a certificate, a BFT algorithm might be invoked by a shard to reach agreement and generate enough signatures. If an input is available, it should be locked after the certificate is generated to prevent another transaction from spending it, i.e., double spending. All availability certificates are supposed to be sent to every input and output shard related to the transaction. In the commit phase, shards that receive all related certificates could verify if the transaction is valid by checking if all inputs are available. If there is at least one input that is unavailable, then the transaction is invalid, and the previously locked inputs should be unlocked. If all inputs are available, the inputs should be removed from the corresponding unspent transaction outputs (UTXO) pools of related input shards, while the outputs should be created in the output shards.

There exist some problems in current cross-shard processing methods, such as low efficiency and vulnerability to attacks. Omniledger [9] adopts a client-driven 2PC method, where a client is responsible for collecting and transmitting the availability certificates. Chainspace employs a shard-driven 2PC scheme to confirm cross-shard transactions. Both methods could only process a single transaction at a time. In addition, multiple calls of a BFT algorithm are required to commit a transaction. Since a BFT algorithm contains at least two rounds of votes among all shard members (usually there is a committee inside a shard in committee-based sharding blockchains), multiple calls of BFT increase the communication complexity. A more efficient and general method to process cross-shard transactions for sharding blockchains needs to be proposed.

1.1 Our Contributions

In this paper, we make the following contributions.

- **A summary of cross-shard transaction processing methods.** A summary of methods to process cross-shard transactions in sharding blockchains is given. We focus on the 2PC-based methods adopted in most blockchains.
- **A responsive sharding transaction batch processing (RSTBP) scheme.** RSTBP is designed based on the 2PC protocol to process cross-shard transactions in sharding blockchains. In the prepare phase, a large number of inputs of different transactions are processed by the input shard simultaneously by running a BFT algorithm. Then an availability certificate for each input is generated by the input shard and sent to related shards. In the commit phase, transactions are processed at once. The total number of BFT calls is reduced by a factor of at least Q_{tx} which is the number of transactions in a set.
- **A responsive sharding transaction shard-batch processing (RSTSBP) scheme.** We make adaptions to RSTBP and design the RSTSBP scheme to improve the processing efficiency of sharding transactions. In the prepare phase, a Merkel tree is constructed with every shard as a leaf node. The Merkel tree root is committed by the BFT algorithm as a part of the availability certificate. For every related shard, there is an availability certificate that contains a committed root, a leaf node, and a hash path. The number of hash values in the hash path is reduced to $\log_2^m$ compared to $\log_2^q$ in RSTBP where m denotes the number of shards and q represents the number of total inputs. The inter-shard communication complexity is reduced.

2 Preliminaries

In this section, we give the adversary model and transaction model in this paper. Besides, the concept of responsiveness is introduced.

2.1 Adversary Model

We assume that "honest committee" holds for every shard such that a BFT algorithm could operate normally. For some partially synchronous BFT algorithms such as PBFT [10], HotStuff [11], "honest committee" means that $n \geq 3f + 1$ holds in a committee where n denotes the total number of committee members and f is the number of malicious nodes. Other synchronous BFT algorithms [12] require $n \geq 2f + 1$. Besides, "honest committee" guarantees the correct execution of BFT, achieving the consistency and liveness properties. Consistency means that honest committee members output an identical log. Liveness ensures that any transaction submitted could be processed eventually in a certain time.

2.2 Transaction Model

Transaction models could be divided into UTXO model and account model.

- **UTXO Model.** UTXO model is the most common model in the blockchain systems, which is used by Bitcoin [1], Litecoin [13], and Zerocash [14], etc. A transaction takes in one or several unspent transaction outputs as its inputs, and creates new outputs. If the transaction is contained in a block and committed as valid, the old outputs are spent and removed from the UTXO, and new unspent transaction outputs are created. The participating nodes, especially miners, maintain a UTXO pool locally.
- **Account Model.** Account model is adopted in Ethereum [15] and Monoxide [16], etc. A transaction is based on money transfer between different accounts. If the balance in an input account is enough to cover the transaction input value, then the transaction is regarded as valid.

2.3 Responsiveness

Responsiveness means that the transaction confirmation time is only related to the actual network delay. While it is not influenced by the upper bound of the network delay. Responsiveness is an important indicator to evaluate the transaction confirmation delay of a blockchain system.

3 Cross-shard Transaction Processing Overview

In this section, we summarize basic methods to process cross-shard transactions. We focus on the situations where there is a committee inside each shard running a BFT-style algorithm, which is adopted by most sharding blockchains. As far as we know, the only special sharding blockchain without a committee is Monoxide [16], which uses relay transactions to process cross-shard transactions.

Basic 2PC. Most cross-shard transaction processing methods are designed on top of the 2PC protocol which contains a prepare phase and a commit phase. In a 2PC protocol, there is a coordinator who is responsible for collecting availability certificates of inputs and transmitting them among the related participating shards. In the prepare phase, a coordinator collects certificates to prove that the inputs of a transaction are available from different shards. A such proof is usually generated by a shard through running BFT to reach agreement on if the inputs are available. So the proof might be some signatures or a single signature aggregated. Meanwhile, the inputs should be locked to prevent themselves from being spent by other transactions. Then in the commit phase, the coordinator sends all availability certificates of inputs to all related shards, including input and output shards. If all inputs are available, then the transaction is regarded as valid and committed in related shards. The inputs should be spent and outputs should be created. Else, i.e., at least one input is unavailable (locked or already spent), the transaction is invalid. The previous locked inputs should be unlocked.

According to the role of a coordinator, we divide current cross-shard processing methods into client-driven basic 2PC and shard-driven basic 2PC.

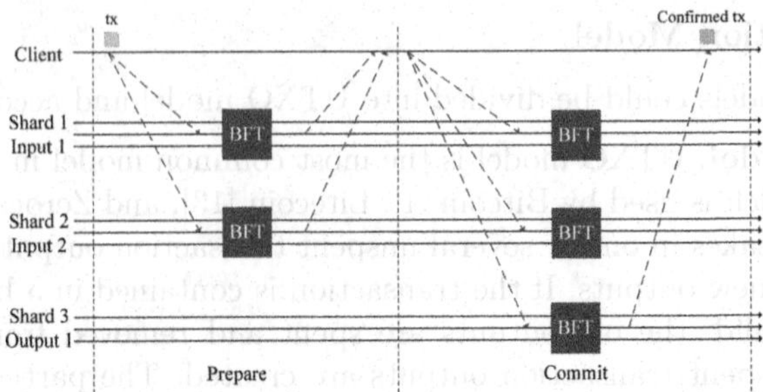

Fig. 1. The flowchart of client-driven basic 2PC.

3.1 Client-Driven Basic 2PC

Client-drive basic 2PC is represented by Omniledger. The basic process is shown in Fig. 1. A client is responsible for collecting proofs in the prepare phase and transmitting them to related shards in the commit phase. Note that in the prepare phase, if the proof of a input is generated by a single leader instead of a committee running BFT, then a malicious leader might provide false proofs or fail to respond. In addition, letting a client act as a coordinator could lead to an input being locked permanently if the client fails to send the corresponding proof to related input shards. The burden on a client is increased as well.

3.2 Shard-Driven Basic 2PC

In a shard-driven 2PC protocol, one or more shards play the role of coordinator. The flowchart is shown in Fig. 2. The burden on a client is released. It just submits a transaction and waits for a response. In the prepare phase, the availability certificates of inputs are generated by input shards. In the commit phase, a valid transaction will be accepted in all input and output shards. The confirmation information of the transaction will be sent to the client.

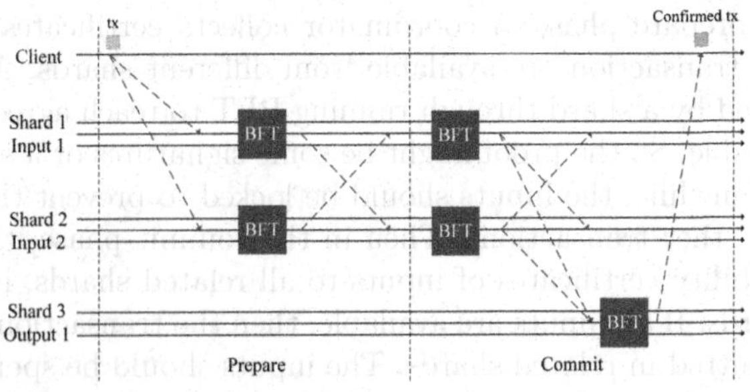

Fig. 2. The flowchart of shard-driven basic 2PC.

4 Responsive Sharding Transaction Batch Processing

In this section, we describe our responsive sharding transaction batch processing scheme, i.e., RSTBP.

RSTBP contains a prepare phase and a commit phase. The concrete processing steps are as follows.

Prepare. In the prepare phase, every shard plays the role of an input shard. The flowchart of the prepare phase is shown in Fig. 3.

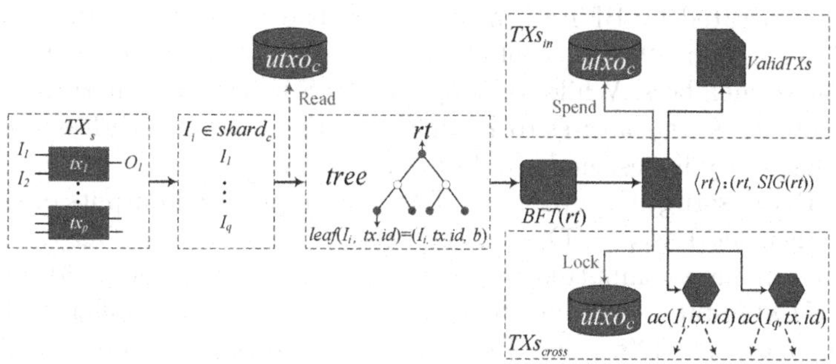

Fig. 3. The flowchart of the prepare phase.

1. **Merkel Tree Construction.** A leader of a shard S_c is denoted as L_c. $c \in \{1, \cdots, m\}$ is the sequence number of shards. L_c first receives a set of transactions from the environment $\mathcal{Z}$ or the clients, which is denoted as TXs. TXs contains p transactions, which might have multiple inputs. Then L_c extracts all inputs in TXs, denoted by $I_1, \cdots, I_q$, which are controlled by S_c. L_c constructs a Merkel tree *tree* where a leaf node is set to $\mathsf{leaf}(I_i, \mathsf{tx}.id) = (I_i, \mathsf{tx}.id, b)$. tx is the corresponding transaction to I_i, and $\mathsf{tx}.id$ returns the ID information of tx. b is used to denote the state of I_i.
For every input I_i, L_c queries if it is available in the local UTXO pool utxo_c. If I_i is $\mathsf{available}$, i.e., $\mathsf{utxo}_c[I_i] = 1$, then b is set to 1. Note that if there are multiple transactions using I_i as an input (double spending), then L_c chooses the first received transaction and sets b to be 1, while all other transactions' related b are set to 2. If I_i is locked, i.e., $\mathsf{utxo}_c[I_i] = 2$, then b is set to 2. If I_i does not exist in utxo_c (or already spent), b is set to 0. The root hash of *tree* is denoted as rt.
2. **Merkel Tree Root Commitment.** L_c broadcasts rt and *tree* among the committee members of S_c, and calls the BFT algorithm to reach agreement upon the value rt.
For an honest member of S_c, after receiving the rt and *tree* from L_c, it verifies if *tree* and rt are valid based on the following rules. First, an honest node queries its local utxo_c to verify if the value b is correct in every leaf node. For

every I_i in *tree*, if I_i only appears once, then the value b should be consistent with $\mathsf{utxo}_c[I_i]$, i.e., if $\mathsf{utxo}_c[I_i] = 1$, $b = 1$; if $\mathsf{utxo}_c[I_i] = 2$, $b = 2$; if I_i does not exist, $b = 0$. If there are multiple identical I_i in *tree* and $\mathsf{utxo}_c[I_i] = 1$, there must be only one corresponding b to be 1, and other b are all set to 2. Second, an honest node verifies that if rt equals to be root value of *tree*. If the conditions above are satisfied, then an honest member votes for rt in the BFT algorithm. Else, it could launch a view-change to change a leader.
After two rounds of votes, a valid rt will be committed by the BFT algorithm since honest majority holds in every shard. We use the notion $\langle rt \rangle$ to denote the committed rt. $\langle rt \rangle = (rt, \mathsf{SIG}_c(rt))$ where $\mathsf{SIG}_c(rt)$ is the signature generated by BFT to prove that rt is committed. In the traditional PBFT algorithm [10], SIG_c contains $2f+1$ valid signatures signed by different committee members. While we adopt the BLS threshold signature [17,18] to aggregate $2f+1$ signatures to 1 single signature. $\langle rt \rangle$ is broadcast among S_c such that every honest node could receive it.

3. TXs_{in} **Processing.** For every $\mathsf{tx} \in \mathsf{TXs}$, if all inputs and outputs of tx belong to S_c, then set $\mathsf{TXs}_{in} := \mathsf{TXs}_{in} \cup \{\mathsf{tx}\}$; else, set $\mathsf{TXs}_{cross} := \mathsf{TXs}_{cross} \cup \{\mathsf{tx}\}$. After receiving a committed $\langle rt \rangle$, every committee member could update its local UTXO state according to the content of *tree*. For different kinds of transactions, i.e., TXs_{in} and TXs_{cross}, the processing methods are different. For TXs_{in}, UTXO and the valid transaction pool ValTXs need to be updated.
   (a) **UTXO Update (Spend).** For a $\mathsf{tx} \in \mathsf{TXs}_{in}$, if the values b in the *tree* belonging to all inputs of tx are marked as 1, and the input sum equals to the output sum, then tx is regarded as a valid transaction and committed to be $\mathsf{accepted}$. So for every input that satisfy $I_i \in \{\mathsf{tx}\} \cap \mathsf{utxo}_c$, I_i needs to be removed from utxo_c, i.e., I_i is spent. Else, i.e., at least one b of tx is not 1, or the input sum does not equal to the output sum, then tx is a invalid transaction and is committed to be $\mathsf{rejected}$. In this case, the UTXO state does not need to be updated.
   (b) ValTXs **Update.** The valid transaction pool ValTXs is used to store all valid transactions, including valid TXs_{in} and TXs_{cross} (a cross shard transaction is only written to its output shard). After a tx is regarded as a valid intra-shard transaction, tx is added to ValTXs, waiting to be written on a block through the BFT algorithm.
4. TXs_{cross} **Processing.** For a cross-shard transaction, all committee members update their UTXO state according to the committed $\langle rt \rangle$ and its *tree*. Then a proof that an input is available needs to be generated and sent to corresponding shards, which is done by a leader of the input shard.
   (a) **UTXO Update (Lock).** After receiving the committed $\langle rt \rangle$ and related *tree*, committee members check the value b of every input. For an input I_i which belongs to a cross-shard transaction, if its b is 1, then the local UTXO state of I_i is changed from 1 to 2 for locked, i.e., $\mathsf{utxo}_c[I_i] = 2$. Else, nothing is changed.
   (b) **Availability Certificate Construction.** A leader of an input shard is responsible for constructing the availability certificates for every input of

cross-shard transactions. $ac(I_i, \mathsf{tx}.id)$ is adopted to represent the availability certificate of I_i. For every input I_i of a cross-shard transaction in the *tree*, $ac(I_i, \mathsf{tx}.id)$ is constructed as $(\langle rt\rangle, \mathsf{leaf}(I_i, \mathsf{tx}.id), \mathsf{hpath}(I_i, \mathsf{tx}.id))$ where $\mathsf{leaf}(I_i, \mathsf{tx}.id) = (I_i, \mathsf{tx}.id, b)$ is exactly the leaf node in *tree* and $\mathsf{hpath}(I_i, \mathsf{tx}.id)$ denotes the hash path to prove that $\mathsf{leaf}(I_i, \mathsf{tx}.id)$ is a valid leaf node of the root value $\langle rt\rangle$. For a Merkel tree that includes q leaf nodes, the number of hash values in $\mathsf{hpath}(I_i, \mathsf{tx}.id)$ is $\log_2^q$. Note that the value b in a $ac(I_i, \mathsf{tx}.id)$ could be $1, 2$ or 0, that is, regardless of whether I_i is available, its availability certificate must be constructed and sent to the corresponding shards.

(c) **Availability Certificate Transmission.** After generating an ac for every input I_i, a leader sends the $ac(I_i, \mathsf{tx}.id)$ to related shards. Namely, L_c first finds the transaction tx corresponding to I_i, then confirms the shards to which the other inputs and outputs of tx belong, and sends $ac(I_i, \mathsf{tx}.id)$ to all these shards.

Commit. In the commit phase, all shards related to a specific transaction decide to accept or reject the transaction according to the availability certificates receiving in the prepare phase. As an input shard, the UTXO state needs to be updated, while as an output shard, valid transactions are put into ValTXs. The flowchart of the commit phase is shown in Fig. 4.

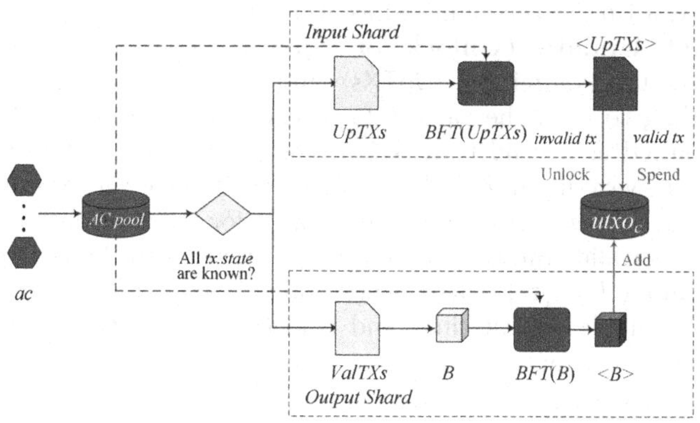

Fig. 4. The flowchart of the commit phase.

1. **Availability Certificate Processing.** After receiving an ac, a verification must be carried out to verify its correctness. In order to cut down the communication complexity among shards, a leader in every shard is required to receive the availability certificates, i.e., ac is transmitted among shard leaders. We use $ac(I_j, \mathsf{tx}.id)$ to denote a received ac for an input I_j. $ac(I_j, \mathsf{tx}.id)$ could be parsed as $(\langle rt\rangle, \mathsf{leaf}(I_j, \mathsf{tx}.id), \mathsf{hpath}(I_j, \mathsf{tx}.id))$. If $\langle rt\rangle$ is valid and $\mathsf{leaf}(I_j, \mathsf{tx}.id)$ is a valid leaf node, then $ac(I_j, \mathsf{tx}.id)$ is regarded as a valid one

and put into an availability certificate pool AC. When all input states are known for a cross-shard tx, the processing methods for input and output shards are different.

2. **Input Shard Processing.** In an input shard, a leader puts all transactions whose input states are known into a update transaction pool UpTXs. Then UpTXs is committed by the BFT algorithm and every committee member updates its local UTXO states according to UpTXs.
   (a) **Transaction Validation and UpTXs Formation.** For a transaction tx, if all its input states are known, i.e., all ac are received, then tx is added into UpTXs. Note that for a tx, if every b of every input equals to 1, then tx is a valid cross-shard transaction. Else, tx is an invalid one. Both valid and invalid cross-shard transactions are put into UpTXs with the marks "valid" or "invalid".
   (b) **UpTXs Commitment.** If the number of transactions in UpTXs reaches Q_{tx}, or T_{tx} time has passed since the last committed UpTXs, the leader broadcasts UpTXs and the availability certificate pool AC, and initiates the BFT algorithm to reach agreement on UpTXs. An honest committee member first verifies if every $ac(I_j, \mathsf{tx}.id)$ in AC is valid, where the verification rule is the same as before, i.e., $\langle rt \rangle$ is valid and $\mathsf{leaf}(I_j, \mathsf{tx}.id)$ is a valid leaf node. Then an honest member verifies if the mark of every transaction in UpTXs is correct based on the information in AC. If all the conditions are satisfied, an honest member votes for UpTXs in BFT. Finally, a UpTXs is committed as ⟨UpTXs⟩.
   (c) **UTXO Update (Unlock or Spend).** Every committee member receives the committed ⟨UpTXs⟩ and updates its local UTXO state according to it. If the value b for every input of tx in AC is 1, then tx is regarded as a valid transaction, i.e., tx is committed to be accepted. And for every input $I_i \in \{\mathsf{tx}\} \cap \mathsf{utxo}_c$, I_i is removed from utxo_c, which means I_i is spent. Else, i.e., at least one b for an input of tx is not 1, then tx is an invalid transaction, i.e., tx is committed to be rejected. And for every input $I_i \in \{\mathsf{tx}\} \cap \mathsf{utxo}_c$ whose value b is set to 1 in the original $tree$ (I_i is available at that time, and is locked later), $\mathsf{utxo}_c[I_i]$ is set from 2 for locked to 1 for available. That is, I_i is unlocked.
3. **Output Shard Processing.** In an output shard, a leader also receives different ac from other shard leaders and puts them into the availability certificate pool AC.
   (a) **Transaction Validation and ValTXs Update.** For a transaction tx, if all its input states are known and the b of every input equals to 1, then tx is treated as a valid cross-shard transaction, committed to be accepted, and added into the valid transaction pool ValTXs, i.e., $\mathsf{ValTXs} := \mathsf{ValTXs} \cup \{\mathsf{tx}\}$. Else, tx is regarded as invalid and rejected.
   (b) **Block Generation.** When the number of transaction in ValTXs reaches a certain value Q_{tx}, or T_{tx} time has passed since the last committed block $B_{\ell-1}$, a new block B_ℓ is constructed as (str, ValTXs) where str is the hash value of the last block, i.e., $str = \mathsf{H}(B_{\ell-1})$. $\mathsf{H}(\cdot)$ is used to compute a hash

value of an input. Then the leader broadcasts B_ℓ and its related AC, and calls the BFT algorithm to commit B_ℓ.
An honest committee member will check the validity of B_ℓ by two conditions. First, for every input I_i of every tx in ValTXs, the corresponding $ac(I_i, \mathsf{tx}.id)$ in AC is valid and the input state b in $ac(I_i, \mathsf{tx}.id)$ is 1. Second, str equals to $\mathsf{H}(B_{\ell-1})$. If the conditions are satisfied, then an honest member votes for the block B_ℓ. The agreement property and the liveness property of the BFT algorithm ensure that a committed $\langle B_\ell \rangle$ will be confirmed.

(c) **UTXO Update (Add).** After receiving a committed block $\langle B_\ell \rangle$, the local UTXO state could be updated for a shard member. As an output shard of a transaction tx which is committed in $\langle B_\ell \rangle$, corresponding outputs need to be created in the local utxo_c.

5 Responsive Sharding Transaction Shard-Batch Processing

In this section, we give the construction details of the responsive sharding transaction shard-batch processing scheme, i.e., RSTSBP.

In the RSTBP scheme described above, for every transaction input, there is a corresponding availability certificate ac in need to prove the state of the input. And every ac contains a committed root value, a leaf node of the Merkel tree, and a hash path to prove that the leaf node is a valid one. If there are q inputs in a Merkel tree, then a hash path hpath of an input I_i contains $\log_2^q$ hash values. The message complexity increases as q becomes larger. As a result, the communication complexity among shards will increase. In order to reduce the communication complexity and improve processing efficiency, we make further improvements to the RSTBP scheme. The processing method is as follows.

Prepare

1. **Merkel Tree Construction.** When constructing a Merkel tree, instead of using every input as a leaf node, each shard is used as an index of a leaf node to generate a Merkel tree. Assume that there are m shards in total, then a Merkel tree has a fixed number of m leaf nodes denoted as $S_1, \cdots, S_m$. For every leaf node $S_{c'}$, $\mathsf{leaf}(S_{c'})$ is set to be $(S_{c'}.id, [(I_1, \mathsf{tx}.id, b), \cdots, (I_{q_i}, \mathsf{tx}.id, b)])$ where $I_1, \cdots, I_{q_i}$ are the inputs related to the shard $S_{c'}$. That is, for the transaction tx to which I_i belongs, $S_{c'}$ is an input or output shard (or both) of tx. $S_{c'}.id$ returns the ID number of shard $S_{c'}$. Note that for the current shard S_c, there is also a leaf node $\mathsf{leaf}(S_c)$ which contains all intra-shard transaction inputs of S_c. For every I_i, the value b is obtained from the same way as that in RSTBP.
2. **Merkel Tree Root Commitment.** This step is identical to that of RSTBP.
3. TXs_{in} **Processing.** Note that in the Merkel tree $tree$, there are m leaf nodes, including a node of the current shard, denoted by $\mathsf{leaf}(S_c)$. The input states of intra-shard transactions in S_c are all contained in $\mathsf{leaf}(S_c)$. So the "UTXO Update (Spend)" and "ValTXs Update" steps are done similar to RSTBP according to the information in $\mathsf{leaf}(S_c)$ and the committed $\langle rt \rangle$.

4. TXs_{cross} **Processing.**
   (a) **UTXO Update (Lock).** This step is identical to that of RSTBP.
   (b) **Availability Certificate Construction.** There is only one availability certificate for each shard $S_{c'}$ where $c' \in \{1, \cdots, m\}/\{c\}$, i.e., $ac(S_{c'}) = (\langle rt \rangle, \mathsf{leaf}(S_{c'}), \mathsf{hpath}(S_{c'}))$. $\mathsf{hpath}(S_{c'})$ includes the hash values to prove that $\mathsf{leaf}(S_{c'})$ is a valid node of *tree*. The number of hash values in $\mathsf{hpath}(S_{c'})$ is only $\log_2^m$. Note that for the current shard S_c, there is no need to construct a ac, so there are $m - 1$ availability certificates.
   The specific construction process of a Merkel tree is shown in Fig. 5. In the figure, we assume that there are 8 shards, i.e., $m = 8$. For the shard S_1, the hash path in $ac(S_1)$ contains the hash values of S_2, S_{34} and S_{5678} that are green nodes in the figure.
   Availability Certificate Transmission. There is an $ac(S_{c'})$ generated for each shard $S_{c'}$. So a leader of shard S_c sends each ac to the other shards, i.e., sends $ac(S_1), \cdots, ac(S_m)$ to shard $S_1, \cdots, S_m$, respectively.

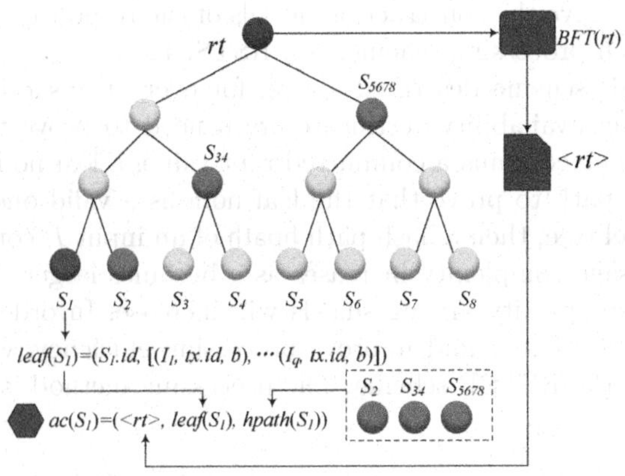

Fig. 5. The example of a Merkel tree in RSTSBP.

Commit

1. **Availability Certificate Processing.** A leader in every shard S_c receives $ac(S_c)$ from other shards. Suppose that $ac(S_c)$ is sent by the shard leader of $S_{c'}$, then $ac(S_c)$ could be parsed as $(\langle rt \rangle, \mathsf{leaf}(S_c), \mathsf{hpath}(S_c))$. $\langle rt \rangle$ is $(rt, \mathsf{SIG}_{c'}(rt))$ where $\mathsf{SIG}_{c'}(rt)$ is a threshold signature generated by $2f + 1$ members of shard $S_{c'}$ and could be verified by the group public key of shard $S_{c'}$. $\mathsf{leaf}(S_c)$ could be parsed as $(S_c.id, [(I_1, \mathsf{tx}.id, b), \cdots, (I_{q_i}, \mathsf{tx}.id, b)])$. If $\mathsf{SIG}_{c'}(rt)$ is a valid threshold signature and $\mathsf{leaf}(S_c)$ is a valid leaf node of $\langle rt \rangle$, $ac(S_c)$ is treated as a valid one and put into the AC pool.
2. **Input Shard Processing.** This step is almost the same as that of RSTBP, except that the availability certificates of multiple inputs are in a single leaf

node. So in the "UpTXs Commitment" step, honest committee members only need to verify several $ac(S_c)$ sent from different shards. The other operations are identical to those of RSTBP.

3. **Output Shard Processing.** The operations in this step are the same as those of RSTBP.

6 Security Analysis

In this section, we give the security analysis of the RSTBP and RSTSBP schemes.

For a cross-shard transaction scheme, we propose and prove the security properties that should be satisfied. Note that we only focus on the security analysis of the cross-shard transaction processing method, while the whole analysis of a sharding blockchain could be based on the works such as [19,20].

Definition 1 (Consistency). For any input I_i, it could only be spent by a single transaction (no double spending happens). Furthermore, for any transaction tx, if it is valid or invalid, then it is committed to be accepted or rejected in all related input and output shards, respectively.

Definition 2 (Liveness). For any input I_i, its UTXO state does not remain locked permanently. Furthermore, for any transaction tx, after it is submitted, it will be committed to be accepted or rejected in a certain time.

Theorem 1 (Consistency). *RSTBP and RSTSBP both satisfy the consistency property as per Definition 1.*

Proof (Proof of Theorem 1). RSTBP and RSTSBP are both designed based on the 2PC mechanism. The main differences between them are the construction method of a Merkel tree and the transmission way of availability certificates, which do not influence the system security.

Suppose there are two different transaction tx and tx' that intend to spend a same input I_i, then at most one transaction will be committed as a valid one. If tx and tx' are in the same transaction set, then according to the rule in the "Merkel Tree Construction" step, only one b of the two leaf nodes $\mathsf{leaf}(I_i, \mathsf{tx}.id)$ and $\mathsf{leaf}(I_i, \mathsf{tx}'.id)$ is set to 1, while the other is set to 2 (in RSTSBP, it is the same except the form of a leaf node). Whether tx and tx' are intra-shard or cross-shard transactions, only the transaction with a $b = 1$ is possible to be committed as a valid one. If tx and tx' are in different transaction sets, for example, tx comes before tx'. Then a Merkel tree containing $\mathsf{leaf}(I_i, \mathsf{tx}.id) = (I_i, \mathsf{tx}.id, 1)$ as a leaf node will be committed by the BFT algorithm. After receiving the committed $\langle rt \rangle$, every honest member sets the local UTXO to be $\mathsf{utxo}_c[I_i] = 2$ for locked. At this time, if tx' is submitted, the corresponding leaf node is set to $\mathsf{leaf}(I_i, \mathsf{tx}'.id) = (I_i, \mathsf{tx}'.id, 2)$ since I_i is locked. So tx' will not be committed by any shard. After a certain time, if tx is committed as valid, then I_i will be removed from the UTXO. If tx' is uploaded at this time, $\mathsf{leaf}(I_i, \mathsf{tx}'.id) = (I_i, \mathsf{tx}'.id, 0)$ since I_i is spent. tx' will still be committed as invalid.

For a cross-shard transaction tx, if it is accepted or refused in a shard, then it is committed to be accepted or refused respectively in all related shards. This is because the unforgeability property of the availability certificate. For every related shards, honest majority ensures the correct execution of the BFT algorithm. The Merkel tree root is committed by BFT with a threshold signature as a proof. An adversary could not forge a valid threshold signature since he controls limited nodes. As a result, he could not forge any availability certificate. So for every related shard, members receive the same *ac* for tx. Hence, the final verification results of tx are consistent in all related shards.

Theorem 2 (Liveness). *RSTBP and RSTSBP both satisfy the liveness property as per Definition 2.*

Proof (Proof of Theorem 2). For an input I_i, if it is locked, then it must be used by some cross-shard transaction tx. The other inputs of tx might be managed by other shards. When related *ac* is not broadcast by a leader after a certain time, the other honest members could contact the honest members in related shards for the *ac*. Then a view-change could be launched to change the malicious leader. Based on the liveness property of the BFT algorithm, the *ac* of other inputs are sure to be generated and sent to all related shards. So I_i will be either spent or unlocked to be available according to whether tx is a valid transaction.

Liveness also holds for a transaction tx. For all inputs of tx, every related input shard will generate the corresponding availability certificate and send it to other related shards. Therefore, any transaction submitted will be committed as accepted or rejected in every related shard after a certain time.

7 Performance Analysis

We compare RSTBP and RSTSBP with previous schemes, i.e., client-driven basic 2PC represented by Omniledger, and leader-driven basic 2PC such as Chainspace. The comparison is shown in Table 1.

Table 1. Comparison of RSTBP and RSTSBP with other cross-shard transaction processing schemes.

System	Total BFT calls	Message number for a shard	Message length for a shard
Client-driven basic 2PC	$(2k+1)n$	nk	$\ell_{\mathsf{Input}} + \ell_{\mathsf{txid}} + \ell_{\mathsf{SIG}}$
Leader-driven basic 2PC	$(2k+1)n$	nk	$\ell_{\mathsf{Input}} + \ell_{\mathsf{txid}} + \ell_{\mathsf{SIG}}$
RSTBP	$\frac{kn}{q} + \frac{(k+1)n}{Q_{tx}}$	nk	$\ell_{\mathsf{Input}} + \ell_{\mathsf{txid}} + \ell_{\mathsf{SIG}} + \ell_{\mathsf{rt}} + \log_2^q \ell_{\mathsf{hash}}$
RSTSBP	$\frac{kn}{q} + \frac{(k+1)n}{Q_{tx}}$	$(m-1)$	$\frac{nk}{m}(\ell_{\mathsf{Input}} + \ell_{\mathsf{txid}}) + \ell_{\mathsf{SIG}} + \ell_{\mathsf{rt}} + \log_2^m \ell_{\mathsf{hash}}$

Assume there are n transactions. For every transaction, the number of inputs are k on average, while the number of outputs is set to 1, just like most blockchains adopting the UTXO model. The number of inputs in *tree* is q.

The number of transactions in UpTXs and ValTXs are both Q_{tx}. The number of shards is m. The "Total BFT Calls" denotes the number of calling BFT algorithms in order to process n transactions. In the basic 2PC schemes, the number of BFT calls is $2k+1$ for a single transaction. While in RSTBP and RSTSBP, in the prepare phase, the number of BFT calls in every input shard is reduced by a factor of q, and in the commit phase, the number is decreased by a factor of Q_{tx}. The "Message Number for a Shard" represents the number of proof messages that a shard needs to generate. In the basic 2PC and RSTBP scheme, an input shard is required to generate a proof for every single input, so the number is $n(k-1+1)=nk$. While in RSTSBP, the availability certificates are packed according to shards. The "Message Length for a shard" denotes the length of a single proof produced by a shard. In the basic 2PC scheme, the proof, i.e., availability certificate, contains the input address, its related transaction ID information, and $2f+1$ signatures (or 1 single signature if multi-signature is employed) from the BFT algorithm. In RSTBP, the availability certificate $ac(I_i)=(\langle rt\rangle, \mathsf{leaf}(I_i), \mathsf{hpath}(I_i))$ contains a root value of a Merkel tree, and $\log_2^q$ hash values to prove a leaf node is a valid one. ℓ_{hash} denotes the length of a hash value. In RSTSBP, a certificate $ac(S_i)=(\langle rt\rangle, \mathsf{leaf}(S_i), \mathsf{hpath}(S_i))$ includes nk/m inputs in $\mathsf{leaf}(S_i)$, and the number of hash values in $\mathsf{hpath}(S_i)$ is reduced to $\log_2^m$. q and Q_{tx} is usually on the order of hundreds or even thousands. m is the number of shards, such as 16 or 32, which could be regarded as a constant.

8 Discussion

A shard leader is responsible for collecting votes from shard members, generating availability certificates, and transmitting these messages to other shard leaders. When a leader is malicious or does not respond in a certain time, honest members of a same shard could launch a view-change mechanism to replace the old leader with a new random leader.

Note that for the intra-shard consensus, the BLS threshold signature [17] could be adopted in the BFT algorithm to cut down message complexity inside a shard. Meanwhile, the BFT threshold signature could be used to generated an unbiased and public verifiable randomness. Such a randomness could be employed to select a leader randomly in a shard to prevent an adversary from controlling a leader in advance.

9 Related Work

Classical Cross-shard Transaction Processing Schemes. Many classical cross-shard schemes have been proposed, while they have some limitations. Omniledger [9] is designed based on Elastico [21], adopting a client-driven cross-shard transaction processing method. However, a malicious leader could provide a false proof-of-acceptance to cause inconsistency among shards. Moreover, the detailed construction of a proof-of-rejection is not given. Collecting and transmitting proofs increase the burden on clients. RapidChain [7] splits a cross-shard transaction into multiple single-input single-output transactions and commits them in order. However, this enlarges the total number of transactions, resulting in an increased processing and storage burden on the entire network. Besides, the details about how to generate a transaction of a specific shard are not given. Chainspace [22] combines 2PC with BFT, while transactions could only be processed one by one, leading to a large number of BFT calls. Monoxide [16] adopts an account model and proposes atomic transfer using a relay transaction to process a cross-shard transaction.

Improvements and Analysis of Cross-shard Schemes. In view of the limitations of the classical cross-shard schemes, some improvements and analysis are proposed. Sonnino, Bano, and Al-Bassam et al. propose a replay attack against the cross-shard transaction processing method in Chainspace. Then the Byzcuit protocol is designed where a transaction manager is responsible for message collecting and transmitting. Zamyatin, Al-Bassam, and Zindros et al. [23] study the communication across distributed ledgers and claim that communication across chains is as hard as fair change.

10 Conclusion

In this paper, RSTBP and RSTSBP are proposed as cross-shard transaction processing methods for sharding blockchains. The two schemes are designed to decrease the number of BFT calls and cut down message complexity among shards. The processing methods could be applied to most sharding blockchains that adopt a UTXO transaction model.

Acknowledgment. This paper is supported by the National Key R&D Program of China through project 2017YFB1400702 and 2017YFB0802500, the National Cryptography Development Fund through project MMJJ20170106, the Natural Science Foundation of China through projects 61932014, 61972018, 61972019, 61932011, 61772538, 61672083, 61532021, 61472429, 91646203, 61402029, 61972017, 61972310, the foundation of Science and Technology on Information Assurance Laboratory through project 61421120305162112006, the China Scholarship Council through project 201906020015.

References

1. Nakamoto, S., et al.: Bitcoin: a peer-to-peer electronic cash system (2008). https://bitcoin.org/bitcoin.pdf
2. Bano, S., Al-Bassam, M., Danezis, G.: The road to scalable blockchain designs. Login **42**(4), 31–36 (2017)
3. Conti, M., Kumar, E.S., Lal, C., Ruj, S.: A survey on security and privacy issues of bitcoin. IEEE Commun. Surv. Tutor. **20**(4), 3416–3452 (2018)
4. Liu, Y., Liu, J., Zhang, Z., Xu, T., Yu, H.: Overview on consensus mechanism of blockchain technology. J. Cryptologic Res. **6**(4), 395–432 (2019)
5. Avarikioti, G., Kokoris-Kogias, E., Wattenhofer, R.: Divide and scale: formalization of distributed ledger sharding protocols (2019). CoRR abs/1910.10434

6. Liu, Y., Liu, J., Zhang, Z., Yu, H.: A fair selection protocol for committee-based permissionless blockchains. Comput. Secur. **91**, 101718 (2020)
7. Zamani, M., Movahedi, M., Raykova, M.: RapidChain: scaling blockchain via full sharding. In: CCS 2018, pp. 931–948 (2018)
8. Bernstein, P.A., Hadzilacos, V., Goodman, N.: Concurrency Control and Recovery in Database Systems. Addison-Wesley, Boston (1987)
9. Kokoris-Kogias, E., Jovanovic, P., Gasser, L., Gailly, N., Syta, E., Ford, B.: OmniLedger: a secure, scale-out, decentralized ledger via sharding. In: SP 2018, pp. 583–598 (2018)
10. Castro, M., Liskov, B.: Practical Byzantine fault tolerance. In: OSDI 1999, pp. 173–186 (1999)
11. Yin, M., Malkhi, D., Reiter, M.K., Golan-Gueta, G., Abraham, I.: HotStuff: BFT consensus with linearity and responsiveness. In: PODC 2019, pp. 347–356 (2019)
12. Ren, L., Nayak, K., Abraham, I., Devadas, S.: Practical synchronous Byzantine consensus. IACR Cryptology ePrint Archive 2017, 307 (2017)
13. Charlie, L.: [ANN] litecoin - a lite version of bitcoin. launched! (2011). https://bitcointalk.org/index.php?topic=47417
14. Hopwood, D., Bowe, S., Hornby, T., Wilcox, N.: Zcash protocol specification (2016). https://zips.z.cash/protocol/protocol.pdf
15. Buterin, V.: A next-generation smart contract and decentralized application platform (2014). https://whitepaperdatabase.com/wp-content/uploads/2017/09/Ethereum-ETH-whitepaper.pdf
16. Wang, J., Wang, H.: Monoxide: scale out blockchains with asynchronous consensus zones. In: NSDI 2019, pp. 95–112 (2019)
17. Boneh, D., Boyen, X.: Short signatures without random oracles. In: Cachin, C., Camenisch, J.L. (eds.) EUROCRYPT 2004. LNCS, vol. 3027, pp. 56–73. Springer, Heidelberg (2004). https://doi.org/10.1007/978-3-540-24676-3_4
18. Boldyreva, A.: Threshold signatures, multisignatures and blind signatures based on the Gap-Diffie-Hellman-group signature scheme. In: Desmedt, Y.G. (ed.) PKC 2003. LNCS, vol. 2567, pp. 31–46. Springer, Heidelberg (2003). https://doi.org/10.1007/3-540-36288-6_3
19. Garay, J., Kiayias, A., Leonardos, N.: The bitcoin backbone protocol: analysis and applications. In: Oswald, E., Fischlin, M. (eds.) EUROCRYPT 2015. LNCS, vol. 9057, pp. 281–310. Springer, Heidelberg (2015). https://doi.org/10.1007/978-3-662-46803-6_10
20. Pass, R., Seeman, L., Shelat, A.: Analysis of the blockchain protocol in asynchronous networks. In: Coron, J.-S., Nielsen, J.B. (eds.) EUROCRYPT 2017. LNCS, vol. 10211, pp. 643–673. Springer, Cham (2017). https://doi.org/10.1007/978-3-319-56614-6_22
21. Luu, L., Narayanan, V., Zheng, C., Baweja, K., Gilbert, S., Saxena, P.: A secure sharding protocol for open blockchains. In: ACM SIGSAC 2016, pp. 17–30 (2016)
22. Al-Bassam, M., Sonnino, A., Bano, S., Hrycyszyn, D., Danezis, G.: Chainspace: a sharded smart contracts platform. In: NDSS 2018, pp. 18–21 (2018)
23. Zamyatin, A., et al.: SoK: communication across distributed ledgers. IACR Cryptology ePrint Archive 2019, 1128 (2019)

Lexicon-Enhanced Transformer with Pointing for Domains Specific Generative Question Answering

Jingying Yang[1,2(✉)], Xianghua Fu[2(✉)], Shuxin Wang[2], and Wenhao Xie[1,2]

[1] College of Computer Science and Software Engineering, Shenzhen University, Shenzhen, China
yjygo1103@163.com, lesamourai@163.com

[2] College of Big Data and Internet, Shenzhen Technology University, Shenzhen, China
{fuxianghua,wangshuxin}@sztu.edu.cn

Abstract. Aiming at the problem of inaccurate generation caused by the lack of external knowledge in the generative automatic question answering system, we propose a new answer generation model (LEP-Transformer) that integrates domain lexicon and copy mechanism, which can enable the Transformer to effectively deal with the long-distance dependence of different text granularity and have the ability to reproduce the details of the facts when generating answers. And the experimental results on two different datasets show that the model can alleviate this problem and has ability to model short text and long text sequences simultaneously.

Keywords: Transformer · Domain lexicon · Copy mechanism · Knowledge fusion structure

1 Introduction

Questions in the generative question and answering task contain little effective information. Without enough valid information, the model cannot understand the intention of the question well, and therefore cannot generate an accurate answer [1]. Keyword information in the dataset has certain knowledge representation [6], So fusing domain keyword knowledge information can alleviate the questions lack of valid information.

The problem of effectively fusing external knowledge information into the traditional generative model can be regarded as finding a way to effectively deal with long-term memory problem. Some research work uses RNN [1] or LSTM [7, 8] to capture the long-term structure of the sequences. The RNN model can be trained to predict the next word or group of words to be output after reading the word stream, but the memory obtained by the hidden layer state and weight encoding is too small, and the knowledge is compressed into a dense vector without sufficient separation, so The model cannot accurately remember the facts of the past [9]. The LSTM-based models solve this problem by locking the past state of the network with local memory units [10]. Other research work uses attention mechanism to deal with long-distance dependence. Bahdanau et al. [2] proposed an

M. Qiu (Ed.): ICA3PP 2020, LNCS 12454, pp. 340–354, 2020.
https://doi.org/10.1007/978-3-030-60248-2_23

alignment mechanism to learn the alignment relationship between each target word and related words in the source sentence. Each time the model generates a target word, it will soft search the most relevant set of positions in the source sentence. Seo et al. [11] proposed the BiDAF model, which uses a cyclic model to process sequential inputs and uses attention mechanisms to deal with long-term interactions. Vaswani et al. [5] created the Transformer model and proposed a multi-head attention mechanism that can connect all positions of the sequence with only a constant number of operations, making the model easier to learn Long-term dependencies. If the input information also contains a series of discrete words, these words and the sequential sequence cannot directly form a positional relationship, Using the attention mechanism alone is not able to complete the generative question and answering task well. Other research work uses memory networks [9, 10, 12, 13], which use a large memory to store knowledge, and model the action of the neural network to read and write storage. The memory network proposed by Weston et al. [9] can combine inference learning strategies with readable and writable memory components. The end-to-end memory network model proposed by Sukhbaatar et al. [10] is a continuous form of memory network, which improves model performance by performing multiple calculation steps (hops) on long-term memory. An et al. [12] uses a specific memory to store keyword embedding.

In order to solve the above problem, we propose a lexicon-enhanced transformer with pointing for domains specific generative question answering (LEP-Transformer). The model is divided into two parts: knowledge encoder and knowledge decoder.

To summarize, our contributions lie in the following three points:

- We designed a knowledge fusion module containing two knowledge memory units, can enhance the long-term memory ability of the Transformer encoder for different text granularities.
- Using Transformer's self-attention mechanism to design a new copy mechanism can alleviate word repetition problems and have the ability to reproduce facts and details, and reduce the gap between the model and the question when generating answers.
- We conduct experiments on the medical answering dataset and the stock question answering dataset. Through the analysis of the experimental results, we found that the integration of domain lexicon knowledge can indeed alleviate the lack of external knowledge of the automatic question answering system.

Our article describes the problem to be solved in the first section of the introduction, the second section describes the related work, the third section introduces the structure and various parts of the model, the fourth section verifies the performance of the model through experiments, and the fifth section gives the summary.

2 Related Works

The work closed to ours is An's KM-BiLSTM [12], but there are a few differences: (i) Oriented task: KM-BiLSTM is used to complete the answer selection task, and our LEP-Transfomer is used to complete the answer generation task. (ii) Method of extracting keywords: An uses the TF-IDF algorithm [15] to extract keywords in the datasets.

The keywords obtained in this way will contain some high-frequency pseudo-words such as "YES", some domain keywords may be over-segmented, resulting in inaccurate results. We directly use the vocabulary in the domain lexicons, because they contain all important information words. (iii) Basic model: KM-BiLSTM use Bi-LSTM [14], LEP-Transformer uses Transformer [5]. The self-attention mechanism of Transformer can change the dependence calculation between any two characters in the sequence into a constant, it is better handled in the long-distance dependence problem in the sentence. (iv) Knowledge fusion method: An only designed a knowledge memory unit in KM-BiLSTM to store knowledge information, in which each piece of knowledge (keyword) has only a fixed embedded form, questions are processed by Bi-LSTM will be lost part of the original information. In the knowledge fusion module of LEP-Transformer, Knowledge is stored in two different memory blocks, each piece of knowledge has two different embedding forms. Compared with the pre-trained fixed embedding form, the updated word embedding form will be closer to the original semantic information of the word. (v) Ability to reproduce facts: The LEP-Transformer model can reproduce the factual details in the question during the generation stage by pointing to copy words from the source text [16], while effectively alleviating the problem of out-of-vocabulary words. But this aspect is not involved in An's work.

3 Model

We propose LEP-Transformer that adds a knowledge fusion module to enhance Transformer's long-term memory ability for different text granularities. As shown in Fig. 1, LEP-Transformer includes knowledge encoder (a) and knowledge decoder (b).

3.1 Knowledge Encoder

The knowledge encoder is composed of knowledge fusion module and Transformer encoder module.

Knowledge Fusion Module. Considering that the original question (sentence level) and domain keywords (word level) are different text granularity, if simply splicing their embedded form and as input to Transformer encoder, the effect is not ideal, which may be related to position embedding [17]. Therefore, we design a knowledge fusion module (Knowledge Fusion module in Fig. 1) to store domain keyword knowledge information and add information to the representation of question.

The reason for designing two different forms of word embedding is that word embedding is constantly updated during the model training process. When the question embedding and the keyword embedding in the input knowledge memory unit are related calculations, then enter the output knowledge memory unit that the keyword embedding has been updated, and the information contained in the updated word embedding is more suitable for the original semantic information of the keyword.

First, given the set of keywords $\{k_i\}$, $i = 1, \ldots, n$, n is the number of keywords. As the Fig. 2 shows, in the input knowledge memory unit of the knowledge fusion module, each keyword k_i is converted into a corresponding input vector m_i in continuous space.

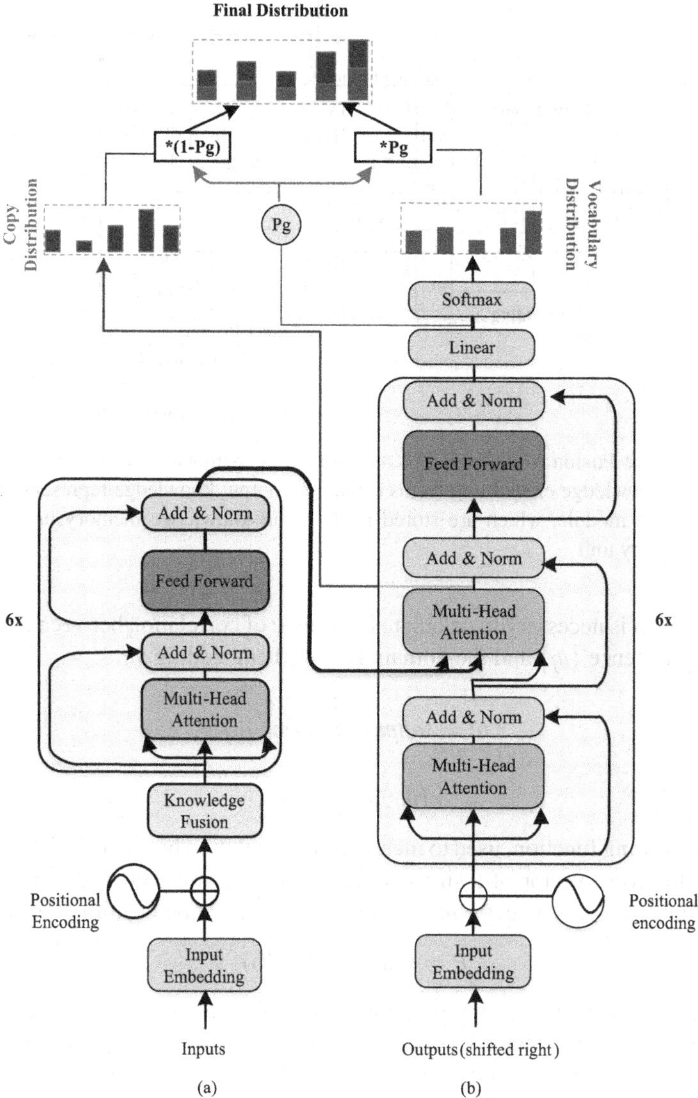

Fig. 1. LEP-Transformer.

Questions q is embedded as a sequence of vectors $\{x_i\}, j = 1, \ldots, T$. which T is the length of the question. Then use Transformer's position coding method [5] to embed the position pos_j into the question vector sequence to obtain a vector sequence containing position information:

$$u_j = x_j + pos_j \tag{1}$$

All m_i and u_j form the vector matrix A and vector matrix B respectively. Specifically, in the embedding space, in order to obtain knowledge information related to the question

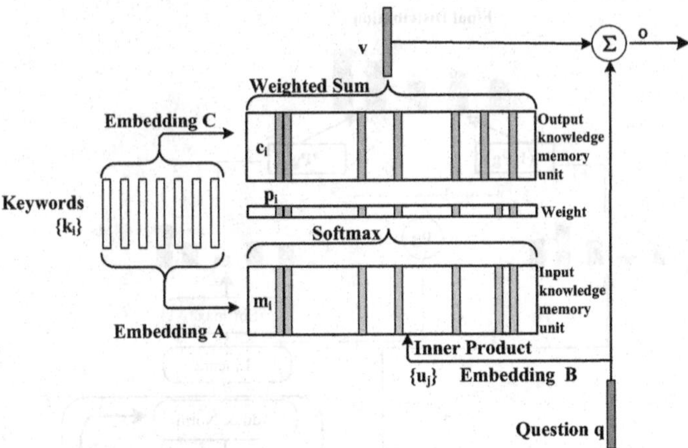

Fig. 2. Knowledge Fusion Module. Based on the idea of memory storage in Sukhbaatar [10], we design two knowledge embedding forms (input and output knowledge representation) in the knowledge fusion module, which are stored in the input knowledge memory unit and output knowledge memory unit

representation, it is necessary to calculate the degree of correlation between the question embedding sequence $\{u_j\}$ and the domain keyword embedding m_i:

$$p_j = softmax \sum_{j=1}^{T} e(m_j, u_j) \tag{2}$$

$$e(m_i, u_j) = u_j m_i \tag{3}$$

$e(\bullet)$ is a scoring function, used to measure the correlation between the keyword word vector and the word vector of a single word in the question sequence. Softmax is the activation function. p_i is a vector representation, all the p_i composition matrix P:

$$P = softmax(e(A, B)) \tag{4}$$

The weight information P_i represented by each piece of knowledge information and the original question is obtained. Next, in the output knowledge representation (Fig. 2 output knowledge memory unit), each keyword k_i is converted into the corresponding output vector c_i in the continuous space, and the knowledge vector is obtained by weighting and summing it:

$$v_i = \sum_i p_i c_i \tag{5}$$

All c_i form a matrix C, the corresponding matrix form is:

$$V = PC \tag{6}$$

Finally, by integrating the original question vector and the knowledge vector related to the question, a new fusion knowledge question expression can be obtained:

$$O = V + B \tag{7}$$

Transformer Encoder Module. Through the formula (7), we can get the question expression O of the fusion keyword knowledge information, and then input it into the Transformer's stack encoder [5] for encoding. The final output is noted as M.

3.2 Knowledge Decoder

The knowledge decoder is composed of The Transformer decoding module and the copy mechanism module.

Transformer Decoder Module. Assuming that the output y_{t-1} of the Transformer decoder is at the previous moment, so for the current moment t, y_{t-1} will be used as the input of the masked multi-attention layer in the first sublayer of the Transformer decoder. It is assumed that the mask attention layer will output a vector matrix Q, then Q and M will be input into the multi-head attention layer of the first sublayer of the Transformer decoder together (M will be used as the vector matrix K and vector matrix V in the self-attention mechanism). The Transformer decoder finally outputs a vector matrix S. The vector matrix S passes through the linear layer and the softmax layer to generate the vocabulary distribution P_v. The calculation formula is as follows:

$$P_v = softmax(VS + b) \tag{8}$$

Where V and b are parameters, and their values can be obtained through training. P_v is the probability distribution of all vocabularies in the vocabulary, from which the probability of the predicted word w can be obtained:

$$P(w) = P_V(w) \tag{9}$$

During the training phase, the loss function at time step t is as follows:

$$loss_t = -logP\left(w_t^{'}\right) \tag{10}$$

Therefore, the total loss of the entire sequence is:

$$Loss = \frac{1}{T}\sum_{t=0}^{T} loss_t \tag{11}$$

Copy Mechanism Module. Our copy mechanism is similar to the copy mechanism of See's pointer generator network [16]. We use the multi-head attention distribution as the copy distribution. The copy distribution obtained in this way is simpler and more efficient. The copy distribution calculation formula is as follows:

$$A = softmax\left(\frac{1}{\mathrm{h}}\sum_{\mathrm{i}=1}^{\mathrm{h}}\frac{\mathrm{Q_i K_i^T}}{\sqrt{\mathrm{d_k}}}\right) \tag{12}$$

The vector matrix Q and the vector matrix K can be obtained through the multi-attention layer of the Transformer decoder, Q_i and K_i is the *ith* part divided by Q and K, which h refers to the number of matrices that divide the matrix Q and the matrix K at the same time. The decoder can find the position of the next word through the probability distribution A. Similar to the practice of the pointer generator network, we introduce $P_g \in [0, 1]$ that the word in the vocabulary is selected with the probability of P_g in the decoding stage, and the word in the question fusion knowledge in the domain is selected with the probability of $1 - P_g$. The calculation formula of P_g is as follows:

$$P_g = \sigma(W_s S + b_s) \tag{13}$$

The values of W_s and b_s can be obtained through training. The probability distribution after the introduction of the copy mechanism is as follows:

$$P(w) = p_g P(w) + \left(1 - p_g\right) \sum_{i:w_i=w} A_i^t \tag{14}$$

From this, the total loss function of the LEP-Transformer model is as follows:

$$Loss = \frac{1}{T} \sum_{t=0}^{T} -\log\left(p_g P(w) + \left(1 - p_g\right) \sum_{i:w_i=w} A_i^t\right) \tag{15}$$

4 Experiments

4.1 Data and Evaluation

Medical Question Answering Dataset (Referred to as Medical). This dataset is provide by Yan [18]. We make a preliminary analysis of the distribution of the data, and drop a small amount of data. The processed Medical dataset contains 1,000K, 5K, and 5K pairs for training, validation, and testing, respectively.

Stock Question Answering Dataset (Referred to as Stock). This dataset is provide by Tu [13]. We make a preliminary analysis of the distribution of the data, and drop a small amount of data. As a result, the processed Stock dataset includes 182.557K, 1K, 0.5K pairs for training, validation, and testing, respectively.

Medical Lexicon. Because no domain lexicon can cover the entire data set well, we intend to build a domain dictionary that can cover the data set. First of all, we obtain the data on the medical health website[1], and use these data as the training corpus of this article. Then we use the longest matching algorithm based on segmentation unit [19] to extract related vocabulary, collect and sort the thesaurus related to the medical field from the Internet: Sogou medical lexicon[2], Baidu medical lexicon[3], Chinese medical lexicon[4], and combined with the extracted domain vocabulary to get the medical field dictionary.

[1] http://3g.xywy.com/.
[2] https://pinyin.sogou.com/dict/.
[3] https://shurufa.baidu.com/dict/.
[4] http://www.hankcs.com/nlp/corpus/tens-of-millions-of-giant-chinese-word-library-share.html.

Stock Lexicon. Due to the large number of stock websites and the complicated content, which is not consistent with the **Stock**, we only screens the current mainstream stock domain thesaurus to obtain five stock thesauruses related to the Stock (of which the stock thesaurus 1-3 are the same source as the **Medical lexicon**): stock lexicon 1, stock lexicon 2, stock lexicon 3, stock lexicon 4[5], and stock lexicon 5[6], and integrate them as the stock domain lexicon.

Evaluation. We use automatic evaluation metrics and human evaluation metric to evaluate the generated results. For automatic evaluation metrics, we use the BLEU [20] on text level and topic similarity: Greedy Matching, Embedding Average, Vector Extrema [21] on theme level to evaluate the generated results. For human evaluation metric, we use human evaluation of the fluency, accuracy and diversity [22] to evaluate the generated answers.

4.2 Baselines

We compare the proposed LEP-Transformer with the following baseline models:

ACNM [1]: The first model to apply RNN-based Seq2seq framework to dialogue generation tasks.

RNNsearch [2]: The model combining attention mechanism with Bi-LSTM-based Seq2seq framework.

DIAL-LV [3]: Combining latent variables with standard Gaussian priors and decoders in the seq2seq model based on Bi-RNN.

Adver-REGS [4]: An adversarial reinforcement learning model consisting of a generator and a discriminator : an RNN-based discriminator is used to evaluate the sequence, and the learning of the generator based on the Seq2seq model is guided by reinforcement learning to generate answers.

Transformer [5]: Transaction model based entirely on attention mechanism.

4.3 Training Details

We trained our models on one machine with 2 NVIDIA P100 GPUs. For **Medical**, the maximum length is 100 words for per generated sentence, LEP-Transformer's hidden size is 512 for encoder and decoder, vocabulary size is 4,329, and batch size is 64. We use Adam optimizer with the initial learning rate 0.001. In training, the epochs are 8. For **Stock**, the maximum length is 250 words for per generated sentence, the vocabulary size is 3777, the epochs are 30. In order to verify the effectiveness of the domain lexicon and copy mechanism, we designed the LE-Transformer model and the P-Transformer model respectively.

4.4 Experimental Results

On two Chinese datasets, as can be seen from the Table 1, Transformer has the highest score in the baseline models. In the Transformer-based models, LE-Transformer's

[5] http://cdict.qq.pinyin.cn/v1/.

[6] https://www.5118.com/ciku/detail/71.

Table 1. Different model's BLEU score.

Dataset	Model	BLEU-1	BLEU-2	BLEU-3	BLEU-4
Medical	ANCM	14.43	24.32	26.33	25.19
	RNNsearch	16.37	25.93	27.41	27.01
	DIAL-LV	17.27	27.42	28.86	27.20
	Adver-REGS	17.64	25.92	26.73	25.00
	Transformer	20.22	29.64	30.36	28.26
	LE-Transformer	27.65	34.50	33.50	30.35
	P-Transformer	20.83	29.90	30.45	28.27
	LEP-Transformer	**29.80**	**35.81**	**34.28**	**30.90**
Stock	ANCM	13.84	15.71	15.63	14.53
	RNNsearch	12.67	19.78	18.04	17.81
	DIAL-LV	13.12	21.48	23.50	22.71
	Adver-REGS	**20.84**	20.46	17.65	17.65
	Transformer	10.86	19.55	22.18	21.79
	LE-Transformer	12.63	20.58	24.31	24.29
	P-Transformer	12.44	19.69	21.99	21.44
	LEP-Transformer	15.93	**21.54**	**25.50**	**25.90**

BLEU score is higher than Transformer, which shows that the introduction of domain lexicon can better complete the generational question and answer tasks in the medical and stock fields. P-Transformer's BLEU score is higher than Transformer, showing the copy mechanism can directly copy keywords from the source question and domain keyword. LEP-Transformer's BLEU score is higher than LE-Transformer and P-Transformer, this is because LEP-Transformer combines the advantages of LE-Transformer model and P-Transformer model. Compared to tabulated Transformer,the LEP-Transformer model obtained 29.80's BLEU-1 score, 35.81's BLEU-2 score, 34.28's BLEU-3 score and 30.90's BLEU-4 score, has increased by 47%, 21%, 13% and 9% respectively. Although BLEU-1 is not the highest in Stock, compared to the standard Transformer model, the LEP-Transformer model has obtained 15.93's BLEU-1 score, 21.54's BLEU-2 score, 25.50's BLEU-3 score, and 25.90's BLEU-4 score, which is improved 46%, 10%, 15%, 18%. The results show that the LEP-Transformer can improve the accuracy of generating answers.

It can be found from Table 2 that the similarity scores of all models are very high, which are basically close to full marks. We conduct research from the dataset, vocabulary and word vectors separately to find out the causes for the high similarity of topics:

1. Dataset: By observing the corpus, we found that the answers to many questions start with "Hello, based on your description, basically understand the situation", statements like this account for a large part of the entire answer, the real core keywords

Table 2. The results of the embedding-based topic similarity.

Dataset	Model	Greedy matching	Embedding average	Vector extrema
Medical	ACNM	94.390	99.565	97.701
	RNNsearch	94.438	99.534	97.891
	DIAL-LV	95.160	99.674	98.237
	Adver-REGS	94.613	99.204	96.537
	Transformer	95.188	99.726	98.083
	LE-Transformer	96.047	**99.772**	98.455
	P-Transformer	93.934	99.413	96.038
	LEP-Transformer	**96.343**	99.765	**98.486**
Stock	ACNM	92.456	98.829	94.422
	RNNsearch	93.379	98.833	94.541
	DIAL-LV	93.869	99.392	95.340
	Adver-REGS	92.933	98.450	94.762
	Transformer	92.978	98.558	95.379
	LE-Transformer	94.832	**99.452**	96.717
	P-Transformer	92.934	99.413	96.038
	LEP-Transformer	**95.734**	99.441	**96.906**

such as "medical treatment with phlegm" in **Medical** occupy a small proportion. Since many models have learned such patterns, many of the answers they generate start with this form, but the keywords are not predicted. Topic similarity comprehensively considers the similarity of all words in a sentence, and keywords account for a very small part of the content of the answer, so the corpus itself will also affect the topic similarity score.

2. Vocabulary: The model's training vocabulary obtains from the training set. Although **Medical** has more than one million question and answer pairs, nearly 500 MB, indicating that most of the text in the dataset is repeated. When the model is sampling, it is likely that the same word will be repeatedly selected and generated. Similar to "Hello, according to your description" and other answers that do not give specific suggestions, and these words continue to appear in the reference answers, it may cause the topic similarity scores to become higher.
3. Word vectors: In the testing stage, we directly use the Bert model [23] trained Chinese word vectors, each word vector and sentence vector have a dimension of 768; while in the training stage, the word vectors follow the model Simultaneous training, different word vectors used in training and testing may also cause similarity deviation.

We use the artificial evaluation proposed by Shen et al. [22] to evaluate the model (Table 3). We randomly select 300 question and answer pairs from the test corpus and

Table 3. The results of human evaluation of sentences generated by different models.

Dataset	Model	F(%)	A(%)	D(%)
Medical	ACNM	32	34	28.5
	RNNsearch	47	35.5	31.5
	DIAL-LV	48	21	44.5
	Adver-REGS	69.5	15	19
	Transformer	66	54	51.5
	LE-Transformer	67	56	57
	P-Transformer	69.5	55.5	52.5
	LEP-Transformer	**71**	**60**	**61**
Stock	ACNM	69	32	48
	RNNsearch	70	34	49
	DIAL-LV	60	41	52
	Adver-REGS	76	22	48
	Transformer	84	47	46
	LE-Transformer	83	50	54
	P-Transformer	82	46	48
	LEP-Transformer	**84**	**53**	**55**

uses 8 different models to generate responses. And invited 10 graduate students with background in related fields, let them judge whether the generated answer is credible from three aspects of fluency (F), accuracy (A) and diversity (D).

On Medical: The Transformer model scored highest among the traditional Seq2seq models on three small evaluation indicators. (1) Fluency: LEP-Transformer combines the advantages of LE-Transformer and P-Transformer, and obtains 71% fluency, which is 8% higher than the standard Transformer model; (2) Accuracy: Compared with the standard Transformer model, the LEP-Transformer model has achieved 60% accuracy, an increase of 10%; (3) Diversity: Compared to the standard Transformer model, LEP-Transformer has a 61% diversity, an increase of 18%.

On **Stock**: Transformer has the highest score in the baseline models in Fluency and Accuracy. (1) Fluency: All models can achieve basic fluency; (2) Accuracy: Compared with the standard Transformer model, the LEP-Transformer model has achieved 53% accuracy, an increase of 13%; (3) Diversity: Compared with the highest DIAL-LV, the diversity of the LEP-Transformer model is 55%, an increase of 5%.

From the above analysis, it can be seen that the introduction of external knowledge can alleviate the lack of external knowledge of the generative automatic question and answer task and enhance the understanding of questions and improve the accuracy of generation.

Table 4 shows the answers generated by different models on **Medical,** the situation is similar for different models on **Stock**. Two true answers are given. Blue bold font for

Table 4. Analysis of the generated samples.

Question: My baby is 10 days after birth, he felt a little short of breath, nasal congestion, sputum, a little milk spitting, choking, but in a good mental state, what's going on with my baby, please?
GroudTruth1: Hello, you baby may be **cold**, if cold usually cough, expectoration, fever. Guidance: You can use **pediatric aminophenol huangnamin granules** to treat, **drink plenty of water**, pay attention to **keep warm** is very important, **if necessary, use infusion for treatment.**
GroudTruth2: Hello, this situation considers **upper respiratory tract infections**, upper respiratory tract infections generally have nasal congestion, children are relatively small, there may be poor breathing, spitting milk and other symptoms.
ANCM: Hello, according to your description, the baby's situation is caused by a cold, you can give the baby to take children's cold particles, children's cold particles, children's cold particles, children's cold particles, children's cold particles, drink more water, eat more fruit, drink water and eat more fruits. <unk>
RNNsearch: Hello hello, according to your description, consider the baby baby's **snoring**, you can still pay attention to when you can give your baby. Baby babies are recommended UNK.
DIAL-LV: **Your situation is normal**, don't worry, don't worry about advice: hello, I suggest you go to the hospital to do a test, **see if there is inflammation, if it is inflammation, can be treated with antibiotics**.
Adver-REGS: Hello, your situation is a certain risk, I suggest you go to the hospital to check. Suggestion: Hello, I suggest you go to the hospital for a comprehensive check to see if there is a problem. **If it is calcium deficiency, you can take calcium treatment**.
Transformer: The child has the above symptoms and is considered to be caused by **respiratory infections**. He can take **pediatric aminophenol huangnamin granules** and **ribavirin granules**.
LE-Transformer: If the child has the above-mentioned conditions and is considered to be caused by **an upper respiratory tract infection**, he can take **pediatric aminophenol huangnamin granules**.
P-Transformer: Hello, it may consider **colds**, **upper respiratory tract infections** with inflammation, and the above symptoms caused by wind and cold. You can take **pediatric cold granules** and **cephalosporin granules** for treatment and observation to avoid cold and wind. **If necessary, go to the hospital for further examination and treatment**.
LEP-Transformer: Hello, according to the situation you described, if the baby has **sputum** when **the nose is blocked**, the initial consideration may be caused by **respiratory infection**. It is recommended that you can give your baby **pediatric aminophenol huangnamin granules** and give your baby **more water**.

disease names and treatment recommendations related to real-life answers. Purple bold font is new words generated by the model. The orange squiggly font indicates the word that is repeatedly generated. The green font is OOV words.

Traditional Seq2seq models repeatedly generate words, like ACNM's answer repeatedly generated "children's cold particles", etc., it's a reason for not being smooth. Transformer has generated a new word "ribavirin granules". By referring to the information, this drug can also treat colds, but it is aimed at adults and is not applicable here. P-Transformer was successfully copied to the "upper respiratory tract infection" in the question, and the new word "cephalosporin particles" was generated. By consulting the information, this drug "cephalosporin particles" can be used to treat the respiratory system caused by sensitive bacteria, both adults and children Can be taken. Both the LE-Transformer and LEP-Transformer models can correctly generate the disease "upper respiratory tract infection", and the generated recommended drug "Pediatric aminophenol huangnamin granules" is also correct, which shows that the introduction of domain lexicon knowledge can improve the generation model performance.

4.5 Effectiveness Analysis of Knowledge Fusion Structure

Taking the **Medical** as an example, compare the BLEU scores of the following three models: (1) LEP-Transformer-1-hard:using the knowledge fusion module of KM-BiLSTM [12]); (2) LEP-Transformer-1-soft: Using a memory to store Multi-layer perceptrons measure similarity; (3) LEP-Transformer: Including two memory stores different word vectors of keywords, use multi-layer perceptrons to measure similarity. The results are shown in Table 5:

Table 5. BLEU score results obtained using different knowledge fusion modules.

Dataset	Model	BLEU-1	BLEU-2	BLEU-3	BLEU-4
Medical	LEP-Transformer-1-hard	14.90	25.51	27.52	26.23
	LEP-Transformer-1-soft	23.10	31.32	31.30	28.81
	LEP-Transformer	**29.80**	**35.81**	**34.81**	**30.90**

As can be seen from the Table 5, the LEP-Transformer's BLEU score is higher than LEP-Transformer-1-hard and LEP-Transformer-1-soft, indicating that using the updated word embedding form for related calculations can further improve the generated content Accuracy.

5 Conclusion and Future Work

Due to the lack of knowledge in the generative question-and-answer model, we propose LEP-Transformer that combines domain lexicon and copy mechanism. The knowledge fusion structure with two knowledge memory units can enhance the long-term memory ability of the Transformer. And using the copy mechanism to copy the questions and vocabulary in the domain dictionary can shorten the gap between the answer and the question. We also conducted experiments on the medical Q & A dataset and the stock Q & A dataset to verify the effectiveness of integrating the lexicon knowledge in the field to alleviate the lack of external knowledge in the automatic question answering system. In the future, we can consider designing more complex and efficient fusion structures to better complete the generative question-and-answer tasks.

Acknowledgements. This research is supported by the Scientific Research Platforms and Projects in Universities in Guangdong Province under Grants 2019KTSCX204.

References

1. Vinyals, O., Le, Q.V.: A neural conversational model (2015)
2. Bahdanau, D, Cho, K., Bengio, Y.: Neural machine translation by jointly learning to align and translate. In: 3rd International Conference on Learning Representations, ICLR 2015 (2015)
3. Cao, K., Clark, S.: Latent variable dialogue models and their diversity. In: Proceedings of the 15th Conference of the European Chapter of the Association for Computational Linguistics: volume 2, Short Papers, pp. 182–187
4. Li, J., Monroe, W., Shi, T., et al.: Adversarial learning for neural dialogue generation. In: Proceedings of the 2017 Conference on Empirical Methods in Natural Language Processing, pp. 2157–2169 (2017)
5. Vaswani, A., Shazeer, N., Parmar, N., et al.: Attention is all you need. In: Advances in Neural Information Processing Systems, pp. 5998–6008 (2017)
6. Furnas, G.W., Landauer, T.K., Gomez, L.M., et al.: Human factors and behavioral science: statistical semantics: analysis of the potential performance of key-word information systems. Bell Syst. Tech. J. **62**(6), 1753–1806 (1983)
7. Hochreiter, S., Schmidhuber, J.: Long short-term memory. Neural Comput. **9**(8), 1735–1780 (1997)
8. Chung, J., Gulcehre, C., Cho, K., et al.: Empirical evaluation of gated recurrent neural networks on sequence modeling. In: NIPS 2014 Workshop on Deep Learning, December 2014 (2014)
9. Weston, J., Chopra, S., Bordes, A.: Memory networks (2014)
10. Sukhbaatar, S., Weston, J., Fergus, R.: End-to-end memory networks. In: Advances in Neural Information Processing Systems, pp. 2440–2448 (2015)
11. Seo, M., Kembhavi, A., Farhadi, A., et al.: Bidirectional attention flow for machine comprehension (2016)
12. An, W., Chen, Q., Yang, Y., He, L.: Knowledge memory based LSTM model for answer selection. In: Liu, D., Xie, S., Li, Y., Zhao, D., El-Alfy, E.S. (eds.) International Conference on Neural Information Processing, pp. 34–42. Springer, Cham (2017). https://doi.org/10.1007/978-3-319-70096-0_4
13. Tu, Z., Jiang, Y., Liu, X., et al.: Generative stock question answering. arXiv preprint arXiv: 1804.07942 (2018)
14. Hou, W., Nie, Y.: Seq2seq-attention question answering model (2017)
15. Salton, G., Buckley, C.: Term-weighting approaches in automatic text retrieval. Inf. Process. Manage. **24**(5), 513–523 (1988)
16. See, A., Liu, P.J., Manning, C.D.: Get to the point: summarization with pointer-generator networks. In: Proceedings of the 55th Annual Meeting of the Association for Computational Linguistics (Volume 1: Long Papers), pp. 1073–1083 (2017)

17. Shaw, P., Uszkoreit, J., Vaswani, A.: Self-attention with relative position representations. In: Proceedings of NAACL-HLT, pp. 464–468 (2018)
18. Yan, G., Li, J.: Mobile medical question and answer system with improved char-level based convolution neural network and sparse auto encoder. In: Proceedings of the 2019 Asia Pacific Information Technology Conference, pp. 20–24 (2019)
19. Sun, X., Zheng, Q., Wang, C., et al.: Method for generating domain dictionary based on raw corpus (2005)
20. Papineni, K., Roukos, S., Ward, T., et al.: BLEU: a method for automatic evaluation of machine translation. In: Proceedings of the 40th Annual Meeting on Association for Computational Linguistics. Association for Computational Linguistics, pp. 311–318 (2002)
21. Liu, C.W., Lowe, R., Serban, I.V., et al.: How NOT to evaluate your dialogue system: an empirical study of unsupervised evaluation metrics for dialogue response generation. In: Proceedings of the 2016 Conference on Empirical Methods in Natural Language Processing, pp. 2122–2132 (2016)
22. Shen, X., Su, H., Niu, S., et al.: Improving variational encoder-decoders in dialogue generation. In: Thirty-Second AAAI Conference on Artificial Intelligence (2018)
23. Devlin, J., Chang, M.W., Lee, K., et al.: BERT: pre-training of deep bidirectional transformers for language understanding. In: Proceedings of the 2019 Conference of the North American Chapter of the Association for Computational Linguistics: Human Language Technologies, Volume 1 (Long and Short Papers), pp. 4171–4186 (2019)

Design of Smart Home System Based on Collaborative Edge Computing and Cloud Computing

Qiangfei Ma[1], Hua Huang[1](✉), Wentao Zhang[1], and Meikang Qiu[2]

[1] School of Information Engineering, Jingdezhen Ceramic Institute, Jingdezhen, China
{1920045006,1920045003}@stu.jci.edu.cn, huanghua@jci.edu.cn
[2] Department of Computer Science, Texas A&M University-Commerce, Texas 75428, USA
Meikang.Qiu@tamuc.edu

Abstract. With the rapid development of the Internet, more and more smart devices are beginning to enter people's daily lives. The intelligentization of home appliances has gradually become a trend in home networking. How to construct and manage smart homes reasonably system has become the focus of research. Considering the needs and characteristics of smart homes, this paper combines edge computing and cloud computing in intelligent home scenarios to design a smart home system. To improve the system's reliability and real-time performance, we propose a cloud and edge collaborative processing algorithm strategy and use Kubernetes container management platform technology to deploy the system uniformly. Experimental results showed that the system has higher operating efficiency than the system using cloud computing as the center and single edge computing as the center, which can further ensure the smart homes real-time requirements.

Keywords: Smart home · Edge computing · Cloud computing · Container · Algorithm strategy

1 Introduction

The rapid development of the Internet has propelled the Internet of Things into a new era. Based on the Internet of Things concept proposed by the Internet of Things, it has quietly entered thousands of households [1]. People begin to have higher and higher requirements for the comfort, convenience, and safety of home life, which also promotes the continuous development of home appliances in the direction of more and more intelligent. According to the latest Cisco research report, there will be 28 billion devices connected to the Internet by 2022, and IoT devices will also reach 13.7 billion, The amount of global equipment production data will increase from 218ZB in 2016 to 847ZB in 2021, which shows that many intelligent devices will need connecting to the Internet in the future. However, with the increasing number of smart devices connected to the Internet, it will inevitably increase the cloud operations' burden. The centralized processing model of cloud computing will have some deficiencies in the Internet of Everything.

M. Qiu (Ed.): ICA3PP 2020, LNCS 12454, pp. 355–366, 2020.
https://doi.org/10.1007/978-3-030-60248-2_24

The purpose of edge computing is to process the data closer to the terminal device [2]. The downstream of edge computing represents cloud services, and the upstream represents services of the Internet of Everything. Edge computing is a new type of computing model. It provides computing, storage, and network resources for the data generated and collected by terminal devices on the physical segment or data source's network side. At the same time, it can also deal with big data accumulated on the edge side. Provide better support and services to solve the problems of excessive bandwidth and energy consumption caused by many devices accessing the cloud. Because edge computing can process data autonomously, data can be processed locally without uploading to the cloud, which can significantly alleviate the problem of insufficient cloud network bandwidth, which also makes the traditional cloud computing center processing mode begin to edge computing The distributed processing model is transformed. In the past cloud architecture model, the data is processed in the cloud data center. It can provide users with reliable data processing and storage and provide services such as high flexibility, scalability, and on-demand deployment. With the increasing number of devices connected to the Internet and the explosive growth of data generated by machines, cloud computing is beginning to be insufficient to meet people's needs. The proposal of edge computing can solve this problem.

Considering the needs and characteristics of smart homes, applying edge computing to smart home systems is a new model of the Internet of Things. By setting up a private network inside the house, smart home devices are connected to the private network, and manage, store and distribute the data generated and collected by the terminal equipment in the internal environment. The emergence of edge computing is not to replace cloud computing but to combine with cloud computing. Edge computing has changed the traditional cloud computing data processing model, accelerating the development of the field of smart home, the concept of interconnection of everything began to slow from a concept slowly turned into reality, edge computing can solve the problems of real-time, data privacy, and high task latency in smart homes. Therefore, the combination of edge computing and cloud computing in intelligent home systems has significant research value.

In this paper, we mainly take the application of smart home as the central premise. On the one hand, consider the real-time performance of edge computing; On the other hand, consider the computing and storage capabilities of cloud computing, we designed a smart home system that combines cloud-edge computing. Starting from the real-time and reliability of data processing, We designed an algorithm strategy combining cloud-side computing. By combining existing technologies, we have built an experimental platform with cloud hosts and personal computers and finally carried out the system Instance verification.

Specifically, our contributions are summarized as follows:

- We proposed a three-tier architecture model, including cloud, edge, and terminal, and apply cloud-edge computing to smart home systems.
- We designed a collaborative cloud-edge computing algorithm strategy, mainly to improve the real-time and reliability of the entire system.

– We use the Kubernetes container management platform technology to deploy the system uniformly. Through many simulation experiments, the results show that the system model and algorithm strategy we proposed are effective.

The rest of the paper is organized as follows. Section 2 discusses the research background and related work. Section 3 introduces the system architecture and algorithm strategy proposed in this paper. Section 4 builds the system proposed in this paper. Section 5 tests and evaluates the built system. Section 6 concludes.

2 Related Works

The smart home concept originated very early, but there has been no specific architectural case to achieve it. It was not until 1984 that the United Technologies Building System applied the concept of informationization and integration of Building equipment to the CityPlaceBuilding in Hartford, Connecticut, the United States. From then on, it opened the prelude to the application of smart homes in concrete buildings. Smart homes connect various devices in the house through the Internet of Things, such as audio and video equipment, lighting systems, air conditioning control, security systems, digital audio, video systems, and network home appliances, realizing intelligent management of various devices in the house. Smart home development is mainly divided into three stages: single access of smart devices, interconnection access of smart devices and artificial intelligence applications. At present, it is in the transitional stage from interconnection to artificial intelligence, mainly by combining smart devices and digital technology, using home network environment, such as wireless network, Zigbee and other technologies to achieve short-distance, high-speed communication, and then Unified management of these intelligent devices through cloud computing, and provide customized services according to user needs.

In recent years, there have also been many kinds of research on smart homes, and how to improve users' comfort is the focus of research. In [4], Han proposed a new smart home energy management system based on IEEE802.15.4 and ZigBee standards. The system can integrate various physical sensor information with the support of an active sensor network with sensors and actuator components. Control various household appliances. In [5], AR proposed an energy management system for smart homes, using off-the-shelf business intelligence (BI) and big data analysis software packages to better manage energy consumption and meet consumer needs. In [6], Babou proposes Home Edge Computing (HEC): a new three-tier edge computing architecture that provides data storage and processing in close proximity to the users.

There has been a lot of researches on cloud-edge computing. For cloud-edge coexistence, the method of data processing between the two is a research focus. In [7], Dautov introduces a distributed hierarchical data fusion architecture for the IoT networks, consisting of edge devices, network and communications units, and cloud platforms. According to the proposed approach, different data sources are combined at each level of the IoT hierarchy to produce timely and accurate results by utilizing intermediate nodes' computational capabilities. In [8], Thai introduces a general architecture of cloud edge computing that was proposed, which aims to provide vertical and horizontal offloading between service nodes. In [9], Wu considers the heterogeneity of edge and central cloud servers in the offloading destination selection. To jointly optimize the system utility and the bandwidth allocation for each mobile device, a Distributed Deep learning-driven Task Offloading (DDTO) algorithm is proposed to generate near-optimal offloading decisions over the mobile devices, edge cloud server and central cloud server. In [10], Wang presents a tensor-based cloud-edge computing framework that mainly includes the cloud and edge planes. The cloud plane is used to process large-scale, long-term, global data; The edge plane is used to process small-scale, short-term, local data. However, these solutions analyzed above require very high hardware configurations and cannot be deployed and applied quickly. This article mainly considers edge computing to smart homes as a research premise and cooperates with edge computing and cloud computing to quickly process terminal tasks.

3 System Model Architecture and Algorithm Design

3.1 Three-Tier System Architecture of the System

The overall system architecture proposed in this paper includes cloud, edge and terminal. The following will give a brief description of these three parts:

Cloud layer: The cloud layer is composed of multiple high-performance servers; it has powerful storage and computing capabilities and can perform super large-scale calculations. The cloud layer can effectively schedule and manage edge nodes through the control strategy and provide better services for users.

Edge layer: The edge layer is located between the cloud and the terminal layer, and generally consists of a large number of edge nodes, such as gateways, base stations, routers, switches, personal computers, and other devices. These edge nodes can calculate and store the data collected and generated by the terminal equipment. Because these edge nodes are relatively close to the terminal device, it can provide real-time services, which can better meet users' needs. The edge node can also pre-process the data from the terminal device, and then upload the pre-processed data to the cloud, easing the pressure on the cloud and reducing the core network traffic.

Terminal layer: The terminal layer is composed of various Internet of Things devices, such as smart cameras, smart temperature regulators, smart access control, intelligent robots, etc. Generally, complex data is not calculated on the terminal device. This is to extend the terminal device's service time because the terminal device is only responsible for generating and collecting data and uploading these data to the upper layer, which calculates and saves.

The entire system's workflow is that the terminal device first generates and collects data and then uploads it to the edge layer. The edge layer is responsible for calculating and storing these data, managing and scheduling the edge device, and the edge layer is also responsible for uploading part of the data to the cloud. The cloud is responsible for the management and scheduling of the overall system. Simultaneously, when the edge layer encounters tasks that cannot be processed, the cloud will take over these tasks and process it. The cloud also needs to receive the data uploaded by the edge layer, analyze and store these data in order to provide better services to users. The three-tier architecture model of this system design is shown in Fig. 1.

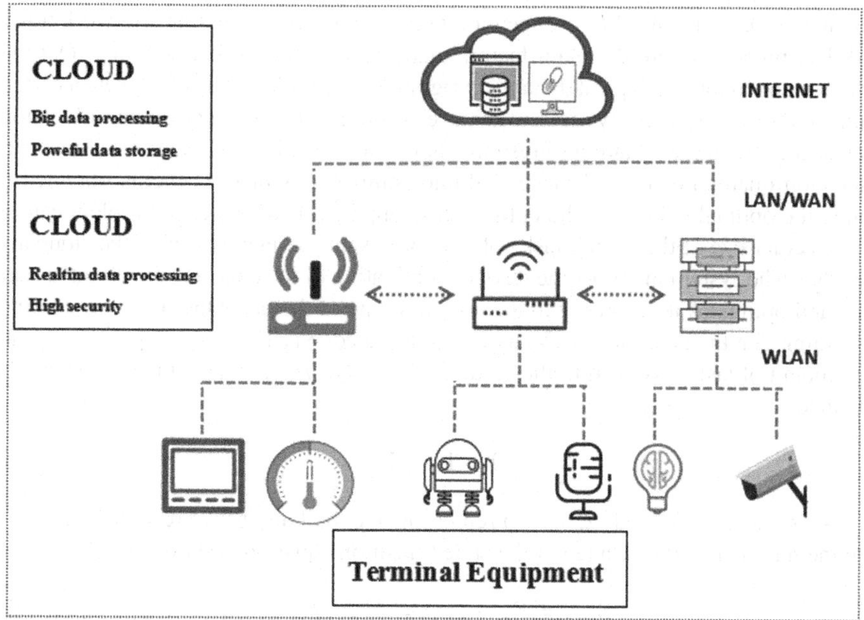

Fig. 1. Three-tier architecture

3.2 Service Management Procedures

The service management program needs to manage the service software and user data, including installing, uninstallation, updating, startup, and suspension of the service software. Server software generally runs on linux systems, and the service environment required by each device is different. We cannot install multiple virtual machines [11]. The solution is to use container technology to isolate the service environment required by each device. This article chooses Docker [12] container to solve this problem. Containers can bundle key application components and achieve portability between different platforms and the cloud. Service providers package services into Docker containers and provide services to users in the form of images, users can use these service software without complicated deployment. The cloud software management center is responsible for the installation and update of local software, when the user needs to use the corresponding service program, the corresponding service is opened in the form of a container, and when the user no longer uses it, the corresponding service is closed in the form of a container. The characteristics of container technology can perfectly match the lightness and portability of edge computing. The platformization of edge computing has become a trend.

3.3 Resource Allocation Optimization Problem

The emergence of edge computing has eased pressure on cloud traffic. Because edge computing is closer to terminal devices, data can be filtered and analyzed at edge nodes, so it is more efficient and more intelligent, but when facing businesses with relatively large amounts of data also need to use the cloud data centre for processing. Network traffic and data processing are the main factors that affect system task scheduling. Optimal resource allocation is a typical dynamic programming problem. It is to allocate a certain number of resources to many users, to achieve the optimal result after resource allocation, then according to the situation of resource allocation, divide the dimension of resource allocation quantitatively, and make multiple comparisons of the benefits obtained to obtain the optimal value. For the optimization object, it is the revenue problem caused by the actual size of the overall task volume for business data allocated to the cloud and the edge. The system monitors the resource utilization of the edge and the cloud in real-time and optimizes it according to the size and type of the task Optimized scheduling. Assuming that the amount of tasks assigned to the edge is $\mathrm{T}_1 = (\mathrm{t}_{11}, t_{12}, \ldots, t_{1n})$, and the amount of tasks assigned to the cloud is $\mathrm{T}_2 = (\mathrm{t}_{21}, t_{22}, \ldots, t_{2n})$, the Eq. (1) can be obtained:

$$\mathrm{T} = \mathrm{T}_1 + \mathrm{T}_2 \tag{1}$$

The state variable $\mathrm{S}_\mathrm{k}(1, 2, \ldots, \mathrm{n})$ represents the total number of resources from the k to the n user. The formula (2) of the state transition equation is shown in:

$$\mathrm{S}_{\mathrm{k}+1} = \mathrm{S}_\mathrm{k} - \mathrm{T}_\mathrm{k}(\mathrm{k} = 1, 2, \ldots, \mathrm{n}) \tag{2}$$

Here we use backward-to-forward recursive method to solve the optimal solution of the problem, as shown in Eq. (3):

$$\begin{cases} \mathrm{f}_\mathrm{k}(\mathrm{S}_\mathrm{k}) = \max\{v_\mathrm{k}(\mathrm{S}_\mathrm{k}, \mathrm{T}_\mathrm{k}) + \mathrm{f}_{\mathrm{k}+1}(\mathrm{S}_{\mathrm{k}+1})\}, k = 1, 2, \ldots, n \\ \mathrm{f}_{\mathrm{k}+1}(\mathrm{S}_{\mathrm{k}+1}) = 0 \end{cases} \tag{3}$$

The index function $v_\mathrm{k}(\mathrm{S}_\mathrm{k}, \mathrm{T}_\mathrm{k})$ represents the level index of the decision T_k with respect to the state variable S_k, and $\mathrm{f}_\mathrm{k}(\mathrm{S}_\mathrm{k})$ represents the maximum profit when the resources of the quantity S_k are allocated to the k to n users.

3.4 Resource Scheduling Algorithm

This paper proposes a resource scheduling algorithm based on edge computing and cloud computing collaborative processing. The main idea is to allocate tasks to the edge for processing when the task volume is relatively small. When the amount of tasks exceeds the threshold set by the system, the system starts to allocate the tasks dynamically and cooperates with the edge and the cloud to process it. Starting from the relevant parameters of the algorithm, we adjust the parameters through multiple iterations to achieve the optimal solution and set the threshold for dynamic programming of related tasks. To ensure that the task can be processed in a short time, the design of the algorithm is mainly to use a greedy strategy under local search in the program [13], using the utilization rate of the virtual machine in the edge server as the allocation choice of the greedy strategy, by updating the edge in real-time Utilization list of virtual machines, giving priority to assigning tasks to virtual machines with low utilization. The specific design of the algorithm is shown below.

Algorithm 1 Resource management

```
Input: 1) Business request queue L1
       2) Edge host resource utilization rate (el, e2, e3....en) and cloud host re- source
          utilization rate (c1,c2,c3....cn)
       3)Algorithm threshold Val
Output: Service distribution queue L2
while L1 ≠ Ø do
    E1 ← e1 + e2 + e3 + ... + en; C1 ← c1 + c2 + c3 + ... + cn
    if E1 < Val then
        Distribute services to edge hosts
    else
        Build resource allocation list
        Update service distribution queue L2
        Distribute services to corresponding edge hosts and cloud hosts
    end if
        Update edge host resource utilization rate E1 and cloud host resource uti-
        lization rate C1
end while
return Service Distribution List L2
```

4 System Construction

4.1 Container Cluster Management

This article selects Kubernetes container management platform for deployment. Kubernetes is a brand-new leading solution of distributed architecture based on container technology. It is based on Docker technology and provides a series of complete functions such as deployment and operation, resource scheduling, service discovery and dynamic scaling for containerized applications, which improves large-scale convenience of container cluster management [15]. Kubernetes is a complete distributed system support platform with complete cluster management capabilities, multiple expansion and multi-level security protection and access mechanisms, multi-tenant application support capabilities, transparent service registration and discovery mechanisms, and built-in intelligent load balancer, powerful fault discovery and self-repair capabilities, service rolling upgrade and online expansion capabilities, scalable automatic resource scheduling mechanism, and multi-granular resource quota management capabilities. At the same time, Kubernetes provides comprehensive management tools covering all aspects including development, deployment testing, operation and maintenance monitoring. Figure 2 shows a typical Kubernetes architecture diagram.

4.2 Realization of the System

This system chooses one cloud computing server and two personal computers to build. The cloud host server has 16 logical processors and a memory size of 32G; both personal computers have four core processors and the memory is 8G. These two personal

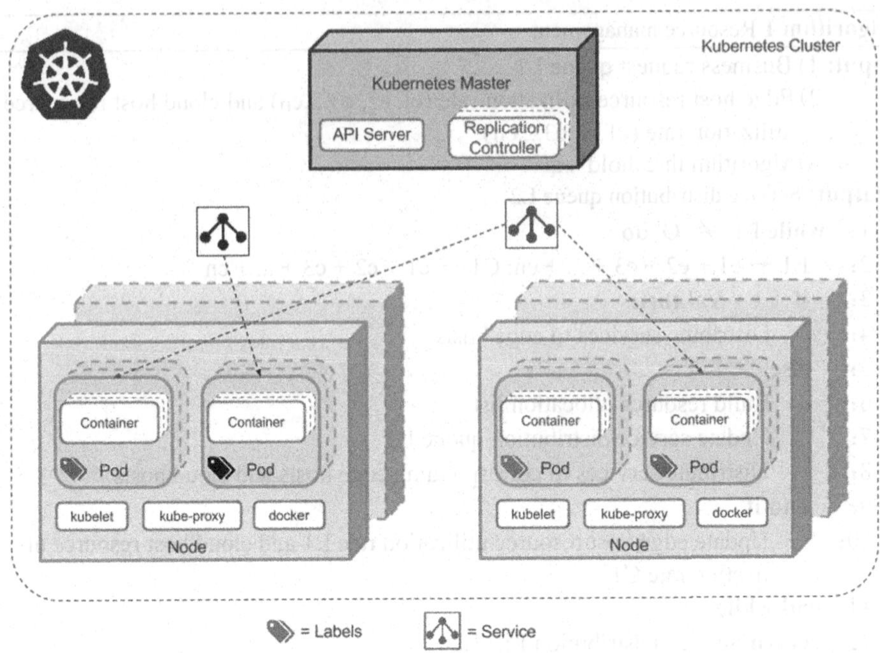

Fig. 2. Kubernetes architecture

computers simulate the system's edge server and coordinator, respectively. First, deploy this computer that simulates the coordinator, and deploy the edge and cloud coordination algorithms designed by this system on the coordinator. The data services generated by the terminal device are sent to the Api Server via kubectl, the Api Server receives the client's request and stores the resource content in the database etcd, the Co-troller component (including Scheduler, replication, endpoint) monitors and responds to resource changes. ReplicaSet checks the database changes and creates the desired number of Pod instances. The Scheduler checks the database changes again and finds that it has not been assigned to a specific node Pod, and then distribute Pod to the nodes that can run them according to the algorithm strategy designed in this article, and update the database records to record Pod distribution, kubeproxy runs on each host in the cluster, manages network communication, service discovery, and load balancing. When there is data sent to the host, it will be routed to the correct pod or container. For the data sent on the host, it can discover the remote server based on the requested address and process the data correctly. In some cases, a round-robin scheduling algorithm is used to send requests to multiple instances in the cluster. In this paper, two-node clusters are deployed on the edge server and the cloud computing server. The node cluster of the edge server is relatively close to the coordinator. To ensure the real-time performance of the terminal device, the coordinator first allocates the service when the data service is relatively small. Give the node on the edge server. The cloud server is far away from the coordinator and has strong computing power. When the data service exceeds the threshold set by the edge server, the coordinator dynamically adjusts according to the algorithm strategy to optimally allocate the data service to the edge and cloud servers. The realization of the system is shown in Fig. 3.

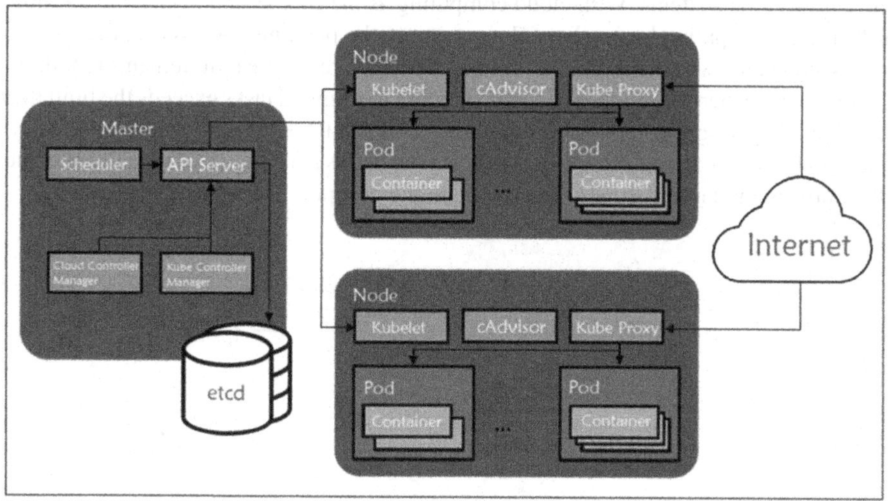

Fig. 3. The overall implementation of the system

5 Experiment

In the current smart home system architecture, there are three main data processing modes, namely cloud computing as the core, single edge computing as the core, and collaborative edge and cloud as the core. This paper selects two different terminal applications, one is to collect different environmental information through sensors, such as infrared temperature sensors, humidity sensors, voice switches, etc. that are often used in homes, and the other is to read images through cameras. Different terminal tasks will produce different business requirements. The sensor generally generates a small amount of data, but it is more sensitive and needs to be processed in real-time. The amount of data generated by the camera is relatively large, and the delay requirement is not so high, but it requires more powerful computing power to deal with it. This paper selects 0 to 100 terminal applications to test and evaluate the collaborative edge and cloud systems. We respectively examine the failure rate and average time to complete the task from the system.

5.1 Compare the Failure Rate in the Three Modes

From the success rate and failure rate of completing terminal tasks, we can observe the ability of the system structure to deliver applications continuously. Therefore, this article mainly compares the failure rates of terminal tasks in the cloud, edge, and collaborative cloud-edge modes. As shown in Fig. 4 (a) and (b) show the situation where the tasks of the sensor and camera are completed in three different system modes, respectively. In general, with the increase of terminal tasks, the collaborative cloud and edge system has excellent performance, that is, its failure rate is much lower than the cloud-centric and edge-centric systems. With cloud computing as the core system, when the terminal tasks increase rapidly, due to channel congestion, the pending business data cannot be processed in time, so it will cause those tasks with high real-time requirements to fail. As a core system, when the number of a large number of terminal tasks exceeds the limit that the edge computing can carry, the computing power of the edge node will drop sharply. The collaborative edge and cloud system designed in this paper will dynamically allocate when processing large business data to obtain the best possible distribution results.

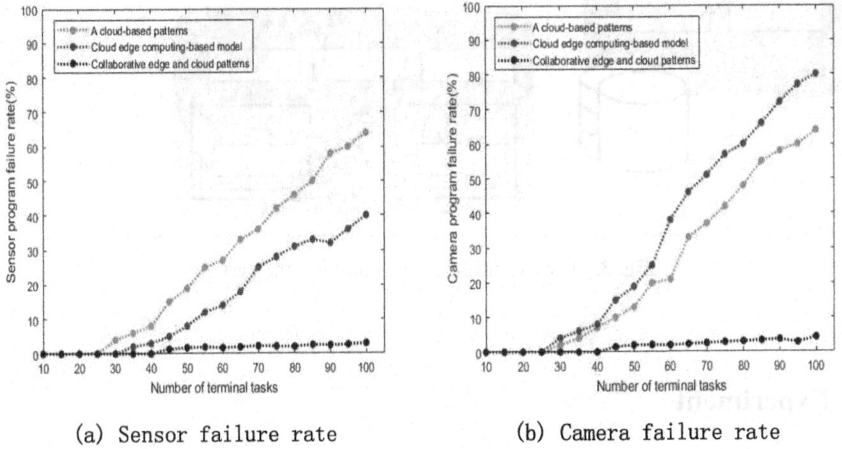

(a) Sensor failure rate (b) Camera failure rate

Fig. 4. Sensor and camera failure rates in three modes

5.2 Compare the Average Completion Time in the Three Modes

From the average time to complete the terminal task can further investigate the performance of the system. A system with a short average time represents relatively high real-time performance. It can quickly process terminal tasks, thereby providing users with more convenient services. This article mainly compares the average completion time of terminal tasks in the cloud, edge, and collaborative modes. As shown in Fig. 5 (a) and (b) represent the average completion time of the sensor and camera tasks in the three system modes, respectively. From the figure, we can observe that the processing time of the terminal task of the collaborative edge and cloud system is lower than the other two system modes, and this situation will become more and more obvious as the terminal task increases. When the terminal task increases rapidly, due to the limitation of network bandwidth, data cannot be quickly uploaded to the cloud, which will cause the average time of the system to process the terminal task. When the number of terminal tasks increases beyond the threshold that the edge computing can handle, the computing power of the edge computing nodes will drop sharply, resulting in long processing time for the terminal tasks. In the face of the above two situations, the system with collaborative edge and cloud can optimally allocate terminal tasks through the algorithm designed in this paper, so that the average completion time of the task is smaller.

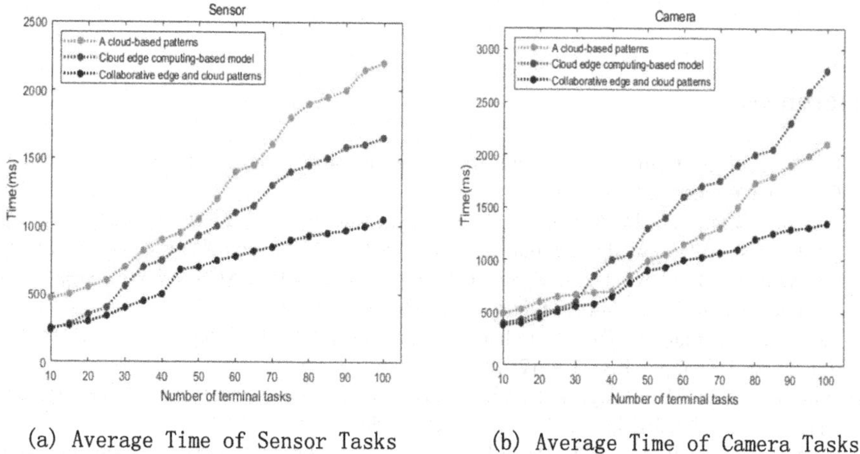

(a) Average Time of Sensor Tasks (b) Average Time of Camera Tasks

Fig. 5. Average time comparison of processing tasks

6 Conclusion

The purpose of this study is to combine edge computing and cloud computing into the smart home system, and provide users with more convenient services through their collaborative work. First of all, we analyzed whether edge computing is feasible to solve the problem of the rapid increase in Internet data volume. Secondly, we analyzed the application of collaborative edge and cloud computing design patterns in the smart home system and proposed a three-tier architecture model. Finally, we proposed an algorithm to improve the performance of the system, and then built a collaborative edge and cloud computing smart home system to conduct simulation experiments on the system. The experimental results verified that the method proposed in this paper could indeed improve the overall performance of the smart home system. The smart home system that collaborates with the edge and the cloud can quickly respond to terminal tasks, ensuring the real-time requirements of the smart home and protecting the privacy within the home. As the next step, we will further study the usability and security aspects of the system in this paper, and at the same time refer to the development trends of edge computing and cloud computing to design more efficient smart home systems.

Acknowledgment. This work is supported by the National Key Research and Development Plan of China under Grant No. 2016YFB0501801, National Natural Science Foundation of China under Grant No. 61170026, the National Standard Research Project under Grant No.2016BZYJ-WG7 – 001,the Key Research and Development Plan of Jiang Xi province under Grant No. 20171ACE50022 and the Natural Science Foundation of Jiang Xi province under Grant No.

20171BAB202011, the science aánd technology research project of Jiang Xi Education Department under Grant Nos. GJJ180730,GJJ1807 – 27,GJJ181520, and the Science and Technology project of Jingdezhen under Grant Nos. 20182GYZD011 – 01, 20192GYZD008 – 01, 2019 GYZD008 – 03,and the Open Research Project of the State Key Laboratory of Industrial Control Technology, Zhejiang University, China (No.ICT 20025).

References

1. Liu, Q., Cui, L., Haiqiang, C.: Key technologies and applications of the internet of things. Computer Science (2010)
2. Babou, C.S.M., Fall, D., Kashihara, S., Niang, I., Kadobayashi, Y.: Home edge computing (HEC): design of a new edge computing technology for achieving ultra-low latency. In: Liu, S., Tekinerdogan, B., Aoyama, M., Zhang, L.J. (eds.) EDGE 2018. LNCS, vol. 10973, pp. 3–17. Springer, Cham (2018). https://doi.org/10.1007/978-3-319-94340-4_1
3. Fang, J., Guangjian, Z.: The research and design on smart home system based on internet of things. Int. J. Comput. Eng. 1(4) (2016)
4. Han, D.M., Lim, J.H.: Design and implementation of smart home energy management systems based on zigbee. IEEE Trans. Consum. Electron. **56**(3), 1417–1425 (2010)
5. Al-Ali, A.R., Zualkernan, I.A., Rashid, M., et al.: A smart home energy management system using IoT and big data analytics approach. IEEE Trans. Consum. Electron. **63**(4), 426–434 (2018)
6. Babou, C.S.M., Fall, D., Kashihara, S., Niang, I., Kadobayashi, Y.: Home edge computing (HEC): design of a new edge computing technology for achieving ultra-low latency. In: Liu, S., Tekinerdogan, B., Aoyama, M., Zhang, L.J. (eds.) EDGE 2018. LNCS, vol. 10973, pp. 3–17. Springer, Cham (2018). https://doi.org/10.1007/978-3-319-94340-4_1
7. Dautov, R., Distefano, S.: Three-level hierarchical data fusion through the IoT, edge, and cloud computing. International Conference on Internet of Things & Machine Learning, 1–5 ACM (2017)
8. Thai, M.T., Lin, Y.D., Lai, Y.C., et al.: Workload and capacity optimization for cloud-edge computing systems with vertical and horizontal offloading. IEEE Trans. Netw. Serv. Manag. **17**(1), 227–238 (2020)
9. Wu, H., Zhang, Z., Guan, C., et al.: Collaborate edge and cloud computing with distributed deep learning for smart city internet of things. IEEE Internet Things J. **7**, 8099–8110 (2020)
10. Wang, X., Yang, L.T., Xie, X., et al.: A cloud-edge computing framework for cyber-physical-social services. IEEE Commun. Mag. **55**(11), 80–85 (2017)
11. Goldberg, R.P.: Survey of virtual machine research. Computer **7**(6), 34–45 (1974)
12. Wan, X., Guan, X., Wang, T., Bai, G., Choi, B.Y.: Application deployment using microservice and docker containers: framework and optimization. J. Netw. Comput. Appl. **119**, 97–109 (2018)
13. Du, D.Z., Ko, K.I., Hu, X.: Design and analysis of approximation algorithms. Springer optimization and its applications. Greedy Strategy, vol. 62. Springer, New York (2012)
14. Pang, Yu.: Application research of docker technology in software development process. Commun. Manag. Technol. **000**(002), 45–47 (2019)

Classification of Depression Based on Local Binary Pattern and Singular Spectrum Analysis

Lijuan Duan[1,2,3], Hongli Liu[1,2,3], Huifeng Duan[1,2,3], Yuanhua Qiao[4(✉)], and Changming Wang[5]

[1] Faculty of Information Technology, Beijing University of Technology, Beijing, China
[2] Beijing Key Laboratory of Trusted Computing, Beijing, China
[3] National Engineering Laboratory for Critical Technologies of Information Security Classified Protection, Beijing 100124, China
[4] College of Applied Science, Beijing University of Technology, Beijing, China
qiaoyuanhua@bjut.edu.cn
[5] Beijing Anding Hospital, Capital Medical University, Beijing 100124, China

Abstract. Depression is a common mental disease characterized by significant sadness and feeling blue all the time. At present, most classifications and predictions of depression rely on different characteristics. Comparing with the previous work, we use local binary pattern (LBP) and signal singular spectrum analysis (SSA) technology to extract features from the original signal. Firstly, the LBP signal is obtained by encoding the segmented signal. Then, we use SSA to decompose and reconstruct the LBP signal to remove noise and divide the frequency band. Finally, we feed the data of each frequency band to K-nearest neighbor (KNN), decision tree (DT), support vector machine (SVM) and extreme learning machine (ELM) for classification. The experimental results show that LBP and SSA features achieve the best classification effect on SVM, and the accuracy of beta band is the highest with 99.24% accuracy, 99.34% sensitivity and 99.12% specificity respectively.

Keywords: Electroencephalography · Local Binary Pattern · Singular Spectrum Analysis

1 Introduction

Depression was a mental disease characterized by low mood, loss of interest, lack of attention, and even suicidal thoughts [1], which turned to be the second most harmful disease for human health in the world in 2020. However, the traditional treatment of depression mainly relied on the dialogue and scale assessment between doctors and patients, which was highly subjective and easily leads to misdiagnosis. Therefore, it was particularly important to find an objective and effective diagnostic method, which can not only help doctors diagnose the disease quickly, but also save the time for patients to cure.

A large number of studies about depression from physiological data are obtained through functional magnetic resonance imaging (fMRI), which used magnetic resonance

M. Qiu (Ed.): ICA3PP 2020, LNCS 12454, pp. 367–380, 2020.
https://doi.org/10.1007/978-3-030-60248-2_25

imaging to measure neuronal activity caused by the change of blood dynamics. FMRI can aid in the diagnosis of depression and allows for multifunctional tradeoffs between SNR, spatial-temporal resolution and field of view. It achieved millimeter spatial resolution and generated great functional localization capability. EEG signals record and describe the electrical transformation of brain activity, which was the specific response of brain nerve cells in the cerebral cortex or the surface layer of the scalp [14]. EEG sampling process was simple and of low cost, and the temporal resolution is of millisecond level. It can not only quickly identify the dynamic changes of the spontaneous brain nerves or cognitive activity and different cognitive processing, but also made real-time record of brain activity such as depression [2], epilepsy [3], epileptic seizure prediction [4], Alzheimer's disease [5], mild cognitive impairment (MCI) [6], Parkinson's disease [7], Creutzfeldt-Jakob disease [8], sleep disorders [9], schizophrenia [10], anxiety disorders [11] and brain-computer interface (BCI) [12]. Among them, resting state electroencephalogram (rsEEG) revealed the activity of brain network, which can be used to evaluate nervous system [13]. Therefore, rsEEG was used in this study because it contributes to the auxiliary diagnosis and recognition of depression.

The main features of EEG in depression study were usually focused on absolute and relative power distribution, power spectrum asymmetry between right and left hemispheres and within hemispheres, and coherence of electrodes. Fachner et al. [15] used music therapy to find that patients under antidepressant treatment had significantly higher absolute work values in the alpha and theta bands in the frontotemporal. Iznak et al. [16] also found similar results that the absolute power value was decreasing in beta band. It was suggested that with the remission of depressive symptoms, the EEG signal showed that the slow wave rhythm increases while the fast wave rhythm decreases. Knott et al. [17] analyzed the power spectrum asymmetry between the right and left hemispheres and within hemispheres of EEG signals in patients with depression, and found that the power spectrum asymmetry between the right and left hemispheres was significantly higher than that of normal people in alpha band. To some extent, it indicated that the activity in the left hemisphere of patients with depression was enhanced. In both hemispheres, the power spectrum asymmetry of beta band has decreased. The alpha band asymmetry in patients also decreased in different conditions in the left and right hemispheres. This supported that rsEEG alpha activity in the right frontal lobe was stronger than alpha activity in the left frontal lobe [18]. However, the stability of frontal alpha asymmetry varied between groups and was not reliable at different retesting stages in patients [19].

Most of the research methods of EEG signals are to extract the features of the pre-processed EEG signals directly, so as to obtain more information. In practice, there are also a large number of features in the sampling points of EEG signals. Therefore, from the perspective of sampling points, we proposed that the representative features are extracted by decomposition and reconstruction of local binary pattern (LBP) signal which are realized by singular spectrum analysis (SSA). Firstly, the pre-processed signal is segmented. Then the SSA is used to decompose and reconstruct the multi-electrode segmenting signal, so as to remove noise and divide frequency band. Finally, the data of each frequency band is fed into KNN, DT and ELM for classification. The experimental results show that the best classification effect is obtained on SVM classifier, and the accuracy of the beta band is the highest.

In future research, we need a lot of data from hospitals, and all the data need to be protected. Differential privacy is a relatively strong privacy protection technology, which can be supported by block chain. Therefore, block chain technology needs to be investigated in depth. The data on the block chain can be stored directly in storage or on cloud servers assigned to the users. However, the sensitive data stored in cloud servers may be accessed by cloud computing operations, which greatly increases user anxiety and reduces the acceptability of cloud computing in many areas. Therefore, a security aware and efficient distributed Storage (SAEDS) model is proposed [21], which effectively separates files and stores data separately in a distributed cloud server to avoid cloud service operators directly obtaining data. It is mainly supported by Secure and Efficient Data Distribution (SED2) algorithm and Efficient Data Consolidation (EDCon) algorithm. Then data needs to be protected by introducing the encryption algorithm with good confidentiality. Dynamic Data Encryption Strategy [22] is proposed to maximize the privacy protection scope within the required execution time and enhance the privacy protection. A data protection method combining the selective encryption (SE) with fragmentation and scatter storage is proposed [23], we divide the uncertain data into three segments of different protection levels using the reversible discrete wavelet transform (DWT). Then they are scattered to different storage areas with different levels of protection for the end user's data by preventing possible leaks in the cloud. Therefore, the proposed method saves expensive, private, secure storage space and achieves high level of protection. The effectiveness of the algorithm is verified by optimizing the task deployment between the CPU and the general purpose graphics processing Unit (GPGPU).

The main structure of this paper is as follows: The first part introduces the background of depression and EEG signal. The second part is the main framework and experimental process, then the third part describes the data set used in the experiment, the fourth part is mainly the analysis and comparison of experimental results, and the last part is the summary.

The main contributions are as follows: Proposing a feature extraction method based on LBP and SSA, and the best effect is obtained on SVM classifier so far.

2 Proposed Model

The overall framework of this paper is shown in Fig. 1 as follows. It is divided into three parts: data processing, EEG feature extraction and classification.

2.1 Data Processing

The experiment was conducted using an undocumented data set, which was obtained from the Clinical Research Center for Mental Disorders, Anding Hospital affiliated to Beijing Capital Medical University. The collection process was approved by the Ethics Committee of Beijing Anding Hospital. All participants were informed of the situation and signed a written consent form. The experimental collection device was the 64-channel 10–20 system EEG collection system produced by Brain Products, as shown

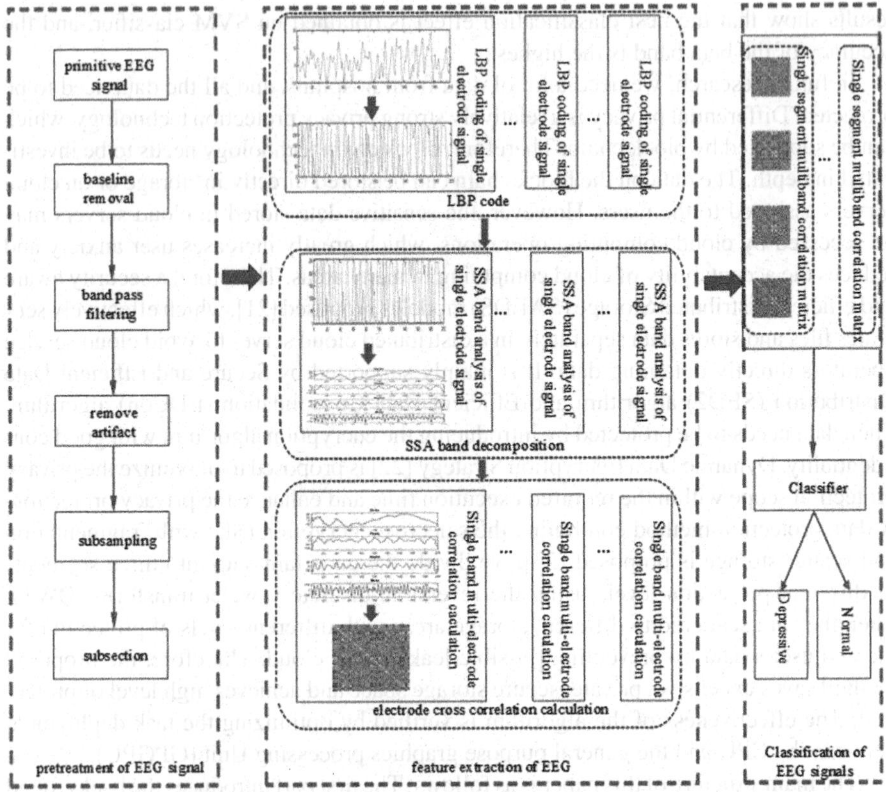

Fig. 1. Depression recognition framework based on LBP and SSA features

in Fig. 2 below. The online reference electricity was FCz, and the grounding electricity was AFz. The resistance of all electrodes was reduced to below 10 K, and the sampling rate was 100 Hz. During the recording of resting EEG signals, the subjects sit quietly and still with their eyes closed. The collected data were processed using the EEGLAB toolbox. The main steps were baseline removal, band-pass filtering, artifact removal and sub-sampling.

2.2 Feature Extraction

In this paper, LBP and SSA are mainly used to extract the features of EEG signals. The concept of LBP was introduced by Ojala et al. [23] in their study of texture measurement. Its initial version is based on a 3 × 3 pixel image block. In order to generate the local weight of the center pixel, the center pixel value depends on the adjacent 8 pixels, which will generate an 8-bit binary code whose decimal value is called the LBP value of the specific center pixel. Taking advantage of these features of LBP, Chatlani and Soraghan. [24] proposed 1D-local binary patterns on the basis of 2D-LBP for detecting acoustic signal activity on non-stationary speech signals. The basic operation of 1D-LBP is very similar to the texture operator. However, it analyzes the neighborhood of samples in

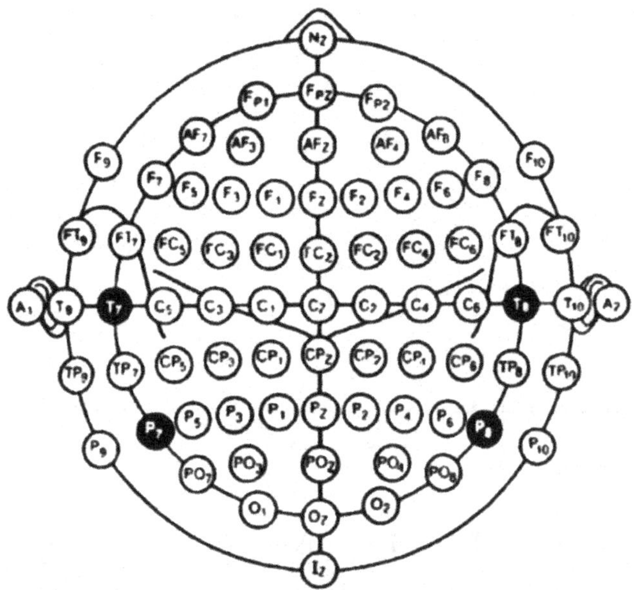

Fig. 2. The distribution of the electrodes in the acquisition system

time series data sequentially. For each data sample in the signal, the binary code is generated by threshing its value with that of the central sample. This process is iterative implemented throughout the signal.

As shown in Fig. 3(a), the number of equally divided sampling points selected $P = 9$, the four sampling points before the central sampling point (P_C) (P_0, P_1, P_2 and P_3) and the four sampling points after (P_4, P_5, P_6 and P_7). As shown in Fig. 3(b), amplitude $P = \{P_0, P_1, P_2, P_3, P_4, P_5, P_6, P_7\}$ of all adjacent sampling points was compared with the amplitude of the central sampling point (P_C). According to the formula to calculate the LBP value, 8 adjacent sampling points were threshold PC to generate binary code. If the P_i amplitude of the adjacent sampling point is greater than or equal to the amplitude of the central sampling point, then the P_i value is 1; otherwise, it is 0. Thus, a neighborhood binary is formed whose decimal value represents the LBP value of the central sampling point P_C.

SSA technology is mainly used to get rid of noise to obtain more pure EEG signals. As it does not need to choose basis function in advance, it is more flexible to reduce noise for nonlinear and non-stationary time series. Track matrix is constructed by observing the time series to extract the different composition of the original time series signal, and by decomposing the trajectory matrix, the characteristic quantity is separated from the original signal, then the characteristic quantity is used to reconstruct the signal, so as to reduce the noise or remove other unimportant signals.

After SSA processing, each segment of the signal was divided into four frequency bands of alpha, beta, theta and delta. Pearson correlation coefficients between 64 electrode signals in different frequency bands were calculated, which represented the degree

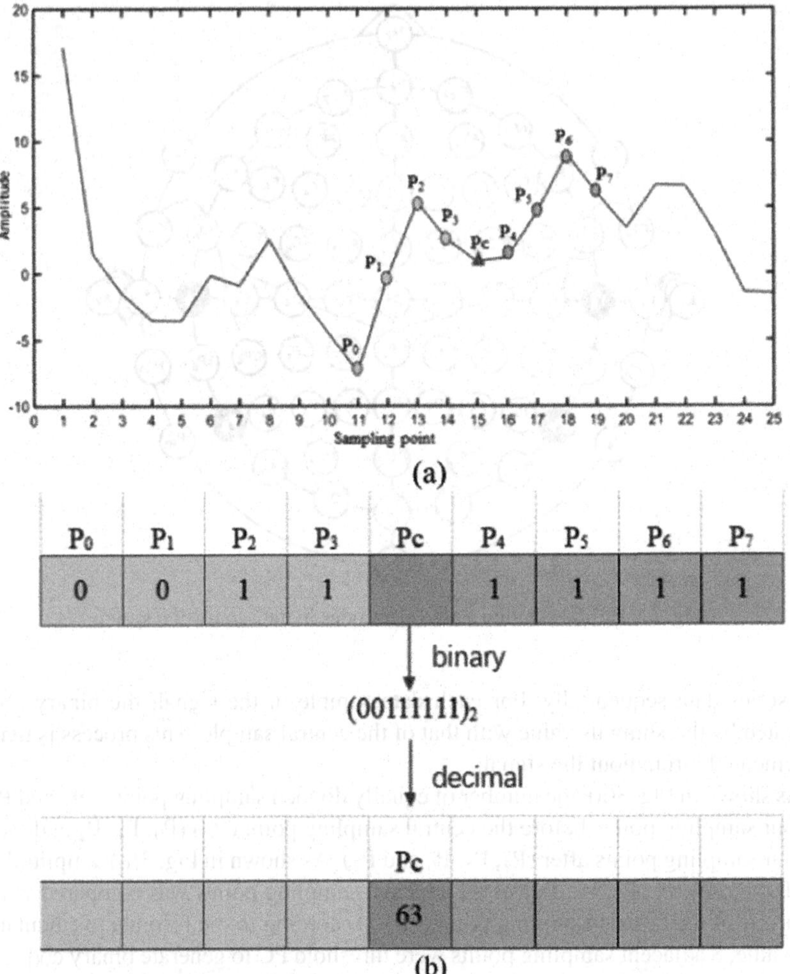

Fig. 3. Implementation steps of LBP encoding. (a) segment of sample EEG signal (b) LBP calculation process of sample points of EEG signal

of connection between electrodes, and then the correlation matrix was used as input fed into the classifier for experiments.

In short, LBP is a local information extraction technique with the advantage of rotation invariability and gray invariability. The center point information is expressed with the information of the surrounding 8 points, which make the information of the center point more richer and contain a lot of features. However, the information may be redundant as the feature is of high dimension, which makes the calculation more complex and time-consuming. We use SSA to separate the characteristic quantity from the noise signals, and the experimental results is well.

2.3 Classifier

After the EEG signal processing and feature extraction, the feature matrix were send to classifier. The classifier used KNN, DT, SVM and ELM in this experiment. KNN algorithm is a classification algorithm based on sample instances. In recent years, KNN algorithm has been greatly improved. Li et al. [25] used EEG to apply KNN algorithm to facial expression recognition, and the results showed that the classification accuracy of EEG wave band increased with the increase of EEG channels, while the classification accuracy of EEG wave γ band, β band, α band and θ band data decreased in turn. SVM is a kind of generalized linear classifier that performs binary classification of data according to supervised learning. The basic principle is to find the optimal decision surface in space and distribute different categories of data on both sides of the decision surface to achieve classification. DT [26] adopts tree structure and classifies based on characteristic performance. ELM is a special single hidden layer feedback neural network (SLFNs) with only one hidden node layer [27]. Later, it was extended to the general single hidden layer feed-forward neural network, whose hidden nodes were similar to neurons [28].

3 Experiments and Results

There were 16 patients and 16 normal controls in this study. The data of all objects was divided into equally-sized fragments, totaling 3758 pieces. Among them, the number of fragments were 2056 in the patient group and 1702 in the control group. The accuracy of training results was tested by 10 fold cross validation. There were 375 test sets and 3383 training sets.

In order to make full use of the data and reduce the randomness of the experimental results, we adopted 10-fold cross validation. The fragments of all subjects were divided into ten subsets as data sets. In each experiment, one subset was used for testing and the rest for training. The average of the results of the ten experiments was then taken as the final result.

3.1 Evaluation Metrics

Accuracy, specificity, sensitivity, precision and F1 score were used to determine the performance of a specific classification model. They were defined as follows:

The accuracy of classification was the ability of a network to differentiate the patients and healthy cases correctly.

$$\text{Accuracy} = \frac{TP + TN}{TP + FN + FP + TN} \tag{1}$$

The sensitivity, recall, hit rate, or true positive rate of classification was the ability of the network to determine the patient's cases correctly.

$$\text{Sensitivity} = \frac{TP}{TP + FN} \tag{2}$$

The specificity, selectivity, or true negative rate of classification was the ability of the network to determine the healthy cases correctly.

$$\text{Specificity} = \frac{TN}{FP + TN} \tag{3}$$

F1 score can be regarded as a weighted average of model accuracy and recall rate, considering both accuracy and recall rates of classification models.

$$\text{F1-score} = \frac{2 * TP}{2 * TP + FN + FP} \tag{4}$$

Where condition positive (P) represents the number of depressed cases, condition negative (N) represents number of healthy cases, true positive (TP) represents depressed cases correctly identified as depressed, False positive (FP) represents healthy cases incorrectly identified as depressed, true negative (TN) represents healthy cases correctly identified as healthy, and false negative (FN) represents depressed cases incorrectly identified as healthy.

3.2 Experimental Process

Processing the original EEG signal includes removing artifact, filtering, et al. As fixed length sliding window was used to segment each processed EEG data. After that, SSA method was used to decompose and reconstruct segmented signal. Each electrode data of each segment of the signal was decomposed into four sub-frequency bands, namely alpha, beta, theta and delta. Then the electrode correlation matrix of each frequency band was calculated according to the decomposition results of each sub-signal, and the characteristic matrix with dimensions of 4 * 64 * 64 was formed. Finally, the feature matrix of all samples was sent to the classifier to obtain the final classification result. The following figure shows the electrode correlation matrix for a certain frequency band (Fig. 4).

3.3 Classifier Parameter

In a large number of studies related to EEG signals, SVM, KNN and DT are widely used algorithms. In this experiment, ELM classifier and the above three classifiers are used to complete the classification of depressed patients and normal people. The parameters of the four classifiers are shown in Table 1.

3.4 Classification Results

In order to evaluate the detection and classification of MDD EEG signals by this framework, several experiments were carried out, the experimental results at different frequency band were shown in the Fig. 5(a), Fig. 5(b), Fig. 5(c), Fig. 5(d), Fig. 5(e).

Comparing of the following five figures suggests that all the classifiers performed best in the beta band is almost 100%. This indicates that in the beta band, not only the total number of features is quite abundant but also patients with depression are more

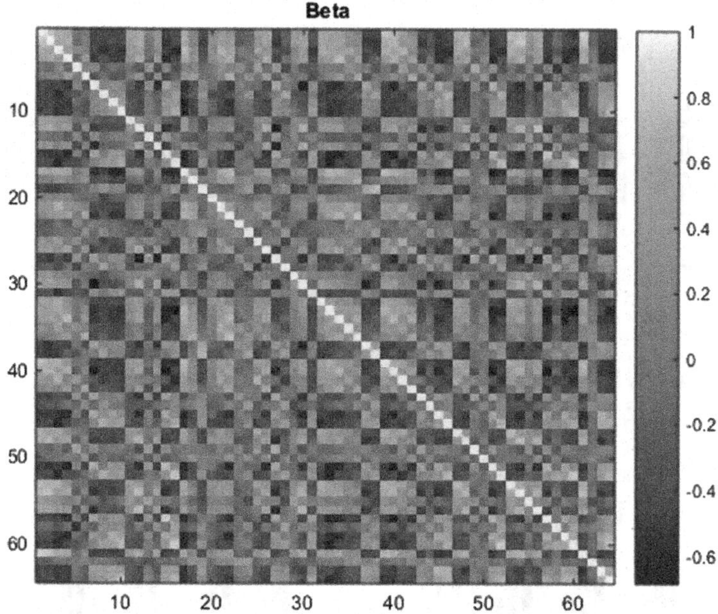

Fig. 4. Electrode correlation matrix for beta band

discriminative, which can prove that the EEG signals of depressed patients in this band are significantly different from those of the normal control group. The classification is worst in the delta band.

Studies have found that people's minds are tense, sensitive to the surrounding environment, and the brain is prone to fatigue when EEG signals are in the beta band, which is very similar to the symptoms of depression patient. In this experiment, no matter

Table 1. Classifier parameters

Classifier	Parameter	Parameter value
KNN	n_neighbors	7
DT	Default parameter	
SVM	Kernal function	Poly
	Degree	2
	Gamma	0.5
	max_iter	30000
ELM	Elm_type	1
	ActivationFunction	sig

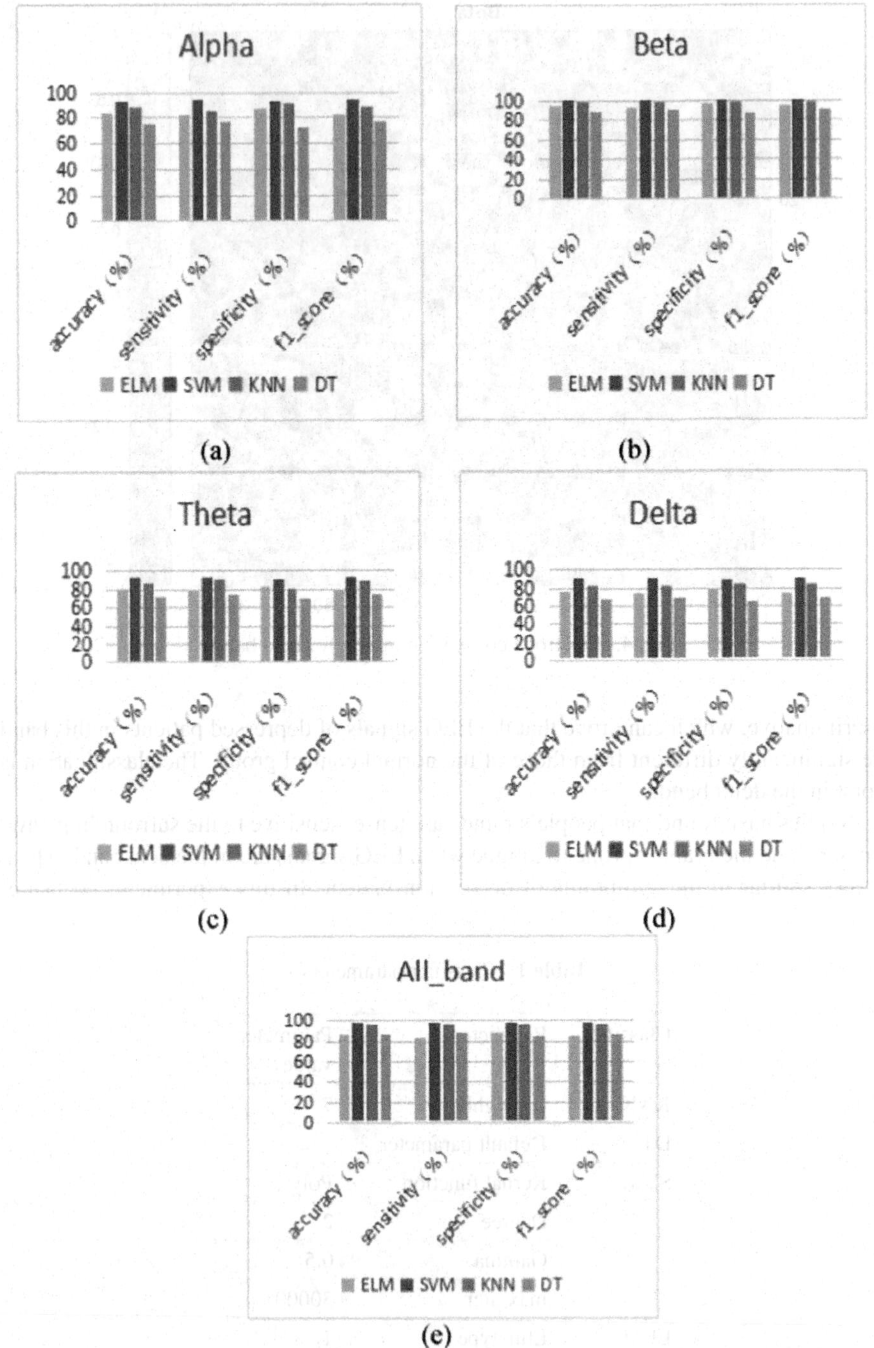

Fig. 5. Experimental results at different frequency band. (a) Alpha band (b) Beta band (c) Theta band (d) Delta band (e) All band

which classifier, it is in the beta band to achieve the best classification effect. This is sufficient to illustrate the validity and rationality of the features we have chosen.

By comparing the performance of each classifier, we found that SVM classification effect is better than the other methods. The classification results are shown in the following Table 2.

Table 2. SVM classification results

Band	Accuracy	Sensitivity	Specificity	f1_score
Alpha	93.59	93.72	93.43	94.1
Beta	**99.24**	**99.34**	**99.12**	**99.3**
Theta	92.89	94.07	91.65	93.59
Delta	90.75	91.68	89.64	91.53
All_band	98.08	98.37	97.74	98.24

As can be seen from Table 2, SVM classifier has the best classification effect in beta band and the accuracy of beta band is the highest with 99.24% accuracy, 99.34% sensitivity and 99.12% specificity respectively.

The experimental results of this paper are compared with recent classification studies of depression, as shown in Table 3 below:

Table 3. Comparison of different methods in the detection of depression

Author	Year	Database(MDD/Healthy)	Feature	Analysis method	Performance		
					Accuracy	Sensitivity	Specificity
Achary [29]	2015	15/15	Nonlinear feature	SVM	**98.00**	97.00	**98.50**
Mumtaz [30]	2017	34/30	Wavelet transform	LR	87.5	95	80
Mumtaz [31]	2017	33/30	Interhemispheric Alpha Asymmetry	SVM	98.4	96.66	100
Acharya [32]	2018	15/15	The left and right hemisphere	CNN	Left: 93.54 Right: 95.9	Left: 91.89 Right: 94.99	Left: 95.18 Right: 96.00
Ay B [33]	2019	15/15		CNN-LSTM	Left: 97.66 Right: 99.12	Left: 97.67 Right: 99.11	–
Our work	–	16/16	Cross correlation LBP-SSA in beta band	SVM	**99.24**	**99.34**	**99.12**

Compared with Achary et al. [29], it indicates that the features extracted by LBP and SSA are better than Achary et al. nonlinear combination features improving the identification. Compared with Mumtaz et al. [31], the difference in beta band between patients with depression and the control group is more significant. By comparing with Mumtaz et al. [30], we found that searching decision surface is more efficient than regression for the recognition of depression EEG signals. We do not compare with Acharya et al. [32] because we selected the whole brain EEG signals.

On the whole, we have the best classification using the features extracted by LBP and SSA on the SVM. Compared with other experiments, our data set is relatively small, but it performs well, which verify the validity and practicability of the extracted features.

4 Conclusion

In this paper, a novel method for depression recognition based on local binary coding and singular spectrum analysis is proposed. Firstly, the processed and segmented EEG signals are coded by LBP to realize feature extraction and obtain LBP signals. Then, the SSA method is applied to realize the decomposition and reconstruction of LBP signal, and obtain the decomposition of multi-frequency band. Then, the correlation of multi-electrode decomposition results of each frequency band is calculated, and the electrode correlation matrix of multi-frequency band is finally gained. In the end, the feature matrix of all samples is sent to the classifier to get the final classification result. The effectiveness of this method in depression identification was verified on the Beijing Anding Hospital EEG data set. The experimental results show that LBP and SSA features can achieve good classification effect on SVM classifier. It is verified that beta band signal is an important indicator in the study of EEG signals in depression, and the combination of LBP and SSA can be used as a new detection method for depression. In addition, we used the shallow neural network ELM because of the amount of data and achieved relatively good results.

Acknowledgement. The research is partially sponsored by the National Natural Science Foundation of China (No. 61672070), the Beijing Municipal Natural Science Foundation (No. 4202025, 4162058), the Beijing Municipal Education Commission Project (No. KM201910005008, No. KM201911232003), and the Beijing Innovation Center for Future Chips (No. KYJJ2018004).

References

1. Sarin, K., Punyaapriya, P., Sethi, S., et al.: Depression and hopelessness in institutionalized elderly: a societal concern. Open J. Depress. **05**(3), 21–27 (2016)
2. Acharya, U.R., Sudarshan, V.K., Adeli, H., Santhosh, J., et al.: Computer-aided diagnosis of depression using EEG signals. Eur. Neurol. **73**(5–6), 329–336 (2015)
3. Cogan, D., Birjandtalab, J., Nourani, M., et al.: Multi-biosignal analysis for epileptic seizure monitoring. Int. J. Neural Syst. **27**, 345–355 (2016)
4. Direito, B., Teixeira, C.A., Sales, F., Castelo-Branco, M., Dourado, A.: A realistic seizure prediction study based on multiclass SVM. Int. J. Neural Syst. **27**(3) (2017)
5. Morabito, F.C., et al.: A longitudinal EEG study of Alzheimer's disease progression based on a complex network approach. Int. J. Neural Syst. **25**(2)(2015)
6. Mammone, N., et al.: Permutation disalignment index as an indirect, EEG-based, measure of brain connectivity in MCI and AD patients. Int. J. Neural Syst. **27**(5) (2017)
7. Yuvaraj, R., et al.: Brain functional connectivity patterns for emotional state classification in Parkinson's disease patients without dementia. Behav. Brain Res. **298**, 248–260 (2016)

8. Morabito, F.C.: Deep learning representation from electroencephalography of early-stage Creutzfeld-Jakob disease and features for differentiation from rapidly progressive dementia. Int. J. Neural Syst. **27**(2) (2017)
9. Bruder, J.C., Duemepelmann, M., Piza, D.L., Mader, M., Schulze-Bonhage, A., Jacobs-Le Van, J.: Physiological ripples associated with sleep spindles differ in waveform morphology from epileptic ripples. Int. J. Neural Syst. **27**(7) (2017)
10. Akar, S.A., Kara, S., Latifoğlu, F., Bilgiç, V.: Analysis of the complexity measures in the EEG of Schizophrenia patients. Int. J. Neural Syst. **26**(2) (2016)
11. Tonoyan, Y., Looney, D., Mandic, D.P., van Hulle, M.M.: Discriminating multiple emotional states from EEG using a data-adaptive, multiscale information-theoretic approach. Int. J. Neural Syst. **26**(2) (2016)
12. Sereshkeh, A.R., Trott, R., Bricout, A., Chua, T.: Online EEG classification of covert speech for brain-computer interfacing. Int. J. Neural Syst. **27**(8) (2017)
13. Tóth, B., File, B., et al.: EEG network connectivity changes in mild cognitive impairment—preliminary results. Int. J. Psychophysiol. **92**(1), 1–7 (2014)
14. Kaiming, W., et al.: Nonlinear research and diagnosis model construction of EEG signal in depression. Thesis of Beijing University of Technology (2015)
15. Fachner, J., Gold, C., Erkkil, J.: Music therapy modulates frontotemporal activity in rest-EEG in depression clients. Brain Topogr. **26**(2), 338–354 (2013). https://doi.org/10.1007/s10548-012-0254-x
16. Iznak, A.F., Iznak, E.V., et al.: Changes in EEG and reaction times during the treatment of apathetic depression. Neurosci. Behav. Physiol. **43**(1), 79–83 (2013). https://doi.org/10.1007/s11055-012-9694-8
17. Vemer, K., Colleen, M., Sideny, K., et al.: EEG power, frequency, asymmetry and coherence in male depression. Psychiatry Res. Neuroimaging Sect. **106**, 123–140 (2001)
18. Henriques, J.B., Davidson, R.J.: Regional brain electrical asymmetries discriminate between previously depressed and healthy control subjects. Abnorm. Psychol. **99**, 22–31 (1990)
19. Debener, S., Beauducel, A., Nessler, D., Brocke, B., Heilemann, H., Kayser, J.: Is resting anterior EEG alpha asymmetry a trait marker for depression? Neuropsychobiology **41**, 31–37 (2000)
20. Gai, K., Qiu, M., Zhao, H.: Security-aware efficient mass distributed storage approach for cloud systems in big data. In: 2016 IEEE 2nd International Conference on Big Data Security on Cloud (BigDataSecurity), IEEE International Conference on High Performance and Smart Computing (HPSC), and IEEE International Conference on Intelligent Data and Security (IDS), New York, NY, pp. 140–145 (2016)
21. Gai, K., Qiu, M., Zhao, H.: Privacy-Preserving Data Encryption Strategy for Big Data in Mobile Cloud Computing. IEEE Trans. Big Data 2705807 (2017)
22. Qiu, H., Noura, H., Qiu, M.: A user-centric data protection method for cloud storage based on invertible DWT. IEEE Trans. Cloud Comput. 2911679 (2019)
23. Ojala, T., Pietikäinen, M., Harwood, D.: A comparative study of texture measures with classification based on feature distributions. Pattern Recogn. **29**(1), 51–59 (1996)
24. Chatlani, N., Soraghan, J.J.: Local binary patterns for 1-D signal processing. In: 18th European Signal Processing Conference (EUSIPCO-2010), 23–27 August, Aalborg, Denmark, pp. 95–99 (2010)
25. Mi, L., Hongpei, X., Xingwang, L., et al.: Emotion recognition from multichannel EEG signals using K-nearest neighbor classification. Technol Health Care **26**(S1), 509–519 (2018)
26. Shuo, H.J., et al.: Study on the characteristics of sleep EEG in depressive population. The thesis of Lanzhou University (2019)
27. Huang, G.B., Zhu, Q.Y., Siew, C.K.: Extreme learning machine: a new learning scheme of feedforward networks with random hidden nodes. IEEE Trans. Neural Netw. **17**, 879–892 (2006)

28. Huang, G.B., Zhu, Q.Y., Siew, C.K.: Extreme learning machine: theory and applications. Neurocomputing **70**, 489–501 (2006)
29. Acharya, U.R., Sudarshan, V.K., Adeli, H., et al.: A novel depression diagnosis index using nonlinear features in EEG signals. Eur. Neurol. **74**(1–2), 79–83 (2015)
30. Mumtaz, W., Xia, L., et al.: A wavelet-based technique to predict treatment outcome for major depressive disorder. PLoS One **12**(2), e0171409 (2017)
31. Mumtaz, W., Yasin, M.A.M., et al.: Electroencephalogram (EEG)-based computer-aided technique to diagnose major depressive disorder (MDD). Biomed. Signal Process. Control **31**, 108–115 (2017)
32. Ay, B., Yldrm, Z., Talo, M., et al.: Automated depression detection using deep representation and sequence learning with EEG signals. J. Med. Syst. **43**(7), 205 (2019)
33. Acharya, U.R., Oh, S.L., Hagiwara, Y., et al.: Automated EEG-based screening of depression using deep convolutional neural network. Comput. Methods Programs Biomed. **161**, 103–113 (2018). S0169260718301494

Cloud Allocation and Consolidation Based on a Scalability Metric

Tarek Menouer[1(✉)], Amina Khedimi[2], Christophe Cérin[2], and Congfeng Jiang[3]

[1] Umanis Research & Innovation, 7 Rue Paul Vaillant Couturier, 92300 Levallois-Perret, France
tmenouer@umanis.com

[2] Université Sorbonne Paris Nord, LIPN/CNRS UMR 7030, 93430 Villetaneuse, France
{Khedimi,christophe.cerin}@lipn.univ-paris13.fr

[3] Hangzhou Dianzi University, Hangzhou, Zhejiang 310018, China
cjiang@hdu.edu.cn

Abstract. In this paper, we present a new allocation and resource consolidation system based on a scalability metric. According to cloud computing principles, the end users rent computing and big data analytic services with a pay-as-you-go cost model. However, when users' data size increases or when the application stresses the memory or requires more computing power, they need to scale their rental to achieve approximately the same performance, such as task completion time and normalized system throughput. In this paper, we propose to delegate the responsibility to scale-up and scale-out the cloud system to a new component of the cloud orchestrator. The decision is taken on a metric that quantifies the scalability of the cloud system consistently under different system expansion configurations. The scalability metric is defined as a ratio between the new and the current situation, over the size of the i-th workload in terms of CPU cores and a performance metric. The considered metrics are the waiting time in the queue and the average resource (cores) usage rate during task execution. To validate our approach, we conduct experiments by emulation, on a real platform. The experimental results demonstrate the validity of the proposed general automatic strategies for cloud allocation and consolidation.

Keywords: Cloud computing · Resource management and scheduling · Performance of cloud systems · Consolidation of servers · Scalability

1 Introduction

Cloud computing is an opportunity for big changes in the ICT field, that offers an efficient and flexible service delivery. The main objective of cloud computing is to provide the desired services to each user efficiently, according to high capacity for both compute and storage resources. As the amount of data and the number

M. Qiu (Ed.): ICA3PP 2020, LNCS 12454, pp. 381–395, 2020.
https://doi.org/10.1007/978-3-030-60248-2_26

of tasks increase, users expect the system to scale and the performance of the system to increase accordingly. Hence, the most important thing to control is the scalability of the cloud system which should not be degraded.

Regarding the request scheduling and resources allocation problems, the scalability should not decrease even if the number of users' requests increases. One technique to control the cloud system is to consider strategies and technologies that can reduce the number of the geographic distribution of servers owned and managed by a cloud provider. This strategy is called consolidation, either it reduces the communication overhead between servers or it balances the workload among all available nodes.

Unfortunately, there is no effective way to help end-users to choose between the direction of scale-up and scale-out to keep the cloud system balanced, i.e. the number of running servers is not too high nor too low whereas the user satisfaction is still preserved.

We propose, to control the scale-up and scale-out through an automatic process, implemented in the cloud orchestrator. This component automatically makes a reasonable choice without the end-users intervention, and it is also coupled with the dynamic computation of resources that need to be attached with any request. Summarizing, the overall framework selects automatically one request from the queue of requests according to a multi-criteria algorithm, then computes the number of resources required for the selected request, allocates it to a server, and finally consolidates the server based on a scalability metric.

The latter quantifies the scalability of the cloud system consistently under different system expansion configurations. Most of the available methods measure this metric by calculating the speed-up ratio and/or the throughput of the system, which ignores the metrics that end-users and service providers care about, such as resource consumption and cost.

In our case, the scalability metric is defined as a ratio between the new and the current situation. This metric has two variables, the size of the i-th workload which is defined in terms of CPU cores, while the performance metric is defined in terms of waiting time in the queue and the average resource (cores) usage rate during task execution.

Finally, note that for the sake of simplicity, we reason about the number of cores that make up the cloud platform, but alternative choices can be made without fundamentally questioning our model. This paper finds its roots in [22] where a methodology for solving the general problem of container scheduling is proposed. The methodology we promote is decomposed according to the following steps:

1. Selection of one container from a set of containers saved in a global queue;
2. Computation, on the fly, of the resources which must be allocated to the container;
3. Container placement;
4. Server consolidation.

This paper introduces a new method for step 4 according to the general scheme. The server consolidation problem is revisited according to a scalability metric and also takes into account the number of resources allocated to the server. The combinatorial problem for step 4 is solved by PROMETHEE (*Preference Ranking Organization METHod for Enrichment Evaluations*) multi-criteria decision making method [3,16,23].

Our paper identifies the issues as follows. Given a queue of user requests, find an allocation of the requests on physical nodes such that user satisfaction is maximized, the number of the allocated resources is minimized and the scalability metric is controlled. In the literature, different combinations of work are proposed for the optimal allocation of resources and their utilization. The main focus of our study is the analysis of the consolidation scenarios on different virtual machines by considering a traditional performance metric, such as throughput and latency, which are not sufficient to effectively measure the performance. However, there is a little work that analyses the scalability.

Compared to other studies, our approach optimize the resource allocation through a new consolidation heuristic that takes into account the scalability metric of the cloud system. Therefore, resource consolidation will be done in such a way that the scalability is not degraded, by making a (locally)-optimal usage of resources, thanks to PROMETHEE multi-criteria algorithm.

The remainder of this paper is organized as follows. In Sect. 2, we present some related work. Section 3 describes the principle of the PROMETHEE multi-criteria algorithm. Section 4 describes the request selection mechanism, the resource allocation approach, and the scalability measurement model. Section 5 presents the principle of the proposed consolidation approach. Furthermore, experimental evaluations are presented in Sect. 6. Finally, the conclusion of this paper is presented in Sect. 7.

2 Related Work

In the literature, many problems of resource allocation, or placement of user's requests refer to the same class of scheduling problems. Most of these problems are *NP*-hard [30]. In the following, we propose to introduce a selection of relevant works on scheduling for cloud computing, on Service Level Agreement (SLA) in the cloud context, on consolidation and on scalability, which are all the core concepts in our paper.

In this section we consider isolated problems related to resource allocation whereas our proposal is a global approach, mixing considerations of SLA, consolidation, dynamic computation of resources, and placement. Our work is unique and difficult to rank regarding the state-of-the-art. This fact is also true for the experimental section. The evaluation work does not compare the proposed solution with any other related work because we have taken and motivated very different options of the scheduling process for the overall organization. The reader is rather invited to examine the hypothesis to objectively evaluate the authors' contributions.

Summarizing, it is hard to explain what would be the difference between existing server consolidation algorithms which are used in this proposal because our consolidation mechanism depends on previous steps of the overall framework. The trend in the works we are going to mention now is to consider separate problems, without any global approach.

2.1 Scheduling and Cloud Computing

In [17], authors consider a form of consolidation strategy with a focus on the interference between collocated Virtual Machines (VMs) in data centers. Collocation may result in poor performance because of the contention on the shared resources. They devise efficient VM assignment and scheduling strategies in which they consider optimizing both the operational cost of the data center and the performance degradation of the running applications. Our work is also focussing on consolidation but it does not support any strategy regarding the collocation problem.

In [24], authors consider the problem where total resource consumption and allocation (within and across applications in the infrastructure) are driven by incremental gains in quality-of-execution (QoE), which relates the resources allocated to an application and the performance of the application can extract from having those resources. This is quite different from our approach since, in our work, we make no assumption on the application requirements.

2.2 Service Level Agreement Issues

In [25], authors envisioned a partial utility-driven scheduling schema that allows more VMs to be allocated without the schema. In this study, the authors propose to schedule CPU processing capacity to VMs using an algorithm that strives to account for users and providers potentially opposing interests. The proposal brings benefits to providers, regarding revenue and resource utilization, allowing for more revenue per resource allocated. This is also an issue and objective in our paper but, in our case, the benefits are a consequence of the user choices.

In [26], authors envisioned self-management of services that automatically manages QoS requirements of cloud users thus helping the cloud providers in achieving the Service Level Agreement (SLA) and avoiding SLA violations. This is another view of the problem encountered in cloud environments, regarding uncertainty and dispersion of resources.

2.3 Consolidation Heuristics

In the context of cloud computing, consolidation is one method to address energy reduction issues by improving the utilization of resources. Dynamic consolidation of virtual machines (VMs) are enabled by live migration. Dynamic VM consolidation consists to find a balance between two basic processes: migrating VMs from underutilized hosts to minimize the number of active hosts and offloading VMs from hosts when those become overloaded to avoid performance degradation, thus a violation of the QoS requirements.

Regarding the consolidation context, authors in [4] introduce an architecture and implementation of OpenStack Neat which is an open-source software framework for distributed dynamic VM consolidation in cloud data centers based on the OpenStack platform. The proposed framework applies VM consolidation processes by invoking public API's of OpenStack.

In [31], authors present a comprehensive study of the state-of-the-art VM placement and consolidation techniques used in the green cloud which focusses on improving energy efficiency. In this study a detailed comparison is presented, revealing pitfalls and suggests improvisation methods in this direction. This study covers additional concerns related to the overhead caused by continuous live migration of VMs during consolidation, due to the continuously changing resource demands. Our focus is not on this performance metric but the quality of the placement, meaning that we are interested in the performance metric in terms of nodes utilization and CPU/cores distribution regarding the technique for consolidation.

In [8], authors propose the AVVMC VM consolidation scheme that focuses on balanced resource utilization of servers across different computing resources (CPU, memory, and network I/O) to minimize power consumption and resource wastage. The underlying multi-objective problem is captured with an adaptation and integration of the Ant Colony Optimization (ACO) meta-heuristic. The work is done through simulation.

In [6], authors made a focus on Bag-of-Tasks jobs and proposed an algorithm and heuristics that adaptively predict the number of hosts to be allocated, so that the maximum speed-up can be obtained while respecting a given predefined budget. The proposed solution simulated against real and theoretical workloads, allows obtaining speed-ups in line with the number of allocated hosts, while being charged less than the predefined budget.

In [32] a neural network-based adaptive selection of VM consolidation algorithms is proposed, which adaptively chooses appropriate algorithm according to cloud providers goal priority and environment parameters. They propose a Dynamic VM (Virtual Machine) consolidation algorithm to reduce energy consumption through VM migration.

Authors in [18] also propose a Dynamic consolidation of Virtual Machines (VMs). They use the Cloud Service Users (CSUs) provided information to minimize the energy consumption of the Cloud Data Centers (CDC). They exemplify the fact that if VMs are dynamically consolidated based on the time when a VM can be removed from CDC - useful information to be received from respective CSU, then more physical machines can be turned into a sleep state, yielding lower energy consumption.

In [2] authors present a taxonomy with a new classification for load balancing and server consolidation, such as migration overhead, hardware threshold, network traffic, and reliability. They review the literature on load balancing, server consolidation and present a ready reference taxonomy on the most efficient algorithms that achieve load balancing and server consolidation.

2.4 Scalability Metrics and Analysis

In [10,15] authors put throughput, QoS and other indicators into a radar diagram, and uses the area of the radar map to measure the scalability of the SaaS applications.

Other researchers [7,9,20] have also attempted to measure the performance of cloud services. Regarding more specifically IO or data transfers, Sun [27] proposed *isospeed* to measure the scalability of the system. In this context, the *Isoefficiency* notion is proposed by Grama et al. [11,12] to measure the scalability of parallel systems. *Isoefficiency* is commonly used to measure the combined parallel structure and algorithms.

These traditional scalability evaluation methods are not applied to cloud environments. Scalability, resiliency, and efficiency are three important metrics to measure the performance of cloud-based services [1]. To study the scalability of applications on the cloud platform, Gunther et al. [14] proposed a general model, USL (Universal Scalability Law), which uses the speedup ratio to measure the scalability of the system.

In [29] authors proposed a new metric, *PRR*, to measure the scalability of SaaS applications, which aims at measuring the scalability when varying workloads.

3 PROMETHEE Algorithm

In this paper, the PROMETHEE method is used in both the request selection step and in the consolidation step. The PROMETHEE algorithm has been proven to provide a good compromise between several qualitative and quantitative criteria. It allows comparing objects between them pair by pair, along with different criteria for each object. All criteria of objects are evaluated according to two functions: (i) Minimization; and (ii) Maximization. That means each criterion can be minimized or maximized. However, the use of the PROMETHEE algorithm requires two pieces of information for each criterion: a weight and a preference function. The preference function characterizes the difference for a criterion between the evaluations obtained by two possible objects into a preference degree ranging from 0 to 1. In [16], six basic preference functions have been proposed. In this work, we use the usual preference functions described in the following. To summarize, the PROMETHEE algorithm is composed of four steps [28]:

1. It computes the value of the preference degree for each pair of possible requests ($request_a$ and $request_b$) and for each criterion. Let $g_j(request_a)$ be the value of a criterion j for $request_a$. We note $d_j(request_a, request_b)$ the difference of value of a criterion j for $request_a$ and $request_b$;

$$d_j(request_a, request_b) = g_j(request_a) - g_j(request_b)$$

   $P_j(request_a, request_b)$ is the value of the preference degree of a criterion j for $request_a$ and $request_b$.

The preference function used to compute these preference degrees is defined such as:

$$P_j(d_j) = \begin{cases} 0 & d_j \leq 0 \\ 1 & d_j > 0 \end{cases}$$

2. It computes a global preference index for each pair of possible requests. Let C be the set of considered criteria (qualitative and/or quantitative) and w_j the weight associated to the criterion j. The global preference index for a pair of possible $request_a$ and $request_b$ is calculated as follows:

$$\pi(request_a, request_b) = \sum_{j \in C} W_j \times P_j(request_a, request_b)$$

3. It computes the positive outranking flow $\phi^+(request_a)$ and the negative outranking flow $\phi^-(request_a)$ for each possible request. Let A be the set of requests with size of n. The positive and negative outranking flow of requests are calculated by the following formula:

$$\phi^+(request_a) = \frac{1}{n-1} \sum_{x \in A} \pi(request_a, x)$$

and

$$\phi^-(request_a) = \frac{1}{n-1} \sum_{x \in A} \pi(x, request_a)$$

4. It uses the outranking flows to establish a complete ranking between the requests. The ranking is based on the net outranking flows $\phi(request_a)$ which is calculated as follows: $\phi(request_a) = \phi^+(request_a) - \phi^-(request_a)$. In our work, the first requests returned by PROMETHEE algorithm are requests with the highest net outranking values.

4 Problem Statement and Solution

The aim of our work is to answer the following problem: *Given a set of requests submitted online by users with SLA's classes in a cloud infrastructure, which request must be selected firstly, how many resources must be allocated to each request, and which node must be used to execute the selected request in order to control the scalability of the cloud system and to optimize the schedule?*

To explicitly present our work, we take the example of a user who rents computing services with a pay-as-you-go cost model regarding his SLAs classes. The user expects to have a better quality of services in terms of task completion time and normalized system throughput, especially if he decides to pay for a very expensive server. When the user's data size increases, the cloud needs to scale the rental service to achieve approximately the same performance as before. Therefore, the user, who is not necessarily an expert in cloud technologies, is faced with a multi-criteria optimization problem.

Moreover, for the sake of simplicity, we propose two SLA classes, namely a qualitative and a quantitative one. The qualitative SLA class describes the user needs in terms of satisfaction time, for instance some users like to have a solution as soon as possible, whereas others may not have a constraint on the response time. However, the quantitative SLA class describes the user needs in terms of the required resources. Some users like to execute a request with a large number of cores to reduce the long computing time for their requests, others need few cores to execute their requests which requires a minimum computing time. Each time the system receives a new submitted request, it is saved in a global queue, then it goes through the four steps shown below:

1. **Request selection**: select the first request to be executed from the queue of requests;
2. **Resource computing**: decide the number of cores that must be allocated for the selected request,
3. **Request assignment**: decide the node that will execute the selected request.
4. **Consolidation**: calculate the scalability metric of all nodes in the system to decide the scale-up or scale-out.

As said previously in Sect. 1, this paper extends the ideas for step 4 (consolidation). The consolidation is revisited according to a scalability metric and also takes into account the number of resources allocated to nodes. More information about our consolidation approach is presented in the following.

4.1 Scalability Measurement Model

Scalability in the context of cloud computing can be defined as the ability to handle resource consolidation to meet business demands optimally. Usually, the speed-up ratio and throughput are calculated to measure the scalability of the system [1], which ignores the metrics that users care about, like resource consumption and cost.

In our paper, we define the scalability metric as a ratio between the new and the current situation, over the size of the i-th workload in terms of CPU cores and a performance metric in terms of waiting time in the queue and the average resource (cores) usage rate during task execution.

In Eq. (1), we define the metric P, which represents the performance resource rate. In Eq. (1), T represents the total time spent by the task, in seconds, and C represents the average resource consumption of the task during execution. The resource consumption here means the average amount of resources that are consumed during the execution of the task. The resource type considered in this paper is the CPU.

We separate the T and C in Eq. (1) into different components Eq. (2) and Eq. (3). In Eq. (2), T consists of two parts: T_q and T_e. T_q is the waiting time in the queue, and T_e is the task execution time.

$$P = \frac{1}{T \times C} \tag{1}$$

$$T = T_q + T_e \tag{2}$$

$$C = R \times U \tag{3}$$

C is calculated using Eq. (3), where R represents the total amount of resources, and U represents the average resource usage rate during task execution.

In this paper, we propose Eq. (4) to calculate the scalability of the system. S_i in Eq. (4) represents the scalability of the system under the *i-th* workload. Let n represents the total number of workloads and i represents the *i-th* workload, then $D(i)$ represents the size of the *i-th* workload before the platform expansion and $D_a(i)$ represents the size of the *i-th* workload after the platform expansion. In Eq. (4), $P(i)$ represents the performance resource rate P before the platform expansion, and $P_a(i)$ represents the performance resource rate P after the platform expansion of the *i-th* workload.

$$S_i = \frac{D_a(i) \times P_a(i)}{D(i) \times P(i)} (i = 1, 2, 3, \cdots, n) \tag{4}$$

If the value of S_i is close to 1 or greater than 1, it indicates that the system has good scalability. When S_i is greater than 1, it represents super-linear scalability, and when S_i is equal to 1, it represents linear scalability. When the value of S_i is close to 0, the scalability of the system will get worse.

Due to the difference in the scalability of the system under different workloads, this paper uses the variance (V) of S_i to measure the scalability offset under multiple workloads. As shown in Eq. (5), n represents the number of workloads, and S_i represents the scalability of the system under the *i-th* workload. When the value of V is getting closer to 0, the scalability will become more stable.

$$V = E[(S_i - \frac{1}{n}\sum_{i=1}^{n} S_i)^2](i = 1, 2, 3, \cdots, n) \tag{5}$$

4.2 Requests Selection Based on an Economic Model

The economic model that we use in this paper, as well as the opportunistic requests scheduling model, were initially proposed in [22]. The economic model is based on two SLA classes: qualitative and quantitative. The quantitative SLA class represents the number of resources requested by the user, while the qualitative represents time satisfaction. Each SLA class has three services (Premium, Advanced and Best effort). We can use different interpretations for each QoS internal to an SLA. For instance, the *Premium* service in the qualitative class is designed for users who have many requests to be executed and always have strict deadlines to consider. Each time a user submits a request, he wants to get the solution back quickly without considering the price of the operation. Please, refer to [22] for more examples of users' requests.

All submitted requests are stored in the request queue with pairs of priorities $P_{i,1}$ (priority of the qualitative SLA class) and $P_{i,2}$ (priority of the quantitative SLA class). The PROMETHEE (*Preference Ranking Organization METHod for Enrichment Evaluations*) [23] multi-criteria decision making method is applied to select the first request that must be executed ($Request_*$). The PROMETHEE algorithm is presented in Sect. 3.

Considering all requests saved in the queue ($Request_1, \cdots, Request_p$) with their only quantitative priorities ($P_{1,2}, P_{2,2}, \cdots, P_{p,2}$) and the number of waiting cores (R) in all nodes of the infrastructure, the system calculates the number of cores (C_i) that must be allocated to $Request_i$ using formula (6) and the scalability of all nodes is calculated using formula (4).

$$C_i = \frac{P_{i,2} * R}{\sum_{j=0}^{p} P_{j,2}} \tag{6}$$

To assign a request to a node, the bin packing strategy is applied [22]. The principle consists to assign a $request_i$ to the $node_j$ which has the least available resources.

5 Consolidation Algorithm

According to the previous definitions, the consolidation mechanism is introduced in Algorithm 1. The principle consists firstly to test if a total load of active nodes is greater than the value of the max load metric, in this case, the node with the best compromise between the biggest number of cores and the best value for the scalability metric from the waiting nodes becomes an active node. Otherwise, if no waiting node exists, the node that has a good compromise between the biggest number of cores and the best value for the scalability metric from the stopped nodes now becomes an active node.

If no stop node and no waiting node exist, the system cannot add a node because all nodes of the infrastructure are used as active nodes. If a total load of active nodes is smaller than the value of the max load metric, the algorithm selects the node from active nodes which has a good compromise between the smallest number of cores and the worst value of scalability metric, then it changes its state to waiting node. Nodes with no working core, in waiting or active states will be stopped.

We note that each time we select the node with a good compromise between the number of cores and the best/worst value for the scalability metric we use the PROMETHEE multi-criteria decision aid system [23] that we have introduced above, in Sect. 3.

6 Experimental Evaluation

6.1 Scenario and Settings

In this section, we introduce experiments in using our system to check if it meets our expectations. For these experiments, we propose to use the Grid'5000 platform [13], an experimental large-scale testbed for distributed computing in

Algorithm 1. Consolidation Heuristic

Require: T, Total load of active nodes
 if T > *max load* metric **then**
  if $\exists$ at least one waiting node **then**
   Choose the node M_x having a good compromise between the biggest number of cores and the best value for the scalability metric (PROMETHEE in action)
   Add M_x as an active node
  else
   if $\exists$ at least one stop node **then**
    Choose the node M_x having a good compromise between the biggest number of cores and the best value of scalability metric (PROMETHEE in action)
    Add M_x as an active node
   end if
  end if
 else
  if T < *min load* metric **then**
   Choose from active nodes the node M_x having a good compromise between the smallest number of cores and the worst value of scalability metric (PROMETHEE in action)
   Add M_x as a waiting node
  end if
 end if
 if $\exists$ at least one node M_x active or waiting with all inactive cores **then**
  Add M_x as stop node
 end if

France. For the Grid'5000 platform we booked an infrastructure composed of 94 computing cores, distributed over 4 nodes. Each node has 24 cores.

To calculate the scalability factor, we use the general formula given in (7) and taking into consideration the following special case:

- All the containers are submitted simultaneously, therefore, the waiting time T_q will be equal to 0.
- There is just one workload, so the size of the workload will simply represent the number of containers. All containers are homogeneous in terms of execution time (5 min) and CPUs number (8 cores).

$$S_i = \frac{D_a \times P_a}{D \times P} \tag{7}$$

Finally, the scalability factor will be given as follows. We start by calculating the performance metric using formula 1 where T represents the execution time. As all containers are homogeneous, C represents the resource consumption in terms of CPU number. Now, the scalability of the node i is given by formula 7 where:

- D_a represents the number of the submitted containers after the scale;
- P_a represents the performance after the scale;
- D represents the number of the submitted containers before the scale;
- P represents the performance before the scale.

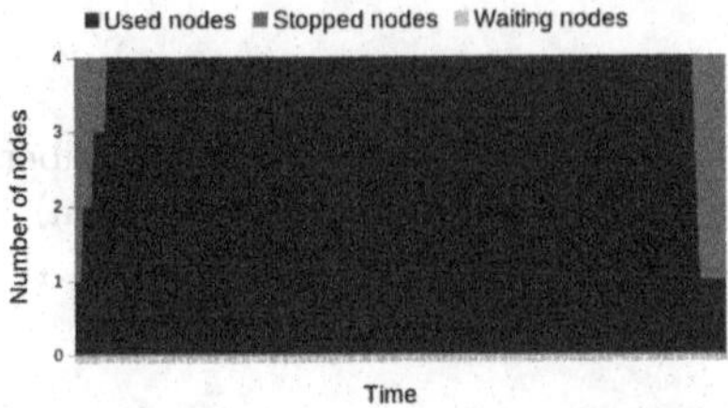

Fig. 1. Variation of active, waiting and stopped nodes to execute 12 requests submitted at the same time

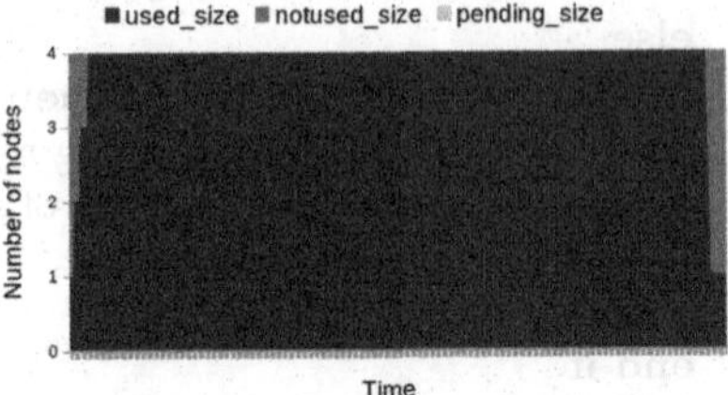

Fig. 2. Variation of active, waiting and stopped nodes to execute 24 requests at the same time

Figure 1 shows the variation of the number of active, stopped, and waiting nodes for the submission of 12 requests at the same time. Each submitted request represent a Linux Container (LXC). Figure 2 shows the variation of the number of active, stopped, and waiting nodes for the submission of 24 requests at the same time. We note from Figs. 1 and 2 that when the system load is high, all nodes of the infrastructure are used. However, if the system load is low or medium, the number of active, stopped and waiting nodes is varied. That's a good outcome for the system behavior.

In this context, we do not have any waiting nodes because all containers are submitted at the same time. When containers finish their execution in a node, the node switches directly to a stopped node as it is presented in Algorithm 1.

Table 1. Comparison between the overall computation time consumed without consolidation and with consolidation

Containers submission	Overall computation time consumed (s)	
	Without consolidation	With consolidation
12 containers	2508.28	1976.07
24 containers	3821.16	3396.38

Table 1 shows the overall computation time consumed without consolidation and with consolidation. We note that our consolidation approach reduces the overall computing time. This result shows the effectiveness of our approach in term of computing time.

6.2 Discussion

As the consolidation heuristic is based on PROMETHEE, we may wonder if this method scales since PROMETHEE is based on a sorting step. In [5,21] we propose a family of accelerated methods for multi-criteria decisions algorithms. The key idea was to compute an approximation of the result, that we experimentally evaluated according to quality metrics.

Let n be the number of nodes and p the number of requests in the queue, the sequential time complexity is given by the sum of the time complexity of each step. We get $\mathcal{O}((p \log p) + (p + n) + n + (n \log n))$ which is in $\mathcal{O}(n \log n)$. Note that, in this calculation, step 3 is implemented with a search of the most loaded server, among n nodes. We also use the time complexity of the traditional PROMETHEE sequential algorithm, which is given by the initial sorting step performed by the method.

7 Conclusion

In this paper, we present a new cloud allocation and consolidation based on the scalability metric. The proposed solution allows delegating the responsibility to scale-up and scale-out the cloud system to a new component of cloud orchestrator. The decision is taken on a metric that quantifies the scalability of the cloud system consistently under different system expansion configurations.

As a first perspective, we propose to implement our approach inside the Kubernetes scheduler framework [19]. Kubernetes is an open-source system for automating deployment, scaling, and management of containerized applications.

As a complementary perspective, we propose to use learning techniques in the consolidation mechanism to automatically adapt the number of active nodes.

In our approach, we consider that each time a request is assigned to a node it will be executed without problems. As a third perspective, we suggest to work on the problem of fault tolerance in case that a node fails in the cloud infrastructure. In this situation, we offer to use a smart replica approach of requests.

Acknowledgments.. We thank the Grid5000 team for their help to use the testbed. Grid5000 is supported by a scientific interest group (GIS) hosted by INRIA and including CNRS, RENATER and several Universities as well as other organizations.

References

1. Ahmad, A.A.-S., Andras, P.: Measuring the scalability of cloud-based software services. In: 2018 IEEE World Congress on Services (SERVICES), pp. 5–6. IEEE (2018)
2. Ala'Anzy, M., Othman, M.: Load balancing and server consolidation in cloud computing environments: a meta-study. IEEE Access **7**, 141868–141887 (2019)
3. Behzadian, M., Kazemzadeh, R., Albadvi, A., Aghdasi, M.: PROMETHEE: a comprehensive literature review on methodologies and applications. Eur. J. Oper. Res. **200**(1), 198–215 (2010)

4. Beloglazov, A., Buyya, R.: OpenStack Neat: a framework for dynamic and energy-efficient consolidation of virtual machines in OpenStack clouds. Concurr. Comput. Pract. Exp. **27**(5), 1310–1333 (2015)
5. Cérin, C., Menouer, T., Lebbah, M.: Accelerating the computation of multi-objectives scheduling solutions for cloud computing. In: 8th IEEE International Symposium on Cloud and Service Computing, SC2 2018, Paris, France, 18–21 November 2018, pp. 49–56. IEEE (2018)
6. de Oliveira e Silva, J.N., Veiga, L., Ferreira, P.: A^2HA - automatic and adaptive host allocation in utility computing for bag-of-tasks. J. Internet Serv. Appl. **2**(2), 171–185 (2011)
7. Elmubarak, S.A., Yousif, A., Bashir, M.B.: Performance based ranking model for cloud SaaS services. Int. J. Inf. Technol. Comput. Sci. **9**(1), 65–71 (2017)
8. Ferdaus, M.H., Murshed, M., Calheiros, R.N., Buyya, R.: Virtual machine consolidation in cloud data centers using ACO metaheuristic. In: Silva, F., Dutra, I., Santos Costa, V. (eds.) Euro-Par 2014. LNCS, vol. 8632, pp. 306–317. Springer, Cham (2014). https://doi.org/10.1007/978-3-319-09873-9_26
9. Gao, J., et al.: A cloud-based TaaS infrastructure with tools for SaaS validation, performance and scalability evaluation. In: 4th IEEE International Conference on Cloud Computing Technology and Science Proceedings, pp. 464–471. IEEE (2012)
10. Gao, J., Pattabhiraman, P., Bai, X., Tsai, W.-T.: SaaS performance and scalability evaluation in clouds. In: Proceedings of 2011 IEEE 6th International Symposium on Service Oriented System (SOSE), pp. 61–71. IEEE (2011)
11. Grama, A., Gupta, A., Kumar, V.: Isoefficiency function: a scalability metric for parallel algorithms and architectures. IEEE Trans. Parallel Distrib. Syst. **4**(8), 02 (1996)
12. Grama, A.Y., Gupta, A., Kumar, V.: Isoefficiency: measuring the scalability of parallel algorithms and architectures. IEEE Parallel Distrib. Technol. Syst. Appl. **1**(3), 12–21 (1993)
13. Grid500: Grid5000. https://www.grid5000.fr/. Accessed 08 May 2020
14. Gunther, N., Puglia, P., Tomasette, K.: Hadoop superlinear scalability. Queue **13**(5), 20 (2015)
15. Hwang, K., Bai, X., Shi, Y., Li, M., Chen, W.-G., Wu, Y.: Cloud performance modeling with benchmark evaluation of elastic scaling strategies. IEEE Trans. Parallel Distrib. Syst. **27**(1), 130–143 (2015)
16. Brans, J.-P., Mareschal, B.: PROMETHEE methods - multiple criteria decision analysis: state of the art surveys. In: International Series in Operations Research & Management Science, vol. 78. Springer, New York (2005). https://www.springer.com/gp/book/9780387230818
17. Jin, X., Zhang, F., Wang, L., Hu, S., Zhou, B., Liu, Z.: Joint optimization of operational cost and performance interference in cloud data centers. IEEE Trans. Cloud Comput. **5**(4), 697–711 (2017)
18. Khan, M.A., Paplinski, A.P., Khan, A.M., Murshed, M., Buyya, R.: Exploiting user provided information in dynamic consolidation of virtual machines to minimize energy consumption of cloud data centers. In: 2018 Third International Conference on Fog and Mobile Edge Computing (FMEC), pp. 105–114, April 2018
19. kubernetes: kubernetes. https://kubernetes.io/. Accessed 08 May 2020
20. Meena, M., Bharadi, V.A.: Performance analysis of cloud based software as a service (SaaS) model on public and hybrid cloud. In: 2016 Symposium on Colossal Data Analysis and Networking (CDAN), pp. 1–6. IEEE (2016)

21. Menouer, T., Cérin, C., Darmon, P.: Accelerated PROMETHEE algorithm based on dimensionality reduction. In: Hsu, C.-H., Kallel, S., Lan, K.-C., Zheng, Z. (eds.) IOV 2019. LNCS, vol. 11894, pp. 190–203. Springer, Cham (2020). https://doi.org/10.1007/978-3-030-38651-1_17
22. Menouer, T., Cérin, C., Hsu, C.-H.R.: Opportunistic scheduling and resources consolidation system based on a new economic model. J. Supercomput. (2020)
23. Deshmukh, S.C.: Preference ranking organization method of enrichment evaluation (PROMETHEE). Int. J. Eng. Sci. Inven. **2**, 28–34 (2013)
24. Simão, J., Veiga, L.: QoE-JVM: an adaptive and resource-aware Java runtime for cloud computing. In: Meersman, R., et al. (eds.) OTM 2012. LNCS, vol. 7566, pp. 566–583. Springer, Heidelberg (2012). https://doi.org/10.1007/978-3-642-33615-7_8
25. Simão, J., Veiga, L.: Partial utility-driven scheduling for flexible SLA and pricing arbitration in clouds. IEEE Trans. Cloud Comput. **4**(4), 467–480 (2016)
26. Singh, S., Chana, I., Buyya, R.: STAR: SLA-aware autonomic management of cloud resources. IEEE Trans. Cloud Comput., p. 1 (2018)
27. Sun, X.-H., Rover, D.T.: Scalability of parallel algorithm-machine combinations. IEEE Trans. Parallel Distrib. Syst. **5**(6), 599–613 (1994)
28. Taillandier, P., Stinckwich, S.: Using the PROMETHEE multi-criteria decision making method to define new exploration strategies for rescue robots. In: 2011 IEEE International Symposium on Safety, Security, and Rescue Robotics, pp. 321–326, November 2011
29. Tsai, W.-T., Huang, Y., Shao, Q.: Testing the scalability of SaaS applications. In: 2011 IEEE International Conference on Service-Oriented Computing and Applications (SOCA), pp. 1–4. IEEE (2011)
30. Ullman, J.: Np-complete scheduling problems. J. Comput. Syst. Sci. **10**(3), 384–393 (1975)
31. Usmani, Z., Singh, S.: A survey of virtual machine placement techniques in a cloud data center. Procedia Comput. Sci. **78**, 491–498 (2016). 1st International Conference on Information Security & Privacy 2015
32. Witanto, J.N., Lim, H., Atiquzzaman, M.: Adaptive selection of dynamic VM consolidation algorithm using neural network for cloud resource management. Future Gener. Comput. Syst. **87**, 35–42 (2018)

Adversarial Attacks on Deep Learning Models of Computer Vision: A Survey

Jia Ding and Zhiwu Xu(✉)

College of Computer Science and Software Engineering, Shenzhen University, Shenzhen, China
2269242754@qq.com, xuzhiwu@szu.edu.cn

Abstract. Deep learning plays a significant role in academic and commercial fields. However, deep neural networks are vulnerable to *adversarial attacks*, which limits its applications in safety-critical areas, such as autonomous driving, surveillance, and drones and robotics. Due to the rapid development of adversarial examples in computer vision, many novel and interesting adversarial attacks are not covered by existing surveys and could not be categorized according to existing taxonomies. In this paper, we present an improved taxonomy for adversarial attacks, which subsumes existing taxonomies, and investigate and summarize the latest attacks in computer vision comprehensively with respect to the improved taxonomy. Finally, We also discuss some potential research directions.

Keywords: Deep learning · Adversarial attacks · Black-box attack · White-box attack · Machine learning

1 Introduction

As a major branch of machine learning, deep learning [23] has always been a popular research direction in the artificial intelligence community. It can solve the classification problems that are difficult or even impossible to solve in a relatively short time, and has many applications in academic and commercial fields, such as computer vision [1,38], speech recognition [44,48], natural language processing [41], malware detection [18], autonomous vehicles [36], network security [17], surveillance [31], drones and robotics [11,28], and so on. Moreover, deep learning has become the preferred choice in computer vision, which plays a major role in our daily lives, after Krizhevsky et al.'s work in 2012 [20]. Thus we focus on computer vision in this paper.

Although deep learning can perform a wide variety of hard tasks with remarkable accuracies, especially in computer vision, Szegedy et al. [42] discovered that the robustness of neural networks encounters a major challenge when adding imperceptible non-random perturbation to input in the context of image classification. They firstly defined the perturbed examples with the ability to misclassify the classifiers as *adversarial examples* and the imperceptible perturbations

M. Qiu (Ed.): ICA3PP 2020, LNCS 12454, pp. 396–408, 2020.
https://doi.org/10.1007/978-3-030-60248-2_27

to images as *adversarial attacks* [42]. This phenomenon implies that deep neural networks are vulnerable to *adversarial attacks*, which limits its applications in safety-critical areas, such as autonomous driving, surveillance, and drones and robotics, and could cause huge economic losses.

After the findings of Szegedy et al., lots of researchers realize the importance of adversarial examples for neural networks, as they are essential to the robustness of the neural networks in some sense. As a result, adversarial examples have become a hot research field in recent years, and many approaches for generating adversarial examples have been proposed. And there is also some review work [1,9,38,49], which gives a comprehensive survey on adversarial attacks in computer vision of that time. However, due to the rapid development of adversarial examples, many novel and interesting approaches for adversarial attacks have been proposed recently, which are not covered by existing review work. Moreover, some adversarial attacks are hard to categorize with respect to existing attack taxonomies presented in [38,49], which indicates existing taxonomies may not be suitable for the latest attacks.

This paper aims to briefly review the latest interesting attacks in computer vision, and revise existing taxonomies for adversarial attacks. More specifically, we first present an improved taxonomy for adversarial attacks, which combines existing taxonomies in [38,49] with a brand-new category, namely, functional-based attacks [22]. Then we explore different approaches, including the classic ones and the latest ones, for generating adversarial examples by taking advantage of the attributes of adversarial examples, such as transferability, and the attributes of images, such as geometric transformation invariance. Finally, in light of the development of adversarial attacks, we also discuss some potential research directions.

Our main contributions are summarized as follows:

- An improved taxonomy for adversarial attacks is presented, including a brand-new category that has never been mentioned in previous work.
- The state-of-the-art approaches for generating adversarial examples are explored, according to the improved taxonomy.

The remainder of this paper will be organized as follows. We introduce some definitions of terms in Sect. 2 and give some related work in Sect. 3. In Sect. 4, we present the improved taxonomy and review the classic attacks and the state-of-the-art attacks. In Sect. 5, we discuss the potential research directions for researchers. We conclude this paper in Sect. 6.

2 Definitions of Terms

In this section, we introduce the technical terms used in the adversarial attacks literature.

Adversarial example: the input with small perturbations that can misclassify the classifier. In the application scene of computer vision, the image with carefully prepared perturbations noise that can make the classifier misclassification.

Adversarial perturbation: the noise data that is capable to change original images to adversarial examples.

Black-box attack: the attackers know nothing about the architecture, training parameters, and defense methods of the attacked model, and can only interact with the model through the input and output.

White-box attack: in contrast to the black-box attack, the attackers master everything about the model and the defense schemes should be public to attackers. At present, most attack approaches are white-box.

Gray-box attack: between black-box attack and white-box attack, only a part of the model is understood. For example, the attackers get only the output probability of the model, or know only the model structure without the parameters.

Untargeted attack: the attackers only need to make the target model misclassify, but do not specify which category is misclassified.

Targeted attack: the attackers specify a certain category, so that the target model not only misclassify the sample but also need to be misclassified into the specified category. It is more difficult to achieve targeted attacks than untargeted attacks.

Transferability: transferability refers to the effective adversarial examples for one model, and still effective for other models.

Geometric transformation invariance: the target in the image can be successfully identified whether it is translated, rotated, or zoomed, or even under different lighting conditions and viewing angles, such as translation invariance and scale invariance.

3 Related Work

This section presents some related work of surveys on adversarial attacks.

There are some review work [1,9,38,49] on adversarial attacks in computer vision so far. Fawzi et al. [9] discussed the robustness of deep networks to a diverse set of perturbations that may affect the samples in practice, including adversarial perturbations, random noise, and geometric transformations. Serban et al. [38] provided a complete characterization of the phenomenon of adversarial examples, summarized more than 20 kinds of attacks at that time by dividing the attack approaches into four categories: (*i*) attacks based on optimization methods, (*ii*) attacks based on sensitive features, (*iii*) attacks based on geometric transformations, and (*iv*) attacks based on generative models. Akhtar et al. [1] reviewed the adversarial attacks at that time for the task of image classification and beyond classification, and introduced some adversarial attacks in the real world. Zhou et al. [49] summarized the latest attack approaches at that time and divided them into four categories: (*i*) gradient-based attack, (*ii*) score-based attack, (*iii*) transfer-based attack, and (*iv*) decision-based attack.

However, because of the popularity of adversarial attacks, after the latest review work [49], many novel and interesting approaches for adversarial attacks have been proposed recently. Moreover, some adversarial attacks are hard to categorize with respect to existing attack taxonomies presented in [38,49], which indicates existing taxonomies may not be suitable for the latest attacks. To supplement the latest development and revise existing taxonomies, in this paper we give an improved taxonomy, which subsumes existing taxonomies in [38,49], and summarize the latest attack approaches according to the improved taxonomy.

In addition to computer vision, there are some surveys on adversarial attacks in other areas, such as (vector) graphs [3], speech recognition [14], autonomous driving [36], and malware detection [27]. Chen et al. [3] investigated and summarized the existing works on graph adversarial learning tasks systemically. Hu et al. [14] provided a concise overview of adversarial examples for speech recognition. Ren et al. [36] systematically studied the safety threats surrounding autonomous driving from the perspectives of perception, navigation and control. Martins et al. [27] explored applications of adversarial machine learning to intrusion and malware detection.

4 Adversarial Attacks

In this section, we first give an improved taxonomy for attack approaches, and then explore the attack approaches according to this taxonomy. As the classic attack approaches have been summarized in existing work [1,9,38,49], we only give a brief review on these classic approaches in this section. In other words, we focus on the latest attack approaches (*i.e.*, those are not covered by existing surveys [1,9,38,49]), each of which is marked with its abbreviated name in bold.

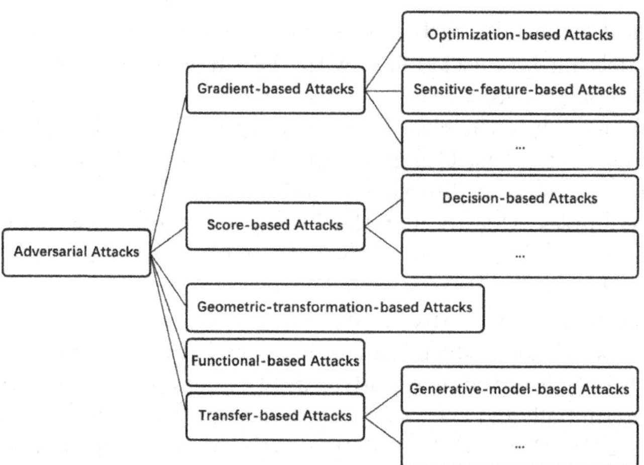

Fig. 1. The improved taxonomy for adversarial attacks

4.1 An Improved Taxonomy

There have been many interesting approaches of adversarial attacks in computer vision. To investigate and summarize them comprehensively, we present a taxonomy for them.

Based on existing taxonomies [38,49], we propose an improved taxonomy for adversarial attacks, which is given in Fig. 1. In detail, we classify the adversarial attacks into five categories, that is, (*i*) gradient-based attacks (GA) (from [49]), (*ii*) score-based attacks (SA) (from [49]), (*iii*) geometric-transformation-based attacks (GTA) (from [38]), (*iv*) functional-based attacks (FA) (a brand-new category), and (*v*) transfer-based attacks (TA) (from [49]). Moreover, we categorize optimization-based attacks and sensitive-feature-based attacks from [38] into gradient-based attacks, as both of them are almost based on gradient but with different objectives, and generative-model-based attacks from [38] into transfer-based attacks, as it is stated in [49] that generative-model-based attacks are a subclass of transfer-based attacks. The decision-based attacks from [49] are treated as a special case of score-based attacks, due to the fact that the decisions are always made according to the scores. In conclusion, our taxonomy subsumes both the taxonomies in [38] and [49].

According to our taxonomy, we summarize different attacks, including the classic ones and the latest ones (in bold), in Table 1. All these attacks are reviewed in the following.

4.2 Gradient-Based Attacks

Gradient-based attacks perturb the images in the direction of the gradient, so that the model can be misclassified with the smallest perturbation. They are mainly white-box attacks.

We briefly introduce the classic gradient-based attacks. In 2014, Goodfellow et al. [12] proposed the Fast Gradient Sign Method (FGSM), applying small perturbations in the gradient direction to maximize the loss function to generate adversarial examples. Due to that the FGSM algorithm only involves a single gradient update and that a single update is sometimes not enough to attack successfully, Kurakin et al. [21] proposed the Iterative Fast Gradient Sign Method (I-FGSM) based on FGSM. After that, Madry et al. [26] proposed the Projected Gradient Descent (PGD), which is a more powerful gradient attack than I-FGSM and FGSM. It initializes the search adversarial examples at random points within the allowed norm ball, and then runs the Basic Iterative Method (BIM) multiple iterations. In 2018, Dong et al. [6] proposed the Momentum Iterative Fast Gradient Sign Method (MI-FGSM), which integrates momentum into the iterative attack and leads to a higher attack success rate and transfer-ability than other gradient-based methods for adversarial examples.

Besides attacks based on FGSM, there are some other gradient-based attacks. In 2016, Moosavi et al. proposed the DeepFool [29], which can generate adversarial examples that are very close to the minimum perturbation, so it can be used as a measure of the robustness of the classifier. Later, Moosavi-Dezfooli et al. [30]

Table 1. Catalog of adversarial attacks.

Attack	Category	White-box/ Black-box	Targeted/ Untargeted	Specific/ Universal	Iterative/ One-shot	Year
FGSM [12]	GA	White-box	Targeted	Image specific	One-shot	2015
I-FGSM [21]	GA	White-box	Targeted	Image specific	Iterative	2017
PGD [26]	GA	White-box	Untargeted	Image specific	Iterative	2018
MI-FGSM [6]	GA	White-box	Untargeted	Image specific	Iterative	2018
DeepFool [29]	GA	White-box	Untargeted	Image specific	Iterative	2016
UAP [30]	GA	White-box	Untargeted	Universal	Iterative	2017
C&W [2]	GA	White-box	Targeted	Image specific	Iterative	2017
LogBarrier [10]	GA	White-box	Untargeted	Image specific	Iterative	2019
NI-FGSM [24]	GA	White-box	Untargeted	Image specific	Iterative	2020
ZOO [4]	SA	Black-box	Untargeted	Image specific	Iterative	2017
OPA [39]	SA	Black-box	Untargeted	Image specific	Iterative	2019
AutoZOOM [43]	SA	Black-box	Untargeted	Image specific	Iterative	2019
CornerSearch [5]	SA	Black-box	Untargeted	Image specific	Iterative	2019
BayesOpt [37]	SA	Black-box	Untargeted	Image specific	Iterative	2020
ManiFool [19]	GTA	White-box	Untargeted	Image specific	Iterative	2018
Xiao et al. [46]	GTA	White-box	Untargeted	Image specific	Iterative	2018
DIM [47]	GTA	White-box	Untargeted	Image specific	Iterative	2019
TI [7]	GTA	White-box	Untargeted	Image specific	Iterative	2019
SIM [24]	GTA	White-box	Untargeted	Image specific	Iterative	2020
ReColorAdv [22]	FA	White-box	Untargeted	Image specific	One-shot	2019
Substitute [34]	TA	Black-box	Targeted	Image specific	Iterative	2017
Ensemble [25]	TA	Black-box	Targeted	Image specific	Iterative	2019
ILA [15]	TA	Black-box	Targeted	Image specific	Iterative	2019
TREMB [16]	TA	Black-box	Targeted	Image specific	Iterative	2020

proposed attacks using Universal Adversarial Perturbations (UAP), which is an extension of DeepFool. In 2017, Carlini et al. [2] proposed the Carlini & Wagner attack (C&W), which optimizes the distances from adversarial examples to benign examples and is an optimization-based method.

LogBarrier Attack. Recently, Finlay et al. [10] proposed a new type of adversarial attack (called LogBarrier attack) based on optimization. Different from existing approaches that use training loss functions to achieve misclassification, LogBarrier uses the best practices in the optimization literature to solve "how to make a benign sample transform into adversarial sample", wherein the well-known logarithmic barrier method [33] is used to design a new untargeted attack. The LogBarrier attack performs well on common data sets, and in images that require large perturbation for misclassification, LogBarrier attacks always have an advantage over other adversarial attacks.

NI-FGSM. Lin et al. [24] proposed the Nesterov Iterative Fast Gradient Sign Method (NI-FGSM)[1], which aims to adapt Nesterov accelerated gradient into the iterative attacks so as to effectively look ahead and improve the transferability

[1] The other method SIM proposed by Lin et al. [24] is discussed in Sect. 4.4.

of adversarial examples, according to the fact that Nesterov accelerated gradient method [32] is superior to the momentum for conventionally optimization method [40]. Experiments show that it can effectively improve the transferability of adversarial examples.

4.3 Score-Based Attacks

Score-based attacks are black box approaches that rely only on predicted scores, such as category probability or logarithm. Conceptually, these attacks use numerically estimated gradient predictions.

In 2017, inspired by C&W attacks [2], Chen et al. [4] proposed the Zeroth Order Optimization (ZOO), which directly estimates the gradient of the target model to generate adversarial examples. In 2019, Vargas et al. [39] proposed the One Pixel Attack (OPA), wherein only one pixel can be modified at a time. As a result, less adversarial information is required. And thus it can deceive more types of networks, because of the inherent function of differential evolution.

AutoZOOM. To address the major drawback of existing black-box attacks, that is, the need for excessive model queries, Tu et al. [43] proposed a generic framework, the Autoencoder-based Zeroth Order Optimization Method (AutoZOOM), for query-efficient black-box attacks. AutoZOOM has two novel building blocks towards efficient black-box attacks: (*i*) an adaptive random gradient estimation strategy to balance query counts and distortion, and (*ii*) an autoencoder that is either trained offline with unlabeled data or a bilinear resizing operation for attack acceleration.

CornerSearch. Despite highly sparse adversarial attacks have a great impact on neural networks, the perturbations imposed on images by sparse attacks are easily noticeable due to their characteristics. To prevent the subtle perturbations of highly sparse adversarial attacks from being detected, Croce et al. [5] proposed a fresh black-box attack. They use locally adaptive component wise constraints to minimize the difference between the modified image and the original image, which enables to limit pixel perturbation to occur only in areas of high variance and to reduce the number of pixels that need to be modified. Experiments show that their score-based l_0-attack CornerSearch needs the least pixels to complete the task.

BayesOpt Attack. The existing black-box attacks are based on the substitute model, gradient estimation or genetic algorithm. The number of queries they need is usually very large. For projects that need to control costs in real life or items that have a limited number of query numbers, these approaches are obviously not applicable. Therefore, Ru et al. [37] proposed a new gradient-free black-box attack, which uses Bayesian Optimization (BayesOpt) in combination with Bayesian model selection to optimise over the adversarial perturbation and the optimal degree of search space dimension reduction. Experiments show that, in the constraint of l_∞-norm, BayesOpt adversarial attack can achieve a considerable success rate with a model query of about 2 to 5 times, compared with the latest black-box attacks.

4.4 Geometric-Transformation-Based Attacks

Geometric-transformation-based attacks transform targets in the images via geometric transformation (*e.g.*, rotating or zooming) to generate adversarial examples. According to geometric transformation invariance, no matter how geometrically transformed its input image is, the classifier for image classification tasks should produce the same output. Algorithms based on geometric transformation invariances often work together with algorithms based on gradients.

Engstrom et al. [8] showed that only simple transformations, namely rotations and translations, are sufficient to fool DNN. Kanbak et al. [19] proposed ManiFool, an approach to find small worst-case geometrical transformations of images. Xiao et al. [46] proposed to spatially transform the image, that is, to change the geometry of the scene, while keeping the original appearance.

DIM. Xie et al. [47] a Diverse Inputs Method (DIM) to improve the transferability of adversarial examples. Inspired by the data augmentation strategy [13], DIM randomly applies a set of label-preserving transformations (e.g., resizing, cropping and rotating) to training images and feeds the transformed images into the classifier for gradient calculation. DIM can be combined with the momentum-based method (such as MI-FGSM [6]) to further improve the transferability. By evaluating DIM against top defense solutions and official baselines from NIPS 2017 adversarial competition, the enhanced attack M-DI^2-FGSM reaches an average success rate of 73.0%, which outperforms the top-1 attack submission in the NIPS competition by a large margin of 6.6%.

TI. Dong et al. [7] proposed a Translation-Invariant (TI) attack method to generate more transferable adversarial examples against the defense models. TI optimizes the adversarial samples by using a set of translated images, making the adversarial samples less sensitive to the distinguished regions of the white-box model being attacked, and thus the transferability of the adversarial samples is increased. To improve the efficiency of attacks, TI can be implemented by convolving the gradient at the untranslated image with a pre-defined kernel. Dong et al. [7] also showed that TI-DIM, the combination of TI and DIM [47], performed best on common data sets.

SIM. Besides NI-FGSM (see Sect. 4.2), Lin et al. [24] also proposed another method to improve the transferability of adversarial examples, that is, the Scale-Invariant attack Method (SIM). SIM utilizes the scale-invariant property of the model to achieve model augmentation and can generate adversarial samples which are more transferable than other black box attacks. Combining NI-FGSM and SIM, SINI-FGSM is a powerful attack with higher transferability. Experiments show that SINI-FGSM can break other powerful defense mechanisms.

4.5 Functional Adversarial Attacks

Unlike standard l_p-ball attacks, functional adversarial attacks allow only a single function, which is called *the perturbation function*, to be used to perturb input features to generate an adversarial example. Functional adversarial attacks are in some ways more restrictive because features cannot be perturbed individually.

ReColorAdv. Laidlaw et al. [22] proposed ReColorAdv, a functional adversarial attack on pixel colors. ReColorAdv generates adversarial examples to fool image classifiers by uniformly changing colors of an input image. More specifically, ReColorAdv uses a flexibly parameterized function f to map each pixel color c in the input to a new pixel color $f(c)$ in an adversarial example. Combining functional adversarial attacks with existing attacks that use the l_p-norm can greatly increase the attack capability and allow the model to modify the input locally, individually, and overall. Experiments show that the combination of ReColorAdv and other attacks (*e.g.*, Xiao et al.'s work [46]) can produce the strongest attack at present.

4.6 Transfer-Based Attacks

Transfer-based attacks do not rely on model information, but need information about training data. This is a way to transition between black-box attacks and white-box ones.

In 2014, Szegedy et al. [42] firstly proposed the concept of adversarial examples, in the same time they observed that adversarial examples generated for one model can be effectively transferred to other models regardless of architecture, which is named by model-transferability. And later, Papernot et al. [35] explored deeply this property. In 2017, Papernot et al. [34] proposed a transfer-based attack, called Substitute in this paper, which trains a local model to substitute for the target DNN, using inputs synthetically generated by an adversary and labeled by the target DNN. In 2017, Liu et al. [25] proposed a novel strategy (called Ensemble here) to generate transferable adversarial images using an ensemble of multiple models, which enables a large portion of targeted adversarial examples to transfer among multiple models for the first time.

ILA. In order to enhance the transferability of black box attacks, Huang et al. [15] proposed the Intermediate Level Attack (ILA), which fine-tunes the existing adversarial examples and increases the perturbations on the pre-designated layer of the model to achieve high transferability. ILA is a framework with the goal of enhancing transferability by increasing projection onto the *Best Transfer Direction*. Two variants of ILA, namely ILAP and ILAF, are proposed in [15], differing in their definition of the loss function L.

TREMBA. Unlike previous attack methods, which training substitute models with data augmentation to mimic the behavior of the target model, Huang et al. [16] proposed a method called TRansferable EMbedding based Black-box Attack (TREMBA), which combines transfer-based attack and scored-based attack, wherein transfer-based attack is used to improve query efficiency, while scored-based attack is to increase success rate. TREMBA contains two steps: the first step is to train an encoder-decoder to generate adversarial perturbations for the source network with a low-dimensional embedding space, and the second step is to apply NES (Natural Evolution Strategy) [45] to the low-dimensional embedding space of the pretrained generator to search adversarial examples for the target network. Compared with other black box attacks, the success rate of TREMBA is increased by about 10%, and the number of queries is reduced by more than 50%.

5 Future Directions

In this section, we discuss some potential research directions.

Explanation. Lots of approaches can generate adversarial examples effectively and efficiently. But why classifier makes a misclassification on these adversarial examples? Few work gives a systematic study on this problem. Moreover, are these adversarial examples helpful to explain the classifiers, that is, can we extract some negative but useful knowledge from adversarial examples? This is also an interesting problem. On the other hand, as the geometric-transformation-based attacks, we could take advantage of the domain knowledge or the knowledge that is extracted from models if possible to guide the adversarial attacks to get more effective adversarial examples, via the perturbation functions [22]?

Ensemble. It has been shown in existing work [7,16,22] that some attacks can be combined together. Indeed, some categories of our taxonomy are orthogonal, such as gradient-based attack and geometric-transformation-based attack. Similar to ensemble learning, how to combine different attacks to get a more powerful attack is worth investigating.

Transferability. As stated in [38], transferability is inherent to models that "learn feature representations that preserve non-robust properties of the input space". If models were to learn different features, it would have not been possible to transfer adversarial examples. Therefore, when designing a model, it is better to consider transferability.

Robustness. Finally, one goal of adversarial examples is to improve the robustness of DNN. This is also a fundamental problem in the community and deserves special attention.

6 Conclusion

In this paper, we have presented an improved taxonomy for adversarial attacks in computer vision. Then according to this taxonomy, we have investigated and summarized different adversarial attacks, including the classic ones and the latest ones. Through the investigation, some future directions are discussed. This paper is expected to provide guidance on adversarial attacks for researchers and engineers in computer vision and other areas.

Acknowledgements. This work was partially supported by the National Natural Science Foundation of China under Grants No. 61972260, 61772347, 61836005; Guangdong Basic and Applied Basic Research Foundation under Grant No. 2019A1515011577; and Guangdong Science and Technology Department under Grant No. 2018B010107004.

References

1. Akhtar, N., Mian, A.: Threat of adversarial attacks on deep learning in computer vision: a survey. IEEE Access **6**, 14410–14430 (2018)
2. Carlini, N., Wagner, D.: Towards evaluating the robustness of neural networks. In: 2017 IEEE Symposium on Security and Privacy (SP), pp. 39–57. IEEE (2017)
3. Chen, L., et al.: A survey of adversarial learning on graphs. arXiv preprint arXiv:2003.05730 (2020)
4. Chen, P., Zhang, H., Sharma, Y., Yi, J., Hsieh, C.: ZOO: zeroth order optimization based black-box attacks to deep neural networks without training substitute models. In: Proceedings of the 10th ACM Workshop on Artificial Intelligence and Security (AISec@CCS 2017), pp. 15–26 (2017)
5. Croce, F., Hein, M.: Sparse and imperceivable adversarial attacks. In: Proceedings of the IEEE International Conference on Computer Vision, pp. 4724–4732 (2019)
6. Dong, Y., et al.: Boosting adversarial attacks with momentum. In: 2018 IEEE Conference on Computer Vision and Pattern Recognition (CVPR), pp. 9185–9193 (2018)
7. Dong, Y., Pang, T., Su, H., Zhu, J.: Evading defenses to transferable adversarial examples by translation-invariant attacks. In: IEEE Conference on Computer Vision and Pattern Recognition (CVPR 2019), pp. 4312–4321 (2019)
8. Engstrom, L., Tsipras, D., Schmidt, L., Madry, A.: A rotation and a translation suffice: fooling CNNs with simple transformations. CoRR abs/1712.02779 (2017)
9. Fawzi, A., Moosavi-Dezfooli, S., Frossard, P.: The robustness of deep networks: a geometrical perspective. IEEE Signal Process. Mag. **34**(6), 50–62 (2017)
10. Finlay, C., Pooladian, A., Oberman, A.M.: The logbarrier adversarial attack: making effective use of decision boundary information. In: 2019 IEEE/CVF International Conference on Computer Vision (ICCV 2019), pp. 4861–4869 (2019)
11. Giusti, A., et al.: A machine learning approach to visual perception of forest trails for mobile robots. IEEE Robot. Autom. Lett. **1**(2), 661–667 (2016)
12. Goodfellow, I.J., Shlens, J., Szegedy, C.: Explaining and harnessing adversarial examples. In: 3rd International Conference on Learning Representations (ICLR 2015) (2015)
13. He, K., Zhang, X., Ren, S., Sun, J.: Identity mappings in deep residual networks. In: Leibe, B., Matas, J., Sebe, N., Welling, M. (eds.) ECCV 2016. LNCS, vol. 9908, pp. 630–645. Springer, Cham (2016). https://doi.org/10.1007/978-3-319-46493-0_38
14. Hu, S., Shang, X., Qin, Z., Li, M., Wang, Q., Wang, C.: Adversarial examples for automatic speech recognition: attacks and countermeasures. IEEE Commun. Mag. **57**(10), 120–126 (2019)
15. Huang, Q., Katsman, I., He, H., Gu, Z., Belongie, S., Lim, S.: Enhancing adversarial example transferability with an intermediate level attack. In: 2019 IEEE/CVF International Conference on Computer Vision (ICCV 2019), pp. 4732–4741 (2019)
16. Huang, Z., Zhang, T.: Black-box adversarial attack with transferable model-based embedding. In: 8th International Conference on Learning Representations (ICLR 2020) (2020)
17. Ibitoye, O., Abou-Khamis, R., Matrawy, A., Shafiq, M.O.: The threat of adversarial attacks on machine learning in network security-a survey. arXiv preprint arXiv:1911.02621 (2019)
18. John, T.S., Thomas, T.: Adversarial attacks and defenses in malware detection classifiers. In: Handbook of Research on Cloud Computing and Big Data Applications in IoT, pp. 127–150. IGI global (2019)

19. Kanbak, C., Moosavi-Dezfooli, S., Frossard, P.: Geometric robustness of deep networks: analysis and improvement. In: 2018 IEEE Conference on Computer Vision and Pattern Recognition (CVPR 2018), pp. 4441–4449. IEEE Computer Society (2018)
20. Krizhevsky, A., Sutskever, I., Hinton, G.E.: ImageNet classification with deep convolutional neural networks. In: Advances in Neural Information Processing Systems 25: 26th Annual Conference on Neural Information Processing Systems 2012, pp. 1106–1114 (2012)
21. Kurakin, A., Goodfellow, I.J., Bengio, S.: Adversarial examples in the physical world. In: 5th International Conference on Learning Representations (ICLR 2017). OpenReview.net (2017)
22. Laidlaw, C., Feizi, S.: Functional adversarial attacks. In: Advances in Neural Information Processing Systems 32: Annual Conference on Neural Information Processing Systems 2019 (NeurIPS 2019), pp. 10408–10418 (2019)
23. LeCun, Y., Bengio, Y., Hinton, G.: Deep learning. Nature **521**, 436–44 (2015). https://doi.org/10.1038/nature1453910.1038/nature1453910.1038/nature14539
24. Lin, J., Song, C., He, K., Wang, L., Hopcroft, J.E.: Nesterov accelerated gradient and scale invariance for adversarial attacks. In: 8th International Conference on Learning Representations (ICLR 2020). OpenReview.net (2020)
25. Liu, Y., Chen, X., Liu, C., Song, D.: Delving into transferable adversarial examples and black-box attacks. In: 5th International Conference on Learning Representations (ICLR 2017) (2017)
26. Madry, A., Makelov, A., Schmidt, L., Tsipras, D., Vladu, A.: Towards deep learning models resistant to adversarial attacks. In: 6th International Conference on Learning Representations (ICLR 2018) (2018)
27. Martins, N., Cruz, J.M., Cruz, T., Abreu, P.H.: Adversarial machine learning applied to intrusion and malware scenarios: a systematic review. IEEE Access **8**, 35403–35419 (2020)
28. Mnih, V., et al.: Human-level control through deep reinforcement learning. Nature **518**(7540), 529–533 (2015)
29. Moosavi-Dezfooli, S., Fawzi, A., Frossard, P.: DeepFool: a simple and accurate method to fool deep neural networks. In: 2016 IEEE Conference on Computer Vision and Pattern Recognition (CVPR), pp. 2574–2582 (2016)
30. Moosavidezfooli, S., Fawzi, A., Fawzi, O., Frossard, P.: Universal adversarial perturbations. In: 2017 IEEE Conference on Computer Vision and Pattern Recognition (CVPR), pp. 86–94 (2017)
31. Najafabadi, M.M., Villanustre, F., Khoshgoftaar, T.M., Seliya, N., Wald, R., Muharemagic, E.: Deep learning applications and challenges in big data analytics. J. Big Data **2**(1), 1–21 (2015). https://doi.org/10.1186/s40537-014-0007-7
32. Nesterov, Y.: A method for unconstrained convex minimization problem with the rate of convergence $o(1/k^2)$. Doklady AN USSR **269**, 543–547 (1983)
33. Nocedal, J., Wright, S.J.: Numerical Optimization, 2nd edn. Springer, New York (2006). https://doi.org/10.1007/978-0-387-40065-5
34. Papernot, N., McDaniel, P., Goodfellow, I., Jha, S., Celik, Z.B., Swami, A.: Practical black-box attacks against machine learning. In: Proceedings of the 2017 ACM on Asia Conference on Computer and Communications Security, pp. 506–519 (2017)
35. Papernot, N., McDaniel, P.D., Goodfellow, I.J.: Transferability in machine learning: from phenomena to black-box attacks using adversarial samples. CoRR abs/1605.07277 (2016)
36. Ren, K., Wang, Q., Wang, C., Qin, Z., Lin, X.: The security of autonomous driving: threats, defenses, and future directions. Proc. IEEE **108**(2), 357–372 (2019)

37. Ru, B., Cobb, A., Blaas, A., Gal, Y.: BayesOpt adversarial attack. In: 8th International Conference on Learning Representations (ICLR 2020) (2020)
38. Serban, A.C., Poll, E., Visser, J.: Adversarial examples-a complete characterisation of the phenomenon. arXiv preprint arXiv:1810.01185 (2018)
39. Su, J., Vargas, D.V., Sakurai, K.: One pixel attack for fooling deep neural networks. IEEE Trans. Evol. Comput. **23**(5), 828–841 (2019)
40. Sutskever, I., Martens, J., Dahl, G., Hinton, G.: On the importance of initialization and momentum in deep learning. In: International Conference on Machine Learning, pp. 1139–1147 (2013)
41. Sutskever, I., Vinyals, O., Le, Q.V.: Sequence to sequence learning with neural networks. In: Advances in Neural Information Processing Systems, pp. 3104–3112 (2014)
42. Szegedy, C., Zaremba, W., Sutskever, I., Bruna, J., Erhan, D., Goodfellow, I., Fergus, R.: Intriguing properties of neural networks. In: 2nd International Conference on Learning Representations (ICLR 2014) (2014)
43. Tu, C., et al.: AutoZOOM: autoencoder-based zeroth order optimization method for attacking black-box neural networks. In: The Thirty-Third AAAI Conference on Artificial Intelligence (AAAI 2019), The Thirty-First Innovative Applications of Artificial Intelligence Conference (IAAI 2019), The Ninth AAAI Symposium on Educational Advances in Artificial Intelligence (EAAI 2019), pp. 742–749. AAAI Press (2019)
44. Vaswani, A., et al.: Attention is all you need. In: Advances in Neural Information Processing Systems, pp. 5998–6008 (2017)
45. Wierstra, D., Schaul, T., Glasmachers, T., Sun, Y., Peters, J., Schmidhuber, J.: Natural evolution strategies. J. Mach. Learn. Res. **15**(1), 949–980 (2014)
46. Xiao, C., Zhu, J., Li, B., He, W., Liu, M., Song, D.: Spatially transformed adversarial examples. In: 6th International Conference on Learning Representations (ICLR 2018). OpenReview.net (2018)
47. Xie, C., Zhang, Z., Zhou, Y., Bai, S., Wang, J., Ren, Z., Yuille, A.L.: Improving transferability of adversarial examples with input diversity. In: IEEE Conference on Computer Vision and Pattern Recognition (CVPR 2019), pp. 2730–2739. Computer Vision Foundation/IEEE (2019)
48. Zhang, Z., Geiger, J., Pohjalainen, J., Mousa, A.E.D., Jin, W., Schuller, B.: Deep learning for environmentally robust speech recognition: an overview of recent developments. ACM Trans. Intell. Syst. Technol. (TIST) **9**(5), 1–28 (2018)
49. Zhou, Y., Han, M., Liu, L., He, J., Gao, X.: The adversarial attacks threats on computer vision: a survey. In: 2019 IEEE 16th International Conference on Mobile Ad Hoc and Sensor Systems Workshops (MASSW), pp. 25–30. IEEE (2019)

FleetChain: A Secure Scalable and Responsive Blockchain Achieving Optimal Sharding

Yizhong Liu[1,3], Jianwei Liu[2], Dawei Li[2(✉)], Hui Yu[1], and Qianhong Wu[2]

[1] School of Electronic and Information Engineering, Beihang University, Beijing, China
{liuyizhong,yhsteven}@buaa.edu.cn

[2] School of Cyber Science and Technology, Beihang University, Beijing, China
{liujianwei,lidawei,qianhong.wu}@buaa.edu.cn

[3] Department of Computer Science, University of Copenhagen, Copenhagen, Denmark

Abstract. Sharding blockchains are promising in improving transaction throughput and achieving network scalability. Intra-shard consensus and cross-shard communication are two essential parts for almost every kind of sharding blockchain. However, some security problems still exist in current sharding solutions such as replay attacks, and there is still room for improvement in efficiency.

In this paper, we propose FleetChain, a secure and scalable sharding blockchain. First, we make modification of the original BLS multi-signature scheme to a robust (t, u)-multi-signature protocol supporting further aggregation, which could shorten vote messages. Second, a leader-stable fast Byzantine fault tolerance (FBFT) protocol is designed for efficient intra-shard consensus, combining pipeline technology and multi-signature. FBFT is specially designed for sharding blockchains, with the ability to process different types of proposals that might be transactions or transaction inputs. Third, a responsive sharding transaction processing (RSTP) protocol is given, which greatly improves the processing efficiency of cross-shard transactions by using multi-signature aggregation. FleetChain employs a star network in both intra-shard and cross-shard communication, achieving responsiveness when confirming transactions. In addition, FleetChain achieves optimal sharding with a scaling factor of $O(n/\log n)$ where n denotes the total number of participating nodes.

Keywords: Sharding blockchains · Byzantine fault tolerance · Multi-signature aggregation · Scalability · Responsiveness

1 Introduction

Nowadays, the blockchain technology has been developing rapidly ever since it was first proposed in Bitcoin [1]. Blockchains offer the advantages of decentralization, data transparency, and temper-resistant transaction histories with great

M. Qiu (Ed.): ICA3PP 2020, LNCS 12454, pp. 409–425, 2020.
https://doi.org/10.1007/978-3-030-60248-2_28

application potential in various fields [2], such as finance, Internet of Things, and supply chain management. However, there are currently some problems that limit the development and application of blockchains, the most critical of which are security and performance issues [3].

There are various security problems in current blockchain systems, such as double spending [4] and eclipse attacks [5] on general blockchains, selfish mining [6], stubborn mining [7,8] and block withholding attacks [9] on proof-of-work (PoW) blockchains. Meanwhile, there are works dedicated to the analysis of blockchain security, such as [10,11], which lay a good foundation for blockchain security analysis. Designing a secure blockchain protocol and proving its security is the key and hotspot of current research.

Regarding performance, the most important indicators are transaction throughput and confirmation latency. Methods of improving throughput could be divided into on-chain and off-chain ones. Off-chain methods use payment channels [12], state channels [13] and other technologies to support some small-value transactions. On-chain methods mainly combine classical state machine replication algorithms with blockchains to design new consensus protocols. Sharding blockchain is one of the most promising technology to achieve scalability. Network nodes are divided into multiple shards, where each shard processes and maintains its own blockchain.

There are still many problems to be solved in sharding blockchains for them to be applied in practice [14]. Most intra-shard consensus rely on a Byzantine fault tolerant (BFT) protocol, where the processing efficiency is low while communication complexity is high. At the same time, current solutions for cross-shard communication are vulnerable to different attacks, like Denial of Service (DoS) in Omniledger [15], or replay attack [16] against Chainspace [17]. So we aim to propose a secure sharding blockchain through designing an efficient intra-shard consensus algorithm and cross-shard communication mechanism.

1.1 Our Contributions

In this paper, we make the following contributions.

A Robust (t,u) Multi-signature Protocol with Public Key Aggregation Using Proof-of-Possession (PoP). We make modification of the BLS multi-signature scheme in [18] to a robust (t, u)-`MSP-PoP` protocol. By adding the maintenance of a valid public key list, a signature share generation interface, and the verification of an aggregation public key, our protocol is secure and achieves robustness. (t, u)-`AMSP-PoP` is also robust and secure which adds multi-signature aggregation on the basis of (t, u)-`MSP-PoP`. Besides, we combine PoP with PoW for the first time, realizing a consistent view of a public key list among all nodes without a trusted party. Both of them could be applied easily in a committee without a complicated distributed key generation phase.

A Leader-Stable Fast Byzantine Fault Tolerance Protocol. We design a fast Byzantine fault tolerance protocol utilizing (t, u)-`MSP-PoP` to cut down message complexity inside shards. Besides, the pipelined technology is adopted to improve processing efficiency. FBFT uses a normal phase ($O(u)$) to process proposals and a view-change phase ($O(u^2)$) to change leaders. FBFT could be used alone or embedded in other protocols that require a intra-shard consensus. FBFT is specially designed for sharding blockchains to handle different types of proposals that could be transactions or transaction inputs.

A Responsive Sharding Transaction Processing Protocol. We propose a responsive sharding transaction processing (RSTP) protocol based on the classical two phase commit (2PC) protocol. RSTP uses (t, u)-AMSP-PoP, and treats an output shard leader as a coordinator. RSTP utilizes FBFT inside every shard to provide input availability certification and process transactions simultaneously.

A Secure and Scalable FleetChain Achieving Optimal Sharding. FBFT and RSTP make up FleetChain which employs a star network in both intra-shard and cross-shard communication. FleetChain achieves responsiveness when confirming transactions, i.e., confirmation delay is only related to the actual network delay. The computation, communication, and storage complexity [19] of FleetChain are all $O(n/m)$, achieving an optimal scaling factor of $O(n/\log n)$.

Comparison of FleetChain with other sharding blockchains is shown in Table 1. The specific meaning of the parameters is given in Sect. 3.2. For a detailed description of the complexity calculation, please see Sect. 6.2.

Table 1. Comparison of FleetChain with other sharding blockchains.

System		Elastico	Omniledger	RapidChain	Chainspace	FleetChain
Intra-shard	Network model	Partial sync	Partial sync	Sync	Partial sync	Partial sync
	Adversary model	$3f+1$	$3f+1$	$2f+1$	$3f+1$	$3f+1$
	Consensus algorithm	PBFT	Omnicon	Sync BFT	PBFT	FBFT
	Communication complexity	$O(u^2)$	$O(u)$	$O(u^2)$	$O(u^2)$	$O(u)$
Cross-shard	Basic algorithm	–	2PC	Split	2PC	2PC
	Coordinator	–	Client-driven	Shard-driven	Shard-driven	Shard-driven
	Malicious leader resistance	✗	✗	✗	✓	✓
Communication complexity ω_m		$\Theta(n)$	$O(k_{all}\frac{n}{m})$	$O(k_{all}\frac{n}{m})$	$O(k_{all}\frac{n}{m})$	$O(\frac{n}{m})$
Computation complexity ω_c		$\Theta(n)$	$O(k_{in}\frac{n}{m})$	$O(k_{in}\frac{n}{m})$	$O(k_{in}\frac{n}{m})$	$O(\frac{n}{m})$
Storage complexity ω_s		$\Theta(n)$	$O(\frac{n}{m})$	$O(k_{in}\frac{n}{m})$	$O(\frac{n}{m})$	$O(\frac{n}{m})$
Responsiveness		✗	✗	✗	✓	✓
Scalability		✗	✓	✓	✓	✓

1.2 Paper Organization

We introduce the background in Sect. 2. In Sect. 3, the system model and notations are shown. A multi-signature protocol is described in Sect. 4. Intra-shard consensus including a FBFT protocol and its security analysis is given in Sect. 5. Then RSTP is illustrated in Sect. 6. Related works are described in Sect. 7. Finally, we conclude this paper in Sect. 8.

2 Background

In this section, we introduce the background, namely the classification, features and key components of sharding blockchains.

Classification of Sharding Blockchains. Based on the underlying structure, sharding blockchains could be divided into instant and eventual ones. The difference is that in instant ones, there exists a committee running some intra-shard consensus within each shard to handle transactions and generate blocks. In this way, the generation of each block is immediately confirmed, achieving strong consistency. Eventual sharding blockchains, like Monoxide [20], do not use a committee in each shard, while still rely on PoW or proof-of-stake (PoS) to generate blocks directly. In this case, a certain number of blocks at the end of a blockchain must be removed to confirm transactions, which is known as weak consistency.

Features of Sharding Blockchains. In general, sharding blockchains have the following three characteristics. The first one is communication sharding. Participating nodes are divided into different shards where nodes in each shard only need internal communication most of the time. The second one is computation sharding, which means that each shard is only responsible for processing its corresponding transactions, so transactions are handled by different shards in parallel. When the number of nodes in the network increases, more shards could be added to achieve scalability. The third one is storage sharding, i.e., nodes of different shards only need to store their corresponding transactions. Each shard is only responsible for maintaining its own blockchain exclusive to this shard.

Key Components of Sharding Blockchains. There are several important components of sharding blockchains: node selection, node assignment, intra-shard consensus, cross-shard communication and shard reconfiguration. For permissionless sharding blockchains, node selection, node assignment, and shard reconfiguration are important to defend against Sybil Attack [21] and guarantee a target honest fraction inside every committee. While intra-shard consensus and cross-shard communication are essential parts in every sharding blockchain. In this paper, we mainly analyze and design these two parts.

3 System Model and Notations

In this section, we give our system model that consists of network and adversary model, and notations which are useful in later protocol construction.

3.1 System Model

Network Model. We assume a partially synchronous network. The actual network delay is δ and there is an upper bound Δ of the network delay. Since the concrete value of Δ is unknown, Δ could not be used directly in the design of the protocol.

Adversary Model. We assume $u = 3f+1$ in a committee where f is the number of malicious nodes and u denotes the total number of nodes. An adversary $\mathcal{A}$ could control all malicious nodes perfectly. The corruption process is mild, where it takes some time for $\mathcal{A}$ to control an honest node. $\mathcal{A}$ is responsible for transferring network messages. $\mathcal{A}$ could delay and reorder messages from honest nodes, yet constraint to the Δ limit. $\mathcal{A}$ can not forge a signature of an honest node.

3.2 Notations

There are m shards, namely $shard_1, \cdots, shard_m$ where $c \in \{1, \cdots, m\}$ denotes the shard sequence number. $utxo_c$ denotes the UTXO pool of $shard_c$. In each shard, there is a committee C_c that runs a intra-shard consensus FBFT to process transactions. PKlist_c denotes the public key list of committee C_c. u denotes the number of nodes in a committee while n represents the total number of nodes that participate the protocol. u, m and n satisfy $u = \frac{n}{m}$. We use k_{in} and k_{all} to indicate the average number of input shards and total shards (input shards plus output shards) that a transaction spans on average, respectively (used in Table 1). While we use $\mathsf{tx} = (I^1, \cdots, I^k, O^1, \cdots, O^{k'})$ to denote a specific transaction with k inputs and k' outputs (used in Sect. 6). Each shard maintains a ledger log including committed transactions.

4 BLS Multi-signature and Aggregation

In this section, we describe our robust (t, u)-MSP-PoP and (t, u)-AMSP-PoP scheme. The former one is useful in our FBFT protocol for intra-shard communication, while the latter is suitable for cross-shard communication.

4.1 Robust t Out of u Multi-signature with Public Key Aggregation

(t, u)-MSP-PoP (Fig. 1) makes three modification of the scheme in [18]: the first point is about a public key list PKlist used to resist rogue key attacks, we combine PoP and PoW to provide a consistent view of PKlist for all nodes without a trusted party. Since this change involves BFT, we introduce it in Sect. 5.1. Here we use a certifying authority (CA) to provide a public key registration [22]. The second one is separating share signature and multi-signature combination for easier calling in FBFT. The third one is that when verifying a multi-signature, a verifier should first verify the legality of the aggregation public key $apkt$ to prevent the leader from generating a multi-signature with less than t signature shares. A leader in our protocol is equivalent to a combiner in previous works. In (t, u)-MSP-PoP, we assume a leader to be honest, where in later FBFT protocol, a view-change mechanism could be adopted to change a malicious leader.

As a signature scheme, (t, u)-MSP-PoP only provides several algorithm interfaces which could be called by FBFT, while message interaction between nodes is not reflected in it. We add these message interaction in the following description.

Assume hash functions $\mathsf{H}_0 : \{0,1\}^* \to \mathbb{G}_2$ and $\mathsf{H}_1 : \{0,1\}^* \to \mathbb{G}_1$.

- **Parameter Generation.** $\mathsf{ParaGen}(\kappa)$: *// (for each participant P_i)*
  Sets up a bilinear group $(q, \mathbb{G}_1, \mathbb{G}_2, \mathbb{G}_t, e, g_1, g_2) \leftarrow \mathcal{G}(\kappa)$, sets $par \leftarrow (q, \mathbb{G}_1, \mathbb{G}_2, \mathbb{G}_t, e, g_1, g_2)$, and outputs par.
- **Key Generation.** $\mathsf{KeyGen}(par)$: *// (for each participant P_i)*
  Computes: $sk_i \xleftarrow{\$} \mathbb{Z}_q$, $pk_i \leftarrow g_2^{sk_i}$, $\pi_i \leftarrow \mathsf{H}_1(pk_i)^{sk_i}$. Then P_i sends (pk_i, π_i) to the CA. CA broadcasts $\mathsf{PKlist} = \{pk_1, pk_2, \cdots, pk_u\}$. Outputs (pk_i, sk_i).
- **Share Signature.** $\mathsf{ShareSign}(par, sk_i, m)$: *// (for each participant P_i)*
  Computes $s_i \leftarrow \mathsf{H}_0(m)^{sk_i}$. Outputs (s_i, m).
- **Multi-Signature Combination.** $\mathsf{Combine}(par, \mathsf{PKlist}, m, \{s_i, pk_i\}_{|t|})$: *// (for a leader P_l)*
  Sets $\mathsf{SIGlist} \leftarrow \varnothing$.
  For every s_i, if $(pk_i \in \mathsf{PKlist}) \wedge (e(s_i, g_2) \stackrel{?}{=} e(\mathsf{H}_0(m), pk_i))$: $\mathsf{SIGlist} \leftarrow \mathsf{SIGlist} \cup \{(s_i, pk_i)\}$.
  If $|\mathsf{SIGlist}| = t$, computes: $\sigma \leftarrow \prod_{\{s_i|(s_i,-)\in\mathsf{SIGlist}\}} s_i$, $apkt \leftarrow \prod_{\{pk_i|(-,pk_i)\in\mathsf{SIGlist}\}} pk_i$, $T \leftarrow [b_1, b_2, \cdots, b_u]$ where u elements in T correspond to u pk in PKlist, and for every $i = 1$ to u, if $(-, pk_i) \in \mathsf{SIGlist}$, sets $b_i \leftarrow 1$.
  Outputs $(\sigma, apkt, T, m)$.
- **Multi-Signature Verification.** $\mathsf{MulVer}(par, \mathsf{PKlist}, \sigma, apkt, T, m)$: *// (for a verifier P_v)*
  Verifies if $apkt \stackrel{?}{=} \prod_{\{i|T[i]=1\}} \mathsf{PKlist}[i]$. If it is, then verifies if $e(\sigma, g_2) \stackrel{?}{=} e(\mathsf{H}_0(m), apkt)$. If it is, outputs 1; otherwise, outputs 0.

Fig. 1. The robust (t, u)-MSP-PoP scheme.

To defend against the rogue public key attack proposed in [22], (t, u)-MSP-PoP requires every participant to generate a proof-of-possession π_i of his public key pk_i and send (pk_i, π_i) to the CA in the key generation phase. CA generates a valid public key list PKlist which is useful in the multi-signature verification phase since only pk on the list is regarded as a valid one to generate a $apkt$.

In the share signature phase, every node P_i computes a signature share s_i on a message m, and sends $(s_i, \mathsf{H}_0(m))$ instead of (s_i, m) to the leader to reduce the message length. In the multi-signature combination phase, a leader first verifies a signature share, then uses valid shares to compute a valid multi-signature σ and an aggregation public key $apkt$. Besides, a bitmap T indicating whose public key is used should be broadcast for other nodes to verify the validity of $apkt$. The leader P_l broadcasts $(\sigma, apkt, T, \mathsf{H}_0(m))$ instead of $(\sigma, apkt, T, m)$.

About multi-signature verification, a verifier computes if $apkt$ is generated by t valid pk on PKlist. This is to prevent a leader from launching a rogue public key attack, and guarantee that a leader could not use less than t signature shares to forge a (t, u)-multi-signature. This verification is crucial to the safety of the FBFT protocol where t is set to be $2f + 1$. $\mathcal{A}$ is able to control at most f nodes, he could not forge a valid $(2f + 1, u)$ multi-signature.

Since (t, u)-MSP-PoP just adds some verification to the original multi-signature protocol, completeness and unforgeability follow from the original protocol where concrete proof is referred to [18,22]. Completeness means that a legally generated multi-signature is sure to pass the verification. Unforgeability in (t, u)-MSP-PoP means that $\mathcal{A}$ controlling at most f ($f < t$) nodes could not forge a (t, u)-multi-signature on a message m that has not been signed by honest nodes. To make the multi-signature protocol better applied in FBFT, we define a robustness property.

Definition 1 (Robustness). *A multi-signature scheme is said to satisfy robustness if for any m, a valid (t, u)-multi-signature is sure to be generated.*

Theorem 1. *Assume that $\mathcal{A}$ could control at most f nodes where $f \leq u-t$, then the (t, u)-MSP-PoP scheme as per Fig. 1 satisfies robustness as per Definition 1.*

4.2 Multi-signature Aggregation

The robust (t, u)-AMSP-PoP scheme is shown in Fig. 2. In the multi-signature aggregation phase, a leader P_l is required to verify a multi-signature using MulVer(). Only valid multi-signatures are used to compute an aggregated signature Σ.

MulAgVer() takes in par, a public key and message set $\{apkt_i, T_i, m_i\}_{i=1}^k$, a signature Σ as inputs. For a node to verify an aggregated multi-signature, he should first verify the validity of each $apkt_i$. So he should keep a valid PKlist for every multi-signature group. In sharding blockchains, this means that a node should keep public key lists of other related shards.

- **Multi-Signature Aggregation.** $\mathsf{MulAgSign}(par, \{\mathsf{PKlist}\}_{i=1}^k, \{\sigma_i, apkt_i, T_i, m_i\}_{i=1}^k)$: // (for a leader P_l)
  Verifies if for every $i = 1$ to k, $\mathsf{MulVer}(par, \mathsf{PKlist}_i, \sigma_i, apkt_i, T_i, m_i) \stackrel{?}{=} 1$.
  If it is, computes $\Sigma \leftarrow \prod_{i=1}^k \sigma_i$ and outputs Σ.
- **Aggregated-Multi-Signature Verification.** $\mathsf{MulAgVer}(par, \{\mathsf{PKlist}\}_{i=1}^k, \Sigma, \{apkt_i, T_i, m_i\}_{i=1}^k)$: // (for a verifier P_v)
  For every $i = 1$ to k, verifies if: $apkt_i \stackrel{?}{=} \prod_{\{j|T_i[j]=1\}} \mathsf{PKlist}_i[j]$.
  If it is, then verifies if $e(\Sigma, g_2) \stackrel{?}{=} \prod_{i=1}^k e(\mathsf{H}_0(m_i), apkt_i)$.
  If it is, outputs 1; otherwise, outputs 0.

Fig. 2. The robust (t, u)-AMSP-PoP scheme.

Note that completeness, unforgeability, and robustness hold for protocol (t, u)-AMSP-PoP. The definition and proof are based on that of (t, u)-MSP-PoP, so we do not discuss it in detail here.

5 Intra-Shard Consensus

In this section, we describe our intra-shard consensus, leader-stable fast Byzantine fault tolerance. FBFT adopts the $(2f+1, 3f+1)$-MSP-PoP multi-signature to cut down the message length and pipelined technology to improve efficiency.

5.1 Regarding Sybil Attack and a Consistent View of PKlist

(t, u)-MSP-PoP exists as a multi-signature scheme alone, yet when it is applied to intra-shard consensus, the following two problems arise. One is that $\mathcal{A}$ might launch Sybil Attack by creating a great amount of public keys. The other is about PKlist. To ensure that nodes clearly know which public keys are involved in the aggregation when verifying a *apkt*, all nodes need to have a consistent view of PKlist. We make the following improvements to handle the above issues and realize a perfect combination of (t, u)-MSP-PoP and a sharding blockchain.

Permissioned Networks. In a permissioned network, a trusted CA is employed like in [22]. CA acts as a trusted third party to provide identity authentication. A node first sends its public key and proof (pk_i, π_i) to a CA who adds it to the PKlist if both the identity and public key are valid. When there are enough nodes, CA publishes PKlist, so that all nodes have a consistent view of PKlist.

Permissionless Networks. In a permissionless network, FleetChain could rely on PoW and a reference committee to defend against Sybil Attack and complete the proof-of-possession simultaneously. FleetChain extends the fair selection protocol proposed in [23] with two improvements.

A node who wants to participate the protocol tries to find a PoW solution using $h = \mathsf{H}(\xi, nonce, pk_i)$ where ξ is the current puzzle, $nonce \in \{0,1\}^\lambda$ denotes a potential solution, pk_i is the node's public key, and $\lambda \in \mathbb{N}$ is a security parameter. If $h < D$ where D is a target difficulty value, a node is required to send $(h, nonce, pk_i, \pi_i)$ (in the original scheme, node only sends $(h, nonce, pk_i)$) to members of a reference committee C_R. In addition to verifying if $h = \mathsf{H}(\xi, nonce, pk_i)$ and $h < D$, a reference committee member adds a verification on the public key: $e(\pi_i, g_2) = e(\mathsf{H}_1(pk_i), pk_i)$. If it is, sets $\mathsf{PKlist} \leftarrow \mathsf{PKlist} \cup \{pk_i\}$. When the number of nodes reaches a target, C_R runs a BFT algorithm to confirm PKlist and broadcasts it. In this way, all nodes obtain a consistent view of PKlist. Note that the reference committee itself could be generated through this fair selection protocol. Besides, in a sharding blockchain, the selection of multiple committees could be done by running multiple fair selection protocols sequentially. For a detailed security proof, please refer to [23].

Through the above measures, every node has a same view on m public key list $\mathsf{PKlist}_1, \mathsf{PKlist}_2, \cdots, \mathsf{PKlist}_m$ for m committees. On the basis of ensuring security, FleetChain focuses on intra-shard consensus and cross-shard communication, not paying much attention to issues such as committee configuration.

5.2 Leader-Stable Fast Byzantine Fault Tolerance

FBFT is specially designed for sharding blockchains, where the point is to process transactions and input availability in parallel. For transactions, only legal ones are proposed and processed. For input availability, members reach agreement on whether an input is available, and append a Boolean value b to indicate whether the input could be spent. FBFT could be modified into a general BFT protocol through simple changes, i.e., removing b and the processing of input availability.

As shown in Algorithm 1, FBFT employs a normal phase to process requests from clients. The tuple (v, r) is used to denote some round r of view v. P_l and $P_{l'}$ denote the leader in view v and $v+1$, respectively, which is decided by a round-robin method. req is a request that might be some transaction tx, or some input I to be confirmed. p denotes a proposal. MSC is a multi-signature certification including $(\sigma, apkt, T, \mathsf{H}(p))$.

IsValid(), as shown in Algorithm 2, is a predict to verify if the request is a valid one, depending on the type of req. For a req whose type is "tx", IsValid() verifies the legality using relative $\mathcal{TC}$. Only valid transactions are able to pass through the verification of IsValid() and processed by FBFT. Illegal transactions get no vote from nodes. If $req.type$ equals to "input", IsValid() queries the availability state in the local UTXO pool. IsValid() returns 1 or 0 for available or unavailable, respectively. FBFT sets $req||b$ as a new proposal p_r. Different from transactions, no matter whether an input state is available, it will be processed by FBFT.

When the leader of the protocol is offline due to network problems or behaves maliciously, FBFT relies on view-change to step into a new view where a new leader is in charge. Concrete operations are illustrated in view-change phase.

We give the security definition of FBFT. Due to space limit, the detailed proofs of Theorem 2 and Theorem 3 is omitted.

Definition 2 (A secure FBFT). FBFT *is said to be secure if and only if it satisfies the following properties.*

- *Agreement. For any two honest nodes, if they output two committed* p *and* p' *in the same* (v, r), *then* $p = p'$.
- *Liveness. After a valid request req is submitted, there must be a corresponding* $p = req||b$ *committed and output by all honest nodes after a certain time.*

Theorem 2. FBFT *satisfies the agreement property as per Definition 2.*

Theorem 3. FBFT *satisfies the liveness property as per Definition 2.*

6 Cross-shard Communication

In this section, we describe our RSTP protocol. RSTP could efficiently process intra-shard and cross-shard transactions through invoking FBFT.

6.1 Responsive Sharding Transaction Processing

Cross-shard communication is as hard as fair exchange, as described in [24]. Due to the existence of a large fraction of cross-shard transactions, the processing of these transactions is important for sharding blockchains. The most commonly used and safest processing method is the 2PC method. In the prepare phase, each input shard is required to verify the input availability and lock the input,

Algorithm 1: FBFT (fast Byzantine fault tolerance)

Input: a request *req*.
Output: a committed tuple (MSC, p).
▷ **Normal Phase:**
for any (v, r),
▷ as a **leader** P_l:
  sets $p_r \leftarrow \perp$;
  on receiving a *req*,
    if *req.type* = "tx", then:
      if $\mathsf{IsValid}(req) = 1$, then sets $b \leftarrow 1$, $p_r \leftarrow req||b$; else, ignores it; // *(only valid transactions are processed by* FBFT*)*
    if *req.type* = "input", then:
      sets $b \leftarrow \mathsf{IsValid}(req)$, $p_r \leftarrow req||b$; // *(an input is sure to be processed by* FBFT *no matter if it is available)*
  on receiving $2f + 1$ valid vote messages m_v:
    parses them as $\{(\text{"vote"}, \mathsf{H}(p_{r-1}), s_i)\}_{|2f+1|}$;
    computes $(\sigma, apkt, T, p_{r-1}) \leftarrow \mathsf{Combine}(par, \mathsf{PKlist}, p_{r-1}, \{s_i, pk_i\}_{|2f+1|})$;
    sets $\mathsf{MSC}_{r-1} \leftarrow (\sigma, apkt, T, \mathsf{H}(p_{r-1}))$;
    constructs $m_{np} \leftarrow (\text{"new-proposal"}, p_r, \mathsf{MSC}_{r-1})$ and broadcasts m_{np}.
▷ as a **normal node** P_i or a **leader**:
  on receiving a new proposal message m_{np}:
    parses m_{np}, then gets $\mathsf{MSC}_{r-1} = (\sigma, apkt, T, h)$ and local $p_{r-1} = req||b$;
    if $(h \stackrel{?}{=} \mathsf{H}(p_{r-1})) \wedge (\mathsf{MulVer}(par, \mathsf{PKlist}, \sigma, apkt, T, p_{r-1}) \stackrel{?}{=} 1)$, then:
      sets $p_{r-1}.state \leftarrow$ prepared, $p_{r-2}.state \leftarrow$ committed; // *(executes state change if the multi-signature is valid)*
      outputs $(\mathsf{MSC}_{r-2}, p_{r-2})$; // *(*MSC *serves as a proof for a committed p)*
      if $\mathsf{IsValid}(req) \stackrel{?}{=} b$, mark m_{np} as "valid", then: // *(useful in view-change)*
        if P_i has not voted in (v, r), then:
          $s_i \leftarrow \mathsf{ShareSign}(par, sk_i, p_r)$; $m_v \leftarrow (\text{"vote"}, \mathsf{H}(p_r), s_i)$;
          sends m_v to P_l and starts a timer *timeout*.
▷ **View-Change Phase**:
▷ as a **normal node** P_i:
  in (v, r), if $((timeout \text{ ends}) \wedge (\text{receives no "valid" } m_{np} \text{ from } P_l)) \vee (\text{receives conflicting } m_{np} \text{ and } m'_{np} \text{ both signed by } P_l)$, then:
    sets $r' \leftarrow \max_{\{r|p_r=\text{prepared}\}}(r)$ // *(r' is the round number of the most recently prepared p_r)*
    constructs $m_{vc} \leftarrow (\text{"view-change"}, v + 1, r', p_{r'}, \mathsf{MSC}_{r'})$;
    sends m_{vc} to $P_{l'}$.
▷ as a new **leader** $P_{l'}$:
  sets $\mathcal{VC} \leftarrow \emptyset$; sets p_0 exactly like setting p_r in the normal phase;
  on receiving a valid view-change messages m_{vc}:
    sets $\mathcal{VC} \leftarrow \mathcal{VC} \cup \{(r', p_{r'}, \mathsf{MSC}_{r'})\}$; // *($\mathcal{VC}$ is the certification for the new leader to launch a view-change)*
    if $|\mathcal{VC}| = 2f + 1$, then:
    lets Φ denote a tuple in $\mathcal{VC}$; sets $(r^*, p_{r^*}, \mathsf{MSC}_{r^*}) \leftarrow \operatorname{argmax}_{\Phi \in \mathcal{VC}}(\Phi.r)$;
      constructs $m_{nv} \leftarrow (\text{"new-view"}, v + 1, p_{r^*}, \mathsf{MSC}_{r^*}, \mathcal{VC}, p_0)$;
      broadcasts m_{nv}.

Algorithm 2: The predict IsValid

```
Input: a request req.
Output: {0,1}.
if req.type =? "tx":
    if tx is a intra-shard transaction:
        if every input of tx is available, then:
            returns 1, and for every I^i ∈ shard_c, removes I^i from utxo_c;
        else, returns 0.
    else, i.e., tx is a cross-shard transaction:
        finds the corresponding (Σ, TC) related to tx.id;
        parses TC as {(apkt^i, T^i, p^i)}_|k| and every p^i as I^i||H(tx.id)||b^i;
        sets {PKlist_c}_|k| to be all k input shard public key list of tx;
        if (MulAgVer(par, {PKlist_c}_|k|, Σ, TC) =? 1) ∧ (for every i = 1 to k, b^i = 1):
            returns 1; else returns 0.
else, i.e., req.type =? "input", parses req as I^i||H(tx.id), and verifies:
    if utxo_c[I^i].state =? available:
        returns 1 and sets utxo_c[I^i].state ← locked; // (lock an input)
    else returns 0.
```

while in the commit phase, the transaction is committed and the previously locked input is unlocked. When verifying the input availability, if it is only a leader who makes the judgment and returns the result, then a malicious leader might censor the transaction or provide a wrong result. Hence, to ensure the correctness of cross-shard transactions, it is required that a committee instead of a single leader to run BFT to provide an input availability certification.

The concrete RSTP protocol is shown in Algorithm 3 and Fig. 3. In RSTP, the operations of input and output shard are separate, while the actual situation is that for a transaction, a certain shard might be both an input shard and an output shard. Note that when we call $\mathsf{FBFT}(I^i||\mathsf{H}(\mathsf{tx}.id))$, we require the concatenation of the input address I^i and the hash value of tx's ID to prevent an adversary from launching a replay attack against sharding blockchains, while the details of this kind of attack is analyzed in [16].

We give an definition for a sharding blockchain to be secure in Definition 3. Then it is described that RSTP satisfies consistency in Theorem 4, and liveness in Theorem 5. The detailed proofs are omitted due to space limit.

Definition 3 (Secure RSTP). *RSTP is said to be secure if the following properties are satisfied.*

- *Consistency.*
  - *Common prefix in one shard. For any two honest nodes of a same shard who output* log *and* log′ *at time t and t', respectively, it holds that either* $\mathsf{log} \prec \mathsf{log}'$ *or* $\mathsf{log}' \prec \mathsf{log}$. *"$\prec$" means "is a prefix of".*

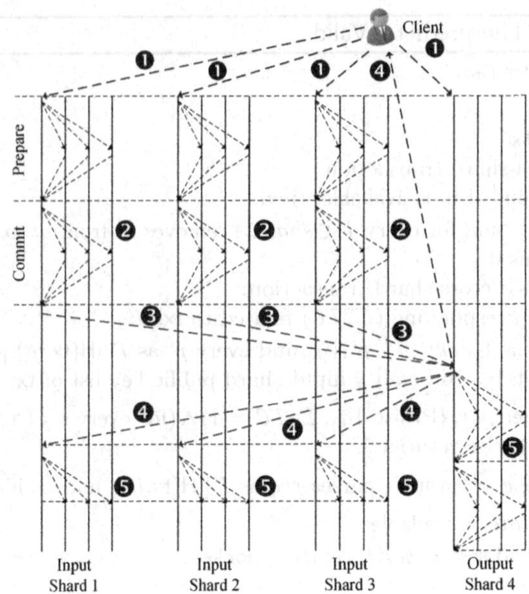

Fig. 3. The transaction processing in RSTP where $k = 3, k' = 1$. 1. A client submits a tx to all related shards' leaders; 2. All input shards run FBFT to verify the input availability of tx; 3. Every input shard's leader sends the result with a multi-signature (MSC^i, p^i) to the first output shard leader; 4. The output shard leader computes an aggregated multi-signature, and sends $(\Sigma, \mathcal{TC})$ to the client as well as all input shards' leaders. 5. Output shard runs FBFT to write tx onto log, while input shards update their UTXO states according to $(\Sigma, \mathcal{TC})$.

- *No conflict between shards. For any two honest nodes of two different shards who output* log *and* log′*, and any two* tx ∈ log *and* tx′ ∈ log′*, if* tx ≠ tx′*, then* tx ∩ tx′ = ∅.

– *Liveness. If a valid* tx *is submitted by a client, then after a certain time,* tx *must appear in the* log *output by all honest nodes of all output shards.*

Theorem 4. RSTP *satisfies consistency as per Definition 3.*

Theorem 5. RSTP *satisfies liveness as per Definition 3.*

6.2 Regarding Optimal Sharding

We follow the model in [19] to analyze the sharding extent of FleetChain, including computational, communication and storage complexity, and scaling factor.

Computational Complexity ω_c: during a protocol execution, the total number of verification times of all parties. In Omniledger, for every transaction, a proof-of-acceptance contains k_{in} signatures which is linear to the number of input shards.

Algorithm 3: RSTP (responsive sharding transaction processing)

▷ **Input Shard Operations:**
▷ as a **leader** of **an input** $shard_c$, $c \in \{1, \cdots, m\}$:
  on receiving a cross-shard tx $= (I^1, \cdots, I^k, O^1, \cdots, O^{k'})$:
    for every $i = 1$ to k, if $I^i \in shard_c$, calls $\mathsf{FBFT}(I^i||\mathsf{H}(\mathsf{tx}.id))$.
  on receiving (MSC^i, p^i) from FBFT:
    sends (MSC^i, p^i) to the shard leader of O^1.
  on receiving $(\Sigma, \mathcal{TC})$, broadcasts $(\Sigma, \mathcal{TC})$ inside $shard_c$;
▷ as a **normal node** or a **leader** of an **input** $shard_c$:
  on receiving $(\Sigma, \mathcal{TC})$:
    parses $\mathcal{TC}$ as $\{(apkt^i, T^i, p^i)\}_{|k|}$, and parses every p^i as $I^i||\mathsf{H}(\mathsf{tx}.id)||b^i$.
    if $\mathsf{MulAgVer}(par, \{\mathsf{PKlist}_c\}_{|k|}, \Sigma, \mathcal{TC}) \stackrel{?}{=} 1$, then:
      if for $i=1$ to k, $b^i \stackrel{?}{=} 1$ then for every $I^i \in shard_c$, removes I^i from $utxo_c$;
      else for every $I^i \in shard_c$, if its related $b^i \stackrel{?}{=} 1$, then:
        sets $utxo_c[I^i].state \leftarrow$ available. *// (unlocks UTXO states)*
    else, ignores it.
▷ **Output Shard Operations:**
▷ as a **leader** of an **output shard**:
  on receiving (MSC^i, p^i) from $shard_c$ for a same tx.id: *//(leader of O^1)*
    parses MSC^i as $(\sigma^i, apkt^i, T^i, h^i)$, and parses p^i as $I^i||\mathsf{H}(\mathsf{tx}.id)||b^i$;
    if $((I^i \in shard_c) \wedge (h^i \stackrel{?}{=} \mathsf{H}(p^i)) \wedge (\mathsf{MulVer}(par, \mathsf{PKlist}_c, \sigma^i, apkt^i, T^i, p^i) \stackrel{?}{=} 1)$:
      sets $\mathcal{TC} \leftarrow \mathcal{TC} \cup \{(apkt^i, T^i, p^i)\}$. *// (collects all input multi-signatures)*
  if $|\mathcal{TC}| = k$: *// (aggregates them and forwards it to relevant shards)*
    sets $\{\mathsf{PKlist}_c\}_{|k|}$ to be all k input shard public key list of tx;
    computes $\Sigma \leftarrow \mathsf{MulAgSign}(par, \{\mathsf{PKlist}_c\}_{|k|}, \{\sigma^i, apkt^i, T^i, p^i\}_{\{i|p^i \in \mathcal{TC}\}})$;
    sends $(\Sigma, \mathcal{TC})$ to all leaders of the input and output shards of tx;
    calls FBFT(tx) to commit tx.
  on receiving $(\Sigma, \mathcal{TC})$: *// (other output shards handle cross-shard tx)*
    parses $\mathcal{TC}$ as $\{(apkt^i, T^i, p^i)\}_{|k|}$ and every p^i as $I^i||\mathsf{H}(\mathsf{tx}.id)||b^i$;
    sets $\{\mathsf{PKlist}_c\}_{|k|}$ to be all k input shard public key lists of tx;
    if $(\mathsf{MulAgVer}(par, \{\mathsf{PKlist}_c\}_{|k|}, \Sigma, \mathcal{TC}) \stackrel{?}{=} 1) \wedge$ (for every $i=1$ to k, $b^i = 1$):
      broadcasts $(\Sigma, \mathcal{TC})$ to shard members and calls FBFT(tx) to commit tx.
  on receiving an intra-shard tx: *// (directly processes intra-shard tx)*
    broadcasts tx to shard members, and calls FBFT(tx).
▷ as a **normal node** or a **leader** of an **output shard**:
  on receiving committed (MSC, p) from FBFT:
    parses p as tx$||b$, sets local log $\leftarrow$ log$||$tx, and outputs log.

In a committee, there are $\frac{n}{m}$ members. For a cross-shard tx, all $\frac{n}{m}$ members have to verify k_{in} signatures. So the communication overhead is $O(k_{in}\frac{n}{m})$. The case is also true for RapidChain and Chainspace. In FleetChain, k_{in} multi-signatures are aggregated into a single signature, hence ω_c equals to $O(\frac{n}{m})$.

Communication Complexity ω_m: the average number of messages per participant in a protocol execution. The calculation of ω_m is similar to ω_c, except that to commit a cross-shard transaction, k_{all} shards should participate the cross-shard communication. Hence, ω_m equals to $O(k_{all}\frac{n}{m})$ for Omniledger, RapidChain and Chainspace, while $\omega_m = O(\frac{n}{m})$ for FleetChain since an output shard behaves as a coordinator and multi-signature aggregation is employed.

Storage Complexity ω_s: the number of stable transactions included in every node's local ledger over the total number of transactions. In Omniledger, Chainspace and FleetChain, every shard only needs to store its own blockchain, so the storage complexity is $O(\frac{n}{m})$. While the number of transactions created by RapidChain is k_{in} times larger than the original one. So it is $O(k_{in}\frac{n}{m})$ for RapidChain.

Scaling Factor. The scaling factor for a sharding blockchain is $\Lambda = \frac{n}{\max(\omega_c, \omega_m, \omega_s)}$. The scaling factor for FleetChain could then be calculated as $\Lambda = \frac{n}{O(\frac{n}{m})} = \Omega(m) = O(\frac{n}{\log n})$. The specific proof of the last equation $m = O(\frac{n}{\log n})$ is referred to [19]. Intuitively, to guarantee that there are enough honest nodes in each committee, the members of each committee can not be less than a certain value, that is, the number m of committees should not exceed a certain value.

7 Related Work

Byzantine Fault Tolerance. Castro and Liskov [25] propose the practical Byzantine fault tolerance (PBFT) protocol. PBFT allows a certain number of Byzantine nodes in the network. These nodes might behave arbitrarily in the process of consensus while honest nodes follow the protocol. PBFT achieves agreement through three phases, namely pre-prepare, prepare and commit, in the case that the fraction of malicious nodes does not exceed 1/3 of the total number of nodes. When a primary node fails to process a request in time, the other backup nodes initiate a view-change phase. HotStuff [26] utilizes three-round votes to integrate view-change into general phase, realizing a linear view-change to reduce the communication complexity. In addition, HotStuff uses threshold signature to combine votes into a single signature. Scalable Byzantine fault tolerance [27] uses a leader as a collector of the signature and is compatible with smart contracts.

Instant Sharding Blockchains. Elastico [28] first applies sharding technology to the blockchain, yet it does not support atomic cross-shard transactions. Omniledger [15] employs an adaption [29] of ByzCoin as an intra-shard consensus algorithm. Omniledger proposes Atomix to process cross-shard transactions where a client is regarded as a coordinator. A malicious client could lock the inputs by not responding to other shards, while a malicious leader could provide a false proof-of-acceptance to cause inconsistency. Besides, the construction of a proof-of-rejection is not given. RapidChain [30] adopts a synchronous Byzantine fault tolerance protocol [31] in its committees which could tolerate no more than

1/2 malicious nodes. As for cross-shard transactions, RapidChain splits such a transaction into multiple single-input single-output transactions and commits them in order. Yet this increases the total number of transactions. Besides, this method is not general, and it is not given how to deal with transactions with multiple outputs. Chainspace [17] proposes S-BAC combining 2PC and BFT to process cross-shard transactions, while cross-shard communication is complicated and is vulnerable to replay attacks. Ostraka [32] uses a node differential environment model, i.e., each node's ability to process transactions is different. Dang et al. [33] use trusted hardware Intel SGX to generate randomness and facilitate the efficiency of BFT.

Eventual Sharding Blockchains. Monoxide [20] proposes atomic transfer using a relay transaction to process a cross-shard transaction. Chu-ko-nu mining is designed to defend against the 1% attack where an adversary focuses his computational power on a single shard.

8 Conclusion

In this paper, we design a new sharding blockchain FleetChain focusing on intra-shard consensus and cross-shard communication. A robust (t, u) BLS multi-signature is first given. Then FBFT is designed to improve transaction processing efficiency, which combines multi-signature with the pipeline technology. In addition, RSTP is proposed to handle cross-shard transactions. FleetChain achieves optimal sharding and responsiveness.

Acknowledgment. The authors would like to thank Prof. Fritz Henglein and Marcos Antonio Vaz Salles for their valuable comments. This paper is supported by the National Key R&D Program of China through project 2017YFB1400702 and 2017YFB0802500, the National Cryptography Development Fund through project MMJJ20170106, the Natural Science Foundation of China through projects 61932014, 61972018, 61972019, 61932011, 61772538, 61672083, 61532021, 61472429, 91646203, 61402029, 61972017, 61972310, the foundation of Science and Technology on Information Assurance Laboratory through project 61421120305162112006, the China Scholarship Council through project 201906020015.

References

1. Nakamoto, S., et al.: Bitcoin: a peer-to-peer electronic cash system (2008). https://bitcoin.org/bitcoin.pdf
2. Bano, S., Al-Bassam, M., Danezis, G.: The road to scalable blockchain designs. Login **42**(4), 31–36 (2017)
3. Conti, M., Kumar, E.S., Lal, C., Ruj, S.: A survey on security and privacy issues of bitcoin. IEEE Commun. Surv. Tutor. **20**(4), 3416–3452 (2018)
4. Bonneau, J.: Why buy when you can rent? Bribery attacks on bitcoin-style consensus. In: FC 2016, pp. 19–26 (2016)
5. Heilman, E., Kendler, A., Zohar, A., Goldberg, S.: Eclipse attacks on bitcoin's peer-to-peer network. In: USENIX Security 2015, pp. 129–144 (2015)

6. Eyal, I.: The miner's dilemma. In: SP 2015, pp. 89–103 (2015)
7. Nayak, K., Kumar, S., Miller, A., Shi, E.: Stubborn mining: generalizing selfish mining and combining with an eclipse attack. In: EuroS&P 2016, pp. 305–320 (2016)
8. Liu, Y., Hei, Y., Xu, T., Liu, J.: An evaluation of uncle block mechanism effect on Ethereum selfish and stubborn mining combined with an eclipse attack. IEEE Access **8**, 17489–17499 (2020)
9. Bag, S., Ruj, S., Sakurai, K.: Bitcoin block withholding attack: analysis and mitigation. IEEE Trans. Inf. Forensics Secur. **12**(8), 1967–1978 (2017)
10. Garay, J., Kiayias, A., Leonardos, N.: The bitcoin backbone protocol: analysis and applications. In: Oswald, E., Fischlin, M. (eds.) EUROCRYPT 2015. LNCS, vol. 9057, pp. 281–310. Springer, Heidelberg (2015). https://doi.org/10.1007/978-3-662-46803-6_10
11. Pass, R., Seeman, L., Shelat, A.: Analysis of the blockchain protocol in asynchronous networks. In: Coron, J.-S., Nielsen, J.B. (eds.) EUROCRYPT 2017. LNCS, vol. 10211, pp. 643–673. Springer, Cham (2017). https://doi.org/10.1007/978-3-319-56614-6_22
12. Poon, J., Dryja, T.: The bitcoin lightning network: scalable off-chain instant payments (2016). https://www.bitcoinlightning.com/wp-content/uploads/2018/03/lightning-network-paper.pdf
13. Dziembowski, S., Eckey, L., Faust, S., Malinowski, D.: Perun: virtual payment hubs over cryptocurrencies. In: SP 2017, pp. 327–344 (2017)
14. Liu, Y., Liu, J., Zhang, Z., Xu, T., Yu, H.: Overview on consensus mechanism of blockchain technology. J. Cryptologic Res. **6**(4), 395–432 (2019)
15. Kokoris-Kogias, E., Jovanovic, P., Gasser, L., Gailly, N., Syta, E., Ford, B.: Omniledger: a secure, scale-out, decentralized ledger via sharding. In: SP 2018, pp. 583–598 (2018)
16. Sonnino, A., Bano, S., Al-Bassam, M., Danezis, G.: Replay attacks and defenses against cross-shard consensus in sharded distributed ledgers (2019). CoRR abs/1901.11218
17. Al-Bassam, M., Sonnino, A., Bano, S., Hrycyszyn, D., Danezis, G.: Chainspace: a sharded smart contracts platform. In: NDSS 2018, pp. 18–21 (2018)
18. Boldyreva, A.: Threshold signatures, multisignatures and blind signatures based on the gap-Diffie-Hellman-group signature scheme. In: Desmedt, Y.G. (ed.) PKC 2003. LNCS, vol. 2567, pp. 31–46. Springer, Heidelberg (2003). https://doi.org/10.1007/3-540-36288-6_3
19. Avarikioti, G., Kokoris-Kogias, E., Wattenhofer, R.: Divide and scale: formalization of distributed ledger sharding protocols (2019). CoRR abs/1910.10434
20. Wang, J., Wang, H.: Monoxide: scale out blockchains with asynchronous consensus zones. In: NSDI 2019, pp. 95–112 (2019)
21. Douceur, J.R.: The sybil attack. In: Druschel, P., Kaashoek, F., Rowstron, A. (eds.) IPTPS 2002. LNCS, vol. 2429, pp. 251–260. Springer, Heidelberg (2002). https://doi.org/10.1007/3-540-45748-8_24
22. Ristenpart, T., Yilek, S.: The power of proofs-of-possession: securing multiparty signatures against rogue-key attacks. In: Naor, M. (ed.) EUROCRYPT 2007. LNCS, vol. 4515, pp. 228–245. Springer, Heidelberg (2007). https://doi.org/10.1007/978-3-540-72540-4_13
23. Liu, Y., Liu, J., Zhang, Z., Yu, H.: A fair selection protocol for committee-based permissionless blockchains. Comput. Secur. **91**, 101718 (2020)
24. Zamyatin, A., et al.: SoK: communication across distributed ledgers. IACR Cryptology ePrint Archive 2019, 1128 (2019)

25. Castro, M., Liskov, B.: Practical Byzantine fault tolerance. In: OSDI 1999, pp. 173–186 (1999)
26. Yin, M., Malkhi, D., Reiter, M.K., Golan-Gueta, G., Abraham, I.: HotStuff: BFT consensus with linearity and responsiveness. In: PODC 2019, pp. 347–356 (2019)
27. Golan-Gueta, G., et al.: SBFT: a scalable and decentralized trust infrastructure. In: DSN 2019, pp. 568–580 (2019)
28. Luu, L., Narayanan, V., Zheng, C., Baweja, K., Gilbert, S., Saxena, P.: A secure sharding protocol for open blockchains. In: ACM SIGSAC 2016, pp. 17–30 (2016)
29. Kokoris-Kogias, E.: Robust and scalable consensus for sharded distributed ledgers. IACR Cryptology ePrint Archive 2019, 676 (2019)
30. Zamani, M., Movahedi, M., Raykova, M.: RapidChain: scaling blockchain via full sharding. In: CCS 2018, pp. 931–948 (2018)
31. Ren, L., Nayak, K., Abraham, I., Devadas, S.: Practical synchronous Byzantine consensus. IACR Cryptology ePrint Archive 2017, 307 (2017)
32. Manuskin, A., Mirkin, M., Eyal, I.: Ostraka: secure blockchain scaling by node sharding (2019). CoRR abs/1907.03331
33. Dang, H., Dinh, T.T.A., Loghin, D., Chang, E., Lin, Q., Ooi, B.C.: Towards scaling blockchain systems via sharding. In: SIGMOD 2019, pp. 123–140 (2019)

DSBFT: A Delegation Based Scalable Byzantine False Tolerance Consensus Mechanism

Yuan Liu[1,2](✉), Zhengpeng Ai[1], Mengmeng Tian[1], Guibing Guo[1], and Linying Jiang[1]

[1] Software College of Northeastern University, Shenyang, China
liuyuan@swc.neu.edu.cn

[2] Key Laboratory of Data Analytics and Optimization for Smart Industry (Northeastern University), Ministry of Education, Shenyang, China

Abstract. Blockchain, a trustless decentralized ledger and cryptography based technology, is a fast-growing computing paradigm, born with Bitcoin. Benefit from its core features, a number of applications have arisen in various application fields, e.g. finance, healthcare, supply chain and so on. A consensus mechanism is a foundational component of a blockchain-based system. The existing mechanisms rare achieve high efficiency and high scalability simultaneously. In this paper, we propose a hybrid consensus mechanism, a delegation based scalable byzantine false tolerant (DSBFT), which combines PoW and BFT to achieve a well-balanced between efficiency and scalability. Our mechanism has been shown to achieve linear scalability and sustain security even when more than 1/3 computing power is byzantine malicious.

Keywords: Blockchain · Byzantine false tolerance · Consensus mechanism · Delegation

1 Introduction

Since the present of BitCoin in 2009, the first decentralized cryptocurrency [11], many altcoins have been generated. By 13 February 2019, the number of cryptocurrency available over the Internet is over 2017 and still keep growing [3], and BitCoin owns the largest market capitalization, taking about 53%. Blockchain, as the core technology supporting Bitcoin and similar altcoins, has been considered to benefit not only finance but also politics, healthcare, social, production industrial and scientific fields [16]. The Blockchain potentially become the next major disruptive technology and worldwide computing paradigm (following the mainframe, PC, Internet, and social networking and mobile phones) to reconfigure all human activities as pervasively as did the Web [15,19].

A blockchain is a distributed ledger that cryptographically links a growing list of records, called blocks, where each block contains a cryptographic hash of the previous block, a timestamp, and transaction data(organized in the way of

M. Qiu (Ed.): ICA3PP 2020, LNCS 12454, pp. 426–440, 2020.
https://doi.org/10.1007/978-3-030-60248-2_29

a Merkle tree) [12]. According to the extent of decentralization, the blockchain systems are divided into three categories, namely permission networks, protected networks, and permissionless networks. There are also researchers who interpret blockchain systems following a two-level architecture, i.e. immutable data organization layer and data deployment and maintenance layer, respectively [18].

With the fast growth for high-quality and high-security services in need, which are delivered by blockchain systems, it presents many critical challenges in designing blockchain protocols. Especially, the effectiveness and efficiency of a blockchain system significantly depends its adopted consensus mechanism, e.g. data throughput, scalability, speed of consensus confirmation, security against maliciously behaving nodes (i.e. Byzantine nodes). According to FLP impossibility theorem [4] and CAP theorem [5], with a constraint network bandwidth, the high scalability and high throughput (efficiency) of consensus protocols cannot be satisfied at the same time. Traditional Byzantine false tolerant (BFT) consensus algorithm has high throughput, but its communication complexity increases exponentially or polynomially increase with the number of the nodes [8,14]. Besides, because of the constraint of a fixed network bandwidth, the number of nodes engaging in a BFT consensus cannot be too large, and nodes can't move in and out dynamically. On the other hand, Proof of work (PoW [11]) based consensus algorithms have good scalability with a communication complexity of consensus being $O(n)$. In PoW, nodes can join or leave the network at will. However, PoW bears low transaction throughput at about seven transactions a second. The performance of a hybrid consensus which combines the above two algorithms could be a good practice, which motivates us to propose a scalable DBFT protocol in this paper.

In our mechanism, the consensus procedure contains five steps. In the first step, a committee is elected based on PoW to avoid identity abuse and then confirmed through publishing their identities within the whole network in the second step. In the third step, the committee achieves the consensus agreement over new created block through BFT based protocol, and each agreed block is then broadcast to the whole network in the fourth step. In the fifth step, the new block is confirmed and updated among the whole network. Our main contributions are summarized as follows.

- A committee election based delegation is proposed in a decentralized PoW manner, where the proper committee size is investigated by considering the proportion of Byzantine malicious nodes.
- A hybrid consensus mechanism DSBFT based on PoW and BFT is proposed, which is potential to achieve well performance in efficiency and scalability in a compatible manner with the existing protocols.
- We show that the proposed mechanism has $O(n)$ scalability and can tolerate more than $1/3$ malicious nodes.

2 Related Works

From the perspective of the design of distributed systems, the consensus is also regarded as state machine replication, which has been extensively studied for more than 30 years [13]. Blockchain systems can be classified into three architectures, namely permission, permissionless, and protected (between permission and permissionless) networks. In permission network, to consider Byzantine nodes, the classical BFT consensus protocols such as PBFT for reaching the consensus among a small group of authenticated nodes. As pointed out in [17], permission consensus protocols rely on a semi-centralized consensus framework and a higher messaging overhead to provide immediate consensus finality and thus high transaction processing throughput. Because the communication complexity of such consensus protocols is at least exponential, they can only be applied in network consensus of small-scale nodes.

On the contrary, in a permissionless blockchain network, a Byzantine agreement in Nakamoto type is achieved by combining a series of cryptographic techniques, for example, cryptographic puzzle systems [1,6], and incentive mechanism design. Permissionless consensus protocols are more suitable for a blockchain network with loose control in synchronizing behaviors of the nodes. In the case of bounded delay with the majority being trustworthy, permissionless consensus protocols provide significantly better scalability but at the cost of lower processing efficiency.

Delegated Byzantine Fault Tolerance (dBFT) is a variant of standard BFT, described in the NEO white paper [20], this fault tolerance algorithm splits clients within nonoverlapped types: bookkeepers and ordinary nodes. Ordinary nodes do not take part in determining the network consensus but vote on which bookkeeper node it wishes to support, which is the meaning of the term "delegated". The bookkeeper nodes that were successfully elected are then participating in the consensus process. We adopt the "delegation" concept in this study and also specify how to set the committee size according to the network status, for example, the proportion of Byzantine computing power, and the network size.

A closely related hybrid consensus mechanism is proposed in [13], which the fast classic permission Byzantine consensus protocols, i.e. Nokamoto PoW, are employed to bootstrap the network consensus and then secured permissionless protocols, i.e. BFT, are utilized to achieve the high scalability. Our work is also a hybrid consensus protocol which joins the ideas of permission and permissionless consensus protocols, however, our protocol is more than a combination of the classic consensus protocols from the following two aspects. First, our mechanism itself is a novel protocol and it can run in the bootstrap stage and sequential stage. Second, it can achieve 1/3 byzantine nodes in proportion by properly setting our model parameters.

3 DSBFT Mechanism

In a blockchain network, a certain number of distributed nodes or users aims to generate new data blocks in a consistent way, through sending and receiving messages among them. The nodes involving in the process of consensus mechanisms are also called "miners", and we will use users and nodes interchangeably in following description. In our system, each node can behavior as a *competitor* or a *follower*, and all the competitors form the *campaign set*. A competitor participates the system consensus by generating one or multiple committee identities. The committee identities exchange messages to achieve the committee agreed block through an improved BFT protocol and then the agreed block is consensus around the whole network. In this way, the distributed nodes finally hold the same block status. After a series of blocks are committed, the campaign set is updated so as to sustain a secure and reliable system in runtime. It is worthy to be noted that our mechanism allows open accessed and nodes can join and leave our system at any time.

The outline of the mechanism is presented as in Fig. 1. Specifically, the consensus goes in epochs. Each epoch can be divided into 5 steps: *committee generation*, *committee confirmation*, *committee consensus*, *network consensus*, *campaign set updation*, and the main workflow is also presented as in Fig. 1. At the beginning of each period, a set of qualified nodes form campaign set and the nodes in this set are eligible to participate in the committee identity generation through solving a PoW puzzle. When a certain size of committee identities is generated, the identities form a network-recognized committee set through the committee confirmation process. After the committee agrees on a block with the improved BFT algorithm, the block is broadcast to the whole network. Each node in the network is able to verify blocks by validating the original signatures. The committee would continue to create new blocks until the number of created blocks reaches a predetermined number or the time reaches a predetermined time constraint. Then the campaign set is updated to continue the next epoch.

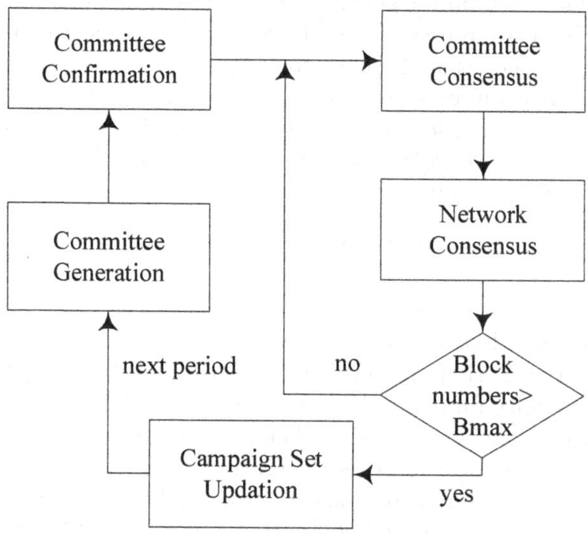

Fig. 1. An overview of the proposed DSBFT

The symbols and concepts we used in the description are summarized as follows. Assume there are N nodes who are actively participating in the network status consistency. In an epoch k, the campaign set is denoted by C_k which contains n_k nodes. The committee identities generated by the campaign set is denoted by D_k which contains m_k identities. Among the n_k nodes, we assume the computing power contributed by Byzantine nodes in the campaign set takes for α_k in proportion.

3.1 Committee Generation

Nodes in the *campaign set* join the committee by solving PoW puzzles. When a node successfully resolves a PoW puzzle, we call the node to create an identity. The node joins the committee through this identity. An identity is a vector consisting of three pieces of information, i.e. $ID_i = (IP_i, PK_i, T_i, Nonce_i)$ where IP_i is the Internet address of the node i who creates the identity; PK_i represents a public key that node i assigns the identity to be recognized by other identities and nodes; T_i is the time when the node i creates this identity; and $Nonce_i$ is the result of PoW puzzle. Through verifying the correctness of $Nonce_i$, any node is able to validate whether the affiliated identity is acceptable or not. The verification process is similar to that in PoW where the calculated hash value with $Nonce_i$ is examined through comparing with the needed difficulty [11].

Specifically, in a certain period k, each node i in C_k generates committee identities through solving a PoW puzzle. The hardness of the PoW is denoted by $D(H_k)$, where H_k being the harness parameter can be adjusted according to the time requirement of committee generation. The hash function in the PoW puzzle is denoted by $\mathcal{H}$. The process of a node generating an identity is exhaustively querying a $Nonce$, such that the hash code $Hash_i$ of the concatenation context is not larger than a target value $D(H_k)$. The concatenation context includes IP_i, PK_i, R_{k-1}(it is a random value serving as a PoW seed left by the previous period $k-1$ to avoid the PoW calculation in advance), T_i and $Nonce_k$. It can be formally described as follows.

$$Hash_i = \mathcal{H}(R_{k-1}||IP_i||PK_i||T_i||Nouce_k) \leq D(H_k) \tag{1}$$

where $D(H_k) = 2^{L-H_k}$ with L being the length of the hash code in bits. To better understand the hardness function, we illustrate it in an example as follows. Assuming that the Hash function generates a 64-bit hexadecimal string and $H_k = 10$, the difficulty function $D(10)$ requires the first 10 bit string to be 0.

The committee member set is a set of created identities in size of m_k, denoted by D_k. The algorithm for committee generation is presented as in Algorithm 1.

In Algorithm 1, the committee identities gradually generated by the nodes in the campaign set until a required committee size m_k is achieved. Then how to set a proper m_k for our algorithm become an important issue. Next, we are to discuss how to set m_k to ensure the security of our mechanism against Byzantine false.

Algorithm 1: Committee Generation in Period k

Set CommitteeSize= m_k, $h = H_k$;
Initialize CommitteeNumber=0,$D_k = \emptyset$;
while *CommitteeNumber≤CommitteeSize* **do**
    foreach $i \in C_k$ **do**
        Computes $Hash_i$ according to Eq.(1);
        if $Hash_i \leq D(h)$ **then**
            CommitteeNumber++;
            Genearate ID_i;
            $ID_i \rightarrow D_k$;% Add ID_i in D_k set

Suppose the Byzantine nodes takes α_k proportionally in the campaign set in period k, and all the nodes have the same probability q in successfully generating identity in a PoW calculation. If a unit of computational power can call the Hash function k times, the number of successes (the number of identities) follows binomial distribution $B(k, p)$. Therefore, the expectations of the number of identities created by normal and malicious computing power are $s_k(1-\alpha_k) \times q$ and $s_k\alpha_k \times q$ respectively, where s_i represents the total computing power of the nodes in the campaign set C_k. Since the created committees are going to get consensus in the third step of our mechanism based on an improved BFT protocol, we require the created identities m_k should contain at most $(m_k - 1)/3$ Byzantine identities [2]. We denote q_m to be the system allowed probability that more than $m_k/3$ $(m_k/3 > (m_k - 1)/3)$ of the created identities are controlled by the Byzantine nodes.

Proposition 1. *Given α_k and q_m, the value m_k is lower bounded by $m_{min} = \frac{2 \ln q_m}{-(1-\alpha_k)(1-\frac{2}{3(1-\alpha_k)})^2}$, in order to ensure that the probability of the Byzantine identities acceding $m_k/3$ is less than q_m.*

Proof. Suppose all the nodes in the *campaign set* start calculating a PoW puzzle at the same time. The Byzantine node takes for α_k in proportion. Then the probability of each identity being created by a byzantine node is α_i, and the probability that each identity is created by a normal node is $1 - \alpha_i$.

Let X_i be a random variable, signaling the i-th identity created by a normal node or by a malicious node. If the i-th identity is created by a normal node, then $X_i = 1$. If the i-th identity is created by a Byzantine node, then $X_i = 0$. So X_i obeys the binomial distribution $B(1, 1-\alpha_i)$. And $X_1, ..., X_n$ are independent with each other. Let $X = \sum_{i=1}^{m_k} X_i$. Then $E[X] = n_k\alpha_i$.

According to the Chernoff bound theory [10], the probability of $X \leq 2m_k/3$ can be obtained.

$$P(X \leq 2m_k/3) = P(X \leq (1-\delta)\mu) \tag{2}$$

where

$$\mu = E[X] = m_k(1 - \alpha_k)$$

and

$$\delta = 1 - \frac{2}{3(1-\alpha_k)}$$

Thus, we can derive that

$$P(X \leq 2m_k/3) \leq (\frac{e^{-\delta}}{(1-\delta)^{(1-\delta)}})^{\mu} \leq e^{-\mu\delta^2/2} \tag{3}$$

Suppose q_m is the upper bound of the system with $m_k/3$ identities controlled by the Byzantine nodes. Thus, the probability that fewer or equal than $2m_k/3$ identities created by normal nodes should also be lower bounded by q_m. So, we let $P(X \leq 2m_k/3) \leq e^{-\mu\delta^2/2} \leq q_m$. We can obtain that

$$m_k \geq \frac{2\ln q_m}{-(1-\alpha_k)(1-\frac{2}{3(1-\alpha_k)})^2} \tag{4}$$

and the minimal value of m_k to be m_{min}:

$$m_{min} = \frac{2\ln q_m}{-(1-\alpha_k)(1-\frac{2}{3(1-\alpha_k)})^2} \tag{5}$$

To quantitatively understand the lower bound of m_{min}, we present the relationship between q_m and m_{min} as in Fig. 2, where the Byzantine computing power proportion α_k set to be 1/16, 1/32, 1/64 respectively. We can observe that m_{min} decreases as q_m grows, but increases as α_k increases. It indicates that, given the same α_k, the number of the committees should be larger if the system requires higher reliability or security. Meanwhile, given the same system reliability or security q_m, the number of the committees is required to be larger when the Byzantine nodes increases.

Please note that α_k is estimated by tools outside system. Besides, the first period in the system is special, we only allow m_i nodes according to α_1 and q_m to join our system to ensure the functionality of our mechanism. After the first period, the follower nodes are allowed to freely join the network. The set of competitor nodes are updated at the end after each period by replacing the nodes with reputable nodes according to m_k.

3.2 Committee Confirmation

After a committee identity is created, the node to which the identity belongs will broadcast it to all the other nodes, so all the nodes in the network can recognize this committee identity. Each node receives this identity simultaneously if all the nodes bear the same communication delay. To let other nodes recognize that an identity has been confirmed by a node, the node broadcasts the received identity again.

When a node receives $m_k/2$ copies of an identity, the identity is treated as "confirmed" among the committee. When the number of confirmed committee

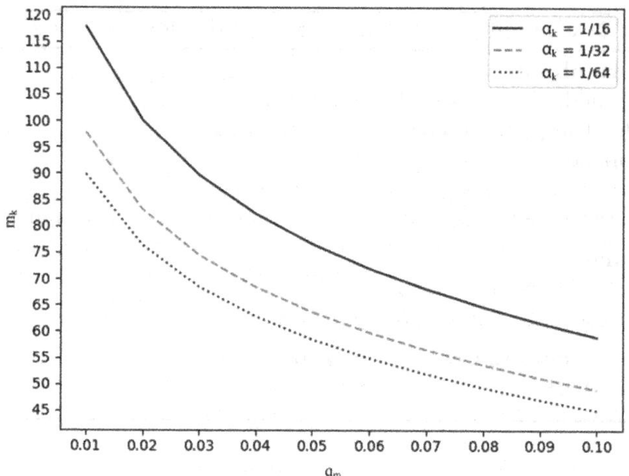

Fig. 2. Relationship between q_m and m_{min}

identities reaches m_k, all the nodes stop creating or broadcasting identities any more.

In the committee confirmation step, the communication complexity of the network is $O(m_k N_k)$ where N_k is the total number of the nodes in the network.

3.3 Committee Consensus and Network Consensus

According to the *committee confirmation* procedure, after the committee is confirmed, the members of the committee can identify one another through their identities. The committee then process to achieve block consensus through the improved version of BFT. The consensus mechanism in this step is similar to PBFT[2], and can tolerant up to $(m_k - 1)/3$ Byzantine nodes, which is guaranteed by setting m_k in the committee generation step.

The m_k committee members are indexed according to the identity creation time which is a component of the identity. The detailed consensus procedure is described in Algorithm 2.

Algorithm 2 contains three phases: pre-prepare, prepare, and commit phases, which is similar to the PBFT proposed in [2]. The main difference is that our committee consensus involves followers which are nodes not included in the committee set, making the original PBFT not applicable in our mechanism.

Specifically, in the pre-prepare phase, a *primary* node is assigned to one of the committee identities sequentially according to their index automatically (the last block $P_n + 1 \bmod m_k$, and the first round is 0). All the other committee nodes are called as *backup* nodes. The primary node, denoted by Pn, is responsible for collecting transactions and packaging them into a block. There are three types of messages, namely pre-prepare, prepare, and commit respectively. In the pre-prepare phase, the primary node creates msg_1, containing

Algorithm 2: Committee Consensus in Period k for a Block BlK_k

```
pre-prepare phase:
  A primary node P_n creates a block BLK_k;
  P_n broadcast msg1(pre-prepare, P_n, n, BLK_k, d);
prepare phase:
  foreach i ∈ D_k verifies msg1.type==pre-prepare do
    i broadcast msg2(prepare, P_n, n, d, i) to all the other nodes in D_k
commit phase:
  foreach i ∈ D_k receives ⌈2m_k/3⌉ + 1 msg2 do
    i broadcast msg3(commit, P_n, n, d, i) to network
  foreach i receives ⌈2m_k/3⌉ + 1 msg3 do
    i accepts BLK_k
```

- pre-prepare—message type,
- P_n—primary node index,
- BLK_k—a newly generated block,
- d—the message digest/hash value of the new block.

In the second phase, called as *prepare* phase. For each backup node i in the committee set D_k receives the msg_1, it then broadcasts a updated message msg_2 (prepare, P_n, n, d, i) to all the other committee identities. To reduce the communication load, the new block BLK_k is not included in msg_2. It is worthy to note that some nodes may fail to send the message as the backup node 3 behaves in Fig. 3.

The third step is the commit phase. When each of the identity in the committee receives msg_2 from $\lceil 2m_k/3 \rceil + 1$ identities, then the node will broadcast msg_3 (commit, P_n, n, d, i) to the whole network, where $\lceil \cdot \rceil$ is the minimal integer greater or equal to the value.

Finally, each node in the network who receives msg_1 and $\lceil 2m_k/3 \rceil + 1$ copies of msg_3 will accept the new block BLK_k. The formal proof of the Byzantine tolerance is similar to that in [2].

To clearly explain the consensus algorithm, we describe the operations in Fig. 3 where the primary node works without failure and one backup node fails to send messages.

When a node accepts a new block, the node then adds the block to its chain. When the primary node fails to complete its duty within a specific time period, the primary node index would be switched to the next.

The consensus process continues to generate new blocks until the predefined number is achieved or a predefined time is over, then an epoch is finished.

3.4 Campaign Set Updation

The committee identities generated by the campaign set should not keep the same due to the fact that the system is changing dynamically. At the end of

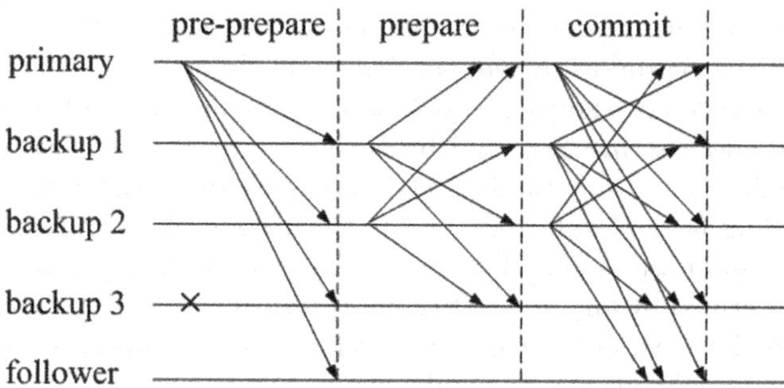

Fig. 3. Operational sequence of the consensus process

each epoch, the campaign set has a chance to be updated. The purpose of the campaign set updation is to increase the fairness between nodes and also increase system security.

Messages received and sent by a node are stored in the message log. According to message log, the behavior of nodes have three types: sending correct messages, not sending messages, and sending fault messages. We then conduct reputation evaluation on nodes through an existing authoritative peer-to-peer network reputation models [7]. The Byzantine computing power proportion α_{k+1} in the *campaign set* can also be estimated.

Based on the reputation values of nodes, the nodes in the campaign set with low reputation values are to be moved out from the set, and a certain number of qualified nodes are added into the set again. We describe the specific procedure as follows.

Algorithm 3: Campaign Set Updation

---

m_{min} is updated according to Proposition 1;
$m_{k+1} = \max\{m_k, m_{min}\}$;
Sort the reputation of nodes in descending order;
Remove the first $m_d = \lceil \beta \times m_{k+1} \rceil$ nodes in campaign set;
Add the highest $m_{k+1} - m_d$ reputation values to campaign set.

First of all, the committee set size m_{min} is recalculated according to Proposition 1. The size of committee of epoch $k+1$ is updated as $m_{k+1} = \max\{m_k, m_{min}\}$. A proportion $\beta(\beta \leq \alpha_{k+1})$ of the campaign nodes with lowest reputation values are replaced by other nodes with high reputation, ensuring that updated committee being m_{k+1}.

To recognize the new campaign set. The nodes of original *campaign set* broadcast the new *campaign set* to the network. If a node receives campaign sets with valid signatures from $m_k/2$ committee identities, it accept the new *campaign set*.

Moreover, the new committees before block consensus need to generate a set of random strings to produce $Rand_k$ serving as the PoW seed. Generating a set of random strings needs two steps. In the first step, each member of the committees generates a random string Str_i and broadcasts it to all nodes in the network. Each node will receive m_k random strings. The node organizes the received strings according to a Merkle tree, and the leaf node of the Merkle tree is the hash of the random strings [9]. The Merkle tree root is the final random $Rand_k$ which can be verified by any nodes. Because $Rand_k$ can only be uncovered at the end of the previous period, it can ensure that all the competitors in campaign set are unable to calculate PoW in advance.

4 Experimental Evaluation

In this section, we carry out several experiments to evaluate the proposed DSBFT consensus mechanism, where a prototype implementation is deployed. The main goal of our evaluation is to quantify the scalability of the mechanism and its throughput as the network size increases.

4.1 Experimental Setting

We develop a prototype system of DSBFT in Python 3.7 programming language, by using a server with a 16-core CPU and 64 GB memory. We visualized distributed nodes through objects, realize the network communication through calls of methods.

In our experiments, we compared the throughput efficiency and scalability of the DSBFT algorithm with PBFT. To test the maximum consensus efficiency, we assume there are no Byzantine nodes in the network. We set the number of *campaign set* to be 52. The expected time to generate the committee is 8s. q_m is set as 0.05. We made two comparisons. In the first case, the consensus mechanisms agree on only one block per epoch. In the second case, the consensus mechanisms agree on ten blocks per epoch. We vary the number of network nodes in each period from 50 to 150.

4.2 Experimental Results

In our experiments, we measure the throughput and scalability by comparing consensus time with the classical PBFT [2]. The experimental results in the two cases are shown in Fig. 4 and Fig. 5 respectively. Because of the instability of network communication, the consensus time of the same scale network will fluctuate in an acceptable small range.

In the first case, we can see from Fig. 4 that the consensus time of each period of DSBFT increases linearly with the increase of network nodes. It is because the smaller the consensus time indicates the greater throughput in the evaluated mechanism. So the throughput of DSBFT decreases linearly with the increase of network node number. When the number of network nodes is fewer than 95,

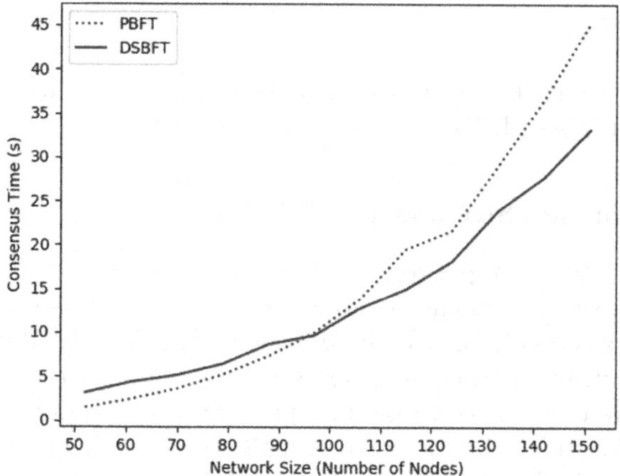

Fig. 4. Consensus time for one block of with different network sizes

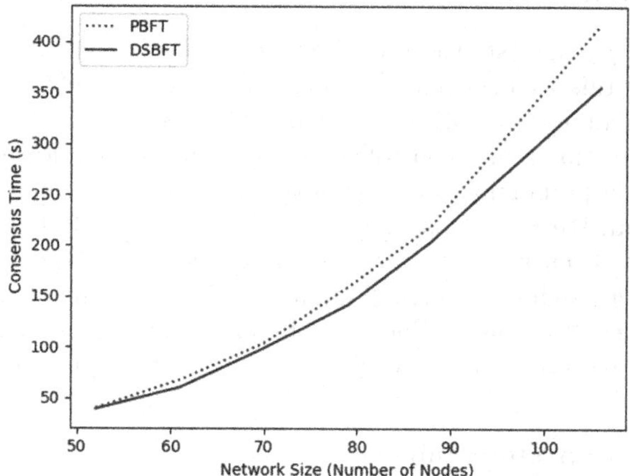

Fig. 5. Consensus time for ten blocks of with different network sizes

the throughput of DSBFT is small than that of DSBFT. When the number of network nodes is greater than 95, the throughput of DSBFT is larger than that of PBFT.

In the second case, the consensus time of each ten blocks are recorded. We can see from Fig. 5 that the throughput of DSBFT is always greater than that of PBFT when the network node size is from 50 to 100. It is because the communication complexity of DSBFT algorithm is linear. The communication complexity of PBFT algorithm is quadratically increase with the network size. When we consider the generation time of the committee to be stable and treat it as a constant time, then the throughput of the whole system will even much better than PBFT as the number of blocks in each epoch grows more than 10.

5 Discussion

In this section, we are to discuss the significant parameters and Byzantine toleration of the proposed DSBFT consensus mechanism.

5.1 The Campaign Size and PoW Difficulty

If the number of the competitors in the campaign set is n_k, the size of the committee identities m_k should be no fewer than n_k. Because if $m_k > n_k$, it is more communication efficient to use BFT algorithm directly. With the increase of n_k, the computing power of *campaign set* will increase. To ensure the time for generating committees in each period is approximately the same, the computing difficulty of PoW should be increased. On the contrary, if n_k becomes smaller, the computing difficulty of PoW should be reduced.

5.2 The System Security Parameter

The parameter q_m suggests the probability that the Byzantine malicious computation power takes more than one-third of the whole network computation power. If the committee identities contain fewer than 1/3 Byzantine malicious ones, the computation power controlled by the Byzantine malicious nodes is also fewer than 1/3 in proportion, because of the same difficulty in generating identities. The committee consensus algorithm is based on the BFT, resulting in the system losing its security when there are more than 1/3 Byzantine identities. The parameter q_m should be a small number close to 0, which can be explained as the probability of an impossible event happens. The value of q_m also stands for the security requirement of the system in a specific application scenario.

5.3 Solution of System Failure

In the case where the Byzantine identities are greater than $m_k/3$ and less than $2m_k/3$ in the committee, the security of blockchain will not be affected when Byzantine nodes choose to send error messages for illegal blocks. Because a follower node will not submit the block until it receives commit messages from more than $2m_k/3$ different identities. When the block proposed by the primary node is illegal, the backup node will not send prepare messages for the block, resulting nodes will not receive more than $2m_k/3$ of commit messages. The illegal block will not be submitted by the normal node. The security of the system can be ensured. When the Byzantine nodes, behaving as the primary node, do not send messages, our consensus algorithm loses availability but remains secure. In this case, the system waits a certain time and continue the consensus by reassigning the next node as the primary node.

5.4 Byzantine Tolerance

The Byzantine computing power ratio α_k can be low in the first period (maybe less than 1/3). However, with the increase of network nodes, Byzantine computing power that system can tolerance increases gradually. Because the campaign set is updated in each round by replacing low reputation nodes with the high reputation nodes. In other words, our system can work properly as long as the 2/3 computation power in campaign set is controlled by honest nodes, regardless of the status of the nodes out of the campaign setting. In this case, even there are Byzantine nodes increase rapidly in the network, the consensus of the normal nodes still can be guaranteed.

6 Conclusion

In this work, we have proposed a consensus mechanism, a delegation based scalable Byzantine tolerant consensus mechanism, abbreviated as DSBFT. In DSBFT, the consensus executives in epoch consisting five steps: committee generation, committee confirmation, committee consensus, network consensus, and campaign set updation, where the committee consensus and network consensus repeats for a certain predefined number in a single epoch. Our mechanism provides a new combinational investigation of multiple consensus mechanisms, where the committee is generated through PoW and committee consensus is achieved through an improved BFT. Experimental results show that the proposed DSBFT can achieve linear scalability and throughput as the network grows.

In futurework, we plan to elaborately evaluate the proposed consensus mechanism through extensive experiments, and also compare our mechanism with other classical ones. Furthermore, we also plan to apply our mechanism in realistic application scenarios to examine its effectiveness.

Acknowledgments. This work is supported in part by the National Natural Science Foundation for Young Scientists of China under Grant No. 61702090 and No. 61702084, and 111 Project (B16009).

References

1. Baldimtsi, F., Kiayias, A., Zacharias, T., Zhang, B.: Indistinguishable proofs of work or knowledge. In: Cheon, J.H., Takagi, T. (eds.) ASIACRYPT 2016. LNCS, vol. 10032, pp. 902–933. Springer, Heidelberg (2016). https://doi.org/10.1007/978-3-662-53890-6_30
2. Castro, M., Liskov, B., et al.: Practical Byzantine fault tolerance. In: OSDI, vol. 99, pp. 173–186 (1999)
3. CoinLore: Cryptocurrency list. https://www.coinlore.com/all_coins. Accessed 01 Feb 2019
4. Fischer, M.J., Lynch, N.A., Paterson, M.S.: Impossibility of distributed consensus with one faulty process. Technical report, Massachusetts Inst of TechCambridge Lab for Computer Science (1982)

5. Gilbert, S., Lynch, N.: Brewer's conjecture and the feasibility of consistent, available, partition-tolerant web services. ACM SIGACT News **33**(2), 51–59 (2002)
6. Goldreich, O.: Zero-knowledge twenty years after its invention. IACR Cryptology ePrint Archive **2002**, 186 (2002)
7. Hendrikx, F., Bubendorfer, K., Chard, R.: Reputation systems: a survey and taxonomy. J. Parallel Distrib. Comput. **75**, 184–197 (2015)
8. Lamport, L., Shostak, R., Pease, M.: The Byzantine generals problem. ACM Trans. Program. Lang. Syst. (TOPLAS) **4**(3), 382–401 (1982)
9. Merkle, R.C.: Protocols for public key cryptosystems. In: 1980 IEEE Symposium on Security and Privacy, p. 122. IEEE (1980)
10. Mitzenmacher, M., Upfal, E.: Probability and Computing: Randomized Algorithms and Probabilistic Analysis. Cambridge University Press, Cambridge (2005)
11. Nakamoto, S.: Bitcoin: a peer-to-peer electronic cash system (2008)
12. Narayanan, A., Bonneau, J., Felten, E., Miller, A., Goldfeder, S.: Bitcoin and Cryptocurrency Technologies: A Comprehensive Introduction. Princeton University Press, Princeton (2016)
13. Pass, R., Shi, E.: Hybrid consensus: efficient consensus in the permissionless model. In: 31st International Symposium on Distributed Computing. Schloss Dagstuhl-Leibniz-Zentrum fuer Informatik (2017)
14. Pease, M., Shostak, R., Lamport, L.: Reaching agreement in the presence of faults. J. ACM (JACM) **27**(2), 228–234 (1980)
15. Swan, M.: Blockchain: Blueprint for a New Economy. O'Reilly Media, Inc., Sebastopol (2015)
16. Tschorsch, F., Scheuermann, B.: Bitcoin and beyond: a technical survey on decentralized digital currencies. IEEE Commun. Surv. Tutor. **18**(3), 2084–2123 (2016)
17. Vukolić, M.: The quest for scalable blockchain fabric: proof-of-work vs. BFT replication. In: Camenisch, J., Kesdoğan, D. (eds.) iNetSec 2015. LNCS, vol. 9591, pp. 112–125. Springer, Cham (2016). https://doi.org/10.1007/978-3-319-39028-4_9
18. Wang, W., Hoang, D.T., Xiong, Z., Niyato, D., Wang, P., Hu, P., Wen, Y.: A survey on consensus mechanisms and mining management in blockchain networks. arXiv preprint arXiv:1805.02707 (2018)
19. Xiao, Y., Zhang, N., Lou, W., Hou, Y.T.: A survey of distributed consensus protocols for blockchain networks. IEEE Commun. Surv. Tutor. **22**(2), 1432–1465 (2020)
20. Zhang, E.: Neo consensus white paper: a Byzantine fault tolerance algorithm for blockchain (2019). https://docs.neo.org/en-us/basic/consensus/whitepaper.html. Accessed 01 Feb 2019

A Privacy-Preserving Approach for Continuous Data Publication

Mengjie Zhang[1(✉)], Xingsheng Zhang[1], Zhijun Chen[1,2(✉)], and Dunhui Yu[1,2]

[1] College of Computer and Information Engineering, Hubei University, Wuhan 430062, China
chenzj@hubu.edu.cn

[2] Education Informationization Engineering and Technology Center Of Hubei Province, Wuhan 430000, China

Abstract. To ensure the privacy protection and improve data availability, addressing the problem of loss of data in continuous data publication. This paper provides a continuous privacy-preserving data publishing algorithm inspired by the slicing algorithm, where the τ-safety condition is applied as an extension of the m-invariance algorithm. The algorithm divides the published data records into several bins horizontally based on the sum of the distances between the attributes corresponding to the records. Then, each bin is divided into multiple columns vertically based on the relationship between attributes. Next, the columns in the same bin are randomly rearranged. Finally, a differential privacy protection algorithm is applied to any remaining records that do not satisfy the L-diversity condition. Application of the proposed algorithm to actual medical data demonstrates that the proposed approach effectively reduces information loss and enhances the retention rate of association relationships compared with a state-of-the-art algorithm.

Keywords: Privacy-preserving data publishing · τ-safety · m-invariance · L-diversity

1 Introduction

Data publication is an increasingly employed information technology that allows for the release of privileged data for widespread public and private use. However, the data aggregated and managed by public and private enterprises include a large volume of private information that cannot be directly published in its original form without inevitably compromising the privacy of individuals. This issue has been addressed by the development of privacy-preserving data publishing (PPDP) technologies such as the k-anonymity [1], L-diversity [2], and t-proximity [3] models. Here, PPDP technologies differ fundamentally from conventional privacy protection techniques, because the purpose of PPDP is to maintain the availability of published data while protecting individuals from the release of sensitive private information. However, most currently available PPDP technologies are applied only to the initially published data, when, in reality, published data is updated continuously after its initial publication. As a result, nefarious actors can link this later data with previously published data to infer private information about individuals,

M. Qiu (Ed.): ICA3PP 2020, LNCS 12454, pp. 441–458, 2020.
https://doi.org/10.1007/978-3-030-60248-2_30

resulting in compromised privacy. Moreover, most currently available PPDP technologies employ algorithms based on generalization. However, while generalization-based algorithms have achieved improved privacy protection relative to earlier approaches, such algorithms have obvious defects, such as large information loss, which reduces data availability. Therefore, a new PPDP approach is needed to ensure the protection of sensitive information with continuous data releases and improved data availability.

2 Related Work and Proposed Approach

A number of PPDP approaches have been developed in recent years. Some typical examples are the Incognito algorithm proposed [4–6] between 2005 and 2012, However, while this approach greatly reduced the loss of important information somewhat, it increased the information loss of the other data. In the same year, a PPDP approach was proposed by Li et al. [7] based on clustering. Later, a PPDP approach based on the semi-division for high-dimensional data was proposed by Gong [8] in 2016. In 2017, a classification anonymity PPDP approach was proposed by Liao et al. [9] based on the weight attribute entropy. While these methods have improved the availability of data to some extent, the effect is not significant. This was addressed by a greedy clustering PPDP approach proposed by Jiang [10]. The algorithm implements equivalence class equalization division, which further improved data availability, but still suffered from a large loss of high-dimensional data.

A number of studies have focused on improving data availability in PPDP. For example, Li et al. [11] proposed a data slicing algorithm based on random rearrange-ment technology that improved data availability by preserving the useful relationships between attributes. Li et al. [12] proposed another data slicing algorithm that divides the data table into bins horizontally and columns vertically to preserve the useful relationships between attributes. Better preservation of the relationships between attributes was obtained by the overlap partitioning anonymous algorithm proposed by Jing [13], which allowed a single attribute to be divided into multiple columns according to a data slicing algorithm. However, the privacy protection provided by this approach is insufficient when the data dimension is high. Later, Rohilla [14] proposed the LDS algorithm to address the poor privacy protection obtained under high-dimensional data by solving the problem of continuous data discretization better than the slicing algorithm. The problems associated with high-dimensional data in PPDP was addressed by the work of Zhang et al. [15], which adopted Bayesian networks to reduce the dimensions of data, and thereby improved the privacy protection performance of the approach. However, the approach still suffers from some shortcomings in conjunction with high-dimensional data.

But, these approaches are applicable only to initially published data, which is not suitable for the generally encountered case of continuously published data [16]. The application of PPDP to continuous data releases has been addressed in many studies. For example, Xiao and Tao [17] proposed the m-invariance algorithm, which used the technique of falsifying records to ensure that any quality improvement (QI) group in which a record exists in a different data release version has the same signature. However, this algorithm easily leads to privacy issues. This privacy protection problem associated with re-insertion was addressed in the τ-safety model proposed by Anjum et al. [18], which negatively impacts data availability. This loss in data availability was addressed by the SDUPPA algorithm proposed by Liao et al. [19] using random replacement techniques based on the premise of satisfying the m-invariance model. In addition, Min [20] proposed the m-slicing algorithm, which combines the m-invariance and data slicing algorithms, and not only reduces the amount of information loss, but also helps to preserve the relationships between useful attributes. However, these previously developed PPDP approaches for continuously updated data offer considerable scope for improving the privacy protection and data availability provided.

The present work addresses these issues by proposing a PPDP approach denoted as privacy protection based on local partitioning (PPLP) for continuous data publication. Our goal is to improve data availability while ensuring privacy protection by requiring that the published data meet the k-anonymity, L-diversity, and extended □τ-safety conditions as an extension of the m-invariance algorithm. First, the individual data releases are divided horizontally into several bins according to the distance between records, where the distance between two records is defined as the sum of the distances between their corresponding attributes, and the signatures of any records in the data published at each time are made consistent. Then, since the relationships between attributes change as the attribute value ranges change, local vertical partitioning is used to retain the more useful attribute relationships by dividing each bin vertically into several columns according to the relationships between attribute values. Then, a random rearranging technology is adopted for ensuring data anonymity. Finally, differential privacy protection is applied to better balance privacy protection and data availability for remaining data that do not satisfy the L-diversity condition. Our proposed approach ensures privacy protection while significantly reducing data information loss and improving the retention rate of the association relationships. The superior privacy protection performance of the proposed PPLP algorithm is validated by comparisons with the performances of the m-invariance algorithm and m-slicing algorithm based on the multiply released patient diagnosis records of a hospital.

3 Related Concepts

Assume that a data table T to be published has been assembled with attributes $A = \{A_1, A_2, \cdots, A_i, \cdots, A_n, S\}$, which include a total of n attributes, such as the name, identification (ID) number, age, and zip code, as well as a single sensitive attribute n. Then, we define T_i^* as the anonymous data table of the ith release, and the attributes in set A are organized according to their relationships based on the following ten definitions. Single sensitive attributes often appear in continuously published data, so this article mainly introduces the application of single sensitive attributes.

Definition 1 (generalized hierarchical trees, *TSet*). Any node in a tree that has a parent-child relationship with another node in a tree satisfies the generalization relationship, and the tree is denoted as a generalized hierarchical tree. Here, the kth node in the jth layer of the generalized hierarchy tree of attribute A_i is denoted as A_{ijk}, and its value range is $D(A_{ijk})$. The symbol "<" is employed to represent the generalization relationship between attributes. If the node A_{ijk} and the mth node in the lth layer ($l = j + 1$) of the generalized hierarchy tree of attribute A_i (i.e., $A_{ij\text{m}}$) have a parent-child relationship, then

$D(A_{ijm}) < D(A_{ijk})$. In addition, $node_i$ represents any node in the generalized hierarchical tree of attribute A_i. Generalized hierarchical trees of age and zip code attributes are illustrated in Fig. 1.

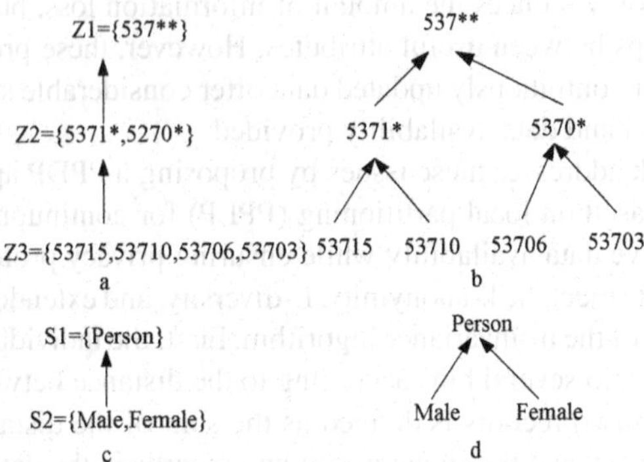

Fig. 1. Generalized hierarchical tree of the age (a,b) and zip code(c,d)

Definition 2 (generalized lattice g_i). If existing vertices of any edges in a graph satisfy the generalization relationship, and the vertex is a set of nodes with different attributes in generalized hierarchical trees, this graph is denoted as a generalized lattice. A generalized lattice g_i is formed of i generalized hierarchical trees of different attributes. Therefore, any vertex v in g_i is a combination of nodes that come from generalized hierarchical trees with different attributes, namely $v = \{node_1, node_2, \cdots, node_i,\}$. The starting vertex $node_1, v_m$ and ending vertex v_j of any edge $v_m v_j$ in the graph must satisfy the following conditions:

(1) $v_m < v_j$; (2) the compositions of v_m and v_j differ by only a single node.

Definition 3 (bins B). Assuming that a data table T is divided into n bins B_i according to the time of publication, then $\bigcup_{i=0}^{n} B_i = T$ and $\forall k, j, B_k \cap B_j = \emptyset$ $(0 \leq k, j \leq n)$.

Definition 4 (column). Assuming that a bin B is divided into m columns C_i vertically, then $\bigcup_{i=0}^{m} C_i = B$ and $\forall j, k, B_j \cap B_k = \phi$ $(1 \leq j, k \leq m)$.

Definition 5 (constraint on vertex v). The attributes in a record *to* serve as the constraint on vertex v in a generalized lattice, and the values of these attributes also serve as the value range constraint of the attributes in v corresponding to those in t.

Definition 6 (m-uniqueness). If a generalized table includes no less than m records, then the values of the S attributes are different in each QI group in the table, and the generalized table is m-unique.

Definition 7 (m-invariance). A sequence of published relations $T_1^*, T_2^*, \cdots, T_n^*$ (where $n \geq 1$) is m-invariant if the following conditions hold.

1. T_j^* is m-unique for all $j \in [1, n]$.
2. For any record t with a lifespan $[x, y]$, the quasi-identifier groups of record t, denoted as $t.QI_x^*, t.QI_{x+1}^*, \ldots, t.QI_y^*$, have the same signature when t exists within a generalized hosting group at time $j \in [x, y]$.

Definition 8 (exponential noise mechanism [20]). Let a random algorithm M be input as the data set. The output is an entity object $r \in Range$, and $q(D, r)$ is the availability function with a sensitivity Δq. If M selects and outputs r from *Range* with a probability proportional to $exe(\frac{\varepsilon q(D,r)}{2\Delta q})$, where ε is a constant, M provides $\varepsilon-$ differential privacy protection.

Definition 9 (Laplace noise mechanism [21]). Given a dataset D and a function f: $D \rightarrow R^d$ with a sensitivity Δf, and defining $Y \sim Lap(\Delta f / \varepsilon)$ as random noise obeying the Laplace distribution with the scale parameter $\Delta f / \varepsilon$, $M(D) = f(D) + Y$ provides $\varepsilon-$ differential privacy protection.

Description of the Problem

The problem is to improve data availability and reduce information loss while ensuring privacy protection by requiring that the published data meet the k-anonymity, L-diversity, and extended τ-safety conditions as an extension of the m-invariance algorithm.

A sequence of anonymous data releases $T_1^*, T_2^*, \cdots T_n^*$ conform to the τ-safety condition if these tables satisfy the following conditions.

1. T_j^* is m-unique at any time j.
2. The signature of each individual x in an event list $\tau(x)$ must remain the same in each consecutive lifespan $[x]$. Whenever $\tau(x)$ at time i represents a reinsert point r for an individual x, $Sig([x]_i) = Sig([x]_{i-k-1})$, such that the last deletion of x occurs at time $i - k$.

The objective is to minimize the information loss $iloss(T, A)$ during the release process, which can be formally expressed as the following optimization problems:

$$Min(iloss(T, A)) = Min(iloss(T, A_n) + iloss(T, A_c))$$

4 Privacy Protection Algorithm Based on Local Partitioning for Continuous Data Publication

4.1 Outline of the Proposed Algorithm

The general processing of the PPLP algorithm can be given as follows based on the special terms listed in Table 1. First, multi-dimensional and multi-level associations between the attribute values are evaluated by identifying the vertices that are associated with sensitive attributes using the Apriori algorithm and the hierarchical generalization trees, and these attributes are recorded in the sensitive attribute association set *srset*, while the other vertices are recorded in the common association set *rset*. Second, the

Table 1. Symbol definition

Symbol	Definition
A_i	Quasi-identifier attributes
S	Sensitive attribute
$rset$	Non-sensitive attribute association vertex set not containing S
$srset$	Sensitive attribute association set containing S
$sigs$	Signature table
$TSet$	Generalized hierarchical tree
$T_\cap$	Intersection of the historical release data table and the current published data table
T_{new}	Difference between the current published data table and the historical data table
val	Level importance
v	Vertex in generalized lattice g_i consisting of i nodes belonging to generalized hierarchical trees: $nodes$=[$node_1$, $node_2$, L, $node_i$]
B	Bins dividing a data table horizontally according to the distance between records
w	Attribute weight
dw	Dynamic attribute weight
ms	Maximum number of attributes in columns dividing bins vertically
ns	Number of records satisfying the constraint on vertex v
C	Columns dividing the bins B
$t_pA_i^c$	The ith classified attribute of record t_p
$t_pA_i^n$	The ith numerical attribute of record t_p
nt	Number of records in the bin satisfying the vertex v constraint
τ	Event list
r	Reinsert point
∂	Record falsification control parameters
$Davl(v)$	Dynamic importance of the relationship between attributes

data table is divided into bins horizontally based on the distances between records. Then, each bin is divided into multiple columns based on the vertices in $srset$ and $rset$. Third, the algorithm calculates the number of columns in which the number of unique records in the column is greater than or equal to $\frac{k}{2} + 1$, where k is the number of records in the bin. If this value is less than 2, the records in the bin are generalized; otherwise, the records are rearranged randomly. Finally, differential privacy protection is applied to all data that have not been anonymously processed due to their failure to meet the L-diversity condition.

4.2 Main Processing Steps of the Algorithm

The main processing steps of the algorithm is as follow:
Step1: Main Processing Steps of the Algorithm
First, we calculate the degree of support for the nodes in the generalized hierarchical tree of each attribute and delete those nodes whose support is less than a threshold value. This generates a frequent attribute generalization hierarchical tree. Then, we connect the nodes in different frequent attribute generalization hierarchy trees to obtain a generalized lattice. We calculate the degree of support and confidence of the vertices in the generalized lattice. If the support and confidence of any vertex are both greater than or equal to a threshold value, the vertex remains in the generalized lattice according to whether the vertex contains a sensitive attribute. Then, the vertex is added to *srset* if it contains a sensitive attribute; otherwise, it is added to *rset*
Step2: Bin Partitioning
The τ-safety condition is applied to ensure the security of data under conditions of continuous data release. Accordingly, the algorithm maintains a signature table that holds the latest signature for each record. The signature table ensures that the signature for each record is the same in $T_1^*, T_2^*, \cdots, T_n^*$. The amount of information loss is reduced by dividing the records according to a distance constraint, where records with a sufficiently small distance between them are included in the same bin. In addition, the more useful relationships between records are preserved by dividing the records that meet the sensitive attribute column value constraint into a bin. The data of the signature table is structured according to the combination key-value, where the key is the collection of all the quasi-identifier attributes of the record, and the value is the sign of all the different sensitive attributes in the bin. The bin partitioning step can itself be divided into the following steps.
Step3: Bin Segmentation
For each partition, randomly take a record from $T_\cap$, or T_{new} if no record is found in $T_\cap$, and add it to bin B. The signature corresponding to record t in the signature table is then applied as the bin signature. Then, $k-1$ records are added to the bin. These records are obtained from $T_\cap$ if the bin signature is not empty, and the records satisfy the distance constraint, which ensures that they have the same bin signature. Otherwise, the records that satisfy the distance constraint are selected from T_{new}.

As discussed, the distance between a record t and any other record is the sum of the distances between their corresponding attributes. The distance between attributes is divided into the distance between numeric attributes and the distance between classified attributes. Here, $dist\left(t_pA_i^n, t_qA_i^n\right)$ represents the distance between the ith numeric attribute of record t_p and the ith numeric attribute of record t_q, while $dist\left(t_pA_i^c, t_qA_i^c\right)$ represents the distance between the ith classified attribute of t_p and the ith classified attribute of record t_q. These distances are defined as follows:

$$dist\left(t_pA_i^n, t_qA_i^n\right) = \frac{\left|D\left(t_pA_i^n\right) - D\left(t_qA_i^n\right)\right|}{\max\left(TA_i^n\right) - \min\left(TA_i^n\right)} \tag{1}$$

where $D\left(t_pA_i^n\right)$ represents the value of the ith numerical type quasi-identifier attribute of record t_p, $D\left(t_pA_i^n\right) - D\left(t_qA_i^n\right)$ is the absolute value of the difference between the two attribute values, and $\max(TA_i^n)$ and $\min(TA_i^n)$ represent the respective maximum and minimum values in the set of the ith numerical type quasi-identifier attribute of data table T. The formula for the classified attributes is given as follows.

$$dist\left(t_pA_i^c, t_qA_i^c\right) = \frac{len\left(t_pA_i^c, t_qA_i^c\right)}{len\left(t_pA_i^c, root\right) - len\left(t_qA_i^c, root\right)} \tag{2}$$

Here, the distance calculation between the classified quasi-identifier attributes must introduce a generalized hierarchical tree as a basis for measurement. Accordingly, $len\left(t_pA_i^c, t_qA_i^c\right)$ represents the minimum distance between the node corresponding to the ith classified attribute $t_pA_i^c$ of record t_p and the node corresponding to the ith classified attribute $t_qA_i^c$ of record t_q in the generalized hierarchical tree, and the distance between the two nodes is the distance from each of the common minimum ancestor nodes. Finally, $len\left(t_pA_i^c, root\right)$ and $len\left(t_qA_i^c, root\right)$ are the distances to the root node. These two expressions can be combined to obtain the distance $dist\left(t_p, t_q\right)$ between records t_p and t_q, which include d_1 numerical types of quasi-identifier attributes and d_2 classified types of quasi-identifier attributes, as follows.

$$dist\left(t_p, t_q\right) = \sum_{i=1}^{d_1} dist\left(t_pA_i^n, t_qA_i^n\right) + \sum_{i=1}^{d_2} dist\left(t_pA_i^c, t_qA_i^c\right) \tag{3}$$

Step4: Bin Balancing

Any bin B obtained with fewer than k records after conducting bin segmentation indicates that some records with the sensitive attribute corresponding to the bin signature are missing. This represents an unbalanced bin that is addressed by augmenting B with records obtained from T_{new} whose sensitive attributes are consistent with the sensitive attributes of the missing records in the bin signature, and the records satisfy the distance constraint. If B remains unbalanced after this step, falsified records must be added to B to obtain balanced bins. Falsifying a record requires that the algorithm falsify its attributes. For falsifying numeric attributes, we first calculate the average of the current quasi-identifiers in B, and then obtain the falsified value by adding Laplace noise to the average. For falsifying classified attributes, we first identify the $k - |B|$ attribute values that are closest to the existing attribute value in the bin, and then add the existing attribute values as options (that is the Range in the Definition 9). Then, the value of the attribute is selected from the options according to the exponential noise mechanism in Definition 9, and the obtained value is used as the falsified attribute value in B. The use of the exponential noise mechanism requires an explicit form for the availability function $q(D, r)$, which is defined as follows.

$$q(D, r) = \begin{cases} \frac{1}{s} + \frac{(k-b)}{4s \cdot b} & \textit{if } st \in B. \\ \frac{1}{s} - \frac{1}{4s} & \textit{if } st \notin B. \end{cases} \tag{4}$$

Here, s is the number of options, b is the number of records in the original bin before adding the falsified record, which is denoted as $B_\circ$, and st is any of the options.

An example of the bin segmentation and bin balancing processes is given based on the original data tables of the firstly and secondly released patient diagnosis records of a hospital listed in Table 2 and Table 3, respectively. A comparison of Table 2 and Table 3 indicate that $T_{new} = \{Paul\}$. The bin segmentation of Table 2 is given in Table 4, where the signature of bin $B1$ is [endemic typhus, dyspepsia, flu] and the signature of bin $B2$ is [pneumonia, bronchitis, breast cancer]. When the second version of the data table is released and the table is segmented into bins, the records are selected from $T_{\cap}$, whose signatures are consistent with the bin signatures and satisfy the distance constraint. However, we note that bins $B1$ and $B2$ are not balanced. First, the algorithm selects records from T_{new}, and adds those records to the bins to balance them. As shown in Table 5, the addition of the record "Paul" to $B1$ balances the bin. However, bin $B2$ remains unbalanced. Therefore, the falsified record c_1 is added to $B2$.

Table 2. Original patient diagnostic records of the first data release.

Name	Age	Sex	District	Zip code	Disease
Tom	22	M	Connecticut	47902	Flu
David	25	F	Connecticut	47302	Flu
Ray	26	M	Bloomington	47490	Dyspepsia
Lily	27	F	California	47905	Endemic typhus
Hebe	35	M	Washington	47301	Pneumonia
John	40	F	Mississippi	47904	Breast cancer
Ella	49	M	Washington	47301	Bronchitis
Lau	50	F	Alaska	47304	Breast cancer

Table 3. Original patient diagnostic records of the second data release.

Name	Age	Sex	District	Zip code	Disease
David	25	F	Connecticut	47302	Flu
Ray	26	M	Bloomington	47490	Dyspepsia
Lily	27	F	California	47905	Endemic typhus
Paul	30	M	Connecticut	47302	Flu
John	40	F	Mississippi	47904	Breast cancer
Ella	49	M	Washington	47301	Bronchitis
Lau	50	F	Alaska	47304	Breast cancer

Table 4. Example of bin segmentation

Record	Sensitive Attribute	Bin
Ray	Dyspepsia	*B*1
David	Flu	*B*1
	Flu	*B*1
Lily	Endemic typhus	*B*1
John	Breast cancer	*B*2
Ella	Bronchitis	*B*2
	Pneumonia	*B*2
Lau	Breast cancer	*B*2

Table 5. Example of bin balancing

Record	Sensitive Attribute	Bin
Ray	Dyspepsia	*B*1
David	Flu	*B*1
Paul	Flu	*B*1
Lily	Endemic typhus	*B*1
John	Breast cancer	*B*2
Ella	Bronchitis	*B*2
c_1	Pneumonia	*B*2
Lau	Breast cancer	*B*2

Step5: Column Division.

The column division process for each bin is conducted as follows. First, the algorithm traverses each vertex in the sensitive attribute association set *srset*, and calculates the number of records in the bin that satisfy the constraint on each vertex. Then, the dynamic importance of the relationship between attributes $Dval(v)$ is calculated based on the number of records, attribute weights, and the level of the attribute values in the generalized hierarchical tree. This is calculated for each bin according to levels *lev* in the attribute generalization hierarchy tree as follows:

$$val = \frac{lev_c - lev_{root}}{lev_{\max} - lev_{root}} \tag{5}$$

where lev_c is the level of the current attribute, lev_{root} is the level of the root node, and $lev_{\max}$ is the maximum level of the attribute generalization hierarchy tree. This is then employed as follows:

$$Dval(v) = c \times nt \times \sum_{i=1}^{n} w_i val_i + conf(v) \tag{6}$$

where c is a constant in the range (0–1), nt is the number of records in the bin that satisfy the constraint on each vertex, w_i represents the weight of attribute A_i in vertex v, n is the number of attributes in v,val_i is the level importance of A_i, and $conf(v)$ is the confidence of v. The vertex v with the largest $Dval(v)$ is selected, and the attributes corresponding to the node set (which are included in v) are divided into a single column. Second, the algorithm traverses each vertex in the set of non-sensitive attribute associations *rset*, calculates the number of records in each bin that satisfy the constraint on each vertex, and calculates the value of $Dval(v)$ for each vertex. The nodes of *rset* are sorted in descending order according to the values of $Dval(v)$, and the vertices in *rset* are taken in sequence. Then, the attributes are found in the bin corresponding to each node that is taken out, and the attributes that have already been assigned are removed. If the number of remaining attributes is greater than or equal to the minimum number of attributes in a column, these attributes are divided into a single column; otherwise, the division is discarded. Finally, if the number of remaining attributes that are not vertically divided in the bin is greater than the maximum number of attributes in the column *mc*, the remaining attributes are evenly grouped; otherwise, they are directly divided into a column.

Step6: Data Rearrangement or Generalization

Assuming that the attributes in a bin are divided into *c* columns, we calculate the number of columns in which the number of unique records is greater than or equal to $\frac{k}{2}+1$. If the number of these columns is less than 2, the records in the bin are generalized; otherwise, they are randomly rearranged. If the record is still in its original position after rearranging, it will be randomly replaced with other records in the bin. Therefore, the relationship between the attributes in the same column remains unchanged, and the relationship between the attributes in different columns is destroyed, which maintains the highly-related relationship between the attributes.

Step7: Remaining Data Processing

The privacy protection of any remaining data that cannot satisfy the L-diversity condition will be reduced if the records are assigned to a bin that has been partitioned. Meanwhile, unpublished records reduce the availability of data. Therefore, differential privacy protection is applied to the remaining data for a better balance between privacy protection and data availability. However, the dimensions of high-dimensional data must be reduced first. This is conducted as follows. We calculate the *nt* value and the dynamic attribute weight *dw* based on the two-dimensional (2D) multi-level attribute association set in the bin. Then, we reduce the data dimension by selecting attributes based on *nt* and *dw* values. The selection process is conducted as follows. First, the records meeting the distance constraint are divided into a single bin. Then, we traverse each vertex of *srset* and *rset*, and calculate the number of records *nt* that satisfy the constraint on vertex v_i in the bin. Priority is given to attributes in the vertices with the highest value of $nt * val$ (Eq. (5)). The third step calculates the value of *dw* according to the vertices of the 2D multi-level attribute association relationship stored in *rset*. The dynamic attribute weights are important indicators for attribute selection. Because the calculation of *dw* is based on 2D multi-level attribute associations, it is necessary to measure the importance of the same attribute at different levels. Accordingly, *dw* is calculated as follows.

$$dw = val_j\left[w_j - \frac{1}{n}\left(\sum_i^n val_i\frac{nt_i}{b}w_i\right) + \frac{1}{m}\left(\sum_z^m val_z\frac{nt_z}{b}w_z\right)\right] \tag{7}$$

Here, w_j is the attribute weight corresponding to attribute A_j, vertex v_i contains attribute A_j and ingress attribute A_i, nt_i is the number of records in the bin that satisfy the constraint on vertex v_i, b is the number of records in the bin, w_i is the attribute weight of ingress attribute A_i contained in v_i, val_i is the level of the current attribute A_i in the attribute generalization hierarchy tree of A_i. We note that vertex v_z contains attribute A_j and outgoing attribute A_z, and nt_z is the number of records in the bin that satisfy the constraint on vertex v_z, while w_z is the attribute weight of outgoing attribute A_z and val_z is the level value of current attribute A_z in the attribute generalization hierarchy tree of A_i. Finally, the remaining attributes are selected according to the values of *dw*.

The process of dimensionality reduction is illustrated by considering an example dataset consisting of 20,000 records randomly drawn from an existing dataset ("Adult"), which is widely used in the field of privacy protection. Here, it is assumed that bin *B* includes four attributes at the time of publication. Some 2D, multi-level association sets of attributes are identified based on the Apriori algorithm and the attribute generalization hierarchy tree, and they are recorded. Detailed descriptions of the nodes in the vertex are given in Table 6, which include six attributes, such that dimensional reduction is required. The relationships between the four attributes are illustrated visually by the network given in Fig. 4.The bin results are in Table 7. For dimensionality reduction, priority is given to the attributes contained in the vertices in *srset* with the highest $nt * Val$, so occupation and sex are chosen. Next, the *dw* values of the other attributes age, relationship status, race, and education are calculated from Eq. (7) as ((4 – 1)/(5 – 1))*[0.42 – ((3 – 1)/(5 – 1))*0.75*0.35] = 0.185, 1*[0.2 + ½(1*1*0.3 + 1*1*0.3) – 0] = 0.5, 1*[0.3 + 0 – ½(1*0.5*0.35 + 1*1*0.2) – 0] = 0.1125, and 1*[0.35 + (1*0.5*0.3 + ½((3 – 1)/(5 – 1)*(4 – 1)/(5 – 1))*0.75*0.42) – 0] = 0.618, respectively. These values are listed in Table 8 along with the attribute weights *w*. Here, age is important as a new outgoing attribute. In addition, the calculation for the education attribute is complicated by the fact that the vertices v4 and v5 contain the same attribute, but at different levels of the attribute generalization tree. Therefore, these were written separately for the sake of providing a clearer representation. When calculating this *dw*, E3, with a low-level importance, is multiplied by the attribute weight of E3 and its outgoing attribute weight of age. Then, the model regards it as the outgoing degree of education. Accordingly, the final attribute set is {occupation, sex, education, relationship} based on the *dw* values.

Table 6. Detailed descriptions of the nodes in the vertex of the example dataset.

Node	Value	Current Level	Maximum Level
S2	Sex, male	2	2
O4	Occupation, machine-op-inspect, transport-moving, craft-repair	4	5
A4	Age, 30–40	4	5
Re2	Relationship status, husband	2	2
Ra2	Race, white	2	2
E5	Education, HS-grad	5	5
E3	Education, HS-grad, some-college, bachelor's	3	5

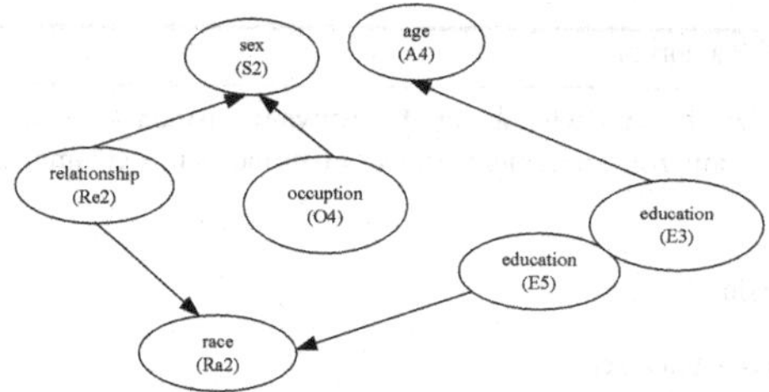

Fig. 2. Association network of the four attributes in the example dataset.

Table 7. Bin results

Age	Education	Sex	Occupation	Relationship status	Race
32	some-college	M	machine-op-inspect	husband	white
31	HS-grad	M	transport-moving	husband	white
32	HS-grad	M	machine-op-inspect	husband	white
36	7th-8th	M	craft-repair	husband	white

Table 8. Attribute weight table

No	Attribute	Weight	Dynamic weight
1	age	0.42	0.185
2	education	0.35	0.618
3	sex	0.3	–
4	occupation	1	–
5	relationship	0.2	0.5;
6	race	0.3	0.1125

Algorithm 1. The PPLP algorithm

Input: table T, $srset$, $rset$, parameter k for k-anonymity, parameter l for L-diversity, maximum number of attributes in a column mc, minimum number of attributes in a column nc, signature table sigs, $TSet$, version, and ∂.

Output: T_i^* for the ith version.

1. Generate the attribute association set.
2. Initialize the sigs of the current version.
3. While $|T|>0$ and T conforms to the L-diversity condition
4. Randomly extract a record t from $T_\cap$, or T_{new} if no record exists in $T_\cap$, and add it to bin B.
5. If version >1, then the signature of $t \leftarrow$ sigs; BinSign $\leftarrow$ signature.
6. While the number of cycles is less than k
7. If BinSign = ϕ, extract the record t_i with the shortest distance to record t of the sensitive attribute, and add t_i to B;
8. If BinSign $\neq \phi$, extract the record t_i whose signature is consistent with the bin signature and distance is closest to t, and add t_i to B;
9. Save the signatures of records in B to current Sigs.
10. End while
11. If BinSign $\neq \phi$ and B is unbalanced, select records from T_{new} to balance the bin.
12. If Bin Sign $\neq \phi$, B is unbalanced, and $|B|<\frac{k}{\partial}$, falsify records until B is balanced; otherwise, abandon the bin division.
13. If $|B|<k$, abandon this B, and return to Step 3.
14. Select vertex v ($v \in srset$) with the largest $Dval(v)$, and merge the attributes corresponding to the node set contained in vertex v into a column.
15. For $v_i \leftarrow rset$
16. Calculate the number of records nt in the bin that satisfy the constraints on vertex v_i and $Dval(v)$;
17. End for
18. Sort vertices in $rset$ in descending order according to $Dval(v)$.
19. For $v_i \leftarrow rset$
20. Find the attributes of each node contained in v_i in bin B, and remove the attributes that have been assigned to other columns.
21. If the number of attributes $> nc$, divide them into columns; otherwise, abandon the partitioning.
22. End for
23. If attributes remain that were not divided into the column in B $> mc$, divide the attributes evenly; otherwise, directly use the remaining attributes in a column.
24. Calculate the number of columns in which the number of unique records is greater than or equal to $\frac{k}{2}+1$. If this number < 2, generalize the records in B; otherwise, randomly rearrange them;
25. End while
26. End while
27. Reduce the dimensionality of remaining high-dimensional records, and add noise to the data after dimension reduction.
28. Update sigs based on current Sigs.
29. Return T_i^*

4.3 Algorithm Analysis

The proposed PPLP algorithm is given in detail as Algorithm 1. We note that the signature table *sigs* is constantly updated in the algorithm. Therefore, conducting bin division according to *sigs* ensures that the signatures of the records in each release version is are consistent, so that the PPLP algorithm satisfies the m-slicing condition. For the records that are re-inserted, the published data satisfy $Sig([t]_i) = Sig([t]_{i-k-1})$ because the signature corresponding to the record before the last deletion is saved in *sigs*. In addition, the point in time $i - k$ is the last time the record is deleted, so the algorithm satisfies the τ-safety condition. Then, differential privacy protection is applied to all remaining data that do not satisfy the L-diversity condition to better balance privacy protection and data availability.

The time complexity of the PPLP algorithm is mainly divided into two components, where the first component is the process of generating the association set, and the second component is the data privacy protection processing. The time complexity of first component is $O(n^2)$. The time complexity of the second component is related to the number of vertices in *srset*, the number of vertices in *rset*, the parameter k of the k- anonymity constraint, and the number n of records in table T. The number of vertices in *srset* and the value of k are much smaller than n, so the time complexity of this component is also $O(n^2)$. Therefore, the total time complexity of the PPLP algorithm is $O(n^2)$.

5 Analysis of the Experimental Results

The performance of the PPLP algorithm proposed in this study was further verified by comparisons with the performance of the m-invariance algorithm from the perspectives of information loss and data availability. The experimental dataset was again derived from the "Adult" dataset. First, records with missing attribute values were removed from the dataset, resulting in a dataset containing a total of 30,162 records. The original data of version 1 consisted of 20,000 records randomly selected from the dataset. The original data of version 2 consisted of the original data of version 1 after deleting 5,000 records randomly and then adding 5,000 new records that were randomly selected from the remaining data in the dataset. The original data of version 3 consisted of the original data of version 2 after randomly deleting 5,000 records, adding 5,000 records randomly selected from the remaining data in the dataset, and reinserting 1,000 records randomly selected from the deleted data in the original data of version 1. The experimental platform was a personal computer with an Intel® Core™ i5 − 2450 M CPU operating at 2.50 GHz, 4.00 GB memory, LITEON T9 (256 GB) hard disk, Windows 10 Professional 64-bit operating system, MySql database system, IntelliJ IDEA2017.2.5 development environment, and the jdk8 operating environment. The algorithm was implemented in Java.

5.1 Information Loss Analysis

The information loss is measured by the amount of information loss in the experiment. The greater the amount of information loss, the more information is lost. The impact of the anonymous parameter *k* on the amounts of information loss suffered by the PPLP and m-invariance algorithms for data versions 1–3 are shown in Fig. 3, respectively. The experimental results for each data version demonstrate that the information loss of the PPLP algorithm is significantly less than that of the m-invariance algorithm with increasing *k*, owing to the adoption of local generalization by the PPLP algorithm.

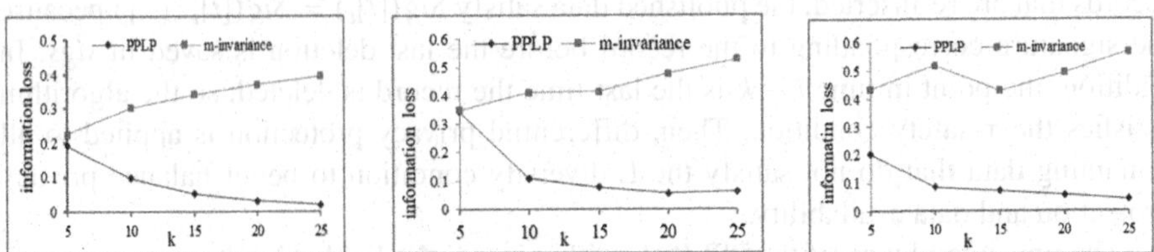

Fig. 3. Information losses suffered by the proposed PPLP and m-invariance algorithms with respect to the anonymous parameter *k* for the three data versions as Version = 1,2,3.

5.2 Data Availability Analysis

The retention rate of the association relationship (AR) is the ratio of the number of association rules retained after PPDP processing to the number of association rules of the original data. The retention rate of the AR was employed to measure the effect of the PPLP and m-invariance algorithms on the reserved-useful AR between the attributes with increasing k, where the Weka Apriori algorithm was employed to mine the association rules. The retention rate of AR results for data versions 1–3 are shown in Fig. 4, respectively. The experimental results show demonstrate that with the increase of the k value, the retention rate of the AR for the PPLP algorithm is generally greater than that of the m-invariance algorithm is greatly reduced. The retention rate of the PPLP algorithm is slightly increased because PPLP is longitudinally divided locally. With increasing k for all data versions, particularly for large values of k. This is because the PPLP algorithm adopts the local longitudinal division of data.

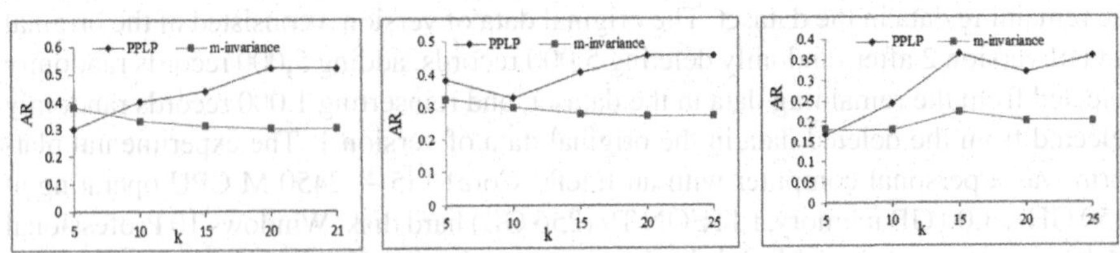

Fig. 4. Retention rate of the association relationship (AR) obtained by the proposed PPLP and m-invariance algorithms with respect to k for the three data versions as Version = 1,2,3.

6 Conclusion

Currently, most PPDP technologies are applied only to the initially published data, when, in reality, published data is updated continuously after its initial publication. As a result, nefarious actors can link this later data with previously published data to infer private information about individuals, resulting in compromised privacy.Therefore, we developed a PPDP approach for continuous data publication inspired by the slicing algorithm, where the τ-safety condition was applied as an extension of the m-invariance algorithm. Experimental results demonstrated that the proposed approach is effective for ensuring data privacy while significantly reducing information loss and increasing the retention rate of the association relationship. The present work examined the case of a single sensitive attribute. However, simply handling conditions of multiple sensitive attributes as a single sensitive attribute will inevitably lead to privacy leakage. Therefore, we intend to consider the case of multiple sensitive attributes in future research.

References

1. Sweeney, L.: K-anonymity: a model for protecting privacy. Int. J. Uncertainty Fuzziness Knowl. Based Syst. **10**(05), 557–570 (2002)
2. Machanavajjhala, A., Kifer, D., Gehrke, J.: L-diversity: privacy beyond k-anonymity. In: 2006 International Conference on Data Engineering, Istanbul, Atlanta, pp. 24–24 IEEE (2006)
3. Li, N., Li, T., Venkatasubramanian, S.: t-closeness: privacy beyond k-anonymity and l-diversity. In: 2007 IEEE International Conference on Data Engineering, Istanbul, Turkey, pp. 106–115 IEEE (2007)
4. Lefevre, K., Dewitt, D.J., Ramakrishnan, R.: Incognito: efficient full-domain K-anonymity. In: 2005 ACM SIGMOD International Conference on Management of Data, Baltimore, Maryland, USA, pp. 49–60 DBLP June 2005
5. Fung, B.C.M., Wang, K., Yu, P.S.: Top-down specialization for information and privacy preservation. In: 2005 International Conference on Data Engineering, Melbourne, pp. 205–216 VIC IEEE Computer Society (2005)
6. Yong, X., Xiaolin, Q., Yitao, Y.: A QI weight-aware approach to privacy preserving publishing data set. J. Comput. Res. Dev. **49**(5), 913–924 (2012)
7. Li, S., Zhu, Y., Chen, G.: Clustering-based alogrithm for data sentitive attributes anonymous protection. J. Comput. Res. Dev. **29**(2), 469–471 (2012)
8. Qiyuan, G.: Research on Data Anonymization Techniques for Data Publishing. Southeast University, NanJing (2016)
9. Liao, J., Jiang, Z., Guo, C.: Classification anonymity alogrithm based on weight attributes entropy. Comput. Sci. **44**(7), 42–46 (2017). (in Chinese)
10. Jiang, H., Zeng, G., Ma, H.: Greedy clustering-anonymity method for privacy preservation of table data-publishing. J. Softw. **28**(2), 341–351 (2017)
11. Li, T., Li, N., Zhang, J.: Slicing: a new approach for privacy preserving data publishing. IEEE Trans. Knowl. Data Eng. **24**(3), 561–574 (2012)
12. Li, T., Li, N., Zhang, J.: Slicing a new approach to privacy preserving data publishing. Int. J. Comput. Trends Technol. **4**(8), 64–78 (2013)
13. Yang, J., Liu, Z., Yue, Y.: A data anonymous method based on overlapping slicing. In: 2014 IEEE International Conference on Computer Supported Cooperative Work in Design, Hsinchu, pp. 124–128. IEEE (2014)
14. Bhardwaj, M., Rohilla, S.: Privacy preserving data publishing through slicing. Am. J. Netw. Commun. **4**(3–1), 45–53 (2015)
15. Zhang, J., Cormode, G., Procopiuc, C.M.: Privbayes private data release via bayesian networks [J]. ACM Trans. Database Syst. (TODS) **42**(4), 25 (2014)
16. Wang, M., Jiang, Z., Yang, H.: T-closeness slicing a new privacy preserving approach for transactional data publishing. Soc. Sci. Electron. Publishing **29**(7), 50–63 (2018)

17. Xiao, X., Tao, Y.: M-invariance: towards privacy preserving re-publication of dynamic datasets (2007)
18. Anjum, A., Raschia, G., Khan, A.: τ -Safety a privacy model for sequential publication with arbitrary updates. Comput. Secur. **66**, 20–39 (2017)
19. Liao, J., Jiang, C., Guo, C.: Data privacy protection based on sensitive attributes dynamic update. In: International Conference on Cloud Computing & Intelligence Systems. IEEE (2016)
20. Wang, M.: Research on the Improved Method of m-invariance Algorithm in Continuous Data Publishing. Jilin University, Changchun (2015)
21. Mcsherry, F., Talwar, K.: Mechanism design via differential privacy. In: IEEE Symposium on Foundations of Computer Science (2007)
22. Dwork, C., McSherry, F., Nissim, K., Smith, A.: Calibrating noise to sensitivity in private data analysis. In: Halevi, S., Rabin, T. (eds.) TCC 2006. LNCS, vol. 3876, pp. 265–284. Springer, Heidelberg (2006). https://doi.org/10.1007/11681878_14

Web Attack Detection Based on User Behaviour Semantics

Yunyi Zhang[1,2], Jintian Lu[1,2], and Shuyuan Jin[1,2](✉)

[1] School of Data and Computer Science, Sun Yat-sen University, Guangzhou, China
zhangyy333@mail2.sysu.edu.cn, ljt45@hotmail.com, jinshuyuan@mail.sysu.edu.cn

[2] Cyberspace Security Research Center, Peng Cheng Laboratory, Shenzhen, China

Abstract. With the development of the Internet and the increased popularity of web applications, the web has become one of the main venues for attackers engaging in cybercrimes. While enjoying the convenience of web applications, consumers also face security problems, such as the leakage of sensitive information and Internet fraud. Security protection mechanisms, such as traditional intrusion detection systems (IDSs) and web application firewalls (WAFs), are becoming incompetent at defending against the new cyber-attacks. In this paper, we propose a web attack detection approach that takes advantage of analysing the malicious intentions hidden in user actions. First, after using the independent user behaviours to build a sequential behaviour model, the proposed approach extracts the hidden malicious intentions of attackers from normal and seemingly normal behaviours utilizing a Long Short-Term Memory (LSTM) network. Then, on the basis of the user intentions, the approach leverages ensemble learning techniques to integrate extra inherent features of abnormal behaviour, resulting in its efficient practicality. The experimental results show the effectiveness of the proposed approach on the CSIC 2010 dataset with 99.87% accuracy.

Keywords: Web attack detection · Web Application Security · Deep learning · HTTP requests · Anomaly detection

1 Introduction

With the development of information and hardware technology [15–19], the Internet has revolutionized the traditional ways of daily life. Web applications support modern lifestyles well, both in life and work, such as telecommuting and recreation. However, the web security situation is not optimistic. For cyber-criminals, web applications are a cake with temptation. First, web applications possess millions of users' private information (e.g., addresses, phone numbers, and names), which are ideal targets for attackers. Second, compromised hosts are used by attackers to distribute malware, which allows attackers to gain access to and control victims' computers.

According to a report Gartner published, web servers have been the main targets of hackers in recent network security incidents. Common web attacks include SQL injections (SQLi), cross site scripting (XSS), XML external entity (XXE), file uploads,

M. Qiu (Ed.): ICA3PP 2020, LNCS 12454, pp. 459–474, 2020.
https://doi.org/10.1007/978-3-030-60248-2_31

command injections, and more. In the report of the Top 10 Web Application Security Risks released by OWASP [1], injection attacks repeatedly rank first.

Most traditional defence mechanisms rely on the rule filters obtained by analysing known attacks. There are two defects:

1) traditional methods require much expert knowledge to build robust rules, and they cannot detect unknown attacks using these rules; and
2) the concern of traditional methods is one user operation, and the context of users' actions is overlooked.

Deep learning has achieved prominent achievements in many fields and provides us with a new research direction. By applying its learning capability, defence mechanisms can reduce the dependence on expert knowledge whenever possible. Detecting web attacks from normal users and attackers using deep learning is challenging, and there are three problems.

1) Data presentation. The way of source data are transformed into effective representations is important in view of the various hidden ways of different attacks.
2) Model update. Detecting web attacks will be an ongoing process. Not only do model updating issues require attention in an actual implementation, but also attacks that aim to detect models need to be guarded against.
3) Data diversity. The ways of attacks are diversified and unpredictable, which result in diverse data, and thus feature selection is not easy.

User interaction is direct communication between costumers and web servers. Through a series of operations, costumers archive their goals on the web application. Each user action contains a specific intention. For example, a user visits the login page and inputs his username and password, or a user inputs "apple" in the search box. It is difficult to identify whether the login is abnormal by a single action in a password guessing attack. When extending the scope of behaviours from one operation to an operating sequence, it becomes clear. Figure 1 illustrates a simple scenario where an attacker attempts to guess a user's password, and a normal user inputs a wrong password. From the perspective of separate actions, it is difficult to identify whether the login is abnormal. However, through extending the scope of behaviours, the attack pattern becomes clear since it contains too many login requests and has attempted a vast number of weak passwords. The main challenge to identifying malicious intentions from users' operating sequences is how to define the sequential behaviour.

To address the aforementioned three problems, this paper proposes a web attack detection approach that identifies the malicious intentions in the background of normal and seemingly normal behaviours. The proposed approach constructs the independent user behaviours as a sequential behaviour model, abstracts complex users' behavioural data as a behaviour model, and then converts it to a vector representation by applying word embedding technology on the premise of preserving the semantics as much as possible. Applying an iterative updating method eliminates the model updating problem and attacks using a deep learning model. Specifically, our work makes the following contributions.

- We propose a method to represent the sequential behaviour. Briefly, we extract some key information from raw data and then split and recombine it to produce the target sequential behaviour.
- The proposed approach can distinguish anomalous actions and normal operations by automatically analysing the hidden intentions in sequential behaviour.
- An iterative updating strategy is applied in the approach to steadily update the detection model and defend against model attacks.

The rest of the paper is organized as follows. Section 2 presents a brief review of the related works. Section 3 describes the details of the proposed approach. The experimental results and discussions are depicted in Sect. 4. Section 5 concludes the paper and discusses the future work.

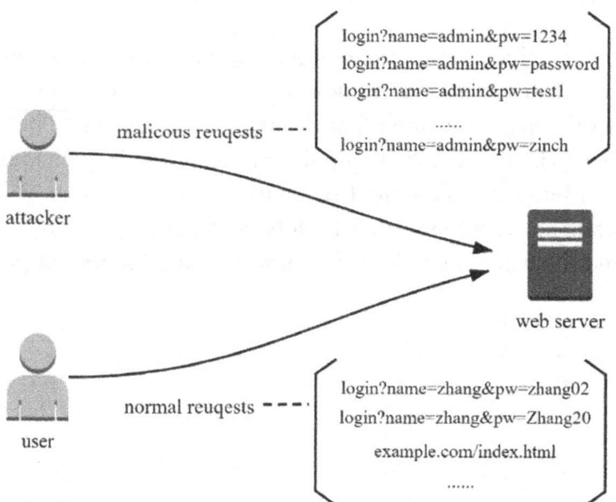

Fig. 1. Scenario: attackers guess a user's password and a normal user login

2 Related Works

Web attack detection has been extensively studied in the last decade. For common attacks, such as XSS and SQLi attacks, Debabrata Kar [2] proposed a novel approach to detect an SQL injection in time by using query transformation and a document similarity measure. A more generic approach is defensive coding practices [3]. However, they are usually difficult. A better option is a Web Application Firewall (WAF), which discovers abnormal operations by monitoring real-time Web traffic. There are some limitations to traditional rules-based WAFs. For example, the maintenance of the rules highly depends on experts' experience and it cannot find new attacks. The powerful data-driven learning of machine learning provides a generalization ability for web attack detection. In [4], Xiaohui Kuang presented the DeepWAF based on deep learning techniques, which uses the currently popular CNN and LSTM models. Ming Zhang [5] uses the same method, a specially designed CNN, to detect Web attacks. Jingxi Liang [6] applied Recurrent Neural Networks (RNNs) to learn the normal request patterns for discriminating between anomalous and normal requests. In general, the effective attack payload is very short. To eliminate the noise of irrelevant background strings, Tianlong Liu [7] proposed a novel Locate- Then-Detect (LTD) system, which detects web threats by using attention-based deep neural networks. For the same purpose, Igino Corona [8] analysed the query data through Hidden Markov Models (HMMs) to induce noise in the training set. To reduce the training costs, Michiaki Ito [9] used a Character-level convolutional neural network (CLCNN) to extract the features of HTTP requests instead of LSTM. To address the threats of the cloud-IoT, Zhihong Tian [10] proposed a web attack detection system that was deployed on edge devices.

3 Methodology

In this section, we describe the overview of the approach and explain every component in detail. Subsect. 3.1 describes the framework of the approach and how the approach takes advantage of the malicious intentions hidden in the sequential behaviour to detect web attacks. Subsect. 3.2 depicts the method for processing the web logs collected from servers and how to build the sequential behaviour model. Subsect. 3.3 introduces the scenario features and the chosen reasons. Finally, Subsect. 3.4 introduces how to identify user intentions and the web attack detection model applied in the proposed approach.

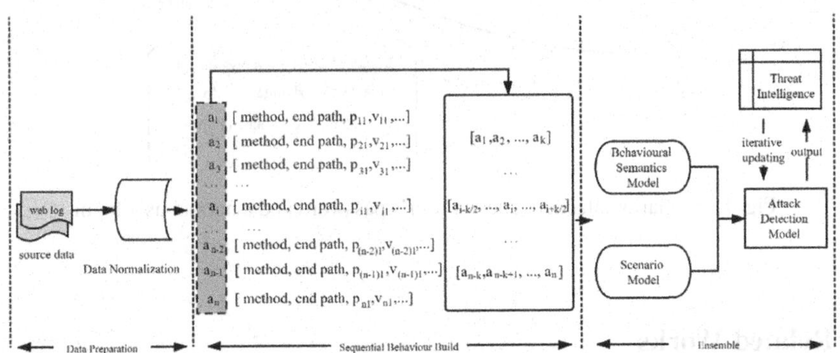

Fig. 2. The overview of approach

3.1 Approach Overview

The approach consists of the Data Preparation, the Sequential Behaviour Build, and the Ensemble. As shown in Fig. 2, the source data, collected from servers, including web logs and system logs, will be first sent to the Data Normalization module to remove invalid and unnecessary data. Then, preliminary processed data will be used to construct independent user behaviours and combine them into sequential behaviours by splitting and recombining them in the Sequential Behaviour Build. Finally, user intentions are extracted from the sequential behaviour and the scenario features from user behaviour will be used simultaneously for detecting attack actions in the Ensemble.

Data Preparation is used to get rid of the irrelevant data in raw logs collected from servers to build a behavioural model. According to the analysis of the target data, we first define the elements that make up the user behaviour, which will be discussed in Subsect. 3.2. Subsequently, the parts containing behavioural elements are preserved and others are dropped.

The Sequential Behaviour Build is designed to convert the behavioural elements obtained from Data Preparation to sequential behaviour. According to the original information of the raw data, the elements are first combined into independent user behaviours. Then, there are two tasks to finish the process. The first task is to split and recombine the independent user behaviours to produce sequential behaviour. Specially, to verify the effect of the sequential behaviour model, we assessed different methods, including the overlapped, successive, and skipped methods, which will be detailed in Subsect. 3.2. The second task is to extract the scenario features, which will be discussed in Subsect. 3.3.

Ensemble not only takes advantage of the user intentions obtained from behavioural semantics analysis, but also applies the scenario features that exist in user actions to detect attacks. In this paper, we utilize LSTM to analyse the behavioural semantics of sequential user behaviours to extract the hidden user intentions, which will be detailed in Subsect. 3.4. Also, the scenario features provide the perspective of independent user actions, supplementing the inadequacy of a user's intentions. Moreover, to update the detection model and mitigate the data pollution problem of the detection model, this paper applies an iterative updating method that is discussed in Subsect. 3.4.

3.2 User Behaviour Model

To identify the intentions in a user's behaviours accurately, we consider the following three problems:

1) how to define user behaviours,
2) how to build sequential behaviours from independent user actions, and
3) how to covert the behavioural model to a vector representation.

For 1), we processed the data of the HTTP DATASET CSIC 2010 [11] that is shown as Fig. 3, including the user request path, parameters, and http header information. We define the user action as a four-tuple $ai = \{method,\ end\ path,\ parameters,\ values\}$. Request methods include POST, GET, and PUT. The end path indicates the target page that disposes of the user action. Due to the URL paths containing too much redundant information, such as domain names, which will lead to a sharp increase in the dimension of our data, we choose the end path. The parameters and values that are the most immediate embodiment of user intentions mainly come from the user's input. Figure 3 presents the four-tuple in web logs.

For 2), as shown in Fig. 4, we first construct the independent user behaviour database for the behavioural elements. Then, as shown in Fig. 5, we use the overlapped, successive, and skipped methods to construct different sequential behaviours. Another factor that must be considered is the length of one sequential behaviour. A sequence may not be

method
end path
```
POST http://localhost:8080/tienda1/publico/anadir.jsp HTTP/1.1
User-Agent: Mozilla/5.0 (compatible; Konqueror/3.5; Linux) KHTML/3.5.8 (like Gecko)
Pragma: no-cache
Cache-control: no-cache
Accept: text/xml,application/xml,application/xhtml+xml,text/html;q=0.9,text/plain;q=0.8,image/png,*/*;q=0.5
Accept-Encoding: x-gzip, x-deflate, gzip, deflate
Accept-Charset: utf-8, utf-8;q=0.5, *;q=0.5
Accept-Language: en
Host: localhost:8080
Cookie: JSESSIONID=3B654D6DF7F1466EE80D7F756B00E5D1
Content-Type: application/x-www-form-urlencoded
Connection: close
Content-Length: 77
```
parameters and values
```
id=2%2F&nombre=Jam%F3n+Ib%E9rico&precio=85&cantidad=49&B1=A%F1adir+al+carrito
```

Fig. 3. The data details of CSIC 2010

able to show the real user intentions once the sequence is too short. Otherwise, it will affect the training efficiency of the model or lead to overfitting.

a_1 GET + index.jsp

a_2 GET + anadir.jsp + [id:1, nombre:Jam%F3n+Ib%E9rico, precio:39, cantidad:41, B1:A%F1adir+al+carrito]

a_3 POST + anadir.jsp + [id:1, nombre:Jam%F3n+Ib%E9rico, precio:39, cantidad:41, B1:A%F1adir+al+carrito]

a_4 GET + autenticar.jsp + [modo:entrar, login:caria, pwd:egipciaca, remember:off, B1:Entrar

a_5 GET + caracteristicas.jsp + [id:2]

Fig. 4. The sample of independent user behaviours

$a_1\ a_2 \cdots a_k\ a_{k+1}\ a_{k+2}$ — overlapped

$a_1\ a_2 \cdots a_k\ a_{k+1}\ a_{k+2} \cdots a_{2k}$ — successive

$a_1\ a_2 \cdots a_k\ a_{k+1}\ a_{k+2} \cdots a_{2k}\ a_{2k+1}$ — skipped

Fig. 5. Three building methods

For 3), to retain the semantics of the behavioural elements, we apply word embedding technology to acquire the behaviour vector. Common word embedding technologies include word2vec [12], doc2vec [13], and GloVe [14].

Word2vec, a popular word embedding technology that converts words to a vector, has been widely applied in NLP studies. The structure of the CBOW (Continuous Bag-of-Words), a word2vec model that predicts the target word according to the context, is shown in Fig. 6. The model takes the one-hot vector $X_i \in R^v$ for every word as its input, uses a hidden layer and outputs an output vector $Y_i \in R^v$. The weight matrix W is the final goal, and we can calculate the vector representation of each word using it.

Doc2vec is an extension of word2vec, which is used to get a document vector. Word2vec contains the original semantics of words, but it lacks semantics for documents or paragraphs. Doc2vec overcomes the problems by adding a paragraph ID in the training stage.

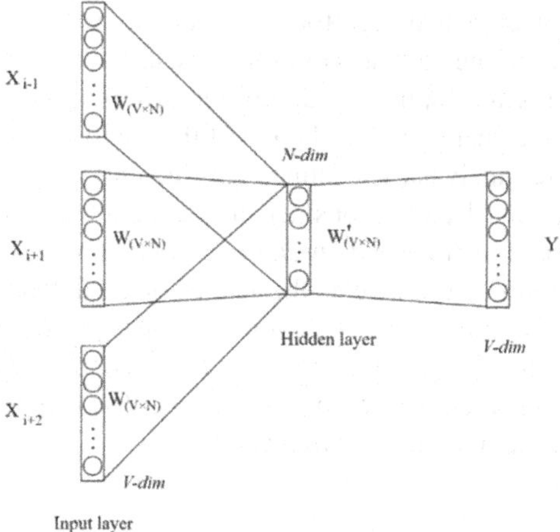

Fig. 6. The CBOW model

The Global Vectors for Word Representation (GloVe) is a word representation method whose basic idea is using count-based global statistics. *Xij* is the number of times word j appears in the context of word i, where *X* refers to the matrix of the word-word co-occurrence counts. In addition, $X_i = \sum_k X_{ik}$ refers to the total number of times any word occurs in the context of the word i. Let $\mathrm{P_{ij}} = p(j|i) = \frac{X_{ij}}{X_i}$ be the probability that word j occurs in the context of the word i. Through the analysis of the co-occurrence probability matrix, there is a relationship between $Ratio = \frac{P_{ik}}{P_{jk}}$ and the correlation among words in Table 1.

Table 1. The correlation among words

Words j and k ratio words i and k	Relevant	Irrelevant
Relevant	Close to 1	Large
Irrelevant	Small	Close to 1

Hence, there is a function $F(w_i, w_j, w_k) = \frac{P_{ik}}{P_{jk}}$, where w_i, w_j, and w_k are the word vectors we need.

3.3 Scenario Features

The scenario features characterize the inherent attributes in user behaviour, and they include two parts: payload presentation analysis and behavioural statistics analysis. Payload presentation analysis assesses the abnormal features in user inputs. The carefully constructed payloads that attackers exploit contain more information about users' intentions. An SQLi attack is a straightforward example, which can take the sensitive information in a database. The payloads of SQLi attacks have some unique features that cannot be found in a normal input. To capture the information, we apply the n-gram technique to calculate the character distribution of the payloads for each behaviour. For the behavioural statistics analysis, we define eight features after analysing the logs, as shown in Table 2. Features 1 and 2 consider the frequency of occurrence for a single parameter or path in a short time, which can demonstrate whether the user replays a particular action. Features 3–6 consider the length of a payload. The length of normal payloads can be determined, but malicious payloads present various distributions for some attacks. Features 7–9 are aimed at the characteristics of attack payloads. For example, the specified characters of feature 8 include "/@()%$-<>?", etc., which are common characteristics in malicious payloads.

Table 2. Behavioural statistics features

Features
1 Maximum occurrence frequency of a single parameter
2 Unique path ratio
3 Maximum payload length
4 Average payload length
5 Variance of the payload length
7 Number of spaces
8 Number of specified characters
9 Number of unprintable characters

3.4 Ensemble

Ensemble learning is a model enhancement technology that combines multiple weak models to get a better strong model. The behavioural semantics model is featureless, and only behavioural semantic representations are taken into account implicitly through the training process. Global patterns in user behaviour and correlations among behaviours can be captured from sequential user behaviours. However, the overall perspectives may inevitably leave out some inherent attributes. Meanwhile, scenario features just supplement the part that the behavioural semantics model does not. To maintain the efficacy of the model with a lower false positive rate (FPR) and false negative rate (FNR), we propose a novel ensemble model that encompasses the behavioural semantics model and scenario model (Fig. 7).

The behavioural semantics model discovers the hidden intentions from sequential user behaviour utilizing LSTM. Compared to the traditional RNN, LSTM adds a cell that stores past information and a memory control mechanism that chooses to forget or

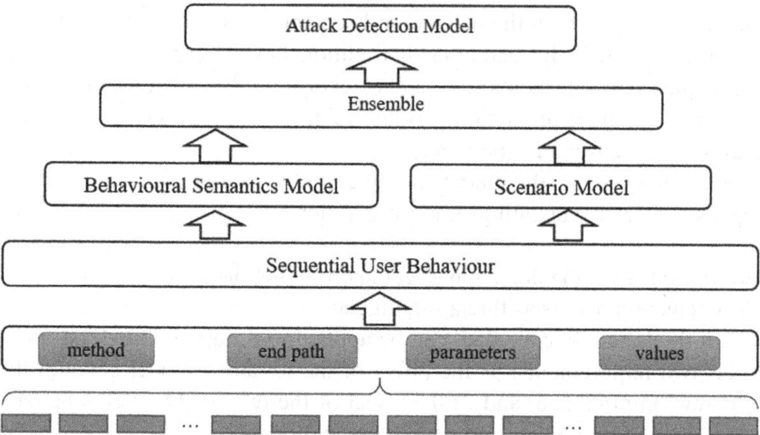

Fig. 7. Ensemble model

store the information. Moreover, it addresses the long-term memory issue of the RNN. Figure 8 shows a typical LSTM cell. There are three gates, including a forget gate, an input gate, and an output gate, to control the cell states. The input of each gate is determined based on the input of the current step x_t and the output of the last step h_{t-1}. The forget gate determines what information to discard, which is realized by a sigmoid function $\delta(z) = 1/(1 + e^z)$. h_{t-1} and x_t are the inputs of the sigmoid function, and the output is a 0–1 vector. For every status in C_{t-1}, 0 means discarded and 1 means reserved. The input gate determines the status that needs to be updated, which gains the candidate status via a tanh layer. The new cell state C_t is updated by the information passed through the forget gate and the input gate. The output gate finishes the calculation of h_t. LSTM has been implemented in semantic related problems with prominent achievements. In this paper, we leverage LSTM to learn the user behavioural semantics to discover user intentions.

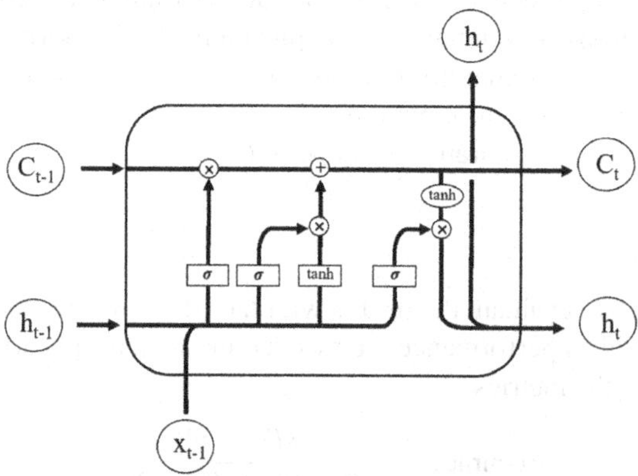

Fig. 8. An LSTM cell

In the training stage, both the behavioural semantics model and the scenario model are first trained separately. Instead of using traditional machine algorithms, the scenario model also applies boosting technology, and we choose XGBoost for the implementation. In the second stage, their predictions on the training data are used in the final model training to solve the ensemble coefficients.

To continuously update the model and mitigate the attacks aimed at the model, this paper applies an iterative updating strategy as follows.

1) Build a baseline for the detection effect and training data.
2) Collect detection results as threat intelligence.
3) The new threat intelligence data can be applied to retrain the model if and only if it meets two requirements: 1) the effect of the retrained model is better than that for the raw training data, and 2) the effect of the retrained model is better for the combined raw training data and new threat intelligence data.

4 Evaluation

The approach is implemented using Keras with TensorFlow as the backend. In this section, we conduct experiments to evaluate the performance of the approach. In particular, we focus on three problems.

P1: How to effectively build a sequential behaviour model.
P2: The influence of different words embedded on the behaviour model.
P3: How to evaluate the effect of the approach.

4.1 Dataset

The HTTP DATASET CSIC 2010 has been widely used in web intrusion detection. CSIC 2010 contains thousands of automatically generated web requests and was developed at the "Information Security Institute" of the CSIC (Spanish Research National Council). It aims to test web attack protection systems. The data were automatically generated by the HTTP traffic targeting an e-Commerce web application. The data set contains 36,000 normal requests and 25,000 anomalous requests. There are various attacks, including SQL injections, buffer overflows, information gathering, file disclosures, CRLF injections, XSS, server-side inclusions, parameter tampering, etc.

4.2 Experiments

In our experiments, we evaluate the results with the F1-score, accuracy, recall and precision, which are typical performance metrics that measure deep learning models. Here are the formulas for the metrics:

$$\text{Accuracy} = \frac{TP + TN}{TP + FP + TN + FN} \tag{1}$$

$$\text{Recall} = \frac{TP}{TP + FN} \tag{2}$$

$$\text{Precision} = \frac{TP}{TP + FP} \tag{3}$$

$$\text{F1-Score} = 2 * \frac{\text{Precision} * \text{Recall}}{\text{Precision} + \text{Recall}} \tag{4}$$

where *TP* refers to true positives, *TN* refers to true negatives, *FP* refers to false positives and *FN* refers to false negatives.

To answer P1, we consider the factors the influence the sequential behaviour model. The first is the window of actions, which influence the length and number of independent user actions. Although a longer action contains more information than a shorter one, it is harder to understand. Table 3 shows the results of different windows, and Fig. 9 is a bar chart for the results. As we can see, the trend is consistent that the detection accuracy improves as the window size increases, and the recall changes most dramatically. Next, the split and recombine method of independent user actions is relevant to the degree of correlation of adjacent sequences. Figure 10(a) presents the bar chart of the results of applying different methods. In addition, Fig. 10(b) shows the accuracy curves in the training stage. The skipped method works better but is not stable. Although the effect of the successive method is slightly worse than others, it is more stable, and its gap can be complemented by combining it with the scenario model.

Table 3. The results of different windows in CSIC 2010

Window	F1-Score	Precision	Accuracy	Recall
1	0.809412	0.992599	0.867614	0.683305
2	0.873423	0.996123	0.90731	0.777636
5	0.948531	0.996934	0.959548	0.90461
10	0.991948	0.996764	0.987179	0.993449
15	0.996972	1	0.997543	0.993961
20	**0.998347**	**1**	**0.998689**	**0.9967**

To answer P2, we vectorize the behavioural data using three methods to train the detection model. Figure 11 shows the differences in the results of the three methods. In terms of the overall effect, doc2vec is more stable than the others. Although the precision of word2vec is higher, the accuracy and recall are both lower than those of doc2vec. Figure 12 represents the trend of accuracy during training for different word embedding technologies. As we can see, the effect and stability of doc2vec are better than those of the others.

To answer P3, we explore the effects of each part of the approach. First, Fig. 13 demonstrates the effects of each part in the approach, including the scenario model, the behavioural semantics model, and the ensemble model. The behavioural semantics model understands user intentions well, but its effects are unstable and the accuracy fluctuates from approximately 88% to 95% for CSIC 2010. After the ensemble, the

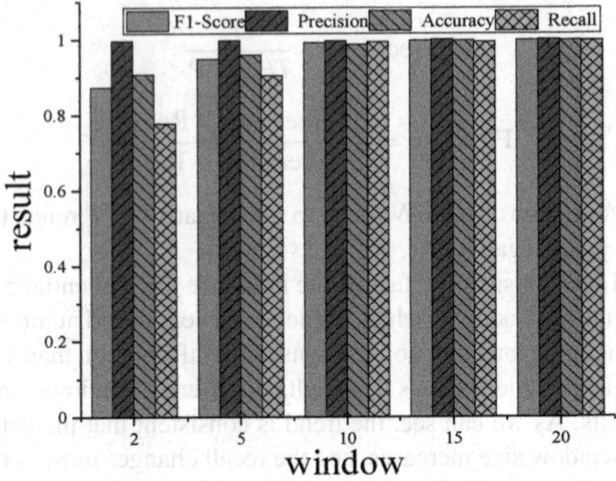

Fig. 9. The results of building sequential behaviour using different windows

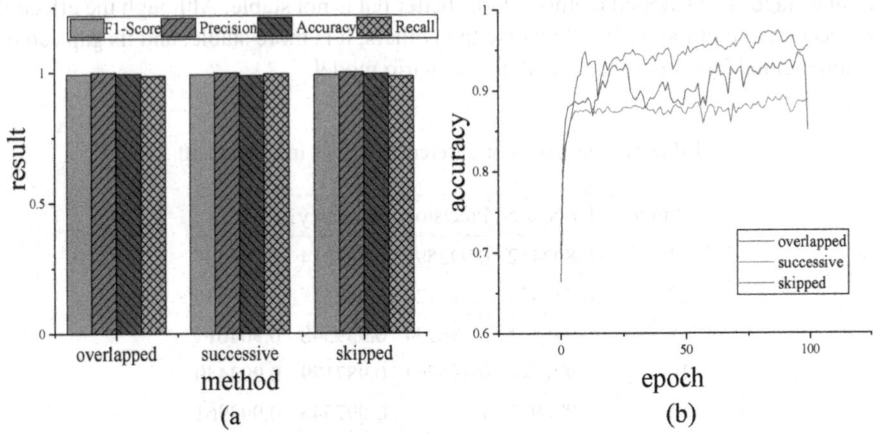

Fig. 10. The effect of different building methods

approach stabilizes at approximately 99.86%. Then, Table 4 compares several existing methods on the CSIC 2010 dataset, which shows that our approach outperforms the other methods.

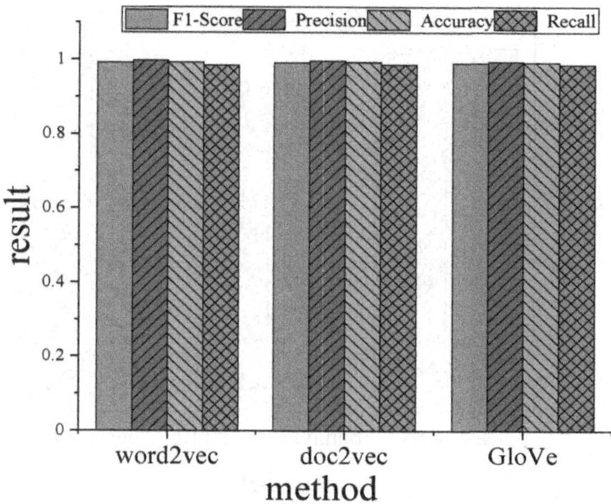

Fig. 11. The differences in the results for three words embedding methods (window = 10, successive)

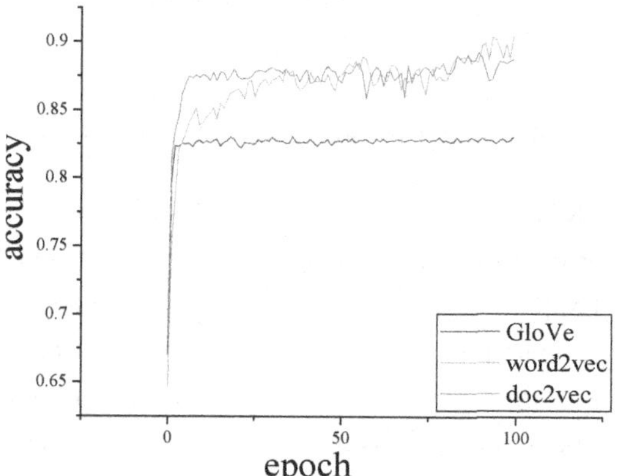

Fig. 12. The trend of the accuracy of different word embedding technologies. (window = 10, successive)

4.3 Discussion

Security Analysis. The proposed approach detects web attacks by identifying the intentions in user behaviours and integrating scenario features, which is more robust than traditional methods. However, there are some possible attacks on it. One possible attack comes from the behavioural model. When building the behavioural model, we require the complete behavioural data for an attacker. If the behavioural data are mixed, the attacker may escape detection by the approach. To address this threat, we can gather as complete of data as possible for each user using multidimensional association analysis to enhance the detectability of user data. Moreover, attacks aimed at models also should be considered. Although the ensemble and iterative updating strategy strengthen the resistance of the approach to them, we should still pay attention to the attacks. The model

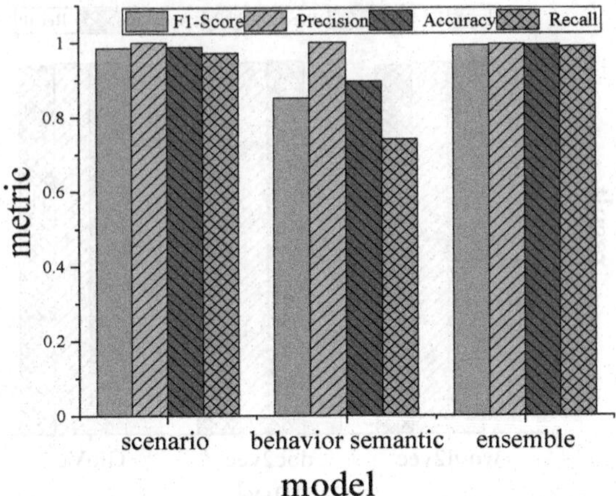

Fig. 13. The effects of each part in the approach

can be protected by ensuring two aspects: reasonable data sampling and the upgrading training process.

Table 4. Comparison on the CSIC2010 dataset

Method	Accuracy	Recall
ModSecurity with CRS [6]	0.5520	0.436
EM [6]	0.7486	0.7516
X-means [6]	0.7493	0.6837
C4.5 [6]	0.9650	0.9914
Model with RNN in [6]	0.8515	0.7403
Model with LSTM in [6]	0.9856	0.9888
Model in [5]	0.9649	0.9335
Model in [9]	0.9881	0.9835
M-ResNet with word2vec in [10]	0.9941	0.9891
Our model	**0.9987**	**0.9967**

5 Conclusion

In this paper, we present a web attack detection approach based on user behaviours, which leverages LSTM to extract hidden user intentions. The approach abstracts the complex data as user behaviours and then constructs the independent user behaviours as a sequential behaviour model. The presented approach not only applies the user intentions extracted from sequential behaviours, but it also adds extra inherent features of abnormal behaviour and leverages ensemble learning techniques, resulting in its efficient practicality. In the experiments, the approach achieves 99.876% detection accuracy on CSIC 2010, which shows its effectiveness in identifying the malicious intentions of attackers.

Acknowledgement. This work was supported by the National Natural Science Foundation of China (Grant No. 61672494), the National key research and development program of China (Grant No. 2018YFB1800705), the Key Research and Development Program for Guangdong Province (Grant No. 2019B010136001).

References

1. OWASP Top Ten https://owasp.org/www-project-top-ten/. Accessed 02 July 2020
2. Kar, D., Panigrahi, S., Sundararajan, S.: SQLiDDS: SQL injection detection using query transformation and document similarity. In: Natarajan, R., Barua, G., Patra, M.R. (eds.) ICDCIT 2015. LNCS, vol. 8956, pp. 377–390. Springer, Cham (2015). https://doi.org/10.1007/978-3-319-14977-6_41
3. Johari, R., Sharma, P.: A survey on web application vulnerabilities (SQLIA, XSS) exploitation and security engine for SQL injection. In: 2012 International Conference on Communication Systems and Network Technologies, Piscataway, pp. 453–458. IEEE (2012)
4. Kuang, X., et al.: DeepWAF: detecting web attacks based on CNN and LSTM models. In: Vaidya, J., Zhang, X., Li, J. (eds.) CSS 2019. LNCS, vol. 11983, pp. 121–136. Springer, Cham (2019). https://doi.org/10.1007/978-3-030-37352-8_11
5. Zhang, M., Xu, B., Bai, S., Lu, S., Lin, Z.: A deep learning method to detect web attacks using a specially designed CNN. In: Liu, D., Xie, S., Li, Y., Zhao, D., El-Alfy, E.-S.M. (eds.) ICONIP 2017. LNCS, vol. 10638, pp. 828–836. Springer, Cham (2017). https://doi.org/10.1007/978-3-319-70139-4_84
6. Liang, J., Zhao, W., Ye, W.: Anomaly-based web attack detection: a deep learning approach. In: Proceedings of the 2017 VI International Conference on Network, Communication and Computing (ICNCC 2017), New York, pp. 80–85. ACM (2017)
7. Liu, T., Qi, Y., Shi, L., et al.: Locate-then-detect: real-time web attack detection via attention-based deep neural networks. In: Twenty-Eighth International Joint Conference on Artificial Intelligence IJCAI-19, Morgan Kaufmann, San Francisco, pp. 4725–4731 (2019)
8. Corona, I., Ariu, D., Giacinto, G.: HMM-Web: a framework for the detection of attacks against web applications. In: 2009 IEEE International Conference on Communications, Piscataway, pp. 1–6. IEEE (2009)
9. Ito, M., Iyatomi, H.: Web application firewall using character-level convolutional neural network. In: 2018 IEEE 14th International Colloquium on Signal Processing & Its Applications (CSPA), Piscataway, pp. 103–106. IEEE (2018)
10. Tian, Z., Luo, C., Qiu, J., Du, X., Guizani, M.: A distributed deep learning system for web attack detection on edge devices. IEEE Trans. Ind. Inform. **16**(3), 1963–1971 (2020)
11. HTTP DATASET CSIC 2010. https://www.isi.csic.es/dataset/. Accessed 02 July 2020
12. Mikolov, T., Chen, K., Corrado, G., Dean, J.: Efficient estimation of word representations in vector space. In: 1st International Conference on Learning Representations, ICLR 2013 (2013)

13. Le, Q., Mikolov, T.: Distributed representations of sentences and documents. In: Proceedings of the 31st International Conference on International Conference on Machine Learning - Volume 32 (ICML 2014), pp. 1188–1196. ACM, New York (2014)
14. Pennington, J., Socher, R., Manning, C.: Glove: global vectors for word representation. In: Proceedings of the 2014 Conference on Empirical Methods in Natural Language Processing, Stroudsburg, pp. 1532–1543. ACL (2014)
15. Qiu, M., Ming, Z., Li, J., Liu, S., Wang, B., Lu, Z.: Three-phase time-aware energy minimization with DVFS and unrolling for chip multiprocessors. J. Syst. Arch. **58**, 439–445 (2012)
16. Qiu, M., Sha, H.M., Liu, M., et al.: Energy minimization with loop fusion and multi-functional-unit scheduling for multidimensional DSP. J. Parallel Distrib. Comput. **68**, 443–455 (2008)
17. Shao, Z., Xue, C., Zhuge, Q., et al.: Security protection and checking in embedded system integration against buffer overflow attacks. IEEE Trans. Comput. **55**, 443–453 (2006)
18. Qiu, M., Chen, Z., Niu, J., et al.: Data allocation for hybrid memory with genetic algorithm. IEEE Trans. Emerg. Top. Comput. **3**, 544–555 (2015)
19. Li, J., Ming, Z., Qiu, M., et al.: Resource allocation robustness in multi-core embedded systems with inaccurate information. J. Syst. Arch. **57**(9), 840–849 (2011)

A Supervised Anonymous Issuance Scheme of Central Bank Digital Currency Based on Blockchain

Wenhao Dai[1,2], Xiaozhuo Gu[1(✉)], and Yajun Teng[1,2]

[1] State Key Laboratory of Information Security, Institute of Information Engineering, CAS, Beijing, China
guxiaozhuo@iie.ac.cn

[2] School of Cyber Security, University of Chinese Academy of Sciences, Beijing, China

Abstract. The emergence of digital currency is seen as a new major revolution in the form of currency. In this revolution, central banks around the world have been focusing on exploring the development and implementation of central bank digital currency (CBDC). Up to now, CBDC is still in the stage of research, and there is no successful experience or precedent worldwide. Existing schemes based on the "indirect CBDC" model, which are deployed on a transparent blockchain, have the problem of financial data leakage of commercial banks. In this paper, we propose a *supervised anonymous issuance* (SAI) scheme of CBDC based on blockchain using a "special" anonymous multi-receiver certificateless signcryption scheme proposed and zero-knowledge proofs, which guarantees the anonymity of commercial banks and the confidentiality of the amount under the supervision of the central bank. Our scheme is secure under some cryptographic assumptions. In the practical instantiation of the SAI scheme, transactions are about 2 KB and take under 6 ms to verify.

Keywords: Central bank digital currency · Supervised anonymity · Blockchain · Zero-knowledge proofs · Certificateless cryptosystem

1 Introduction

In more than a decade, bitcoin [1] has gone from being an obscure curiosity to a household name. In the meantime, hundreds of other cryptocurrencies have emerged in the market. While bitcoin or its sisters unlikely displace sovereign currencies, they have demonstrated the viability of the underlying blockchain technology [2]. However, the promotion of stable cryptocurrencies, such as Libra [3], will have an impact on the sovereign currencies of some countries. In the future, the emergence of a widely accepted and stable cryptocurrency will cause businesses and residents to turn to the cryptocurrency for payment instead of the sovereign currency that continues to depreciate in their country [4]. The

M. Qiu (Ed.): ICA3PP 2020, LNCS 12454, pp. 475–493, 2020.
https://doi.org/10.1007/978-3-030-60248-2_32

implementation of central bank digital currency (CBDC) can reduce cross-border payment fees, improve payment efficiency, restructure the credit system, prevent crime and anti-money laundering, and improve financial inclusion [5–7]. The Bank for International Settlements survey shows that 70% of the countries taking part in the survey in 2018 are or will be participating in the study of CBDC [8]. So far, CBDC is still in the stage of research and exploration, and there is no successful experience or precedent in the world [9].

It is common to divide payments of CBDCs into retail and wholesale segments [10]. Retail payments are relatively low-value transactions such as cheques, credit transfers, direct debits and card payments. These schemes [11–13] aim to implement retail CBDCs. By contrast, wholesale payments, such as RSCoin [14], CADcoin [15] and SGD [16], are large-value and high-priority transactions. One of the reasons for the intense study from central banks is that many central bank-operated wholesale payment systems are at the end of their technological life cycles with inefficient architectures [17]. There are two possible technical architectures of CBDC, "direct CBDC" and "indirect CBDC". In the direct CBDC model, the CBDC represents a direct claim on the central bank, which keeps a record of all balances and updates it with every transaction. However, this model brings the potential adverse impacts on the banking system caused by a shift in deposits from commercial banks to the central bank [18]. In addition, this entails compromises in terms of the payment system's reliability, speed and efficiency. The indirect CBDC model proposed by [19] is also known as the "two-tier CBDC" for its resemblance to the existing two-tier financial system. For consumers, this type of CBDC is not a direct claim on the central bank.Instead, commercial banks, the central bank authorizes to issue CBDCs, handle all communication with retail clients, net payments and send payment messages to other commercial banks and wholesale payment instructions to the central bank. In the indirect CBDC model, wholesale CBDCs are executed on the first tier of the model between commercial banks and the central bank, and retail CBDCs are executed on the second tier between commercial banks and consumers.

Privacy of financial data is highly important due to its sensitive and potentially valuable nature [20]. As far as we know, the first tier of existing schemes in indirect CBDC model is directly deployed on a transparent blockchain, which exposes the information of commercial banks to issue and settle CBDCs. Under the traditional centralized system, there is no such leakage of financial data. The issuance of CBDCs is the first step in CBDCs' life cycle including issuance, settlement and redemption. Therefore, it is necessary to propose an issuance scheme of CBDC with strong privacy guarantees under the supervision of the central bank to solve this financial data leakage problem.

Combining the characteristics of the indirect CBDC model mentioned in [21] and the strong privacy requirements, we propose a supervised anonymous issuance (SAI) scheme of central bank digital currency based on blockchain using a specific anonymous multi-receiver certificateless signcryption scheme and zero-knowledge proofs. Our SAI scheme can be summarized as follows:

- **Issuer anonymity.** A licensed commercial bank can use a new key pair generated by itself to issue a CBDC to ensure that the identity of the issuer is dynamically hidden, and that the amount is always kept encrypted.
- **Supervision of the central bank.** Commercial banks are authorized by the central bank to issue CBDCs. While guaranteeing the anonymity of issuers, the central bank can check online on the blockchain whether the new address issuing CBDCs belongs to a licensed commercial bank and whether the circulation is within the permitted range.
- **Public verifiability.** While ensuring the anonymity of issuers, the public or other commercial banks can verify whether the issuer is one of the licensed commercial banks and whether its circulation is within the allowable range.

The remainder of this paper is organized as follows. Section 2 provides some preliminaries. We define and construct SAI schemes in Sect. 3. Section 4 discusses the concrete instantiation in our scheme. Section 5 shows the performance of our prototype implementation.

2 Preliminaries

The main cryptographic primitives used in this paper are a special kind of *Succinct Non-interactive ARgument of Knowledge* (SNARK) and certificateless-based multi-receiver signcryption schemes. Next, we briefly introduce a *publicly-verifiable preprocessing zeroknowledge* SNARK (zk-SNARK) and the system model of anonymous multi-receiver certificateless signcryption (AMCLS).

2.1 zk-SNARKs

A zk-SNARK is an efficient variant of a zero-knowledge proof of knowledge [22]. We introduce the following definition given in [23].

For a field $\mathbb{F}$, an $\mathbb{F}$-*arithmetic circuit* takes elements in $\mathbb{F}$ as inputs, and its gates output elements in $\mathbb{F}$. We naturally associate a circuit with the function it computes. To model nondeterminism, we consider circuits that have an input $x \in \mathbb{F}^n$ and an auxiliary input $a \in \mathbb{F}^h$, called a *witness*. Arithmetic circuit satisfiability is defined analogously to the boolean case, as follows.

Definition 1. *The arithmetic circuit satisfiability problem of an* $\mathbb{F}$-*arithmetic circuit* $C : \mathbb{F}^n \times \mathbb{F}^h \to \mathbb{F}^l$ *is captured by the relation* $R_C = \{(x, a) \in \mathbb{F}^n \times \mathbb{F}^h : C(x, a) = 0^l\}$*; its language is* $\mathcal{L}_C = \{x \in \mathbb{F}^n : \exists a \in \mathbb{F}^h \text{ s.t. } C(x, a) = 0^l\}$.

Given a field $\mathbb{F}$, a (publicly-verifiable preprocessing) zk-SNARK for $\mathbb{F}$-arithmetic circuit satisfiability is a triple of polynomial-time algorithms $(\mathit{KeyGen}, \mathit{Prove}, \mathit{Verify})$:

- $\mathit{KeyGen}(1^\lambda, C) \to (pk, vk)$. On input a security parameter λ and an $\mathbb{F}$-arithmetic circuit C, *KeyGen* probabilistically samples a proving key *pk* and a verification key *vk*. Both keys are published as public parameters and can be used, any number of times, to prove/verify membership in $\mathcal{L}_C$.

- *Prove*$(pk, x, a) \to \pi$. On input a proving key pk and any $(x, a) \in R_C$, *Prove* outputs a non-interactive proof π for the statement $x \in \mathcal{L}_C$.
- *Verify*$(vk, x, \pi) \to b$. On input a verification key vk, an input x and a proof π, *Verify* outputs $b = 1$ if he is convinced that $x \in \mathcal{L}_C$.

A zk-SNARK satisfies the properties of *completeness, succinctness, soundness* and *perfect zero knowledge.*

2.2 The System Model of AMCLS Scheme

The supervision idea of our scheme is inspired by the public-private key pair generation method of the certificateless public key cryptosystem [24]. A user's public-private key pair is divided into two parts, one part is generated by the key generation center, and the other part is generated by itself. The central bank acts as the key generation center, and uses an anonymous multi-receiver signencryption scheme to supervise the behavior of users.

Selivi *et al.* [25] proposed the first certificateless multi-receiver signencryption scheme. However, the scheme does not guarantee the anonymity of receivers. Islam *et al.* [26] proposed the anonymous multi-receiver certificateless encryption scheme (AMCLE) in 2004. Pang L *et al.* [27] put the signature into AMCLE and proposed the AMCLS scheme. We illustrate the system model of AMCLS scheme that consists of the following algorithms.

- $G_{amcls}(1^\lambda) \to (pk_{master}, sk_{master}, pp_{amcls})$. On input a security parameter λ, G_{amcls} generates a master key pair $(pk_{master}, sk_{master})$ and public parameters pp_{amcls} for AMCLS.
- $P_{amcls}(pp_{amcls}, sk_{master}, ID) \to (pk_{auth}, sk_{auth})$. On input public parameters pp_{amcls}, a master secret key sk_{master}, and user's identity ID, P_{amcls} outputs a partial key pair (pk_{auth}, sk_{auth}).
- $K_{amcls}(pp_{amcls}, pk_{auth}, sk_{auth}) \to (pk_{amcls}, sk_{amcls})$. On input public parameters pp_{amcls} and a partial key pair (pk_{auth}, sk_{auth}), K_{amcls} outputs a complete key pair, where $pk_{amcls} = (pk_{auth}, pk_{pay})$ and $sk_{amcls} = (sk_{auth}, sk_{pay})$.
- $\mathcal{E}_{amcls}(pp_{amcls}, (pk^{sender}_{amcls}, sk^{sender}_{amcls}), pk^1_{amcls}, \ldots, pk^t_{amcls}, m) \to c$. On input public parameters pp_{amcls}, a sender's complete key pair $(pk^{sender}_{amcls}, sk^{sender}_{amcls})$, receivers' public keys $pk^1_{amcls}, \ldots, pk^t_{amcls}$, and a message m, $\mathcal{E}_{amcls}$ outputs a signcryption ciphertext.
- $\mathcal{D}_{amcls}(pp_{amcls}, sk^i_{amcls}, pk^{sender}_{amcls}, c) \to m$. On input public parameters pp_{amcls}, a receiver's secret key sk^i_{amcls}, a sender's public key pk^{sender}_{amcls}, and a signcryption ciphertext c, $\mathcal{D}_{amcls}$ outputs a message m (or $\perp$ if it fails).

An AMCLS scheme used in our scheme should satisfy the security property of *indistinguishability* and *anonymous indistinguishability of certificateless signcryptions against selective multi-ID chosen ciphertext attack* (IND-CLMS-CCA & ANON-CLMS-CCA) and *correctness.*

2.3 Others Building Blocks

We assume familiarity with the definitions of these building blocks; for more details, see [28]. Throughout, λ denotes the security parameter.

Collision-Resistant Hashing. We use a collision-resistant hash function $H : \{0,1\}^* \rightarrow \{0,1\}^{O(\lambda)}$.

Statistically-Hiding Commitment. We use a commitment scheme $COMM$ where the binding property holds computationally, while the hiding property holds statistically. It is denoted $\{COMM_x : \{0,1\}^* \rightarrow \{0,1\}^{O(\lambda)}\}_x$ where x denotes the commitment trapdoor.

Strongly-Unforgeable Digital Signature. We use a digital signature scheme $Sig = (G_{sig}, K_{sig}, S_{sig}, V_{sig})$. The signature scheme Sig satisfies the security property of *strong unforgeability against chosen-message attacks* (SUF-CMA).

Symmetric Encryption. We use a symmetric encryption scheme $Enc = (G_{enc}, K_{enc}, E_{enc}, D_{enc})$. The encryption scheme Enc satisfies the security property of *message indistinguishability under chosen-ciphertext attack.*

3 The SAI Scheme

In this section, we first introduce the data structures and the definition of the SAI scheme. Then, we show how to achieve anonymity, supervision and public verifiability using a special AMCLS and zk-SNARKs. Finally, we analyze the security of our scheme.

3.1 Data Structures

We begin by describing data structures used by the SAI scheme, which are summarized in Fig. 1. The algorithms that use and produce these data structures are introduced in Sect. 3.2.

Basecoin Ledger. Our scheme is applied on top of a ledger-based central bank digital currency; for generality we refer to this base currency as *Basecoin*. At any given time T, all users have access to L_T, the ledger at time T, which is a sequence of transactions. The ledger is append-only.

Address. An issuer has different payment key pairs (sk_{pay}, pk_{pay}) to issue CBDCs, and also has one authentic key pair (sk_{auth}, pk_{auth}), which is generated by the central bank according to the identity information of the issuer. Therefore, the complete secret key $addr_{sk}$ and public key $addr_{pk}$ are represented as (sk_{auth}, sk_{pay}), (pk_{auth}, pk_{pay}), respectively. The payment public key pk_{pay} is published on the blockchain as the public key on the traditional blockchain, but the authentic public key pk_{auth} appears in the form of ciphertext.

Central Bank Digital Currency. We use the CBDC expression in the cryptographic form mentioned in [9] and extend it to fit our scheme. EXP_{CBDC}

equals $\{id, Crypto_{value}, owner, Crypto_{issuer}, Crypto_{key}, ExtSet\}$, where *Crypto* stands for symmetric encryption operation on attribute elements. The expression of CBDC includes a unique identifier *id*, a ciphertext of a denomination $Crypto_{value}$, ownership *owner*, a ciphertext of issuer's identity $Crypto_{issuer}$, a related information of symmetric key $Crypto_{key}$, and extended attribute set *ExtSet*.

Commitment List of Issuers' Balance. The central bank maintains and regularly releases a commitment list of issuers' balance $List = \{cm_i | i = 1, \ldots, n\}$, where n is the total number of licensed commercial banks, on the blockchain. A commitment of issuer's balance is $cm = COMM_{HASH_{head}}(pk_{auth} \| Balance)$, where pk_{auth} is an issuer's authentic public key, $HASH_{head}$ is a hash value of the block that includes issuing transactions, and *Balance* denotes the current balance of the issuer.

New Transactions. Besides *Basecoin* transactions, there are two new types of transactions: *Issue* and *Trust*.

An issue transaction tx_{Issue} is a tuple $(type, EXP_{CBDC}, owner, *)$, where *type* is set as 1 to indicate the type of transaction, EXP_{CBDC} is a new CBDC that belongs to the *owner*, and $*$ denotes other (implementation-dependent) information. The transaction tx_{Issue} records that the new CBDC has been issued by a commercial bank.

A trust transaction tx_{Trust} is a tuple $(type, List^{new}, HASH^{new}_{head}, HASH^{last}_{head}, *)$, where *type* is set as 0, $List^{new}$ is a new commitment list, $HASH^{new}_{head}$ is a hash value of the block pointed to by $List^{new}$, $HASH^{last}_{head}$ is a hash value of the block pointed to by the last commitment list $List^{last}$, and $*$ denotes other (implementation-dependent) information. The public can utilize tx_{Trust} to verify the legitimacy of the issue transactions in the block $HASH^{new}_{head}$.

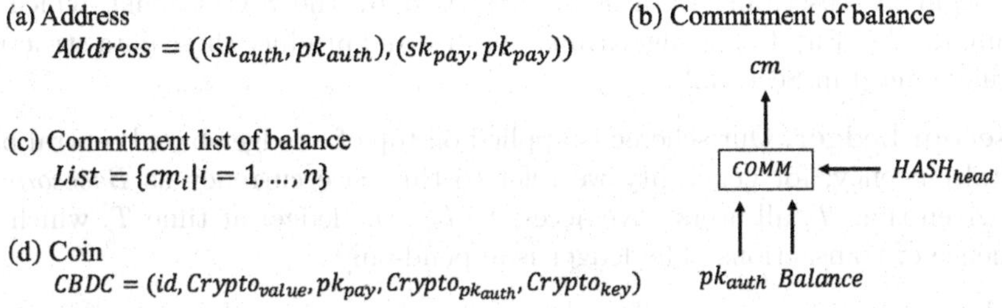

(a) Address
Address = ((sk_auth, pk_auth), (sk_pay, pk_pay))
(b) Commitment of balance
cm
COMM
HASH_head
pk_auth Balance
(c) Commitment list of balance
List = {cm_i | i = 1, ..., n}
(d) Coin
CBDC = (id, Crypto_value, pk_pay, Crypto_pk_auth, Crypto_key)

Fig. 1. (a) Structure of address. (b) Structure of a commitment of issuer's balance. (c) Structure of a commitment list of issuers' balance. (d) Structure of a coin.

3.2 Algorithms in the SAI Scheme

A SAI scheme is a tuple of polynomial-time algorithms

$$(Setup, CreateAddress, Issue, Trust, UpdateBalance, VerifyTransaction)$$

with the following syntax.

System Setup. The algorithm *Setup* generates a list of public parameters and the central bank's address key pair.

Setup
- inputs :
  - security parameter λ
- outputs :
  - central bank's address key pair $(addr^c_{pk}, addr^c_{sk})$
  - public parameters pp

The algorithm *Setup* is executed by the central bank. The resulting public parameters pp are published and made available to all parties. *Setup* is done only once; afterwards, no trusted party is needed.

Creating Addresses. The algorithm *CreateAddress* generates a complete key pair.

CreateAddress
- inputs :
  - public parameters pp
  - central bank's address key pair $(addr^c_{pk}, addr^c_{sk})$
  - identity ID
- outputs :
  - address key pair $(addr_{pk}, addr_{sk})$

The algorithm *CreateAddress* is executed jointly by the central bank and a licensed commercial bank. The resulting $addr_{pk}$ and $addr_{sk}$ equals (pk_{auth}, pk_{pay}), (sk_{auth}, sk_{pay}), respectively. Each licensed commercial bank has only one authentic key pair (pk_{auth}, sk_{auth}) which is generated by the central bank through P_{amcls} (introduced in Sect. 2.2), but can generate at least one payment pair (pk_{pay}, sk_{pay}) by itself.

Issuing CBDCs. The algorithm *Issue* generates a CBDC (of a given value v) and an issue transaction.

Issue
- inputs :
  - public parameters pp
  - coin value v
  - authentic key pair $(pk^{issuer}_{pay}, sk^{issuer}_{pay})$
  - central bank's address key $addr^c_{pk}$
- outputs :
  - coin EXP_{CBDC}
  - issue transaction tx_{Issue}

The algorithm *Issue* is executed by a licensed commercial bank. The commercial bank selects a new payment key pair $(pk^{issuer}_{pay}, sk^{issuer}_{pay})$ to issue a CBDC for achieving anonymity. The output coin EXP_{CBDC} has the ciphertext of a value v to achieving confidentiality; the output issue transaction tx_{Issue} equals $(1, EXP_{CBDC}, pk_{pay}^{issuer}, *)$, where $*$ denotes a signature information.

Trusting Coins. The algorithm *Trust* executed by the central bank generates a trust transaction for the public to verify the legitimacy of all CBDCs issued in a block.

Trust
- inputs :
  - public parameters pp
  - central bank's address key pair $(addr_{pk}^{c}, addr_{sk}^{c})$
  - issuers' authentic public keys $(pk_{auth}^{1}, \dots, pk_{auth}^{n})$
  - last balance list of commercial banks $(Balance_1^{last}, \dots, Balance_n^{last})$
  - new balance list of commercial banks $(Balance_1^{new}, \dots, Balance_n^{new})$
  - last trust transaction tx_{Trust}^{last}
  - new block $HASH_{head}^{new}$
  - issue transactions $(tx_1, \dots, tx_m)$
- outputs :
  - trust transaction tx_{Trust}

The algorithm *Trust* takes as input the center bank's address key pair $(addr_{pk}^{center}, addr_{sk}^{center})$, issuers' authentic public keys $(pk_{auth}^{1}, \dots, pk_{auth}^{n})$, the last trust transaction tx_{Trust}^{last}, the last balance list of commercial banks $(Balance_1^{last}, \dots, Balance_n^{last})$, and a new balance list of commercial banks $(Balance_1^{new}, \dots, Balance_n^{new})$. To ensure that the issue transactions in the block $HASH_{head}^{new}$ are legal, the algorithm *Trust* also takes the issue transactions $(tx_1, \dots, tx_m)$ in the block as input, where m is the total number of issue transactions.

The output tx_{trust}^{new} equals $(0, List^{new}, HASH_{head}^{new}, HASH_{head}^{last}, *)$, where $List^{new}$ is a new commitment list, $HASH_{head}^{new}$ is a hash value of the block pointed to by $List^{new}$, $HASH_{head}^{last}$ is a hash value of the block pointed to by $List^{last}$, and $*$ denotes zero-knowledge proofs π. Crucially, without revealing the identity and balance of issuers, tx_{trust} convinces the public that CBDCs are legal in $HASH_{head}^{new}$.

Updating Balance. The algorithm *UpdateBalance* executed by the central bank verifies issue transactions and updates balance of commercial banks.

UpdateBalance
- inputs :
  - public parameters pp
  - central bank's private key $addr_{sk}^{c}$
  - last balance list of commercial banks $(Balance_1^{last}, \dots, Balance_n^{last})$
  - new block $HASH_{head}^{new}$
  - the current ledger L
- outputs :
  - new balance list of commercial banks $(Balance_1^{new}, \dots, Balance_n^{new})$

The algorithm *UpdateBalance* takes the central bank's secret key $addr^c_{sk}$, the last balance list $(Balance^{last}_1, \ldots, Balance^{last}_n)$, and the block header hash $HASH^{new}_{head}$ as input. In order to calculate the circulation of commercial banks, the algorithm needs to find the issue transactions according to $HASH^{new}_{head}$ in the current ledger L. If one of the issuers is not authorized or there is over-issue in the issue transactions, the algorithm *UpdateBalance* outputs $\perp$; else outputs $(Balance^{new}_1, \ldots, Balance^{new}_n)$.

Verifying Transactions. The algorithm *VerifyTransaction* checks the validity of a transaction.

VerifyTransaction
- inputs :
  - –public parameters pp
  - –trust transaction tx_{Trust}
  - –the current ledger L
- outputs :
  - –bit b, equals 1 if the transaction is valid

The algorithm *VerifyTransaction* is executed by the public. Trust transactions must be verified before being considered well-formed. In practice, for privacy considerations, the public can indirectly verify issue transactions by the trust transaction. The output b equals 1 if the transaction is valid.

3.3 AMCLS for Anonymity and Supervision

We propose a special AMCLS scheme suitable for blockchain, through which we can realize the anonymity of commercial banks and the confidentiality of the amount under the supervision of the central bank. The algorithms $\mathcal{E}_{amcls}$ and $\mathcal{D}_{amcls}$ of the special AMCLS scheme work as follows.

- $\mathcal{E}_{amcls}(pp_{amcls}, (pk^{sender}_{amcls}, sk^{sender}_{amcls}), pk^1_{amcls}, \ldots, pk^t_{amcls}, m) \rightarrow c$. On input public parameters pp_{amcls}, a sender's key pair $(pk^{sender}_{amcls}, sk^{sender}_{amcls})$, receivers' public keys $pk^1_{amcls}, \ldots, pk^t_{amcls}$ and a message m, the sender does the following:
  - Randomly sample a $\beta \in \{0, 1\}^w$, then compute $r = H_1(\beta, pk^{sender}_{auth})$ and $U = r \cdot P$(P derived from pp_{amcls}).
  - For $i = 1, \ldots, t$, compute $F_i = r \cdot pk^i_{pay}$, $K_i = r \cdot pk^i_{auth}$ and $T_i = H_1(K_i, F_i)$.
  - Compute $C_i = H_1(T_i) \| (H_2(T_i) \bigoplus \beta)$ where $\|$ indicates connection operation.
  - Compute symmetric key $sk = H_2(\beta)$, $sk = sk_1 \| sk_2$, $V = E_{enc}(sk_1, m)$ and $\Gamma = E_{enc}(sk_2, pk^{sender}_{auth})$.
  - Sign the message with the sender's secret key as follows:
    $H^{'} = H_1(U, V, \Gamma, pk^{sender}_{auth}, pk^{sender}_{pay}), W = sk^{sender}_{auth} + r * H^{'} + sk^{sender}_{pay} * H^{'}$.
  - Perform the hash operation on ciphertext to ensure the data integrity $\Lambda = H_1(v, pk^{sender}_{auth}, \beta, C_1, \ldots, C_t, V, \Gamma, U, W)$.

- Output ciphertext $c = ((C_1, \ldots, C_t), V, \Gamma, W, U, \Lambda)$.

- $\mathcal{D}_{amcls}(pp_{amcls}, sk^i_{amcls}, pk^{sender}_{pay}, c) \rightarrow (m, pk^{sender}_{auth})$. On input public parameters pp_{amcls}, a receiver's secret key sk^i_{amcls}, the sender's payment public key pk^{sender}_{pay}, and a signcryption ciphertext c, the receiver i does the following:
  - Parse ciphertext $c = ((C_1, \ldots, C_t), V, \Gamma, W, U, \Lambda)$.
  - Compute $K = sk^i_{auth} \cdot U$, $F = sk^i_{pay}$, $T = H_1(K, F)$ and $H_1(T)$.
  - The corresponding C_i can be found by using $H_1(T)$ through $C_i = H_1(T) \| Y$ where Y represents the remaining string by removing $H_1(T)$ from C_i.
  - Recover $\beta' = Y \bigoplus H_2(T)$.
  - Compute $sk' = H_2(\beta')$, $sk' = sk_1' \| sk_2'$, $m' = D_{enc}(sk_1', V)$, $pk'_{auth} = D_{enc}(sk_2', \Gamma)$ and $\Lambda' = H_1(m', pk'_{auth}, \beta', C_1, \ldots, C_t, V, \Gamma, U, W)$.
  - Compute $H' = H_1(U, V, \Gamma, pk'_{auth}, pk^{sender}_{pay})$.
  - If $\Lambda' == \Lambda$ and $pk'_{auth} + U \cdot H' + pk^{sender}_{pay} \cdot H' == P \cdot W$, output the message m and the sender's authentic public key pk'_{auth}; else return $\perp$.

Next, we recall the context motivating **Issue**. When a licensed commercial bank issues a CBDC, a corresponding issue transaction

$$tx_{Issue} = (1, EXP_{CBDC}, pk_{pay}, *)$$

is generated, where EXP_{CBDC} is $\{id, Crypto_{value}, pk_{pay}, Crypto_{issuer}, Crypto_{key}, ExtSet\}$. In our structure, we need to provide a signature in $*$ to resist the transaction malleability attack. Further, tx_{Issue} means that (i) a new payment public key pk_{pay} generated by an issuer possesses the CBDC; (ii) the issuer's identity information is contained in the ciphertext $Crypto_{issuer}$; (iii) the denomination is contained in the ciphertext $Crypto_{value}$; (iv) the key of symmetric encryption is contained in the ciphertext $Crypto_{key}$; (v) the central bank can obtain the symmetric key through $Crypto_{key}$ so that $Crypto_{issuer}$ and $Crypto_{value}$ can be decrypted; (vi) id is unique in the ledger. Figure 2 gives the pseudocode for *Issue*. Subsequently, the central bank can call the algorithm $\mathcal{D}_{amcls}$, which is the subroutine of *UpdateBalance*, to decrypt tx_{Issue} to achieve supervision.

3.4 zk-SNARKs for Public Verifiability

Our construction invokes a zk-SNARK for a specific NP statement, statement **Trust**, and we now define it. We first recall the context motivating **Trust**. When the central bank proves that the issue transactions in the block $HASH^{new}_{head}$ are legal, a corresponding trust transaction

$$tx_{Trust} = (0, List^{new}, HASH^{new}_{head}, HASH^{last}_{head}, *)$$

is generated. $List^{new}$ is represented as $\{cm^{new}_1, \ldots, cm^{new}_n\}$ where cm^{new}_i is $COMM_{HASH^{new}_{head}}(pk^i_{auth} \| Balance^{new}_i)$ and n is the number of licensed commercial banks.

In our construction, we need to provide zero-knowledge proofs π in "$*$" that various conditions were respected by the trust operation. Concretely, tx_{Trust}

Issue

• INPUTS:

- public parameters pp
- coin value $v \in \{0, 1, \ldots, v_{max}\}$
- issuer's authentic key pair $(pk^A_{auth}, sk^A_{auth})$
- central bank public key $addr^c_{pk}$

• OUTPUTS: coin EXP_{CBDC} and issue transaction tx_{Issue}

1. Compute $(addr^A_{pk}, addr^A_{sk}) := K_{amcls}(pp, pk^A_{auth}, sk^A_{auth})$.
2. Parse $addr^A_{pk} = (pk^A_{auth}, pk^A_{pay})$ and $addr^A_{sk} = (sk^A_{auth}, sk^A_{pay})$.
3. Set $c := \mathcal{E}_{amcls}(pp, (addr^A_{pk}, addr^A_{sk}), addr^c_{pk}, addr^A_{pk}, v)$.
4. Parse c as $((C_A, C_c), V, \Gamma, W, U, \Lambda))$.
5. Select a unique id.
6. Set $EXP_{CBDC} := \{id, V, pk^A_{pay}, \Gamma, (C_A, C_c), (W, U, \Lambda)\}$.
7. Compute $\sigma := S_{sig}(sk^A_{pay}, (1, EXP_{CBDC}))$.
8. Set $tx_{Issue} := (1, EXP_{CBDC}, pk^A_{pay}, \sigma)$.
9. Output EXP_{CBDC} and tx_{Issue}.

Fig. 2. Pseudocode for **Issue**.

should demonstrate that (i) $List^{last}$ and $List^{new}$ point to $HASH^{last}_{head}$, $HASH^{new}_{head}$, respectively. (ii) the issuers in this block $HASH^{new}_{head}$ are authorized. (iii) for each issuer i, the new balance $Balance^{new}_i$ is obtained by subtracting the total issuance of the issuer i in the block $HASH^{new}_{head}$ from the last balance $Balance^{last}_i$. (iv) the issue transactions in the block $HASH^{new}_{head}$ are legal and there is no over-issue. Our construction achieves this by including zk-SNARK proofs π_i for the statement **Trust** which checks the above invariants.

For each issuer i, the NP statement is defined as follows (Fig. 3 gives the pseudocode for *Trust*).

- Instances are of the form $x_i = (i, tx_1, \ldots, tx_m, h_{sig}, h, HASH^{last}_{head}, HASH^{new}_{head}, cm^{last}_i, cm^{new}_i)$. Thus, an instance x_i specifies the serial number i of the issuer, the issue transactions $tx_1, \ldots, tx_m$ in the block $HASH^{new}_{head}$, fields h_{sig}, h used for non-malleability, two hashes for the two blocks, and two commitments of issuers' balance.
- Witnesses are of the form $a_i = (pk^i_{auth}, sk^c_{auth}, sk^c_{pay}, Balance^{last}_i, Balance^{new}_i)$, where $(sk^c_{auth}, sk^c_{pay})$ is the secret key of the central bank and pk^i_{auth} is the authentic public key of the issuer i.
- Generate a zero-knowledge proof $\pi_i = Prove(pk_{proof}, x_i, a_i)$.

The algorithm *Verify* for zero-knowledge proofs π is a subroutine of algorithm *VerifyTransaction*. The public can set up the public input x to verify tx_{Trust} in order to indirectly verify issue transactions.

3.5 Security

We informally illustrate the security of the SAI scheme by the following properties: *ledger indistinguishability*, *transaction non-mallebility*, and *mutual restraint*.

Trust
- INPUTS:
- public parameters pp
- issuers' authentic public keys $(pk^1_{auth}, \dots, pk^n_{auth})$
- central bank's address key pair $(addr^c_{pk}, addr^c_{sk})$
- last trust transaction tx^{last}_{Trust}
- block header hash $HASH^{new}_{head}$
- issue transactions $tx_1, \dots, tx_m$ on the block $HASH^{new}_{head}$
- last balance list $(Balance^{last}_1, \dots, Balance^{last}_n)$
- new balance list $(Balance^{new}_1, \dots, Balance^{new}_n)$
- OUTPUTS: trust transaction tx^{new}_{Trust}

1. Parse tx^{last}_{Trust} to get $List^{last}$ and $HASH^{last}_{head}$.
2. Parse $add^c_{pk} = (pk^c_{auth}, pk^c_{pay})$ and $add^c_{sk} = (sk^c_{auth}, sk^c_{pay})$.
3. Compute $h_{sig} := H(pk^c_{pay})$ and $h := H(sk^c_{pay}, h_{sig})$.
2. For each issuer $i \in \{1, \dots, n\}$:
   a) Compute $cm^{new}_i := COMM_{HASH^{new}_{head}}(pk^i_{auth} \| Balance^{new}_i)$.
   b) Set $x_i := (i, tx_1, \dots, tx_m, h_{sig}, h, HASH^{last}_{head}, HASH^{new}_{head}, cm^{last}_i, cm^{new}_i)$.
   c) Set $a_i := (pk^i_{auth}, sk^c_{auth}, sk^c_{pay}, Balance^{last}_i, Balance^{new}_i)$.
   d) Compute $\pi_i := Prove(pk_{Trust}, x_i, a_i)$.
3. Set $m := (x_1, \dots, x_n, \pi_1, \dots, \pi_n)$.
4. Compute $\sigma := S_{sig}(sk^c_{pay}, m)$.
5. Set $List^{new} := (cm^{new}_i, \dots, cm^{new}_n)$.
6. Set $tx^{new}_{Trust} := (0, List^{new}, HASH^{new}_{head}, HASH^{last}_{head}, *)$, where $* := (\pi_1, \dots, \pi_n, pk^c_{pay}, h, \sigma)$.
7. Output tx^{new}_{Trust}.

Fig. 3. Pseudocode for **Trust**.

Each property is formalized as a game between a challenger $\mathcal{C}$ and an adversary $\mathcal{A}$. In each game, $\mathcal{C}$ maintains a SAI scheme oracle $\mathcal{O}^{\mathrm{SAI}}$, providing **Issue** queries to maintain a ledger L for making other banks' issue transactions (i.e. $\mathcal{A}$ specifies identities of previous transactions) and **Insert** queries for inserting its own issue transactions directly. During the oracle, $\mathcal{A}$ makes the input values and learns the resulting transaction.

Ledger Indistinguishability. This property indicates that the ledger reveals no new information to $\mathcal{A}$, even when $\mathcal{A}$ can lead honest commercial banks to perform SAI operations of its choice. This property tells us that: (i) apart from the central bank and the initiator of an issue transaction, no one knows which commercial bank the issue transaction belongs to; (ii) for the public, one can not identify if two issue transactions are issued by the same commercial bank.

More theoretically, $\mathcal{A}$ can hardly distinguish between two ledgers L_0 and L_1 in using two SAI scheme oracles, when the queries are *public consistent*: they have matching type, and are identical in terms of publicly-revealed information.

It can be formalized by an experiment called LIND described below. After initialized, the challenger selects a random bit b and allows $\mathcal{A}$ to issue queries to $\mathcal{O}_0^{\mathrm{SAI}}$ and $\mathcal{O}_1^{\mathrm{SAI}}$, controlling the behavior of honest banks on L_0 and L_1. For **Insert** queries, the modification of the ledger from $\mathcal{C}$ depends on b, which means the queries from $\mathcal{O}_0^{\mathrm{SAI}}$ modifies L_b and from $\mathcal{O}_1^{\mathrm{SAI}}$ modifies L_{1-b}, while of the other queries is normal. At the conclusion of the experiment, $\mathcal{A}$ outputs a guess b', and wins if $b = b'$. Due to perfect zero knowledge of zk-SNARK, ANON-CLMS-CCA of AMCLS, and hiding property of COMM, using hybrid lemma, we conclude $\mathcal{A}$ wins LIND with probablity at most negligibly greater than 1/2.

Transaction Non-malleability. This property requires that no bounded adversary can change any data stored within tx_{Issue} and tx_{Trust}. It prevents attackers from modifying transactions before being added to the ledger. This property tells us that: (i) other commercial banks can not issue in place of the real commercial bank; (ii) no one can fake or tamper a trust transaction initiated by the central bank.

It is formalized by two experiments, $\mathrm{TRNM}_{\mathrm{Issue}}$ and $\mathrm{TRNM}_{\mathrm{Trust}}$, indicating both issue transactions and trust transactions are unextendable. For $\mathrm{TRNM}_{\mathbf{X}}$ experiment ($\mathbf{X}$ is *Issue* or *Trust*, the same below), $\mathcal{A}$ adaptively interacts with an oracle $\mathcal{O}^{\mathrm{SAI}}$ and then outputs an $\mathbf{X}$ transaction $tx'_{\mathbf{X}}$. Letting $\mathcal{T}_{\mathbf{X}}$ be the set of $\mathbf{X}$ transactions generated by $\mathcal{O}^{\mathrm{SAI}}$ in response to $\mathbf{X}$ queries, there exists $tx_{\mathbf{X}} \in \mathcal{T}_{\mathbf{X}}$ such that: (i) $tx'_{\mathbf{X}} \neq tx_{\mathbf{X}}$; (ii) if $\mathbf{X}$ is *Issue*, it can be correctly verified by the central bank, and pk^*_{pay} must be equal to that in $tx_{\mathbf{X}}$; (iii) the trust transaction can be verified; (iv) if $\mathbf{X}$ is *Trust*, the last and new hash in $tx^*_{\mathbf{X}}$ must be equal to those in $tx_{\mathbf{X}}$. In other words, $\mathcal{A}$ wins games if it can change the coin created in a different value or it can modify the balance. Due to SUF-CMA of signature, collision resistant of CRH, correctness of AMCLS, and perfect zero knowledge of zk-SNARK, $\mathcal{A}$ wins TRNM with only negligible probability.

Mutual Restraint. This property means that (i) the commercial bank cannot perform malicious operations such as issuing more than its limit; (ii) the central bank also cannot perform such as using incorrect balances to generate a trust transaction. The first point is decided by the correctness of *UpdateBalance* algorithm, the second by soundness of zk-SNARK.

4 Instantiation

We describe a concrete instantiation of the SAI scheme, aiming at a level of security of 128 bits. Doing so requires concrete choices, described next. Later, in Sect. 5, we will show the performance of our scheme on this basis.

4.1 Instantiation of Building Blocks

H from SHA256. We instantiate H via the SHA256 compression function.

COMM from "special" SHA256. Let $\mathcal{H}$ be the "special" SHA256 compression function, which maps a 512-bit input to a 256-bit output. We instantiate *COMM* via $\mathcal{H}$. As for the commitment scheme *COMM*, we only use it in $cm = COMM_{HASH_{head}}(pk_{auth}\|Balance)$. We use an issuer's authentic address a_{auth} instead of pk_{auth}, where $a_{auth} = H(pk_{auth})$. Due to our instantiation of H, a_{auth} and $HASH_{head}$ are 256 bits. We set coin values to be 128-bit integers (so that, $v_{max} = 2^{128} - 1$ in our implementation), and then compute $cm = COMM_{HASH_{head}}(pk_{auth}\|Balance)$ as $\mathcal{H}(\mathcal{H}(HASH_{head}\|a_{auth})\|0^{128}\|Balance)$.

Instantiating Sig. For the signature scheme *Sig*, we use elliptic curve digital signature algorithm to retain consistency and compatibility with the mainstream cryptocurrencies' source code.

Instantiating Enc. For the encryption scheme *Enc*, we use a XOR operation to simplify the size of the Trust circuit as much as possible, because we guarantee one-time pad in the AMCLS scheme (based on information-theoretic security). If necessary, we can use the lightweight symmetric encryption scheme like Speck [29] to replace XOR.

Instantiating AMCLS. For the specific AMCLS scheme proposed by us, major improvements lie in signcryption and designcryption (see Sect. 3.3). It is worth noting we use SHA256 to instantiate H_1 and SHA512 to instantiate H_2.

Instantiating the NP Statement Trust. For efficiency reasons, we do not check the constraints on the entire ciphertext of AMCLS, but simplify it to the constraints on symmetric ciphertexts of AMCLS (This does not weaken the security of the scheme because the commitment of issuers' balance is chained that a new commitment is calculated from the last commitment).

4.2 Arithmetic Circuit for Trust

Our construction also requires zk-SNARKs relative to the NP statement **Trust**. These are obtained by invoking a zk-SNARK for arithmetic circuit satisfiability on an arithmetic circuit C_{Trust} which verifies the NP statement. In our instantiation, we rely on the implementation [30] for the basic zk-SNARK, and apply it to C_{Trust} described next.

SHA256's Compression Function. The vast majority of the "verification work" in **Trust** is verifying computations of $\mathcal{H}$, the special compression function of SHA256. Thus, we begin by discussing our construction of an arithmetic circuit $C_{\mathcal{H}}$ for verifying SHA256 computations. For this arithmetic circuit, we use the $C_{\mathcal{H}}$ which has been optimized in [23] and has 27,904 arithmetic gates. For the "full" hash verification $a_{auth} = H(pk_{auth})$ and $h = H(sk^{c}_{pay}, h_{sig})$, we transform them to $a_{auth} = \mathcal{H}(m_x\|m_y)$, $h = \mathcal{H}(sk^{c}_{pay}\|h_{sig})$, respectively (Since pk_{auth} removes the header identifier as 512 bits, that is, $pk_{auth} = ID_{pk}\|m_x\|m_y$ where ID_{pk} is 8 bits, and m_x, m_y are 256-bit coordinate information identifying the x and y axes, respectively).

Decryption Function. We need to construct an arithmetic circuit C_D that, given a 256-bit ciphertext and a 256-bit key sk, a plaintext message m is output. We compute $\sum_{i=0}^{255} m_i := \sum_{i=0}^{255}(c_i \bigoplus sk_i)$. C_D requires 2033 gates.

Integer Addition Function. We need to construct an arithmetic circuit C_A that, given 128-bit integers A and B, a 128-bit integer C is output. We compute $\sum_{i=0}^{127} 2^i c_i := \sum_{i=0}^{127} 2^i (a_i + b_i)$. C_A requires 133 gates.

Integer Subtraction Function. We need to construct an arithmetic circuit C_S that, given 128-bit integers A and B, a 128-bit integer C is output. We compute $\sum_{i=0}^{127} 2^i c_i := \sum_{i=0}^{127} 2^i (a_i - b_i)$. C_S requires 137 gates.

Integer Comparison Function. We need to construct an arithmetic circuit C_C that, given 128-bit integers A and B, is satisfied if and only if $A \leq B$. C_C requires 129 gates.

Overall Circuit Sizes. See Table 1 for the size of C_{Trust} (assume that a block contains 20 issue transactions).

Table 1. Size of the circuit C_{Trust}.

Gate count for C_{Trust}	
Ensure $a_{auth} = H(pk_{auth})$ and $h = H(sk^c_{pay}, h_{sig})$	55808
Check computation of $List^{last}$	55936
Check computation of $List^{new}$	55936
Decrypt computation of v in $CBDC$	20330
Decrypt computation of $pk^{'}_{auth}$ in $CBDC$	40660
Ensure that $v_1 + \cdots + v_{20} = Balance^{new}_i - Balance^{last}_i$	2837
Ensure that $0 \leq v_1 + \cdots + v_{20} \leq v_{max}$	2580
Ensure that $Balance^{new}_i \geq 0$	176
Miscellaneous	13056
Total	247319

5 Experiments

To measure the performance of our scheme, we run several experiments on a desktop machine whose environment is Intel Core i7-4770 @ 3.4 GHz and 16 GB of RAM. We implemented the AMCLS scheme applicable to the blockchain based on OpenSSL [31]. Our zk-SNARK for the NP statement **Trust** is obtained by constructing an arithmetic circuit C_{Trust} for verifying **Trust**, and then invoking the generic implementation of zk-SNARK for arithmetic circuit satisfiability.

In Table 2 we report performance characteristics for each of the six SAI scheme algorithms in our implementation (excluding the cost of scanning L, e.g., to find the last trust transaction). Above, we regard the sizes of pp_{sig} as 0 B, because the parameter is "hardcoded" in the libraries we rely on. We obtain that (the number of issuers is expressed as n and the number of issue transactions in the block is expressed as m):

Table 2. Performance of the SAI algorithm. ($N = 10, \sigma \leq 2\%$)

Setup	Time	15.6 s
	Size of pp	44 MiB
	size of pk_{Trust}	44 MiB
	size of vk_{Trust}	64 B
	size of pp_{AMCLS}	294 B
CreateAddress	Time	1.9 ms
	Size of $addr_{pk}$	128 B
	size of pk_{auth} and pk_{pay}	64 ×2 B
	Size of $addr_{sk}$	64 B
	size of sk_{auth} and sk_{pay}	32 ×2 B
Issue	Time	4.8 ms
	Size of $CBDC$	519 B
	size of id	6 B
	size of $owner$	64 B
	size of $Crypto_{vaule}$	32 B
	size of $Crypto_{issuer}$	32 B
	size of $Crypto_{key}$	385 B
	Size of tx_{Issue}	647 B
	size of $CBDC$	519 B
	size of pk_{pay} and σ	64×2 B
Trust	Time	1.5 + 15.7×n s
	Size of tx_{Trust}	224 + 320×n B
	size of $HASH$	2×32 B
	size of $List^{new}$	32×n B
	size of π	288×n B
	size of h	32 B
	size of pk_{pay} and σ	64×2 B
UpdateBalance	Time	0.2 + 1.6×m ms
	Size of the balance list	16×n B
VerifyTransaction	Time	0.5 + 0.7 × n ms

- **Setup** takes about 15.6 s to run; its running time is dominated by the running time of KeyGen on C_{Trust} (**Setup** is run only once). The size of the resulting public parameters pp is dominated by the size of pk_{Trust}.
- **CreateAddress** takes 1.9 ms to run. The size of the resulting address key pair is just a few hundred bytes.
- **Issue** takes 4.8 ms to run. It results in a $CBDC$ of size 519 B and an issue transaction of size 647 B.

- **Trust** takes about $1.5 + 15.7 * n$ s to run. Besides Setup, it is the only "expensive" algorithm to run; as expected, its running time is dominated by *Prove*. It results in a trust transaction of size $224 + 320 * n$ B.
- **UpdateBalance** takes about $0.2 + 1.6 * m$ ms to run. Its running time is dominated by the running time of $\mathcal{D}_{AMCLS}$.
- **VerifyTransaction** takes $0.5 + 0.7 * n$ ms to verify a trust transaction. Its running time is dominated by that of *Verify*, which checks the zk-SNARK proof.

To illustrate the practicality of our scheme, in the same environment, we compared the performance of our scheme and Zcash [23] that provides strong anonymity and confidentiality guarantees. According to EC/DP [9], where there are seven issuers, we set n as 7. The maximum cost of our scheme and Zcash comes from zk-SNARKs, so we compare time and space cost about zk-SNARKs between the two schemes. Comparisons of computation time cost and space cost are shown in Fig. 4 and Fig. 5, respectively. The time cost of our scheme is slightly lower than Zcash, but our scheme pays extra space cost to achieve the supervision feature. The extra space cost is acceptable, because each trust transaction corresponds to a block, that is, the number of trust transactions is approximately equal to the height of blockchain.

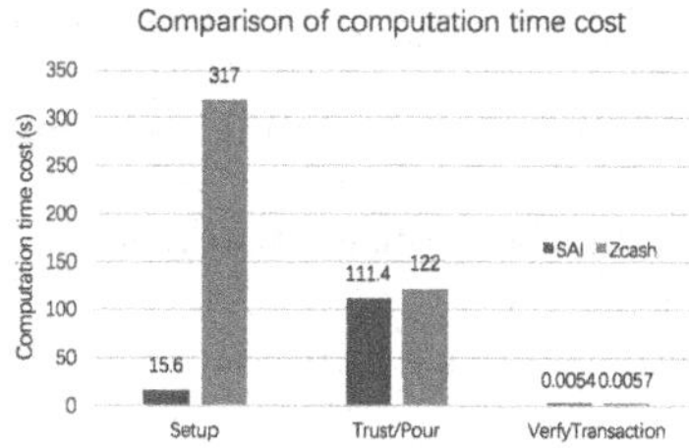

Fig. 4. Comparison of computation time cost for two schemes

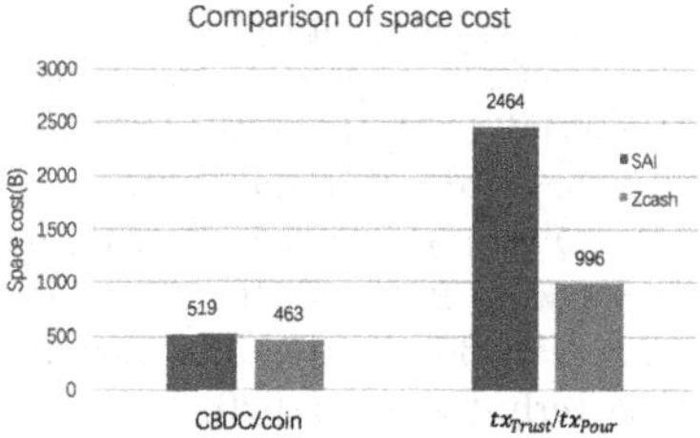

Fig. 5. Comparison of space cost for two schemes

6 Conclusion

In this paper, We propose a supervised anonymous issuance scheme of central bank digital currency based on blockchain to solve the leakage problem of financial data in indirect CBDC model, which can provide stronger privacy of commercial banks under the supervision of the central bank. We realize the commercial bank's anonymity and amount confidentiality through the special AMCLS scheme proposed by us, and public verifiability through zk-SNARKs. The special AMCLS is flexible enough to support a wide range of scenarios, for example, by adding designated receivers of signcryption to achieve cooperation between commercial banks. Anonymous redemption and settlement can be achieved by constructing an inverse process of *Issue* and transfer algorithm of CBDCs, respectively. The security analysis demonstrates that our scheme is secure under some cryptographic assumptions. The overall performance of SAI is comparable to Zcash, indicating that the proposed scheme is practical in terms of deployability.

Acknowledgment. This work was supported by Beijing Municipal Science & Technology Commission (Project Number: Z191100007119004).

References

1. Nakamoto, S.: Bitcoin: A Peer-to-Peer Electronic Cash System (2009). https://bitcoin.org/bitcoin.pdf
2. Bech, M.L., Garratt, R.: Central bank cryptocurrencies. BIS Quarterly Review, September 2017
3. An Introduction to Libra (2019). https://libra.org/en-us/whitepaper
4. Claeys, G., Demertzis, M., Efstathiou, K.: Cryptocurrencies and monetary policy. Bruegel Policy Contribution (2018)
5. Ganne, E.: Can blockchain revolutionize international trade?. World Trade Organization, Geneva (2018)
6. Munro, A.: IBM: we already use Stellar Lumens live, much more in Q2 2018. finder.com.au. (2018). https://www.finder.com.au/ibm-we-already-use-stellar-lumens-live-much-more-in-q2-2018
7. United Nations Capital Development Fund: Building inclusive financial sectors for development. United Nations Publications (2006)
8. Barontini, C., Holden, H.: Proceeding with caution-a survey on central bank digital currency. In: BIS Paper (2019)
9. Yao, Q.: Experimental study on prototype system of central bank digital currency. J. Software **29**(9), 2716–2732 (2018)
10. Auer, R., Böhme, R.: The technology of retail central bank digital currency. BIS Quarterly Review (2020)
11. Koning, J.P.: Fedcoin: a central bank-issued cryptocurrency. R3 report 15 (2016)
12. Ingves, S.: The e-krona and the payments of the future (2018). https://www.bis.org/review/r181115c.pdf
13. Project Sand Dollar: A Bahamas Payments System Modernisation Initiative (2019). https://www.bis.org/review/r190321a.pdf
14. Danezis, G., Meiklejohn, S.: Centrally Banked Cryptocurrencies. arXiv preprint arXiv:1505.06895 (2015)
15. Project Jasper: A Canadian experiment with distributed ledger technology for domestic interbank payments settlement. White paper prepared by Payments Canada (2017)
16. Dalal, D., Yong, S., Lewis, A.: The Future is Here-Project Ubin: SGD on Distributed Ledger. Monetary Authority of Singapore & Deloitte (2017)
17. Ward, O., Rochemont, S.: Understanding Central Bank Digital Currencies (2019)
18. Sayuri, S.: Money and Central Bank Digital Currency (2019). https://www.adb.org/sites/default/files/publication/485856/adbi-wp922.pdf
19. Kumhof, M., Noone, C.: Central bank digital currencies-design principles and balance sheet implications. Bank of England Working Papers (2018)
20. Anton, A.I., Earp, J.B., He, Q., et al.: Financial privacy policies and the need for standardization. IEEE Secur. Privacy **2**(2), 36–45 (2004)

21. Rohan, G.: Regulatory Challenges and Risks for Central Bank Digital Currency. Regulatory Requirements and Economic Impact Working Group (2019)
22. Goldwasser, S., Micali, S., Rackoff, C.: The knowledge complexity of interactive proof systems. SIAM J. Comput. **18**(1), 186–208 (1989)
23. Ben-Sasson, E., Chiesa, A., Garman, C., et al.: Zerocash: decentralized anonymous payments from bitcoin. In: IEEE Symposium on Security and Privacy, pp. 459–474 (2014)
24. Al-Riyami, S.S., Paterson, K.G.: Certificateless public key cryptography. In: Laih, C.-S. (ed.) ASIACRYPT 2003. LNCS, vol. 2894, pp. 452–473. Springer, Heidelberg (2003). https://doi.org/10.1007/978-3-540-40061-5_29
25. Selvi, S.S.D., Vivek, S.S., Shukla, D., Rangan Chandrasekaran, P.: Efficient and provably secure certificateless multi-receiver signcryption. In: Baek, J., Bao, F., Chen, K., Lai, X. (eds.) ProvSec 2008. LNCS, vol. 5324, pp. 52–67. Springer, Heidelberg (2008). https://doi.org/10.1007/978-3-540-88733-1_4
26. Islam, S.H., Khan, M.K., Al-Khouri, A.M.: Anonymous and provably secure certificateless multireceiver encryption without bilinear pairing. Security Commun. Netw. **8**(13), 2214–2231 (2015)
27. Pang, L., Kou, M., Wei, M., et al.: Efficient anonymous certificateless multi-receiver signcryption scheme without bilinear pairings. IEEE Access **6**, 78123–78135 (2018)
28. Katz, J., Lindell, Y.: Introduction to Modern Cryptography. CRC Press, Boca Raton (2014)
29. Beaulieu, R., Shors, D., Smith, J., et al.: The SIMON and SPECK families of lightweight block ciphers. IACR Cryptol. ePrint Arch. 404–449 (2013)
30. Ben-Sasson, E., Chiesa, A., Tromer, E., et al.: Succinct non-interactive zero knowledge for a von Neumann architecture. In: 23rd USENIX Security Symposium (2014)
31. OpenSSL Homepage. https://www.openssl.org/. Accessed 12 Dec 2019

IncreAIBMF: Incremental Learning for Encrypted Mobile Application Identification

Yafei Sang[1,2(✉)], Mao Tian[1,2], Yongzheng Zhang[1,2], Peng Chang[1,2], and Shuyuan Zhao[1,2]

[1] Institute of Information Engineering, Chinese Academy of Sciences, Beijing, China
{sangyafei,tianmao,zhangyongzheng,changpeng,zhaoshuyuan}@iie.ac.cn
[2] School of Cyber Security, University of Chinese Academy of Sciences, Beijing, China

Abstract. Mobile application identification, as the fundamental technique in the field of network security and management, suffers from a critical problem, namely 'encrypted traffic'. The proven methods for encrypted traffic identification have a major drawback, which is new come applications continue to suffer from *catastrophic forgetting*, a dramatic decrease in overall performance when training with new app classes added incrementally. This is due to the current model requiring the entire dataset, consisting of all the samples from the old and the new classes, to update the model. The updating requirement becomes easily unsustainable as the number of apps grows, To address the issue, we propose *IncreAIBMF* framework to learn deep neural networks incrementally, using new apps data and only a small exemplar set corresponding to samples from the old apps. The key idea behind *IncreAIBMF* is an incremental learning framework which possesses new application identification ability by incorporating the cross-distilled loss, which can not only learn the new app classes and also retain the previous knowledge corresponding to the old app classes. Our experiment results show that *IncreAIBMF* achieves 87.3% on Macro Precision, 87.8% on F1 Score and 88.9% on Macro Recall, respectively, on the real-world traces that consists of 50 mobile applications, supports the early prediction, and is robust to the scale of the app classes. Besides, the basic variant of *IncreAIBMF*, AIBMF is superior to the state-of-the-art methods in terms of identification performance.

Keywords: Incremental learning · Mobile application identification · Encrypted traffic analysis

1 Introduction

1.1 Motivation

The rapid development of the mobile Internet has led to a sharp increase in the number of mobile applications. As of December 2018, the number of mobile

M. Qiu (Ed.): ICA3PP 2020, LNCS 12454, pp. 494–508, 2020.
https://doi.org/10.1007/978-3-030-60248-2_33

applications (Apps) in the market monitored in China was 4.49 million [1]. In addition, these mobile applications widely use the Secure Sockets Layer (SSL) or its successor Transport Layer Security (TLS) encryption protocol to communication for network security considerations. This indisputable fact makes that the traditional application identification approaches based on traffic payload content are difficult and ineffective to fingerprint the application from traffic information. Hence, encrypted mobile application identification (EMAI for short), *associating encrypted SSL/TLS flows with the specific mobile applications*, has been a crucial challenge as a fundamental technique for many task scenarios, such as traffic monitoring, QoS guarantees, network anomaly detection and so on.

1.2 Related Work and Their Limitations

The existing EMAI methods can roughly be divided into two categories: *special-field-based* and *state-sequence-based. (1)* Special-field-based methods identify encrypted flow by using plain text protocol fields/chunks transmitted in the process of encryption negotiation, such as certificates [11], Server Name Indication (SNI) [14] etc. These methods have low identification accuracy and can be easily bypassed. *(2)* The basic idea of the state-sequence-based methods is to extract the protocol state information sequence of the SSL/TLS flow and then apply a markov chain or neural network. The SSL/TLS `message type state`[1] information is explored for first-order homogeneous markov chains in [6] and second-order markov chains in [15] . The `payload byte state` information is verified in an end-to-end traffic classification model [19] by extracting the first 784 bytes of a flow or session neglecting any other information, and then using one-dimensional convolution for processing. The `packet size state` information is modeled by employing first-order markov chains in [8] and recurrent neural network based encoder-decoder structure in [9]. While interesting in their own right, these prior arts have two common limitations: *1)* They are single state specific. That is to say, the repeating single specified state sequences between different mobile applications grow sharply as the number of applications increases, which greatly weakens the discrimination capability of the specified state information. *2)* These approaches are data in a fixed category, and no consideration is given to the traffic data of new categories such as new mobile applications. In other words, EMAI problem is the dramatic increase in apps, where new app classes are learned continually. Traditional models require all the samples (corresponding to the old and the new classes) to be trained at training time.

[1] The message type of SSL/TLS denotes the semantic information of exchange packet, and there are 16 categories in one typical session, including Change Cipher Spec(20), Alert(21), Handshake(22), Hello Request(22:0), Client Hello(22:1), Server Hello(22:2), Hello Verify Request(22:3), New Session Ticket(22:4), Certificate(22:11), Server Key Exchange(22:12), Certificate Request(22:13), Server Hello Done(22:14), Certificate Verify(22:15), Client Key Exchange(22:16), Finished(22:17), Application Data(23).

1.3 Our Research

In this work, we present **IncreAIBMF**: **Incre**mental learning for Encrypted Mobile **A**pplication **I**dentification **B**ased on **M**ulti-view **F**eatures; a systematic framework for EMAI task at a per-flow granularity. The key novelty and advantages of *IncreAIBMF* as follows:

- It overcomes the weaken discrimination capability of the specified state problem as the number of applications increases by combining the payload, packet size sequence and message type sequence together and use deep learning architectures to extract valuable information from multiple views.
- It's an incremental learning framework which possesses new application identification ability based on the key points: *1)* using new apps data and only a small representative exemplar set corresponding to samples from the old apps; *2)* applying the cross-distilled loss [4], which combine the cross-entropy loss function [3] to learn the new apps and distillation loss function [5] to retain the previous knowledge corresponding to the old apps.

We evaluate *IncreAIBMF* using 200,000+ flows generated by 50 applications from the Android market-place. Our proposed *IncreAIBMF* remains stable across the incremental app classes (from 25 to 50 in 5 step). Even in the context of 50 apps, *IncreAIBMF* achieves 87.3% on Macro Precision, 87.8% on Macro F1 Score and 88.9% on Macro Recall, respectively.

The rest of this paper is organized as follows. In Sect. 2, we introduce some preliminaries. The basic AIBMF framework is presented in Sect. 3 and the IncreAIBMF framework is illustrated in Sect. 4. Experiments and evaluations are shown in Sect. 5. Section 6 puts forward discussion and conclusion is made in the last section.

2 Preliminaries

2.1 Problem Statement

The encrypted application identification problem is to classify the encrypted traffic into specific applications with the network flows. Formally, it can be described as follows. Assume that there are N observed encrypted flows $F = \{f_1, f_2, \cdots, f_i, \cdots, f_N\}$ and C applications $A = \{a_1, a_2, \cdots, a_j, \cdots, a_C\}$. We aim to build an end-to-end model $\varphi(f_i)$ to predict a label $\hat{a}_j$ that is exactly the real label a_j, such that $i \in \{1, 2, \cdots, N\}$ and $j \in \{1, 2, \cdots, C\}$. An encrypted flow is a set of packets that have the same source Internet Protocol (IP) address, destination IP address, source transport layer port, destination transport layer port, and transport protocol.

2.2 Incremental Learning

The incremental learning is defined as the capability of machine learning architectures to continuously improve the learned model by feeding new data without losing previously learned knowledge. This has been widely studied in the context of problems like image classification and object to solve the *catastrophic forgetting* problem, a phenomenon where the performance on old set of classes degrades dramatically [7,12,16]. Some studies keep a small portion of data belonging to previous tasks and use them to preserve the accuracy on old tasks when dealing with new problems. The exemplar set to store is chosen at random or according to a relevance metric [5,12].

3 Basic AIBMF Framework

Our insights of *AIBMF* into encrypted application identification is three-fold: *(i)* SSL/TLS protocol is essentially a language of communication between mobile app client and server, the specific manifestation of this language is network flow, *i.e.*, a series of packets; *(ii)* Because protocol state machine is an overall abstract description of the network flows, application fingerprints can be constructed from different protocol state information; *(iii)* Although SSL/TLS hides the payload of network flows, side-channel protocol state information is still leaked from encrypted connections. Figure 1 shows the overview of *AIBMF*. Three views are incorporated in our model which are named as packet size (PS) view, packet content type (CT) view and packet payload byte (PB) view respectively. This framework includes three modules: preprocessing, feature learning, and identification.

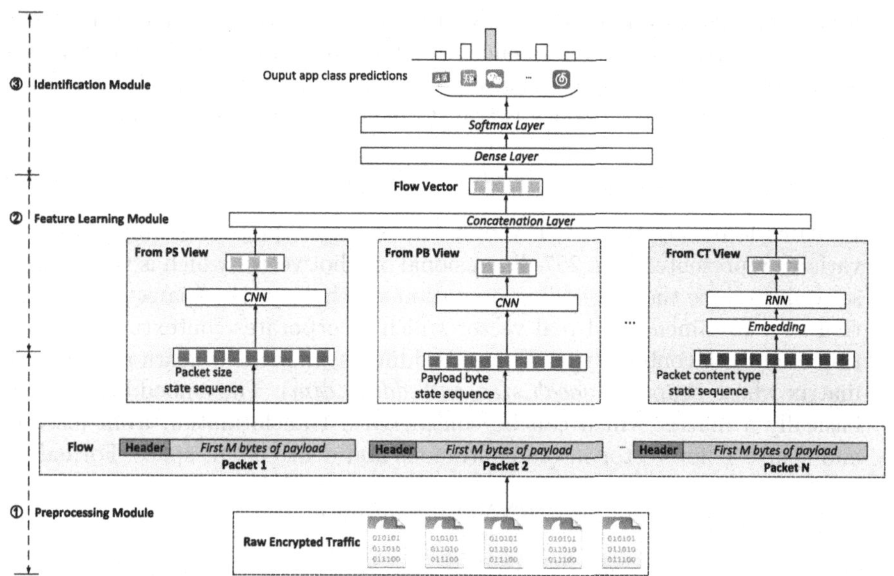

Fig. 1. The overview of the proposed *AIBMF* using multi-view features.

Preprocessing Module: This module first organizes raw network traffic into flows (*i.e.*, the sample objects to be classified). The collection of the whole flows can be indicated as $D = \{\{f_i\}_{i=1}^{i=N}\}_{j=1}^{j=C}$ where f_i means the ith flow. As we already have the flow objects, we can turn to the numerical vector representation of flow. *(i)* For PS view, we form a 16 dimensional vector v_{ps} by computing the packet size of the first 16 packets (padded with −1). *(ii)* For CT view, we form a 16 dimensional vector v_{ct} by extracting the value content type field the first 16 packets (padded with −1), and it ranges from 0 to 255. *(iii)* For PB view, we form a $16 \times 64 = 1024$ dimensional vector v_{pb} by regarding payload byte value as the corresponding ASCII code, and further convert into a integer of 0–255 (padded with −1).

Features Extraction Module: This module focuses on learning high-order abstract feature based on the initial state sequence feature from three different view.

- **Feature learning from PB view:** We use a 1D-CNN network architecture to process the input numerical vector v_{pb}. In the 1D-CNN, two-layer convolutional are used, the number of convolution kernels used in each layer is 64, and the size of the convolution kernel is 1×3, max pooling layer used. Therefore, raw payload byte sequence is transformed into a 64-dimensional vector α.

$$\alpha = \mathit{1D-CNN}(v_{pb}) \tag{1}$$

- **Feature learning from PS view:** The vector v_{ps} is processed by 1D-CNN the same as payload byte sequence, the differences are that the input vector is packet size sequence and the input length is 16 here. After 1D-CNN feature learning, packet size sequence is transformed into a 64-dimensional vector β.

$$\beta = \mathit{1D-CNN}(v_{ps}) \tag{2}$$

- **Feature learning from CT view:** The content type is originally a discrete variable represented by a 257-dimensional one-hot vector which is very sparse, so we introduce the embedding operation which maps the sparse representation to a 16-dimensional real vector which incorporates contextual information about the content type. The embedding layer needs to learn a transform matrix whose shape is $(vocab\_size, embedding\_dim)$. The embedding layer is basically a matrix which can be considered a transformation from discrete and sparse 1-hot-vector into a continuous and dense latent space. Formally,

$$\gamma = RNN(Embedding(v_{ct})) \tag{3}$$

Identification Module: Three different abstract features are concatenated together to form a 192-dimensional vector

$$\delta = \alpha \oplus \beta \oplus \gamma \tag{4}$$

which is then input to the fully connected layer. The output of the fully connected layer, *i.e.*, x, will be used as the input of the softmax regression to identify the flow. Three different abstract features jointly optimize the same category target, jointly calculate the multi-class cross entropy loss, and update the parameters of the three neural networks through the back propagation algorithm [13]. Mathematically, the probability that an output prediction Y is app a_j, is determined by:

$$p(Y = j|x, W, b) = softmax_j(\boldsymbol{W}x + \boldsymbol{b}) \tag{5}$$

where W is a weight matrix between the fully connected layer and the softmax layer, and the b is the bias vector. Then the model's prediction y_{pd} is the class whose probability is maximal:

$$y_{pd} = argmax(p(Y = j|x, W, b)), \forall j \in \{1, 2, 3, ..., C\} \tag{6}$$

Our neural network selects the cross entropy [3] as the loss function.

4 IncreAIBMF Framework

In this section, we describe how we extend the *AIBMF* to form the *IncreAIBMF* framework by utilizing increment learning. The *IncreAIBMF* keeps the first two modules of AIMBF fixed, and replaces the third identification module with distillation module. The key points of *IncreAIBMF* as follows:

- **Distillation Module:** This module is aim to update the decision boundaries of the apps with the cross-distilled loss function (Eq. 7), combines a distillation loss [5], which retains the knowledge from old app classes, with classification loss (using multi-class cross-entropy loss), which learns to classify the new app classes [4]. Specifically, when new apps are trained, we add a new distillation module corresponding to these apps, and connect it to the feature learning module of AIBMF.
- **Representative Memory:** To retain the knowledge acquired from the old mobile apps, we use a `representative memory` that stores the most representative samples from the old apps. The term of `representative memory` refers to a subset with the most representative samples from them is selected and stored in the representative memory, when a new app or set of apps is added to the current model.

Figure 2 shows the overview of our proposed detailed *IncreAIBMF*, consisting of training set constructing phase, training phase, representative memory updating phase. We now describe these phases in detail.

4.1 Training Set Constructing Phase

This phase prepares the training data to be used in the next training phase. The training set is composed of a portion of the representative samples of the old class (exemplars from the old classes stored in the representative memory)

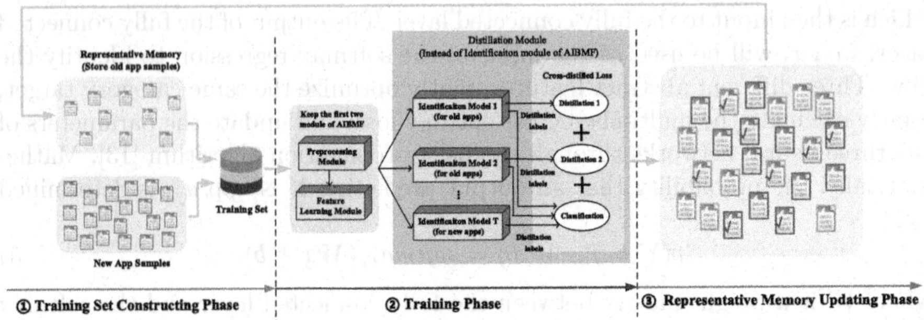

Fig. 2. IncreAIBMF architecture. The icon with a red check mark represents the selected representative sample. (Color figure online)

and the samples of the new app class. As our approach uses two loss functions, *i.e.*, classification and distillation, we need two labels for each sample, associated with the two losses. For classification, we use the one-hot vector which indicates the app appearing in the all app classes. For distillation, we use the `distillation labels` produced by every `identification module` with old app classes. Therefore, we have as many `distillation labels` per sample as `identification module` with old app classes. To better understand, where we are performing the third incremental step of our *IncreAIBMF*. At this point the two `identification module` will operate on old apps, and one `identification module` processes the new apps. When a sample is evaluated, the `distillation labels` produced by the two identification modules with the old apps are used for distillation, and the `distillation labels` produced by the three identification modules are used for classification.

4.2 Training Phase

The training phase takes the training set with its corresponding labels as input for cross-distilled loss computing, and updates all parameters of the entire network architecture. The cross-distilled loss function $J^{loss}(\theta)$ is defined as:

$$J^{loss}(\theta) = C^{loss}(\theta) + \gamma \sum_{t=1}^{T} D_t^{loss}(\theta) \tag{7}$$

where $C^{loss}(\theta)$ is the standard cross-entropy loss applied to samples from the old and new app classes, D_t^{loss} is the distillation loss of the `identification module` t, and T is the total number of `identification module` for the old apps, γ is refereed to ratio that distilled loss in the cross-distilled loss. In our training process, we set it 0.3. The cross-entropy loss $C^{loss}(\theta)$ and the distillation loss $D^{loss}(\theta)$ are defined as:

$$C^{loss}(\theta) = -\frac{1}{N} \sum_{i=1}^{N} \sum_{j=1}^{C} p_i^j \log q_i^j \tag{8}$$

$$D^{loss}(\theta) = -\frac{1}{N}\sum_{i=1}^{N}\sum_{j=1}^{C} u_i^j \log v_i^j \tag{9}$$

where q_i is a score obtained by applying a softmax function, p_i is the ground truth, u_i and v_i are modified versions of p_i and q_i, for the sample i. N is the number of samples, C is the number of app classes.

4.3 Representative Memory Updating Phase

The goal of this phase is to update the sample in the memory, and ensure that the size of the training set does not increase dramatically. That is, in the updated memory, we fixed the memory size to K flows ($K = 100,000$ in this paper), and the number of samples corresponding to each category of application is $\frac{100,000}{C}$, here C indicates the category of the application. Then, we employ the two operations: *(i)* selection of new samples to store: we perform the herding selection strategy [20], which produces a sorted list of samples of one class based on the distance to the mean sample of that class; Given the sorted list of samples, the first n samples of the list are selected. These samples are most representative of the class according to the mean. *(ii)* removal of leftover samples: as the samples are stored in a sorted list, this operation is trivial. The memory unit only needs to remove samples from the end of the sample set of each class. Note that after this operation, the removed samples are never used again.

5 Experimental Evaluation

5.1 Data Collection and Labeling

We collect a traffic dataset based on the SSL/TLS protocol. We execute an app both in several real Android devices and Android emulators. The app is automatically driven by the Android tool, monkeyrunner[2]. we execute the same app in different devices at one time and ensure there is no app running in the background by limiting the permissions of other applications in the android devices. We capture one app's traffic at a time to ensure it is ground truth. To avoid the influence of network environment, we kept capturing traffic by Wireshark[3] in our laboratory in different time manually and automatically during a month. In the early research work [17], we collected 20 popular applications. In this work of this article, we added 30 commonly used applications shown in the Table 1.

5.2 Evaluation Metrics

In order to make a reasonable and effective quantitative evaluation of the identification performance of *IncreAIBMF*, we have used three basic metrics Precision

[2] https://developer.android.com/studio/test/monkeyrunner/index.html.
[3] https://www.wireshark.org/.

Table 1. Data overview of 30 mobile applications

Apps	Developer	Categories	Flows
ALiJianKang	alihealth.cn	Lifestyle	22904
AnJuKe	anjuke.com	Lifestyle	2340
BaiCiZhan	baicizhan.com	Tools	674
BaiHeHunLian	baihe.com	Lifestyle	2452
BeiKeZhangFang	bj.ke.com	Lifestyle	7520
DangDangYueDu	dangdang.com	Books & Reference	2588
DingDing	dingtalk.com	Lifestyle	3468
DingXiang	dxy.cn	Lifestyle	4654
DouBan	douban.com	Social	3998
VigoVideo	huoshan.com	Music & Audio	2924
Keep	www.gotokeep.com	Lifestyle	9646
MiaoPai	miaopai.com	Social	340
csair App	China Southern Airlines	Travel & Local	7656
PinDuoDuo	pinduoduo.com	Shopping	2660
QingTing FM	Music & Audio	Social	2312
QuNaEr	qunar.com	Lifestyle	3294
Soul	soulapp.cn	Communication	4406
TianYa	tianya.cn	Communication	1058
TianYanCha	tianyancha.com	Tools	6146
TongHuaShun	10jqka.com.cn	Finance	5928
Arena of Valor	Tencent	Game	2524
XianYu	2.taobao.com	Shopping	3226
XiaoMiYunDong	huami.com	Lifestyle	2364
XinLangCaiJing	finance.sina.com.cn	Finance	3608
YangShiXinWen	news.cctv.com	Social	1534
YouDaoYunBiJi	note.youdao.com	Tools	2396
ZhangShangShengHuo	cmbchina.com	Finance	4838
ZhiBoBa	zhibo8.cc	Tools	4368
airchina App	Air China	News & Travel & Local	3343
ZuoYeBang	Zybang.com/	Tools	4692
Total	–	–	**129791**

(Pre), Recall (Rec) and F1 Score (F1). The above metrics are described mathematically as follows:

$$Pre = \frac{TP}{TP+FP},\ Rec = \frac{TP}{TP+FN},\ F1 = 2\frac{Pre \times Rec}{Pre + Rec} \tag{10}$$

Here, TP, FP and FN stands for true positive, false positive and false negative, respectively. Furthermore, to evaluate *IncreAIBMF* on multiple app classes, we introduce $MacroPre(MPre)$, $Macro\ Recall(MRec)$, $MacroF1(MF1)$ as the overall metrics, which are defined as follows:

$$MPre = \frac{1}{C}\sum_{i=1}^{C} Pre_i, \; MRec = \frac{1}{C}\sum_{i=1}^{C} Rec_i, \; MF1 = \frac{1}{C}\sum_{i=1}^{C} F1_i \tag{11}$$

where C denotes the number of app classes.

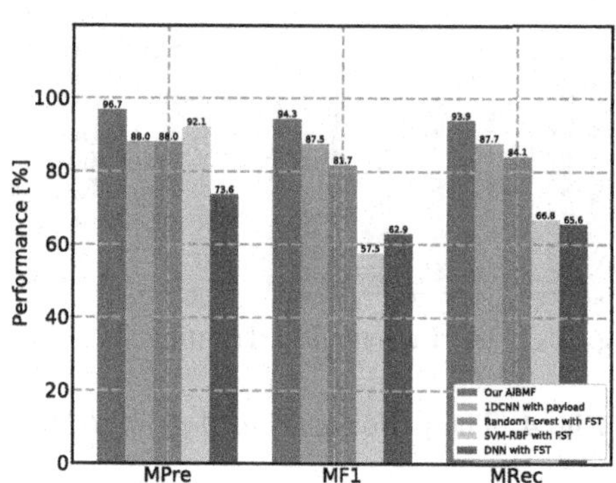

Fig. 3. The AIBMF's evaluation results compared with the previous works

5.3 Basic AIMBF Compared with Existing Approaches

We carefully select a series of previous work on encrypted traffic classification as baselines, some of which are classical and others are state-of-the-art, including 1DCNN with payload byte feature [19], DNN with flow statistic features (FST for short) [18], Random Forest with FST [10], SVM-RBF with FST [2]. We used flow samples from 20 mobile apps and employed 24 flow statistic features, including *the number of packets (1 features), the mean/max/min/median/variance of TTL/window/packet size/client extensions length* ($4\times5 = 20$ *features), the number of client ciphers (1 features), server cipher (1 features)*, for the comparative evaluation experiments. Figure 3 presents the experimental results of different approaches. Compared with the state-of-the-art methods, *AIBMF* performs best in Macro Precision, Macro Recall and Macro F1. The Macro F1 of AIBMF is 6.2%, 9.8%, 27.1%, and 28.3% better than 1DCNN with Payload, RF with FST, SVM-RBF with FST, and DNN with FST, respectively.

5.4 Basic AIMBF Performance with Full Apps Training Data

Figure 4(a) presents the classification accuracy of AIBMF in different scales of app classes. We use $C \in [25, 30, 35, 40, 45, 50]$ app classes respectively from the 50-classes data set (20 old basic app classes and 30 new app classes) to test the performance of AIBMF. Note that, all training samples of all app classes are available in the training phase. As the number of applications increases, the effect of the AIBMF decreases to a certain extent. We can find that AIBMF still performed well, with 95.0% on $MPre.$, 94.2% on $MF1$ and 93.7% on $MRec.$ respectively, even in 50 apps. Hence, AIBMF shows strong robustness to the scale of app classes. However, the time required for training also increases linearly (Fig. 4(c)), which greatly reduces the acquisition efficiency of the good model.

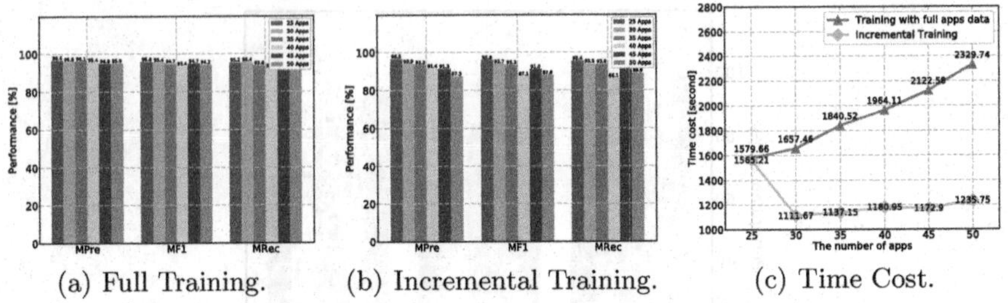

(a) Full Training. (b) Incremental Training. (c) Time Cost.

Fig. 4. The performance of *IncreAIBMF*.

5.5 Basic AIBMF with Fine-Tuning for New App Identification

In the face of the increase in the number of applications, one of the commonly used methods is fine-tuning. The fine-tuning method refers to adding a small number of task-specific parameters based on a trained model. For example, for classification problems, a layer of softmax network is added to the model, and then retrained on the new classes for fine-tune. For the newly collected traffic of 30 applications, we first used the idea of fine-tuning, freezing the parameters of the AIBMF network model trained on the 20 old apps for feature extraction, and only modifying the number of fully connected layers. We added $T \in [5, 10, 15, 20, 25, 30]$ app classes each training step. As shown in Fig. 5, the new model cannot complete the recognition task in either the old category or the new category. We attribute this trouble performance to catastrophic forgetting.

5.6 IncreAIBMF Performance for New App Identification

In this experiment, we use a model trained on 20 applications as the base model, each time adding 5 new applications, and performing experiments according to the sequence of *[25,30,35,40,45,50]* with our proposed *IncreAIBMF*. After each incremental step, the resulting model is evaluated on the test data composed of

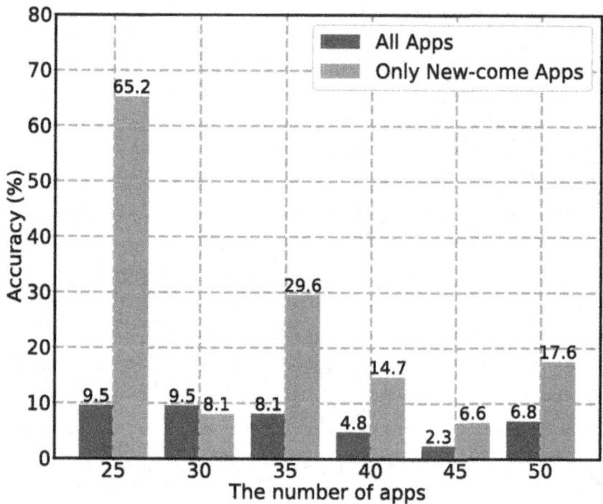

Fig. 5. The performance of *AIBMF* with fine-tuning.

all the training app classes. As shown in Fig. 4(b), it can be observed that the performance of *IncreAIBMF* remains stable across the incremental app classes (from 25 to 50). Even in 50 app classes, *IncreAIBMF* achieves 87.3% on MPre, 87.8% on MF1 and 88.9% on MRec, respectively. We can conclude that our method—*IncreAIBMF* can overcome catastrophic forgetting problem. As shown in Fig. 4(c), we compared the time of full training and incremental learning training. We can find that the time required by the *AIBMF* model increased almost linearly, while the time consumption of *IncreAIBMF* increases very slowly. This is because when we updated the representative memory, the number of training sets is limited to 100,000, and the representative samples was selected after each training process. In order to have an intuitive observation of the experimental results, confusion matrix on the 10,000+ test flow samples of 50 applications is depicted in Fig. 6. The identification performance is almost perfect where most of the mistakes are due to the share of network service among different apps especially for these developed by the same developer (*i.e.* App *XianYu* and *AliJianKang* are both developed by AliBaBa.com and App *XianYu* is wrong recognized as App *AliJiangKang* by *IncreAIBMF*). The recognition for App *TianYa* is also very poor. We checked the composition of the dataset and found that there is little samples for App *TianYa* (for this App is not active now). We attribute this poor result to data imbalance.

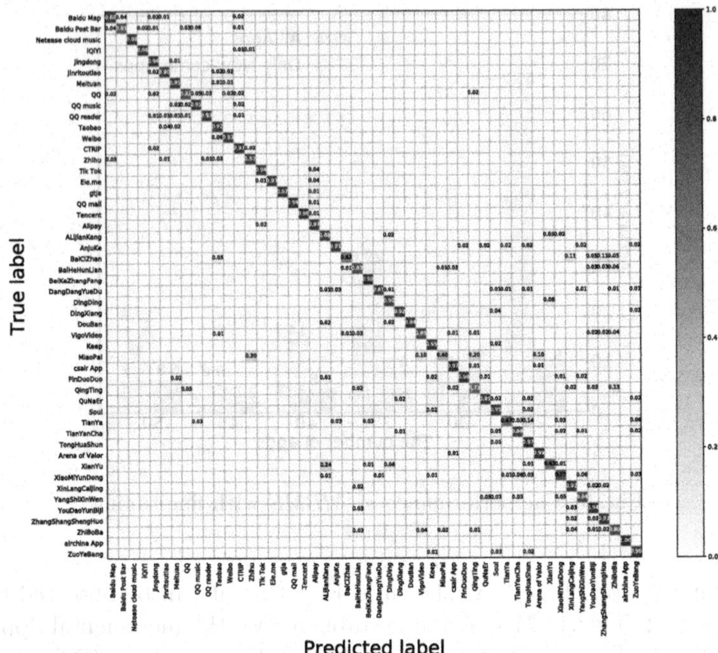

Fig. 6. The confusion matrix of *IncreAIBMF* on the testing set

6 Conclusion and Future Work

This paper proposes an incremental learning framework, called *IncreAIBMF* for encrypted mobile application identification task. Compared with the existing approaches, *IncreAIBMF* possesses new application identification ability by overcoming catastrophic forgetting and training the identification model incrementally. On the basis of this paper, the following work will be carried out: *(i)* We plan to study the impact of app arrival moments on recognition results. We will also compare different growth sequences and compare the effects of the experiments. We also plan to study the impact of the value of γ in the cross-distilled loss and find out how the cross-distilled loss affects the performance *IncreAIBMF*. *(ii)* With the increase of application types, the count of each old application will decrease, and the newly added data will cause data imbalance. We plan to study the impact of data imbalances on identification tasks and find solutions to this problem. *(iii)* We tend to study other SSL/TLS state information and combine them with payload to improve the performance of *IncreAIBMF*.

References

1. The 43th china statistical report on internet develop. Technical report, China Internet Network Information Center, CNNIC (2019)

2. Alshammari, R., Zincir-Heywood, A.N.: Machine learning based encrypted traffic classification: identifying SSH and skype. In: IEEE Symposium on Computational Intelligence for Security and Defense Applications, pp. 1–8. IEEE (2009)
3. Boer, P.T.D., Kroese, D.P., Mannor, S., Rubinstein, R.Y.: A tutorial on the cross-entropy method. Ann. Oper. Res. **134**(1), 19–67 (2005). https://doi.org/10.1007/s10479-005-5724-z
4. Castro, F.M., Marín-Jiménez, M.J., Guil, N., Schmid, C., Alahari, K.: End-to-end incremental learning. In: Ferrari, V., Hebert, M., Sminchisescu, C., Weiss, Y. (eds.) ECCV 2018. LNCS, vol. 11216, pp. 241–257. Springer, Cham (2018). https://doi.org/10.1007/978-3-030-01258-8_15
5. Hinton, G., Vinyals, O., Dean, J.: Distilling the knowledge in a neural network. arXiv preprint arXiv:1503.02531 (2015)
6. Korczyński, M., Duda, A.: Markov chain fingerprinting to classify encrypted traffic. In: IEEE INFOCOM 2014-IEEE Conference on Computer Communications, pp. 781–789. IEEE (2014)
7. Li, Z., Hoiem, D.: Learning without forgetting. IEEE Trans. Pattern Anal. Mach. Intell. **40**(12), 2935–2947 (2017)
8. Liu, C., Cao, Z., Xiong, G., Gou, G., Yiu, S.M., He, L.: MaMPF: encrypted traffic classification based on multi-attribute Markov probability fingerprints. In: IEEE/ACM 26th International Symposium on Quality of Service (IWQoS), pp. 1–10. IEEE (2018)
9. Liu, C., He, L., Xiong, G., Cao, Z., Li, Z.: FS-Net: a flow sequence network for encrypted traffic classification. In: IEEE INFOCOM 2019-IEEE Conference on Computer Communications, pp. 1171–1179. IEEE (2019)
10. Orsolic, I., Pevec, D., Suznjevic, M., Skorin-Kapov, L.: A machine learning approach to classifying YouTube QoE based on encrypted network traffic. Multimedia Tools Appl. **76**(21), 22267–222301 (2017). https://doi.org/10.1007/s11042-017-4728-4
11. Pukkawanna, S., Blanc, G., Garcia-Alfaro, J., Kadobayashi, Y., Debar, H.: Classification of SSL servers based on their SSL handshake for automated security assessment. In: Third International Workshop on Building Analysis Datasets and Gathering Experience Returns for Security (BADGERS), pp. 30–39. IEEE (2014)
12. Rebuffi, S.A., Kolesnikov, A., Sperl, G., Lampert, C.H.: iCaRL: incremental classifier and representation learning. In: Proceedings of the IEEE Conference on Computer Vision and Pattern Recognition, pp. 2001–2010 (2017)
13. Rumelhart, D.E., Hinton, G.E., Williams, R.J.: Learning representations by back-propagating errors. Nature **323**(6088), 533 (1986)
14. Shbair, W.M., Cholez, T., François, J., Chrisment, I.: Improving SNI-based https security monitoring. In: IEEE 36th International Conference on Distributed Computing Systems Workshops (ICDCSW), pp. 72–77. IEEE (2016)
15. Shen, M., Wei, M., Zhu, L., Wang, M.: Classification of encrypted traffic with second-order Markov chains and application attribute bigrams. IEEE Trans. Inf. Forensics Secur. **12**(8), 1830–1843 (2017)
16. Shmelkov, K., Schmid, C., Alahari, K.: Incremental learning of object detectors without catastrophic forgetting. In: Proceedings of the IEEE International Conference on Computer Vision, pp. 3400–3409 (2017)
17. Tian, M., Chang, P., Sang, Y., Zhang, Y., Li, S.: Mobile application identification over https traffic based on multi-view features. In: 26th International Conference on Telecommunications (ICT), pp. 73–79. IEEE (2019)
18. Velan, P., Čermák, M., Čeleda, P., Drašar, M.: A survey of methods for encrypted traffic classification and analysis. Int. J. Network Manage. **25**(5), 355–374 (2015)

19. Wang, W., Zhu, M., Wang, J., Zeng, X., Yang, Z.: End-to-end encrypted traffic classification with one-dimensional convolution neural networks. In: IEEE International Conference on Intelligence and Security Informatics, pp. 43–48 (2017)
20. Welling, M.: Herding dynamical weights to learn. In: Proceedings of the 26th Annual International Conference on Machine Learning, pp. 1121–1128. ACM (2009)

A Multi-level Features Fusion Network for Detecting Obstructive Sleep Apnea Hypopnea Syndrome

Xingfeng Lv[1] and Jinbao Li[2](✉)

[1] School of Computer Science and Technology, Heilongjiang University, Harbin, China
lvxingfeng@hlju.edu.cn
[2] Shandong Artificial Intelligence Institute, Qilu University of Technology (Shandong Academy of Sciences), Jinan, China
jbli@hlju.edu.cn

Abstract. Obstructive Sleep Apnea Hypopnea Syndrome (OSAHS) is the most common sleep disorder. If not treated in time, OSAHS will lead to high blood pressure, heart or cerebrovascular diseases. At present, the commend method is based on deep learning which can extract deep features. But it does not make full use of the influence of shallow features on sleep classification. Therefore, this paper proposes a Multi-Level Features Fusion Network (MLF2N) model to detect OSAHS. MLF2N can learn different levels of features from the different respiratory signals. In this paper, 2056 subjects were classified as Obstructive Sleep Apnea, Hypopnea and Normal sleep on the Multi Ethical Study of Atherosclerosis (MESA) data set using this model. The experimental results show that the accuracy of the three respiratory signals can reach 85.5% through this model.

Keywords: Obstructive Sleep Apnea Hypopnea Syndrome · Multi-level features fusion network · Deep learning · Deep features · Shallow features

1 Introduction

Obstructive Sleep Apnea Hypopnea Syndrome (OSAHS) is one of the wide spread sleep disorders a common disorder affecting at least 49.7% of men and 23.4% of women [7]. The symptoms of this disorder are nocturnal recurrent arousal, excessive daytime sleepiness, memory loss, executive and motor dysfunction. If left untreated, OSAHS could lead to serious health problems, such as stroke and heart disease. It also increases the incidence of traffic accidents [1,16]. OSAHS is usually divided into obstructive sleep apnea (OSA), hypopnea and normal sleep. OSA is caused by upper airway obstruction. The oral and nasal respiratory airflow disappears or decreases significantly, with a decrease of more than 90% compared with the baseline, and the duration is more than 10 s, while

M. Qiu (Ed.): ICA3PP 2020, LNCS 12454, pp. 509–519, 2020.
https://doi.org/10.1007/978-3-030-60248-2_34

thoracic and abdominal respiration still exist. Hypopnea means that the oronasal airflow decreases more than 30% compared with the baseline level during sleep, accompanied by a decrease of oxygen saturation more than 4%, duration greater than 10 s, or the decrease of oronasal air flow was more than 50% from baseline, accompanied by a decrease of oxygen saturation more than 3%, lasting for more than 10 s.

Polysomnography (PSG) is the gold standard for OSAHS detection. But it is a slow and expensive process since it usually requires the patient to be in attendance at a sleep laboratory under the supervision of a specialized technician. It is an uncomfortable experience because the patients need to wear a variety of sensors on their bodies [13]. At present, many studies use a variety of physiological signals to automatically detect OSAHS. In these physiological signals, respiratory signals can directly reflect the breathing situation during sleep [18]. They can be obtained directly from the three parts of nasal airflow, thoracic and abdomen, or indirectly through ECG signals according to certain algorithms [12]. The detection performance of these respiratory signals is different. By comparison, it is found that the respiratory signals obtained directly can be better detect OSAHS [14].

The advent of deep learning and its astonishing progress in various domains have stimulated interest in applying them for detection OSAHS. The power of deep networks lies in their great capability of automatic feature learning from data, thus avoiding the reliance on hand-crafted features. However, there are still some shortcomings in the current methods. First, previous studies have focused on sleep classification by extracting deep features, ignoring the role of shallow features, which provide rich information for different sleep classifications. Second, the respiratory signals obtained from different parts play different roles in detecting sleep classification.

To address the above problems, a Multi-Level Features Fusion Network model (MLF2N) is proposed in this paper, which can effectively utilize the shallow features of different respiratory signals to improve the detection accuracy of OSAHS.

The main contributions of this paper are described as follows:

1) In this paper, a MLF2N model for detecting OSAHS is proposed. By fusing the features of different convolution layers, the shallow and deep feature information of respiratory signals are aggregated, and multi-level features are extracted to improve the detection performance.
2) The amplitude of respiratory signal affects the feature extraction, and the amplitude is adjusted by the expansion factor to improve the detection performance. Through comparative experiments, this paper selects the appropriate expansion factor.
3) Through detailed comparative experiments, it verifies that the different respiratory signals play different roles in detecting OSAHS. This paper demonstrates experimentally good performance on the public available dataset MESA, which is a large sleep dataset with 2056 subjects.

2 Related Works

Various algorithms have been developed for automatically detecting OSAHS events in one or more of the physiological signals originating from an overnight PSG. A common approach is to use machine learning and deep learning methods. Commonly used methods include support vector machine (SVM) [9], logical regression (LR) [4], K Nearest-Neighbor (KNN) and linear discriminant analysis (LDA) and artificial neural network (ANN) [15,17]. Gutierrez et al. [3] extracted spectral features and nonlinear features from nasal airflow respiratory signals, and used AdaBoost ensemble learning algorithm to classify sleep. For severe sleep apnea subjects, the accuracy was 83.3%. Yin Yan Lin et al. [11] used thoracic and abdominal respiratory signals to extract amplitude statistical features and frequency features, and the accuracy reached 81.8% by using SVM classification algorithm. These methods typically start with computing a set of human-engineered features over a certain epoch of the data. For each epoch, how to extract a set of features which can accurately describe the features of OSAHS requires a lot of artificial experience and feature engineering skills. It is still a difficult problem.

In recent years, the interest in using deep learning algorithms to automatically classify raw data of PSG signals has increased. Algorithms such as long short-term memory (LSTM) and convolution neural network (CNN) have been proposed as a good method capable of detecting OSAHS. Van Steenkiste et al. [19] used LSTM to detect normal sleep and sleep apnea according to the temporal correlation of sleep events. Haidar et al. [5] have demonstrated the validity of CNN models to classify the apnea or hypopnea events within sleep PSG recordings, with an accuracy of 77.6% using the nasal flow signal. Incorporating abdominal and thoracic respiratory signals could increase the performance of CNN model to 83.5% [6].

3 Materials and Methods

3.1 Data Set

The data are retrieved from the National Sleep Research Resource (NSRR), which is a new National Heart, Lung, and Blood Institute resource designed to provide big data resources to the sleep research community. The PSG data are available from Multi Ethical Study of Atherosclerosis (MESA) data set. In this data set, there are PSG records of 2056 subjects. Each record included physiological signals such as EEG, respiration signals, ECG and so on. The respiratory signals used in this paper are extracted from nasal thermal sensors, pressure sensors near the mouth, conductive bands around the thoracic and abdominal. These respiration signals were sampled at a frequency of 32 Hz. According to the features of each sleep event, sleep experts marked the start time and duration of OSA and hypopnea events [2], and other monitoring time was the time of normal sleep events.

3.2 Methods

In order to solve the problems of incomplete feature extraction in automatic detection of OSAHS, a MLF2N model is proposed in this paper. The framework is shown in Fig. 1. Firstly, the preprocessing block needs to normalize the respiratory signal, and each record is divided into epochs of 30 s which were manually classified by experts as event of normal sleep, OSA or hypopnea. Secondly, the CNN block with Multi-Level Feature Fusion Network (MLF2N) takes into account different levels of convolution layer to extract features of different frequencies. The shallow features not only contain noise information, but also contain features beneficial to sleep classification [10–13]. The multi-level features are fused by down sampling technology. Finally, the feature vector is transformed into the probability of each sleep event, so as to get OSA (O), normal sleep (N) and hypopnea (H).

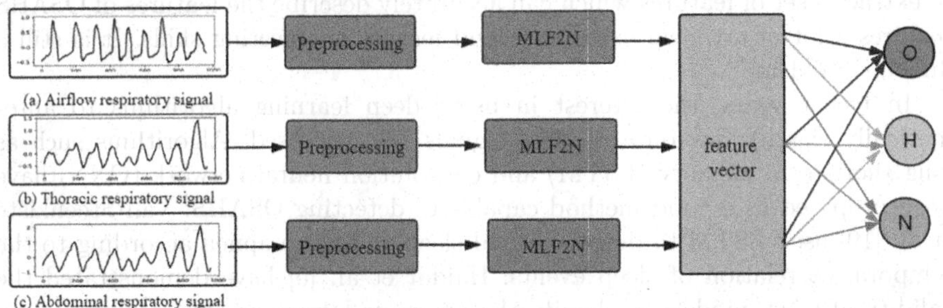

Fig. 1. Structure of OSAHS detection with the MLF2N model.

3.3 Data Preprocessing

Due to the frequent body movements of some subjects, the respiratory signal fluctuates greatly or the signal is missing, resulting in the unstable and incomplete signal collected. At the data preprocessing block, the subjects with significant differences were filtered according to the deviation of each respiratory signal. Secondly, it removed the subjects who slept less than 8 h or included only normal sleep events.

To control the variations in signal strengths due to differences in equipment-settings and inter-subject physiology, each respiratory signal was normalized based on the mean and standard deviation of the normal samples for each subject.

After standardized processing, the amplitude of respiratory signal is small, and the signal changes of different sleep events are not obvious. To overcome this shortcoming, standardized respiratory signals need to be multiplied expansion factor to make it easier to distinguish between different types of sleep events.

The sleep time of each subject was divided into 30s epochs, and 960 data were obtained for each epoch according to the sampling frequency. The sleep events were labeled according to the following rules:

(1) An epoch is labeled as OSA (O) if it includes an obstructive apnea event lasting for more than 10 s.
(2) An epoch is labeled as hypopnea (H) if it includes a hypopnea event lasting for more than 10 s.
(3) An epoch is labeled as normal (N) if it doesn't include any obstructive apnea and hypopnea events, or if include obstructive apnea and hypopnea events with a total duration of less than 10 s.
(4) An epoch was excluded if it included both obstructive apnea and hypopnea events lasting for more than 10 s.

The number of OSA events and hypopnea events is much smaller than that of normal sleep events, which results in a large difference in the number of sleep events. If the data set of unbalanced classification is used to train the model, the model will be biased towards the main classification, resulting in distorted detection results. In order to balance the number of each sleep event, three sleep events of the same number were randomly selected for training and testing of the model.

3.4 Multi-level Feature Fusion Network (MLF2N)

CNN is a feedforward neural network, which is widely used in image recognition, speech recognition and other classification tasks [8,10]. In recent years, it has also been used in physiological signal detection [17,18]. CNN mainly extracts the features of the signal through convolution.

To select the architecture and hyper-parameters of the CNN, this paper evaluated different combinations using random manual search. For every parameter combination, this paper evaluated on the 80% training dataset and selected the parameter combination that gave the best average accuracy over the 10 folds. These parameters are shown in Table 1. This CNN model includes three repeated operations. Each operation includes two convolutions and a max-pooling. The convolution kernel size is (1,3) and (1,2) respectively, stride is 3 and 2, and dropout is 0.1. The pool size of max-pooling layer is (1,2) and the stride is 2. Each convolutional layer has 32 filters and uses Rectified Linear Unit (Relu) activation function.

Different levels of convolution extract features of different frequencies. The shallow convolution contains more high-frequency features and the deep convolution extracts advanced features. Since both shallow and deep features can provide useful identification information for OSAHS detection, this paper adopts the CNN model of multi-level feature fusion to improve the detection performance by fusing the shallow and deep features of respiratory signals during sleep. Model contains six convolution layers and 3 max-pooling layers, and each epoch under different degree of convolution output features through down sampling. Then the

Table 1. Parameters in CNN model.

Layer	Size	Stride	Filter	Activation	Dropout
Conv1	(1,3)	3	32	Relu	0.1
Conv2	(1,2)	2	32	Relu	0.1
Max-pooling	(1,2)	2	32	—	—
Conv3	(1,3)	3	32	Relu	0.1
Conv4	(1,2)	2	32	Relu	0.1
Max-pooling	(1,2)	2	32	—	—
Conv5	(1,3)	3	32	Relu	0.1
Conv6	(1,2)	2	32	Relu	0.1
Max-pooling	(1,2)	2	32	—	—
Flatten	—	—	32	—	—
Dense	—	—	32	Relu	0.1
Output	—	—	3	Sigmoid	—

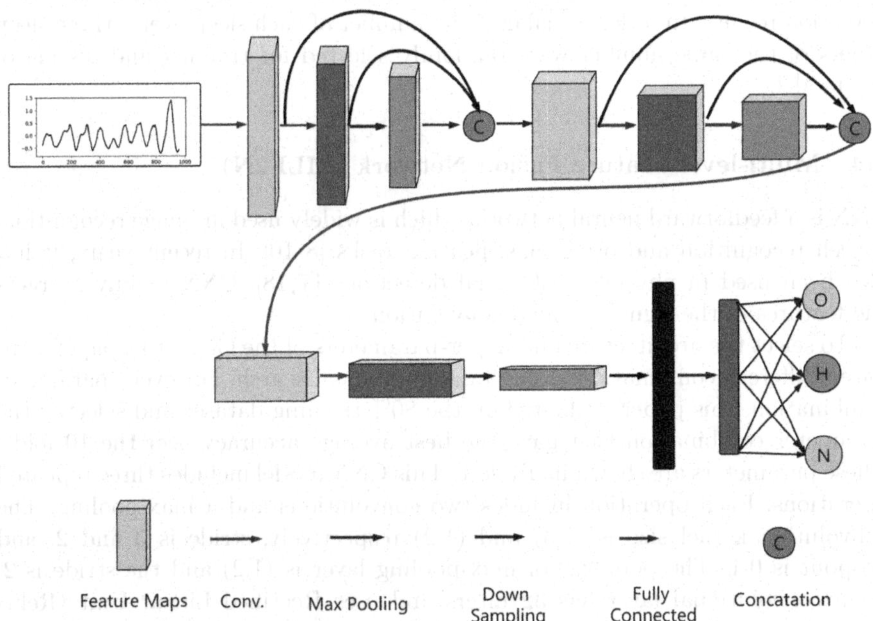

Fig. 2. Multi-level feature fusion network.

concatenation operation is used to concatenate feature together to obtain feature information of different levels and improve information utilization. Finally, output is generated through convolutional layer, max-pooling and full connection layer. The specific process is shown in Fig. 2.

3.5 Performance Evaluation

The main evaluation metrics are the sensitivity, specificity, informedness, accuracy. They are calculated as:

$$Sensitivity = \frac{TP}{TP + FN} \times 100\% \quad (1)$$

$$Specificity = \frac{TN}{TN + FP} \times 100\% \quad (2)$$

$$Informedness = Sensitivity + Specificity - 1 \quad (3)$$

$$Accuracy = \frac{TP + TN}{TP + TN + FN + FP} \times 100\% \quad (4)$$

Where TP is the number of true positives. TN is the number of true negatives. FP is the number of false positives. FN is the number of false negatives. Sensitivity reflects the detection effect of positive samples, specificity reflects the detection effect of negative samples, and informedness is the comprehensive evaluation of positive and negative samples.

4 Experimental Results and Analysis

The MLF2N model was trained and tested on the MESA data set. After preprocessing, 1801 subjects were selected from 2056 subjects. 28036 OSA, hypopnea and normal sleep events were selected from the data set. 80% of the events were randomly selected as the training set and 20% as the test set.

The MLF2N model adopts Adam optimization method and cross entropy as the loss function. The initial learning rate is 1e−3, and the learning rate is reduced to 1e−4 after 50 iterations. A total of 100 epochs of training are conducted, and the size of batch size is 400 sleep events. The training and testing are conducted based on tensorflow framework of Python 3.6, and the hardware is NVIDIA GTX 1080ti GPU.

4.1 The Influence of Respiratory Signal Amplitude

Taking the airflow respiratory signal as an example, the input signal is expanded by different factors, and the accuracy of sleep classification detection obtained is shown in Fig. 3. The accuracy of 75.2% can be obtained by directly inputting the standardized airflow respiratory signal. When the amplitude value is expanded by different multiples, the detection accuracy can be improved in different degrees. From the experimental results, it can be found that the detection result is the best when the amplitude value is expanded by 30 times, and the accuracy can reach 76.4%. Therefore, in the preprocessing, the amplitude of each respiratory signal is expanded by 30 times, and then input into the MLF2N model to detect OSAHS.

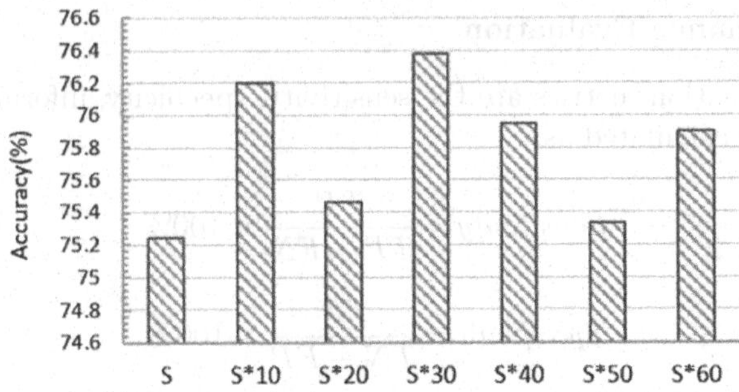

Fig. 3. The influence of different expansion factors on the accuracy.

4.2 The Effect of Different Respiratory Signals

In order to comprehensively evaluate the sensitivity and specificity of different respiratory signals in the detection of OSAHS, informedness was used. The evaluation values of different respiratory signals in detecting three sleep events are shown in Fig. 4.

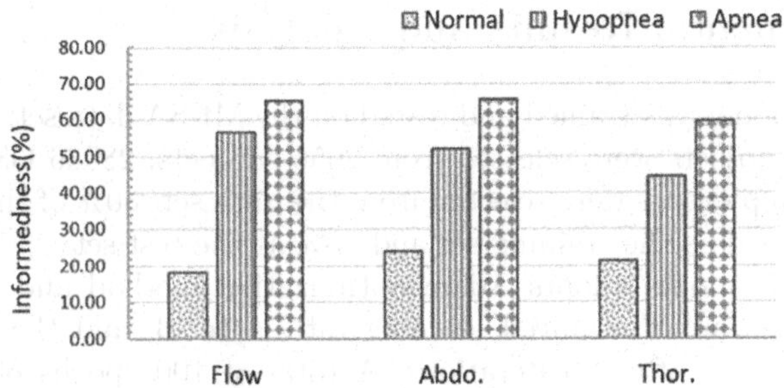

Fig. 4. Informedness of different respiratory signals in sleep classification.

In the detection of hypopnea events, the informedness of airflow respiratory signal was 56.65%, that of abdominal respiratory signal was 51.96%, and that of thoracic respiratory signal was 44.6%. Among these three respiratory signals, airflow respiratory signal provided a higher informedness. In the detection of OSA events, the informedness of abdominal breathing signals was 6.06% higher than that of thoracic respiratory signals. In the detection of normal sleep events, the informedness provided by abdominal respiratory signals was higher than that of other respiratory signals. It can be seen that different respiratory signals are different in the detection of sleep event.

4.3 The Performance of MLF2N

Three respiratory signals were input into typical CNN and the MLF2N respectively. The specificity, sensitivity and accuracy obtained are shown in Table 2.

Flow is the nasal airflow respiratory signal. Thor. is the abbreviation of thoracic respiratory signal. Abdo. is the abbreviation of abdominal respiratory signal.

Table 2. Detection results of different respiratory signals and different block.

Signals	Model Normal	Specificity			Sensitivity			Acc.
		Hypopnea	OSA	Normal	Hypopnea	OSA		
Flow	CNN	45.6	87.6	93.7	72.8	69.0	71.4	76.4
Abdo.	CNN	45.7	86.3	93.0	78.3	65.7	71.6	75.5
Thor.	CNN	45.3	84.0	91.0	76.2	60.6	68.6	71.7
Flow, Thor.	CNN	47.8	87.4	95.9	77.4	79.9	84.8	83.0
Flow, Abdo.	CNN	47.7	87.9	95.9	77.4	79.7	84.5	83.4
Abdo, Thor.	CNN	47.3	87.1	92.9	76.8	73.5	80.7	79.8
Flow, Thor. Abdo.	CNN	47.6	89.8	96.3	81.1	79.4	83.3	84.5
Flow	MLF2N	46.3	86.4	93.1	70.1	72.6	76.0	77.2
Abdo.	MLF2N	47.2	83.3	91.8	70.7	69.9	80.4	76.2
Thor.	MLF2N	46.3	84.0	91.0	74.7	66.3	74.7	74.3
Flow, Thor.	MLF2N	47.6	89.2	95.7	80.0	79.5	83.7	84.0
Flow, Abdo.	MLF2N	47.2	91.0	96.1	83.2	77.0	81.0	84.4
Thor., Abdo.	MLF2N	47.7	86.8	92.8	76.0	74.1	82.1	79.9
Flow, Thor. Abdo.	MLF2N	47.7	91.4	95.7	83.2	79.7	83.8	85.5

When using CNN with MLF2N, the accuracy of three respiratory signals has been improved in varying degrees. The detection accuracy of thoracic respiratory signal has been improved from 71.7% to 74.3%, which is 2.6% higher. For hypopnea and OSA events, the sensitivity is also improved. The sensitivity is not improved in normal sleep events. The main reason is that the shallow features contain little normal sleep identification information. It can be seen that the shallow features of different respiratory signals contain abundant identification information, so the MLF2N improves the detection accuracy.

5 Conclusion

This paper presents a MLF2N network model for the detection of OSA, hypopnea and normal sleep. This network model confirms that the amplitude of different respiratory signals affects the detection performance, and the effect of different respiratory signals on the detection of OSAHS is different. Through the effective fusion of shallow features and advanced features, it can provide more abundant information for sleep detection and improve the OSAHS detection performance. The accuracy is 85.5% for 84108 sleep events on MESA large data set, which is better than other methods.

Acknowledgement. This paper is supported by the Harbin science and technology bureau innovation under Grants No. 2017RAQXJ131, the Basic research project of scientific research operating expenses of Heilongjiang provincial colleges and universities under Grants No. KJCX201815, Heilongjiang Province Natural Science Foundation key project of China under Grant No. ZD2019F003.

References

1. Alex, S., Findley, L.J., Meir, K., Eric, G., Charles, G., Davidson, T.M.: Reducing motor-vehicle collisions, costs, and fatalities by treating obstructive sleep apnea syndrome. Sleep **27**(3), 453–458 (2004)
2. Dean, D.A., et al.: Scaling up scientific discovery in sleep medicine: the national sleep research resource. Sleep **39**(5), 1151–1164 (2016)
3. Gutierrez-Tobal, G.C., Alvarez, D., Del Campo, F., Hornero, R.: Utility of Adaboost to detect sleep apnea-hypopnea syndrome from single-channel airflow. IEEE Trans. Biomed. Eng. **63**(3), 636–646 (2016)
4. Gutierrez-Tobal, G., HorneroRoberto, A.: Linear and nonlinear analysis of airflow recordings to help in sleep apnoea-hypopnoea syndrome diagnosis. Physiol. Meas. **33**(7), 1261–1275 (2012)
5. Haidar, R., Koprinska, I., Jeffries, B.: Sleep apnea event detection from nasal airflow using convolutional neural networks. In: Liu, D., Xie, S., Li, Y., Zhao, D., El-Alfy, E.-S.M. (eds.) ICONIP 2017. LNCS, vol. 10638, pp. 819–827. Springer, Cham (2017). https://doi.org/10.1007/978-3-319-70139-4_83
6. Haidar, R., McCloskey, S., Koprinska, I., Jeffries, B.: Convolutional neural networks on multiple respiratory channels to detect hypopnea and obstructive apnea events. In: 2018 International Joint Conference on Neural Networks, IJCNN 2018, Rio de Janeiro, Brazil, 8–13 July 2018, pp. 1–7. IEEE (2018). https://doi.org/10.1109/IJCNN.2018.8489248
7. Heinzer, R., et al.: Prevalence of sleep-disordered breathing in the general population: the hypnolaus study. Lancet Respir. Med. **3**(4), 310–318 (2015)
8. Hinton, G.E., Srivastava, N., Krizhevsky, A., Sutskever, I., Salakhutdinov, R.: Improving neural networks by preventing co-adaptation of feature detectors. Comput. Sci. **3**(4), 212–223 (2012)
9. Koley, B., Dey, D.: Real-time adaptive apnea and hypopnea event detection methodology for portable sleep apnea monitoring devices. IEEE Trans. Biomed. Eng. **60**(12), 3354–3363 (2013)
10. Krizhevsky, A., Sutskever, I., Hinton, G.E.: Imagenet classification with deep convolutional neural networks. Commun. ACM **60**(6), 84–90 (2017)
11. Lin, Y.Y., Wu, H.T., Hsu, C.A., Huang, P.C., Huang, Y.H., Lo, Y.L.: Sleep apnea detection based on thoracic and abdominal movement signals of wearable piezoelectric bands. IEEE J. Biomed. Health Inf. **21**(6), 1533–1545 (2017)
12. Moody, G.B., Mark, R.G., Bump, M.A., Weinstein, J.S., Goldberger, A.L.: Clinical validation of the ECG-derived respiration (EDR) technique. Comput. Cardiol. **13**(2), 87–92 (1986)
13. Senaratna, C.V., et al.: Prevalence of obstructive sleep apnea in the general population: a systematic review. Sleep Med. Rev. **34**(2), 70–81 (2016)
14. Steenkiste, T.V., et al.: Systematic comparison of respiratory signals for the automated detection of sleep apnea. In: 40th Annual International Conference of the IEEE Engineering in Medicine and Biology Society, EMBC 2018, Honolulu, HI, USA, 18–21 July 2018, pp. 449–452. IEEE (2018). https://doi.org/10.1109/EMBC.2018.8512307
15. Tagluk, M.E., Sezgin, N.: Classification of sleep apnea through sub-band energy of abdominal effort signal using wavelets + neural networks. J. Med. Syst. **34**(6), 1111–1119 (2010). https://doi.org/10.1007/s10916-009-9330-5
16. Tamanna, S., Ullah, M.I.: Sleep apnea and cardiovascular disease. Cardiovasc. J. **8**(2), 143–148 (2016)

17. Thommandram, A., Eklund, J.M., McGregor, C.: Detection of apnoea from respiratory time series data using clinically recognizable features and KNN classification. In: 35th Annual International Conference of the IEEE Engineering in Medicine and Biology Society, EMBC 2013, Osaka, Japan, 3–7 July 2013, pp. 5013–5016. IEEE (2013)
18. Uddin, M.B., Chow, C.M., Su, S.W.: Classification methods to detect sleep apnea in adults based on respiratory and oximetry signals: a systematic review. Physiol. Meas. **39**(3), 03TR01 (2018)
19. Van Steenkiste, T., Groenendaal, W., Deschrijver, D., Dhaene, T.: Automated sleep apnea detection in raw respiratory signals using long short-term memory neural networks. IEEE J. Biomed. Health Inf. **23**(6), 2354–2364 (2018)

BIMP: Blockchain-Based Incentive Mechanism with Privacy Preserving in Location Proof

Zhen Lin[1], Yuchuan Luo[1(✉)], Shaojing Fu[1,2], and Tao Xie[1]

[1] College of Computer, National University of Defense Technology, Changsha, China
linz_cs@163.com, lychuan.cs@gmail.com

[2] Sate Key Laboratory of Cryptology, Beijing, China
shaojing1984@163.com

Abstract. Location based services (LBS), as an important part of people's everyday life, rely on current/historical location information to offer services. However, malicious users may generate some fake information to cheat providers to obtain more profits. Currently, existing schemes directly leverage GPS or a centralized party to claim location information, which may be easily counterfeited or result in leakage of users' privacy. To authenticate the location information, an interactive scheme can be introduced to allow a mobile user as prover to generate proofs of location by exploiting neighboring nodes as witnesses. However, owing to the selfishness of users, it is a great challenge to encourage witnesses to generate proofs. Moreover, witnesses will not be willing to generate proofs unless that their privacy is under well protection. To tackle the above issues, we propose a blockchain-based incentive mechanism with privacy preserving in location proof. First, a novel blockchain system is introduced for interactive location proof to generate secure and trustworthy proofs through the use of distributed ledgers and cryptocurrency. Second, a proof of points protocol is proposed to efficiently reach consensus to build location proof blockchain and incentivise users. Third, motivated by security problem, a novel concept of traceable-detectable-prefix is developed to resist collusion attacks while protecting users' privacy. Finally, theoretical analysis and simulation experiments are provided, which demonstrate that the proposed scheme can provide complete location proof while protecting privacy of users.

Keywords: Blockchain · Location proof (LP) · Proof of points (PoP) · Privacy preserving · Collusion resistant

---

This work is supported by the National Key Research and Development Program of China (No. 2018YFB0204301), NUDT Research Grants (No. ZK19-38), Open Foundation of State Key Laboratory of Cryptology (No: MMKFKT201617), and National Nature Science Foundation of China (grant 61572026, 61872372).

M. Qiu (Ed.): ICA3PP 2020, LNCS 12454, pp. 520–536, 2020.
https://doi.org/10.1007/978-3-030-60248-2_35

1 Introduction

Location based services (LBS) play crucial roles in achieving smart city, which have attracted widespread attention form both academia and industry. For example, LBS for location-based access control [1], detecting and recommending nearby friends [2], or mobile crowdsensing [3] have been presented, among many others. However, with the rapid development of LBS, it becomes an issue of significance to guarantee the authentic geographic location of users. Recently, location proof (LP) has been introduced as an innovative and attractive approach for improving the integrity of location information, which leverages mobile devices to generate proofs in a distributed way [4,5].

LP is a simple primitive that allows mobile devices to prove their locations to services and applications [6]. In other words, a LP is a small piece of meta-data which includes "who, when and where", stating that a user is at a given position at a given time. Other terminologies, such as provenance proof [7], alibi [8] and spatial-temporal provenance (STP) proof [5], have been used for similar concepts. It is demonstrated as an effective and feasible way to use a peer-to-peer approach between mobile devices to generate exact and trustworthy location information [9]. However, LP schemes encounter the following new challenges. On one hand, owing to the selfish nature of users, witnesses may refuse to contribute to save the energy of their devices energy in pursuit of self-interests. This may have a great negative impact on obtaining sufficient LPs. On the other hand, malicious proves or witnesses will threaten the system security by launching various attacks, e.g., proof falsification [10], privacy sniffing [11], collusion attack [12], and wormhole attack [13]. Thus, it is urgently needed to design a secure proof generation and verification scheme in LP system while stimulating witnesses to behave cooperatively to realize trustworthy geographic location and privacy preserving.

Existing works on crowdsensing incentives [14,15], however, cannot be directly applied in LP system. First, due to the heterogeneity and diversity of attacks on participants, the security of LP generation and verification has not been properly addressed in existing works. Second, the risk of privacy disclosure of witnesses during the LP generation process can reduce the incentive of witnesses to cooperate. Third, most of current incentive mechanisms are designed from the perspective of reputation instead of the perspective of contributions of users. Therefore, it is still a vital issue to design a secure incentive scheme to improve the efficiency and security in LP proofs generation and verification.

To acquire LPs in a secure manner, a trusted architecture is needed to support interactions between users and witnesses. Current similar frameworks, such as APPLAUS [16] and STAMP [5], are based on trust evaluation by a centralized entity to detect collusion, raising privacy risks. Fortunately, blockchain provides a novel solution for this question, which records transactions in the decentralized network in a immutable manner [17].

In this paper, to address the aforementioned problems, we propose a blockchain-based incentive mechanism with privacy preserving (BIMP) for location proof, which relies on witnesses to attest the current location of a user and eliminates the dependence on a reliable central authority. In BIMP, mobile users utilize short-range communication devices to collaboratively generate LPs, which are submitted to blockchain by its owner. A verifier can retrieve and verify LPs from blockchain. Afterwards, a novel proof of points (PoP) protocol is developed to reach consensus and stimulate the contributors in LP blockchain. Moreover, a novel structure of LPs is designed for detecting collision attacks while guaranteeing users' privacy from other parties. Finally, theoretical analysis and simulation results are provided to evaluate and demonstrate the efficiency of the proposal by comparisons with other schemes.

In a nutshell, the main contributions of this work are fourfold as follows.

1) First, we propose BIMP, the first blockchain-based incentive mechanism with privacy preserving scheme in LP system. The proposed secure framework based on distributed ledgers keeps the advantages of the existing witness-based LP scheme (distributed, correctness, and collusion-resistant), while addressing the privacy disclosure problem caused by the trusted computation server through blockchain technology.
2) Second, We develop a PoP protocol within the LP blockchain to encourage witnesses against selfish behavior, where the miners are selected based on the total points and hashrate. The points model is designed to arouse the enthusiasm of witnesses based on contribution and credibility computing.
3) Third, We devise a traceable-detectable-prefix structure of LPs that can protect users from prover-witness collisions. Moreover, the privacy problem for users is maximized while guaranteeing the system security.
4) Fourth, extensive simulation experiments demonstrate that the proposed scheme can efficiently stimulate witnesses to help provers with generating LPs while protecting their privacy.

The rest of the paper is organized as follows: Related work is discussed in Sect. 2. Our system model and treatment model are described in Sect. 3. The proposed scheme is presented in Sect. 4. Security analysis against various attacks is formulated in Sect. 5. Performance evaluations are given in Sect. 6. Finally, conclusion and the future work are provided in Sect. 7.

2 Related Work

The notion of location proofs was introduced by Saroiu and Wolman [6]. They proposed a secure mechanism which users can create timestamped location proofs by exchanging their signed public keys with wireless Access Points (APs). Luo and Hengartner [18] developed a trusted third party validation scheme to detect cheating attempts where a prover obtain two LPs of different locations at the same time. Hasan and Burns [7] explored two privacy protection schemes based on Bloom filters and hash chains for guaranteeing the integrity of the chronological order of location proofs.

To reduce dependence on wireless APs and enhance the flexibility of generating location proofs, Zhu and Gao [16] designed a APPLAUS scheme in which co-located mobile devices mutually generate LPs via Bluetooth. Gambs *et al.* [4] investigated the privacy-preserving of location proof system in distributed environments, considering the privacy of verification phase and the correctness of LPs. Wang *et al.* [5] designed a STAMP system and studied the Bussard-Bagga distance-bounding protocol in LP systems to resist wormhole attacks. In addition, an entropy-based trust evaluation approach was employed for P-W collusion detection. However, incentive mechanism and P-W collusion detection issue in location proof are still not fully considered in most of the existing works.

Recently, various works have been reported in leveraging the promising blockchain technology into the location proof scheme. Amoretti *et al.* [19] proposed an blockchain-based proof of location framework to guarantee message integrity and non-repudiation. A hierarchical tree of connected blockchain infrastructure was developed by Dasu *et al.* [20] to improve scalability and security while servers or cellular towers could issue location certifications. Liu *et al.* [21] proposed an efficient data collection and secure sharing scheme in mobile crowdsensing scenarios by combining blockchain technology with deep reinforcement learning. However, the above blockchain systems cannot be directly applied in LP, as security issues are not fully considered, such as privacy issue, collusion and etc.

3 System Model

In this section, the concept of location proof is first introduced, followed by the different entities of our system model. Then, the threat model is elaborated.

Definition 1 (*lp*). *A lp is a one-to-one proof that a user attest another user's current position at that time. In this paper, a prover cooperates with a witness to obtain a lp.*

Definition 2 (Location Proof – LP). *A LP is a proof which includes 'who, when and where', stating that a user is at a given position in time. After collecting at least k lps, a prover can synthesize a LP.*

3.1 Entities

The system model is shown in Fig. 1. There are five different entities in BIMP system: prover, witness, LP blockchain network, verifier, semi-trusted authority (STA).

- **Prover:** A prover is a mobile user willing to obtain a LP to prove his location. And the prover will broadcast a *lp* request to his neighboring nodes through short-range communication technology (e.g. Bluetooth and ZigBee) when he needs a LP. After generating a LP, the prover submits LP obtained to blockchain network.
- **Witness:** A witness is a neighboring user of the prover who agrees to generate a *lp* upon receiving his *lp* request. Notice that, a prover can be a witness and vice versa.

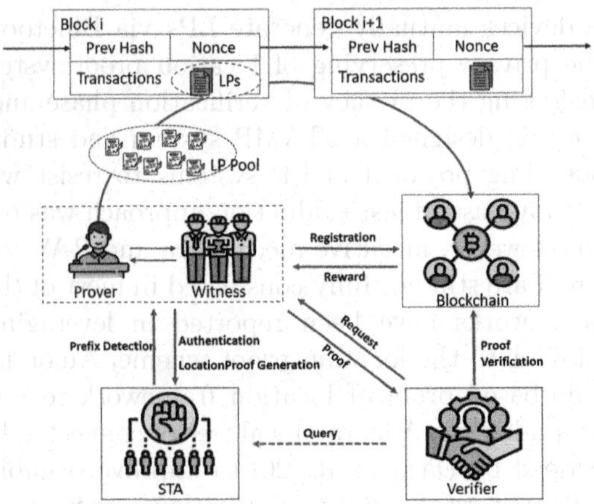

Fig. 1. System model.

- **LP Blockchain:** LP blockchain is a continuously growing sequence of hash-chain blocks, aim to record all the valid LPs during a certain period of time. The users register on blockchain, and all valid LPs are stored on blockchain where the data recorded cannot be altered retroactively. In LP blockchain, each verifier can store the full ledgers to verify. As the storage size of mobile devices is limited, provers and witnesses will store the meta-data of blocks other than the full ledgers for decreasing the cost [22].
- **STA:** The STA is a semi-trusted third-party who is only used to provide authentication service. The STA only knows the correlation of a user's pseudonyms, real identity and device information, but the location information of user is kept secret for it. In special circumstances, the STA have the capacity to retrieve the real identities of the users to ensure the auditable property of system. In addition, STA has power to calculate the point value.
- **Verifier:** A verifier is a third-party organization or user (e.g. bank, police authority, LBS provider...) who is authorized to check a prover's LP.

3.2 Threat Model

In LP system, malicious participants may threaten the network security. Specifically, we define the following malicious behaviors and attackers.

- *Malicious provers:* A malicious prover's goal is obtaining a LP without physically being present at a location. This includes modifying information in a existing LP already endorsed by a witness, or lying to a witness or verifier about the position and time information. Also, a malicious prover tries to learn witnesses' identity information in process of LP generation [5].
- *Malicious witnesses:* This type of witness may repudiate or fool an honest prover and malicious acquire provers' identity information.

- *Malicious verifier:* A malicious verifier obtains a valid LP form a prover and then tries to extract the prover's information without permission. Moreover, a malicious verifier cheat another verifier through pretending to be the owner of this LP.
- *Curious STA:* In this paper, we do not require that the STA is a completely trusted third party. And we assume STA is a semi-honest entity known as honest but curious which always follows the protocol faithfully to perform its functions correctly, i.e., registration, verifying the prefix, etc. However, curious STA tries to infer extra information to outline users' historical track.
- *Collusion:* Two different collusion scenarios are taken into consideration in this paper. (1) P - P collusion. This type of collusion attack has been more commonly known as impersonation fraud or terrorist fraud [23]. More specifically, with the help of a colluding prover $\mathcal{B}$ at position P_B, a prover $\mathcal{A}$ at position P_A tricks a witness $\mathcal{W}$ near $\mathcal{B}$ to obtain a fake LP that $\mathcal{A}$ is at P_B. (2) P - W collusion. A prover $\mathcal{A}$ and colludes with a witness $\mathcal{W}$ to create a LP when $\mathcal{A}$ and $\mathcal{W}$ are not at the claim location or time. To my knowledge, existing solutions do not detect this type of collusion well under the premise of protecting privacy.

4 Proposed BIMP Scheme

In this section, we introduce the proposed scheme, including preliminaries, the PoP protocol, and traceable-detectable-prefix for LP.

4.1 Preliminaries

1) Blockchain and Consensus: Blockchain are widely known by its inherent characteristics of decentralization, anonymity and tamper-resilience [24]. Especially, the data submitted to the blockchain will be exposed to the public, owing to the transparency of blockchain where the whole nodes will replicate the whole chain to ensure consistency. Monetary incentive plays the key role to ensure that most of the consensus nodes participate and follow the rules of blockchain. For example, the 'coinbase reward' in Bitcoin, as a token issuing reward is put forward to incentive the winner in the block puzzle competition and expand the blockchain with the new block. In this paper, we adopt blockchain to jointly address the problems of pseudonym, scalability, data backup, reward and poor synchronization.

2) Commitment and zk-SNARK for Identity Authenticity: A commitment scheme allows one party to 'commit' to a given message while keeping it secret, we use $C(m, r)$ to represent that a message m is committed with a random nonce r in this paper. In our system, we do not require a complex commitment scheme, and we use a based on one-way hashing commitment [25] in our implementation.

Opening a commitment needs m and r which would undermine the privacy we are trying to achieve. Zero-knowledge proof [26] serves the purpose. In this paper,

Table 1. List of notations

Notations	Description
G	G is a generator of the elliptic curve group E
(p, P)	A prover's private key/public key pair on the blockchain
(w_i, W_i)	A witness's private key/public key pair on the blockchain
(a, A)	STA's private key/public key pair on the blockchain
$H(m)$	One-way hashing of message m
$E^K(m)$	Encryption of message m with key K
$\sigma(m)$	A ring signature of message m
$m_1 \vert m_2$	A chain of messages m_1 and m_2
$m_1 \vert\vert m_2$	m_1 is a prefix linking for m_2

the goal of zero-knowledge proofs is to convince a verifier that a prover possesses ownership of a LP, without revealing his ID to the verifier or anyone else. "Zero-knowledge succinct non-interactive argument of knowledge" (zk-SNARK) is used for identity authenticity in our solution. In our implementation, we generate an attestation π_{zk} on a ID P that: the requester P indeed owns a secret key corresponding to a commitment in a valid LP, i.e., $Verify(C_1, \pi_{zk}) \rightarrow Yes/No$. All notations have been summarized in Table 1.

4.2 PoP Protocol

In this study, blockchain technique is exploited in LP system to secure LP generating/verifying process and incentivise witnesses. To efficiently arouse the enthusiasm of witness and reduce the confirmation delay of transaction for reaching consensus, based on the proof of work (PoW) algorithm [27], a novel PoP protocol is presented in this subsection with the following phases: 1) Registering; 2) LP generation; 3) Building blocks; 4) Reward; and 5) LP service requires.

1) Entity Registering: In BIMP, we take advantage of public key cryptography on the blockchain. Users pick a random number a and computes the public key $A = aG$ by the elliptic curve multiplication. (a, A) is the user's private/public key pair registration with blockchain. Here, the public key A can be viewed as the pseudonym of user and can be known by anybody, but the private key a should be kept secretly by the user itself. Moreover, each user, i.e., prover, witness and verifier, becomes legitimate nodes in the LP blockchain after authenticating with STA by binding their real identities with public/private key pairs, e.g.., the MAC address or the business license of verifier. Notice that, STA needs to register on blockchain to get the private/public key pair, and broadcasts its public key to all users.

2) LP Generation: If a man wants to obtain a proof of his current location, he will run this phase as a prover in collaboration with nearby mobile devices to

Fig. 2. Generation Phase of BIMP.

obtain lp. The prover can generate a LP when he has collected k lps from different witnesses at least, and submits it to blockchain. In order to show the intuition of our generation phase, we describe the concrete LP generation architecture in Fig. 2. We depict the figure from four procedures.

a) A prover P anonymously broadcasts a lp request (denoted as $lpReq$) to neighboring witnesses, and a $lpReq$ is constructed as follows:

$$lpReq = C(P, r_p)|L|t|E^A(P) \tag{1}$$

where P is the prover's unique ID (defined in Table I), r_p is a random nonce for the commitment to P. L is the current location, and corresponds to time t. And P is encrypted using public key of STA to protect the prover's ID from being seen by the witness.

b) A witness W_i who is nearby prover P receives a $lpReq$, and decides whether to accept it. If accept, the witness W_i sends an ACK code back to the prover P. Next, we carry out distance bounding protocol between P and W_i.
If the distance bounding protocol return success, the witness W_i will create a lp for prover P. First, witness W_i separates received information to get $E^A(P)$, and constructs a header request (denoted as $HReq$) to STA, as follows:

$$HReq = E^A(W_i, E^A(P)) \tag{2}$$

When STA receives a $HReq$, it has the ability to decrypt and get the ID of two parties. STA is responsible for establishing a detectable-prefix computed by $H(P, W_i)$ for P-W collusion detection. And STA creates a header response (denoted as HRes) as follows and sends it back to witness:

$$HRes_{P,W_i} = H(P, W_i)|Sig^a(H(P, W_i)) \tag{3}$$

where $Sig(\cdot)$ is the signature function.
After receiving the $HRes$ from STA, witness W_i creates the ring signature $\sigma_{W_i}(m, HRes)$ where $m = C(P, r_p)|C(L, r_1)|t$. Finally, the witness generates a lp_{W_i} as follows and sends $lp_{W_i}|m$ to prover P:

$$lp_{W_i} = H(P, W_i)||\sigma_{W_i}(m, HRes_{P,W_i}) \tag{4}$$

c) Suppose prover P receives $lp|m$ from k witnesses (denoted as $lp_{W_1}, \cdots, lp_{W_k}$) and stores associated parameters (i.e., r_p, L, r_1 and t) locally. Up to this point, we can assume that the prover P has created a LP for himself and a LP is constructed as follows:

$$\begin{aligned} LP = & H(P, W_1)|\cdots|H(P, W_k)|| \\ & \sigma_{W_1}(m, HResp_{P,W_1})|\cdots|\sigma_{W_k}(m, HResp_{P,W_k}) \end{aligned} \tag{5}$$

d) Nevertheless, the LP in previous steps is not a valid LP. The prover P must submit it to LP pool, and only the LP recorded on blockchain are considered to be a valid proof. Remark that here we let each prover to generate a new blockchain address for each submitting (i.e. one submitting one address) to enhance anonymity.

3) Building Blocks: A block is a container of an arbitrary subset of LPs. Each prover independently submits LPs to pool during the consensus time interval. And invalid LPs are discarded by traceable-detectable-prefix (defined in Section V-C). All valid LPs are ordered by the timestamp and packaged into a Merkle-tree structure. Then, nodes, i.e., provers and witnesses, build their local blocks concurrently, where each block contains a hash pointer to the prior block to protect the integrity of the LP records. In PoP, we defined that nodes can generate a new block by searching for a solution string, *nonce*, which be concatenated with a given string x to make the hash of them is smaller than a target value. And the difficulty level of the solution searching for each node is adjusted inversely to its points in the network. For simplicity of exposition, let $H(\cdot)$ denote the hash function and $s(\cdot)$ represent the function that returns the difficulty level of the solution searching. Then, we can formally defined that a node can generate a new block by solving the PoP puzzle as follows:

$$H(nonce||x) \leq D(h) \cdot s(points) \tag{6}$$

where $D(h)$ is a difficulty target defined by the system, which can automatically adjust difficulty to target an average fixed number of blocks per hour. The x is based on the candidate block data, i.e., Merkle root, the reference hash pointers, etc. $s(\cdot)$ is monotonically increasing function with respect to the value of points, and can be expressed as followed:

$$s(u_p) = \begin{cases} 1 + \sigma \cdot \frac{u_p}{\sum_{u \in U} u_p}, & u_p \geq 0 \\ 1, & \text{otherwise} \end{cases} \tag{7}$$

where $\sigma > 0$ is the adjustment coefficient, and u_p means the point value of participant in LP blockchain. Note that higher u_p means higher $s(\cdot)$ and lower difficulty level when the $D(h)$ is given. In other word, it is easier to find the solution *nonce* if you have higher point value.

The fastest node who first seeks out answer *nonce* is permitted to broadcast its message to other nodes for verifying. If the auditing is successfully validated by most of the nodes, the new block is appended to their own ledger. With this, the consensus is reached and the valid LPs are stored in blockchain.

4) Reward: After achieving the consensus, the system will update the state variables in the accounts of all nodes in blockchain network and the point value of all participants in LP generating procedure. Then, the system will calculate the corresponding rewards to winner who first seeks out the solution string, *nonce*. Compared to the Nakamoto protocol, the PoP consensus protocol is a combination of contribution and computation-intensive. To win the puzzle solving race, a node needs to achieve the point value and hash querying rate as high as possible. Based on PoP protocol, witnesses prefer to get more points by assisting the provers in generating LPs. Assume that the user's point value is u_p, and u_p is calculated based on the following rules:

$$u_p' = \begin{cases} u_p + \log_2 u_p, & \text{assist in generating} \\ \frac{1}{2} u_p, & \text{collusion attacks} \end{cases} \tag{8}$$

where u_p' is the updated u_p and the initial value of u_p is 5. Notice that, STA is the sole entity of calculating u_p. Obviously, the honest nodes are easier to obtain higher points.

5) LP Service Requests: In this phase, a prover P tries to convince a verifier that he was at a given location (at a given time). In other word, a prover P needs to claim that he has the ownership of a LP existed in LP blockchain to obtain access to services based on location.

a) The verifier V needs the prover P making a claim about his/her location (and time) to acquire the service. And the verifier V send his public key PK_v to the prover P.
b) The prover P creates an LP claim (denoted as LPC) by the following information:

$$LPC = E^{PK_v}(LP|\pi_{zk}|m|L|r_1|t) \tag{9}$$

and then zero-knowledge proof that the prover P is the corresponding to $C(P, r_p)$, proving that he is the master of this LP, and can be expressed as

$$Verify(C_1, \pi_{zk}) \rightarrow Yes/No \tag{10}$$

c) After receiving the prover's LPC, verifier is now responsible for two task: LP verification and collusion detection. First, verifier V decrypts everything in LPC and performs LP verification by the following operations:
   - LP is a valid location proof by querying the LP blockchain.
   - m agrees with m contained in σ_{W_i} of LP.
   - $C(L, r_1)$ can be de-committed with L and r_1.
   - $Verify(C(P, r_p), \pi_{zk})$ returns Yes, meaning that prover P possesses ownership of this LP.
   - Prefix $H(P, W_i)$ agrees with message in $Sig^a(H(P, W_i))$, i.e., $D^A(Sig^a(H(P, W_i))) = H(P, W_i)$. Here, $D^A(\cdot)$ is the decryption function with STA's public key. Afterwards, we conduct P-W collusion detection to link the same prefix LP.

4.3 Traceable-Detectable-Prefix

In LP blockchain, higher point value means a higher probability of winning in the consensus process and receiving the reward. Thus, each witness node is motivated to assist prover in generating LP to increase its point value. However, it is easy for the prover to get a valid LP with fake information by colluding with m witnesses, i.e., P-W collusion. An entropy-based trust model [5] is proposed for such an attack, but this model breach users' privacy.

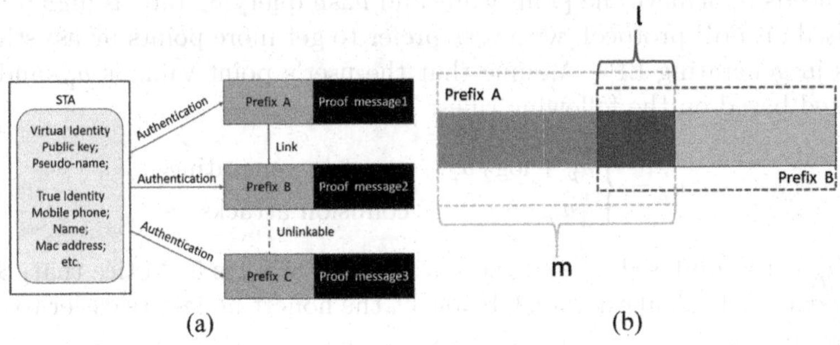

Fig. 3. Anonymous traceable-detectable-prefix scheme.

As a countermeasure against P-W collusion, we proposed a traceable-detectable-prefix structure of LP. The LPs are composed of a traceable-detectable-prefix and the remaining part, as shown in Fig. 3(a). The detectable-prefix is a header computed by hashing the identities of the prover and witness (constructed as $H(P, W)$), which is an invariable for the same prover and witness. In normal cases, the traceable-detectable-prefix ensure the anonymity with data publicly stored in a distributed ledger against any party. Our strong anonymity requires that no one can recognize the real identity from LPs, but we need be alert to collusion attacks. Obviously, through our traceable-detectable-prefix scheme, we can link the same prefix to reveal the collisions without recovering the actual identities. Moreover, traceable-detectable-prefix is also a credential for trust witness to get points reward form STA.

We shall make the followed assumptions on the traceable-detectable-prefix under consideration. A prover may have k LPs and a valid LP need m witnesses to endorse, i.e., a valid LP has m different prefixes. And there might exist l same witnesses for any two LPs. As shown in Fig. 3(b), there are two prefix aggregate A and B corresponding to two LPs respectively, and l same prefixes between them ($0 \leq l \leq m$). Let us compute the probability of prefix linking occurrence P. Obviously, it is impossibility to occur prefix linking on the first LP, i.e., $k = 1$, $P_1 = 0$.

We are now turn to the case $k \neq 1$, and computing the probability P_2:

$$\begin{aligned} P_2 = &C_2^k(C_m^l p^l(1-p)^{m-l} \\ &+ C_m^{l+1} p^{l+1}(1-p)^{m-l-1} + \cdots + C_m^m p^m) \end{aligned} \tag{11}$$

where p is the probability of the event that a witness receives a *lpReq* from prover, and each event is independent.

Consider

$$when \quad \frac{1}{p} \gg \frac{m+1}{l+1} \Rightarrow p \ll \frac{l+1}{m+1},$$

$$\frac{C_m^{l+1} p^{l+1} (1-p)^{m-l-1}}{C_m^l p^l (1-p)^{m-l}} = \frac{m-l}{l+1} \cdot \frac{p}{1-p} \ll 1$$

We can rewrite (11) as follow:

$$P_2 = C_2^k (C_m^l p^l (1-p)^{m-l}) \tag{12}$$

A prover random sends a *lpReq* to surrounding witnesses, so p is inversely proportional to the size of witnesses n. In general, an increase in the number of witnesses n decreases the probability of p. Letting $n \to \infty$ and $p \to 0$, we can ignore influence of $(1-p)^{m-l}$ on P_2. It is easy to find that $O(C_m^l) = O(m^l)$, $O(C_k^2) = O(1)$.

On account of the above remark, we have $P_2{}' = (p \cdot m)^l$. Normally, m and l a constant within 10, it is a simple matter to see that when $p \to 0$, $P_2 \to 0$. In other words, the probability of prefix linking on LPs is rare, for example, $p = 0.01, m = 10, l = 3$, P_2 is equal to 0.001. But in practice the p is even smaller, thus traceable-detectable-prefix do not damage the privacy of the honest witnesses. On the contrary, a malicious prover needs high cost to obtain fake k LPs. Set n' is the number of colluding witnesses for a malicious prover, and n', m and l are related as follows:

$$n' = (k+1)m - \alpha \tag{13}$$

where α is a constant related to m, l and k. If more than 3 same witnesses (i.e., $l = 3$) for two LPs, we shall call it "strong linking". If a LP has been found more than 3 times strong linking, the verifier offers a debriefing to STA that will cut down the point value of every conspirator to punish them by (8), even evict the adversary from BIMP when his points is lower than 1.

5 Security analysis

In this section, we prove that the BIMP scheme can resist malicious attacks to achieve the security and privacy properties.

- *Unforgeability:* A prover must cooperate with nearby witnesses to create a legitimate LP. The location information L is encrypted by commitment, and the witnesses endorse it by ring signature. Assuming a user never gives away his/her private key, a prover must request for k witnesses to create a signature. As long as the size of the malicious witnesses m is less than the number of generating needed, i.e., $k > m$, our BIMP scheme ensures that the prover cannot change the information in a LP. After a valid LP is submitted to blockchain, there is nothing the prover can do about it.

- *Privacy:* First, a prover and a witness cannot find out the real identities of each other. The users creates one-time address for each interaction to remain anonymous. The prover's pseudonyms is encrypted by STA's public key, no one can decrypt it and obtain ID. Second, when generating a LP, a eavesdropper can only learn that the witnesses is around the prover but does not infer the exact position of witnesses. Third, curious STA cannot get any information beyond ID about a prover or witness. Furthermore, the information of a valid LP is uploaded in encrypted form rather than plain text. Therefore, our scheme can protect the privacy of provers and witnesses well.
- *Proof of ownership:* The LP binds to prover's private key, that is to say, who possesses the private key, who is its true master. No one has access to another user's private key, since users do not leaking their private key actively. Furthermore, zero-knowledge proof is used to protect user's private key and judge whether he/she possesses the private key corresponding to ID committed in LP.
- *Unlinkability:* In our scheme, we can guarantee a good unlinkability for each honest user, but malicious conspirators are linked by traceable-detectable-prefix. Normally, no one can distinguish that two LPs, which have different prefixes and submit on blockchain, belong to the same owner, let alone peep what's in LP. Nevertheless, a malicious prover, i.e., P-W collusion, has to satisfy all demand of traceable-detectable-prefix at high costs to keep unlinkability. As the number of collusions increases, the cost of it increases too.

6 Performance Evaluation

In this section, we begin with an introduction to the simulation setup, and then the results are presented.

6.1 Simulation Setup

The simulation is implemented in Python 3.7 with an Intel Core i7-8550U CPU at 1.8 GHz 1.99 GHz and 8G RAM. We use secp256k1 as the elliptic curve algorithm, a Probabilistic Polynomial Time (PPT) algorithm [28] as ring signature scheme and DBPK-Log protocol [29] as distance-bounding protocol. Consider a scenario with 1000 users in a segment of BIMP and every 100 users are divided into ten location areas. All users are walking with a random mobility model in designated territory and 1%–10% witnesses may move from area to area. In each LP generating event, a random prover and m random witnesses are selected among all the users. In addition, we set the difficulty level from 0 to 31 bits to argument the PoP consensus protocol. We run a training phase with 1000–10000 LPs generating events, i.e., the average LP of each user is 10–100. In our experimental tests, we vary some parameters to see their impact on the performance.

6.2 Simulation Results

To evaluate the effectiveness and difficulty of the collusion, we first conduct simulations under different parameters. Figure 4 shows the need number of colluding witnesses at least and the probability of judging collusion at the different boundary setting for "strong linking" l with the increase of LP generating times when $m = 10$, i.e., each valid LP needs 10 witnesses. From Fig. 4(a), it can be seen that a malicious prover needs more colluding witnesses with the increase of number of generating fake LPs to avoid the "strong linking". In addition, from Fig. 4(b), it can be observed that the probability of "strong linking" occurred at $l = 2$ far above the probability at $l = 3$. In fact, it is inevitable that a fraction of traceable-detectable-prefixes are linked when here is not enough witnesses. Besides, combining Fig. 4(a) and Fig. 4(b), it has smaller probability and higher costs of doing evil when setting the boundary of $l = 3$. According to the aforementioned results, we choose $l = 3$ and $m = 10$ because this pair is more in accordance with the actual situation.

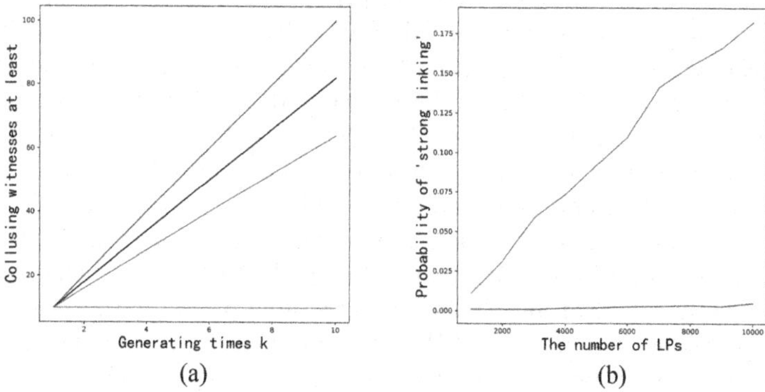

Fig. 4. (a) Collusion difficulty at different setting. (b) Probability of "strong linking" with the increase of number of LPs.

To evaluate the total performance of the proposed scheme, we use *the coefficient of satisfaction* (CS) as our performance metric. CS is inversely proportional to the deviation of the ideality and reality. The top five features with greater impact on CS are "privacy protection", "collusion prevention", "safety employing", "data backup" and "incentive mechanism" in witness-based location proof system. We can define CS as follow:

$$CS = \sum_{j=0}^{5} c_1 \cdot SI_j + c_2 \cdot DI_j \tag{14}$$

where SI_i is the extent of satisfaction, DI is the extent of dissatisfaction corresponding to five features, respectively [30]. If the scheme has some features, and then set $c_1 = 1, c_2 = 0$, else set $c_1 = 0, c_2 = 1$.

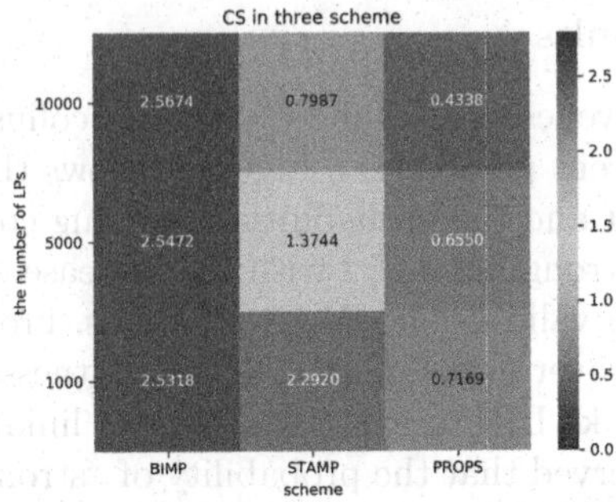

Fig. 5. CS in different scheme.

In Fig. 5, we compare the proposed scheme with STAMP [5] and PROPS [4]. Figure 5 shows the CS in three schemes under different number of LPs. From Fig. 5, we can see that the proposed scheme outperforms the other scheme in terms of CS. In the STAMP scheme, privacy protection at each step is not the optimal due to the entropy-based P-W collusion detection model. In the PROPS scheme, due to the absence of the incentive mechanism and P-W collusion detection, witnesses cannot be stimulated to generate LPs for provers, or in a low intention. And there is a big security risk of using interactive zero-knowledge proof in PROPS. In addition, all LPs are stored in their own mobile devices. Once the devices are lost, all data will be lost. Therefore, in the STAMP scheme and PROPS scheme, the deviation of the demand and the harvest is relatively high, leading to a relatively low CS. In the proposed scheme, thanks to PoP incentive mechanism and traceable-detectable-prefix, witnesses are stimulated to participate in generating LPs for others honestly, which leads to a high CS. Moreover, we can observe that the CS is improving with the LPs increasing in our scheme, owing to PoP protocol that a high point user has more chance to win "mining game", which improves the sense of achievement of participator.

According to the aforementioned results, with the traceable-detectable-prefix and incentive mechanism, users prefer to participate in cooperating to create LPs. In addition, compared with the conventional schemes, the proposed scheme can attain a higher privacy security and an improved CS.

7 Conclusion

In this paper, we proposed a BIMP scheme for LP generation and verification in LBS, to ensure the location information of users is authentic and valid. Specifically, a PoP protocol was proposed to efficiently reach consensus to build LP blockchain. Moreover, the enthusiasm of witnesses was enhanced through PoP protocol, which encouraged a user as a witness to assist prover in generating LP to get a higher point value. In addition, a traceable-detectable-prefix scheme was developed to detect P-W collusion on the premise that protecting users' privacy. Finally, security analysis and simulation results demonstrated that the proposed scheme can efficiently stimulate witnesses to honestly assist prover in generating LP, while preserving privacy between each entity. For future work, we will further investigate the storage problems in the LP generating phase in blockchain.

References

1. Luo, W., Hengartner, U.: Proving your location without giving up your privacy. In: Proceedings of the Eleventh Workshop on Mobile Computing Systems & Applications, pp. 7–12 (2010)
2. Bagci, H., Karagoz., P.: Context-aware friend recommendation for location based social networks using random walk. In: Proceedings of the 25th International Conference Companion on World Wide Web, pp. 531–536 (2016)
3. Liu, W., Yang, Y., Wang, E., Jie, W.: User recruitment for enhancing data inference accuracy in sparse mobile crowdsensing. IEEE Internet Things J. **7**(3), 1802–1814 (2019)
4. Gambs, S., Killijian, M.-O., Roy, M., Traoré, M.: PROPS: a privacy-preserving location proof system. In: IEEE 33rd International Symposium on Reliable Distributed Systems, pp. 1–10. IEEE (2014)
5. Wang, X., Pande, A., Zhu, J., Mohapatra, P.: Stamp: enabling privacy-preserving location proofs for mobile users. IEEE/ACM Trans. Network. **24**(6), 3276–3289 (2016)
6. Saroiu, S., Wolman, A.: Enabling new mobile applications with location proofs. In: Proceedings of the 10th Workshop on Mobile Computing Systems and Applications, pp. 1–6 (2009)
7. Hasan, R., Burns, R.: Where have you been? Secure location provenance for mobile devices. arXiv preprint arXiv:1107.1821 (2011)
8. Davis, B., Chen, H., Franklin, M.: Privacy-preserving alibi systems. In: Proceedings of the 7th ACM Symposium on Information, Computer and Communications Security, pp. 34–35 (2012)
9. Ferreira, J., Pardal, M.L.: Witness-based location proofs for mobile devices. In: IEEE 17th International Symposium on Network Computing and Applications (NCA), pp. 1–4. IEEE (2018)
10. Lee, J.H., Buehrer, R.M.: Characterization and detection of location spoofing attacks. J. Commun. Networks **14**(4), 396–409 (2012)
11. Kasori, K., Sato, F.: Location privacy protection considering the location safety. In: 18th International Conference on Network-Based Information Systems, pp. 140–145. IEEE (2015)
12. Zhu, Z., Cao, G.: Toward privacy preserving and collusion resistance in a location proof updating system. IEEE Trans. Mob. Comput. **12**(1), 51–64 (2011)
13. Hu, Y.-C., Perrig, A., Johnson, D.B.: Wormhole attacks in wireless networks. IEEE J. Sel. Areas Commun. **24**(2), 370–380 (2006)
14. Peng, D., Wu, F., Chen, G.: Pay as how well you do: a quality based incentive mechanism for crowdsensing. In: Proceedings of the 16th ACM International Symposium on Mobile Ad Hoc Networking and Computing, pp. 177–186 (2015)
15. Alswailim, M.A., Hassanein, H.S., Zulkernine, M.: A reputation system to evaluate participants for participatory sensing. In: IEEE Global Communications Conference (GLOBECOM), pp. 1–6. IEEE (2016)

16. Zhu, Z., Cao, G.: Applaus: a privacy-preserving location proof updating system for location-based services. In: 2011 Proceedings IEEE INFOCOM, pp. 1889–1897. IEEE (2011)
17. Kang, J., et al.: Blockchain for secure and efficient data sharing in vehicular edge computing and networks. IEEE Internet Things J. **6**(3), 4660–4670 (2018)
18. Luo, W., Hengartner, U.: Veriplace: a privacy-aware location proof architecture. In: Proceedings of the 18th SIGSPATIAL International Conference on Advances in Geographic Information Systems, pp. 23–32 (2010)
19. Amoretti, M., Brambilla, G., Medioli, F., Zanichelli, F.: Blockchain-based proof of location. In: IEEE International Conference on Software Quality, Reliability and Security Companion (QRS-C), pp. 146–153. IEEE (2018)
20. Dasu, T., Kanza, Y., Srivastava. D.: Unchain your blockchain. In: Proceedings of the Symposium on Foundations and Applications of Blockchain, vol. 1, pp. 16–23 (2018)
21. Liu, C.H., Lin, Q., Wen, S.: Blockchain-enabled data collection and sharing for industrial IoT with deep reinforcement learning. IEEE Trans. Ind. Inf. **15**(6), 3516–3526 (2018)
22. Li, Z., Kang, J., Rong, Yu., Ye, D., Deng, Q., Zhang, Y.: Consortium blockchain for secure energy trading in industrial internet of things. IEEE Trans. Ind. Inf. **14**(8), 3690–3700 (2017)
23. Avoine, G., et al.: Security of distance-bounding: a survey. ACM Comput. Surv. (CSUR) **51**(5), 1–33 (2018)
24. Dinh, T.T.A., Liu, R., Zhang, M., Chen, G., Ooi, B.C., Wang, J.: Untangling blockchain: a data processing view of blockchain systems. IEEE Trans. Knowledge Data Eng. **30**(7), 1366–1385 (2018)
25. Halevi, S., Micali, S.: Practical and provably-secure commitment schemes from collision-free hashing. In: Koblitz, N. (ed.) CRYPTO 1996. LNCS, vol. 1109, pp. 201–215. Springer, Heidelberg (1996). https://doi.org/10.1007/3-540-68697-5_16
26. Goldwasser, S., Micali, S., Rackoff, C.: The knowledge complexity of interactive proof systems. SIAM J. Comput. **18**(1), 186–208 (1989)
27. Wang, W., et al.: A survey on consensus mechanisms and mining strategy management in blockchain networks. IEEE Access **7**, 22328–22370 (2019)
28. Mitchell, J.C.: Probabilistic polynomial-time process calculus and security protocol analysis. In: Sands, D. (ed.) ESOP 2001. LNCS, vol. 2028, pp. 23–29. Springer, Heidelberg (2001). https://doi.org/10.1007/3-540-45309-1_2
29. Bussard, L., Bagga, W.: Distance-bounding proof of knowledge to avoid real-time attacks. In: Sasaki, R., Qing, S., Okamoto, E., Yoshiura, H. (eds.) SEC 2005. IAICT, vol. 181, pp. 223–238. Springer, Boston, MA (2005). https://doi.org/10.1007/0-387-25660-1_15
30. Yao, M.-L., Chuang, M.-C., Hsu, C.-C.: The kano model analysis of features for mobile security applications. Comput. Secur. **78**, 336–346 (2018)

Indoor Positioning and Prediction in Smart Elderly Care: Model, System and Applications

Yufei Liu[1], Xuqi Fang[1], Fengyuan Lu[1], Xuxin Chen[1], and Xinli Huang[1,2](✉)

[1] School of Computer Science and Technology, East China Normal University, Shanghai 200062, China
xlhuang@cs.ecnu.edu.cn

[2] Shanghai Key Laboratory of Multidimensional Information Processing, Shanghai 200062, China

Abstract. The indoor positioning and prediction technologies can be applied to smart elderly-caring application scenarios, helping to discover and reveal irregular life routines or abnormal behavior patterns of the elderly living at home alone, with the aim to predict and prevent the occurrence of emergency or health risks. In this paper, we design and implement an indoor positioning and prediction system, and apply it to smart elderly care. The system consists of two parts: hardware module and software module. The hardware module is responsible for sensing and uploading the location information of the elderly living in the room, while the software module is responsible for receiving and processing the sensing data, modeling and predicting the indoor position of the elderly based on the sensing data, so as to help find the abnormal patterns of the elderly. We also introduce a positioning prediction model that is suitable for processing IoT sensing data. With the model, one can have a more accurate and comprehensive understanding of the regularity and periodicity of the positioning time series for the elderly. To demonstrate the effectiveness and performance gains of our solution, we deploy the system in real world and setup experiments based on the indoor trajectory data collected by sensors. Extensive experimental results show that the model proposed in this paper outperforms its competitors, producing an arresting increase of the positioning prediction accuracy.

Keywords: Internet of Things · Indoor positioning and prediction · System · Positioning prediction model · Smart elderly-caring

1 Introduction

At present, the aging problem has become an important social problem facing and concerned by the whole world. According to United Nations projections, one in six people in the world will be older than 65 by 2050 [1]. With the rapid increase of the elderly population, how to provide sufficient resources and

M. Qiu (Ed.): ICA3PP 2020, LNCS 12454, pp. 537–548, 2020.
https://doi.org/10.1007/978-3-030-60248-2_36

high-quality care services for the elderly have become a big burden for the government and society. To address this problem and ease the pressure on pension institutions, many governments carry out various "ageing-in-place" policies to encourage the elderly to continue to live independently and comfortably in their homes or communities, and this is also in line with the wishes of the most elderly [2]. In order to promote ageing-in-place and provide a safer environment for the elderly, a smart elderly care system is needed in the elderly's home to provide continuous monitoring and analyzing, and to report an emergency or abnormal health event to carers [3].

The smart elderly care systems mainly rely on smart home technology, which monitors the daily activities of the elderly by deploying various types of sensors in their living environment [4]. And their applications include physiological monitoring, functional monitoring, Activities of Daily Living (ADL) monitoring, safety monitoring and emergency response, and social interaction monitoring [5]. Among them, the ADL monitoring of the elderly is the most extensive and valuable research, and the daily activities of the elderly can be analyzed to identify the abnormal behaviors of the elderly, so as to realize early warning. Suryadevara et al. [6] proposed activity monitoring and prediction system for the elderly based on a wireless sensor network. Fleury et al. [7] proposed a SVM-based ADL recognition and classification mechanism by a variety of sensor data. Although the existing proposals could identify and monitor the daily activities of the elderly, they did not take full advantage of the position information of the elderly. In fact, as the daily activities of the elderly are repetitive and periodic, it is easy to model and predict the position of the elderly living alone at home, so as to identify some abnormal behaviors of the elderly to avoid accidents.

In this paper, we focus on the problem of the indoor positioning and prediction in smart elderly care, and manage to introduce new ideas, methods and solutions to locate and predict actual positions of the objective indoors, and to ultimately apply them to smart elderly-caring applications, helping to discover and reveal irregular life routines or abnormal behavior patterns of the elderly living at home alone, with the aim of anticipating and preventing the occurrence of emergency or health risks. The work and main contributions of this paper are summarized as follows:

- We design an indoor positioning and prediction system based on infrared sensors, which can be applied to smart elderly-caring application scenarios. The system consists of two parts: hardware module and software module. The hardware module is responsible for sensing and uploading the location information of the elderly living in the room, while the software module is responsible for receiving and processing the sensing data, modeling and predicting the indoor position of the elderly based on the sensing data, so as to help find the abnormal patterns of the elderly.
- We also propose and incorporate into the system a positioning prediction model, based on the LSTM and Grey model. The model is constructed to utilize the advantages of the LSTM model in dealing with nonlinear time series data of different spans, and the ability of the Grey model in dealing

with incomplete information and in eliminating residual errors generated by the LSTM model, with the aim to further enhance the prediction ability of nonlinear samples in IoT sensing data and improve the prediction accuracy of the conventional models.
- We setup experiments to demonstrate the effectiveness and performance of the positioning prediction model by deploying our indoor positioning and prediction system in a real-world scenario successfully. And the experiment results show that our positioning prediction model based on LSTM model and grey model can effectively correct the residual of the conventional LSTM model, and the prediction results have a better fitting degree with the real action trajectory of the elderly.

The remainder of the paper is organized as follows. We review related work in Sect. 2, and then present the system design in detail in Sect. 3, the model design in Sect. 4, respectively. We describe experimental setup and report the results and analysis in Sect. 5, followed by the conclusion and future work of this paper in the last section.

2 Related Work

A complete positioning and prediction system should be composed of hardware and software, of which the hardware layer is mainly composed of sensors with positioning capability and other hardware, while the software layer provides analysis and prediction capability.

The positioning sensor is the basis of the positioning and prediction system to collect the location information of the elderly, such as Ultra Wide Band (UWB) [8,9], Radio Frequency Identification (RFID) [10], Ultrasonic [11], ZigBee [12], Bluetooth [13], WiFi [14]. However, not all location technologies are suitable for monitoring the location of the elderly. RFID utilizes the characteristics of wireless radio frequency spatial coupling, UWB introduces Time Difference of Arrival (TDOA) and Angle of Arrival (AOA) to locate the terminal, and Ultrasonic uses geometric measurement to locate. They can achieve high precision positioning, but the high cost of the hardware structure make it unaffordable for some elderly and their family. The ZigBee wireless positioning system uses a network of mobile nodes, coordinate reference nodes, and gateways for positioning, that has high communication efficiency and high security but suffers from the short transmission distance and low system stability. Bluetooth and WiFi have the advantages of low power consumption and low cost, they all rely on specific devices such as beacons and signal receivers. Considering the elderly is a vulnerable and sensitive group, carrying the equipment probably makes them feel that their privacy is violated and causes a psychological burden. On the other hand, they are more likely to forget to carry such equipment.

As the core of the system, indoor positioning prediction algorithm provides the function of position modeling and prediction for the elderly. Some researchers try to mine the periodic behaviors of individual users by using their own migration data records. For instance, Kong et al. [15] proposed a Spatial-Temporal

Long-Short Term Memory (ST-LSTM) model which naturally combines spatial-temporal influence into LSTM and employ a hierarchical extension of the proposed ST-LSTM (HST-LSTM) in an encoder-decoder manner which models the contextual historic visit information. Further, Wang et al. [16] proposed a hybrid Markov model for positioning prediction that integrates a long short-term memory model (LSTM). Li et al. [17] discussed the impact of the granularity and duration of stay of target positions, as well as different behavioral features on the prediction accuracy. Yang et al. [18] proposed a novel approach DestPD for destination prediction, which first predicted the most probable future position and then reported the destination based on that. Although these proposals perform well in the prediction of mobility with dense data records, they have high requirements on data quality, requiring long-term movement trajectory of users, which is difficult to obtain in practical applications. The sparsity of data often leads to the occurrence of cold startup problems, or even the failure of prediction.

Most of the proposals are used in business, with little applied to elderly care. Motivated by the aforementioned situations, in this paper, we propose an indoor positioning and prediction system based on infrared sensors, and propose a positioning prediction model based on LSTM-GM model which combines the advantage of LSTM and Grey Model methods to predict the position of elderly people at home.

3 System Design

We design an indoor positioning and prediction system for the elderly who live alone at home. The system is divided into hardware module and software module, where the hardware module is responsible for sensing and uploading the position information of the elderly, while the software module is responsible for storing relevant data, and modeling and predicting the indoor position of the elderly.

3.1 Hardware Module

Due to the limitations of the elderly's specific characteristics, we choose to use the infrared sensors to locate the elderly living alone[19], which can effectively detect human thermal radiation to locate the target. The infrared sensors have the advantages of low cost, easy deployment and being suitable for indoor positioning, which are more applicable to the smart elderly scene. In our system, the sensors are placed in appropriate positions in the rooms, such as toilet, sofa and dining table, etc., to monitor where older people are likely to be active.

Besides, an Internet of Things gateway will be deployed in each room to collect data from the sensors through WiFi. The gateways transmit the collected data to the cloud server via UDP for storage and computation at regular intervals. At the same time, the gateways will also check the WiFi connection of each device in the room and itself at regular intervals. In case of WiFi failure, it will send related error message to the server to inform the monitoring personnel to carry out maintenance. Once the cloud server receives a packet of data, it needs to verify the packet, and reply to the IP address and port of the packet. For invalid data which has illegal data header information, the server will request the gateway to retransmit, and the correct data will be stored in the database. The workflow of the hardware module is shown in the Fig. 1.

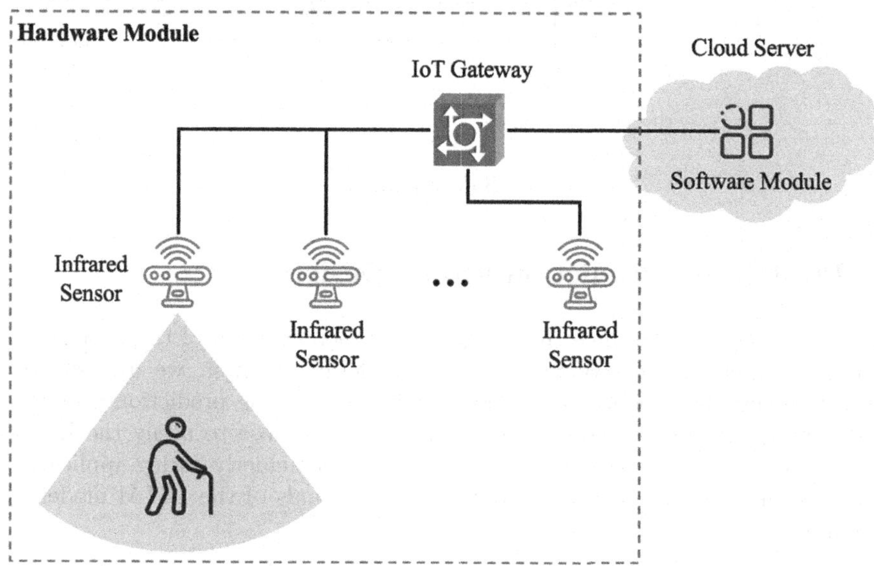

Fig. 1. Hardware module

3.2 Software Module

With the software module, we can predict the position of the elderly based on the stored data collected from the sensors. Firstly, at regular intervals, the raw data are accessed from the database that stores the sensor data collected from the gateway and extract the useful fields from them. After a series of data preprocessing, the data is cleaned to filter out the invalid data such as some field that exceeds a threshold and input into the elderly positioning prediction model for modeling and predicting. The implementation details of the model are described in Sect. 4.

The software module will also carry out real-time monitoring of the position of the elderly, and the current position of the elderly is matched with the prediction model. Once it is found that the position of the elderly is inconsistent with the model, a timer will be started. If the elderly does not move in the position for a long time, the timer will time out, and the warning information will be generated and sent to the corresponding monitoring personnel, so as to prevent accidents in the home where the elderly lives alone. The workflow of the software module is shown in the Fig. 2.

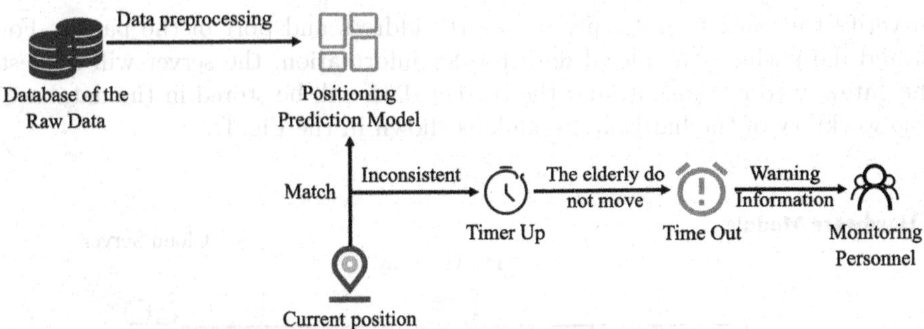

Fig. 2. Software module

4 Positioning Prediction Model Design

In this section, we introduce the design and implementation of the positioning prediction model in the software module in detail. At first, we describe the related definitions and formulate the objective positioning prediction problems of the model. In order to solve the problems, we propose to apply the LSTM model to predict the elderly indoor positions in smart elderly-caring application scenarios, and we also propose to correct the residuals of the LSTM model by grey model.

4.1 Problem Formulation

The purpose of this model is to predict the position of the elderly, helping to detect the abnormal behaviors or irregular life routines of the elderly who live at home alone. So we formally describe the related definitions and formulate the objective positioning prediction problems in this paper:

Definition 1 (Position): Position p is defined as a connected area, representing the position of the elderly at a certain moment. Each position is identified by a unique digital ID.

Definition 2 (Sensor Records): A sensor record r is described as a triple $\{ID_i, V_i, X_i\}$, where ID_i is the number of the sensor, $V_i \in \{0, 1\}$ indicates whether the sensor is in the fired status or not, and X_i is the sensor position. The position triggered indicates the position of the old man at that moment.

Definition 3 (Trajectory): The trajectory $traj = \{p_{t_i}\}_{i=1}^{n}$ represents the position of elderly at time t_i.

Definition 4 (Positioning Prediction Problem): When the set of trajectories at the past n moments $traj_{past} = \{p_1, p_2, \cdots, p_n\}$ of the elderly is given, the positioning prediction task is to predict the set of trajectories $traj_{future} = \{p_{n+1}, p_{n+2}, \cdots, p_m\}$ in the next m moments. Thus the problem that we focus on is to find a function f to improve the accuracy of the predictions $traj_{future}$ as much as possible. The positioning prediction problem is as follows:

$$traj_{future} = f(traj_{past}). \tag{1}$$

4.2 Positioning Prediction Model Based on LSTM and GM

The daily indoor position of the elderly is repetitive, habitual and traceable in time sequence, so we utilize the time series method [20] to convert the elderly indoor position information sensed and collected by the sensors into an effective time series to analyze and predict the position of the elderly. In this paper, we choose the LSTM model [21] which is suitable for predicting the time series with long-term and short-term dependence to predict the time series of the position information of the elderly.

However, in actual design and implementation, we found that although the LSTM model could learn the regularity and periodicity of the elderly position and the results perform well, the residuals that the difference between the measured value and the prediction value still exist in the predicted results. When the residual is large, it will affect the accuracy of the prediction, so it is necessary to correct the residual so as to improve the accuracy of the prediction results. So we propose to correct the residuals of the LSTM model by setting up grey model [22]. The grey model is a prediction method that builds a mathematical model and makes predictions through a small amount of incomplete information, and which processes and quantifies uncertainties by grey mathematics to discover the internal laws in disorderly phenomena.

So the positioning prediction problem proposed above can be further formalized as follow:

$$traj_{future} = LSTM(traj_{past}). \tag{2}$$

Further, we can formulate the correction formula for the prediction results of the LSTM model. The final prediction value $x_t^{'}$ at time t is equal to the difference between the prediction value x_t of the LSTM and the correction value e_t of the grey prediction model:

$$x_t^{'} = x_t - e_t, \tag{3}$$

In our positioning prediction model, we first pre-process the position data of the elderly and convert it into a time series that can be used for prediction. In the training process, the training data of the elderly position information is trained by LSTM, and a preliminary prediction result is calculated. At the same time, the parameters of LSTM will also be optimized for the overall effect of the new model. After getting the results of the preliminary predictions, the residuals generated in the preliminary prediction is calculated by comparing the prediction values by LSTM model with the measured values, and each residual is added to the residual sequence. Then the grey model analyzes the residual sequence at the time t to obtain the residual at the time $t + 1$, and the residual of the predicted value is corrected to get the final predicted value. Algorithm 1 shows the details of our model.

Algorithm 1. LSTM-GM Algorithm

Input:

D_O: position data set, l: size of data set, $D_O^{'}$: LSTM format data set, d_i: the $i-th$ data in the data set, $d_i^{'}$: the i-th data of the LSTM format data set, E: residual sequence, e_i: the i-th residual, d_{t_i}: predicted value at time t, $d_{t_i}^{'}$:corrected the predicted value at time t;

Output:

D: output sequence

Pre-process D_O;
for $i = 1; i < l; i++$ **do**
  $d_i^{'} = \text{reshape}(d_i)$;
  $d_{t_i} = \text{LSTM}(d_i^{'})$;
  $e_i = \text{compare}(d_i, d_i^{'})$;
  $E \leftarrow e_i$;
  $e_(i+1) = \text{GreyModel}(E, d_{t_i})$;
  $d_{t_{i+1}}^{'} = d_{t_{i+1}} - e_(i+1)$;
  $D \leftarrow d_{t_{i+1}}^{'}$;
end for
return D

5 Experimentation and Evaluation

Positioning prediction model is the key of the positioning and prediction system, and its accuracy determines the quality of our system. So in this section, we setup experiments by deploying our system in real-world scenario to compare the accuracy and fitting degree of our positioning prediction model results with the conventional LSTM prediction model results.

5.1 Experimental Setup

We deployed the system successfully inside the house of an old woman living alone and collect the location data of her for three months. Within the scope of her activities, infrared sensors were deployed in 24 different special locations with a unique number, and an IoT gateway was deployed for the entire room to collect sensor data and upload them to the cloud server to store.

In the experiments, we compare the accuracy of the prediction results of the conventional LSTM and our LSTM-GM model. And the models are evaluated from daily level prediction and weekly prediction. We use the root-mean-square error (RMSE) to represent the accuracy of the prediction results of the two models, that is, the square root of the ratio of the differences between the predicted value p and the measured value m and the number of observations n. The calculation formula is as follows:

$$RMSE = \sqrt{\frac{1}{n}\sum_{i=1}^{n}(p_i - m_i)^2} \tag{4}$$

Besides, the parameters of the LSTM model in the conventional LSTM model and the LSTM-GM model should be set to the same. Table 1 lists the effect of the number of hidden neurons in the LSTM model on RMSE in our previous study. It can be seen that the root mean square error is minimal when the number of hidden neurons is 128. So in our experiments, the number of hidden neurons in each layer of *num_units* is set as 128.

Table 1. RMSE of different hidden neurons

Number of hidden neurons	RMSE
4	1.31
16	1.00
32	1.09
64	0.89
128	0.84
256	ERROR

5.2 RSME of Prediction Results

In our experiments, we compare the accuracy of the conventional LSTM model and the proposed positioning prediction algorithm based on LSTM and GM model. And according to the time of the prediction results, the experiments are divided into daily prediction and weekly prediction.

Figure 3 shows the RMSE of the two models in daily prediction, and Fig. 4 shows the RMSE of two models in weekly prediction. It can be seen from Fig. 3 and Fig. 4 that, in the same dimension, the RMSE value of the LSTM-GM model is much lower than that of the LSTM model, and the RMSE value of the weekly prediction higher than that of the daily prediction. This indicates that the proposed model has a higher degree of fit with the real data, but since the time span of weekly prediction is large and the difficulty of prediction is high, the existing RMSE will be higher than the daily prediction. In our model, the residual corresponding to each predicted value is calculated by introducing the grey model, and adds every residual to the residual sequence to correct the residual value of the predicted value, which makes the prediction results perform well and greatly improve the accuracy of the prediction.

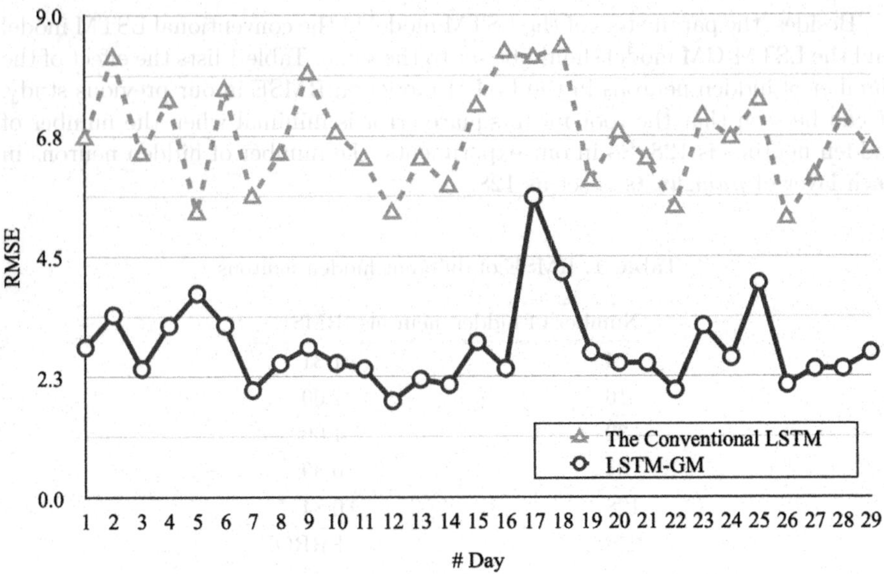

Fig. 3. RMSE of the daily prediction

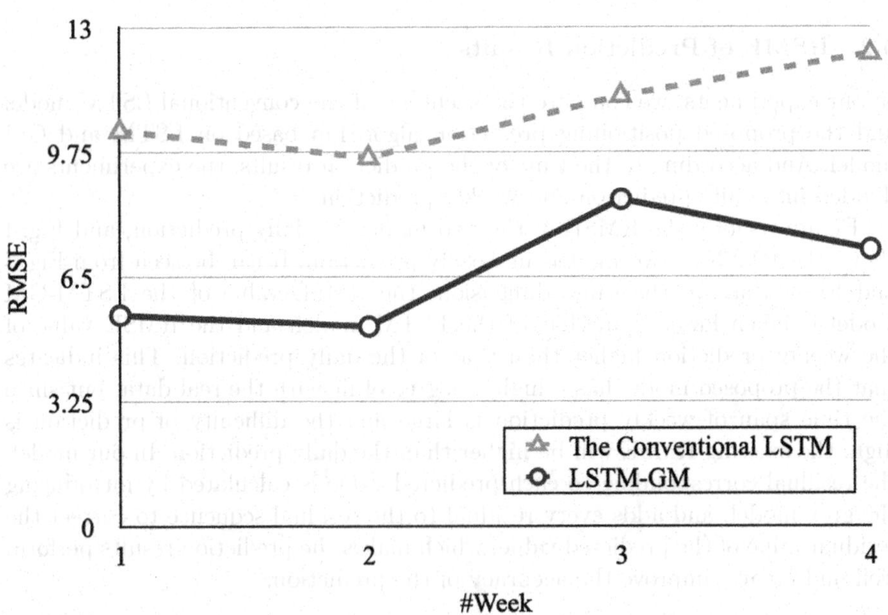

Fig. 4. RMSE of the weekly prediction

6 Conclusions and Future Work

In this paper, we design an indoor positioning and prediction system including hardware module and software module in smart elderly-caring application scenarios, helping to discover and reveal irregular life routines or abnormal behavior patterns of the elderly living at home alone, with the aim to anticipate and prevent the occurrence of emergency or health risks. In the hardware module, the position of the elderly is located based on the infrared sensor, and the data of the elderly is uploaded to the server through the gateway. And the software module is responsible for storing position data, and modeling and predicting the indoor position of the elderly. We also proposed a positioning prediction model based on the LSTM model to predict the position of the elderly who live alone, and the residual error of the model is corrected by the gray model.

To demonstrate the effectiveness and performance gains of our solution, we setup experiments by deploying our system in a real-world scenario successfully. And the experiment results show that our positioning system can effectively correct the residual of the conventional LSTM model, and the positioning prediction model proposed in the paper can also make a more accurate positioning prediction, with a better fitting degree when using real trajectory data collected by the system.

The system and model poposed in this paper can be applied to smart elderly-caring application scenarios, helping to discover and reveal irregular life routines or abnormal behavior patterns of the elderly living at home alone, with the aim to predict and prevent the occurrence of emergency or health risks.

In the future, we plan to carry out our research work in the following two directions: **(i)** consider applying the system to the real-world application scenarios with a large number of sensors deployed in a much more complex indoor environment, to verify the actual performance of the system, and **(ii)** further refine our system, with introducing more types of sensors to enrich the function of the system.

Acknowledgements. We thank the anonymous reviewers for their helpful and constructive comments on our work. This work was supported in part by the National Key Research and Development Plan of China under Grant 2019YFB2012803, in part by the Key Project of Shanghai Science and Technology Innovation Action Plan under Grant 19DZ1100400 and Grant 18511103302, in part by the Key Program of Shanghai Artificial Intelligence Innovation Development Plan under Grant 2018-RGZN-02060, and in part by the Key Project of the "Intelligence plus" Advanced Research Fund of East China Normal University.

References

1. World Population Prospects 2019, June 2019. https://www.un.org/development/desa/publications/world-population-prospects-2019-highlights.html
2. Boldy, D.P., et al.: Older people's decisions regarding 'ageing in place': a Western Australian case study. Australas. J. Ageing **30**(3), 136–142 (2011)

3. Kasteren, Y., Bradford, D., Zhang, Q., Karunanithi, M., Ding, H.: Understanding smart home sensor data for ageing in place through everyday household routines: a mixed method case study. JMIR Mhealth Uhealth **5**(6), e52 (2017)
4. Nathan, V., et al.: A survey on smart homes for aging in place: toward solutions to the specific needs of the elderly. IEEE Signal Process. Mag. **35**(5), 111–119 (2018)
5. Liu, L., Stroulia, E., Nikolaidis, I., Miguel-Cruz, A., Rincon, A.R.: Smart homes and home health monitoring technologies for older adults: a systematic review. Int. J. Med. Inform. **91**, 44–59 (2016)
6. Suryadevara, N.K., Mukhopadhyay, S.C., Wang, R., Rayudu, R.K.: Forecasting the behavior of an elderly using wireless sensors data in a smart home. Eng. Appl. Artif. Intell. **26**(10), 2641–2652 (2013)
7. Fleury, A., Vacher, M., Noury, N.: SVM-based multimodal classification of activities of daily living in health smart homes: sensors algorithms and first experimental results. IEEE Trans. Inf. Technol. Biomed. **14**(2), 274–283 (2010)
8. Chen, P., Kuang, Y., Chen, X.: A UWB/improved PDR integration algorithm applied to dynamic indoor positioning for pedestrians. Sensors **17**(9), 2065 (2017)
9. Paolo, D., Pietra, V., Piras, M., Jabbar, A., Kazim, S.: Indoor positioning using Ultra-wide band (UWB) technologies: positioning accuracies and sensors' performances. In: Proceedings of the 2018 IEEE/ION Position, Location and Navigation Symposium (PLANS), pp. 175–184 (2018)
10. Wang, X., Wang, Y., Zhang, W.: Review of indoor positioning technology based on RFID. Transducer Microsyst. Technol. **28**(2), 1–4 (2009)
11. Hazas, M., Hopper, A.: Broadband ultrasonic location systems for improved indoor positioning. IEEE Trans. Mob. Comput. **5**, 536–547 (2006)
12. Alvarez, Y., Las Heras, F.: ZigBee-based sensor network for indoor location and tracking applications. IEEE Lat. Am. Trans. **14**, 3208–3214 (2016)
13. Devanshi, D., Agrawal, S., Singh, S.: Indoor localization based on bluetooth technology: a brief review. Int. J. Comput. Appl. **97**(8), 31–33 (2014)
14. Yang, C., Shao, H.: WiFi-based indoor positioning. IEEE Commun. Mag. **53**(3), 150–157 (2015)
15. Kong, D., Wu, F.: HST-LSTM: a hierarchical spatial-temporal long-short term memory network for location prediction. In: Proceedings of the Twenty-Seventh International Joint Conference on Artificial Intelligence (2018)
16. Wang, P., Wang, H., Zhang, H., Lu, F., Wu, S.: A hybrid Markov and LSTM model for indoor location prediction. IEEE Access **7**, 185928–185940 (2019)
17. Li, H.: A systematic analysis of fine-grained human mobility prediction with on-device contextual data (2019)
18. Yang, Z., et al.: An efficient destination prediction approach based on future trajectory prediction and transition matrix optimization. IEEE Trans. Knowl. Data Eng. **32**(2), 203–217 (2020)
19. Wu, P., Li, F.: The pyroelectric sensor based system: human tracking and self-calibration scheme. In: 2012 IEEE International Conference on Information Science and Technology, pp. 839–846 (2012)
20. Antunes, C., Arlindo, L.: Temporal data mining-an overview. In: KDD Workshop on Temporal Data Mining, pp. 1–13 (2001)
21. Liu, Y., Su, Z., Li, H., Zhang, Y.: An LSTM based classification method for time series trend forecasting. In: 2019 14th IEEE Conference on Industrial Electronics and Applications (ICIEA), Xi'an, China, pp. 402–406(2019)
22. Deng, J.: Introduction grey system theory. J. Grey Syst. **1**(1), 191–243 (1989)

Research on Stylization Algorithm of Ceramic Decorative Pattern Based on Ceramic Cloud Design Service Platform

Xinxin Liu[1], Hua Huang[1(✉)], Meikang Qiu[2], and Meiqin Liu[3]

[1] School of Information Engineering, Jingdezhen Ceramic Institute, Jingdezhen, China
1920043005@stu.jci.edu.cn, jdz_hh@qq.com

[2] Department of Computer Science, Harrisburg University of Science and Technology, Harrisburg, PA 17101, USA
qiumeikang@yahoo.com

[3] College of Electrical Engineering, Zhejiang University, Hangzhou, Zhejiang, China
liumeiqin@zju.edu.cn

Abstract. Aiming at the problem of limited universality of the ceramic decorative pattern generation method, based on cloud computing and deep learning, a ceramic decorative pattern stylization algorithm based on ceramic cloud design service platform is proposed. This approach adopts the VGG19 network structure to implement the image style transferring algorithm. To evaluate the effect of our approach, eight groups of experiments are conducted to compare the content image and style image, and the effect of parameter modification on the generated image. After a comparative analysis of the experiment, the generated ceramic decorative pattern is very close to the real ceramic decorative pattern, the effect is more realistic, and it is practical.

Keywords: Ceramic decorative pattern · Ceramic cloud design service platform · Deep learning · Image style transferring

1 Introduction

When computer ceramic decorative pattern design [1] has not been widely used, ceramic decorative pattern design is mainly based on traditional manual design, which shows the limitations of traditional manual design in many aspects. With the continuous development of computer technology and craftsmanship, in the traditional workflow, the computer replaces part of manual labor, and in the art of design, computer design replaces part of manual design, repetitive calculation, formulation, and stylized choices and unleashing the human brain makes possible many tasks that are difficult to achieve with traditional operations. In recent years, deep neural networks have continued to develop in practical applications, especially in the field of image processing [2].

Cloud computing [3] is a brand-new business model, known as a "revolutionary computing model". It is a new computing paradigm and a new IT service model. The ceramic cloud design service platform is a professional service platform in the ceramic field

M. Qiu (Ed.): ICA3PP 2020, LNCS 12454, pp. 549–563, 2020.
https://doi.org/10.1007/978-3-030-60248-2_37

created through the cloud computing model and combined with the traditional ceramics industry. The platform is based on the Internet and is a regional base established in major ceramic production areas throughout the country and clustering industrial clusters. The platform is service-oriented and aims at industry characteristics in the ceramics field. It provides online services for effective resources in the ceramics industry, and adopts a cloud model of online and offline coexistence. The ceramic decorative pattern stylization algorithm studied in this paper is based on the ceramic cloud design service platform. It provides an intelligent generation algorithm of ceramic decorative patterns for the platform.

Image style transfer [4, 5] is an image processing method that renders image semantic content with different styles. With the rise of deep learning [6, 7], image style transfer has been further developed, and a series of breakthrough research results have been achieved. Its outstanding style transfer ability has aroused widespread concern in academia and industry, and has important research value. In this paper, it is precisely under such a trend that the stylization algorithm of ceramic decorative patterns based on the ceramic cloud design service platform is proposed.

2 Related Work

2.1 Deep Convolutional Neural Network

As a new field that has developed rapidly for more than ten years, deep learning [6] has received more and more attention from researchers. It has obvious advantages in feature extraction and modeling compared to shallow models. Deep learning is good at mining increasingly abstract feature representations from raw input data, and these representations have good generalization capabilities. It overcomes some problems that were previously considered difficult to solve in artificial intelligence. And with the significant increase in the number of training data sets and the dramatic increase in processing power, it has achieved remarkable results in the areas of object detection [8, 9] and computer vision [10, 11], natural language processing [12, 13], speech recognition [14–16], and semantic analysis [17, 18]. Therefore, it has also promoted the development of artificial intelligence [19]. Convolutional neural networks [20] are one of the most representative neural networks in the field of deep learning technology. Deep convolutional neural network is a kind of feedforward neural network with deep structure with convolution calculation. Nowadays, deep convolutional neural networks have been widely used in the field of image processing [21]. Image style transfer is the use of deep convolutional neural networks to reconstruct the content and style of output images. There are two types of images in image style transfer. One is a style image, usually an artist's work, and the other is a content image, usually a real photo. The style transfer is to transfer the artistic style of the style image to the content image through a deep convolutional neural network, and the specific content such as the object and the shape are not changed.

2.2 Image Style Transfer

In the field of image art style transfer research, Gatys [4, 22, 23] Proposed the method of Neural Style, which simulates the processing of human vision. After training multi-layered convolutional neural network, the computer recognizes and learns the art style, so that it can be used on ordinary photographs. The resulting images are very artistic, and these images are distorted and lose detail. Later, Gatys and others [24] improved their work, strengthened the detail control in style transfer, and could control the degree of style transfer, but did not control the image content. Johnson [25] Proposed the Fast Neural Style method to train a style model, and the GPU usually only needs a few seconds to generate the corresponding style transfer result, which improves the speed of style transfer, but the image generation effect is not improved. Luan [26] Enhanced the work based on Gatys' work and controlled the content details of style transfer [27, 28]. The CNNMRF of Li C [29] Is a combination algorithm of Markov random field model and trained deep convolutional neural network. It matches each input neural region with the most similar region in the style image, reducing inaccurate feature transfer and retaining The specific characteristics of the content image.

This paper mainly uses pre-trained VGG-19 [30] convolutional neural network for feature extraction, and then compares and analyzes through different experiments and modification of different parameters to make the generated image content clearer and the lines smoother.

3 Method Principle

3.1 Image Stylization Principle

Since convolutional neural networks have produced good results in image recognition and classification, Gatys et al. Applied CNN to the study of artistic style features, performed style extraction based on oil painting, and compared the texture of artistic images with real images. Combine to get the result of artistic painting with photo content.

For the network structure of VGG19, each layer is a non-linear convolution kernel, and the complexity of the convolution kernel will also increase as the number of convolution layers increases. For the l layer, the resulting response matrix is expressed as $F^l \in R^{R_l \times M_l}$, where R_l is the number of filters in the l layer; M_l is the size of the filters in the l layer; F^l_{ij} then Represented as the output of the i-th filter at the l-th layer at the position j. After the content image and the style image are input in the network, the style image or the content image needs to be initialized, and the content map and the style map features are reconstructed on the white noise image by the gradient descent method. The content loss function is obtained as the following formula (1):

$$L_{content}(\vec{p}, \vec{x}, l) = \frac{1}{2}\sum_{i,j}\left(F^l_{ij} - p^l_{ij}\right)^2 \quad (1)$$

In formula (1):

$\vec{p}$—input content pattern
$\vec{x}$—results graph
l —convolutional layers
F^l_{ij}—the target graph is under the ith filter of the lth layer and the output at position j
p^l_{ij}—The content map is under the i-th filter of the l layer, and the output at position j

In the feature extraction process of the style map, a white noise image is also used. The Gram matrix of the target image and the style image is calculated for feature matching. $\vec{a}$ represents the style image; $\vec{x}$ represents the final image; A^l and G^l are their characteristic outputs at the l layer. Then the convolution loss of each layer of the two images can be expressed as the following formula (2):

$$E_l = \frac{1}{4N_l^2M_l^2}\sum_{i,j}\left(G_{ij}^l - A_{ij}^l\right)^2 \quad (2)$$

Then the total style loss function can be expressed as formula (3):

$$L_{style}(\vec{a}, \vec{x}) = \sum_{i=0}^{L} E_l W_l \quad (3)$$

In order to obtain a feature image with style image features, Gatys et al. Gave a white noise image, and defined the style loss and content loss as the total loss function of formula (4):

$$L_{total}(\vec{p}, \vec{a}, \vec{x}) = \alpha L_{content}(\vec{p}, \vec{x}) + \beta L_{style}(\vec{a}, \vec{x}) \quad (4)$$

Among them, α and β are weight ratios of the content and style of the target image, respectively.

In the results achieved by Gatys et al., Some of the higher layers of the generated image basically lost the information of the content map, and some of the generated images could not identify the original content image. So this article will discuss the impact of different parameters and make comparisons and analysis.

3.2 Pooling Algorithm

Pooling is mainly to reduce the dimensions of the features extracted by the convolutional neural network, retain the main features, avoid the occurrence of overfitting and improve the calculation efficiency. In neural networks, features are generally downsampled by following the pooling layer after feature extraction using image convolution. There are two common pooling methods: average pooling and maximum pooling.

Average pooling, that is, only averaging the feature points in the domain, Lecun et al. Have introduced detailed pooling and derivation in the paper, as shown in formula (5):

$$\alpha_i \in \{0, 1\}, \alpha_{i,j} = 1 iff j = \underset{k \leq K}{argmin} ||x_i - d_k||_2^2, h_m = \frac{1}{|N_m|}\sum_{ieN_\omega} \alpha_i \quad (5)$$

Maximum pooling, that is, taking the largest feature point in the domain, such as formula (6):

$$\alpha_i = \underset{\alpha}{argmin} L(\alpha, D) \triangleq ||x_i - D\alpha||_2^2 + \lambda||\alpha||_1, h_{m,j} = \underset{ieN_\omega}{max} \alpha_{i,j}, for j = 1, \ldots, K \quad (6)$$

The operation of pooling is very similar to convolution, and the algorithm is different. The internal value is not considered during the pooling process, and only the size of the filter is concerned. In average pooling, the position of the filter size corresponding to the image is averaged for pixels that are not 0. The feature background information obtained by this method is more sensitive, reducing the error of increasing variance of the estimated value, and maximizing pooling. The position corresponding to the size of the filter on the image is taken to the maximum value of the pixel point. The feature texture obtained by this method will be more prominent and the boundary will be clearer.

3.3 Content and Style Feature Weight Influence

Effect on content loss weight α and style loss weight β: After experimenting in the literature [22], it is concluded that the choice of α is between 1 and 10, and the value of β is between 100 and 1000. The style transfer effect is better. In the following experiments, the value of content weight α is fixed at 5 and β is increased from 0 to 1000. According to the results of the experiment, it is concluded that the higher the value of β, the stronger the style, background and The texture of the smooth portion has also increased significantly. When the style weight is fixed at $\beta = 500$ and the value of α is increased from 0 to 10, it is concluded from the experimental results that the smaller the content weight, the less the content information of the output structure map.

4 Experiments and Evaluation

This article makes specific considerations and experiments for the loss of information and low definition of content maps in the realization results of Gatys and others. This article divides the experiments into two categories. The first category is the influence between the style image and the content image. The second category will modify and set certain parameters of the Gatys experiment, and compare the influence of different parameters on the generated image.

4.1 The Influence of Different Style Images and Content Images on Generated Images

In this experiment, it is also divided into 5 small experiments for comparison, which can more clearly and accurately find the circumstances under which the generated image is the best. 5 small experiments mainly compare the same style image and different content image, compare different style image and same content image, compare different style image and different content image, and compare 1 content image and 2 style image, and Comparison between flowers and birds landscape image. You can choose the type of stylized image that the ceramic decorative pattern is most suitable.

- **same style image + different content image.**

Experiments are performed using the same style image and different content images, and the generated images are compared. As shown in Fig. (1), from left to right, it is the style image, the content image, and the generated image. It can be seen from the generated images that the overall style of the content map and the style of the style map

Fig. 1. Comparison of different style images with the same content image

are well migrated. The only disadvantage is that the local lines are not very smooth, and the processing of the background is not ideal.

- different style images + same content images

The comparison of different style images with the same content image is to highlight the influence of style on overall migration. The experiment used nine groups of different style images and the same content image to generate the resulting image. As shown in Fig. (2), from left to right, from top to bottom, the order is 1–9. The 4th and 6th images are all good and the migration effect is the best; for the black and white style images, the 5th generated image is also very good; for the 7th image, the background image is slightly exposed, probably due to the number of iterations. Caused by fitting; for the eighth image, the original style image is mainly line-based, and the image after migration also shows the stylization of the line well, both the content picture and the background are lined. So we can conclude that the choice of style image has a great impact on the final generated image.

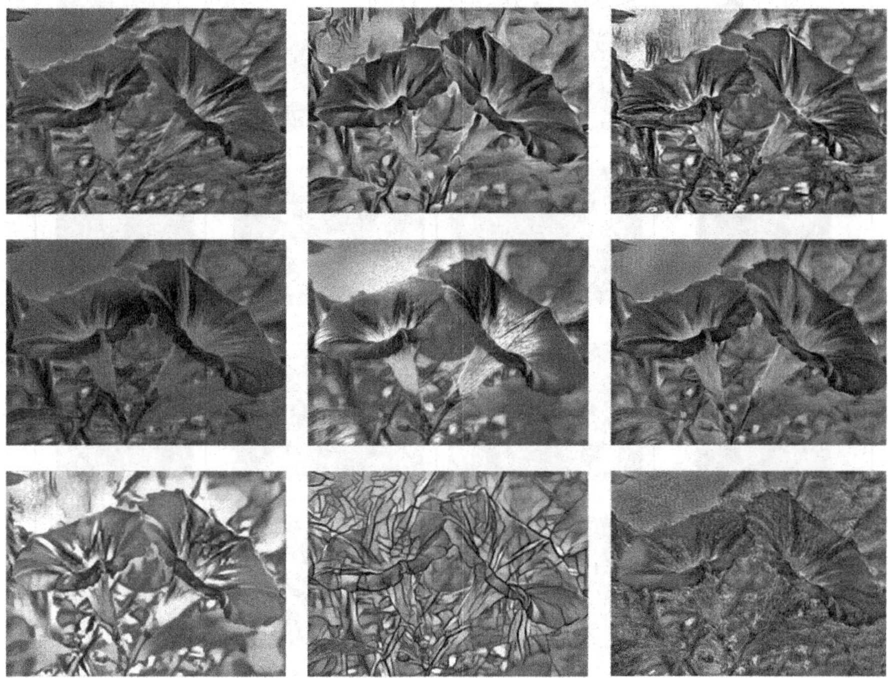

Fig. 2. Comparison of the same content image and different style images

- different style images + different content images

This group of experiments is shown in Fig. (3), which is a combination of (a) and (b) experiments. It can be found from the comprehensive comparison that the same style of ink painting images shows good or not ideal results. Therefore, in the case of the same algorithm, there is a big difference in the selection and matching of content images and style images.

- same content image + 2 style images

For the above experiments, it can be found that the selection and matching of content images and style images will affect the effect of the final generated image. So whether using two style images will eventually improve or worse. As shown in (4), the first group and the second group of images are the result maps generated with one style image, and

Fig. 3. Comprehensive comparison of different content images and different style images

the third group are the result maps generated with the two style images. It can be found that the use of two style images does migrate to the style of the two images and is well expressed, but the lines at the edges and the overall sharpness are significantly reduced. Therefore, the increase of the style image may not bring a good effect to the final result image (Fig. 4).

- flowers, birds, landscape images

Ceramic decorative patterns are mainly composed of patterns of flowers, birds, landscapes, etc. This experiment is to compare these three patterns separately. As shown in Fig. (5), using the same method to transfer the style of animals and landscapes, the effect picture is more vivid than the original content picture, and it also has a stylized effect.

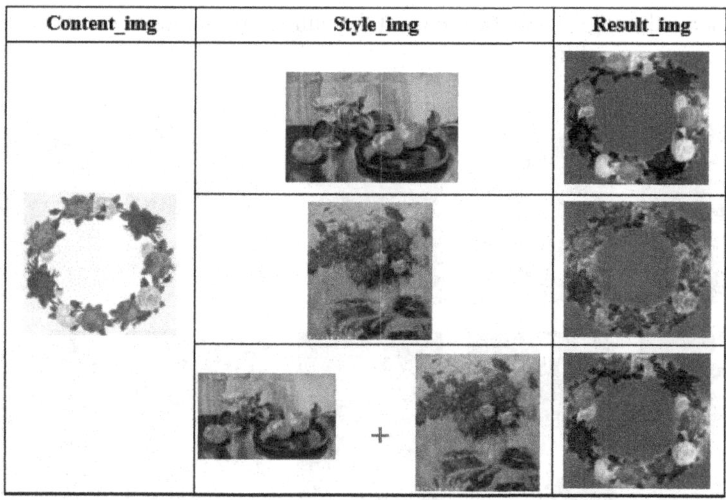

Fig. 4. Comparison of the same content image and two style images

Fig. 5. Comparison of Flowers, Birds, and Landscape Content Images

4.2 Influence of Different Parameters on the Generated Image

- different iterations

The number of iterations is very important for a training. Insufficient or too many iterations will cause fitting problems. This experiment is to study whether the number of iterations of training has an effect on the result graph, and when the number of iterations, the iteration effect is the best. Figure (6) shows the effect of the number of iterations on the generated image. It can be seen that the number of iterations is different and the generated images are different.

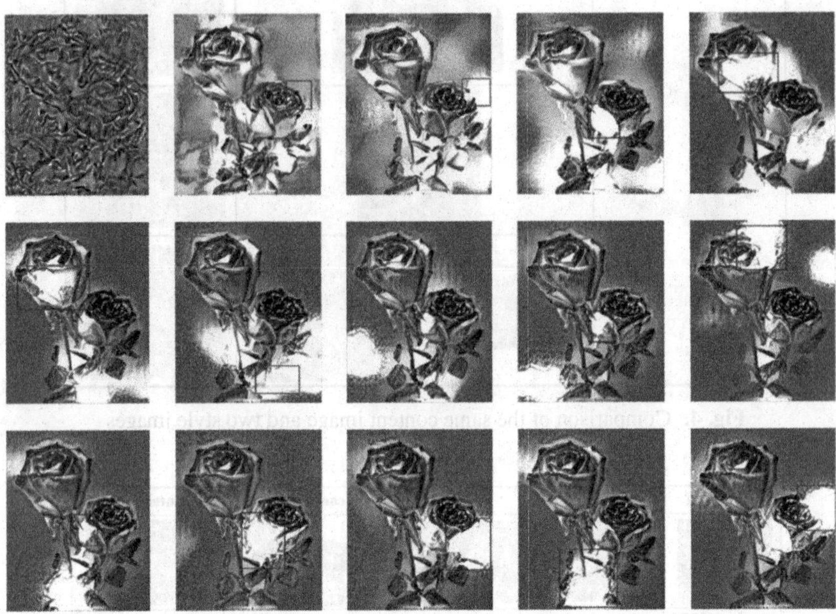

Fig. 6. Impact of the number of iterations on the generated image

Figure (7) shows the change curve of the loss function with the number of iterations. The vertical axis is content loss as the content loss function, style loss is the style loss function, tv loss is the total change regularization weight loss, and total loss is the total loss. The horizontal axis is the number of iterations, which are 100 iterations, 1000 iterations, 2000 iterations, 5000 iterations, 8000 iterations, 10,000 iterations, 20,000 iterations, 30,000 iterations, 40,000 iterations, 50,000 iterations, and 60,000 iterations. Iterations, 70,000 iterations, 80,000 iterations, 90,000 iterations, 100,000 iterations. Experiments show that the best effect of the number of iterations is presented as an approximate cosine curve distribution, which is best at 1000 iterations, 5000–20000 iterations, 70,000–90000 iterations, and so on. In a limited time, training can be performed 5000–20,000 iterations.

- content and style feature weights

As shown in Fig. (8), the study compares the influence of content feature weights and style feature weights on the generated image. When the style feature weights are constant, the value $\beta = 500$, and the content weights are $\alpha = 0$, $\alpha = 0.01$, $\alpha = 0.1$, $\alpha = 1$, $\alpha = 5$, $\alpha = 10$. In the first set of experiments, it can be seen that the content feature has an effect when the content weight is $\alpha = 1$, and the effect is best when $\alpha = 5$. When the content feature weight is constant, the value is $\alpha = 5$, and the style weight is $\beta = 0$,

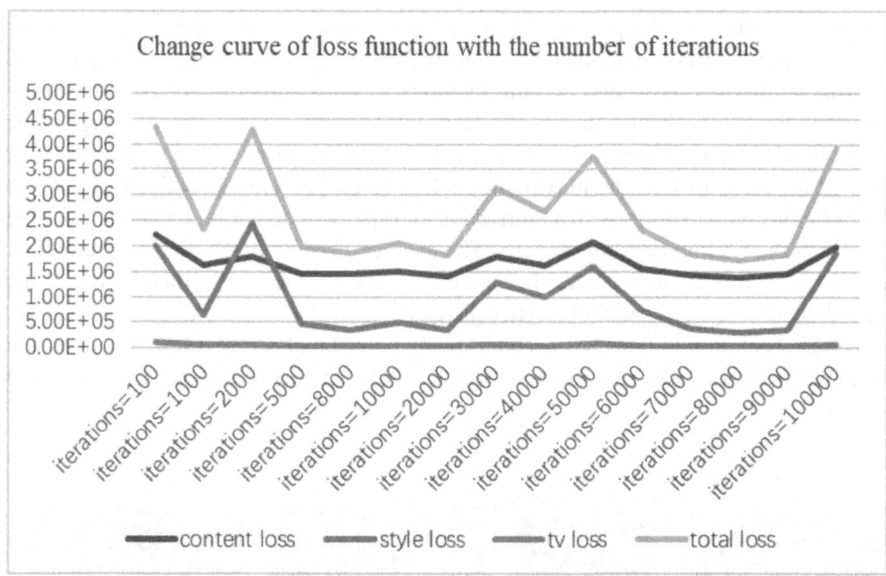

Fig. 7. Change curve of loss function with the number of iterations

β = 1, β = 10, β = 100, β = 500, β = 1000. In the second set of experiments, it can be seen that β = 100 began to have effect on style feature weights, and β = 500 had the best effect.

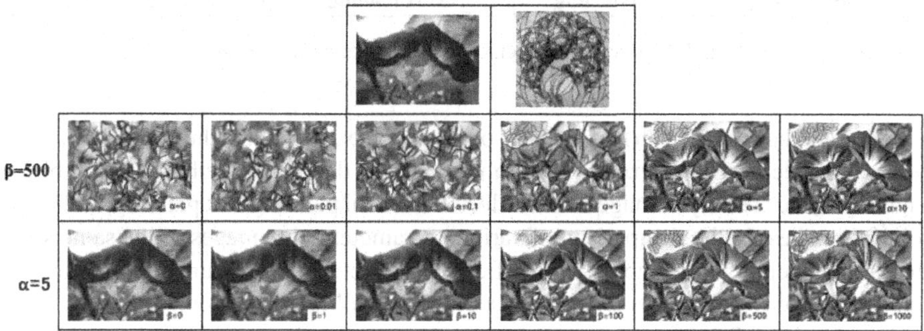

Fig. 8. Comparison of the influence of content feature weight and style feature weight on generated images

- average pooling and maximum pooling

Pooling is mainly to reduce the dimensions of the features extracted by the convolutional neural network, retain the main features, avoid the occurrence of overfitting, and improve the calculation efficiency. In neural networks, features are generally down sampled by following the pooling layer after feature extraction using image convolution. Common pooling methods include average pooling and maximum pooling. This experiment chooses these two pooling methods to study the effect of pooling on the generated image. As shown in Fig. (9), from left to right are the content image, style image, maximum pooling result, average pooling result, and two sets of experiments are selected for comparison. The maximum pooling result has more prominent edges and more obvious lines. Things with a blurred background will be ignored, making the style transfer in the background not obvious. The average pooling result image is smoother, and the background can transfer the features of the original style image even if the blurred details are not prominent. After the analysis of the ceramic decorative pattern, the pattern background does not need too many color block expressions. Therefore, in this paper, we choose the maximum pool in the style transfer of the ceramic decorative pattern.

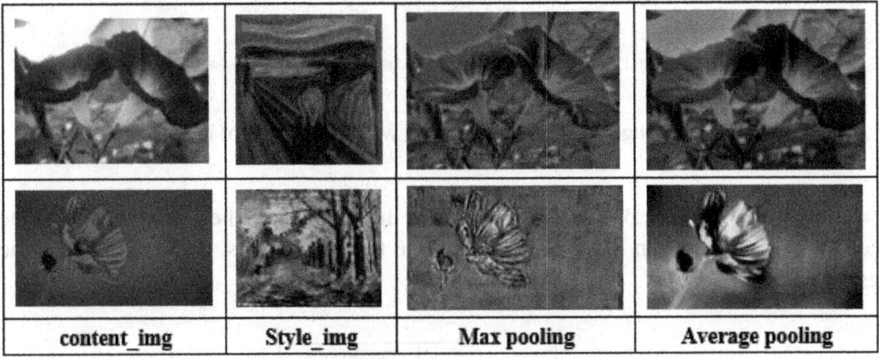

Fig. 9. Effect of different pooling on generated images

Through the comparison and analysis of the above experiments, the content image of the ceramic decorative pattern can be common ceramic decorative such as flowers, birds, landscapes, etc. The style image can be selected from oil painting, ink painting, and more colorful images. In the comparison of various parameters, the images of the same style and the same content are subjected to different iterations, the loss function is different, and the number of iterations is 5000–20000, the overall image presentation is the best. In the selection of pooling, ceramic decorative patterns are more biased towards the algorithm of maximum pooling. In the comparison of content and style feature weights, the content feature weight selection is from $\alpha = 1$ to $\alpha = 10$, and the style feature weight selection is from $\beta = 100$ to $\beta = 1000$. The optimal solution for the stylization algorithm of ceramic decorative patterns based on the ceramic cloud design service platform is to select content images as flowers, birds, landscapes and other images, and style images as oil paintings and ink paintings. In the selection of the parameters, the number of iterations is 8000, the pooling algorithm is the largest pooling, the content feature weight is $\alpha = 5$, and the style feature weight is $\beta = 500$. This choice can make the image show the best effect, and the real ceramic decorative patterns are closer.

5 Conclusion and Contribution

The style transfer algorithm is studied by more and more people. This paper puts forward its own views in the research of ceramic decorative patterns, realizes the stylization algorithm of ceramic decorative patterns based on the ceramic cloud design service platform, and compares and analyzes the difference between content images and style images, as well as different parameters and different iteration times. Comparing the results, it is better to use the appropriate parameter style transfer. The main contents of this paper are as follows:

(1) Carry out detailed research and analysis on ceramic decorative patterns to obtain the main style of ceramic decorative patterns.
(2) Use different styles and different content images for comparison to get the influence of style and content on the generated images.
(3) Get the conclusion that the setting of different parameters, the number of iterations, the use of different pooling methods, and the different weights of content and style features can affect the results of generating images.

This article has achieved certain effects in the style transfer of ceramic decorative, but there are still some differences for real ceramic decorative, and the transfer of details is not in place. I hope to improve it in this aspect. And the paper has the following main contributions:

- Ceramic decorative patterns have high requirements for style images. If the style images are not properly selected and combined, the resulting image effect is not very obvious.
- Under limited conditions, training can be performed for 5000–20000 iterations. In this training interval, the iterative effect basically does not appear to fit.
- Pooling is mainly to reduce the dimensionality of the features extracted by the convolutional neural network and retain the main features. For ceramic decorative patterns, the largest pooling effect should be selected.
- When selecting the content and style feature weights, the content weight $\alpha = 5$ and style weight $\beta = 500$ should be selected.

Acknowledgment. This work is supported by the National Key Research and Development Plan of China under Grant No. 2016YFB0501801, National Natural Science Foundation of China under Grant No. 61170026, the National Standard Research Project under Grant No. 2016BZYJ-WG7-001,the Key Research and Development Plan of Jiang Xi province under Grant No. 20171ACE50022 and the Natural Science Foundation of Jiang Xi province under Grant No. 20171BAB202011, the science and technology research project of Jiang Xi Education Department under Grant Nos. GJJ160906, GJJ171524, GJJ180727. the National Natural Science Foundation of China (No. 61728303) and the Open Research Project of the State Key Laboratory of Industrial Control Technology, Zhejiang University, China (No. ICT20025).

References

1. Yang, L., Zhang, Y., Li, H.: Decorative pattern design of ceramic based on cloud model and fractal art. In: International Conference on Computer-aided Industrial Design & Conceptual Design. IEEE (2008)
2. Huang, T.S., Schreiber, W.F., Tretiak, O.J.: Image processing, vol. 59, no. 11, pp. 1586–1609 (1971)
3. Zheng, M., Luo, L., Jiang, P.Y.: Cloud design service platform and key technology based on semantic Web. Comput. Integr. Manuf. Syst. **18**, 1426–1434 (2012)
4. Gatys, L.A., Ecker, A.S., Bethge, M.: Image style transfer using convolutional neural networks. In: 2016 IEEE Conference on Computer Vision and Pattern Recognition (CVPR) (2016)
5. Yang, F.W., Lin, H.J., Yen, S.H., Wang, C.H.: A study on the convolutional neural algorithm of image style transfer. Int. J. Pattern Recognit. Artif. Intell. (2018)
6. Lecun, Y., Bengio, Y., Hinton, G.: Deep learning. Nature **521**(7553), 436 (2015)
7. Schmidhuber, J.: Deep learning in neural networks: an overview. Neural Netw. **61**, 85–117 (2015)
8. Han, J., Zhang, D., Cheng, G., Liu, N., Xu, D.: Advanced deep-learning techniques for salient and category-specific object detection: a survey. IEEE Signal Process. Mag. **35**(1), 84–100 (2018)
9. Makantasis, K., Karantzalos, K., Doulamis, A., Loupos, K.: Deep learning-based man-made object detection from hyperspectral data. In: Bebis, G., et al. (eds.) ISVC 2015. LNCS, vol. 9474, pp. 717–727. Springer, Cham (2015). https://doi.org/10.1007/978-3-319-27857-5_64
10. Campos, V., Sastre, F., Yagües, M., Bellver, M., Giró-i-Nieto, X., Torres, J.: Distributed training strategies for a computer vision deep learning algorithm on a distributed GPU cluster. Procedia Comput. Sci. **108**, 315–324 (2017)
11. Voulodimos, A., Doulamis, N., Doulamis, A., Protopapadakis, E.: Deep learning for computer vision: a brief review. Comput. Intell. Neurosci. **2018**, 1–13 (2018)
12. Young, T., Hazarika, D., Poria, S., Cambria, E.: Recent trends in deep learning based natural language processing. IEEE Comput. Intell. Mag. **13**(3), 55–75 (2018)
13. Li, H.: Deep learning for natural language processing: advantages and challenges. Natl. Sci. Rev. **5**(1), 28–30 (2017)
14. Huang, J., Kingsbury, B.: Audio-visual deep learning for noise robust speech recognition. In: 2013 IEEE International Conference on Acoustics, Speech and Signal Processing (2013)
15. Zhang, Z., et al.: Deep learning for environmentally robust speech recognition. ACM Trans. Intell. Syst. Technol. **9**(5), 1–28 (2018)
16. Deng, L.: Deep learning: from speech recognition to language and multimodal processing. APSIPA Trans. Signal Inf. Process. **5** (2016)
17. Huang, C.L., Shih, H.C., Chao, C.Y.: Semantic analysis of soccer video using dynamic Bayesian network. IEEE Trans. Multimed. **8**(4), 749–760 (2006)
18. Aerts, D., Czachor, M.: Quantum aspects of semantic analysis and symbolic artificial intelligence. J. Phys. A: Math. Gen. **37**(12), L123–L132 (2003)
19. Russell, S., Dietterich, T., Horvitz, E.: Letter to the editor: research priorities for robust and beneficial artificial intelligence: an open letter. AI Mag. **36**(4) (2015)
20. Krizhevsky, A., Sutskever, I., Hinton, G.E.: ImageNet classification with deep convolutional neural networks. Commun. ACM **60**(6), 84–90 (2017)
21. Vos, B.D.D., Wolterink J.M., Jong, P.A.D., et al.: 2D image classification for 3D anatomy localization: employing deep convolutional neural networks. In: Medical Imaging: Image Processing. International Society for Optics and Photonics (2016)
22. Gatys, L.A., Ecker, A.S., Bethge, M.: A neural algorithm of artistic style. J. Vis. (2015)

23. Gatys, L.A., Ecker, A.S., Bethge, M.: Texture synthesis using convolutional neural networks (2015)
24. Gatys, L.A., Ecker, A.S., Bethge, M., Hertzmann, A., Shechtman, E.: Controlling perceptual factors in neural style transfer, 01 November 2016
25. Johnson, J., Alahi, A., Fei-Fei, L.: Perceptual losses for real-time style transfer and super-resolution. In: Leibe, B., Matas, J., Sebe, N., Welling, M. (eds.) ECCV 2016. LNCS, vol. 9906, pp. 694–711. Springer, Cham (2016). https://doi.org/10.1007/978-3-319-46475-6_43
26. Luan, F., Paris, S., Shechtman, E., Bala, K.: Deep photo style transfer (2017)
27. Chen, L.C., Papandreou, G., Kokkinos, I., Murphy, K., Yuille, A.L.: DeepLab: semantic image segmentation with deep convolutional nets, atrous convolution, and fully connected CRFs. IEEE Trans. Pattern Anal. Mach. Intell. **40**(4), 834–848 (2016)
28. Levin, A., Lischinski, D.: Weiss, Y: A closed-form solution to natural image matting. IEEE Trans. Pattern Anal. Mach. Intell. **30**(2), 228–242 (2007)
29. Li, C., Wand, M.: Combining Markov random fields and convolutional neural networks for image synthesis, 01 January 2016
30. Simonyan, K., Zisserman, A.: Very deep convolutional networks for large-scale image recognition. Comput. Sci. (2014)

Blockchain Consensus Mechanisms and Their Applications in IoT: A Literature Survey

Yujuan Wen[1], Fengyuan Lu[1], Yufei Liu[1], Peijin Cong[1], and Xinli Huang[1,2(✉)]

[1] School of Computer Science and Technology, East China Normal University, Shanghai 200062, China
xlhuang@cs.ecnu.edu.cn

[2] Shanghai Key Laboratory of Multidimensional Information Processing, Shanghai 200062, China

Abstract. With the advance of blockchain technology, blockchain has been used in many fields. As a point-to-point network with characteristics of traceability, decentralization, de-trust and so on, blockchain technology is suitable for deploying Internet of Things networks. It provides the opportunity for the prosperity of IoT networks. As a key component of blockchain, blockchain consensus mechanism is worth studying. In this paper, we systematically survey blockchain consensus mechanisms from the perspective of Internet of Things networks requirements. We first introduce the requirements of consensus mechanisms in IoT networks to help researchers understand the connection of blockchain consensus mechanisms and IoT. Then we divide the blockchain consensus mechanism into four categories, namely, consensus mechanisms for security, consensus mechanisms for scalability, consensus mechanisms for energy saving, and consensus mechanisms for performance improvement. We further analyze those blockchain consensus mechanisms and point out the potential direction of blockchain consensus designs based on our observation. The target of this paper is to provide insights for researchers into the future development of blockchain consensus mechanisms in IoT and encourage more efforts in this field.

Keywords: Blockchain · Consensus mechanisms · Internet of Things · Distributed ledger · PoW

1 Introduction

With the rapid development of Internet of Things (IoT) technology, the scale of Internet of things equipments is experiencing explosive growth. Up to 2020, the number of Internet of Things devices is expected to grow to 50 billion [1]. Though connecting such a large number of global electronic devices to the Internet, IoT transforms the physical world and the digital world from fragmentation to

M. Qiu (Ed.): ICA3PP 2020, LNCS 12454, pp. 564–579, 2020.
https://doi.org/10.1007/978-3-030-60248-2_38

integration, processing between people and things for industry [2–4], finance [5–7], medicine [8–10] and other fields and bringing huge development and economic benefits. Although with a good prospective development, IoT has its own pain points.

The blockchain integrates distributed data storage, point-to-point transmission, consensus mechanism and encryption algorithm to solve the pain points of the Internet of Things. As a typical distributed P2P network, blockchain can provide point-to-point direct interconnection for the IoT to transmit data, rather than through the central processor. Therefore, distributed computing power can process hundreds of millions of transactions, greatly reducing the cost of calculation and storage. As the core component of blockchain, blockchain consensus algorithms grab the attention of researchers in IoT fields.

There are many surveys about blockchain consensus algorithms, however, few of them review the stat-of-art of blockchain consensus algorithm and its application for IoT. Alsunaidi et al. [11] and Nguyen et al. [12] clarified existing blockchain consensus into two categories: proof-based consensus algorithms and voting-based consensus algorithms. However, the number of consensus algorithms analyzed in those surveys is limited. Wang et al. [13] reviewed consensus mechanisms in blockchain networks, but they focused more on mining strategy management rather than consensus mechanisms. Xiao et al. [14] generally sorted consensus protocols into proof-of-work based protocols, proof-of-stake based protocols, and other emerging consensus protocols. They do not consider consensus algorithms for IoT. Cachin et al. [15] and Liu et al. [16] also surveyed blockchain consensus without the consideration of IoT applications. Zoican et al. [17] reviewed blockchain consensus algorithms in IoT, but only mentioned 4 kinds of blockchain consensus algorithms.

Contributions: In this paper, we present a comprehensive survey of blockchain consensus mechanisms and its applications for IoT Networks. We introduce the traditional consensus mechanisms and analyze the requirement of blokchain consensus in a IoT scenario. Then we extract the main idea of various blockchain consensus mechanisms, and classify the consensus mechanisms into four categories from the perspective of IoT requirements:

- consensus mechanisms for security
- consensus mechanisms for scalability
- consensus mechanisms for energy saving
- consensus mechanisms for performance improvement

Furthermore, we discuss some of blockchain applications in IoT networks. Then we point out the future directions for blockchain consensus mechanisms for IoT networks, giving some reference to researchers who are in this field.

Organization: The remainder of this paper is structured as following. Sections 2 introduces traditional consensus mechanisms and analyze the blockchain consensus need of IoT networks. Section 3 divides blockchain consensus mechanisms into four categories according the requirement of IoT networks. Section 4 discusses the application combining IoT with blockchain and indicates the blockchain consensus used in them. Section 5 points out the future directions of the blockchain consensus mechanisms for IoT networks. Section 6 concludes this paper.

2 Traditional Consensus Mechanisms and IoT Requirements for Blockchain Consensus Mechanisms

In the field of distributed consensus, there are already many traditional distributed consensus mechanisms. However, these consensus mechanisms cannot meet the demands of Internet of Things. In this section, we introduce traditional consensus mechanisms and analyze IoT requirements for consensus mechanisms.

2.1 Traditional Consensus Mechanisms

The traditional consensus mechanisms in these paper mainly includes the Paxos family and the Raft.

Pure Paxos Algorithm. Lamport et al. [18] proposed a Paxon parliament's protocol provides a novel way of implementing the state-machine method for distributed systems. Because the protocol is regarded as difficult to understand, Lamport simplified the protocol and proposed the well-known consensus algorithm named Paxos [19]. There are three roles performed in this consensus algorithm, classified as proposers, acceptors, and learners. Paxos is a kind of consistency algorithm with high fault tolerance based on message passing, but it is hard to achieve in engineering. In addition, paxos has derived many variants such as disk paxos [20], fast paxos [21], cheap paxos [22], generalized paxos [23], Bipartisan paxos [24], and HT-paxos [25].

Raft. Ongaro et al. [26] presented Raft which is more understandable than Paxos algorithms to build a practical systems. They separated primary consensus consensual elements like log replication, leader election, and safety, enforcing coherency to a strong degree which reduce the number of state. Raft changes the cluster membership, adopting overlapping majorities to ensure safety. Compared with Paxos, the Raft algorithm has high efficiency and simplicity [27]. Copeland et al. [28] proposed a variant of Raft named Tangaroa. It inherits the advantages of Raft's simplicity and easy to understand, and can maintain security, fault tolerance and activity in the Byzantine context.

Paxos algorithms and Raft are used in many applications and platforms. To sum up, we present the generation time, representative applications of Paxos and Raft in Table 1.

2.2 IoT Requirements for Blockchain Consensus Mechanisms

As mentioned above, there are many traditional consensus mechanisms, but these consensus mechanisms cannot meet the needs of the Internet of Things. In this section, we expound the requirement for consensus mechanisms in IoT

Table 1. Summaries of traditional consensus mechanisms

Consensus algorithms	Years	Cases	Advantages	Disadvantages
Paxos	1998	Chubby [29]	High performance low resource consumption	Difficult to understand achieve in engineering
Raft	2014	ECTD	Easy to understand	Poor performance in high concurrency scenarios

networks from four perspective of security, scalability, energy saving, and performance improvement. The security mainly refers to the fact that honest users can reach a final agreement in an trustless network environment in the presence of an adversary. Scalability refers to whether the performance of network processing transactions can be enhanced with the increase of nodes, and its focus is on the scalability of network processing capabilities. The main consideration of energy saving is the communication complexity between nodes and the computational complexity required by nodes. Performance improvement refers to lower transaction confirmation time, lower latency between nodes in IoT networks, etc.

3 Blockchain Consensus Mechanisms

In this section, based on the needs of IoT networks, we clarify blockchain consensus mechanisms into four categories: blockchain consensus mechanisms based on security, blockchain consensus mechanisms based on scalability, blockchain consensus mechanisms based on energy saving, and blockchain consensus mechanisms based on performance improvement.

3.1 Blockchain Consensus Mechanisms Based on Security

PoW. Nakamoto et al. [30] tackled the problem of double-spending using Bitcoin system. The author designed the peer-to-peer electronic cash system with hash-based proof-of-work (PoW) algorithm. In this system, Every node in Bitcoin network concentrates on figuring out a nounce, and the node figuring out the nounce firstly can obtain the right to pack the transactions into a block and broadcast the block in network. Ren et al. [31] designed a flexible PoW method to solve the problem brought by the usage of application specific integrated circuit (ASIC) machines. Duong et al. [32] proposed **2-hop** which is the combination of PoW and PoS mechanism to secure cryptocurrencies system. Wustrow et al. [33] designed DDoSCoin with a malicious proof-of-DDoS, which can also be a alternative to PoW. Ball et al. [34] designed PoUW using energy to solve Orthogonal Vectors problem.

Proof of Space. Ateniese et al. [35] proposed a novel concept named proof of space (PoSpace) to decrease the DDoS attack. They utilized combinational tools to study space low bounds, developing the corresponding theoretical analysis. Users in proof of space use their CPU0-bound or memory0-bound as a proof of work to get the right of packing a package in blockchain. In addition, proof of space is also called proof of capacity.

ByzCoin. Jovanovic et al. [36] proposed ByzCoin to mitigate attacks like double spending and selfish mining in blockchain. Byzcoin utilizes scalable collective signatures to irreversibly conduct Bitcoin transactions within seconds. It achives consensus by forming hash power-proportionate consensus groups dynamically. Byzcoin can tolerate f fulat nodes within $3f+2$ group members. Kokoris-Kogias et al. [37] improved ByzCoin to ByzCoinX. It optimizes blockchain performance through parallel cross-shard transaction processing.

Proof of Vote. Li et al. [38] observed that Bitcoin-derived consensus mechanisms for public blockchain are insufficient for the deployment scenarios of consortium blockchain. They designed four roles for network participants based-on the key idea of voting campaign and voting mechanism. The four identities in PoV process are butler, butler candidate, commissioner, and ordinary user. The experiment shows that PoV can improve the security, transaction finality, low power consumption of the blockchain and make sure that the blockchain would never fork.

Proof of TEE-Stake. Andreina et al. [39] designed by PoTS (proof of TEE-Stake)to deal the problem of security like nothing at stake attack, long range attck, or the stake grinding attack. PoTS uses a TEE (trusted execution environment) like Intel SGX to assure that each miner can only generate at most one block at each height to limit the increase blockchain height. Furthermore, with TEE, PoTS can prevent posterior corruption by cryptographic techniques. The experimental results show that PoTS balances the security and performance of the POS protocol.

Proof of Authentication. Puthal et al. [40] presented Proof-of-Authentication (PoAh) to take place of Proof-of-Work. It introduces authentication in resource-constrained distributed systems to make the blockchain become specific application and evaluate its sustainability and applicability. Experimental results show that PoAh has a latency in the order of 3 s, while running in limited computer resources. PoAh can also replace existing consensus algorithms, such as proof of stake (PoS) and proof of activity (PoA), for resource and energy constrains infrastructures, such as the IoT [41].

Proof of Trust Negotiation. Feng et al. [42] presented a new consensus algorithm named proof-of-trust-negotiation (PoTN) to tackle the problem of DoS attack and bribery attack happened on the blockchain with fixed miners. They designed negotiation rules and measured the trustworthiness of miners by trust management. Moreover, they select random honest miners by a trusted random selection. The experiment results show that PoTN is of higher efficiency than common consensus algorithms on the block in term of block creation.

Proof-of-Reputation-X. Wang et al. [43] focused the improvement of trusted communication in IIoT to solve the security and efficiency problems of consensus algorithms. The authors proposed a reputation scheme that encourages normal nodes and abnormal nodes both to participate in network collaboration in a good way. The potential of reputation is served as the incentive role in this consensus algorithm. Experimental results show that the proposed scheme can have a good performance at efficiency and safety, resisting Sybil attack, 51% attack and so on.

3.2 Blockchain Consensus Mechanisms for Scalability

Tangle. Popov [44] designed a new consensus mechanism called Tangle, which is DAG (directed acyclic graph)-based consensus mechanism in IOTA. The structure of Tangle is shown as Fig. 1. Each vertex represents a transaction and an edge is a transaction's approval to a previous one [45]. There is no block in Tangle, so there is no limitation of block size. Tangle can do micropayments and ensure data integrity in IoT network, therefore Tangle is suitable for IoT scenario.

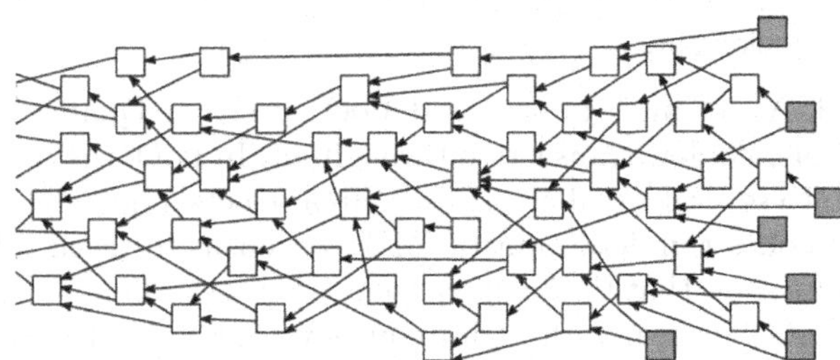

Fig. 1. The tangle network in IOTA

Bitcoin-NG. Eyal et al. [46] handled the problem of low scalability in Bitcoin with a new protocol named Bitcoin-NG. Bitcoin-NG divides traditional Bitcoin mining blocks into two categories: key blocks and micro blocks. Block generation time of Keyblocks is 10 min per block, which is the same as the traditional block. Meanwhile, the block generation time of microblocks is 10 s, in order to improve the transaction speed. The experiment results show that Bitcoin-NG achieves greater throughput and lower latency than Bitcoin while maintaining the premise of Bitcoin trust.

Algorand. Gilad et al. [47] designed a novel crytocurrency named Algorand to improve scalability. Algorand uses the VRF function, combined with the balance ratio of the account, to randomly determine the block generation and voter role. The experiment results show that the throughput of Algorand in one minute is

125 times that of Bitcoin, and there is almost no loss caused by the increase in the number of users.

Proof of Property. Ehmke et al. [48] tackled the problem of an increasing scale of data storage with proof of property. They added a proof of property in each transaction, so other users do not need complete knowledge in data validation process. Proof of property is a kind of lightweight and scalable blockchain protocol, which is suitable for the development of IoT construction.

Stellar Consensus Protocol. Mazieres et al. [49] proposed Stellar Consensus Protocol (SCP) to decrease transaction costs between a heap of closed financial system. SCP enjoys four key attributes: decentralized control, flexible trust, low latency, and asymptotic security. It is a construction of federated Byzantine agreement that can achieve the best flexibility for participants with bad behavior. Like any other Byzantine agreement, SCP has the ability to process transactions quickly, because it verifies through a smaller network of sub-nodes instead of all nodes.

Delegated Proof of Reputation. Do et al. [50] tackled the problem of scalability with DPoR (delegated proof of reputation). DPoR is a improvement of delegated proof of stake, and it takes a reputation ranking system to replace of pure coin staking. In addition, DPoR can resist to loop attack, which guarantee the security of blockchain.

Proof of QoS. Yu et al. [51] proposed PoQ (proof of Quality-of-Service) to address the problem of low transaction throughput. In PoQ. the integral network is partitioned into several small areas. Each area selects a node by QoS and then all the selected nodes run deterministic BFT consensus to reach consensus. The simulation results show that PoQ can realize 9.7K transactions per second (TPS), meaning that PoQ can deal with high-frequency trading scenarios.

FastBFT. Liu et al. [52] designed FastBFT to handle the problem of scalability in traditional Byzantine fault-tolerant (BFT). Existing BFT can only scale to tens of network nodes because of their $O(n^2)$ message complexity, while FastBFT aggregates message in a hardware-based TEE to mitigate the load of message passing. Du et al. [53] designed mixed Byzantine fault tolerance (MBFT), utilizing sharding and layered technology to achieve the improvement of scalability without decrease of security. MBFT divides nodes participating in the negotiation process based on functions of nodes.

Proteus. Jalalzai et al. [54] proposed Proteus to deal with the problem of scalability brought by inherent quadratic message complexity of network replicas. Proteus selects a subset of nodes as a committee to decide the bookkeeper nodes. It improves the quadratic message complexity of traditional BFT mechanism from $O(n^2)$ to $O(cn)$ messages. The experiment results show that the throughput and latency is 2 times compared with PBFT and Bchain.

Proof of Block and Trade. Biswas et al. [55] concentrated on the challenge of scalability and security of IoT blockchain systems. They designed a lightweight consensus algorithm named proof of block and trade (PoBT). It incorporates peers based on the participating number of nodes in a session, which reduces the computational time of peers and improves transaction rates for resource-constrained IoT devices. The experiment shows that the proposed algorithm can reduce the bandwidth required at the critical points of the network.

3.3 Blockchain Consensus Mechanisms for Energy Saving

Proof of Stake. In order to save the energy wasted in mining process, King et al. [56] proposed a peer-to-peer crypto-currency system derived from Bitcoin system. The authors realized the concept of coin age which can facilitate an alternative design named proof-of-stake (PoS). Proof-of-Stake takes place of proof-of-work to make sure the network security. PoS saves computing resource wasted in PoW, however, it suffers from security attacks like nothing at stake attack. Kiayias et al. [57] made an improvement of PoS for security named Ouroboros. In Ouroboros, the bookkeeper is randomly selected by the system, therefore no malicious attacks would occur. Bitshare blockchain [58] makes use of Delegated Proof of Stake (DPoS) to replace the wasteful mining process in proof-of-work algorithm. P4Titan et al. [59] utilized Proof-of-burn in Slimcoin to solve the drawbacks of proof-of-work mining first. Proof-of-burn is the combination of PoS and PoW, derived from PPCoin's and Bitcoin's algorithms.

Tendermint. Kwon [60] designed Tendermint without mining process to solve the problem of energy consumption. Tendermint requires a fixed set of validators who try to reach consensus on a certain block. The consistency of each block is rotated in turn, and each round has a supporter to start the block, after which the verifier decides whether to accept the block or enter the next round of voting.

Proof of Learning. Reeves et al. [61] designed WekaCoin with a novel distributed consensus protocol named proof of learning to decrease the computational power wasted in solving hash-based puzzles. The spent energy of proof of learning in validation process is used to solve tasks of machine learning models. In others words, machine learning competitions is used to validate blockchain transaction. Cheng et al. [62] proposed a novel consensus mechanism called PoDL(proof of deep learning). PoDL recycles a enormous amount of power wasted in mining process to train the task of deep learning, which solve the problem of energy wasting in PoW.

Proof of Exercise. Shoker [63] focused on the problem of energy wasting of PoW and designed a new protocol called proof of exercise. It is an alternative for PoW, using computing power to solve matrix-based problem instead of useless puzzles in PoW.

Proof of Stake Velocity. Ambili [64] presented Proof of Stake Velocity (PoSV) to solve energy wasting problem. PoSV modifies the coin-aging function from the linear relationship of time to an exponential decay function on time, which discourages the coin hoarding of users. Meanwhile, PoSV removes the mining arms race between miners and avoids the menace multipools and ASCIs.

Proof of Contribution. Xue et al. [65] were bent on the problem of the large amount of electricity and expensive mining equipments wasted in proof-of-Work based blockchain. They designed a novel consensus protocols named Proof-of-Contribution (PoC) for cryptocurrency to reduce the energy consumption. It works by rewarding the calculation difficulty of cryptographic puzzle.

Proof of Importance. NEM [66] network first designed a novel consensus algorithm named proof of Importance (PoI). In NEM network, the importance of a node is judged by the number of wallet interactions and currency assets. Unlike power-wasted PoW, powerful equipment is not need to obtain the return of packing a package, which encourages users to trade and maintain on the NEM network actively.

Proof of Search. Shibata [67] designed proof of search protocol to handle with problem of a large amount of energy wasting in mining process of blockchain. In proof of search, nodes assess plenty of candidate solutions and find a good solution to decide the new block added to the chain. The node which get the best candidate solution will get reward. The analysis of this protocol shows that proof of search is less likely to fork compared with PoW.

3.4 Blockchain Consensus Mechanisms for Performance Improvement

Practical Byzantine Fault Tolerance. Castro et al. [68] designed a consensus algorithm named practical byzantine fault tolerance (PBFT) to solve the efficiency of BFT. They reduced the algorithm complexity from exponential to polynomial level. Fang et al. [69] handled with problem of network traffic occurring when number of nodes and transaction increase by optimizing PBFT as selection-based PBFT (SPBFT). They introduce the point mechanism to divide the nodes in networks into two categories as consensus nodes and candidate nodes. Rong et al. [70] proposed a efficient and robust consensus protocol named ERBFT based on BFT state machine protocol. They designed Order-Match (OM) mechanism to order the request of primary and backup nodes. Gao et al. [71] optimized PBFT according to EigenTrust model and named it as T-PBFT. In T-PBFT, nodes is selected by the trust degree, while the trust is evaluated by the transactions between nodes. Wang et al. [72] designed VPBFT by introducing voting mechanism into PBFT.

Proof of Activity. Bentov et al. [73] focused on cost of gaining an advantage over the honest network. They proposed a novel consensus algorithm called Proof-of-Activity (PoA), which is the combination of proof of work and proof of stake. PoA consensus algorithm also makes an improvement on network topology, decreases transaction fees, stimulates nodes to maintain online, and saves energy.

Proof of Luck. Milutinovic et al. [74] designed a novel system with proof of luck (PoL), whose primitives are from proof of work, proof of time, and proof of ownership. PoL utilizes the random number generation of TEE platform to select a consensus leader, providing low-latency transaction validation with deterministic confirmation time. Therefore, the mining process is efficient, and the mining power is saved. The analysis results show that the new protocol can guarantee the liveness and persistence of the blockchain system.

Delegated Byzantine Fault Tolerance. NEO puts forward an improved Byzantine fault-tolerant algorithm named delegated Byzantine fault tolerance [75] to reach a consensus on blockchain. This mechanism allows system to tolerate any number of fault during the lifetime if only the number of faulty nodes is less than $\frac{1}{3}$.

Proof of Elapsed Time. The concept of proof-of-elapsed-time (PoET) was proposed by Intel [76], in order to utilize trusted computing to enforce random waiting times for block construction. While trusted computing component is not perfect and 100% reliable, blockchain networks based on PoET need to tolerate failures of trusted computing components. Chen et al. [77] developed a theoretical framework for evaluating a PoET based blockchain system to improve the performance of blockchain construction.

Proof of Participation and Fee. Fu et al. [78] proposed a novel blockchain consensus called PoPF (proof of participation and fee) to fulfill the demand of enormous and high-frequency transactions in JointCloud Ledger. In PoPF, only candidates have the chance to mine. The candidates are selected by time of participation and fee the participant has paid. The experiment results show that PoPF can balance the unequal computation power of participating nodes and give the contributory users opportunities to be a bookkeeper.

4 Typical Applications of Blockchain Consensus Mechanisms in IoT Networks

IOTA is a novel scalable and feeless distributed ledger, designed to support frictionless data and value transfer. It does help to achieve "Internet of Everything" and exchange value and data between humans and machines. In IOTA network, there is no miners, no blocks. The unit of verification in IOTA network is a transaction rather than a block in general blockchain network. IOTA network is called Tangle shown as Fig. 1. As the volume of transactions increases, the IOTA network will become stronger and consensus will be faster.

The ADEPT. IoT system is jointly created by IBM and Samsung, using blockchain technology to create a decentralized IoT. ADEPT aims to provide optimal transaction security for in IoT System. ADEPT system is based on three protocols: Blockchain, BitTorrent, and TeleHash. It is a proof of concept for a completely decentralized Internet of Things [79].

IoT Chain is committed to becoming a type-safe, functional and flexible blockchain [80]. The specific business model is that IoT Chain charges a processing fee in the transaction process between big data demanders and data providers, and uses platform-specific tokens for payment. IoT Chain uses mobile currency distribution supported by traffic currency to closely integrate information exchange and value transmission on IoT. Furthermore, it uses blockchain as a platform to achieve decentralization and secure encryption of smart products. The bottom layer of IoT Chain is based on DAG technology, using PBFT consensus algorithm [81] to arrive consensus.

5 Future Direction

Through the analysis of the existing consensus algorithms, the design of the future consensus algorithm can be considered from the following aspects:

Security. There are many security attacks like bribery attacks, long range attack, and nothing at stake attack, which is committed to launching attacks on consensus algorithms. Therefore, designing and improving a provably secure blockchain consensus mechanisms is a key point of future direction.

Scalability. Scalability means that a new consensus algorithm should be easy to extend in IoT networks which contain a lot of dynamically changing nodes. When the number of system members and unconfirmed transactions increase, changes in respect of system load and network traffic is considerable. Up to now, partitioning and DAG have been used to increase the scalability block chain consensus algorithm.

Energy Saving. When it comes to energy saving, reducing computing power wasted to compute meaningless computing puzzle is a main consideration in a resource-constrained IoT network, when designing a new consensus algorithms. How to turn meaningless mining energy consumption into meaningful work will be a direction for future blockchain consensus design.

Performance Improvement. Performance improvement means that new consensus mechanisms should be improved in the direction of low latency and high throughput for there are a lot of device in IoT network required to send messages with each other at a actual time.

6 Conclusion

The gist of this survey is to supply researchers a overview of blockchain consensus mechanisms and its applications in IoT networks. We classify the consensus mechanisms into four categories as blockchain consensus mechanisms for security, scalability, energy saving, and performance improvement. Meanwhile, we analyze advantages and disadvantages of these consensus mechanisms. We also point out the future direction of blockchain consensus mechanisms from the perspective of security, scalability, performance improvement, and resource consumption. It is foreseeable that more secure, low energy consumption, high scalability, and more efficient blockchain consensus mechaonclunism will be the pursuit of the future combination of blockchain and Internet of Things.

Acknowledgements. We thank the anonymous reviewers for their helpful and constructive comments on our work. This work was supported in part by the National Key Research and Development Plan of China under Grant 2019YFB2012803, in part by the Key Project of Shanghai Science and Technology Innovation Action Plan under Grant 19DZ1100400 and Grant 18511103302, in part by the Key Program of Shanghai Artificial Intelligence Innovation Development Plan under Grant 2018-RGZN-02060, and in part by the Key Project of the "Intelligence plus" Advanced Research Fund of East China Normal University.

References

1. Ramachandran, P.: Powering the energy industry with the internet of things. PC Quest 46–46 (2014)
2. Jie, W., Yang, D., Liu, X., Yu, Z.: The key technologies and development research of Chinese light industry IoT application. In: China Conference on Wireless Sensor Networks (2013)
3. Lampropoulos, G., Siakas, K.V., Anastasiadis, T.: Internet of Things (IoT) in industry: contemporary application domains, innovative technologies and intelligent manufacturing (2018)
4. Alladi, T., Chamola, V., Parizi, R.M., Choo, K.K.R.: Blockchain applications for industry 4.0 and industrial IoT: a review. IEEE Access **7**, 176935–176951 (2019)
5. Kim, Y., Rue, S., Park, Y.: IoT convergence on finance: fintech trend analysis. Korea Inst. Inf. Technol. Mag. **13**, 45–50 (2015)
6. Cuomo, S., Somma, V.D., Sica, F.: An application of the one-factor Hullwhite model in an IoT financial scenario. Sustain. Cities Soc. **38**, 18–20 (2018)
7. Rubing, H., Yumin, S.: Financial integrated application platform based on BeiDou and quantum security internet of things (2018)
8. Kishor Panda, N., Bhardwaj, S., Bharadwaj, H., Singhvi, R.: IoT based advanced medicine dispenser integrated with an interactive web application. Int. J. Eng. Technol. **7**(4.10), 46–48 (2018)
9. Neto, M.M., Coutinho, E.F., de Oliveira Moreira, L., de Souza, J.N.: Toward blockchain technology in IoT applications: an analysis for e-health applications. In: IFIPIoT (2019)
10. Wu, Y., Zhou, J., Li, J., Liu, J., Li, S., Bai, C.: Application of IoT-based medical diagnosis and treatment in patients with obstructive sleep apnea/hypopnea syndrome in primary hospitals: a preliminary study. Tradit. Med. Mod. Med. **1**(3), 207–212 (2018)
11. Alsunaidi, S.J., Alhaidari, F.A.: A survey of consensus algorithms for blockchain technology. In: 2019 International Conference on Computer and Information Sciences (ICCIS) (2019)
12. Nguyen, G.T., Kim, K.: A survey about consensus algorithms used in blockchain. J. Inf. Process. Syst. **14**(1), 101–128 (2018)
13. Wang, W., et al.: A survey on consensus mechanisms and mining strategy management in blockchain networks. IEEE Access **7**, 22328–22370 (2018)
14. Xiao, Y., Zhang, N., Lou, W., Hou, Y.T.: A survey of distributed consensus protocols for blockchain networks (2019)
15. Cachin, C., Vukolić, M.: Blockchain consensus protocols in the wild (2017)
16. Liu, Y.-Z., Liu, J.-W., Zhang, Z.-Y., Xu, T.-G., Yu, H.: Overview on blockchain consensus mechanisms. J. Cryptol. Res. **6**, 395–432 (2019)

17. Zoican, S., Vochin, M., Zoican, R., Galatchi, D.: Blockchain and consensus algorithms in internet of things. In: 2018 International Symposium on Electronics and Telecommunications (ISETC), pp. 1–4 (2018)
18. Lamport, L.: The part-time parliament. ACM Trans. Comput. Syst. **16**(2), 133–169 (1998)
19. Lamport, L.: Paxos made simple. ACM SIGACT News **32**(4), 18–25 (2001)
20. Gafni, E., Lamport, L.: Disk Paxos. Distrib. Comput. **16**(1), 1–20 (2003)
21. Lamport, L.: Fast Paxos. Distrib. Comput. **19**(2), 79–103 (2006)
22. Lamport, L.B., Massa, M.T.: Cheap Paxos (2007)
23. Pires, M., Ravi, S., Rodrigues, R.: Generalized Paxos made byzantine (and less complex). CoRR abs/1708.07575 (2017). http://arxiv.org/abs/1708.07575
24. Whittaker, M., Giridharan, N., Szekeres, A., Hellerstein, J.M., Stoica, I.: Bipartisan Paxos: a modular state machine replication protocol (2020)
25. Vinit, K., Ajay, A.: HT-Paxos: high throughput state-machine replication protocol for large clustered data centers. Sci. World J. **2015**, 1–13 (2015)
26. Ongaro, D., Ousterhout, J.: In search of an understandable consensus algorithm. In: Proceedings of the 2014 USENIX Conference on USENIX Annual Technical Conference (2014)
27. Huang, D., Ma, X., Zhang, S.: Performance analysis of the raft consensus algorithm for private blockchains. IEEE Trans. Syst. Man Cybern. Syst. **50**, 172–181 (2018)
28. Copeland, C.N., Zhong, H.: Tangaroa: a Byzantine fault tolerant raft (2014)
29. Chubby. http://blogoscoped.com/archive/2008-07-24-n69.html
30. Nakamoto, S.: Bitcoin: a peer-to-peer electronic cash system (2019)
31. Ren, W., Hu, J., Zhu, T., Ren, Y., Choo, K.K.R.: A flexible method to defend against computationally resourceful miners in blockchain proof of work. Inf. Sci. **507**, 161–171 (2020)
32. Duong, T., Fan, L., Zhou, H.S.: 2-hop blockchain: combining proof-of-work and proof-of-stake securely (2016)
33. Wustrow, E., VanderSloot, B.: DDoSCoin: cryptocurrency with a malicious proof-of-work. In: Proceedings of the 10th USENIX Conference on Offensive Technologies, WOOT 2016, pp. 168–177. USENIX Association, Berkeley (2016)
34. Ball, M., Rosen, A., Sabin, M., Vasudevan, P.N.: Proofs of useful work. IACR Cryptol. ePrint Arch. **2017**, 203 (2017)
35. Ateniese, G., Bonacina, I., Faonio, A., Galesi, N.: Proofs of space: when space is of the essence. In: International Conference on Security and Cryptography for Networks (2014)
36. Kokoris-Kogias, E., Jovanovic, P., Gailly, N., Khoffi, I., Gasser, L., Ford, B.: Enhancing bitcoin security and performance with strong consistency via collective signing. In: USENIX Security Symposium (2016)
37. Kokoris-Kogias, E., Jovanovic, P., Gasser, L., Gailly, N., Syta, E., Ford, B.: OmniLedger: a secure, scale-out, decentralized ledger via sharding. In: 2018 IEEE Symposium on Security and Privacy (SP) pp. 583–598 (2018)
38. Li, K., Li, H., Hou, H., Li, K., Chen, Y.: Proof of vote: a high-performance consensus protocol based on vote mechanism consortium blockchain. In: 2017 IEEE 19th International Conference on High Performance Computing and Communications; IEEE 15th International Conference on Smart City; IEEE 3rd International Conference on Data Science and Systems, pp. 466–473 (2017)
39. Andreina, S., Bohli, J.M., Karame, G.O., Li, W., Marson, G.A.: Pots - a secure proof of tee-stake for permissionless blockchains. Cryptology ePrint Archive, Report 2018/1135 (2018). https://eprint.iacr.org/2018/1135

40. Puthal, D., Mohanty, S.P., Nanda, P., Kougianos, E., Das, G.: Proof-of-authentication for scalable blockchain in resource-constrained distributed systems. In: 2019 IEEE International Conference on Consumer Electronics (ICCE), pp. 1–5 (2019)
41. Puthal, D., Mohanty, S.P.: Proof of authentication: IoT-friendly blockchains. IEEE Potentials **38**(1), 26–29 (2019). https://doi.org/10.1109/MPOT.2018.2850541
42. Feng, J., Zhao, X., Lu, G., Zhao, F.: PoTN: a novel blockchain consensus protocol with proof-of-trust negotiation in distributed IoT networks. In: Proceedings of the 2nd International ACM Workshop on Security and Privacy for the Internet-of-Things, pp. 32–37. Association for Computing Machinery, New York (2019)
43. Wang, E.K., Liang, Z., Chen, C.M., Kumari, S., Khan, M.K.: PoRX: a reputation incentive scheme for blockchain consensus of IIoT. Future Gener. Comput. Syst. **102**, 140–151 (2020)
44. Popov, S.: The tangle (2015)
45. Zhao, L., Yu, J.: Evaluating DAG-based blockchains for IoT. In: 2019 18th IEEE International Conference On Trust, Security And Privacy in Computing and Communications/13th IEEE International Conference on Big Data Science and Engineering (TrustCom/BigDataSE), pp. 507–513 (2019)
46. Eyal, I., Gencer, A.E., Sirer, E.G., Van Renesse, R.: Bitcoin-NG: a scalable blockchain protocol. In: Proceedings of the 13th Usenix Conference on Networked Systems Design and Implementation, NSDI 2016, pp. 45–59. USENIX Association, Berkeley (2016)
47. Gilad, Y., Hemo, R., Micali, S., Vlachos, G., Zeldovich, N.: Algorand: scaling byzantine agreements for cryptocurrencies. In: Proceedings of the 26th Symposium on Operating Systems Principles, SOSP 2017, pp. 51–68. Association for Computing Machinery, New York (2017)
48. Ehmke, C., Wessling, F., Friedrich, C.M.: Proof-of-property - a lightweight and scalable blockchain protocol. In: IEEE/ACM International Workshop on Emerging Trends in Software Engineering for Blockchain (2018)
49. Mazieres, D., Polu, S., Barry, N., Mccaleb, J., Losa, G.: The stellar consensus protocol (SCP) (2018)
50. Do, T., Nguyen, T., Pham, H.: Delegated proof of reputation: a novel blockchain consensus. In: Proceedings of the 2019 International Electronics Communication Conference, IECC 2019, pp. 90–98. Association for Computing Machinery (2019)
51. Yu, B., Liu, J., Nepal, S., Yu, J., Rimba, P.: Proof-of-QoS: QoS based blockchain consensus protocol. Comput. Secur. **87**, 101580 (2019)
52. Liu, J., Li, W., Karame, G.O., Asokan, N.: Scalable byzantine consensus via hardware-assisted secret sharing. IEEE Trans. Comput. **68**, 139–151 (2019)
53. Du, M., Chen, Q.J., Ma, X.: MBFT: a new consensus algorithm for consortium blockchain. IEEE Access **8**, 87665–87675 (2020)
54. Jalalzai, M.M., Busch, C., Richard, G.G.: Proteus: a scalable BFT consensus protocol for blockchains. In: 2019 IEEE International Conference on Blockchain (Blockchain), pp. 308–313 (2019)
55. Biswas, S., Li, F., Maharjan, S.: PoBT: a lightweight consensus algorithm for scalable IoT business blockchain. IEEE Internet Things J. **7**(3), 2343–2355 (2020)
56. King, S., Nadal, S.: PPCoin: peer-to-peer crypto-currency with proof-of-stake (2012)
57. Kiayias, A., Russell, A., David, B., Oliynykov, R.: Ouroboros: a provably secure proof-of-stake blockchain protocol. In: Katz, J., Shacham, H. (eds.) CRYPTO 2017. LNCS, vol. 10401, pp. 357–388. Springer, Cham (2017). https://doi.org/10.1007/978-3-319-63688-7_12

58. The BitShares (2014). https://www.bitshares.foundation/papers/BitSharesBlockchain.pdf
59. P4Titan: Slimcoin: a peer-to-peer crypto-currency with proof-of-burn (2014). https://github.com/slimcoin-project/slimcoin-project.github.io/raw/master/whitepaperSLM.pdf
60. Kwon, J.: Tendermint: consensus without mining (2014). https://tendermint.com/static/docs/tendermint.pdf
61. Bravo-Marquez, F., Reeves, S., Ugarte, M.: Proof-of-learning: a blockchain consensus mechanism based on machine learning competitions. In: 2019 IEEE International Conference on Decentralized Applications and Infrastructures (DAPPCON) (2019)
62. Chang, H., Chen, L., Li, B., Shi, Y., Jung, T.: Energy-recycling blockchain with proof-of-deep-learning. In: 2019 IEEE International Conference on Blockchain and Cryptocurrency (ICBC) (2019)
63. Shoker, A.: Sustainable blockchain through proof of exercise. In: 2017 IEEE 16th International Symposium on Network Computing and Applications (NCA), pp. 1–9 (2017)
64. Ambili, K.N., Sindhu, M., Sethumadhavan, M.: On federated and proof of validation based consensus algorithms in blockchain. In: IOP Conference Series: Materials Science and Engineering, vol. 225, p. 012198, August 2017. https://doi.org/10.1088/1757-899x/225/1/012198
65. Xue, T., Yuan, Y., Ahmed, Z., Moniz, K., Cao, G., Cong, W.: Proof of contribution: a modification of proof of work to increase mining efficiency. In: IEEE Computer Software & Applications Conference (2018)
66. NEM technical reference (2018). https://nem.io/wp-content/themes/nem/files/NEM_techRef.pdf
67. Shibata, N.: Proof-of-search: combining blockchain consensus formation with solving optimization problems. IEEE Access **7**, 172994–173006 (2019)
68. Castro, M., Liskov, B.: Practical byzantine fault tolerance and proactive recovery. ACM Trans. Comput. Syst. **20**(4), 398–461 (2002)
69. Weiwei, F., Ziyue, W., Huili, S., Yunpeng, W., Yi, D.: An optimized PBFT consensus algorithm for blockchain. J. Beijing Jiaotong Univ. **43**, 58 (2019)
70. Rong, Y., Zhang, J., Bian, J., Wu, W.: ERBFT: efficient and robust byzantine fault tolerance. In: 2019 IEEE 21st International Conference on High Performance Computing and Communications; IEEE 17th International Conference on Smart City; IEEE 5th International Conference on Data Science and Systems (HPCC/SmartCity/DSS), pp. 265–272 (2019)
71. Gao, S., Yu, T., Zhu, J., Cai, W.J.: T-PBFT: an EigenTrust-based practical byzantine fault tolerance consensus algorithm. China Commun. **16**, 111–123 (2019)
72. Haiyong, W., Kaixuan, G., Qiqing, P.: Byzantine fault tolerance consensus algorithm based on voting mechanism. J. Comput. Appl. **36**, 1766–1771 (2019)
73. Bentov, I., Lee, C., Rosenfeld, M., Mizrahi, A.: Proof of activity: extending bitcoin's proof of work via proof of stake. ACM Sigmetrics Perform. Eval. Rev. **42**(3), 34–37 (2014)
74. Milutinovic, M., He, W., Wu, H., Kanwal, M.: Proof of luck: an efficient blockchain consensus protocol. In: Proceedings of the 1st Workshop on System Software for Trusted Execution. SysTEX 2016, Association for Computing Machinery (2016)
75. Neo whitepaper (2016). https://docs.neo.org/docs/zh-cn/basic/whitepaper.html
76. Sawtooth with PoET-SGX. https://sawtooth.hyperledger.org/docs/core/nightly/1-1/introduction.html

77. Chen, L., Xu, L., Shah, N., Gao, Z., Lu, Y., Shi, W.: On security analysis of proof-of-elapsed-time (PoET). In: Spirakis, P., Tsigas, P. (eds.) SSS 2017. LNCS, vol. 10616, pp. 282–297. Springer, Cham (2017). https://doi.org/10.1007/978-3-319-69084-1_19
78. Fu, X., Wang, H., Shi, P., Mi, H.: PoPF: a consensus algorithm for JCLedger. In: 2018 IEEE Symposium on Service-Oriented System Engineering (SOSE), pp. 204–209 (2018)
79. Windley, P.J.: IBM's adept project: rebooting the internet of things (2015)
80. Iotchain (2019). https://github.com/iot-block/iotchain
81. IoT chain: a high-security lite IoT OS (2017). https://www.chainwhy.com/upload/default/20180613/038d009d2747d379f343f3aee991a401.pdf

Towards a Secure Communication of Data in IoT Networks: A Technical Research Report

Bismark Tei Asare[1,2,3,4](✉), Kester Quist-Aphetsi[1,2,3], and Laurent Nana[4]

[1] Computer Science Department, Ghana Technology University, Accra, Ghana
kquist-aphetsi@gtuc.edu.gh
[2] Cyber Security Division, CRITAC, Cape Coast, Ghana
[3] Directorate of Information Assurance and Intelligence Research, CRITAC, Cape Coast, Ghana
[4] Univ Brest, Lab-STICC, CNRS, UMR 6285, 29200 Brest, France
{bismark.asare,laurent.nana}@univ-brest.fr

Abstract. The past decade has seen a sharp rise in the number and research efforts towards improving optimization, efficiency, security and privacy challenges of internet-of-things (IoT) networks. Internet-of-Things (IoT) and its related frameworks have brought about convenience and improved lifestyle through the use of automation services that have afforded individuals, businesses and Governments the unique opportunities to expand their control and efficiently handle growth. Although recent IoT network architecture have attempted to incorporate cloud components to assist in storing sensed data from edge devices in order to create a much larger storage space for these sensed data and also make them readily accessible and available on demand; issues of security, trust and privacy challenges in IoT network architectures that incorporate the cloud have not been satisfactorily addressed. There is the need for a cloud based IoT network architecture that provides a good balance and fair trade-off between security, efficiency and trust to facilitate the transfer of data between edge nodes and the cloud to lessen the storage workload on the edge nodes to make them efficient. We aim to address the challenges of ensuring an efficient and secured approach in offloading the validated messages from edge nodes and the cloud by proposing an architecture that uses blockchain-based cryptographic techniques to allow IoT devices to be discovered each other to reach an agreement in the validation of messages on devices and the cloud.

Keywords: Internet-of-Things · Stellar Consensus Protocol · Blockchain · Cloud-Computing · Signature schemes

1 Introduction

The Internet-of-Things (IoT) provides a unique platform for ordinary objects (devices) to have connectivity capabilities in order to allow them to connect with other objects where they can collect sensitive data, process and share the information among themselves using the internet. The critical issues of security, trust and privacy challenges of IoT networks have not been fully addressed. Academia and industry continue to

M. Qiu (Ed.): ICA3PP 2020, LNCS 12454, pp. 580–591, 2020.
https://doi.org/10.1007/978-3-030-60248-2_39

heighten efforts to address the security, trust and privacy challenges of IoT networks. Sensor data collected by an edge device could be tapped and modified according to the requirements of a wire tapper before finally being communicated to an edge node. These edge devices could also be manipulated such that sensor data communicate collected data to an unintended recipient. Device access control management through authentication and authorization is a critical part of security. Unfortunately, most studies do not include access control management protocols in dealing with solutions that comprehensively address security of IoT networks.

Fundamental IoT models have the end user, devices or sensors, and cloud whereby privacy and security policies are employed to ensure secured and reliable data. These systems use the trusted third-party systems that rely on a centralized system for generating certificates and authenticating users and devices. Such a system introduces a single point of failure which makes the system rather unreliable and susceptible to tampering of messages. In a classical IoT network, devices are authenticated using access credentials like username and passwords or other access tokens. The architecture for those networks involved a protocol to help establish connection between the edge de- vices (wireless sensors, actuators, and computer devices) that served as entry point into enterprise service providers and edge nodes (devices that provide intelligence to collect, measure and connect to an internet gateway). The edge devices sense data from its immediate environment using Message Queuing Telemetry Transport protocol (MQTT) from the client to a broker; a server for message validation, trans-formation and routing as depicted in Fig. 1.

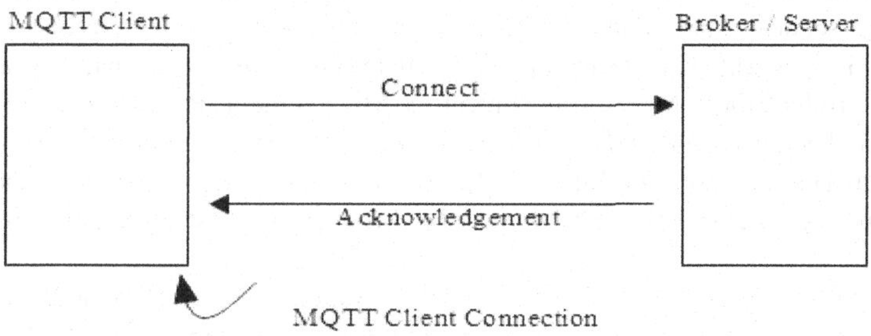

Fig. 1. MQTT connection

The client-broker architecture above does not provide much security and privacy of data. Other architectures also introduced decentralized systems for the authentication of devices and users to assist with access control and management.

The use of blockchain smart contract and cryptographic-based key management protocols assist in dealing with classical password attacks like dictionary, brute force, rule-based, hash-injection, man-in-the-middle, wire-sniffing, replay attacks that most password-based authentication systems face. An appropriate digital signature scheme for constrained devices and internet of things (IoT) will ensure security that preserves the privacy of both messages and devices in the network. Since IoT networks sense and interpret large volume of data, any architecture that will ensure an efficient yet secure way of shedding off excess validated message from edge nodes with the cloud would be desirable. The architecture adopted in this article introduced security, trust, an efficient and robust way of shedding the overhead loads on edge nodes with the cloud. It also used hash-based cryptography to provide device-to-device verification and secure access control management to guarantee a secure mechanism for communication of data between the local IoT devices and the cloud. The rest of the article is organized as follows. Section 2 presents blockchain concepts as well as their fundamental principles for ensuring data security. Section 3 discusses related work on blockchain-based internet of Things architecture. Section 4 describes the architecture and workflow within the architecture. Section 5 discusses the advantages and limitations of the proposed architecture. Section 6 concludes the paper.

2 Description of Concepts and Fundamental Principles for Blockchain for Ensuring Data Security

2.1 Description of Concepts

Internet-of-Things (IoT) is a network consisting of sensing devices or objects that have connectivity features for collecting data from their immediate environments by sensing, measuring, interpreting, connecting, analyzing, predicting and communicating the data using the internet. The interconnection between these devices enable interactivity, collaboration and intelligence. These devices include sensors, actuators, diodes and other electronic components and circuitry that are found in cars, thermo-stats, alarm clocks and washing machines. Industrial and home automation, fire alarm systems, smart meters, wearables, security surveillance systems, industrial cooling systems are some examples of IoT networks. Categories of IoT applications cover areas such as: Safety and Security, Product Flow Monitoring, Inventory Management, Packaging Optimization, Logistic and Supply Chain Optimization, Smart Cities, Smart and Precision Agri- culture, Smart Health Care. The Internet of things possess-es the magic wand for solving real business challenges and playing a critical role in the fourth industrial revolution [1, 2].

Edge Node is also known as gateway node or edge communication node. It supplies the intelligence for sensing, measuring, interpreting and communicating data to other nodes using an internet gateway to the cloud. Thus, edge nodes represent an end portal or gateway for communication with other nodes using the internet to the cloud. Examples of edge node include smartphones and routers [3, 4].

Blockchain is an open ledger with a time-stamp group of cryptographically chained blocks where the state of all validated transactions are stored on active nodes. These trans- actions are traceable, transparent and also decentralized. The participating nodes are cryptographically joined using cryptographic primitives based on mathematical foundations [5, 6]. Blockchain provides a lot of prospects for the provision of se-cured data communication in IoT network due to the rich cryptographic security features that it possesses [7]. The cryptographic security in blockchain allows for the pseudonymous identities, data integrity and authentication, decentralized architecture, fault-tolerance features in resolving the privacy and security lapses in classical architectures for IoT networks. Blockchain based cryptographic architecture has the potential to guarantee an improved security framework for secure data communication in IoT networks. There have been a number of fields where blockchain has shown a great promise of optimization and enhanced security using the smart contract. Protection for intellectual property, electronic voting, identity management systems, healthcare records management systems, and electronic funds payments systems are a few of the domain or fields were smart contract can be useful [8].

Stellar Consensus Protocol (SCP) is a blockchain-based consensus protocol that uses an open membership system for accurately agreeing to the order in recording validated transactions. Thus, it provides a leaderless computing network to obtain consensus on decisions relating to how entries are appended to the open ledger in a decentralized way that operates based on federated byzantine agreement system. The Stellar Consensus Protocol assures low latency, flexible trust, asymptotic security with a decentralized control. The consensus protocol deviates from traditional byzantine fault-tolerant agreement protocols that uses Proof-of-work, Proof-of-stake to guarantee a network that provides a resilient, dependable, and secure system that ensures an uninterrupted access to data anywhere, anytime and on any device [9–11].

2.2 Fundamental Principles for Blockchain for Ensuring Data Security

Security of network systems could be addressed using authentication, authorization, and identification. Safety principles underpinning any network infrastructure hinges on these fundamental concepts, and for the purposes of establishing a case for the choice of the proposed architecture the concepts have been explained. Concepts namely: Privacy, Availability, Integrity, Confidentiality of data, Authentication, Nonrepudiation.

Privacy: The quality of a computer system to protect sensitive information such that only users with the right and privilege to access and use the information do so. Blockchain uses cryptographic schemes and hash functions to protect the privacy of users as well as the messages communicated on such systems.

Availability: The assurance of data to continue to be available at a required time from data storage. Thus, Data availability is the quality of data storage system that de-scribes its ability to guarantee that data is readily accessible, handy, and available when and where required and at all times to authorized users even in situations of disturbances of a system process. Blockchain uses decentralized storage of data on all active nodes where the current state of data (validated transaction) is duplicated and stored on all nodes in the network.

Integrity: The quality of a message that assures that the accuracy and consistency of a message has not been altered in transit, except for those authorized to make those changes. Data integrity answers for the completeness, accuracy, consistency and security of a message over its entire lifecycle. Cryptographic primitives such as hash functions are used for verifying integrity of data.

Confidentiality: It concerns itself with the quality of protecting messages using cryptographic mechanisms from being accessed by unauthorized users. Blockchain uses hash functions and other cryptographic primitives to hide user identities whiles providing protection for ensuring data or message confidentiality.

Authentication: The quality of cryptographic protocol in Blockchain for ensuring the genuineness or validity of a user, device or process. Several techniques such as: passwords, biometrics are used to achieve authentication. Blockchain ensures Authentication, Authorization and Auditing by undertaking identity verification for devices and users, using public key infrastructure scheme where users whose public keys were used for encrypting transactions can use their private keys to decrypt messages. All transactions are public and auditable.

Nonrepudiation: The quality of a database system that concerns itself with the assurance that a user cannot deny or dispute the authorship, authenticity of initialization or validity of a transaction. In Blockchain, all transactions are hashed using the serial number of users' node and are timestamped as well.

3 Related Work

In [12], the authors proposed an authentication framework between IoT devices and active nodes in a blockchain network to guarantee security for IoT network. The block- chain-based security architecture provided a security protection against replay attacks, Denial-of-Service (DOS) attacks, enforced identity authentication, data integrity, enhanced access controls using the digital fingerprint: unique primitive for secret key storage and authentication using the identification numbers on chips thus, the physically unclonable function (PUF) of IoT devices to offer hardware security. The PUF is light weight, introducing no overhead load or delays to the architecture.

The authors in [13] implemented a highly scalable digital signature scheme that provided a good balance between security, execution time and memory requirements. The Merkle Signature scheme: a digital signature scheme was implemented on an 8-bit smart card microprocessor. The algorithms used were relatively short but produced a faster verification time. The architecture operated on three key algorithms: Pair Generation, Signature Generation; and Signature Verification.

In [14], Blockchain was adopted in monitoring mission critical IoT applications. The authors made the case where smart contracts were used to verify and provide the unique platform for monitoring the asset lifecycle of these critical IoT applications. The blockchain-based IoT architecture in that research had three clusters: Applications, IoT platform, IoT devices and gateway in which the Hyperledger composer. A smart contract application was used to monitor mission-critical systems in real time.

In [15], decentralized security mechanism for detecting security attacks that assured a strong defense against cyberattacks was adopted and used. The architecture in that article used a Software Defined Networking (SDN) that operated in an open, decentralized and distributed ledger in monitoring and detecting attacks at the fog node and edge nodes.

In [16] the authors adopted a blockchain based technology to secure a smart home system that presented a secured, scalable and efficient architecture. Blockchain technology has shown to be a reliable management technique for data validation and asset management in which internet of things networks can benefit from. The authors in [17] proposed for a decentralized architecture for the management and controlling of access of devices in an IoT network.

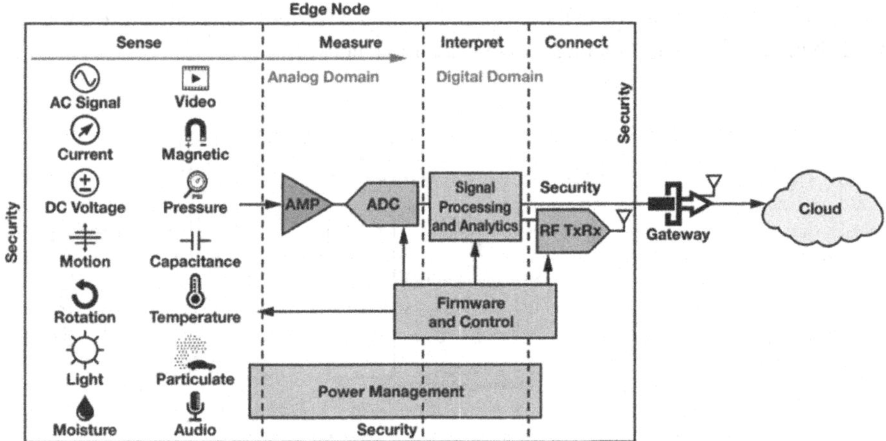

Fig. 2. Security architecture for edge nodes [15]

Other researchers proposed architectures that relied on blockchain-based cryptographic techniques for other categories of internet of things in which the devices or nodes were moderately strong in terms power, processing speed, and storage capacity. Those cryptographic schemes although strong were rather computationally complex for those devices [18]. The architecture proposed in [4] included cloud-based components but the cloud together with its associated privacy and trust challenges were not factored into the general architecture of the systems of the IoT network. Thus, there were separate individual security mechanisms adopted for the edge devices as well as the cloud. The architecture specifically concentrated efforts at providing security for the edge node. A secured edge node in itself cannot guarantee the overall security for an IoT network that will be using the internet to share sensed data. For the security of such architecture to be achieved, there will be the need for other means to be exploited to address the specific cloud security and privacy challenges. Any such security mechanism will introduce additional overheads and affect the overall efficiency of the security mechanism.

The authors in [17] proposed an architecture that was sensitive to energy and other resource demands within an IoT network particularly for factory automation and management of energy in a sustainable way. The architecture specifically adopted an approach that helped with the energy management for the expansion of production systems. A blockchain-based architectural framework for internet of things using the physically unclonable function (PUF) helped in the authentication of devices and active nodes towards validating transactions. Here hardware fingerprints resulting from the use of PUF were produced to assist with data source authentication. A decentralized digital ledger assisted in providing protection against tampering of data. The fault-tolerant consensus and smart contract in blockchain were also used by the authors to include cloud services in the architecture for device and data authentication [18]. The PUF implemented within the cloud helped with the enrollment and verification phases of IoT devices and data. Cryptographic operations were adopted to enhance data integrity, and cloud authentication for IoT networks. Figure 3 describes an architecture for using blockchain to control and manage access in an IoT network.

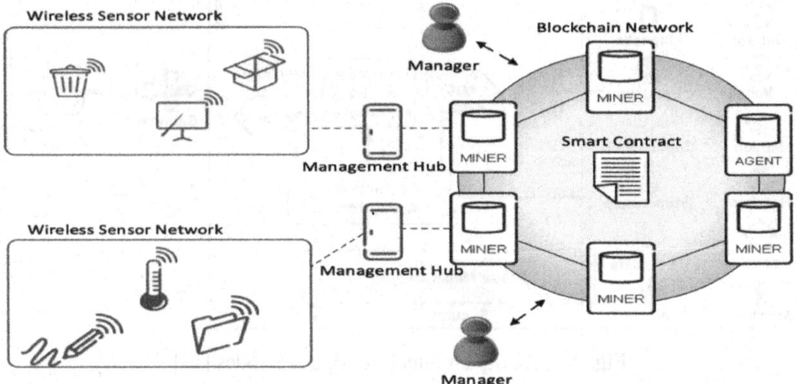

Fig. 3. Blockchain-based architecture for IoT [17]

In [15] the authors made a case for a protocol that allowed for blockchain platform to assist in recording contracts. The paper described a new protocol for confirming contractual documents. In the protocol, each contractor gets the protection from privacy invasion from blockchain by the provision of hash function and digital signatures. An implementation of the IoT architecture proposed is shown in Fig. 2. It incorporated cloud-based technology to secure a smart home system.

4 Architecture and Workflow

4.1 The Blockchain-Based IoT Architecture

Figure 4 outlines the IoT architecture adopted for assisting the storage and sharing of verifiable data and Fig. 5 illustrates data workflow between an edge device and the cloud, passing through an edge node.

4.2 Description of the Architecture

The architecture adopted a protocol that assisted in the exchange of messages between local IoT devices and the cloud in message transfer such that: Blockchain smart contract and cryptographic schemes were used for verification of messages as well as device-to-device authentication between the local edge nodes and the cloud. In a de-centralized and distributed fashion, a sender and recipient agree to share authenticated message using a blockchain consensus such that:

Physical device -to-Cloud Transaction: The Address of the destination, the hash of the previous transaction, the information of the sender (signature, the hash value of the sender's address), timestamp of the trans- action was cryptographically hashed and

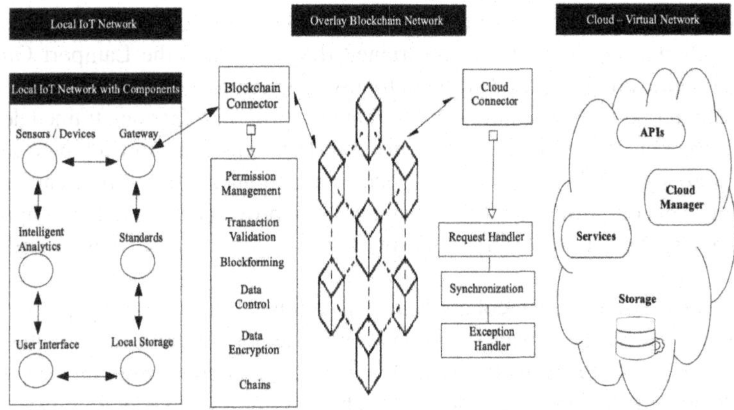

Fig. 4. Block diagram for architecture [19]

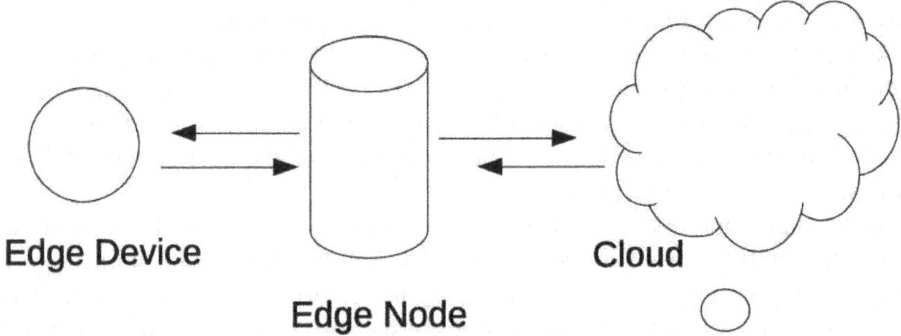

Fig. 5. Data workflow [20]

digitally signed. The recipient of the transaction generates a new transaction using the received transaction request as basis to arrive at a consensus using the stellar consensus protocol. The transactions are cryptographically chained. The most recent validated data that is stored on the edge node is transferred to the cloud using smart contract.

The architecture above integrated smart contract as well as hash-based cryptography scheme to provide device-to-device verification as well as a secured mechanism for communication of data between the local IoT devices and the cloud, a sure measure for shedding off validated data to the cloud from the edge nodes making them efficient in the verification of data from edge devices. To maintain a relatively small key size storage in the cloud for faster and efficient storage and retrieval of data between the edge nodes and the cloud, an indexable and dynamic storage structure that contained the most recently used key pair was made accessible. The key pair was stored in the Certificate Authority a decentralized edge node that formed part of the blockchain overlay network.

The algorithm was based on a one-time signature scheme such that the public-key of any node was used only once in the cryptographic hashing operation. Although the classical Lamport one-time signature scheme produces a strong and secured hashing, it always maintains a larger key size that is not conducive for internet of things network since most of the devices are resource constrained. The goal for a light weighted digital signature scheme appropriate for constrained devices, made the Lamport One-Time Signature scheme not the desired protocol for use. The challenge was for a rather efficient yet lightweight one-time signature scheme that is appropriate for con-strained de- vices, the Winternitz One-Time Signature Scheme (WOTSS) was there-fore adopted and used. The signature scheme helped with key generation, signature generation and signature verification. This one-time signature scheme is an improvement upon the Lamport One-Time Signature Scheme (LOTSS), that produced a larger size of public and private key.

The stellar consensus protocol used an open membership by adopting quorum slicing approach in a decentralized way. Decisions were based on quorums–the situation where each individual participant freely selected among its neighbours whom to trust towards arriving at agreements to validate messages. The sets of nodes that facilitated agreement in the stellar consensus protocol formed the quorum slice. A quorum is a set of nodes from all active nodes on the network which has the capacity to arrive at an agreement. Thus, a type of blockchain-based cryptographic approach that employed open-membership for active nodes for the formation of quorum to facilitate agree-ment.

Actions by these active participants in a permissioned blockchain were regulated by executing a control mechanism (smart contract). The architecture supported self-verifying data (that is data that cannot be easily altered at the server end because they have been digitally signed). All active nodes that formed the quorum agreed through consensus in validating messages.

The configuration setting for each node is tuned to facilitate trust. Nodes with same configuration setting form a quorum. These give way for quorum slice formation. The ability of nodes to select trust in such permissioned environment ensures the decentralization within the network. The architecture consists of a set of edge nodes (servers) in addition to arbitrary number of clients.

4.3 Signature Scheme

The one-time signature scheme has these three components: Key generation, Signature generation and Signature verification.

The Winternitz One-Time Signature Scheme (WOTSS) is a cryptographic secure hash function such that a function, where: $H : \{0, 1\}^* \rightarrow \{0, 1\}^s$, H – Hash function and S – the length of the results of the hash function in bits. The one-time signature has the following characteristics: “preimage resistant”, “second preimage resistant” and “collision resistant”.

Preimage Resistant: A hash function H is preimage resistant if it is hard to find any m for a given h with h = H(m).

Second Preimage Resistance: A hash function H is second preimage resistant if it is hard to find for a given m_1 any m_2 with $H(m_1) = H(m_2)$.

Collision Resistant: A hash function H is collision resistant if it is hard to find a pair of m_1 and m_2 with $H(m_1) = H(m_2)$.

The signature function was the ideal cryptographic secure hash function because it conceals its algorithm to enable it solve preimage resistance and second preimage resistance problem more efficiently than a brute force attack. An attacker in a brute force situation, chooses m randomly until $h = H(m).H(m)$ can have 2^s different results.

Each result represents a common probability P_h with $P_h = 1/(2^s)$. An attacker will be required to averagely pick $2^s/2$ different inputs m_2, until the attacker finds $h = H(m)$. An attacker is faced with same challenge of picking on average $2^s/2$ different inputs m_2 until an expression for $H(m_1) = H(m_2)$ is found. The attack against preimage and second image resistance has a complexity such that $1/2 * 2^s = 0(2^s)$.

In Fig. 6, the structure and functional working of the winternitz one-time signature scheme is presented.

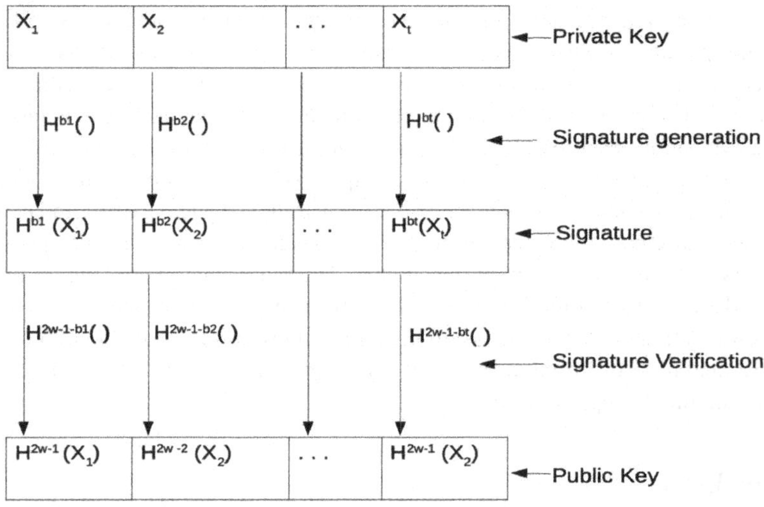

Fig. 6. Winternitz one-time signature scheme [21]

Key Generation
Let $\{0, 1\}^* \rightarrow \{0, 1\}^s$ be the cryptographic hash function. where $h \in \mathbb{N}$ and supposed 2^k signatures are to be generated that are verifiable with one another. A parameter w, with $w \in \mathbb{N}$, is chosen and $t = \left[\frac{s}{w}\right] + \left[(\log_2\left[\frac{s}{w}\right] + 1 + w)/w\right]$ gets calculated. The computational time is dependent of the parameter size of w.

Signature Generation
Let $M = m_1 \ldots m_s \in \{0, 1\}$ be the message to be signed $X_1, ..., X_t$ the private key, and $w$and t. The message is split up into $\lceil s/w \rfloor$blocks $b_1, \ldots, b_{\lceil \frac{s}{w} \rceil}$.

Signature Verification
To verify the signature $sig = (sig_1 || \ldots sig_t)$ for a given message $M = \{0, 1\}^s$ the

parameters $b_1, \ldots, b_t$ are computed first. For $i = 1, \ldots, t$, $sig_i' = H^{2^w-1-b^i}(sig_i)$ is calculated.

$$sig_i' = H^{2^w-1-b^i}(sig_i) = H^{2^w-1-b^i}\left(H^{b_1}(X_i)\right) = H^{2^w-1}(X_i) = Y_i.$$

Hence if $Y' = Sig_1'||\ldots||Sig_t'$ equals $Y = H(Y_1||\ldots||Y_t)$ then the signature is valid or vice versa.

5 Discussion

The stellar consensus protocol helped with formation of trust clusters for the validation and exchange of messages between the local IoT network and the cloud in which device-to-device authentication and message sharing was made possible. Using the protocol, multiple agents agreed to on an output value in an open membership fashion. The trust clusters were achieved using an intactness algorithm in the stellar consensus protocol. The protocol relied on the quorum slices of federated participants which adopted a dynamic set of participants in a decentralized way towards the formation of clusters. The choice of the dynamic membership of participants appropriately addressed the possibilities of some of the active resource-constrained devices exiting and new ones joining the participation cluster due to challenges such as power or other processing constraints that these devices might encounter. The block-chain-based stellar protocol offered a decentralized control, flexible trust, low latency for arriving at a consensus for message validation, and above all asymptotic security which is a combination of digital signatures and hash functions with longer crypto-graphic keys that operated in a rather efficient run time for optimum security.

6 Conclusion

The blockchain technology adopted by the architecture was to ensure a tamper-proof distributed ledger that did not use a centralized authentication system but rather a decentralized mechanism for consensus for the authentication of validated messages. The blockchain-based security mechanism implemented at the network and device level layers facilitated a secure communication of data in the network. At the network layer, an appropriate blockchain-based cryptographic function was adopted to manage authentication, authorization and accounting as well as control the network connectivity and handle the information flow in the IoT network; At the device layer, a fault-tolerant decentralized consensus mechanism (stellar protocol) and smart contract assisted in the management of device-to-device communication in a rather secured manner. The proposed architecture introduced security and trust in the IoT network, achieving a non-repudiation of event records. It also supported a robust way of shedding the validation overhead of transaction between edge nodes and the cloud. The architecture provided a fair balance for an efficient, fast and secured algorithm that used hash functions for secure communication of data within the IoT network and adequately addressed privacy, trust, integrity, confidentiality, and security challenges in the network.

References

1. IoT Applications | Internet Of Things Examples | Real World IoT Examples | Edureka. https://www.edureka.co/blog/iot-applications/. Accessed 16 Jun 2020
2. Jesus, E.F., Chicarino, V.R.L., De Albuquerque, C.V.N., Rocha, A.A.D.A.: A survey of how to use blockchain to secure Internet of Things and the stalker attack. Secur. Commun. Networks **2018** (2018)
3. Shi, W., Cao, J., Zhang, Q., Li, Y., Xu, L.: Edge computing: vision and challenges. IEEE Internet Things J. **3**, 637–646 (2016)
4. Bevers, I.: Intelligence at the edge part 1: the edge node (2018714), pp. 1–6 (2019)
5. Watanabe, H., Fujimura, S., Nakadaira, A., Miyazaki, Y., Akutsu, A., Kishigami, J.: Blockchain contract: securing a blockchain applied to smart contracts. In: 2016 IEEE International Conference on Consumer Electronics, ICCE 2016, pp. 467–468 (2016)
6. Dorri, A., Kanhere, S.S., Jurdak, R.: Towards an optimized blockchain for IoT. In: Proceedings - 2017 IEEE/ACM 2nd International Conference on Internet-of-Things Design and Implementation, IoTDI 2017 (part of CPS Week) (2017)
7. Makhdoom, I., Abolhasan, M., Abbas, H., Ni, W.: Blockchain's adoption in IoT: the challenges, and a way forward. J. Netw. Comput. Appl. **125**, 251–279 (2019)
8. Di Francesco Maesa, D., Mori, P.: Blockchain 3.0 applications survey. J. Parallel Distrib. Comput. **138**, 99–114 (2020)
9. Mazieres, D.: The stellar consensus protocol: a federated model for internet level consensus. J. Am. Chem. Soc. **120**(42), 11022–11023 (2016)
10. Gaul, A., Khoffi, I., Liesen, J., Stüber, T.: Mathematical analysis and algorithms for federated byzantine agreement systems, pp. 1–42 (2019)
11. Cachin, C., Vukolić, M.: Blockchain consensus protocols in the wild (2017)
12. Patil, A.S., Hamza, R., Yan, H., Hassan, A., Li, J.: Blockchain-PUF-based secure authentication protocol for Internet of Things. In: Wen, S., Zomaya, A., Yang, L.T. (eds.) ICA3PP 2019. LNCS, vol. 11945, pp. 331–338. Springer, Cham (2020). https://doi.org/10.1007/978-3-030-38961-1_29
13. Rohde, S., Eisenbarth, T., Dahmen, E., Buchmann, J., Paar, C.: Fast hash-based signatures on constrained devices. In: Grimaud, G., Standaert, F.-X. (eds.) CARDIS 2008. LNCS, vol. 5189, pp. 104–117. Springer, Heidelberg (2008). https://doi.org/10.1007/978-3-540-85893-5_8
14. Hammoudeh, M., Ghafir, I., Bounceur, A., Rawlinson, T.: Continuous monitoring in mission-critical applications using the internet of things and blockchain. In: ACM International Conference Proceeding Series (2019)
15. Rathore, S., Wook Kwon, B., Park, J.H.: BlockSecIoTNet: blockchain-based decentralized security architecture for IoT network. J. Netw. Comput. Appl. **143**(2018), 167–177 (2019)
16. Singh, S., Ra, I.H., Meng, W., Kaur, M., Cho, G.H.: SH-BlockCC: a secure and efficient Internet of Things smart home architecture based on cloud computing and blockchain technology. Int. J. Distrib. Sens. Networks **15**(4), 1550147719844159 (2019)
17. Novo, O.: Blockchain meets IoT: an architecture for scalable access management in IoT. IEEE Internet Things J. **5**, 1184–1195 (2018)
18. Stenum Czepluch, J., Zangenberg Lollike, N., Malone, S.O.: The use of block chain technology in different application domains, May 2015
19. Wang, G., Shi, Z., Nixon, M., Han, S.: ChainSplitter: towards blockchain-based industrial IoT architecture for supporting hierarchical storage, pp. 166–175 (2020)
20. Ahmad, S., Afzal, M.M.: Sustainable communication networks. In: Deployment of Fog and Edge Computing in IoT for Cyber-Physical Infrastructures in the 5G Era, pp. 351–359 (2019)
21. Becker, G.: Merkle signature schemes, merkletrees and their cryptanalysis.pdf (2008)

Efficient Thermography Guided Learning for Breast Cancer Detection

Vishwas Rajashekar, Ishaan Lagwankar, Durga Prasad S N, and Rahul Nagpal(✉)

Department of Computer Science and Engineering, PES University, Bangalore, Karnataka, India
{vishwasrajashekar,ishaankiranl,durgaprasad}@pesu.pes.edu, rahulnagpal@pes.edu

Abstract. Early-stage breast cancer detection is often thwarted due to privacy concerns, the need for regular scanning, among other factors, thereby severely reducing the survival rate of patients. Thermography is an emerging low cost, portable, non-invasive, and privacy-sensitive technique for early-stage breast cancer detection gaining popularity over the traditional mammography based technique that requires expert intervention in a lab setup. Earlier proposals for machine learning augmented thermography for early-stage breast cancer detection suffer from precision as well as performance challenges. We developed a novel voting based machine learning model with on the fly parallel retraining using the Dask library. Experimental evaluation reveals that our novel high-performance thermography based learning technique brings up the accuracy of early-stage life-saving breast cancer detection to 93%.

Keywords: Cancer · Thermography · Performance · Parallelism · Learning · Dask

1 Introduction

Breast Cancer is a prevalent disease that is notoriously hard to detect. However, early-stage detection is directly correlated to the survival rate. Experienced radiologists and doctors scrutinize mammography scans more recently aided by machine learning models to arrive at a diagnosis. However, the radiology based technique is invasive and requires specialized X-ray machines in a carefully orchestrated diagnostic setup. The complete protocol to diagnose as described by The National Breast Cancer Foundation[1] involves mammography followed by ultrasound followed by MRI and finally a Biopsy for a conclusive diagnosis. Recent advancements in thermal imaging for breast cancer detection have enabled faster conclusions compared to mammography. The thermography technique has the added advantage of being non-invasive, non-contact, not

[1] National Breast Cancer; Breast Cancer Diagnosis. https://www.nationalbreastcancer.org/breast-cancer-diagnosis/, accessed 2020-03.

M. Qiu (Ed.): ICA3PP 2020, LNCS 12454, pp. 592–600, 2020.
https://doi.org/10.1007/978-3-030-60248-2_40

harmful, painless, privacy-sensitive, low-cost as well as portable only requiring thermal cameras [8]. Moreover, Thermography techniques can detect early-stage vascular changes in even the dense breast tissues with implants and is agnostic to any hormonal or menstrual changes, critical to the successful treatment and consequently, survival of patients. This paper proposes our novel and efficient technique that constructs, trains, and evaluates a model that automatically diagnoses from thermography images for further review by medical experts. Earlier proposals in the context of machine learning augmented thermography for early-stage breast cancer detection suffer from high maintenance time and minimal use of parallelism. We perform efficient task and pipeline parallel processing using the lightweight, scalable, easy-to-use, and efficient Dask Python library on clusters of computers leveraging lazy execution and in-memory computation which is superior as compared to Spark and Hadoop. Our main contributions in this paper are follows:

1. We propose our novel, efficient, and scalable thermography based learning algorithm to detect early-stage breast cancer. It demonstrates the highest accuracy for this life-threatening disease among already known techniques to the best of our knowledge.
2. We have developed a Dask based framework for distributed processing that enables much-needed no-maintenance, efficient and instantaneous retraining to improve accuracy.
3. We perform a detailed experimental evaluation and analysis of the results of proposed and earlier techniques that demonstrates the superiority of the proposed methodology.

The rest of the paper is organized as follows. We present related work in Sect. 2 followed by our methodology and algorithm in Sect. 3. Experimental evaluation with a detailed analysis of results is in Sect. 4. We conclude in Sect. 5 with a mention of future directions.

2 Related Work

This section presents earlier work related to thermal imaging coupled with neural-network-based detection of breast cancer. Many techniques use convolutional neural networks (CNN) with modules such as ResNet50, SeRes-Net50, Inception among others. J. Zuluaga-Gomez et al. [11] outline a CNN-based methodology for detecting breast cancer using a novel CNN-CAD methodology for the DMR-IR database. They create baseline CNN models to replicate the results of other models first, followed by building the CNN algorithm, with a hyperparameters optimization algorithm based on a tree parzen estimator, to increase performance. Arena et al. [2] enlist the benefits of thermography over the classical methods for cancer diagnoses. They tested a weighted algorithm in 109 tissue proven cases that use a six-feature classification (threshold, nipple, areola, global asymmetry, and hotspot). Schaefer et al. [9] perform a fuzzy logic classification algorithm, with an accuracy of 80%, explaining that statistical feature analysis is a key source of information in achieving high accuracy. Krawczyk et al. [4] propose an ensemble approach using clustering.

The majority of computer-aided diagnosis studies related to IR imaging techniques for breast cancer diagnosis employ the database from [10]. It consists of 1140 images from 57 patients, where 38 carried anomalies, and 19 were healthy women from Brazil, with each patient having a sequence set of 20 images. This comprised the DMR-IR database the same as used in our work. Rajendra et al. [1] built an SVM algorithm for the automatic classification of normal and malignant breast cancer, with an accuracy of 88.1%. Mambou et al. [6] described a method to use deep neural networks and SVMs for breast cancer, with camera awareness and physical modeling of the breast. Whereas earlier techniques are focused on overall performance and complexity, our ensemble learning architecture allows for better and more accurate predictions of 93% for the first time for this deadly disease than a complex CNN that takes a long time to retrain with newer examples.

3 Methodology

In this section, we thoroughly describe our model and methodology to breast cancer diagnosis based on thermographic images. The Dask programming paradigm is applied to the use case of detecting breast cancer, using an ensemble of Dask parallel workers executing parallel models, working on Tensorflow engines running convolutional neural networks trained on different subsets of the data.

The algorithm structure is defined in a way to introduce maximum efficiency in compute and reusability whenever possible. First, raw images from the DMR-IR dataset are preprocessed by a filter to extract features. Next, the features extracted are fed to different workers running in parallel on a cluster to extract votes providing an index for the malignancy of a given breast subject followed by the prediction of dominant value based on voting. Learners with an erroneous outcome are retrained on the corresponding sample point for later inclusion while the other workers function normally providing no downtime in execution.

Preprocessing. The first stage prepares the images for the machine learning classifier by cropping the Region of Interest (ROI), to reduce any interference from thermal signatures of other areas such as the neck. The images in the dataset are filtered to ensure that only frontal images are used for the learning process. The final input into the model is a $224 \times 224 \times 3$ vector with all the three channels as shown in Fig. 1.

Parallel Swarm Optimised CNN Search. The preprocessed images are fed to a set of weak CNNs for the voting framework. A swarm architecture model is used to allow weak learners to search through various hypotheses using velocity frameworks. A particle swarm optimizer searches a complex hyperplane space for minimum specifications and stimulates thousands of workers searching this space. This search is parallelized [5] to allow better interaction among workers.

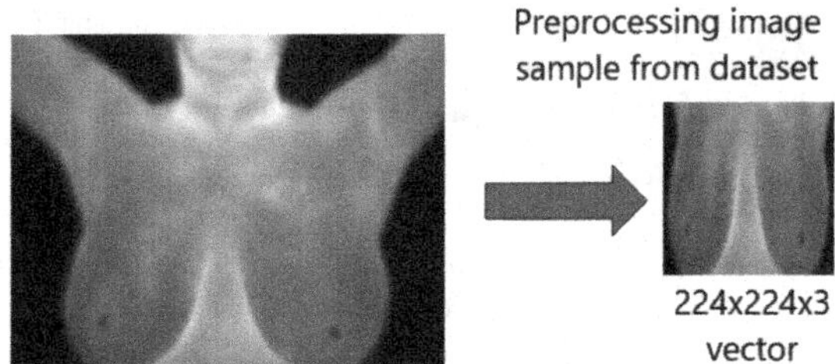

Fig. 1. Preprocessing of images

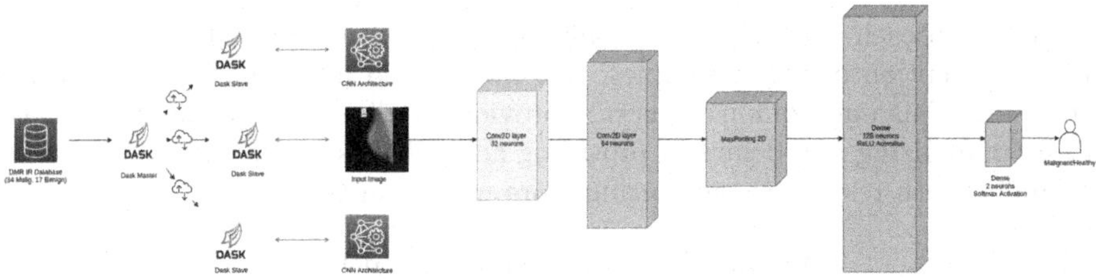

Fig. 2. Architecture overview

Convolutional Neural Network (CNN) Build and Training Architecture. This stage builds the CNN architecture as shown in Fig. 2, with specification in Table 1 taking inputs as native grayscale images, cropped to enhance the features of the tumorous region detected by the image processing algorithm.

Parallelizing Training. The model runs in parallel on a Dask cluster of machines detailed in the next section in the two phases namely the Voting phase and the Retraining Phase. The voting-based architecture allows a voting/consensus-based framework to validate whether a tumorous growth is benign or malignant. The ensemble consists of a combination of weak learner models with different architecture specifications for voting on different features inferred by the networks. Each learner is weighted by its retraining history with the assumption that the retraining boosts the performance of the model. This voting allows a more precise prediction for detecting the malignant parts of breast cancer positive patients, allowing more sensitivity for positive cases.

Retraining Architecture. One of the salient and distinctive features of our approach is that the model is retrained on the fly for the new cases with a unique localized approach. As depicted in Fig. 3, the models with the wrong result on previously unseen inputs are retrained in parallel, with one model per worker, while the other models operate at full functionality. This allows no maintenance time and allows on the fly retraining with no delay attributed to our Dask based approach. It is pertinent to mention here that in our approach, the hyperparameters are adapted only for the learner with the incorrect prediction. As the

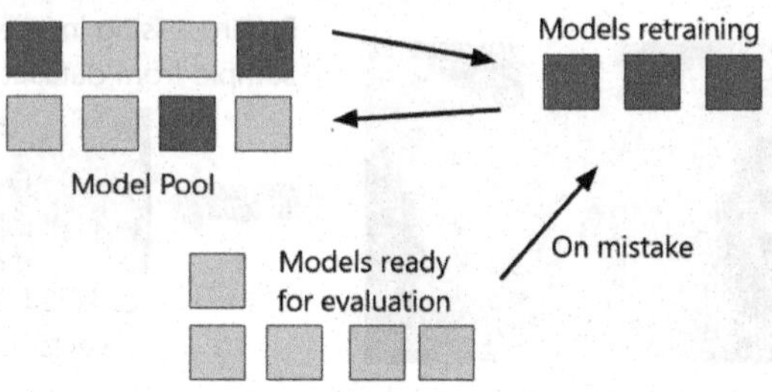

Fig. 3. Retraining of models from pool

retraining on each worker happens only on one architecture build to allow concurrent execution of newer samples, this leads to no maintenance hence faster retraining. This autonomous structure provides resilience and robustness, drastically minimizing any delays in providing accurate results as well as providing scalability with respect to supporting more models in the pool.

Table 1. Description of optimal model layers

Layer	Output shape	Parameters
Convolution 2D	(None, 222, 222, 32)	896
Convolution 2D	(None, 220, 220, 64)	18496
MaxPooling 2D	(None, 110, 110, 64)	0
Dropout	(None, 110, 110, 64)	0
Flatten	(None, 774400)	0
Dense	(None, 128)	99123328
Dropout	(None, 128)	0
Dense	(None, 2)	258

4 Experimental Evaluation

In this work, we used a publicly available DMR-IR database which follows various protocols enlisted in [7] and [3], to ensure consistency and quality. It consists of a population of 57 patients, aged between 21 and 80 years, 19 of these patients are healthy and 37 present malignancies. The diagnostic has been confirmed by mammographic tests, ultrasounds, and biopsies. The thermal images are captured with a FLIR thermal camera model SC620, which has a sensitivity lesser than 0.04 °C. Each infrared image has a dimension of 640 $\times$ 480 pixels, with the software creating two types of files: (i) a heat-map file; (ii) a matrix with 640 $\times$ 480 points. Each patient goes under a thermal stress period for decreasing the breast surface temperature and twenty image sequences are captured every 5 min. Several standards are also enforced in the acquisition of this data.[2]

[2] The DMR-IR dataset used in this work is publicly available at http://visual.ic.uff.br/dmi.

Our approach safeguards the privacy of every patient with a non-invasive, non-contact procedure to get the thermographic image.

We used Dask Python Library over Hadoop, it allows retraining on the fly at a much lower cost, allowing lightweight and quick compute for results on the machine learning pipelines using a distributed asymmetric compute cluster of 4 workers out of which one acts as a master node with an Intel Core i7 7700 @ 3.6 GHz 4 cores 8 threads and 16 GB of RAM; and 3 worker nodes with Intel Core i5 4440 @ 3.1 GHz. The machine with Intel Core i7 8750H @ 2.2 GHz 6 cores with 16 GB RAM is used to run the serial models for benchmarking and comparison.

The metrics inferred from our results are used against the other models used for the same approach under similar conditions, and since none of these models is inherently parallel, the serial versions are considered versus our parallel architecture on three machines. As inferred from Table 2, our model performs significantly better over more data samples provided, with retraining built into the framework. Our work was able to beat the standard VGG16 and SeResNet50 networks for image classification with Dask Tensorflow, allowing us better accuracy and results for our predictions. Our model also boasts an **instantaneous retraining time, which is the key novelty in our approach**. Figure 4 contrasts our results individually with said models, depicting accuracy and performance when these models were trained on the standalone machine, as a proof of concept presentation.

Table 2. Accuracy metrics

Model	Accuracy	F1 score	Precision	ROC/AUC	Time per epoch
SeResNet18	0.90	0.91	0.91	0.90	30
SeResNet34	0.86	0.86	0.91	0.86	35
SeResNet50	0.82	0.81	0.85	0.82	42
ResNet50	0.79	0.77	0.90	0.80	30
VGG16	0.90	0.89	0.85	0.90	22
InceptionV3	0.80	0.80	0.82	0.80	**21**
InceptionV2	0.65	0.72	**0.93**	0.72	44
Xception	0.90	0.89	0.89	0.90	30
Our model	**0.93**	**0.91**	0.87	**0.94**	67

The ROC/AUC is also depicted in Fig. 4b, with our model having the highest ROC/AUC value of 0.94. This is due to the serial nature of their execution whereas our model is trained to work in a distributed fashion leading to performance gain close to 4.5% resulting in fewer false positives and false negatives by our model as depicted in Fig. 4c. It shows that our model efficiently predicts lesser false positives, i.e. interpret a sick person as healthy fewer times than the two closest competing models. It is pertinent to mention here that even small improvement is very significant in this context given the life threatening consequences of delayed detection thereby our model is demonstrably more reliable.

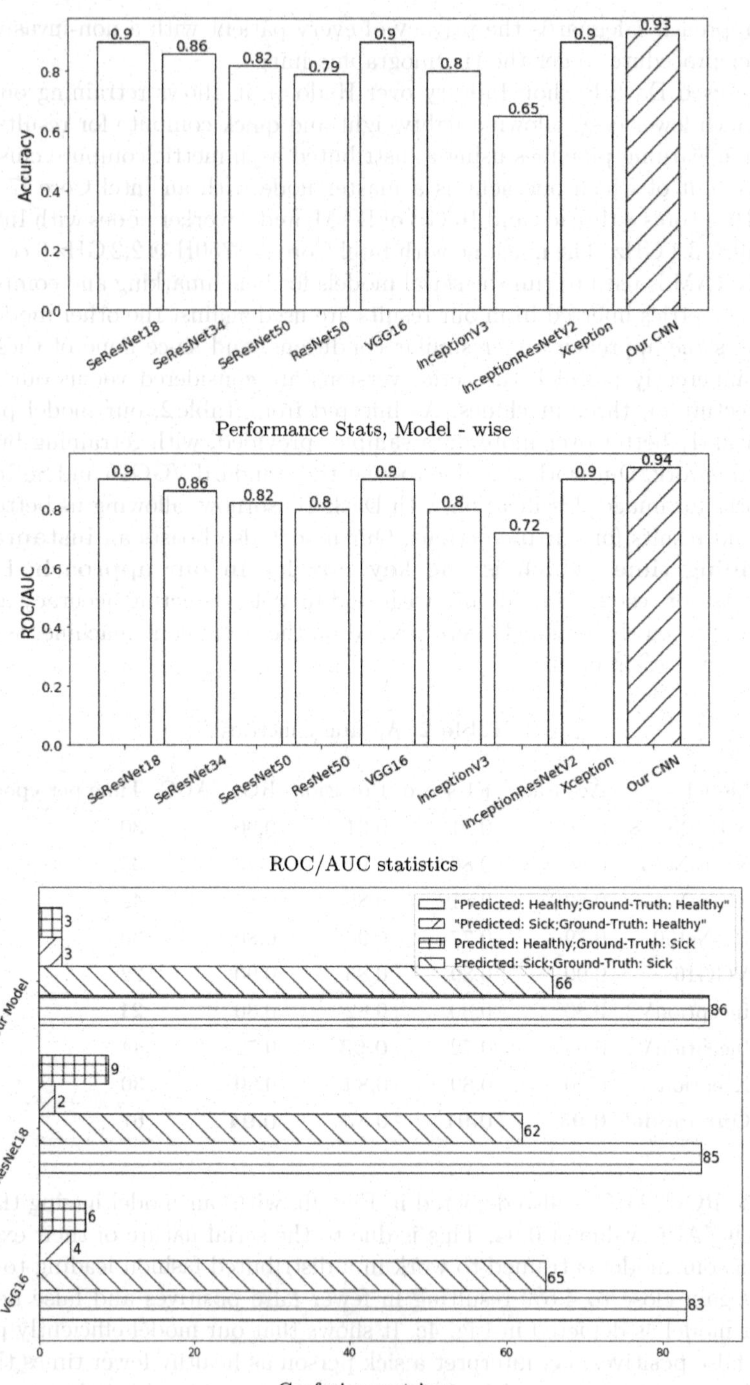

Performance Stats, Model - wise

ROC/AUC statistics

Confusion matrices

Fig. 4. Evaluation results

5 Conclusions and Future Directions

Thermography is an emerging low cost, portable, non-invasive, and privacy-sensitive technique for early-stage breast cancer detection key to survival. We developed a novel voting based machine learning model with on the fly parallel retraining using the Dask python library. Experimental evaluation reveals that our novel and simple high-performance thermography based learning technique delivers better results at higher confidence levels using consensus-based learning and adversarial learning as a single pipeline, thereby bringing up the accuracy of early-stage life-saving breast cancer detection to 93% which is the highest among all known techniques. We are working towards a mobile-based solution with a special thermal add-on camera. Thermography guided learning for efficient and accurate detection of various other diseases such as other various types of tumors, cancers, neuromusculoskeletal disorders, thyroid gland abnormalities are the few other future directions of interest.

References

1. Acharya, U.R., Ng, E.Y.K., Tan, J.H., Sree, S.V.: Thermography based breast cancer detection using texture features and support vector machine. J. Med. Syst. **36**(3), 1503–1510 (2012)
2. Arena, F., Barone, C., DiCicco, T.M.: Use of digital infrared imaging in enhanced breast cancer detection and monitoring of the clinical response to treatment. In: Proceedings of the 25th Annual International Conference of the IEEE Engineering in Medicine and Biology Society (IEEE Cat. No. 03CH37439), vol. 2, pp. 1129–1132 (2003)
3. Kandlikar, S.G., et al.: Infrared imaging technology for breast cancer detection-current status, protocols and new directions. Int. J. Heat Mass Transf. **108**, 2303–2320 (2017)
4. Krawczyk, B., Schaefer, G., Zhu, S.Y.: Breast cancer identification based on thermal analysis and a clustering and selection classification ensemble. In: Imamura, K., Usui, S., Shirao, T., Kasamatsu, T., Schwabe, L., Zhong, N. (eds.) BHI 2013. LNCS (LNAI), vol. 8211, pp. 256–265. Springer, Cham (2013). https://doi.org/10.1007/978-3-319-02753-1_26
5. Lorenzo, P.R.: Particle swarm optimization for hyper-parameter selection in deep neural networks. In: Proceedings of the Genetic and Evolutionary Computation Conference, pp. 481–488. ACM (2017)
6. Mambou, S.J., Maresova, P., Krejcar, O., Selamat, A., Kuca, K.: Breast cancer detection using infrared thermal imaging and a deep learning model. Sensors **18**(9), 2799 (2018)
7. Ng, E.K.: A review of thermography as promising non-invasive detection modality for breast tumor. Int. J. Therm. Sci. **48**(5), 849–859 (2009)
8. Niramai: Thermalytix. https://niramai.com/
9. Schaefer, G., Závišek, M., Nakashima, T.: Thermography based breast cancer analysis using statistical features and fuzzy classification. Pattern Recogn. **42**, 1133–1137 (2009). https://doi.org/10.1016/j.patcog.2008.08.007

10. Zuluaga-Gomez, J., Zerhouni, N., Al Masry, Z., Devalland, C., Varnier, C.: A survey of breast cancer screening techniques: thermography and electrical impedance tomography. J. Med. Eng. Technol. **43**(5), 305–322 (2019)
11. Zuluaga-Gomez, J., Masry, Z.A., Benaggoune, K., Meraghni, S., Zerhouni, N.: A CNN-based methodology for breast cancer diagnosis using thermal images (2019)

PTangle: A Parallel Detector for Unverified Blockchain Transactions

Ashish Christopher Victor, Akhilarka Jayanthi, Atul Anand Gopalakrishnan, and Rahul Nagpal(✉)

Department of Computer Science and Engineering, PES University, Bangalore, KA, India
ashish.chrivic@gmail.com, akhiljay99@gmail.com, reachme.atul@gmail.com, rahulnagpal@pes.edu

Abstract. Tangle is a novel directed acyclic graph (DAG)-based distributed ledger preferred over traditional linear ledgers in blockchain applications because of better transaction throughput. Earlier techniques have mostly focused on comparing the performance of graph chains over linear chains and incorporating the Markov Chain Monte Carlo process in probabilistic traversals to detect unverified transactions in DAG chains. In this paper, we present a parallel detection method for unverified transactions. Experimental evaluation of the proposed parallel technique demonstrates a significant, scalable average speed-up of close to 70%, and a peak speed-up of approximately 73% for a large number of transactions.

Keywords: Blockchain · DAG · Tangle · Markov Chain Monte Carlo · Parallelism

1 Introduction

Blockchain has gained widespread popularity from a simple transaction medium to being the core of many computing and allied domains such as IoT, cybersecurity, decentralized applications, supply chain management (SCM), and digital identity, among many others. In a blockchain application, many users compete to solve a computationally complex problem and the winner gets cryptocurrency. This process becomes more and more complex as the ledger grows, and consequently, the time taken to add a transaction to the blockchain increases. This problem is particularly pronounced in linear ledgers because entire blocks of transactions are verified together. DAG-based ledgers avoid this problem.

Tangle is a DAG-based non-linear ledger built by the IOTA foundation [1]. In the Tangle ledger, vertices represent transactions, and every new transaction needs to verify two other transactions, consequently adding two new edges to the DAG. The core Tangle algorithm is finding an optimal point of attachment for new transactions in the DAG, thereby verifying previously unapproved transactions popularly named tips. The very first transaction is called the genesis. Figure 1 shows tips as grey nodes.

M. Qiu (Ed.): ICA3PP 2020, LNCS 12454, pp. 601–608, 2020.
https://doi.org/10.1007/978-3-030-60248-2_41

A Poisson Point Process [2] is used to model the arrival of transactions with λ as the rate of incoming transactions. Whereas a large value of λ results in the simultaneous arrival of multiple transactions all seeing only the genesis, a small λ degenerates the DAG to a linear chain. Usually, a delay factor h is used to make incoming transactions temporarily invisible to existing transactions.

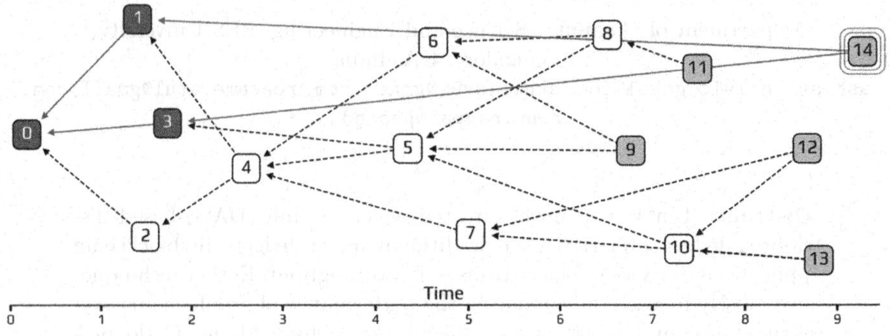

Fig. 1. The Tangle

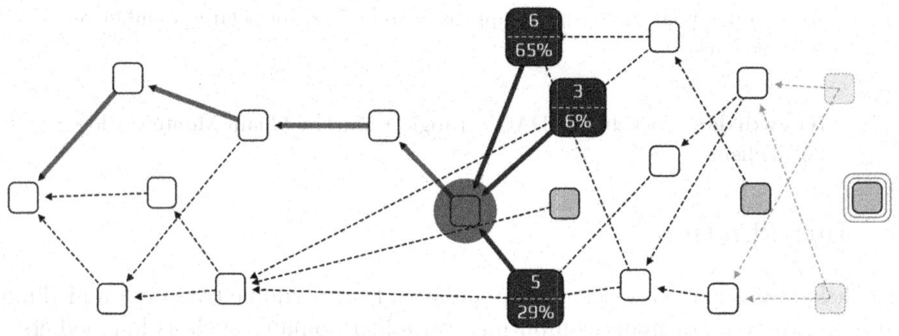

Fig. 2. The weighted random walk decided by MCMC

A lazy tip is defined as a transaction that approves older transactions instead of newer transactions. It is computationally wasteful to re-verify old transactions. In Fig. 1, transaction 14 is a lazy tip. Cumulative weights are used to denote the importance of transactions to solve the lazy tip problem. It is more probable to walk towards a more important transaction than a less important one. The cumulative weight is defined by one more than the number of approvers of a transaction. Since using a purely weighted approach will result in hot spots of tips throughout the graph, a new variable α is used to introduce some randomness by giving importance to the cumulative weight of a transaction. The Markov Chain Monte Carlo (MCMC) process [3] uses α to estimate the probability in each step of the traversal. Figure 2 shows the probability of walking towards a node using this weight factor.

All too often, a new transaction arrives necessitating the process of attachment and verification, thereby frequently triggering the expensive weighted random walk as explained above. We observed that this process can be parallelized in the interest of better scalability.

In this paper, we present our parallel algorithm and implementation to identify unapproved transactions with significant performance benefits. Our main contributions are as follows:

1. We propose a novel parallel algorithm for identifying unapproved transactions in the non-linear DAG-based Tangle ledger.
2. We have implemented our serial and parallel algorithm and performed a thorough performance evaluation that demonstrates a consistent, scalable, and significant performance gain close to 70% on an average for a large number of transactions.

The rest of the paper is organized as follows. We explain related work in Sect. 2, followed by our methodology and parallel algorithm in Sect. 3. Experimental evaluation is presented in Sect. 4. We conclude in Sect. 5 with future directions of this work.

2 Related Work

Bitcoin [4] is a digital currency experiment that drove the blockchain revolution. Its underlying structure is a decentralized distributed ledger that has been used for a plethora of other applications leading to the evolution of blockchain. Blockchain is extensively used in banking and insurance because of its immutability and security. It is used in the supply chain and logistics sector to track food items and manufacturing parts. Distributed ledgers are used in IoT applications to keep track of hundreds of thousands of connected smart devices. Blockchain technology is also used for social good in the fields of education, privacy preservation, and promotion of ethical trade, among others [5].

To build on the model of bitcoin's cash-like tokens, a second-generation blockchain called Ethereum [6] introduced "smart contracts" [7], which allowed loans and bonds to be represented in the blockchain. Both Bitcoin and Ethereum are public, permissionless blockchains. As the name suggests, no permission is required to become part of this blockchain network. Transaction information stored on permissionless blockchains can be validated by anybody. To gain more control over the blockchain, permissioned or private blockchains were introduced. The owner of the blockchain can dictate the network's structure and issue software updates [8]. IBM pioneered permissioned blockchains by implementing the private blockchain Fabric, which is now part of the Hyperledger Consortium [9].

Miners operate vast data centers to provide security in exchange for cryptocurrency in the context of "Proof of Work", which requires time-consuming verification of huge blocks of transactions, leading to the development of scalable non-linear blockchain systems such as Tangle [1], Dagcoin [10], Graphchain [11], Dexon [12], and SPECTRE [13].

Dagcoin uses an algorithm to find the "best parent", starting from unverified transactions to the genesis. Graphchain uses a Resource Description Framework (RDF) graph [14] and extends blockchain functionality to it instead of modifying the blockchain to accommodate graph features [15]. Dexon uses a Byzantine agreement protocol similar to Algorand for consensus [16] and achieves a very high transaction throughput. SPECTRE uses a DAG that allows frequent, concurrent mining without knowledge of the propagation delay and the main chain in the ledger.

It is pertinent to note that none of the earlier techniques combine Bayesian Statistical models like Tangle, developed by the IOTA organization. Tangle has walkers that perform a weighted random walk throughout the graph using a probability-based MCMC process to arrive at unverified tips. This probabilistic approach prevents a large number of transactions from verifying only one tip and prevents verification of older tips in the DAG. The tip selection algorithm is unique to Tangle. We, for the first time in this work, present the parallel algorithm for tip selection in Tangle.

3 Algorithm and Methodology

3.1 Serial Algorithm

The Markov Chain Monte Carlo process plays a key role in selecting tips by probabilistically arriving at the next node in the traversal. In our implementation as outlined in Algorithm 1, the input is the DAG consisting of all previous transactions, and the function Randomchoice() chooses nodes randomly from the graph based on the bounds. The bounds are arbitrarily chosen [1]. The number of nodes chosen is equal to the number of walkers spawned.

Algorithm 2 outlines the weighted random walk algorithm with the input as the current node and the entire graph. The output is two unapproved transactions that will be verified by the new node. H_{node} is the cumulative weight for a particular node. The function Probability() based on Eq. 1 [1], calculates the transition probability from transaction x to transaction y, when y approves x. RandomNode() chooses one node from this set of probable nodes.

Transition probability: x→y   Cumulative weight of node y

$$\underbrace{P_{xy}}_{} = exp(-\alpha(H_x - H_y)) \underbrace{\Big(\sum_{z:z->x} exp(-\alpha(H_x - H_z))\Big)^{-1}}_{\text{All transactions z that directly approve x}} \quad (1)$$

Cumulative weight of node x   All transactions z that directly approve x

Algorithm 1: Markov Chain Monte Carlo

```
   Input: graph
1  numwalkers = 2;
2  lowerBound = maximum(0, (graph.noOfNodes - 20.0) * graph.arrivalRate);
3  upperBound = maximum(1, (graph.noOfNodes - 10.0) * graph.arrivalRate);
4  particles = Randomchoice(graph.nodelist[lowerBound:upperBound],
     numwalkers);
5  for each node in particles do
6      RandomWeightedWalk(node, graph);
7  end
```

In lines 5 through 7, we find an unapproved transaction starting from *node*.

Algorithm 2: Weighted Random Walk

```
   Input: currentNode, graph
   Output: unapprovedList
1  repeat
2      nextsetofnodes = DirectlyApproveCurrentNode(graph, node);
3      if α >0 then
4          weightlist = [];
5          sum = 0;
6          for each nextnode in nextsetofnodes do
7              weightlist.add(H_nextnode);
8              sum += exp(α(H_currentNode - H_nextnode));
9          end
10          probs = [];
11         for each weight in weightlist do
12             probs.add(exp(α(weight - H_currentNode)) / sum);
13         end
14         currentNode = RandomChoice(nextsetofnodes, probs);
15     end
16 until currentNode not approved & length(unapprovedList) < 2;
17 if currentNode not in unapprovedList then
18     unapprovedList.add(currentNode);
19 end
```

Here, *weightlist* holds the weights for the next set of connected nodes from the current node. In lines 6-13, we calculate the probability using Equation 1 and choose a node.

3.2 Parallelization

We have taken a three-stage approach to parallelize the algorithm as explained below.

Parallelization Stage 1. In this stage, we used a modified MCMC process to enable parallel execution of the weighted random walk. We used threading functionalities and other synchronization methods like shared lists and locks provided by the PyMP library to collect tips from the weighted random walk. Each walker collects one tip. We increased the number of walkers to boost speed-up. However, on testing, we found the speed-up to be unsatisfactory. On further analysis, we found that running many walkers on fewer threads led to context switching and bad work sharing.

Parallelization Stage 2. To prevent re-computation of the weight of each node by multiple threads, we used a shared list so that subsequent threads can use the weights computed by previous threads when accessing the same node in the graph. We modified the number of walkers being spawned to achieve a one to one mapping of threads. All walkers run in parallel and there is no concurrent execution. As soon as two tips are collected, we stop the execution of subsequent threads. Thus, stage 2 mapped each walker to a different thread improving performance over stage 1. We exploited further potential for speed-up in stage 3.

Parallelization Stage 3. We spawned more than two walkers to collect only two tips in stage 2 and stopped the execution of threads once we collect two tips, resulting in wastage of the initial work done by the remaining threads. We addressed this in stage 3, where each walker now runs independently from all other walkers and collects a tip. Each tip is accumulated for future use, even if there is no new incoming transaction. All walkers are utilized optimally, and more than one transaction can be added to the graph at a time. Thereafter we use four walkers in parallel to attach two transactions and verify two sets of tips concurrently to get optimal results.

4 Experimental Evaluation

4.1 Setup

To parallelize the tip selection algorithm, we used PyMP and Numba. Numba is an open-source JIT compiler that translates a subset of Python and NumPy code into fast machine-level code using the industry-standard LLVM compiler framework. PyMP is a package that brings OpenMP-like functionality to Python. We ran both the serial and parallel implementations of the Tangle ledger on an AMD Ryzen 3900X CPU, with 12 cores and 32 GB of main memory. We divided the benchmarking into two classes namely, for a small number of transactions i.e. $N < 20{,}000$ and for a large number of transactions i.e. $N \geq 20{,}000$ transactions.

4.2 Results

For $N < 20{,}000$ transactions, we incremented transaction count by 2K each time and recorded the performance improvement of the parallel version over the serial version as depicted in Fig. 3, observing an average speed-up close to 40% with a peak speed-up of 65%. Figure 4 shows performance improvement for $N \geq 20{,}000$ in increments of 20K, which is consistent and close to 70% on an average with a peak observed speed-up of 73%. With an increasing number of transactions, the speed-up also improved very significantly. The proposed parallel algorithm scales consistently with a larger number of transactions modulo a few, very minor fluctuations attributed to non-determinism and asymmetry.

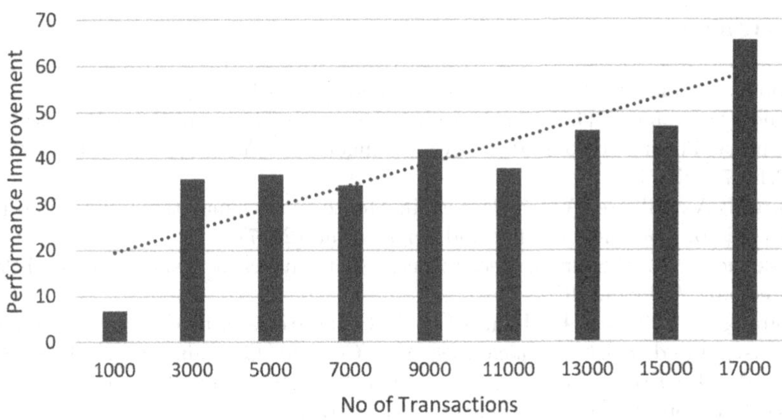

Fig. 3. Performance improvement for small N

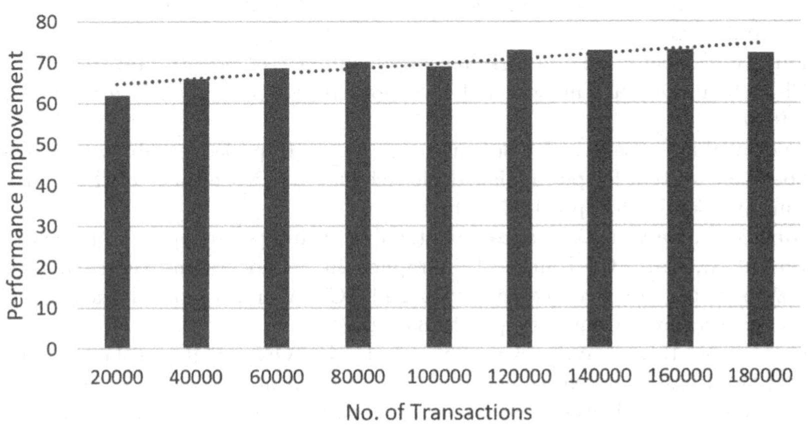

Fig. 4. Performance improvement for large N

5 Conclusion and Future Work

In this paper, we presented and evaluated our novel, multistage, parallel, and scalable algorithm for transaction verification in Tangle, a DAG-based non-linear ledger using PyMP and Numba. Experimental evaluation demonstrates that our novel parallel algorithm scales very well and gives a consistent average speed-up of close to 70% over the serial algorithm. In the future, we plan to focus on further improving performance by batching tip detection for lazy execution with bookkeeping.

References

1. Popov, S.: The tangle (2016). https://iota.org/IOTAWhitepaper.pdf
2. Daley, D.J., Jones, D.V.: An Introduction to the Theory of Point Processes: Elementary Theory of Point Processes. Springer, New York (2003). https://doi.org/10.1007/b97277
3. Berg, B.A., Billoire, A.: Markov chain Monte Carlo simulations. In: Wiley Encyclopedia of Computer Science and Engineering (2007)
4. Nakamoto, S.: Bitcoin: a peer-to-peer electronic cash system (2008). https://bitcoin.org/bitcoin.pdf
5. Narang, S., Chandra, P., Jain, S., et al.: Foundations of blockchain technology for industrial and societal applications. Adv. Comput. Commun. **2**, 32–51 (2018)
6. Wood, G.: Ethereum: a secure decentralised generalised transaction ledger. Ethereum project yellow paper, pp. 1–32 (2014)
7. Bartoletti, M., Pompianu, L.: An empirical analysis of smart contracts: platforms, applications, and design patterns. In: Brenner, M., et al. (eds.) FC 2017. LNCS, vol. 10323, pp. 494–509. Springer, Cham (2017). https://doi.org/10.1007/978-3-319-70278-0_31
8. Lai, R., Chuen, D.L.: Blockchain-from public to private. In: Handbook of Blockchain, Digital Finance, and Inclusion, vol. 2, pp. 145–177. Academic Press (2018)
9. Androulaki, E., Barger, A., Bortnikov, V., et al.: Hyperledger fabric: a distributed operating system for permissioned blockchains. In: Proceedings of the Thirteenth EuroSys Conference, pp. 1–15 (2018)
10. Ribero, Y., Raissar, D.: Dagcoin whitepaper. Whitepaper, pp. 1–71 (2018)
11. Boyen, X., Carr, C., Haines, T.: Graphchain: a blockchain-free scalable decentralised ledger. In: Proceedings of the 2nd ACM Workshop on Blockchains, Cryptocurrencies, and Contracts, pp. 21–33 (2018)
12. Chen, T.Y., Huang, W.N., Kuo, P.C., et al.: DEXON: a highly scalable, decentralized DAG-based consensus algorithm. arXiv preprint arXiv:1811.07525 (2018)
13. Sompolinsky, Y., Lewenberg, Y., Zohar, A.: SPECTRE: a fast and scalable cryptocurrency protocol. IACR Cryptol. ePrint Arch. (2016)
14. Manola, F., Miller, E., McBride, B.: RDF primer. W3C Recommendation (2004)
15. Sopek, M., Gradzki, P., Kosowski, W., et al.: GraphChain: a distributed database with explicit semantics and chained RDF graphs. In: Companion Proceedings of the The Web Conference, pp. 1171–1178 (2018)
16. Gilad, Y., Hemo, R., Micali, S., et al.: Algorand: scaling byzantine agreements for cryptocurrencies. In: Proceedings of the 26th Symposium on Operating Systems Principles, pp. 51–68 (2017)

DOS-GAN: A Distributed Over-Sampling Method Based on Generative Adversarial Networks for Distributed Class-Imbalance Learning

Hongtao Guan[1(✉)], Xingkong Ma[2], and Siqi Shen[1]

[1] Science and Technology on Parallel and Distributed Laboratory, National University of Defense Technology, Changsha, Hunan, People's Republic of China
{guanhongtao07,shensiqi}@nudt.edu.cn

[2] College of Computer, National University of Defense Technology, Changsha, Hunan, People's Republic of China
maxingkong@nudt.edu.cn

Abstract. Class-imbalance Learning is one of the hot research issues in machine learning. In the practical application of distributed class-imbalance learning, data continues to arrive, which often leads to class-imbalance situations. The imbalance problem in the distributed scenario is particular: the imbalanced state of different nodes may be complementary. The imbalanced states of different nodes may be complementary. Using this complementary relationship to do oversampling to change the imbalanced state is a valuable method. However, the data island limits data sharing in this case between the nodes. To this end, we propose DOS-GAN, which can take turns to use the data of one same class data on multiple nodes to train the global GAN model, and then use this GAN generator to oversampling the class without the original data being exchanged. Extensive experiments confirm that DOS-GAN outperforms the combination of traditional methods and achieves classification accuracy closes to the method of data aggregating.

Keywords: Class-imbalance learning · Distributed learning · Generative adversarial networks · Data isolated islands

1 Introduction

Class-imbalance learning problems [11] is one of the hot issues in machine learning. It happened in the classification problem. A data set is considered to be class-imbalanced if it contains different proportions of samples of different classes. There is a large gap in the number of samples between classes. The study of the imbalance problem is very meaningful because many real-world classification applications naturally have imbalance problems. For example, in the task of credit card user behavior classification, the number of malicious behaviors

M. Qiu (Ed.): ICA3PP 2020, LNCS 12454, pp. 609–622, 2020.
https://doi.org/10.1007/978-3-030-60248-2_42

detected is much lower than normal behaviors. In addition, there are some imbalances that come from the process of data generation and collection. For example, in the classification task of pneumonia cases, the number of COVID-19 case samples will be far less than other common pneumonia cases, because it has just appeared.

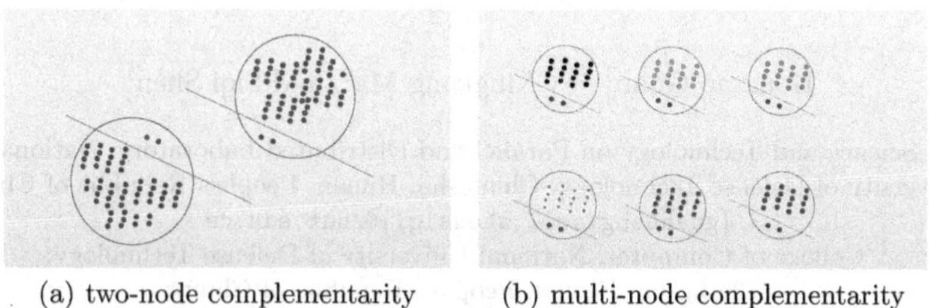

(a) two-node complementarity (b) multi-node complementarity

Fig. 1. Two common complementary situations

With the rapid development of the Internet and big data technology [25], data analysis has played an important role in many real-world applications. Whether it is companies, research institutions, or government agencies and are constantly collecting data for analysis [17,18]. However, most organizations have a limited amount of data, which means that using their own data for analysis may not be able to produce universal results. Therefore, there is a willingness to conduct joint data analysis between organizations to improve the effectiveness of data analysis. Therefore, the joint data analysis brings a new breakthrough point to class-imbalance learning: the minority classes among organizations may be complementary with each other. Figure 1 presents two common situations of complementarity. Figure 1a shows the Min-Maj complementarity between two nodes. The minority class on one node is the majority class on the other nodes. Figure 1b is the Min-Min complementation among multiple nodes, which means that one class is a minority class on each node. However, due to a large number of nodes, the number of this class in the entire system is enough for data analysis. If the joint data analysis can be carried out smoothly between organizations, the class-imbalance problem in the two above cases can be converted into a class-balance problem, which can be solved by using an efficient and accurate classifier.

However, the joint analysis of the data is not trivial. The data contains the user's privacy or business secrets, so it is often inconvenient to be shared with partners [24]. We call this situation data isolate island, where data between organizations cannot be shared. The traditional method of conjoint analysis is that the cooperative organizations directly aggregate the data together for analysis, that is not allowable in data isolate island situation.

The generative models [19] can summarize the distribution information of the samples and is very suitable as an intermediary to complete the information exchange. Instead of sharing any data samples between nodes, information is exchanged by sharing model parameters, which is ideal for data isolated island scenarios. DCIGAN [7] use GAN as an intermediary to change information between multi nodes in the distributed class-imbalance learning situation. However, if DCIGAN is applied to the problem of class-imbalance, there will be privacy protection issues. If the number of minority samples is too small, after a lot of GAN training, the pseudo samples generated by the generator will be too close to the real samples, thus exposing the privacy of the true samples.

So far, we have found that to solve the joint imbalance training problem in the data island scenario, and we need to find an intermediary to exchange information. This intermediary should be able to contain the division information of the minority class of all nodes instead of a single node. To this end, we proposed a distributed class-imbalance oversampling method based on Generative Adversarial Networks: DOSGAN. It could be used to solve the distributed class-imbalance oversampling problem in the data isolated island scenario, especially when different nodes carry complementary classification information of minority class. DOSGAN consists of two parts: a rotationally GAN training method (RGAN) and an oversampling method based on GAN generator (GANOS). Especially, we provide the following contributions in this paper:

1. We find a way that comprehensively uses the distribution information of all nodes to oversample a minority classes data in class-imbalance isolated island scenario.
2. We propose a rotationally GAN training method RGAN, which comprehensively uses the information on each node to train a global GAN without exchanging data.
3. We propose a method to comprehensively use the information of multiple nodes to do filtering work, without exchanging data

We evaluate the proposed model on 6 multi-class benchmark datasets of multiple different application areas tasks, including image classification, network attack classification, and human activity recognition, etc. Extensive empirical results show that: (i) DOSGAN significantly outperforms the other 4 distributed class-imbalance oversample methods modified from state-of-the-art learning methods on 6 real-world data sets under different minority classes and node amount settings; (ii) DOSGAN gets performance most closed to the method that aggregates all data.

The rest of the paper proceeds as follows. In Sect. 2, we discuss related work. In Sect. 3, we present the model of DOSGAN and introduce the key techniques. In Sect. 4, we present the experiment environment and analyze the experimental results. In Sect. 5, we conclude the paper.

2 Related Works

To the best of our knowledge, no work has been proposed to solve the class-imbalance problem in the data isolated islands scenario. To this end, we review inspiring related methods for our method from three aspects: data isolated islands class-imbalance learning, distributed learning, and generative adversarial networks.

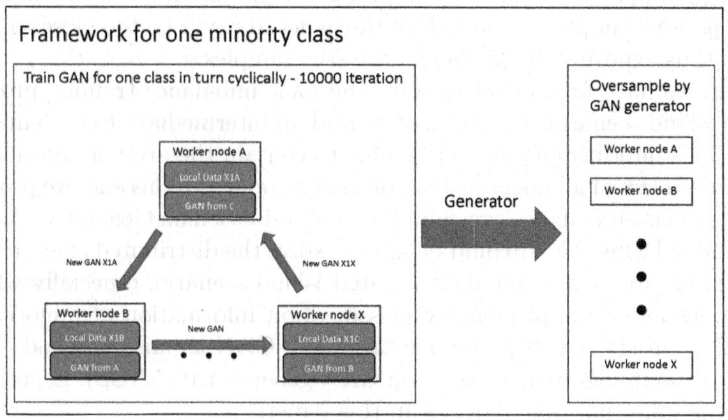

Fig. 2. DOSGAN method framework

2.1 Class-Imbalance Learning

In general, there are two types of strategies for solving class-imbalances. One is the data level method, and the other is the algorithm level method. The data level method is to use oversampling or undersampling to make the originally imbalanced sample set balanced. Algorithm-level methods usually use some additional weights to modify the training process, so that the model tends to classify the minority class correctly.

Data-Level Methods. SMOTE (Synthetic Minority Oversampling Technique) [2] performs oversampling by synthesizing minority samples between the nearest neighbors to avoid the over-fitting problem of random oversampling. Border-line SMOTE [9] is an improvement to the SMOTE algorithm. It enhances the value of synthetic samples by preferentially selecting samples near the classification boundary as base samples. Similar work includes SVM SMOTE [22]. ADASYN [10] improves the learning of data distribution in the following two ways: (1) reduces the deviation caused by the imbalance of categories; (2) adaptively transfers the classification decision boundary to difficult examples.

Algorithm-Level Methods. Some strategies choose an algorithm that is relatively insensitive to data-imbalance. Such as the tree model [14], Ensemble [10], and so on. Cost-sensitive learning [3,4] gives different classes different cost. When the classifier is trained, it will make a larger penalty for the samples with fewer misclassifications, forcing the final classifier to have a higher recognition rate for the positive samples.

However, all methods above need aggregating all instances of new classes to train the model, which is not suitable for the scenario of data isolated island.

2.2 Data Isolated Islands Distributed Learning

Federation Learning. Federated Learning [13,16] is a machine learning mechanism in a data island scenario, through which data exchange can be avoided to perform joint training problems. However, the existing work without federal learning can solve the problem of class-imbalance oversampling. This article is a solution for class-imbalance oversampling in the context of the federal learning classroom method.

Secure Multi-party Computation (SMC). This is a type of distributed security computing method that uses encryption mechanisms. They use encryption operator [23] to ensure that multiple parties calculate and protect privacy securely. There is currently work to encrypt existing machine learning models using SMC to protect privacy, such as support vector machines [5], K-means [12], Bayesian network [26]. However, due to the lack of flexibility of this method, such work has not been applied to the problem of class-imbalance learning.

Ensemble-Based Methods. Ensemble [20] is an intuitive joint learning method. The ensemble obtained by combining the classifiers trained by multiple nodes through weighting and other operations is often better than the classifier obtained by a single node in the test. However, in the class-imbalance scenario, if the classification performance of the classifier on each node is poor (due to the class-imbalance on each node), the ensemble is also difficult to obtain better results.

2.3 Generative Adversarial Networks

GAN [6] is the current hotspot model for deep learning because it can generate realistic data in application fields such as face generation. Many studies are constantly pursuing higher GAN training quality to generate more realistic data. In Least Square GAN [15], the author uses least squares to replace the cross-entropy loss in the original GAN as a loss function. In Wasserstein GAN (WGAN) [1], the author uses the bulldozer (EM) distance as a loss function to optimize, thus avoiding the problem of mode collapse of the original GAN. WGAN-GP [8] is a further improvement in WGAN. It adds gradient penalty to the loss function and obtains higher quality generated data than WGAN.

In summary, none of the above work can deal with class-imbalance learning problems without data sharing. To the best of our knowledge, we are the first to try to solve such a problem.

3 Distributed Class-Imbalance Generative Adversarial Networks

In this chapter, we first introduced the basic principles of DOSGAN. After that, we described the basic principles of two functional modules of DOSGAN: Rotational training GAN and generator oversampling method.

3.1 Overview of DOSGAN

The basic idea of DOSGAN is to use GAN generators to maintain the information of historical data.

DOSGAN consists of two steps. The first step is to train GAN in turn. For each minority class, all the distributed nodes which have samples of this class cyclically train the GAN model of this class. Until the global GAN model of this class is obtained. In the second step, the GAN generator is used to perform oversampling. The frame is shown in Fig. 2. This method is applied to a distributed environment that includes multiple nodes. All the nodes cyclically execute multiple GAN training algorithms for a minority class, and finally get a global GAN model through a large number of iterations, which contains Sample information of this type on all nodes in the entire system. We then use these trained GAN generators to oversample the minority classes of each node. Finally, the data of each node is balanced, thereby improving the classification accuracy of the system.

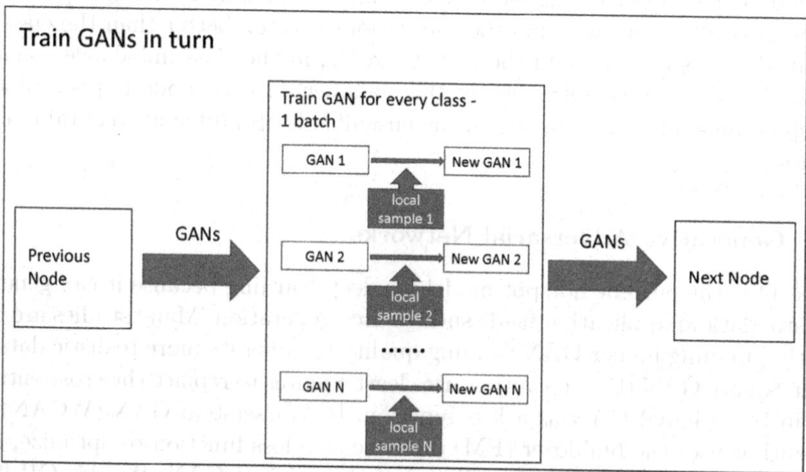

Fig. 3. GAN training on 1 node

3.2 Recurrent GAN (RGAN)

In this section, we discuss how to train GANs in turn. We know that training GAN requires a large number of iterations. However, if training on a node multiple times, then GAN will overfit the samples on this node. Therefore, if we stay too many times on a single node during the round-robin training, the final GAN model only tends to store the sample information of the last node, which is not the result we want. What we want is that the final GAN can contain the sample information on all nodes, which requires a strategy to take turns training, which is to minimize the number of training on a single node in a loop. We have found through experiments that the smaller the number of batches trained on a single node, the better the effect, that is, the final GAN generated samples are more representative of real samples. As shown in Table 1, we tried to use 1 batch, 10 batches, and 100 batches as the number of training cycles in a loop on a single node, and the 1 batch method works best. Oversampling with such a comprehensive GAN generator works better.

The parameter passing rules between nodes are like this. Before the training starts, we number all the nodes: 0, 1, 2... n. When numbering, you can consider the communication cost between nodes. Here we do not discuss the numbering rules but focus on the logic of the algorithm. The trigger condition for $Node_i$ to perform GAN training is to receive the new GAN parameters and corresponding labels passed from the previous $Node_{i-1}$. $Node_i$ uses this parameter to initialize the GAN model after receiving it and then uses the local samples of the corresponding label to train the GAN. After training 1 batch, the node gets a new GAN model, and then passes the GAN parameters and corresponding labels to the next $Node_{i+1}$. If i is the last node, pass the parameter to $Node_0$. The process is shown in Fig. 3.

We take three nodes as an example to illustrate how to perform iterative rotation training for one minority class. As shown in Fig. 2, first select a node A randomly, initialize a set of GAN parameters randomly on the node, use the sample of the node to train the GAN, the training process only executes 1 batch. Then the GAN parameters are saved locally and passed to the next Node B. Node B receives the GAN parameter as a trigger immediately uses the received value as the GAN initialization parameter, and then uses the local sample to update the GAN model. Node B also stops by only executing 1 batch, then saves the parameters, and passes the GAN parameters to its next node C. When node C receives the GAN parameter, C will perform the same process as B, and then save the GAN parameter and pass it to A. At this point, a loop is completed.

3.3 Over-Sampling with GAN Generator (GANOS)

Generating Process. After updating the GAN, each node will save a copy of the latest generator parameters locally, and it will be updated synchronously when the next gan with the same label is passed. When we need to do data analysis, such as classification, we can use the GAN generator corresponding to the minority class to generate the corresponding pseudo samples. The number of pseudo-samples generated corresponds to the gap between the majority and minority samples. After oversampling, the sample size of the minority and the majority will be close to balance, which improves the performance of data analysis.

Such oversampling takes full account of global sample information, not just local samples. The information is more comprehensive. The following experiment shows that it is better than using the local oversampling method.

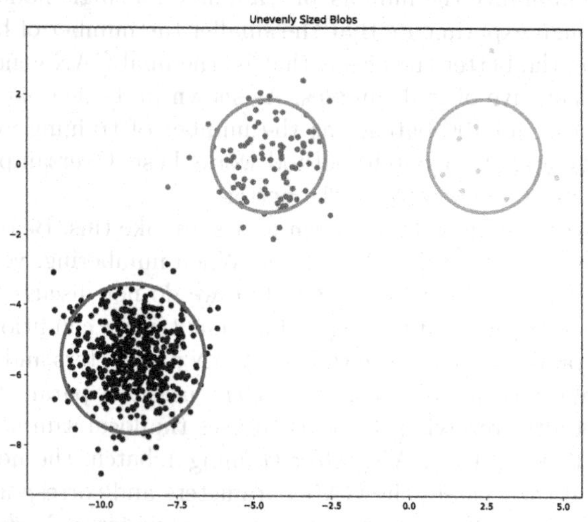

Fig. 4. Mode redundancy filtered

Filter Process. During the experiment, we find that WGAN-GP can save the distribution of training data better than other common GAN models. However, As shown in Fig. 4, the distribution of generated data is often a little more divergent than the distribution of the original data. We call this phenomenon mode redundancy [7]. To eliminate mode redundancy, we used an anomaly detection algorithm to filter the generated pseudo samples and discard the 5% pseudo sample with the highest anomaly score so that the distribution of the remaining data is as close as possible to the distribution of the global minority samples. The filter process is shown in Fig. 5. It is worth noting that in the environment of this article, the number of samples of the local minority class is very small, and it is not suitable to directly use these samples for training anomaly detection models. To this end, we use the generated pseudo samples to train the anomaly detection model, and then use it to filter 5% pseudo samples. The anomaly detection model is One-Class SVM [21]. This is different from the filter in [7], its filter is directly made using local samples. The complete process of generating and filtering data is shown in Fig. 5.

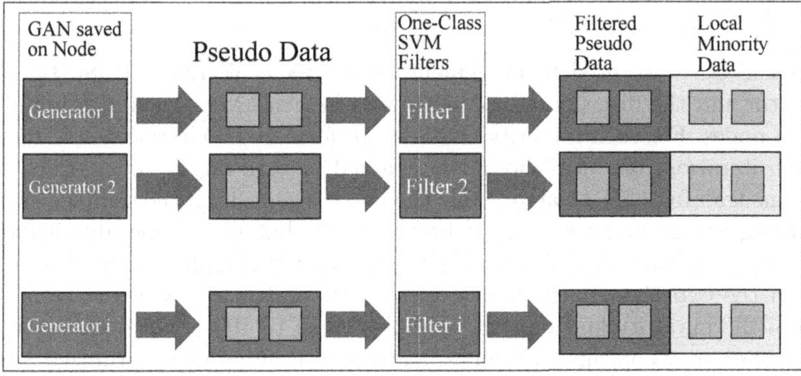

Fig. 5. Generating and filtering process on 1 node

4 Experimental Evaluation

The basic GAN we used to implement DOSGAN is WGAN-GP [8]. Both the generator and the discriminator contain two hidden layers, each containing 128 neuron nodes. Moreover, we also proposed 4 comparison methods to show the effect of the proposed method. During the experiment, both the numbers and labels of known classes and new classes are random. For each iteration, the optimization method is performed 500 times.

4.1 Datasets

We conduct experiments on 6 multi-class benchmark datasets of multiple different application areas, such as images, network security, etc. They are Mnist[1], Kdd99_6c[2], Sensorless drive diagnosis (SDD)[3], Statlog (Landsat Satellite) Data Set(Satellite)[4], Human Activity Recognition Using Smartphones Data Set (HAR)[5] and STL-10 dataset (STL10)[6]. These data sets were originally balanced. In order to facilitate testing, we randomly specify the majority and minority classes during the training process. Then let the ratio of the majority class and the minority class be set to 100:1, that is, the sample of the minority class uses 1% of them randomly. When testing, we directly used a balanced data set to test, so we can directly use the accuracy rate as the evaluation standard. The summary of these datasets is shown in Table 1.

---

[1] http://yann.lecun.com/exdb/mnist.

[2] https://archive.ics.uci.edu/ml/datasets/kdd+cup+1999+data.

[3] https://archive.ics.uci.edu/ml/datasets/dataset+for+sensorless+drive+diagnosis.

[4] https://archive.ics.uci.edu/ml/datasets/Statlog+(Landsat+Satellite).

[5] https://archive.ics.uci.edu/ml/datasets/human+activity+recognition+using+smartphones.

[6] https://cs.stanford.edu/~acoates/stl10/.

4.2 Programming Settings

The basic GAN we used to implement DOSGAN is WGAN-GP [8]. Both the generator and the discriminator contain two hidden layers, each containing 128 neuron nodes. For each minority class, we perform 10,000 iterations of training. In each iteration, one GAN would be trained in turn on all nodes. Each node performs training only 1 batch, while the batch-size is 32. If the samples amount of minority class in some node is less than the batch-size, the algorithm will randomly copy some samples so that the number of samples reaches the batch-size. In the experiment, the labels of majority and minority are random. The ratio of the majority and minority samples is set to 100:1. For each batch, the GAN generator is optimized 3 times, and the discriminator is optimized 1 time.

Table 1. Dataset information

Name	Dimension	Type	Original train amount	Test amount	Class amount
Mnist	784	Numeric	60000	10000	10
Kdd99_6c	41	Hybrid	42000	18000	6
SDD	49	Numeric	40956	17552	11
Satellite	36	Numeric	4435	2000	6
HAR	561	Numeric	7352	2947	6
STL10	27648	Numeric	10400	2600	10

In this article, we used two performance measures to compare various methods. One is overall accuracy (OvACC). That is, the classification accuracy of all data in the entire test set. The other is the average accuracy (AvACC). AvACC is a special evaluation standard for multi-class imbalance problems. It means the average accuracy of each class. We assign the same weight to each class. After obtaining the accuracy of each category separately, we obtain the final AvACC result by averaging:

$$AvACC = \frac{\sum_{i=1}^{N} ACC_i}{N} \tag{1}$$

In each combination, we repeated the experiment 10 times and reported the mean and standard deviation of the performance measures of the classifier trained after oversampling.

4.3 Comparison Methods

We compared our method with 4 comparison methods, which are intuitive and effective oversampling methods.

Table 2. The average OvACC and standard deviation of the five methods over multiple data sets on a distributed imbalance environment with 2, 4, 8 minority classes.

Minority classes amount	Dataset	NC	SMOTE	ENSEMBLE	DOSGAN	DA
Two minority classes	Mnist	.4391 ± .0137	.4418 ± .0198	.5695 ± .0823	**.7613 ± .0284**	**.7573 ± .0212**
	Kdd99_6c	.5573 ± .0443	.5507 ± .0301	.6428 ± .0248	**.9514 ± .0044**	**.9498 ± .0064**
	SDD	.5525 ± .0353	.5520 ± .0231	.5847 ± .0054	**.7646 ± .0238**	**.6421 ± .0550**
	Satellite	.6025 ± .0054	.6060 ± .0051	.6821 ± .0452	**.7696 ± .0125**	**.7705 ± .0100**
	HAR	.5495 ± .0115	.5539 ± .0118	.6990 ± .0295	**.8535 ± .0224**	**.8909 ± .0216**
	STL10	.6456 ± .0080	.6484 ± .0108	.5023 ± .0050	**.8950 ± .0132**	**.9251 ± .0034**
Four minority classes	Mnist	.6028 ± .0145	.6010 ± .0144	.5881 ± .0395	**.7872 ± .0142**	**.7599 ± .0255**
	Kdd99_6c	.8393 ± .0315	.8457 ± .0417	.8760 ± .0837	**.9559 ± .0022**	**.9586 ± .0059**
	SDD	.5882 ± .0101	**5894 ± .0102**	.5725 ± .0461	5889 ± .0623	**.6066 ± .0250**
	Satellite	.7745 ± .0055	.7747 ± .0064	.7912 ± .0226	**.7852 ± .0103**	**.7822 ± .0131**
	HAR	.5661 ± .0186	.5722 ± .0312	.5880 ± .0583	**.8234 ± .0242**	**.7576 ± .0195**
	STL10	.6029 ± .0064	.6040 ± .0084	.6941 ± .0073	**.8888 ± .0169**	**.9106 ± .0039**
Eight minority classes	Mnist	.7688 ± .0053	.7683 ± .0071	.8203 ± .0112	**.8254 ± .0095**	**.8293 ± .0136**
	SDD	.6830 ± .0451	.6778 ± .0417	.6577 ± .0895	**.7070 ± .0425**	**.7187 ± .0556**
	STL10	.8188 ± .0033	.8155 ± .0052	.8885 ± .0056	**.9006 ± .0052**	**.9190 ± .0041**

Native Copy (NC). On each node, we directly increase the samples of the minority class by copying them until the number of samples of the minority class reaches the same as the number of samples of the majority class.

Synthetic Minority Over-Sampling on Every Node (SMOTE). This is a variant of the oversampling method SMOTE [2]. SMOTE synthesizes new minority samples. The synthesis strategy is to randomly select a sample Y from its nearest neighbor for each minority sample X, and then randomly select a point on the line between X and Y as the newly synthesized minority samples.

SMOTE and Ensemble (ENSEMBLE). This strategy is a variant of SMOTE in distributed environment. In order to be able to fully consider the common information on multiple nodes. We classify the classifiers trained by the single node smote method as an ensemble and use them as global classifiers.

Data Aggregation (DA). This method directly aggregates the data distributed on different nodes together to train the global classifier, which is not in line with the needs of data isolated islands. We only use DA here as the goal of our design algorithm. Therefore, we let the algorithm behave as close as possible to the performance of DA.

4.4 Results

We tested three situations: the number of minorities is 2, 4, and 8, respectively. In each case, the labels of the multi-class and the small class, the number of nodes, and the node where the minority class appears are all randomly combined.

Overall Accuracy (OvACC). Table 2 shows a comparison of the average and standard deviation of the Overall Accuracy of all methods in a variety of sequential and quantitative combinations for all data set. The experimental results show that DOSGAN performs better in most minority class amount and minority appearance on all node combinations. DOSGAN is not the best in only one scene. In that scene, DOSGAN's performance is only slightly behind SMOTE, and very close to the method of Data Aggregation.

Average Accuracy (AvACC). Table 3 shows a comparison of the average and standard deviation of the Average Accuracy of all methods in a variety of sequential and quantitative combinations for all data set. The experimental results show that DOSGAN performs better with the average accuracy of all minority class amount and minority appearance on all node combinations. Among all methods, the accuracy of DOSGAN is closest to that of the Data Aggregation method.

Table 3. The average AvACC and standard deviation of the five methods over multiple data sets on a distributed imbalance environment with 2, 4, 8 minority classes.

Minority classes amount	Dataset	NC	SMOTE	ENSEMBLE	DOSGAN	DA
Two minority classes	Mnist	.5802 ± .0100	.5821 ± .0148	.5975 ± .0133	**.8418 ± .0154**	**.8386 ± .0166**
	Kdd99_6c	.6498 ± .0298	.6455 ± .0202	.7573 ± .0163	**.9636 ± .0032**	**.9623 ± .0047**
	SDD	.6628 ± .0262	.6623 ± .0172	.6857 ± .0041	**.7719 ± .0321**	**.7607 ± .0340**
	Satellite	.6750 ± .0075	.6798 ± .0076	.6821 ± .0452	**.7589 ± .0096**	**.7593 ± .0081**
	HAR	.6090 ± .0120	.6135 ± .0122	.7501 ± .0218	**.8760 ± .0194**	**.9086 ± .0172**
	STL10	.7374 ± .0083	.7201 ± .0175	.8208 ± .0083	**.9200 ± .0077**	**.9284 ± .0032**
Four minority classes	Mnist	.6433 ± .0131	.6417 ± .0129	.6811 ± .0238	**.8078 ± .0132**	**.7847 ± .0212**
	Kdd99_6c	.9027 ± .0186	.9065 ± .0247	.9287 ± .0199	**.9700 ± .0013**	**.9707 ± .0053**
	SDD	.5917 ± .0101	.5908 ± .0073	.5880 ± .0066	**.6002 ± .0313**	**.5988 ± .0320**
	Satellite	.7809 ± .0201	.7830 ± .0121	.7912 ± .0226	**.7970 ± .0106**	**.7930 ± .0135**
	HAR	.6631 ± .0116	.6709 ± .0265	.7233 ± .0189	**.7856 ± .0285**	**.7555 ± .0232**
	STL10	.6380 ± .0059	.6424 ± .0060	.7739 ± .0063	**.9036 ± .0148**	**.9241 ± .0030**
Eight minority classes	Mnist	.8013 ± .0059	.8015 ± .0081	.7795 ± .0131	**.8307 ± .0091**	**.8254 ± .0140**
	SDD	.6890 ± .0381	.6880 ± .0355	.6821 ± .0653	**.7172 ± .0308**	**.7310 ± .0283**
	STL10	.8570 ± .0027	.8514 ± .0028	.8670 ± .0114	**.8988 ± .0055**	**.9182 ± .0041**

5 Conclusion

In this paper, we propose DOSGAN: a distributed class-imbalance oversampling method based on Generative Adversarial Networks. DOSGAN enables multiple nodes to collectively train one GAN for the minority class without sharing data. The main idea of DOSGAN is to use global GAN generators to keep The probability distribution of minority classes. It exchanges information by passing generator parameters between nodes to replace the direct exchange of data. Through this mechanism to train GANs in turn, the global GAN contains the subclass information of all the classes on all nodes. Finally, DOSGAN uses GAN generators to generate pseudo samples for oversampling. Extensive experiments validate that DOSGAN can successfully address distributed class-imbalance oversampling without data sharing.

Acknowledgment. The authors would like to thank the anonymous reviewers for their valuable comments.

References

1. Arjovsky, M., Chintala, S., Bottou, L.: Wasserstein GAN. arXiv preprint arXiv:1701.07875 (2017)
2. Chawla, N.V., Bowyer, K.W., Hall, L.O., Kegelmeyer, W.P.: SMOTE: synthetic minority over-sampling technique. J. Artif. Intell. Res. **16**, 321–357 (2002). https://doi.org/10.1613/jair.953
3. Domingos, P.: MetaCost: a general method for making classifiers cost-sensitive. In: Proceedings of the Fifth ACM SIGKDD International Conference on Knowledge Discovery and Data Mining, pp. 155–164 (1999)
4. Fan, W., Stolfo, S.J., Zhang, J., Chan, P.K.: AdaCost: misclassification cost-sensitive boosting. In: ICML, vol. 99, pp. 97–105 (1999)
5. González-Serrano, F.J., Navia-Vázquez, Á., Amor-Martín, A.: Training support vector machines with privacy-protected data. Pattern Recogn. **72**, 93–107 (2017). https://doi.org/10.1016/j.patcog.2017.06.016
6. Goodfellow, I., et al.: Generative adversarial nets. In: Advances in Neural Information Processing Systems, pp. 2672–2680 (2014)
7. Guan, H., Wang, Y., Ma, X., Li, Y.: DCIGAN: a distributed class-incremental learning method based on generative adversarial networks. In: 2019 IEEE International Conference on Parallel Distributed Processing with Applications, Big Data Cloud Computing, Sustainable Computing Communications, Social Computing Networking (ISPA/BDCloud/SocialCom/SustainCom), pp. 768–775 (2019)
8. Gulrajani, I., Ahmed, F., Arjovsky, M., Dumoulin, V., Courville, A.C.: Improved training of Wasserstein GANs. In: Advances in Neural Information Processing Systems, vol. 3, pp. 5767–5777 (2017)
9. Han, H., Wang, W.-Y., Mao, B.-H.: Borderline-SMOTE: a new over-sampling method in imbalanced data sets learning. In: Huang, D.-S., Zhang, X.-P., Huang, G.-B. (eds.) ICIC 2005. LNCS, vol. 3644, pp. 878–887. Springer, Heidelberg (2005). https://doi.org/10.1007/11538059_91

10. He, H., Bai, Y., Garcia, E.A., Li, S.: ADASYN: adaptive synthetic sampling approach for imbalanced learning. In: 2008 IEEE International Joint Conference on Neural Networks (IEEE World Congress on Computational Intelligence), pp. 1322–1328. IEEE (2008)
11. He, H., Garcia, E.A.: Learning from imbalanced data. IEEE Trans. Knowl. Data Eng. **21**(9), 1263–1284 (2009)
12. Jagannathan, G., Wright, R.N.: Privacy-preserving distributed k-means clustering over arbitrarily partitioned data. In: Proceedings of the Eleventh ACM SIGKDD International Conference on Knowledge Discovery in Data Mining, pp. 593–599. ACM (2005)
13. Konecný, J., McMahan, H.B., Yu, F.X., Richtárik, P., Suresh, A.T., Bacon, D.: Federated learning: strategies for improving communication efficiency. CoRR abs/1610.05492 (2016). http://arxiv.org/abs/1610.05492
14. Liu, X.Y., Wu, J., Zhou, Z.H.: Exploratory undersampling for class-imbalance learning. IEEE Trans. Syst. Man Cybern. B (Cybernetics) **39**(2), 539–550 (2008)
15. Mao, X., Li, Q., Xie, H., Lau, R.Y., Wang, Z., Paul Smolley, S.: Least squares generative adversarial networks. In: Proceedings of the IEEE International Conference on Computer Vision, pp. 2794–2802 (2017)
16. McMahan, H.B., Moore, E., Ramage, D., Arcas, B.A.: Federated learning of deep networks using model averaging. CoRR abs/1602.05629 (2016). http://arxiv.org/abs/1602.05629
17. Ming, Y., Zhao, Y., Wu, C., Li, K., Yin, J.: Distributed and asynchronous stochastic gradient descent with variance reduction. Neurocomputing **281**, 27–36 (2018)
18. Ming, Y., Zhu, E., Wang, M., Ye, Y., Liu, X., Yin, J.: DMP-ELMs: data and model parallel extreme learning machines for large-scale learning tasks. Neurocomputing **320**, 85–97 (2018)
19. Ng, A.Y., Jordan, M.I.: On discriminative vs. generative classifiers: a comparison of logistic regression and naive Bayes. In: Advances in Neural Information Processing Systems, pp. 841–848 (2002)
20. Rokach, L.: Ensemble-based classifiers. Artif. Intell. Rev. **33**(1–2), 1–39 (2009). https://doi.org/10.1007/s10462-009-9124-7
21. Schölkopf, B., Williamson, R., Smola, A., Shawe-Taylor, J., Platt, J.: Support vector method for novelty detection. In: Proceedings of the 12th International Conference on Neural Information Processing Systems, NIPS 1999, pp. 582–588. MIT Press, Cambridge (1999)
22. Tang, Y., Zhang, Y.Q., Chawla, N.V., Krasser, S.: SVMs modeling for highly imbalanced classification. IEEE Trans. Syst. Man Cybern. B (Cybern.) **39**(1), 281–288 (2008)
23. Teo, S.G., Cao, J., Lee, V.C.: DAG: a general model for privacy-preserving data mining. IEEE Trans. Knowl. Data Eng. 1–1 (2018). https://doi.org/10.1109/tkde.2018.2880743
24. Wang, Y., Ma, X.: A general scalable and elastic content-based publish/subscribe service. IEEE Trans. Parallel Distrib. Syst. **26**(8), 2100–2113 (2015). https://doi.org/10.1109/TPDS.2014.2346759
25. Wang, Y., Pei, X., Ma, X., Xu, F.: TA-Update: an adaptive update scheme with tree-structured transmission in erasure-coded storage systems. IEEE Trans. Parallel Distrib. Syst. **29**(8), 1893–1906 (2018). https://doi.org/10.1109/TPDS.2017.2717981
26. Yang, Z., Wright, R.: Privacy-preserving computation of Bayesian networks on vertically partitioned data. IEEE Trans. Knowl. Data Eng. **18**(9), 1253–1264 (2006). https://doi.org/10.1109/tkde.2006.147

Effective Sentiment Analysis for Multimodal Review Data on the Web

Peiquan Jin[1(✉)], Jianchuan Li[1], Lin Mu[1], Jingren Zhou[1], and Jie Zhao[2(✉)]

[1] University of Science and Technology of China, Hefei 230027, Anhui, China
jpq@ustc.edu.cn

[2] School of Business, Anhui University, Hefei 230601, Anhui, China
zhaojie@ahu.edu.cn

Abstract. Multimodal review data on the web mainly involve videos, audios, and texts, which have been the major form of review data on E-business, social networks, etc. Extracting the sentimental opinions of the multimodal review data on the web is helpful for many web-based services, such as online customer services, product recommendation, and personalized web search. However, the multiple modalities of the multimodal review data on the web introduces new challenges for sentiment analysis. The key issue is how to make fusion of multimodal data to improve the effectiveness of sentiment analysis. In this paper, we present a novel two-staged self-attention-based approach for multimodal sentiment analysis. At the first stage, we perform an inter-utterance learning by using a self-attention neural network. At the second stage, we conduct a cross-model fusion to integrate information from different modalities. To the best of our knowledge, this is the first work that utilizes the self-attention network to achieve inter-utterance sentiment analysis for multimodal data. We conduct experiments on two public datasets and compare our proposal with several state-of-the-art methods. The results suggest the effectiveness and robustness of our proposal.

Keywords: Multimodal data · Reviews · Sentiment analysis · Neural network

1 Introduction

Nowadays, multimodal data has been a major form of data on the web. For example, a typical web news page may include videos, audios, and texts; a blog or a tweet may also include short videos and texts [1]. The term "multimodal" means that multiple data representations are used in the data, and in this paper, we mainly focus on three modalities including video, audio, and text, which are typical data representations on current web-based applications.

The sentiment analysis of web data has been a widely-studied topic [2]. Sentiment analysis aims to detect the sentimental polarity, e.g., positive or negative, which is represented by the underlying data. Sentiment analysis is helpful for many web-based applications such as online reviews analysis, product recommendation, and personalized search. For example, in E-commerce platforms, vendors can know the public opinion toward a

M. Qiu (Ed.): ICA3PP 2020, LNCS 12454, pp. 623–638, 2020.
https://doi.org/10.1007/978-3-030-60248-2_43

new product by analyzing the sentimental polarity of online reviews, which can be taken as the basis of market-promoting decision.

However, traditional sentiment analysis has several limitations. Early works were only towards one kind of data, e.g., textual data or video data. This is not enough for the sentiment analysis of multimodal data on the web. Recently, there were some studies focusing on multimodal sentiment analysis [3–6], but most of them only performed cross-model interactions and neglected the inter-utterance interactions. Here, an utterance is a segment of data separated from another utterance by obvious pauses or sudden silence. For video data, an utterance is a video clip, while for audio data, an utterance is a voice segment. For text data, an utterance is a paragraph or a series of words. Generally, different utterances may include different sentimental polarities. Thus, it is necessary to detect the polarity of each utterance and conduct an inter-utterance interaction to detect the sentimental polarity of data.

In this paper, aiming to improve the performance of the sentiment analysis for multimodal data on the web, we propose a novel framework that is based on self-attention neural networks. The basic idea of our approach is a two-staged fusion framework, including an inter-utterance fusion and a cross-model fusion:

(1). *Inter-Utterance Learning.* We first measure the inter-utterance interactions of each modality by using a self-attention network [3, 4], which differs from previous works that did not consider the inter-utterance interactions. Generally, sequential utterances within a video may contain contextual correlations with each other. On the other hand, an utterance may also relate to another utterance that is not close to it. As a result, in this study, we propose to utilize the inter-utterance learning to improve the performance of sentiment analysis. The self-attention network originated from machine translation, which aimed to translate a sequence of words into another language by modeling the contextual information of words. Motivated by this, we utilize the self-attention network to model the inter-utterance interactions in this study. Our experimental results will show that this idea can improve the effectiveness of sentiment analysis.

(2). *Cross-Model Fusion.* Then, we perform a cross-model interaction to fuse the classification results from different modalities to get the final result of multimodal sentiment analysis. This is mainly because that one modality of data can only contain part of sentimental information. By using the inter-modality fusion, we can augment the sentimental information, yielding a higher effectiveness of sentiment analysis. Figure 1 shows an example of cross-model interactions. If a speaker says something which is hard to tell positive or negative, we can check his/her facial expressions (video) or the texts he/she posted to know the exact sentiment. As each modality can reveal different kinds of sentimental information, the cross-model interaction is expected to augment sentiment analysis by a data fusion. However, how to realize cross-model fusion is a challenging issue. An intuitive solution is to concatenate multimodal vectors and input them to some classification model, but overfitting will happen due to the influence of noisy features. In our work, we propose to calculate the attention matrix of three modalities to make the cross-model fusion.

Briefly, the differences of this study from existing works and the unique contributions of the paper can be summarized as follows:

(1). We propose a two-staged fusion framework to extract sentimental polarity of multimodal data on the web. It consists of an inter-utterance learning stage and a cross-model fusion stage. With this design, we can capture the sentimental information inside the utterance series of each modality as well as that across different modalities.

(2). We present a self-attention network to perform inter-utterance learning. The self-attention network is employed to extract the contextual information among different utterances. Compared with previous models, our method can extract contextual information between two utterances that are far from each other. To the best of our knowledge, this work is also the first one that utilizes self-attention networks in multimodal sentiment analysis.

(3). We conduct experiments on two real multimodal datasets and compare our work with three state-of-the-art methods. The results show that our proposal achieves higher accuracy of sentiment analysis than all the competitors.

2 Related Work

Sentiment analysis is an important research area in natural language processing (NLP). In recent years, researchers have begun to integrate verbal and nonverbal behavior into sentiment analysis. Basically, there are three kinds of models that are related to multimodal sentiment analysis.

Concatenation of Multimodal Vectors. The first category of models takes the concatenation of multimodal vectors as a single modal to perform multimodal sentiment analysis. This concatenated modal will be used as input to a learning model, such as Hidden Conditional Random Fields (HCRFs) [9], Support Vector Machines (SVMs) [8], Hidden Markov Models (HMMs) [10], and Convolutional Neural Network (CNN) [23]. In the last decade, with the boosting of deep learning algorithms, Recurrent Neural Networks (RNN), or its variants, i.e., Long-short Term Memory (LSTM) [11] have been widely used for sequence modeling. Chen et al. [27] proposed to use SVM, random forest, and logistic regression to implement late fusion of multimodal features. Majumder et al. [6] proposed a hierarchical fusion approach with context modeling that used fully-connected layers to combine unimodal vectors and used RNN to extract context-aware features. These concatenation methods have achieved success to some extent for modeling multimodal problems. However, this type of models will introduce the over-fitting problem, especially when a small number of training samples are used.

Learned Multimodal Fusion. The second category of models aims to learn a temporal representation for each of the different modalities, but they do not consider the time sequences of utterance. Conventional multimodal learning approaches, such as multiple kernel learning [12, 29], sub-space learning [28], subspace clustering [24], and multi-view learning [13] were used for multimodal representations. Other approaches trained individual models for each modality, and used decision voting [14, 25], tensor products [15], or deep neural networks to make a fusion. Poria et al. [5] proposed a context-dependent model (Simple-LSTM) based on LSTM to extract contextual utterance-level features for concatenated multimodal features. Sun et al. [26] proposed a fusion network that adopted the weighted product rule for fusing the audio and image recognition results in the decision layer. Those methods used the average features that were computed over time but ignored the temporal correlations among utterances.

Attention-Based Multimodal Fusion. The third category of models used the attention mechanism to find cross-model interactions over time and modality-specific interactions among utterances. Poria et al. [7] proposed a contextual attention-based LSTM network (CAT-LSTM) to model the contextual relationship among utterances and prioritized the important contextual information for classification. They also introduced an attention-based fusion mechanism to emphasize the informative modalities for multimodal fusion. Ghosal et al. [16] proposed a novel method that employed a recurrent neural network based multimodal attention framework for sentiment prediction. They computed pairwise-attentions on various combinations of three modalities, which were concatenated with individual modalities for final classification. Zadeh et al. [17] proposed a neural architecture called the Memory Fusion Network (MFN) that relied on the Delta-memory Attention Network (DMAN). This attention network implemented a coefficient assignment technique on the concatenation of LSTM memories, which assigned high coefficients to the dimensions that jointly formed a cross-modal interaction. Zadeh et al. [18] proposed a multi-attention recurrent network (MARN) for understanding human communication. They adopted a neural component called Multi-Attention Block (MAB) to discover the interactions between modalities through time. Rahman et al. [19] injected multimodal information within the input space of Bidirectional Encoder Representations from Transformers (BERT) for modeling multimodal languages. They employed an attention gating mechanism to create a shift vector by controlling the influence of acoustic and visual features with respect to the input embedding.

Differences of This Study from Existing Works. This study differs from the existing categories of multimodal sentiment analysis in several aspects. First, differing from all the three previous models, we propose to combine the inter-utterance learning with the cross-model fusion, resulting in a two-staged multimodal sentiment analysis framework. Second, differing from the learned multimodal fusion, our approach considers the temporal correlation among utterances, which can capture the sentimental influence between utterances that are distributed far in videos, audios, or texts. Third, differing from the attention-based models that use soft attention to calculate weight vectors for modalities, we propose to calculate the attention matrix of three modalities to make the cross-model fusion.

3 Two-Staged Multimodal Sentiment Analysis

3.1 Preliminary

In this section, we first present some preliminaries that are necessary to describe our algorithms. In this paper, we focus on three modalities, i.e., video, audio, and text. We assume that multimodal data are represented by videos with embedded audios, and texts can be embedded in videos or presented separately along videos. A typical example of such multimodal data is a YouTube video, in which video clips represent the video

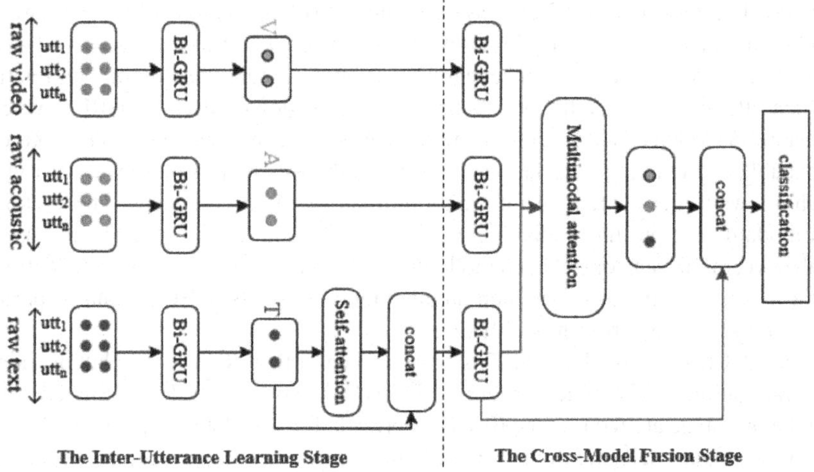

Fig. 1. The two-staged framework for multimodal sentiment analysis

modality and the audio information inside video clips are the audio modality. The caption and textual descriptions of the YouTube video can be regarded as the text modality.

Generally, a video can be defined as a series of utterances.

$$V_j = \left[u_{j,1}\, u_{j,2}\, u_{j,3}, \ldots, u_{j,i} \ldots u_{j,L_j}\right] \tag{1}$$

Here, the ith utterance in video v_j is noted as $u_{j,i}$ and the number of utterances in v_j is L_j. In each video, we obtain features of three modalities: textual, audio, visual noted as {t, a, v}, which will be used to label each utterance with the sentiment of the speaker. Our point of view is that not all modalities are equally useful in classifying, e.g., textual features can describe the words of the speaker accurately, while audio features with loud background noise can give poor information. We also assume that, all utterances in a video are related, in the similar way that all words in a sentence are related, the contextual information between utterance $u_{j,i}$ and the other utterances $[u_{j,k} | \forall k \leq L_j, k \neq i]$ can make a difference in the classification. Assume there are C classes of sentiment in a dataset and m videos in a dataset, the goal of multimodal sentiment analysis can be defined by (2).

$$y_{j,i}(t, a, v) \rightarrow c \in \{1, \ldots, C\}, s.t.1 \leq j \leq m, 1 \leq i \leq L_j \tag{2}$$

3.2 The Two-Staged Framework

We propose a two-staged framework to extract sentiment polarities from multimodal data on the web. Figure 1 shows the overall architecture, which includes an inter-utterance learning stage and a cross-model fusion stage. At the first stage, we handle raw unimodal features for context-sensitive feature representations, in which the self-attention network will be employed. At the second stage, we fuse the results from each modality by computing attention matrix and output the final classification result.

Stage 1: Inter-Utterance Learning. Existing works did not consider the temporal correlations between utterances. In this paper, we present a self-attention-network-based approach to perform inter-utterance learning, so as to capture the temporal correlations between utterances. Specially, we choose the Bi-directional GRU (Bi-GRU), which is similar to LSTM but costs less computation, as the self-attention network. The GRU cells are capable of modeling correlations between utterances that are far away in a video, which allows to get contextual information of surrounding utterances. The vectors of each modality are put into individual GRU networks and the dense layers are used to equalize vector dimensions of each modality for the second stage of cross-model fusion. All the vectors of utterances are then inputted into another Bi-GRU module to perform the second stage, i.e., cross-model fusion.

Stage 2. Cross-Model Fusion. This stage also has a Bi-GRU network but its training and computation is independent from the Bi-GRU at the first stage. The outputted vectors from the first stage are fed into the Bi-GRU network. For text data, we put textual vectors into an additional self-attention step and concatenate generated vectors with textual vectors as the input vectors for the Bi-GRU network. This process is to further enhance the informativeness of text vectors. However, we do not add the self-attention process to audios and videos, because the feature matrices of them are much sparse, which is not friendly to the self-attention process. Then, all the vectors from the three modalities are learned in the multimodal attention module, which generates a vector. As the text vector is more important in sentiment analysis, we concatenate the textual vector learned at the first stage with the generated attention vector for the final classification.

Algorithm 1 describes the overall training process of the two-staged multimodal sentiment analysis. The symbols involved in the algorithm are described in Table 1. The testing process is the same as the training process, except that we use a test dataset in the testing process.

3.3 Self-attention-Based Inter-utterance Learning

Aiming to model the inter-utterance correlations for sentiment analysis, in this paper, we present a self-attention-based approach. The attention mechanism can help deep neural networks focus on the most important parts of the input data, motivating by the way of human brains responding to audio sound and visual images. Attention has been proven useful in computer vision, NLP tasks, and other fields. For example, Vaswani et al. [3] proposed to replace RNNs with self-attention, yielding high performance in language modeling and machine translation. An attention function is defined in the attention mechanism to map the output with the query as well as a set of key-value pairs. Self-attention means computing a representation of a single sequence by correlating the utterances at different positions in the sequence. In this study, utterances can be regarded as a time-ordered single sequence, and self-attention allows to model the dependencies among the utterances within the sequence. Figure 2 shows an example of the self-attention in our study, which indicates that a target utterance (utterance3) is related to other utterances appearing in the same sequence and such correlations are reflected into the representation of the target utterance.

Table 1. Some Notations for Algorithm 1

Weight	Bias
$W_p, W_r, W, U_p, U_r, U \in R^{d \times k}$	$b_p, b_r, b \in R^d$
$W_z, W_f \in R^{m \times d}$	$b_z, b_f \in R^m$
$W_{sft} \in R^{c \times m}$	$b_{sft} \in R^c$

d = dimension of hidden units; k = dimensions of the input vectors;
m = number of utterances in a video; c = number of sentiment classes.

Algorithm 1. Training

Input: *text x, audio x', video x'' as the input data*
Output: *predictions of sentiment*
[The first stage]
Train GRU with X, X', X''
for $i : [1, M]$ **do**
$Z_i \leftarrow$ getGRUFeatures(X_i)
$Z_i' \leftarrow$ getGRUFeatures(X_i')
$Z_i'' \leftarrow$ getGRUFeatures(X_i'')
$A \leftarrow$ self-attention(Z)
for $i : [1, M]$ **do**
for $j : [1, L_i]$ **do**
$Z_{i,j} \leftarrow (Z_{i,j} || A_{i,j})$ *//concatenation*
[The second stage]
for $i : [1, M]$ **do**
$Z_i \leftarrow$ getGRUFeatures(Z_i)
$Z_i' \leftarrow$ getGRUFeatures(Z_i')
$Z_i'' \leftarrow$ getGRUFeatures(Z_i'')
$A^* \leftarrow$ multimodal-attention(Z, Z', Z'')
for $i : [1, M]$ **do**
for $j : [1, L_i]$ **do**
$Z_{i,j}^* \leftarrow (A_{i,j,0}^* || A_{i,j,1}^* || A_{i,j,2}^* || Z_{i,j})$ *//concatenation*
for $i : [1, M]$ **do**
for $j : [1, L_i]$ **do**
$f_{i,j} \leftarrow ReLU(W_f Z_{i,j}^* + b_f)$ *//dense layer*
$predcitions_{i,j} \leftarrow softmax(W_{sft} f_{i,j} + b_{sft})$
return *predictions*
End Training

Procedure getGRUFeatures(X_i)

$Z_i \leftarrow \emptyset$

for $t : [1, L_i]$ **do**

$p_t = \sigma(W_p \cdot x_t + U_p \cdot h_{t-1} + b_p)$

$r_t = \sigma(W_r \cdot x_t + U_r \cdot h_{t-1} + b_r)$

$\tilde{h}_t = tanh(W * x_t + U * h_{t-1} + b)$

$h_t = (1 - p_t) * h_{t-1} + p_t * \tilde{h}_t$ *//GRU output*

$z_t \leftarrow ReLU(W_z h_t + b_z)$ *//dense layer*

$Z_i \leftarrow Z_i \cup z_t$

return Z_i

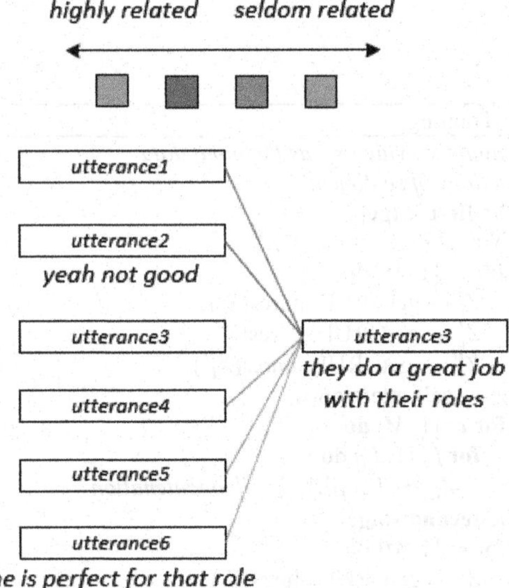

Fig. 2. An example of self-attention (the impact of other utterances to utterance3 is reflected in the representation of utterance3).

While three modalities are involved in this study, we only consider to use the self-attention mechanism for the text modality. This is mainly because the video and audio modalities are not suitable for the self-attention mechanism. For example, the facial features in videos are supposed to reflect speaker sentiment. However, a frown eyebrow may be regarded as sadness. Thus, if we correlate each facial emotion with all utterances in a video, additional noise will be introduced, which will worsen the effectiveness of sentiment analysis. On the other hand, previous work in language modeling has shown that the contextual information in a text sequence contributes a lot to the semantics of a specific text segment. Therefore, it is reasonable to introduce the impact of other text utterances through the self-attention mechanism.

Particularly, we have the textual feature vectors of d dimensions for each utterance, noted by X_t. To implement self-attention, we have three types of vectors, namely the vector of query, keys, and values, which are denoted as q_i, k_i, and v_i for the ith utterance.

These three vectors will be used to generate the output vector. We use a compatibility function of the query with the corresponding key to compute the weight assigned to each value, and accumulate the weighted sum of all values as the output, denoted as Z. Specially, we create a query vector, a key vector and a value vector for each utterance by multiplying the input vector x_i by three parameter matrices trained during the training process, denoted as W^Q, W^K, and W^V. The vectors Q, K, and V are with fewer dimensions (denoted as d_q, d_k, and d_v) than the input vector. Thus, if there are L utterances in a video, the sizes of matrices Q, K, and V are $L*d_q$, $L*d_k$, and $L*d_v$, respectively. And the sizes of matrices W^Q, W^K, and W^V are $d*d_q$, $d*d_k$, and $d*d_v$, respectively. Algorithm 2 describes the overall computation for the self-attention inter-utterance learning process, where we use scaled dot-product as the attention function.

Algorithm 2. Self-attention of inter-utterance learning

Input: *feature vectors X of N utterances*
Output: attention vectors

initialize weight matrices W_Q, W_K, W_V;
for i : [1, N] **do**
  $q_i = x_i \cdot W^Q$;
  $k_i = x_i \cdot W^K$;
  $v_i = x_i \cdot W^V$;
$M = \frac{Q \cdot K^T}{\sqrt{d_k}}$;
$W^*(i',j') = \frac{e^{M(i',j')}}{\sum_{k'=1}^{N} e^{M(i',k')}} for\ i',j' = 1, \ldots, N$;
for i : [1, N] **do**
  for j : [1, N] **do**
    $z_i = z_i + W^*(i,j) * v_i$
return $Z = [z_1, \ldots, z_N]$

3.4 Cross-model Fusion

In the multimodal sentiment analysis, we note that each modality does not contribute equally to the classification results. Inspired by this observation, we adopt the attention mechanism to conduct cross-model fusion so that we can leverage the multimodal information in sentiment analysis.

Note that the self-attention mechanism we discussed in the previous section may be a bad choice for cross-model fusion because vectors of different modalities are from different embedding spaces and cannot be regarded as a single sequence. Hence, we adopt the multimodal attention framework [16] to fuse cross-model information. In particular, for each utterance, we have a textual vector X_t, an audio vector X_a, and a video vector X_v. To make these vectors share the same dimension d, we send the raw features of the three modalities to the dense layers consisting of the same count of units. Let $X_{u_q} = [X_t, X_a, X_v] \in R^{3 \times d}$ be the information matrix for the qth utterance, we can calculate the attention matrix A_{u_q} with the following formulas.

$$M_{u_q} = X_{u_q} \cdot X_{u_q}^T \tag{3}$$

$$N_{u_q}(i,j) = \frac{e^{M_{u_q}(i,j)}}{\sum_{k=1}^{3} e^{M_{u_q}(i,k)}} \, for \; i,j = 1,2,3 \tag{4}$$

$$O_{u_q} = N_{u_q} \cdot X_{u_q} \tag{5}$$

$$A_{u_q} = O_{u_q} \odot X_{u_q} \tag{6}$$

Here, the symbol • stands for matrix multiplication, and $\odot$ stands for element wise multiplication. After this computation process, we concatenate the attention matrix A_{u_q} with the textual vectors X_t for the dense layers. The reason why we do not concatenate A_{u_q} with vectors of all the three modalities is that text contains significantly more important information than audio and video, which is already involved in A_{u_q}.

The concatenated vectors are then fed into a dense layer and finally into a SoftMax layer to calculate the distributed probabilities of sentiment labels. The sentiment label with the highest probability will be outputted as the sentiment polarity.

4 Performance Evaluation

We conduct experiments on a computer with an Intel core i5-4460 CPU @3.20 GHz, 16 GB DRAM, and a 512 GB SSD. All algorithms were implemented by Python 3.5. All the training of models was conducted on TensorFlow 1.14.

4.1 Settings

We evaluate all algorithms on two public web datasets, namely the MOSI dataset provided by Carnegie Mellon University and the IEMOCAP dataset provided by the University of Southern California.

MOSI. The first dataset, Multimodal Corpus of Sentiment Intensity (MOSI for short)[1], is a multimodal dataset containing a collection of 2199 opinion video clips from 93 YouTube movie-review videos, which can be regarded as a typical dataset of multimodal data on the web. Each opinion video is annotated with sentiment in the range between +3 (strong positive) to −3 (strong negative) annotated by 5 people. The dataset is rigorously annotated with textual labels for subjectivity, sentiment intensity, per-frame and per-opinion annotated visual features, and per-milliseconds annotated audio features. For comparison with state-of-the-art methods, we use the average of these five annotations as the sentiment polarity and classify sentiment into positive or negative. We use the first 62 videos as the train/validation set and the rest 31 videos as the test set. Consequently, the train set contains 1447 opinion video clips while the test set has 752 ones.

IEMOCAP. The second dataset, IEMOCAP[2], contains the acts of 10 speakers in a two-way dialogue in English. There are eight categorical labels in the dataset, namely

[1] http://multicomp.cs.cmu.edu/resources/cmu-mosi-dataset/.
[2] https://sail.usc.edu/iemocap/.

anger, happiness, sadness, neutral, excitement, frustration, fear, surprise, and other, but we only retain the first four types in the comparisons of experimental results. There are 120 videos in the training set, 31 videos in the test set. The train and test folds contain 4290 and 1208 utterances respectively.

4.2 Training

Feature extraction is not our contribution in this paper, we use utterances-level features provided in [5]. Textual features are extracted from a convolutional neural network [20] where each utterance is represented as the concatenation of vectors of the constituent words processed with the publicly available word2vec vectors. Audio features are extracted with an open-source software called openSMILE [21] at 30 Hz frame-rate and a sliding window of 100 ms. Visual features are obtained through 3D convolutional neural network [22]. In the MOSI dataset, dimensions of utterance-level features are 100, 100, and 73 for text, video, and audio, respectively. In the IEMOCAP dataset, the dimensions of utterance-level features are 100, 512, and 100 for text, video, and audio, respectively.

The training of the Bi-GRU network is performed using the categorical cross-entropy on each utterance's SoftMax output per video.

$$\text{loss} = -\frac{1}{\left(\sum_{1}^{M} L_i\right)} \sum_{i=1}^{M} \sum_{j=1}^{L_i} \sum_{c=1}^{C} y_{i,c}^{j} log_2\left(\tilde{y}_{i,c}^{j}\right) + \lambda\|\theta\|^2 \tag{7}$$

where $y_{i,c}^{j}$ = original output of class c, $\tilde{y}_{i,c}^{j}$ = predicted output for j^{th} utterance of ith video. λ is the L2 regularization term and θ is the parameter set $\{W_p, b_p, W_r, b_r, W, b, U_p, U_r, U\}$.

In our experiments, we pad videos with dummy utterances to enable batch processing. We also use bit-masks to mitigate noise created from dummy utterances. As for optimizer, we use Adam optimizer which has better robustness than SGD optimizer and train for 50 epochs. We set the learning rate = 0.0001 and the batch size = 20. To test the robustness of our model, we conduct an experiment to verify the impact of epochs on the performance in the training process, where the MOSI dataset is used. As Fig. 3 shows, the precision, recall, and F1-measure are relatively stable when the number of epochs is over 10, showing that our model can be robust for multimodal sentiment analysis if we train the morel for more than 10 epochs.

To avoid overfitting during the training process, we set the dropout of the Bi-GRU layer to 0.4 and the dropout of the dense layer to 0.2. In addition, we use the *ReLu* activation function in the dense layer and the *softmax* activation to calculate the final classification result.

4.3 Results of Sentiment Analysis

Multimodal Sentiment Analysis. We first compare our method with three state-of-the-art approaches that were proposed for multimodal sentiment analysis, which are Simple-LSTM, Hierarchical RNN, and CAT-LSTM. These three competitors are representatives of the three categories of multimodal sentiment analysis that we introduced in the related work section.

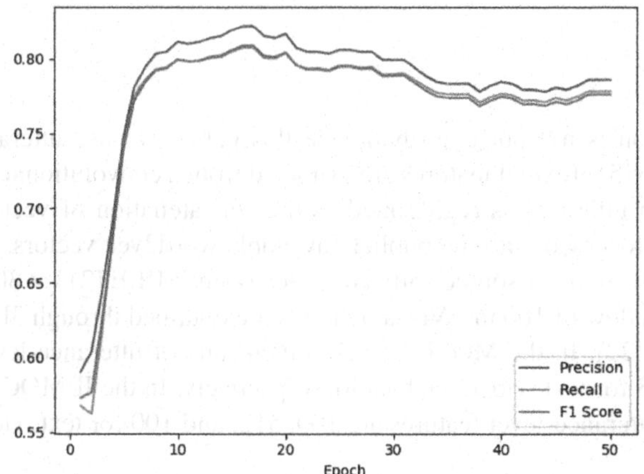

Fig. 3. Impact of epochs on the performance in the training process.

(1) Simple-LSTM [5]. This method was proposed at ACL 2017 and is a representative of the learned multimodal fusion approaches. It aims to learn a temporal representation for each of the different modalities and the LSTM model was employed to learn the contextual utterance-level features, which were used to generate the concatenated multimodal features. However, this method does not consider the temporal correlations among utterances, which differs from our design.

(2) Hierarchical RNN [6]. This method was presented by Majumder et al. in 2018. It takes the concatenation of multimodal vectors as a single model to perform multimodal sentiment analysis. Specially, it employed A hierarchical fusion approach with context modeling and RNN to extract context-aware features. However, it does not consider the inter-utterance correlations and does not distinguish the influences of different modalities on sentiment analysis. Thus, our design is expected to outperform it as we do consider the inter-utterance learning as well as the different impacts of modalities on sentiment polarifying.

(3) CAT-LSTM [7]. This method is an attention-based multimodal fusion approach, which was proposed at ICDM 2017. It used the attention mechanism to find cross-model interactions over time and modality-specific interactions among utterances. In particular, a contextual attention-based LSTM network (CAT-LSTM) was designed to model the contextual relationship among utterances. An attention-based fusion mechanism was also presented to emphasize the informative modalities for multimodal fusion. The major difference between our design and CAT-LSTM is that we employ the self-attention mechanism rather than the soft attention. In addition, we also distinguish the impact of three modalities in the multimodal attention layer.

The comparative results of multimodal sentiment analysis are shown in Table 2. In this experiment, we tested the accuracy of all methods in the setting of three modalities and two modalities. Our approach outperformed all the three competitors on both datasets and on both tri-model and bi-model settings, showing that our method can make better use of the multimodal information in the sentiment analysis. Specially, the higher accuracy of our method compared with that of Simple-LSTM indicated that the self-attention-based inter-utterance learning is helpful to the improvement of the sentiment analysis. Note that when only video and audio modalities are considered in the experiment, our method achieved higher performance in all settings except the comparison with CAT-LSTM on the MOSI dataset. This is mainly because in our design, we only employ the self-attention mechanism in the text modality. To this end, if we remove the text modality, our method is expected to have comparable performance as the CAT-LSTM, which also adopts the attention-based multimodal fusion idea. Therefore, we need to pay more attention to the text modality in multimodal sentiment analysis.

Unimodal Sentiment Analysis. Self-attention can also be used to extract unimodal inter-utterance information for better accuracy of unimodal sentiment analysis. Therefore, in this experiment, we evaluate our method on unimodal sentiment analysis.

As the CAT-LSTM model cannot be modified for unimodal sentiment analysis, we implement the contextual RNN model [16] as another baseline. For our method, we simply implement the self-attention mechanism for all the three modalities including video, audio, and text. This experiment aims to measure the advantage of the self-attention mechanism in unimodal sentiment analysis.

Table 3 shows the results of all methods when running on single modality data. our unimodal self-attention model yields better performance than all baseline methods. It shows that the self-attention mechanism can outperform LSTM-based contextual models like Simple-LSTM and CAT-LSTM because it can retain more contextual information among utterances. Thus, it is a better choice to consider the self-attention mechanism in multimodal data analysis.

Table 2. Accuracy of Multimodal Sentiment Analysis.

Modality	MOSI (%)				IEMOCAP (%)			
Text, Video, Audio	Simple -LSTM [5]	Hierarchical RNN [6]	CAT -LSTM [7]	Ours	Simple -LSTM [5]	Hierarchical RNN [6]	CAT -LSTM [7]	Ours
T+V	80.2	79.3	79.9	**80.5**	75.4	75.9	77.1	**77.3**
T+A	79.3	79.1	80.1	**80.2**	75.6	76.1	77.4	**77.8**
V+A	62.1	58.8	62.9	**62.5**	68.9	69.5	69.7	**69.8**
T+V+A	80.3	80.0	81.3	**81.5**	76.1	76.5	78.2	**78.5**

Performance of Other Neural Networks. In the implantation of our method, we used Bi-GRU as the basic neural network. As there are other possible networks that can be

Table 3. Accuracy of unimodal sentiment analysis

Modality	MOSI (%)				IEMOCAP (%)			
Text, Video, Audio	Simple -LSTM [5]	CAT -LSTM [7]	Contextual RNN [16]	Ours	Simple -LSTM [5]	CAT -LSTM [7]	Contextual RNN [16]	Ours
T	78.1	79.1	80.1	**80.2**	73.6	73.9	74.0	**74.2**
V	55.8	55.5	63.7	**64.1**	53.2	52.8	54.1	**53.9**
A	60.3	60.1	62.1	**63.6**	57.1	57.5	55.3	**58.2**

Table 4. Accuracy comparison of different networks

	Bi-LSTM	LSTM	Bi-GRU	GRU
MOSI	80.8%	78.7%	81.5%	79.0%
IEMOCAP	77.6%	76.7%	78.5%	76.3%

used as alternatives, in this experiment, we evaluate the performance of our method when other networks are used. We compare four networks, including the unidirectional GRU, the bidirectional GRU (Bi-GRU), the unidirectional LSTM, and the bidirectional LSTM (Bi-LSTM). The results are shown in Table 4. We can see that the bidirectional networks outperform the unidirectional networks. This is because the bidirectional networks consider the information of two directions (preceding and following) in the utterance sequence. LSTM and GRU share the ability of modeling long-range dependencies, but GRU performs slightly better than LSTM. On both datasets, Bi-GRU gets the best performance among all networks.

5 Conclusions

In this paper, we presented a novel framework for the sentiment analysis of multimodal review data on the web. Particularly, we propose a two-staged method that includes an inter-utterance learning stage and a cross-model fusion stage. Differing from previous solutions, we consider the inter-utterance correlations in the sequence of each modality and introduce the self-attention mechanism to learn the inter-utterance influence. Combined with the following cross-model fusion, our method can achieve higher performance than existing approaches. We conducted experiments on two public web datasets and compared our proposal with three methods for multimodal sentiment analysis. The results showed that our method got the best accuracy of multimodal sentiment analysis. We also evaluated our design on unimodal sentiment analysis and on different neural networks. The results suggested the robustness of our proposal.

In the future, we will integrate multimodal sentiment analysis with event evolution [30, 31] and develop a prototype to monitor sentiment evolution of events.

Acknowledgement. This study is supported by the National Key Research and Development Program of China (2018YFB0704404) and the National Science Foundation of China (61672479).

References

1. Zhou, J., Jin, P., Zhao, J.: Sentiment analysis of online reviews with a hierarchical attention network. In: SEKE, pp. 429–434 (2020)
2. Zheng, L., Jin, P., Zhao, J., Yue, L.: Multi-dimensional sentiment analysis for large-scale E-commerce reviews. In: Decker, H., Lhotská, L., Link, S., Spies, M., Wagner, Roland R. (eds.) DEXA 2014. LNCS, vol. 8645, pp. 449–463. Springer, Cham (2014). https://doi.org/10.1007/978-3-319-10085-2_41
3. Vaswani, A., Shazeer, N., Parmar, N., et al.: Attention is all you need. In: NIPS, pp. 5998–6008 (2017)
4. Truong, Q., Lauw, H.: VistaNet: visual aspect attention network for multimodal sentiment analysis. In: AAAI, pp. 305–312 (2019)
5. Poria, S., Cambria, E., Hazarika, D., et al.: Context-dependent sentiment analysis in user-generated videos. In: ACL, pp. 873–883 (2017)
6. Majumder, N., Hazarika, D., Gelbukh, A., et al.: Multimodal sentiment analysis using hierarchical fusion with context modeling. Knowl.-Based Syst. **161**, 124–133 (2018)
7. Poria, S., Cambria, E., Hazarika, D., et al.: Multi-level multiple attentions for contextual multimodal sentiment analysis. In: ICDM, pp. 1033–1038 (2017)
8. Rozgic, V., Ananthakrishnan, S., Saleem, S., et al.: Ensemble of SVM trees for multimodal emotion recognition. In: APSIPA ASC 2012, pp. 1–4 (2012)
9. Quattoni, A., Wang, S., Morency, L., et al.: Hidden conditional random fields. IEEE Trans. Pattern Anal. Mach. Intell. **29**(10), 1848–1852 (2007)
10. Morency, L., Mihalcea, R., Doshi, P.: Towards multimodal sentiment analysis: harvesting opinions from the web. In: ICMI, pp. 169–176 (2011)
11. Hochreiter, S., Schmidhuber, J.: Long short-term memory. Neural Comput. **9**(8), 1735–1780 (1997)
12. Poria, S., Cambria, E., Gelbukh, A.: Deep convolutional neural network textual features and multiple kernel learning for utterance-level multimodal sentiment analysis. In: EMNLP, pp. 2539–2544 (2015)
13. Li, J., Zhang, B., Lu, G., Zhang, D.: Generative multi-view and multi-feature learning for classification. Inf. Fusion **45**, 215–226 (2019)
14. Nojavanasghari, B., Gopinath, D., Koushik, J., et al.: Deep multimodal fusion for persuasiveness prediction. In: ICMI, pp. 284–288 (2016)
15. Zadeh, A., Chen, M., Poria, S., et al.: Tensor fusion network for multimodal sentiment analysis. In: EMNLP, pp. 1114–1125 (2017)
16. Ghosal, D., Akhtar, M., Chauhan, D., et al.: Contextual inter-model attention for multimodal sentiment analysis. In: EMNLP, pp. 3454–3466 (2018)
17. Zadeh, A., Liang, P., Mazumder, N., et al.: Memory fusion network for multi-view sequential learning. In: AAAI, pp. 5634–5641 (2018)
18. Zadeh, A., Liang, P., Poria, S., et al.: Multi-attention recurrent network for human communication comprehension. In: AAAI, pp. 5642–5649 (2018)
19. Rahman, W., Hasan, M.K., Zadeh, A., et al.: M-BERT: injecting multimodal information in the BERT structure. arXiv preprint arXiv:1908.05787 (2019)
20. Karpathy, A., Toderici, G., Shetty, S., et al.: Large-scale video classification with convolutional neural networks. In: CVPR, pp. 1725–1732 (2014)

21. Eyben, F., Wöllmer, M., Schuller, B.: Opensmile: the munich versatile and fast open-source audio feature extractor. In: ACM Multimedia, pp. 1459–1462 (2010)
22. Ji, S., Xu, W., Yang, M., et al.: 3D convolutional neural networks for human action recognition. IEEE Trans. Pattern Anal. Mach. Intell. **35**(1), 221–231 (2013)
23. Mishra, A., Dey, K., Bhattacharyya, P.: Learning cognitive features from gaze data for sentiment and sarcasm classification using convolutional neural network. In: ACL, pp. 377–387 (2017)
24. Zhang, C., Fu, H., Hu, Q., et al.: Generalized latent multi-view subspace clustering. IEEE Trans. Pattern Anal. Mach. Intell. **42**(1), 86–99 (2020)
25. Schuller, B.: Recognizing affect from linguistic information in 3D continuous space. IEEE Trans. Affect. Comput. **2**(4), 192–205 (2011)
26. Sun, B., Li, L., Zhou, G, et al.: Combining multimodal features within a fusion network for emotion recognition in the wild. In: ICMI, pp. 497–502 (2015)
27. Chen, S., Li, X., Jin, Q., et al.: Video emotion recognition in the wild based on fusion of multimodal features. In: ICMI, pp. 494–500 (2016)
28. Gaurav, R., Verma, M., Shukla, K.: Informed multimodal latent subspace learning via supervised matrix factorization. In: ICVGIP, pp. 36:1–36:8 (2016)
29. Poria, S., Peng, H., Hussian, A., et al.: Ensemble application of convolutional neural networks and multiple kernel learning for multimodal sentiment analysis. Neurocomputing **261**, 217–230 (2017)
30. Mu, L., Jin, P., Zheng, L., Chen, E.-H.: EventSys: tracking event evolution on microblogging platforms. In: Pei, J., Manolopoulos, Y., Sadiq, S., Li, J. (eds.) DASFAA 2018. LNCS, vol. 10828, pp. 797–801. Springer, Cham (2018). https://doi.org/10.1007/978-3-319-91458-9_51
31. Mu, L., Jin, P., Zheng, L., Chen, E., Yue, L.: Lifecycle-based event detection from microblogs. In: WWW, pp. 283–290 (2018)

An Energy-Efficient AES Encryption Algorithm Based on Memristor Switch

Danghui Wang[1,2,3](✉), Chen Yue[1], Ze Tian[4,5], Ru Han[1,2], and Lu Zhang[1]

[1] School of Computer Science, Northwestern Polytechnical University, Xi'an, China
{wangdh,hanru}@nwpu.edu.cn,
{yuechen,2015201697}@mail.nwpu.edu.cn

[2] Engineering Research Center of Embedded System Integration Ministry of Education, Xi'an, China

[3] National Engineering Laboratory for Integrated Aero-Space-Ground-Ocean Big Data Application Technology, Xi'an, China

[4] AVIC Computing Technique Research Institute, Xi'an 710068, China
tarmz@126.com

[5] Key Laboratory of Aviation Science and Technology on Integrated Circuit and Micro-system Design, Xi'an 710068, China

Abstract. AES algorithm has a large amount of data migration between storage side and computing side. Especially on mobile devices, where the data migration demand is expensive, however, the existing Von Neumann architecture cannot support such demand due to the separation of storage and computation. Memristor, a new device, which integrates storage and computing together, is expected to solve the problem with low power and short delay of AES. The XOR logic operation plays centric rule in AES for carry addition and data encryption. However, the current researches on the logic operation of memristors only stays at basic logic (AND, OR, NOT) level, and the XOR logic operations constructed by these basic logics generally have problems such as too large area, long calculation sequence and complicated control. Therefore, this paper provides a new way to design energy efficient XOR logic and uses it to implement adder and AES encryption algorithm. The results show that the XOR logic method in this paper uses the least number of memristors and the shortest calculation sequence length comparing with the two popular baselines. The XOR logic based on memristor effectively decreases the power consumption and delay by at most 99% and 50% respectively.

Keywords: Memristor switch · XOR logic · Crossbar · Adder · AES

1 Introduction

The research of Advanced Encryption Standard (AES) algorithm has become a hot spot [1], and the algorithm has been widely used in the field of information security. But now it encounters problems of power consumption and delay [2]. AES algorithm has a large amount of data migration between storage side and computing side. Especially

M. Qiu (Ed.): ICA3PP 2020, LNCS 12454, pp. 639–653, 2020.
https://doi.org/10.1007/978-3-030-60248-2_44

on mobile side, where the data migration demand is expensive, the existing Von Neumann architecture cannot support its demand because of the separation of storage and computation [3]. Memristor, a new device that integrates storage and computing [4], is expected to solve the problems with low power and short delay of AES. In addition, this combination of storage and calculation structure is helpful to improve the resistance to aggression of AES.

As an emerging nanoelectronic device, memristor has a memory function [5]. Its resistance value change is affected by the charge passing through it, and it has the function of data storage and logic operation. On the one hand, as a memory [6], memristor has various excellent characteristics such as high integration, low power consumption, non-volatile, and compatibility with CMOS technology [7]. On the other hand, memristor can also perform logical operations [8]. The memory fusion feature of memristor can reduce the data transfer process greatly to improve the bandwidth and effectively solve the problem of "storage wall" [9].

There has been plenty of researches into memristor logic, including Implication Logic [10] and MAGIC [11]. They all use the characteristics of memristor storage and calculation, but there are still some shortcomings in solving the problems of energy consumption and delay.

Implication Logic. Memristor can implement logical operations through complete logic-imply and negation, which is called Implication Logic [12]. Any logic can be expressed in the form of the two basic logics and their combination. This logic operation method breaks traditional circuit structure based on logic gates. Implication Logic shows the characteristics of memristor storage and computing integration and provides a new direction for memristor logic operation. But its complicated control and long calculation sequence have led to problems such as high time cost and high power consumption.

MAGIC. In order to solve the problems caused by Implication Logic, Memristor-Aided Logic (MAGIC) has been proposed in recent years [11]. Unlike Implication Logic, MAGIC does not require a complicated control voltage sequence, and a logic gate can be realized by applying a simple voltage pulse. Although this circuit is simpler than Implication Logic in implementing simple logic, it still has problems such as excessively long calculation sequences when implementing more complex logic operations.

XOR logic operation has a significant meaning in computer science, and it is the basis for many applications such as carry addition and data encryption. However, current logic operation research of memristors only stays at the level of basic logic. XOR logic operation method constructed using these basic logic is very complicated. So this paper designs a new operation method to implement an efficient XOR logic circuit to improve calculation performance of many applications such as adder and data encryption which called "XOR logic operation based on memristor switch". This article gives its circuit structure, calculation sequence and constraint conditions. Compared with Implication Logic and MAGIC, the method designed in this paper uses a small number of memristors and produces a shorter calculation sequence length when perform XOR operations. We use XOR logic based on memristor switch to design circuit of AES encryption algorithm. It makes full use of feature of storage and computing integration to encrypt data directly in memory, reduce the process of data transfer, and effectively improve access bandwidth and throughput rate.

We use the designed XOR logic operation method to achieve encryption and decryption operations, design and optimize crossbar circuit for XOR logic operations in MixColumns, AddRoundKey, and KeyExpansion in AES algorithm. The experimental environment of this article is based on Virtuoso platform of Hspice and Cadence, and memristor VTEAM model [13] is used to simulate the designed XOR logic circuit. The results show that, in terms of time cost of MixColumns, AddRoundKey, and KeyExpansion, memristor switch-based design method is 21.4%, 15.4%, and 15.4% of implication-based logic design method respectively, and is 50%, 40%, 40% higher than MAGIC-based design method respectively.

The main contributions of this paper are:

- We redesigned XOR logic on memristor for faster and less power consumption.
- We have implemented a more concise adder with this new structure.
- We applied the new adder and XOR logic to AES circuit, and obtained better benefits than MAGIC and implication logic respectively.

2 Background and Motivation

2.1 Background

XOR operations are almost everywhere in AES encryption algorithm. XOR is widely used to "disorganize" data [14]. Among them, MixColumns, AddRoundKey and KeyExpansion are the most used XOR operations [15]. Chapter 4 will introduce them in detail, including calculation method and corresponding circuit design.

MixColumns. By means of linear function, the arithmetic property of GF (2^8) in finite field is replaced by XOR logic operation.

AddRoundKey. Perform XOR logic operations on matrix with RoundKey (RoundKey generated by KeyExpansion algorithm).

KeyExpansion. Generate RoundKey used in each round of encryption operation.

Calculation and storage of common AES algorithm are completed in different units, a large amount of data needs to be moved, which increases power consumption and generates additional delays [16]. Memristor can effectively integrate the two functions of storage and calculation. Therefore, this paper uses XOR logic based on memristor switch to design AES algorithm circuit, and makes full use of its storage and computing integration characteristics to directly encrypt data in memory, reduce the process of data movement, and effectively improve memory access bandwidth and throughput rate, while improving attack resistance of encryption operations.

2.2 Motivation

From the above introduction, we can know that XOR in AES encryption algorithm accounts for a large and very important proportion. Memristor integrated with storage and calculation can directly encrypt data in memory, reduce the process of data movement and effectively improve memory access bandwidth and throughput. However, there is no direct and efficient XOR logic operation method based on memristor, and XOR logic operation using existing basic logic is very complicated. XOR logic operation based on Implication Logic is complicated and expensive, and XOR logic operation based on MAGIC has a complex structure and many controls in crossbar array. Therefore, it is necessary to design an efficient XOR logic circuit to improve the overall performance of many applications such as AES encryption algorithm, etc.

We designed a simplified XOR logic based on memristor switch to solve this problem, and when we applied this structure to AES, the time cost and power consumption of AES algorithm to complete one round was significantly reduced compared to Implication Logic and MAGIC.

3 XOR Logic Structure Based on Memristor Switch

Memristors are usually studied in the structure of crossbar. Crossbar is made up of two sets of nanowires intersecting vertically, where an overlap occurs at vertical intersection. By placing memristor material in the overlap, a large number of memristors can be placed in a parallel plane to form a memristor crossbar, as shown in Fig. 1. By applying different voltages to the two ends of memristor array, reading and writing of memristor value can be realized.

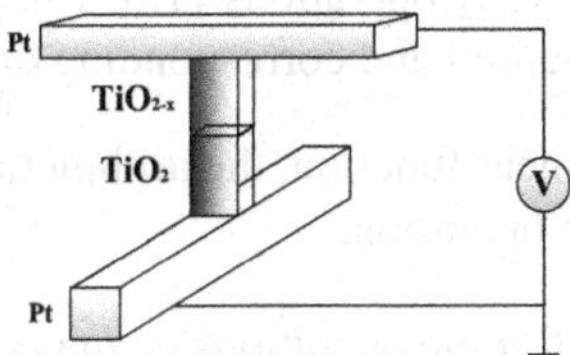

Fig. 1. Schematic diagram of memristor crossbar

In order to avoid Sneak Path, all of memristor memory cells used in this article are One Transistor One Resistor (1T1R) type, which is formed by adding a selection transistor to memory cell of memristor (for the convenience of explanation and highlighting memristor, this article has omitted the MOS structure in 1T1R type in some figures, but in actual research and manufacturing, the MOS structure cannot be ignored). Each memory cell of 1T1R type is composed of a MOS tube and a memristor. When a memory cell needs to be read or written, the transistor of the cell is turned on, and the other transistors are in the off state.

3.1 XOR Logic Circuit Structure Based on Memristor Switch

Since memristor crossbar originally contains a selection tube, in order to achieve efficient XOR logic operations, we use selection tube to appropriately select the input data to obtain the output result of XOR logic. We call this method a "XOR logic operation based on memristor switch".

The XOR logic operation based on memristor switch consists of three memristors and their corresponding selection switches, as shown in Fig. 2. Among them, two memristors (in_1, $\overline{in_1}$) with opposite resistance states are connected in parallel as an input value of XOR logic operation, and two selection switches (in_2, $\overline{in_2}$) are connected in series with the former two memristors respectively as the other of XOR logic operation. One input value, the third memristor (out) as output state.

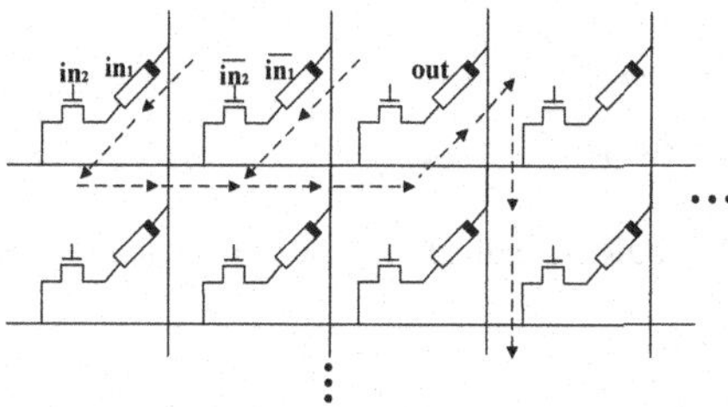

Fig. 2. XOR circuit diagram based on memristor switch

Before logic operation, the three memristors are initialized first, that is, the first operation value of XOR logic operation is written into memristor in_1. Memristor $\overline{in_1}$ is the opposite state value, and the memristor out is initialized to logic "0" means high resistance state. When XOR logic operation is performed, the second operand of XOR logic operation will control selection tube to determine which branch will perform effective work. The operation is implemented by applying a pulse voltage V_0 at logic gate. The equivalent circuit of XOR logic operation based on memristor switch is shown in Fig. 3.

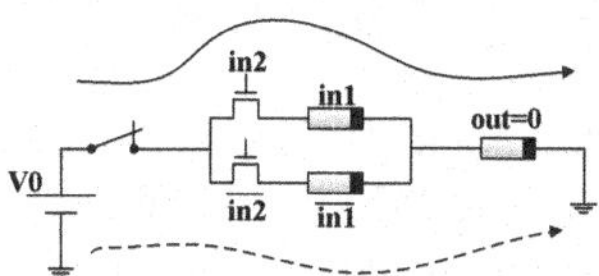

Fig. 3. Schematic diagram of XOR branch selection based on memristor switch

In XOR logic operation based on memristor switch, the states of memristor and selector switch are shown in Fig. 4. in_1 and in_2 are two operands, where in_1 is memristor, in_2 is MOS switch, out_{init} is initial resistance state of the output memristor, out is final calculated state of the output memristor. The initial state of the output memristor is high resistance state R_{off}. When the memristor in_1 is in high resistance state and the MOS tube in_1 is on, the output memristor calculation result is in high resistance state, that is, logic "0"; when the memristor in_1 is high resistance state, when the MOS tube in_1 is off, the calculation result of the output memristor is in low resistance state, that is, logic "1"; when the memristor in_1 is in low resistance state and the MOS tube in_1 is on, the calculation result of the output memristor is in low resistance state, that is, logic

"1"; when the memristor in_1 is in low resistance state and the MOS tube in_1 is off, the calculation result of the output memristor is high resistance state, that is, logic "0". It can be seen that only need to apply a simple voltage excitation, memristor can realize XOR logic operation.

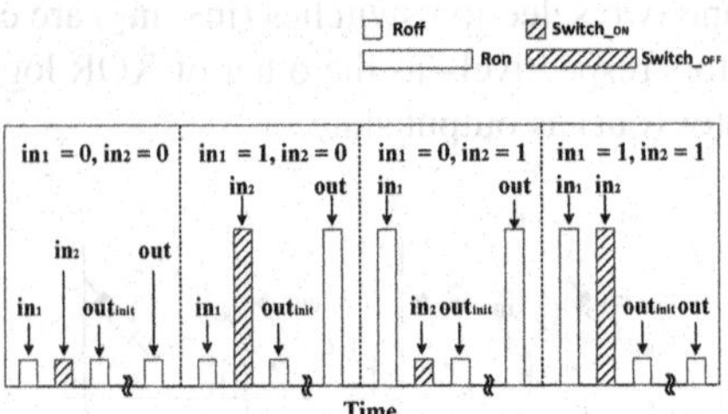

Fig. 4. XOR state diagram based on switch memristor

3.2 XOR Logic Constraints Based on Memristor Switch

It is assumed that the threshold voltage of memristor is Vopen and Vclose, the conducting impedance and turning off impedance of memristor are respectively R_{on} and R_{off}, and the conducting resistance of MOS tube is R_{mos}. Excitation pulse voltage is designed with the voltage of the output memristor out above and below the threshold voltage V_0. When the operand in_1 and in_2 are both high resistance state logic "0" or the operand in_1 and in_2 are both low resistance state logic "1", the output memristor out voltage is lower than the threshold voltage V_{close}, and the excitation voltage should meet the formula 1.

$$V_0 < 2 \cdot \left(\frac{R_{mos}}{R_{off}} + 1\right) \cdot Vclose \tag{1}$$

Consider the low-resistance state logic "1", in_2 is the high-resistance state logic "0" or in_1 is the high-resistance state logic "0", in_2 is the low-resistance state logic "1", the voltage of the output memristor out exceeds the threshold voltage V_{close}, so the minimum voltage needs to satisfy the formula 2.

$$V_0 > \left(\frac{2R_{mos} + R_{on}}{R_{off}} + 1\right) \cdot Vclose \tag{2}$$

In summary, we can get the pulse excitation voltage based on XOR logic of memristor switch V_0 which needs to satisfy formula 3.

$$\left(\frac{2R_{mos} + R_{on}}{R_{off}} + 1\right) \cdot Vclose < V_0 < 2 \cdot \left(\frac{R_{mos}}{R_{off}} + 1\right) \cdot Vclose \tag{3}$$

3.3 Evaluation of XOR Logic Operation Under Three Design Methods

Under the premise of 1-bit XOR logic operation, the difference between Implication Logic and MAGIC and memristor switch in terms of the number of memristors used and the calculated sequence length are shown in Table 1. It can be seen from the table that 1-bit XOR logic operation based on Implication Logic requires at least 5 memristors, and the shortest calculation sequence length is 13; 1-bit XOR logic operation based on MAGIC requires at least 6 memristors, the shortest calculation sequence length is 5; 1-bit XOR logic operation based on Memristor Switch requires at least 3 memristors, and the calculation sequence is 2. Obviously, the calculation sequence of 1-bit XOR logic operation based on Implication Logic is too long, resulting in a large time cost. 1-bit XOR logic calculation sequence based on MAGIC is reduced, but the number of memristors used increases. In comparison, XOR logic design based on memristor switch uses the least number of memristors which is 60% of Implication Logic design and 50% of MAGIC design. Besides, XOR logic design based on memristor switch shows the shortest calculated sequence length which is 15.4% of Implication Logic design and 40% of MAGIC design. To sum up, XOR logic design based on memristor switch improves computational efficiency and has obvious advantages.

Table 1. Comparison of three different 1-bit XOR logic operation design methods

Design method	Number of memristors used	Calculate sequence length
Implication logic	5	13
MAGIC	6	5
Our memristor switch	3	2

3.4 Adder Based on Memristor Switch XOR Logic

The processes of task processing in a computer system are logical operations and arithmetic operations. The basis of these arithmetic operations is addition. Therefore, in traditional computer systems, the design of adder is crucial. The traditional adder design uses CMOS as basic circuit element, calculation and storage are divided into two components. But memristor can integrate calculation and storage together, thereby improving overall performance of the system. Hence, designing an efficient memristor-based adder has important theoretical significance and application value.

Because of the operations generated by the carry can be completed by one XOR logic, two AND logic and one OR logic. In order to implement adder efficiently, we need to implement logic operations in memristor array. As is shown in Fig. 5.

The circuit structure of a full-adder is shown in Fig. 6. M_{11}, M_{12}, M_{13}, M_{14}, M_{15}, M_{22}, M_{24} and M_{25} are memristors, S_{11}, S_{12}, S_{13}, S_{14}, S_{15}, S_{22}, S_{24} and S_{25} are corresponding control switch MOS tubes. Among them, the memristor M_{11} means addend A_i, the switch S_{11} means addend, and the switch S_{13} means low carry. The calculation result of the sum is stored in memristor M_{15}, and the calculation result of the carry is stored in

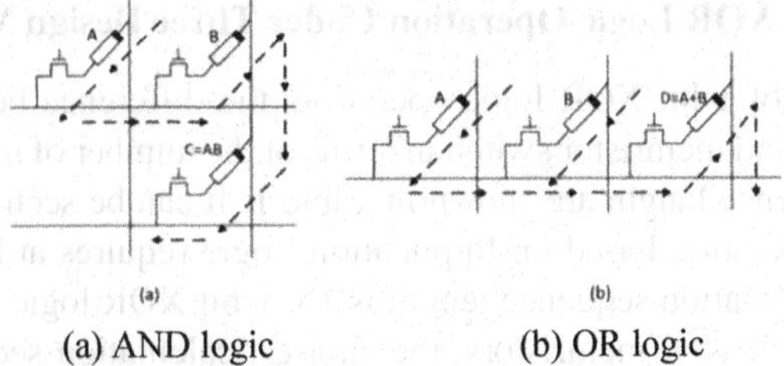

(a) AND logic (b) OR logic

Fig. 5. AND logic and OR logic circuit diagram in memristor array

memristor M_{25}. It can be seen that the 1-bit full-adder based on memristor switch only needs 8 memristors.

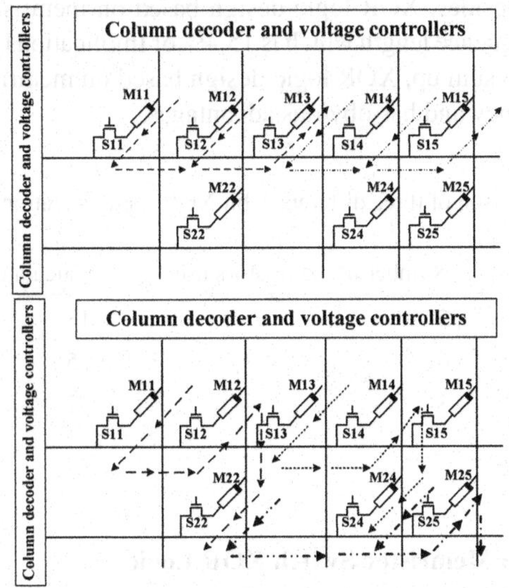

Fig. 6. 1-bit full-adder circuit diagram based on memristor switch

The addition operation flow is as follows:

The first step, memristor M_{11} and M_{12} denote A_i and $\overline{A_i}$ respectively, switch S_{11} and S_{12} denote B_i and $\overline{B_i}$ respectively, that is, $M_{11} = A_i$, $M_{12} = \overline{M_{11}} = \overline{A_i}$, $S_{11} = B_i$ and $S_{12} = \overline{B_i}$.

The second step, the result of XOR logic of addend A_i and B_i is stored in memristor M_{13}, that is, $M_{13} = A_iB_i$.

The third step, the memristor M_{13} performs inversion operation, and the result is stored in memristor M_{14}, that is, $M_{14} = \overline{M_{13}} = \overline{A_i \oplus B_i}$.

The fourth step, the result of XOR logic between C_{i-1} and $A_i \oplus B_i$ is stored in memristor M_{15}, which is the sum S_i, $M_{15} = A_i \oplus B_i \oplus C_{i-1} = S_i$.

The fifth step, assign the memristor M_{12} to addend B_i, $M_{12} = B_i$.

The sixth step, the memristor M_{11} and M_{12} perform AND logic operation, that is, the addend A_i and B_i perform AND logic operation, and the result is stored in memristor M_{22}, $M_{22} = M_{11} \cdot M_{12} = A_i \cdot B_i$.

The seventh step, assign carry C_{i-1} to memristor M_{14}, $M_{14} = C_{i-1}$.

The eighth step, memristor M_{13} and M_{14} perform AND logic operation, and the result is stored in memristor M_{24}, that is, $M_{24} = M_{13} \cdot M_{14} = (A_i \oplus B_i) \cdot C_{i-1}$.

The ninth step, perform an OR logic operation between memristor M_{22} and M_{24}, and the result which means carry C_i is stored in memristor M_{25}, that is, $M_{25} = M_{22} + M_{24} = C_i$.

Figure 7 shows the structure diagram of Serial Carry Adder based on memristor switch XOR logic operation. Traditional adder cannot save intermediate result in calculation process, but the adder based on designed memristor array can store data or perform logical operations as input values. Each carry generation will be stored in memristor, so that a n-bit adder can be integrated into one component. Meanwhile, due to the particularity of memristor crossbar structure, it can support the same row or column to perform a certain logic operation at the same time. This saves a lot of time and area costs.

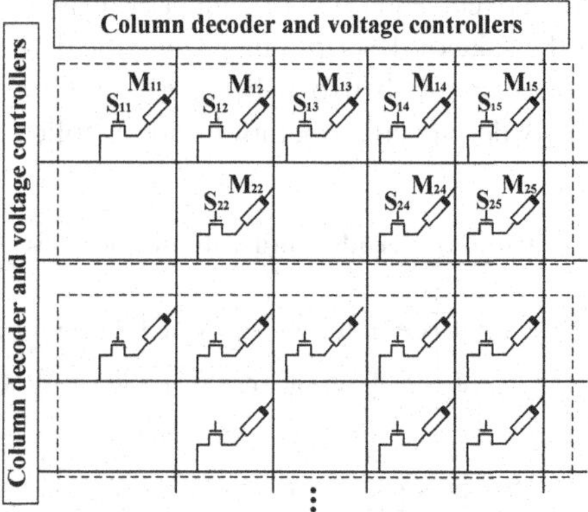

Fig. 7. Structure diagram of Serial Carry Adder based on memristor switch

4 AES Structure Based on Memristor Switch XOR Logic

In AES algorithm, 128 bits are used as a block for data grouping, and the key size is also 128 bits. Through a series of block-based operation transformations, the plaintext is converted into ciphertext with the same size. In encryption transformation process, a certain number of operations need to be performed, each of which is called a round. AES uses 10, 12 or 14 rounds of encryption. This article will take a 128-bit key and 10 rounds of encryption as an example, which is AES-128. The other two ideas are the same, but the number of rounds will increase accordingly.

AES mainly includes four processes: SubBytes, ShiftRows, MixColumns and AddRoundKey. SubBytes uses the form of a lookup table to replace each byte with S box to corresponding byte; ShiftRows performs a corresponding cyclic shift for each row in the matrix; MixColumns is the substitution of arithmetic properties on finite field GF(2^8) by means of linear functions, and are completed by XOR logic operations; AddRoundKey is XOR logic operation of the matrix and RoundKey (RoundKey is generated by KeyExpansion algorithm).

4.1 MixColumns Based on Memristor Switch XOR Logic

MixColumns is a replacement for arithmetic property of finite field GF (2^8). According to the arithmetic rule of GF (2^8) in finite field, if a byte is multiplied by 1, the result is consistent with current byte; Multiplying a byte by 2 results in moving the byte to the left by one bit. If the maximum bit of the byte is 1, the shifted result needs to be equal to 1BH, which is 00011011 XOR; A byte multiplied by 3 can be broken down to the sum of the byte multiplied by 1 and multiplied by 2 according to the allocation rate of multiplication, which is $03 \cdot S_{00} = (01 \oplus 02) \cdot S_{00} = S_{00} \oplus (02 \cdot S_{00})$.

The specific calculation process of MixColumns based on memristor switch is divided into six steps. The circuit of AES algorithm based on memristor switch XOR logic is shown in Fig. 8. In the figure, horizontal is the different bits of 8-bit data, and the operation is the same between each row of data. The array is calculated step by step from top to bottom. For the sake of description, assume the data as follows: $S_{00} = A = a_7a_6a_5a_4a_3a_2a_1a_0$, $S_{11} = B = b_7b_6b_5b_4b_3b_2b_1b_0$, $S_{22} = C = c_7c_6c_5c_4c_3c_2c_1c_0$ and $S_{33} = D = d_7d_6d_5d_4d_3d_2d_1d_0$. Specific operations are as follows:

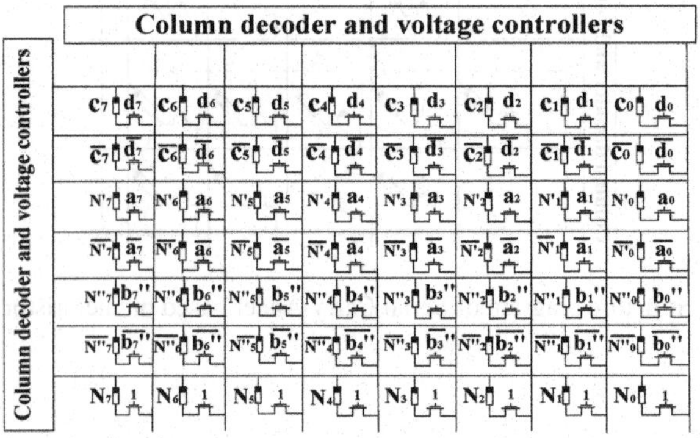

Fig. 8. Circuit diagram of MixColumns

The first step. Invert the bit of C to get $\overline{C}$; At the same time, the highest bit a_7 of A is judged, when the value of a_7 is 0, $A' = a_6a_5a_4a_3a_2a_1a_00$, and when the value of a_7 is 1, $A' = a_6a_5a_4a_3a_2a_1a_00 \oplus 00011101 = a_6a_5a_4\overline{a}_3\overline{a}_2\overline{a}_1a_01$; At the same time, the highest bit b_7 of B is judged, when b_7 is 0, $B' = b_6b_5b_4b_3b_2b_1b_00$, when b_7 is 1, $B' = b_6b_5b_4b_3b_2b_1b_00 \oplus 00011101 = b_6b_5b_4\overline{b}_3\overline{b}_2\overline{b}_1b_01$.

The second step. Carry out bitwise OR for C and D, and the result is denoted as N′, $N' = C \oplus D$. At the same time, XOR B′ with B and get the result B″, that is, $B'' = B' \oplus B$.

The third step. Reverse N′ bitwise and get $\overline{N'}$, $\overline{N'} = \overline{C \oplus D}$.

The fourth step. Perform a bitwise XOR between N' and A', and record the result as N'', that is, $N'' = C \oplus D \oplus A'$.

The fifth step. Perform bitwise inversion of N'' to get $\overline{N''}$, that is, $\overline{N''} = \overline{C \oplus D \oplus A'}$.

The sixth step, Perform a bitwise XOR between N'' and B'', and record the result as N, that is, $N = C \oplus D \oplus A' \oplus B''$.

4.2 AddRoundKey Based on Memristor Switch XOR Logic

The function of AddRoundKey is to perform XOR logical operation between the matrix and Roundkey. The specific methods are described and analyzed in the next section. The main process of Roundkey addition operation is also XOR logic operation.

Since any number is zero with its own XOR result, Roundkey addition during decryption is the same as the encryption process. AES algorithm based on memristor switch XOR logic Roundkey addition operation circuit is shown in Fig. 9, horizontal represents each bit of 8-bit data, each row is an XOR logical operation unit.

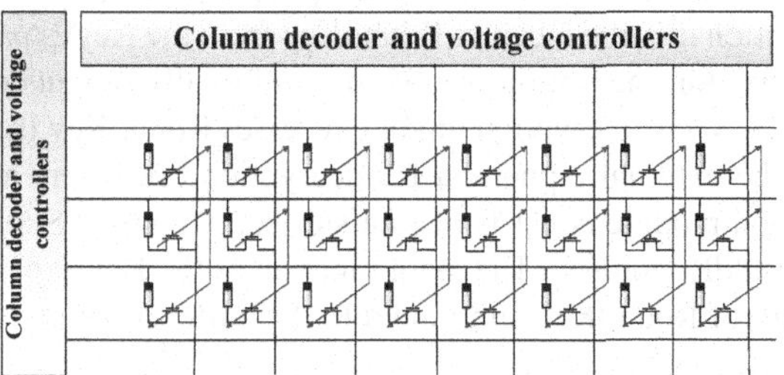

Fig. 9. Circuit diagram of AddRoundKey

4.3 KeyExpansion Based on Memristor Switch XOR Logic

KeyExpansion is mainly to expand initial key to generate RoundKey used in each round of encryption. The circuit diagram of KeyExpansion operation based on memristor switch XOR logic is similar to the circuit diagrams in Sect. 4.2 and 4.3, except that it requires more rows. As before, horizontal rows represent each bit of 8-bit data, and each column is an XOR logic operation unit. The result of XOR logic operation in the fourth part will also be used as the input of XOR logic operation in the fifth part.

4.4 Circuit Design of AES Algorithm Based on Memristor Switch XOR Logic

The AES algorithm is mainly composed of two parts: encryption and KeyExpansion. Among them, the encryption part mainly completes the encryption function, which is composed of four sub-parts, namely SubBytes, ShiftRows, MixColumns and AddRoundKey. The main function of the KeyExpansion part is to complete the output of RoundKey. The system structure of AES algorithm is shown in Fig. 10.

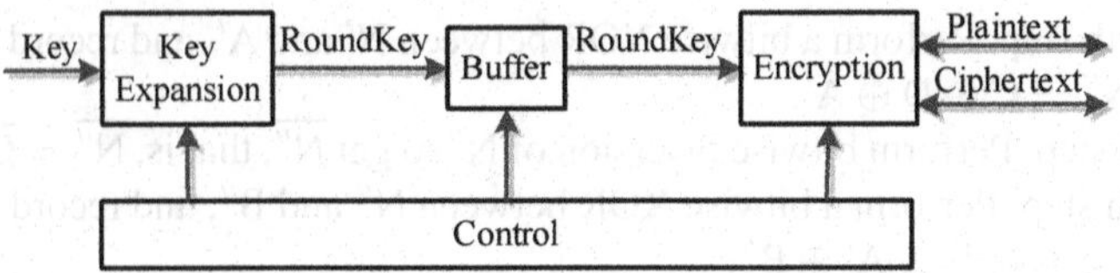

Fig. 10. AES system structure diagram

AES is divided into four modules, namely Buffer, Control, Encryption and Key-Expansion. The main function of the buffer module is to receive RoundKey from the KeyExpansion module and temporarily store it, and then output RoundKey to the encryption module during each round of encryption of AES algorithm. The main function of the control module is to judge the current state of each sub-module and provide control signals. Because of the decryption process and the encryption process are similar, and the decryption process is the reverse operation of the encryption process, so the encryption and decryption modules are also referred to as the encryption module. The encryption module mainly completes encryption and decryption functions. During the encryption process, the encryption module receives RoundKey from the KeyExpansion module and the external plaintext for encryption operations, and finally outputs ciphertext. During the decryption process, the encryption module receives RoundKey from the KeyExpansion module and the external ciphertext for decryption operation, and finally outputs plaintext. Encryption module consists of four sub-parts, namely SubBytes, ShiftRows, MixColumns and AddRoundKey. The main function of the KeyExpansion module is to complete the output of RoundKey and send it to the buffer module.

5 Evaluation

5.1 Setup

The experiment in this article is based on the VTEAM model of memristor. The model is implemented with Verilog-A in the SPICE environment. It can simulate the behavior of memristor controlled by voltage and set threshold voltage. Some parameters of the experimental model used in this article are shown in Table 2 (in order to use the model under different design methods, some parameters are set in this article). In the table, k_{on}, k_{off}, α_{on} and α_{off} are constants, V_{open} and V_{close} are threshold voltage, R_{on} and R_{off} are memristor impedance, x_{on} and x_{off} are the boundary values of memristor length.

The experimental environment of this article is based on the Virtuoso platform of Hspice and Cadence, and the memristor VTEAM model is used to simulate and verify the designed XOR logic circuit. Figure 11 shows the Spice simulation diagram of memristor when V_O is equal to 1. As can be seen from the figure, memristor completes state reversal at 0.4 ns.

In this paper, the model circuit simulation and digital circuit design as well as synthesis are adopted for AES algorithm operation module and encryption control module respectively, and the corresponding experimental results are obtained.

Table 2. Partial parameters of the memristor VTEAM model

Parameter	Value
k_{on}	−216.2 m/s
k_{off}	30.51 m/s
$V_{T,ON}$	−1.5 V
$V_{T,OFF}$	0.6 V
x_{on}	0
x_{off}	3 nm
α_{on}	4
α_{off}	4
R_{on}	1 kΩ
R_{off}	300 kΩ

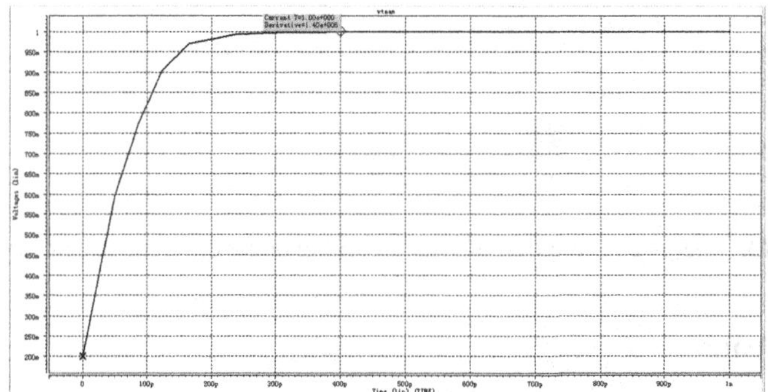

Fig. 11. Spice Simulation diagram of memristor ($V_0 = 1$ V)

5.2 Time Cost

This section compares the time cost required for AES algorithm to complete a round under different design methods, as shown in Fig. 12. It can be seen that AES algorithm based on memristor switch XOR logic has obvious advantages in time cost. In terms of the time cost of MixColumns, AddRoundKey, and KeyExpansion, XOR logic based on memristor switch design method is 21.4%, 15.4%, and 15.4% of Implication Logic design method respectively. Compared with MAGIC, the performance of MixColumns, AddRoundKey, and KeyExpansion are respectively increased by 50.0%, 40.0%, and 40.0%.

5.3 Energy Consumption

This section compares the energy consumption required by AES algorithm versus that of different design methods, as shown in Fig. 13, AES algorithm based on memristor switch XOR logic has obvious advantages in energy consumption.

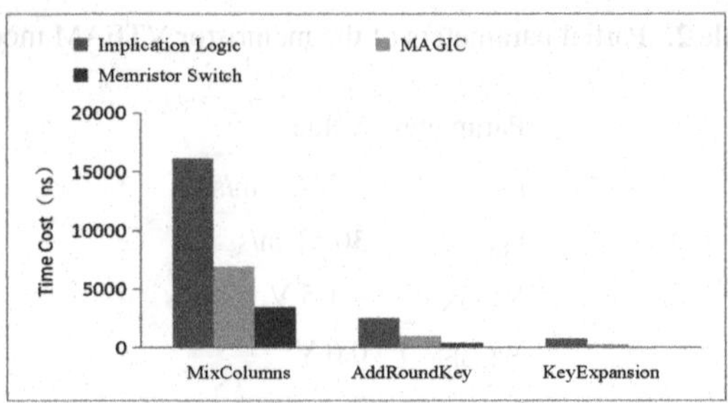

Fig. 12. Time cost of AES algorithm under different design methods

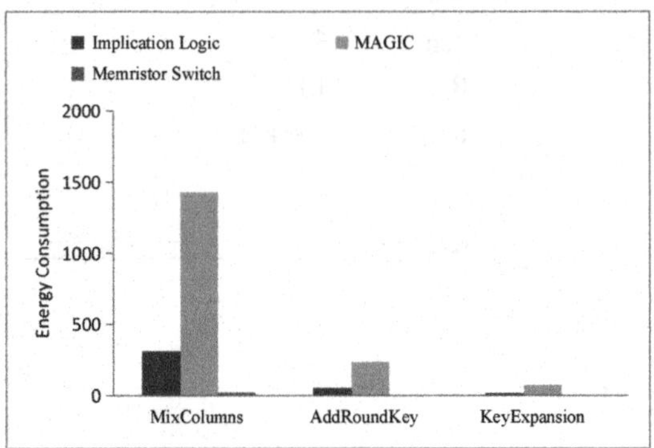

Fig. 13. Energy consumption of AES algorithm under different design methods

6 Conclusion

Memristor has shown great potential in fields such as AES encryption algorithm. Based on the fusion of memristor storage and operation, this paper focuses on the complexity of existing logic operations and the importance of XOR logic in actual circuits. The following researches have been done:

1. This paper designs a new operation method to realize efficient XOR logic, and calls it "XOR logic operation based on memristor switch". This article gives its circuit structure, calculation sequence and constraint conditions, and provides theoretical support for subsequent research.
2. Use the designed XOR logic operation method to realize full-adder, and expand on crossbar array, so as to realize adder and improve the overall operation efficiency of adder.
3. Use the designed XOR logic operation method to realize encryption and decryption operations, and take AES algorithm as an example to perform XOR logic operations in SubBytes, ShiftRows, MixColumns, AddRoundKey and KeyExpansion crossbar array circuit design.

References

1. Murphy, S.: The advanced encryption standard (AES). Inf. Secur. Tech. Rep. **4**(4), 12–17 (1999)
2. Shao, F., Chang, Z., Zhang, Y.: AES encryption algorithm based on the high performance computing of GPU. In: Second International Conference on Communication Software and Networks, Singapore, 2010, pp. 588–590 (2010). https://doi.org/10.1109/iccsn.2010.124
3. Shute, M., Hennessy, J.L., Patterson, D.A.: Computer Architecture: A Quantitative Approach. Morgan Kaufmann Publishers Inc., California (1990). ISBN 1-55860-069-8. £24.95. Elsevier 24(1-2), 1993
4. Shao, F., Chang, Z., Zhang, Y.: AES encryption algorithm based on the high performance computing of GPU. In: 2010 Second International Conference on Communication Software and Networks, Singapore, pp. 588–590 (2010). https://doi.org/10.1109/ICCSN.2010.124
5. Rose, G.S., Rajendran, J., Manem, H., Karri, R., Pino, R.E.: Leveraging memristive systems in the construction of digital logic circuits. Proc. IEEE **100**(6), 2033–2049 (2012). https://doi.org/10.1109/JPROC.2011.2167489
6. Ho, Y., Huang, G.M., Li, P.: Nonvolatile memristor memory: device characteristics and design implications. In: 2009 IEEE/ACM International Conference on Computer-Aided Design - Digest of Technical Papers, San Jose, CA, pp. 485–490
7. Mazumder, P.: Memristor-based RRAM with applications. Sci. China (Inf. Sci.) **55**(06), 1446–1460 (2012)
8. Raja, T., Mourad, S.: Digital logic implementation in memristor-based crossbars. In: International Conference on Communications, Circuits and Systems, Milpitas, CA, pp. 939–943 (2009). https://doi.org/10.1109/icccas.2009.5250374
9. Kim, H., Sah, M.P., Yang, C., Chua, L.O.: Memristor-based multilevel memory. In: 2010 12th International Workshop on Cellular Nanoscale Networks and their Applications (CNNA 2010), Berkeley, CA, pp. 1–6 (2010). https://doi.org/10.1109/cnna.2010.5430320
10. Kvatinsky, S., Kolodny, A., Weiser, U.C., Friedman, E.G.: Memristor-based IMPLY logic design procedure. In: IEEE 29th International Conference on Computer Design (ICCD), Amherst, MA, pp. 142–147 (2011). https://doi.org/10.1109/iccd.2011.6081389
11. Kvatinsky, S.: MAGIC—Memristor-aided logic. IEEE Trans. Circuits Syst. II Express Briefs **61**(11), 895–899 (2014). https://doi.org/10.1109/TCSII.2014.2357292
12. Borghetti, J., Snider, G.S., Kuekes, P.J., Yang, J.J., Stewart, D.R., Williams, R.S.: 'Memristive' switches enable 'stateful' logic operations via material implication. Nature **464**(7290), 873–876 (2010)
13. Kvatinsky, S., Ramadan, M., Friedman, E.G., Kolodny, A.: VTEAM: a general model for voltage-controlled memristors. IEEE Trans. Circuits Syst. II Express Briefs **62**(8), 786–790 (2015). https://doi.org/10.1109/TCSII.2015.2433536
14. Shao, F., Chang, Z., Zhang, Y.: AES encryption algorithm based on the high performance computing of GPU. In: 2010 Second International Conference on Communication Software and Networks, Singapore, pp. 588–590 (2010). https://doi.org/10.1109/iccsn.2010.124
15. Chen, D., Qing, D., Wang, D.: AES key expansion algorithm based on 2D logistic mapping. In: 2012 Fifth International Workshop on Chaos-fractals Theories and Applications, Dalian, pp. 207–211 (2012). https://doi.org/10.1109/iwcfta.2012.81
16. Rao, F., Tan, J.: Energy consumption research of AES encryption algorithm in ZigBee. In: International Conference on Cyberspace Technology (CCT 2014), Beijing, pp. 1–6 (2014). https://doi.org/10.1049/cp.2014.1330

Digital Currency Investment Strategy Framework Based on Ranking

Chuangchuang Dai[1,2], Xueying Yang[1,2], Meikang Qiu[3], Xiaobing Guo[4(✉)], Zhonghua Lu[1,2(✉)], and Beifang Niu[1,2(✉)]

[1] Computer Network Information Center of the Chinese Academy of Sciences, Beijing, China
{dcc,zhlu,bniu}@sccas.cn, yangxueying@cnic.cn
[2] University of the Chinese Academy of Sciences, Beijing, China
[3] Department of Computer Science, Harrisburg University of Science and Technology, Harrisburg, PA 17101, USA
qiumeikang@yahoo.com
[4] The Lenovo Research, Beijing, China
guoxba@lenovo.com

Abstract. Quantitative research of digital currency is important in the financial sector. Existing quantitative research mainly focuses on pairs trade, multifactor models, and investment portfolios. Investment portfolios refer to the allocation of funds to different types of financial products to minimise investment risk when expected returns can be obtained, or maximise returns on investment when investment risks are controllable. Herein, we propose a ranking-based digital currency investment strategy framework for investment portfolios in the digital currency market. The framework mainly involves selecting digital currency attributes, preprocessing historical data, exporting the ranking model of the investment portfolio strategy, and parameter optimisation.

Keywords: Digital currency · Bitcoin · Investment portfolio · Parameter optimisation

1 Introduction

In April 2020, the global digital currency field underwent vigorous development, and Digital Currency Electronic Payment and Libra announced milestones in digital currency [1]. As of the end of 2017, the total value of the entire digital currency market reached a staggering USD 572.48 billion, and 1,381 types of digital currencies exist worldwide [2]. Many researchers have applied quantitative stock market research to the digital currency market and achieved results that were better than that of the true return rate of digital currencies [3]. Su tested more than 40 typically used stock-market factors in the digital currency market. After performing both single- and multifactor regressions, 16 factors were selected to construct a trading strategy. The profitability of the trading strategy proved the effective of all those 16 factors in the currency market [4].

The prices in the digital currency market and stock market are determined by demand. Stock price depends more on a company's fundamentals. However, the price of digital

M. Qiu (Ed.): ICA3PP 2020, LNCS 12454, pp. 654–662, 2020.
https://doi.org/10.1007/978-3-030-60248-2_45

currency is based more on human behaviour and emotions, and a strong relationship exists between different digital currencies. Whether the traditional investment portfolio method (i.e. investing funds at a certain ratio) is applicable to digital currencies has become a new challenging topic. Therefore, we herein propose a ranking-based digital currency portfolio framework.

The remainder of this paper is organised as follows: Sect. 2 describes the characteristics of digital currencies compared with other financial products and the selection of monetary attributes. Section 3 details investment portfolios. Section 4 introduces the ranking-based digital currency investment strategy framework proposed herein. The final section provides a summary of this study.

2 Characteristics of Digital Currencies and Selection of Monetary Attributes

2.1 Characteristics of Digital Currencies

Digital currency is a product of integrating Internet technology and finance. Compared with traditional banknotes, digital currency does not rely on monetary entities and can provide security features, such as unforgeability and tamper-resistance through cryptography [5]. Generally, digital currencies can be classified into two categories: private and legal digital currencies. Currently, most digital currencies circulating in the market are private digital currencies. Unlike stock products, digital currencies can be sold short and traded uninterrupted 24 h a day; furthermore, they are highly correlated with one another.

Private digital currencies can be categorised into centralised and decentralised digital currencies based on whether there is a central node exists to control each link (e.g. issuance and circulation) in the currency lifecycle. The characteristics of these two types of digital currencies are introduced below.

To further enhance the security and privacy of transactions, Chaum developed an electronic currency that can simulate physical coins and banknotes and then proposed the first centralised digital currency system—e-cash [6]. Centralised digital currencies typically exhibit the following characteristics [7]: 1) non-repeatability, i.e. digital currency can only be consumed once and double spending can be detected easily; 2) anonymity, i.e. the privacy of legitimate users is protected; 3) unforgeability, i.e. no individual can forge digital currencies that can be paid and circulated; 4) transferability, i.e. the ownership of a digital currency does not require the participation of a third party when being transferred, and neither can it be traced during the transfer process. In addition, because centralised digital currencies allow banks to participate in the currency issuance and circulation links, they are characterised by non-repeatability and unforgeability. However, they generally pose various problems, such as non-transferability, indivisibility, and non-aggregatability, which are not conducive to large transactions.

Decentralised digital currencies use peer-to-peer networks. Users are allowed to join the network to participate in the issuance, exchange, and circulation of currencies. Users in the system can create multiple transaction addresses to protect the privacy of their personal information. Decentralised digital currencies typically exhibit three characteristics. First, they are pseudonymous. Users do not reveal their true identities when using digital currencies. The second characteristic is unlinkability [8]. The unlinkability of users' transaction addresses means that external attackers cannot link different transaction addresses of the same user. The unlinkability of transaction inputs and outputs means that external attackers cannot link the sender and receiver of a transaction. The last characteristic is divisibility and aggregatability. Digital currencies have the same change and combined payment functions as physical currencies [9]. Price curve of Bitcoin cash (BCH) is shown in Fig. 1.

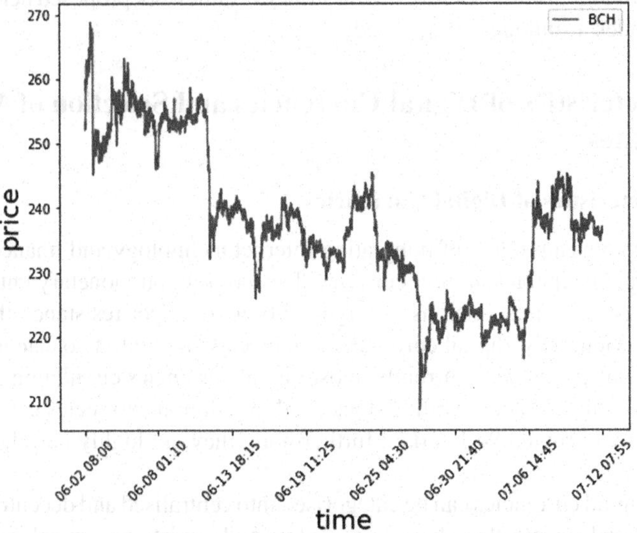

Fig. 1. Price curve of Bitcoin cash (BCH)

2.2 Selection of Monetary Attributes

The appropriate selection of monetary attributes (not necessarily of a significant quantity) can yield high returns on investment (ROIs). Selecting redundant monetary attributes will increase the computing amount and algorithm complexity, thereby causing unnecessary waste of computing resources. In normal cases, many attributes may be involved in completing a learning task. In most cases, however, only part of the attributes is required to make a reasonable and correct inference. Removal of weakly correlated attributes can reduce the difficulty of the learning task but important related attributes must be retained [10]. The importance score measures the value of each property in the algorithm's learning task. The higher the score, the greater is the importance, and the more important it will become in the learning task.

The importance of attributes is calculated in XGBoost [11] for a single decision tree by the amount that each attribute split point improves the performance measure; furthermore, it is weighted by the node that records the number of observations. Finally, the results of an attribute in all of the decision trees are weighted, summed, and averaged to obtain the final importance score. The monetary data used herein comprises different attributes, such as the highest price, lowest price, opening price, closing price, and trading volume. We randomly selected the data of six types of digital currencies and ranked the importance of their attributes. According to the rankings, we obtained the top three attributes: the closing, highest, and lowest prices.

3 Investment Portfolios

The initial idea of investment portfolios is relatively simple, i.e. "Do not put all your eggs in one basket". The more assets an investor selects, the lower are the risks faced by an investor. With the continuous expansion of the cognitive field of investors, popular ideas of investment portfolios have formed gradually. The way by which fund proportions are allocated and the selection of investment portfolio size have become topics of considerable interest to researchers.

Portfolio management is a decision-making process in which a certain amount of funds is continuously reallocated to different financial products to maximise returns and diversify risks. Traditional asset portfolio management methods can be classified into four types: follow-the-winner [12], follow-the-loser [13] pattern matching [14], and meta-learning [15]. The former two methods are based on pre-established models and requires the use of machine learning to determine certain parameters. These two methods do not offer stable performance and will render better results only in certain markets. Pattern matching is used for predicting the market distribution of the next stage based on historical samples; furthermore, it can maximise the investment portfolio based on historical samples. Meta-learning combines multiple strategies to obtain a more stable effect.

Markowitz [16], a Nobel Laureate in Economics, first propounded the mean-variance (MV) portfolio model, which measures asset risks and returns using the expectation and variance of the rate of returns. Using the MV model, he searched for efficient boundaries to obtain the optimal portfolio. However, this model regards fluctuations as risks, which means that fluctuations above the mean are risks as well. This approach does not conform to the thinking of real investors. In addition, the variance of the model is the average variance, in which events with small probabilities are disregarded. Based on the existing portfolio construction methods, researchers have conducted numerous extended studies. For example, Xu et al. [17] proposed an interval quadratic programming model for the construction of securities portfolios based on the traditional MV model for market conditions, where short selling is not allowed. Seyedhossein et al. [18] used the artificial bee colony and hybrid harmony search algorithms to obtain effective investment portfolios. This method is more effective for obtaining the optimal decision for returns and risks. Zhao et al. [19] leveraged the coefficient of variation method to improve standardisation in the factor analysis method, calculated the optimal investment ratio, and performed analysis and simulation in the Chinese market. Portfolio construction methods mainly involve optimising the solution method and reducing the computing amount. However, with the rapid development of computer technology, the once complicated computing process can be replaced by quantum computers. In addition, portfolio construction methods are susceptible to various external factors, such as the economic environment in

China and abroad, the international economic environment, the diversification of investment products, and particularly the returns on the investment before selection. If the previous return is positive, investors may change the preference weight to a large value, and vice versa. Therefore, it is particularly important to select the appropriate portfolio construction methods.

Many analytical studies related to investment portfolios have been conducted, most of which only focus on returns and risks. Simple and direct investment portfolios can no longer satisfy the changing needs of investors [20]. Therefore, an investment portfolio should contain two or more factors and comprehensively consider investors' attitudes towards risks and multiple evaluation indices [21]. For example, Xue designed a trading strategy based on the statistical results of mathematical models. She used ranking and scoring methods to filter effective factors, established an analytic hierarchy process weighting model, and designed a multifactor quantitative strategy. This method provides moderate reference for the majority of investors and is conducive to improving the level of investment management [22].

4 Digital Currency Portfolio Framework

4.1 Pre-processing of Historical Data

In the real world, most data are incomplete or inconsistent. Satisfactory results cannot be obtained by directly using these data even if the model is optimal. The data quality determines the upper limits of the algorithm results. Data preprocessing refers to the necessary processing (review, filtering, and ranking) of the acquired data. Methods such as data cleaning, integration, transformation, and reduction are typically used.

1Token organises and encapsulates the application programming interfaces (APIs) of different exchanges and provides investors with a unified API. Investors can use the unified API provided by 1Token and the corresponding exchange token to trade on the exchange. In this study, data were obtained as samples from the 1Token exchange every 5 min through the API of historical transaction data, among which six mainstream currencies (i.e. BTC, ETH, BSV, EOS, BCH, and XRP) were selected as investment targets. After analysing the data distribution characteristics, we segmented the sample data into a training and a test dataset.

4.2 Ranking-Based Model

Since the MV model was first proposed in 1952, research on investment portfolio decision-making has developed rapidly. Such research mainly focuses on theoretical analysis and method application. Investment portfolio theory has provided valuable theoretical support for reducing investment risks. Technical indices include the Sharpe ratio, mean/CVaR, mean/VaR, stochastic KDJ, ASI, BOLL, BRAR, and RSI.

In the results achieved by Gatys et al., Some of the higher layers of the generated image basically lost the information of the content map, and some of the generated images could not identify the original content image. So this article will discuss the impact of different parameters and make comparisons and analysis. The Sharpe ratio, proposed by William Sharpe in 1996, is an index measuring fund performance. It is based on the capital market line and uses the standard deviation of the fund investment portfolio as a risk measure. It is calculated using the equation $S_p = \sigma_p^{-1}\left(\overline{R_p} - \overline{R_f}\right)$, where S_p represents the Sharpe ratio, $\overline{R_p}$ the average return rate of a single fund in the sample period, $\overline{R_f}$ the riskless interest rate, and σ_p^{-1} the standard deviation of the return rate of a single fund during the sample period [23]. The Sharpe ratio represents the excess return that a fund can obtain when undertaking one unit of the total risk. The performances of funds are typically evaluated or ranked according to the Sharpe ratios of the funds. Top-ranked funds have better performances and are often purchased by fund investors. However, under the Sharpe ratio ranking, the fund performance is not sustainable, i.e. top-ranked funds in the previous period perform poorly in the later period. However, the trading frequency used in the proposed strategy will compensate for this shortcoming in the digital currency market.

The VaR [24] is a new risk indicator proposed by P. Jorion in 1997. It refers to the maximum possible loss of a certain financial asset or investment portfolio in a specific future time period (e.g. one day, one week, or 10 days) or under a certain confidence level in cases of normal market fluctuations. VaR is a statistical index for financial institutions to measure the possible loss of a securities portfolio under normal market fluctuations. This measure can ensure (with a high probability) that the maximum possible loss in the future does not exceed a certain value [25]. The efficient frontier of the mean-VaR investment portfolio is a subset of the MV efficient frontier. This means that the mean-VaR portfolio excludes some of the investment portfolios with lower expected return rates in the efficient frontier. Therefore, the mean-VaR investment portfolio narrows the range of investors' options and improves the speed and efficiency of investment decision-making [26].

Unlike the VaR, the CVaR refers to the conditional average of loss exceeding the VaR. Known as the average excess loss as well, the CVaR represents the average level of excess loss and reflects the average potential loss that may occur when the loss exceeds the VaR threshold. It can reflect the potential value at risk more efficiently than the VaR. The CVaR is superior over the VaR owing particularly to its unique subadditivity. In addition, the optimal portfolio model based on the CVaR can be transformed into a linear programming problem, thereby affording convenience to model solving. Hence, the CVaR is widely used in various fields that require risk control [27]. In the mean-VaR model, the mean or expected value is used to measure investment returns, whereas the VaR value is used to measure risks. The goal of investors using this model is still to maximise returns or minimise risks [28].

The scoring standard R is set as the weighted sum of the Sharpe ratio, mean/CVaR, and mean/VaR according to three variable parameters. We applied the buy/hold/sell trading strategy to evaluate the performance of the digital currency investment strategy [29]. At each trading time t + 24 h during the test period, we used the digital currency investment strategy to simulate trading as follows:

(1) At eight o'clock in the morning of the designated day, the trader uses scoring standard methods such as R to obtain the performance ranking, i.e. the performance

ranking of each digital currency. The trader buys the highest-scoring digital currency (i.e. the top-ranked digital currency).

(2) At trading time t + 24 h, the trader sells the digital currency bought at trading time t. Considering the possibility of short selling, we further adjusted the ranking model as follows: (a) At eight o'clock in the morning of the designated day, the trader uses scoring standard methods such as R to obtain the performance ranking, i.e. the performance ranking of each digital currency. The trader buys the highest-scoring digital currency A (i.e. the top-ranked digital currency) and sells the lowest-scoring digital currency B (i.e. the lowest-ranked digital currency B) simultaneously [30]. (b) At trading time t + 8 h, the trader sells the digital currency A bought at trading time t and buys the digital currency B bought at trading time t [31].

4.3 Calculation of ROIs

For the test dataset, we adopted the grid search method to optimise the parameters of the scoring standard R with parameter weighting and employed the task-level parallel method to accelerate the calculation. To obtain an excess return on digital currency investment, we used the test data from the previous section to calculate the cumulative ROI to report the performance of the model [32]. The ROI was calculated by adding up the return rates of the selected digital currencies every 24 h. The greater the ROI value, the better is the performance of the model. We calculated the cumulative ROI based on the following simple assumptions:

(1) The trader spends the same amount (e.g. USD50,000) on each trading day.
(2) The market always has sufficient liquidity such that the closing price of the buy order at time t is filled. The selling price is the closing price at time t + 24 h.
(3) Trading costs are disregarded.

5 Conclusion and Contribution

An investment strategy framework in the digital currency market was proposed herein. Under this framework, we applied investment portfolio indices (i.e. the Sharpe ratio, mean/CVaR, and mean/VaR) in the stock market to construct a ranking-based investment strategy and used the grid search method to optimise the parameters. This framework can be applied easily. In the future, we will consider adding new technical indices to the investment strategy of digital currencies and seek improved parameter optimisation methods.

Acknowledgment. This work is supported by the National Key R&D Program of China, under Grant No 2018YFB0203903 and Technological Achievements of Qinghai Province, China under Grant No. 2016-SF-127. The author Chuang-Chuang Dai would like to thank all the members of the Computer Network Information Center, Chinese Academy of Science, Beijing. In addition, professor Guangwu Liu in City University of Hong Kong and Rui Tao in Academy of Mathematics and Systems Science Chinese Academy of Sciences greatly helped the completion of the thesis. The authors wish it to be known that, in their opinion, the first 2 authors should be regarded as Joint First Authors.

References

1. Long, B.: Political economy analysis of global digital currency competition: public digital RMB DC/EP and private digital dollar Libra. J. Northeast Univ. Financ. Econ. 1–20 (2020). http://kns.cnki.net/kcms/detail/21.1414.F.20200708.1137.002.html
2. Song, W.: Research on digital currency portfolio strategy. Nanjing University of Information Technology (2019)
3. Wu, C.: GBR and other six regression models to study the multi factor arbitrage of digital currency. Huazhong University of Science and Technology (2019)
4. Su, C.: Research on the application of multi factor model in digital money market. University of Electronic Science and Technology (2019)
5. Fu, S., Xu, H., Li, P., Ma, T.: Research on anonymity of digital currency. Acta Sinica Sinica **42**(5), 1045–1062
6. Chaum, D.L.: Blind signatures for untraceable payments. In: International Cryptology Conference (2008)
7. Tian, Z., Li, M., Qiu, M., Sun, Y., Su, S.: Block-DEF: a secure digital evidence framework using blockchain. Inf. Sci. **491**, 151–165 (2019)
8. Narayanan, A., Bonneau, J., Felten, E.W.: Bitcoin and Cryptocurrency Technologies: A Comprehensive Introduction. Princeton University Press, Princeton (2016)
9. Okamoto, T., Ohta, K.: Universal electronic cash. In: Feigenbaum, J. (ed.) CRYPTO 1991. LNCS, vol. 576, pp. 324–337. Springer, Heidelberg (1992). https://doi.org/10.1007/3-540-46766-1_27
10. Qiu, M., Ming, Z., Li, J., Liu, S., Wang, B., Lu, Z.: Three-phase time-aware energy minimization with DVFS and unrolling for chip multiprocessors. J. Syst. Archit. **58**(10), 439–445 (2012)
11. Zhang, N.: Research on e-cash payment system and its key technologies. Information Engineering University of PLA (2005)
12. Zwirlein, T.J., Reddy, V.K., Doyle, B.: Follow the winner: periodic investing strategies. J. Financ. Plann. **8**(4) (1995)
13. Jeong, T., Kim, K.: Effectiveness of F-SCORE on the loser following online portfolio strategy in the Korean value stocks portfolio. Am. J. Theoret. Appl. Bus. **5**(1), 1–13 (2019)
14. Qiu, M., et al.: Data allocation for hybrid memory with genetic algorithm. IEEE Trans. Emerg. Top. Comput. **3**(4), 544–554 (2015)
15. Finn, C., Abbeel, P., Levine, S.: Model-agnostic meta-learning for fast adaptation of deep networks. arXiv preprint arXiv:1703.03400 (2017)
16. Markowitz, H.: Portfolio selection. J. Financ. **7**(1), 77–91 (1952)
17. Xu, X., He, F.: Interval quadratic programming of portfolio without short selling. China Manage. Sci. (3), 57–62 (2012)
18. Seyedhosseini, S.M., Esfahani, M.J., Ghaffari, M.: A novel hybrid algorithm based on a harmony search and artificial bee colony for solving a portfolio optimization problem using a mean-semi variance approach. J. Central South Univ. **23**(1), 181–188 (2016)
19. Zhao, J., Yang, Y., Zhao, L.: Research on portfolio problem based on improved factor analysis. Pract. Underst. Math. **2**, 44–49 (2015)
20. Gu, R.: A series of risk portfolio models with multiple indexes and its application. Yunnan University of Finance and Economics (2020)
21. Shao, Z., Xue, C., Zhuge, Q., Qiu, M., Xiao, B., Sha, E.H.M.: Security protection and checking for embedded system integration against buffer overflow attacks via hardware/software. IEEE Trans. Comput. **55**(4), 443–453 (2006)
22. Xue, H.: Research on multi factor stock selection strategy based on Shanghai and Shenzhen 300 component stocks. Hebei University of Economics and Trade (2020)

23. Jiao, D.: Research on fund performance evaluation based on improved sharp ratio. Financ. Econ. **20**, 81–83 (2018)
24. Jorion, P.: Value at Risk: The New Benchmark for Controlling Market Risk. Irwin Professional Publishing, Chicago (1997)
25. Kang, Z., Li, Z.: CVaR robust mean CVaR portfolio model and its solution. J. Oper. Res. **21**(01), 1–12 (2017)
26. Chen, M., Zhao, X.: Mean var portfolio optimization based on declu algorithm. Software **39**(10), 79–86 (2018)
27. Song, X., Han, L.: Research on currency allocation of foreign exchange reserve based on mean CVaR model. J. Beijing Univ. Aeronautics Astronaut. (Soc. Sci. Ed.) **25**(02), 82–87 (2012)
28. Lu, D.: Portfolio strategy analysis based on mean crvar model. Shandong University of Finance and Economics (2018)
29. Qiu, H., Noura, H., Qiu, M.: A user-centric data protection method for cloud storage based on invertible DWT. IEEE Trans. Cloud Comput. 1–1 (2019)
30. Gai, K., Qiu, M., Zhao, H.: Security-aware efficient mass distributed storage approach for cloud systems in big data. In: 2016 IEEE 2nd International Conference on Big Data Security on Cloud (BigDataSecurity), IEEE International Conference on High Performance and Smart Computing (HPSC), and IEEE International Conference on Intelligent Data and Security (IDS). IEEE, pp. 140–145 (2016)
31. Qiu, M., Xue, C., Sha, H.M., et al.: Voltage assignment with guaranteed probability satisfying timing constraint for real-time multiproceesor DSP. J. VLSI Sig. Proc. **46**(1), 55–73 (2007)
32. Gai, K., Qiu, M., Zhao, H.: Privacy-preserving data encryption strategy for big data in mobile cloud computing. IEEE Trans. Big Data 1–1 (2017)

Authentication Study for Brain-Based Computer Interfaces Using Music Stimulations

Sukun Li[1(✉)] and Meikang Qiu[2]

[1] Department of Mathematics and Computer Science, Adelphi University, Garden City, NY 11530, USA
sli@adelphi.edu

[2] Department of Computer Science, Texas A&M University-Commerce, Commerce, TX 75428, USA
Meikang.Qiu@tamuc.edu

Abstract. Electroencephalography (EEG) brain signals have been used in a number of pattern recognition studies. Due to the high degree of uniqueness and inherent security, EEG-based brain signals have been considered as a potential pattern for biometrics, especially for continuous authentication. Although previous research has been conducted with different visual stimuli, music stimuli for brain signal authentication are rarely considered. One unresolved question is whether these brain signals that change over time will affect the authentication results. An EEG database with 16 volunteers has been employed in the analysis to observe any changes in brain wave patterns as well as changes in authentication rates. The data were regularly collected over a period of four months while exposed to three different genres of music.

1 Introduction

Biometrics provide an effective way of realizing user authentication. Biometrics refers to the automatic recognition of individuals through their unique and measurable physiological or behavioral characteristics. Compared with the knowledge-based methods like passwords and tokens [1,2], the authentication of biometrics offers a more convenient and safer way to authenticate people by being an inherent quality of the individual. Conventional biometric techniques refer to fingerprint, face, voice recognition, etc. However, not all biometric technologies provide the same level of security: fingerprints may remain on surfaces, faces can be photographed, and voices can be recorded. In many ways this biometric information is somehow "public" as it can be captured without the knowledge and replicated to bypass whatever security measures are in place. In addition to security issues [3,4], most conventional biometrics are not suitable, or at least are inadequate, for continuous authentication as the acquisition of data requires the user to be interrupted during normal operations.

Electroencephalography (EEG) has been considered as a potential option for reliable user authentication. Previous research in both early neurophysiologic

M. Qiu (Ed.): ICA3PP 2020, LNCS 12454, pp. 663–675, 2020.
https://doi.org/10.1007/978-3-030-60248-2_46

studies and biometric studies have already pointed out that EEG signals provide relevant information about individual differences. Research indicated that by controlling the cognitive state of users through specific stimulation modes, user identification can be achieved based on the differential responses of different users to the stimulation. Brain-Computer Interface (BCI) based EEG biometric systems also have the added benefit of allowing for high-security and continuous authentication. The potential to combine with other human modality devices and mix reality applications [5,6] make this novel biometric become adaptable for not only the general people but also beneficial for people with disability.

The majority of previous studies in EEG biometrics conducted make use of either resting states or visual-based stimuli to evoke a response from subjects. Research rarely considered auditory stimuli to evoke signals for EEG-based authentication. Besides, the EEG data generated by the brain usually changes over time, which leads to higher experimental and computational costs for data collection and analysis. Although studies have shown that it is possible to correctly predict the user identification and verification with an accuracy up to 99.7% [7], the changes in EEG signals over time and whether collected samples can be reused in the future as continuous identification has not been thoroughly investigated. Therefore, our research attempts to fill the gap with a novel authentication framework.

The goal of this study is to expand upon the EEG based authentication system initially developed in the previous research. We analyze the EEG database, which collected under multi music stimulation modality and over four months with nine experiment weeks. The first week of recorded brain signals will then be used to train a machine learning classifier, using the recently established model [8] for feature extraction. Signals collected during subsequent weeks from the same volunteers will then be classified using this model in order to record any potential change in classification rates. The experiments also evaluated the changes in brainwave patterns that could suggest at what point an EEG sample is no longer suitable for authentication. Based on the best of our current knowledge, our research is the first one to conduct multi music stimulation with a multi-dialogue long-term session in the field of EEG-based biometrics.

The rest of the paper is organized as follows: Sect. 2 outlines the related works of EEG-based biometrics. Section 3 describes the methodology, including experimental music stimuli modality, the data preparation process, as well as the feature extraction model and classification. Section 4 presents experiments, and discusses the obtained experimental results and future works. Section 5 is the conclusion.

2 Literature Review

Event-related potential (ERP) is a common methods for EEG-based study. The brain signals generated in the brain structures in response to specific events or stimuli [9]. Despite recent interest in this area of research, stimulus tends to fall mainly into merely three categories: Rest, Text, and Images or Video.

Resting state stimulus with eyes-closed and eyes-open is a baseline in the study of EEG for biometrics. Although that classification result reached a relatively high accuracy rate in [7,10], which demonstrates the potential of the use of EEG as biometrics, the resting stage EEG signals is not enough to be applied for real-world applications. Since in real cases, people usually not only stay resting when interacting with computers or other devices.

Researchers at Nanyang Technological University performing classification tests on 109 subjects in resting states were able to achieve an accuracy of 99.7%for beta-band in frequency (15–30 Hz) [7]. Rahman and Gavrilova found in their study of text-evoked brain signals that alpha brainwaves were the most effective with their K-Nearest Neighbor (KNN) and Backpropagation Neural Network (BPNN) models [11]. Another text stimuli study, performed at Tianjin University, was able to achieve over 99% accuracy in their classification of EEG data with the use of a Convolutional Neural Network (CNN) [12]. Although they go on to stress that the model lacks the ability to train data over time and that more research is required to test a CNN's ability to handle new EEG data to authenticate past users. Koike-Akino et al. completed a study that made use of Zener Cards [13] as a form of digital image stimulus. 25 volunteers were tasked with selecting one of five cards on the screen and counting the number of times their card appeared. This data set in conjunction with Principal Component Analysis (PCA) as feature extraction and Quadratic Discriminant Analysis (QDA) as classifier reached 96.7% accuracy [14]. A smaller study of only four subjects attempted to "investigate the efficacy of self-related visual stimuli" [15] by showing subjects pictures of different faces. With a classifier based on Pearson's Correlation Coefficient (PCC) they were able to finish this study with an average accuracy of 87.5%.

The research of biometrics based on music stimulation to evoke EEG signals is a novel area. The EEG-based applied research can be summarized as visual stimuli in the large majority of cases. Other sensory stimuli, such as smell [16], auditory [17], or touch stimuli are rarely considered. Recent studies have shown that Music, in particular, has a well-documented effect on brain waves with EEG measurement. Straticiuc et al., who performed "a preliminary analysis of music on human brainwaves" found that music resulted in an increase in alpha waves and a dramatic decrease in beta waves [17], but no attempt at classifying subjects was made. In addition of the EEG-based applied research, prior research in the field of brain-based biometrics has done much to establish three major benefits to using brain signals over other types of biometric data. Namely, that brain signals are unique among individuals, very difficult to circumvent, and the features extracted have a high degree of permanence [18]. However, as the authors of [18] point out most of the research focused on the optimization of the enrollment and authentication but rarely evaluated the quantifying permanence of EEG data and the stimuli that can be sustainably induced for authentication systems [19]. In [20], researchers indicated that quantify the effects of aging on brain signals and they found that the authentication rates could dropped from 100% to 70% over the length of time when using a Random Forest classifier.

Inspired by the previous studies, this research attempts to fill the gap with different types of multi music stimulation to evoke brain signals for a multi-dialogue long-term evaluation in EEG-based authentication studies. The following sections describe the details.

3 Methodology

According to a prior study for EEG channel selection [21] and with consideration for the practical application needs of authentication via EEG measurement, the Opening Ultracortex Mark IV with Cyton board [22] is used. We selected eight channels for data analysis, of which six channels (T3, T4, C3, C4, Cp5, Cp6) were selected proximity the auditory cortex, and two of them were used as general channels in the frontal and occipital regions (Fz and Oz). The default amplifier 250 Hz sampling frequency and 2 references (A1, and A2) located on the earlobe. The EEG channels positions is illustrated in Fig. 1 accordance with the international 10–20 system.

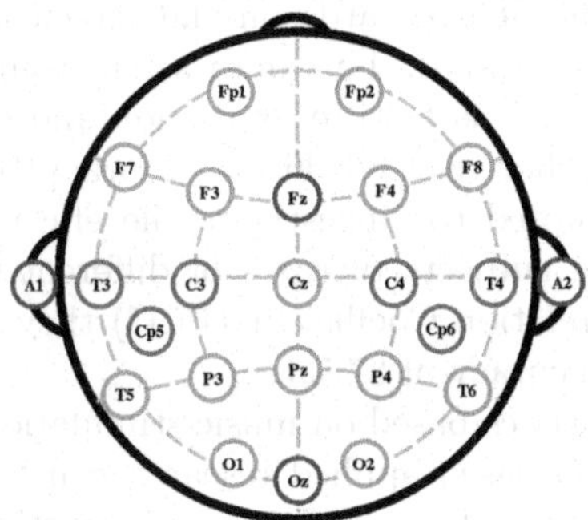

Fig. 1. EEG electrode positions: left (C3, T3, and Cp5), vertex (Cz and Fz), and right (C4, T4, and Cp6), with 2 references (A1 and A2)

The music track is carefully designed as a specific modality with three different types of music. Subjects participated in a 5-min session, wherein they listened to each of the three different types of music stimuli. The time frame of the experimental session is illustrated in Fig. 2. The first 60 s is for resting, then 60 s of classical (M1) music, 30 s of rest, 60 s of jazz (M2) music, 30 s of rest, and 60 s of electronic (M3) music. The time during music stimulus (M1 to M3) are considered as the valid data for user authentication purposes.

3.1 EEG Data Acquisition and Prepossessing

The EEG data was continuously collected from 16 healthy participants at multiple sessions. The average age of the participants was 26 ± 3 years old, including

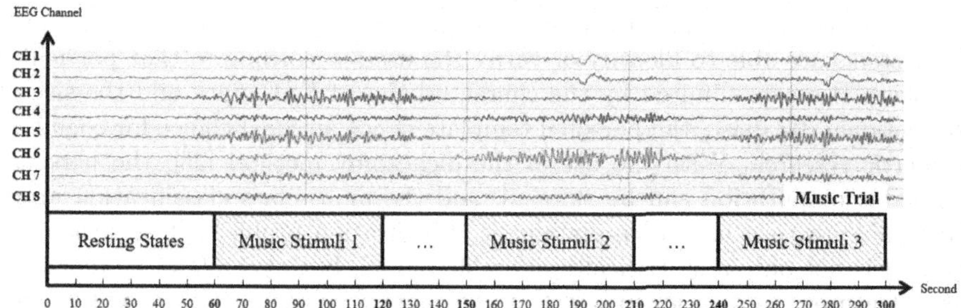

Fig. 2. The 5-min music trial procedure for EEG evoked stimuli.

43.8% females and 56.3% males. Before each day of the data collection session, the participants received 7 h of sleep on average. During the experiment, participants participated in EEG data collection once a week and reported their current health status before taking measurements. Because the study period was several months, 5 of the participants only participated in part of the data collection (6 weeks). The rest of them participated in all the sessions.

The purpose of this study is to explore the feasibility of EEG authentications with music stimulations, hence the data preprocessing steps have been simplified as much as possible. A necessary signal processing procedure is used to prepossess the raw EEG data. A notch filter is applied to remove background noise from power lines follow with a band-pass filter. During the collection process, some channels were challenging to make a comfortable connection with the subject or even could become unconnected. Therefore those 'bad channels' have to removed from the collection during the preprocessing.

3.2 Feature Extraction and Classification

EEG data can be represented from three different domain, and the Dynamic Histogram Measurement (DHM) is one of highly effective method for analyzing the time-frequency properties of EEG data [8]. It involves computing a signal data based on the frequency distribution (standard histogram, frequency domain) and taking slope computing of the frequencies (dynamic histograms, time-frequency domain). Figure 3 illustrates the idea of the histogram feature extraction process for a data set [8].

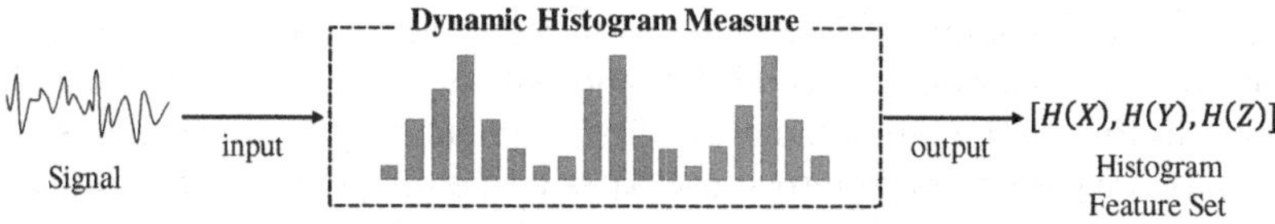

Fig. 3. The Dynamic Histogram Measurement algorithm

Biometric systems usually require to design personal templates, in which the systems are able to identify or verify the users according to that particular template. However, limited by the characteristics of EEG data and the size of conventional database, the personal template is not an ideal choice for cognitive biometric research. With the use of the dichotomy method [23], the personal template computation can be skipped, and the multi-class classification problem is able to turn into a simple binary classification problem. The dichotomy method is not only a statistically inferable model, but also allows nonhomogeneous features. As mentioned in [21], the dichotomy method is more suitable for the EEG based biometrics research because the human brain produces natural signal features, and they are not homogeneously displayed. Therefore, before passing the original features into classifiers, we applied a dichotomy transformation on extracted vectors by pair up them with a distance calculation.

3.3 Algorithm Design

In this section, we give descriptions of our proposed algorithm. The main algorithms support our authentication system which is the Dynamic Histogram Measurement (DHM) and Segmented Dynamic Histogram Transformation (SDHT) algorithm. The sections below explain the detailed mechanism of the algorithm.

Dynamic Histogram Measurement (DHM) algorithm is design for signal feature extraction with dynamic computes the inter difference of the input sequence and output a feature set including histogram information of the input sequence. Pseudo codes of DHM algorithm are shown in Algorithm 1. We describe the main steps of Algorithm 1 in the following statement:

1. Input subsequences X
2. Dynamic computes the slopes of each input subsequence X and obtained slope sequences, Y and Z.
   (a) Input subsequence: $X = \{x_1, \ldots, x_n\}$ with n number of samples.
   (b) Obtained Y sequence: Absolute distance between adjacent samples. $Y = \{y_1, y_2, ...y_k\}$ is the first slope sequence with k samples, where $k = n - 1$.
   (c) Obtained Z sequence: $Z = \{z_1, z_2, ...z_{k-1}\}$ as the second slope sequence with $k - 1$ samples.
3. Histogram measure $h(f)$ on the subsequences X, Y, and Z in order to obtain frequency of the data samples within successive numerical intervals, respectively.
4. Generate histogram feature set H

Segmented Dynamic Histogram Transformation (SDHT) algorithm is inspired by the dichotomy method [23] and enhance with the DHM algorithm. As shows in Fig. 4, the input signal are first segment into small segments. When a non-stationary data is cut into small parts based on its time information, and each part is assumed as relatively stationary data. Then the segments are individually pass though the DHM to calculate their dynamic histogram information. Finally pass through a feature domain transformation function.

Algorithm 1: Dynamic Histogram Measurement

```
Require: X_{i,1~k}
Ensure : H
Input X_{i,1~k}
for each X_{i,1~k} do
    for j ← k to 1 do
        if j > 2 then
            calculate the seconde slope Z_{j-1};
        end
        calculate the first slope Y_j;
    end
    measuer histogram frequency h(f) for sequences X, Y, Z
end
Generate H_{1~d} = [h(X), h(Y), h(Z)];
Output H
```

Pseudo codes of SDHT algorithm are shown in Algorithm 2. We describe the main steps of Algorithm 2 in the following statement:

1. Input sequences D
2. Segment D with window function and overlapping in to k subsequence $\{X_1, X_2, ...X_k\}$
3. If a subsequences is match execute the DHM model.
4. Transform the histogram feature from a original domain to distance domain by computing from each pair of the features.
   (a) Intra distance label as 1
   (b) Inter distance label as 0
5. Generate the combination feature set F_{com}

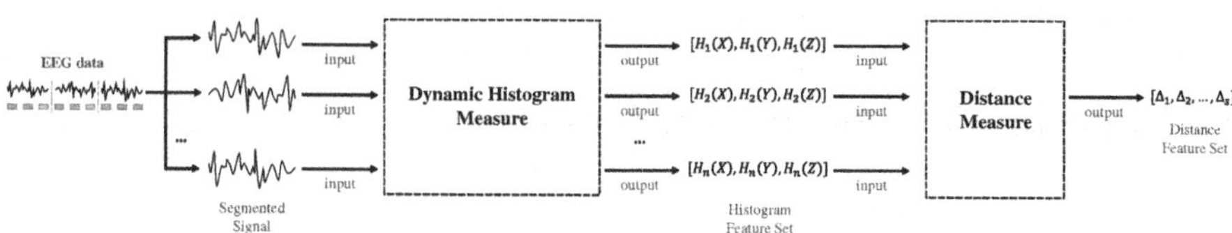

Fig. 4. Illustration of the feature extraction progress in Segmented Dynamic Histogram Transformation algorithm

According to the Algorithm 2, these features are transformed from a feature domain to a feature distance domain by computing from each pair of the features. The δ can be the certain respective distance measure depending on the type of feature but here the Euclidean distance is used.

Algorithm 2: Segmented Dynamic Histogram Transformation (SDHT) algorithm

```
Require: X_{1~n,1~p}, l_{1~n}, W, m
Ensure : F_com
Input X_{1~n,1~p}, t_{1~n}, m
for i ← 1 to n do
    segment X with window W and overlapping;
    if X_i ∈ X then
        | Execute DHM Algorithm /* Algorithm 1 */
    end
end
for i ← 1 to m do
    randomly select H_x ;
    randomly select another H_y where l_x = l_y ;
    w_i = δ(H_x, H_y) ;
end
for i ← 1 to m do
    randomly select H_x ;
    randomly select another H_y where l_x ≠ l_y ;
    b_i = δ(H_x, H_y) ;
end
Obtain the values of F_com
Output F_com
```

3.4 Classification

The features are trained in a Support Vector Machine (SVM) algorithm with a linear kernel, and its accuracy through 10 cross-validations. We majorly focused on the frequency 4 Hz to 45 Hz in this study, and the data are downsampled 128 Hz to reduce computing complexity. The results will be given and discussed in the next section.

4 Experiments and Result Discussions

This section describes the details of the experimental and presents the results that were obtained after performing various experiments on the collected dataset.

As describes in Sect. 3, we selected eight channels for data analysis (T3, T4, C3, C4, Cp5, Cp6, Fz, and Oz), which are near the auditory cortex area and two in the frontal and occipital region. During the experiments, the collected EEG data also segmented based on the music track time series, and the time range for M1 is [60:120], M2 is [150:210], and M3 is [240:300]. Each sequence was then evaluated for the authentication performance of individual channels. A notch filter 60 Hz is then applied to remove background noise from power lines. Follow by a downsampled step to reduce the data size 200 Hz to 128 Hz. All the performance has been evaluated in using 10-fold cross validation and the first week of the EEG data has been used in training the subsequent weeks are used for testing. The performance of the system was also evaluated by individual EEG channels and the music stimuli types. In this study, 16 subjects volunteered to participate in data collection, but due to the longer time period and some personal reasons, 6 of them partially withdrew from the experiment halfway. Since incomplete data sets may affect some experimental data results, we decided to separate them for experiment and discussion.

Table 1. EER for M1 experiments which have 16 subjects participant over the first 6 weeks

Channels	W1	W2	W3	W4	W5	W6	Average
T3	0.06	0.06	0.08	0.07	0.08	0.06	0.07
T4	0.07	0.05	0.06	0.07	0.08	0.07	0.06
C3	0.07	0.07	0.10	0.04	0.10	0.09	0.08
C4	0.05	0.07	0.04	0.06	0.06	0.08	0.06
Cp5	0.03	0.05	0.10	0.05	0.08	0.03	0.06
Cp6	0.06	0.03	0.05	0.04	0.07	0.04	0.05

The Equal Error Rate (EER) values from each signal channel are calculated and given in the following Tables, which include all 16 subjects. Overall, different types of music stimuli get relatively consistent average EER values and the median value is 0.06. In Table 1, the channel T4, C4, and Cp6 shown better performance then the other channels. These channels are located on the scalp in the right cerebral cortex. The interest part is for M3 types (Electronic music) the EER results are exactly the opposite. Table 3 shown the authentication performance with electronic music. The channel T3, C4, and Cp6 shown relative better performance than the other channels (Table 2).

Table 2. EER for M2 experiments which have 16 subjects participant over the first 6 weeks

Channels	W1	W2	W3	W4	W5	W6	Average
T3	0.06	0.04	0.09	0.08	0.07	0.04	0.06
T4	0.04	0.05	0.11	0.12	0.09	0.11	0.09
C3	0.05	0.07	0.09	0.10	0.09	0.09	0.08
C4	0.04	0.04	0.08	0.08	0.07	0.08	0.07
Cp5	0.03	0.06	0.11	0.11	0.10	0.09	0.08
Cp6	0.03	0.03	0.05	0.05	0.06	0.06	0.05

Table 3. EER for M3 experiments which have 16 subjects participant over the first 6 weeks

Channels	W1	W2	W3	W4	W5	W6	Average
T3	0.06	0.06	0.07	0.08	0.05	0.06	0.06
T4	0.06	0.11	0.13	0.04	0.06	0.13	0.09
C3	0.06	0.07	0.09	0.11	0.03	0.08	0.07
C4	0.05	0.08	0.08	0.10	0.06	0.11	0.08
Cp5	0.04	0.06	0.10	0.09	0.08	0.07	0.07
Cp6	0.04	0.10	0.10	0.09	0.07	0.07	0.08

Table 4 given accuracy values for single-channel modalities for 11 subjects across weeks 7–9 for the remaining three weeks, only 11 subjects were available to continue. The best accuracy rate achieved was 98.02%, achieved on Week 9 on the Cp5 channel. Figure 5 shows the average accuracy values across all channels from session one to nine. The first six weeks of experimentation were carried out with 16 subjects. These overall results show that as users become

Table 4. Accuracy for single-channel modalities for 11 subjects across weeks 7–9

Week	Channels	Accuracy (%)			
		M1	M2	M3	Average
Week 7	T3	76.14	84.70	84.90	81.91
	T4	89.12	83.49	89.30	87.30
	C3	87.20	87.15	91.70	88.68
	C4	84.75	86.48	88.28	86.50
	Cp5	87.49	90.17	85.40	87.69
	Cp6	86.19	90.10	92.21	89.50
Week 8	T3	84.90	88.15	95.44	89.50
	T4	87.59	83.69	95.05	88.78
	C3	92.62	94.07	89.81	92.17
	C4	80.88	91.12	94.65	88.88
	Cp5	91.39	90.70	96.15	92.75
	Cp6	83.27	82.50	86.04	83.94
Week 9	T3	85.94	88.93	86.48	87.12
	T4	85.75	86.11	90.34	87.40
	C3	91.18	91.65	94.02	92.28
	C4	93.00	94.65	89.41	92.35
	Cp5	98.02	96.71	92.05	95.59
	Cp6	88.51	87.20	82.75	86.15

more accustomed to the musical stimuli over an extended period of time is also light on trends inherent to this type of stimulus or authentication model.

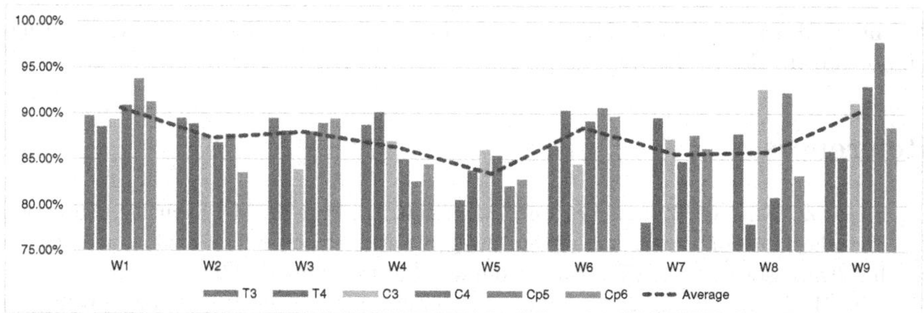

Fig. 5. The average accuracy for single-channel modality for session 1 to session 9 during four months

In this experiment, we also found that the verification results decreased significantly in the fourth and fifth weeks of the experiment then began to rebound. Will it be because of other external factors that affect the user's brain? Or may because they are familiar with the music and no longer concentrate during the collection? We tried to find the correlation and compared the EEG dataset with the meta dataset. The metadata set mainly records the health status of the subjects on the day of data collection, including caffeine, sleep, alertness, and other relative information. Through analysis, we found that most of the participants have relatively high-level self-alertness and low caffeine intake. The difference between EEG and traditional biometrics is the participation of human cognition. Whether different musical stimuli or different emotions will affect the classification results is an interesting question.

Although the achieved results in this paper look promising, many issues need to be addressed in future work. As mentioned in the previous section, the research was designed as an experiment for up to 4 months and planned to analyze the data once a week. Even though we arranged the research as much as possible, only nine consecutive effective data collection sessions were performed. In future research, extending the collection cycle and expanding the number of subjects is our first major work.

5 Conclusions

This study explored the use of different musical stimuli to authenticate EEG-based biometrics. We propose a novel algorithm to test the feasibility of the brain-computer interface system with music stimulation for user authentication. This approach was capable of performing user authentication across nine sessions. These results show that the musically evoked response carries participant discriminating features, which can be potentially employed as a biometric using Brain-Computer Interface devices. Analysis of authentication rates as users become more accustomed to the musical stimuli over an extended period of time is also light on trends inherent to this type of stimulus or authentication model.

Acknowledgment. Thank Mr. Leonard Marino and Ms. Vineetha Alluri for their support on the data collection.

References

1. Shao, Z., Xue, C., Zhuge, Q., Qiu, M., Xiao, B., Sha, E.-M.: Security protection and checking for embedded system integration against buffer overflow attacks via hardware/software. IEEE Trans. Comput. **55**(4), 443–453 (2006)
2. Qiu, H., Noura, H., Qiu, M., Ming, Z., Memmi, G.: A user-centric data protection method for cloud storage based on invertible DWT. IEEE Trans. Cloud Comput., 1 (2019)
3. Gai, K., Qiu, M., Zhao, H.: Privacy-preserving data encryption strategy for big data in mobile cloud computing. IEEE Trans. Big Data, 1 (2017)
4. Gai, K., Qiu, M., Zhao, H.: Security-aware efficient mass distributed storage approach for cloud systems in big data. In: IEEE 2nd International Conference on Big Data Security on Cloud (BigDataSecurity), IEEE International Conference on High Performance and Smart Computing (HPSC), and IEEE International Conference on Intelligent Data and Security (IDS), pp. 140–145 (2016)
5. Li, S., Leider, A., Qiu, M., Gai, K., Liu, M.: Brain-based computer interfaces in virtual reality. In: IEEE 4th International Conference on Cyber Security and Cloud Computing (CSCloud), pp. 300–305 (2017)
6. Li, S., Savaliya, S., Marino, L., Leider, A.M., Tappert, C.C.: Brain signal authentication for human-computer interaction in virtual reality. In: IEEE International Conference on Computational Science and Engineering (CSE) and IEEE International Conference on Embedded and Ubiquitous Computing (EUC), pp. 115–120 (2019)
7. Thomas, K.P., Vinod, A.P.: Biometric identification of persons using sample entropy features of EEG during rest state. In: IEEE International Conference on Systems, Man, and Cybernetics (SMC), pp. 003 487–003 492 (2016)
8. Li, S.: Dynamic histogram for time-series brain signal classification. Ph.D. dissertation, Pace University (2019)
9. Blackwood, D.H.R., Muir, W.J.: Cognitive brain potentials and their application. Br. J. Psychiatry **157**(S9), 96–101 (1990)
10. La Rocca, D., et al.: Human brain distinctiveness based on EEG spectral coherence connectivity. IEEE Trans. Biomed. Eng. **61**(9), 2406–2412 (2014)
11. Rahman, M.W., Gavrilova, M.: Overt mental stimuli of brain signal for person identification. In: IEEE International Conference on Cyberworlds (CW), pp. 197–203 (2016)
12. Di, Y., An, X., Liu, S., He, F., Ming, D.: Using convolutional neural networks for identification based on EEG signals. In: IEEE 10th International Conference on Intelligent Human-Machine Systems and Cybernetics (IHMSC), vol. 2, pp. 119–122 (2018)
13. Kennedy, J.L.: The visual cues from the backs of the ESP cards. J. Psychol. **6**(1), 149–153 (1938)

14. Koike-Akino, T., et al.: High-accuracy user identification using EEG biometrics. In: 38th Annual International Conference of the IEEE Engineering in Medicine and Biology Society (EMBC), pp. 854–858 (2016)
15. Thomas, K.P., Vinod, A., et al.: EEG-based biometrie authentication using self-referential visual stimuli. In: IEEE International Conference on Systems, Man, and Cybernetics (SMC), pp. 3048–3053 (2017)
16. Kotas, R., Ciota, Z.: Olfactory event-related potentials recordings analysis based on modified eeg registration system. In: IEEE 21st International Conference Mixed Design of Integrated Circuits and Systems (MIXDES), pp. 512–516 (2014)
17. Straticiuc, V., Nicolae, I., Strungaru, R., Vasile, T., Băjenaru, O., Ungureanu, G.: A preliminary study on the effects of music on human brainwaves. In: IEEE 8th International Conference on Electronics, Computers and Artificial Intelligence (ECAI), pp. 1–4 (2016)
18. Höller, Y., Uhl, A.: Do EEG-biometric templates threaten user privacy?. In: 6th ACM Workshop on Information Hiding and Multimedia Security, pp. 31–42 (2018)
19. Li, S., Marino, L., Alluri, V.: Music stimuli for EEG-based user authentication. In: The Thirty-Third International Flairs Conference (2020)
20. Kaur, B., Kumar, P., Roy, P.P., Singh, D.: Impact of ageing on EEG based biometric systems. In: IEEE 4th IAPR Asian Conference on Pattern Recognition (ACPR), pp. 459–464 (2017)
21. Li, S., Cha, S.-H., Tappert, C.C.: Biometric distinctiveness of brain signals based on EEG. In: IEEE 9th International Conference on Biometrics Theory, Applications and Systems (BTAS), pp. 1–6 (2019)
22. OpenBCI: Software Screenshot (2019) https://openbci.com/
23. Srihari, S.N., Cha, S.-H., Arora, H., Lee, S.: Individuality of handwriting. J. Forensic Sci. **47**(4), 1–17 (2002)

An Ensemble Learning Approach to Detect Malwares Based on Static Information

Lin Chen[1(✉)], Huahui Lv[2], Kai Fan[2], Hang Yang[2], Xiaoyun Kuang[1], Aidong Xu[1], and Siliang Suo[1]

[1] Electric Power Research Institute, CSG, Guangzhou 510663, China
chenlin_temp@163.com
[2] China Southern Power Grid Co., Ltd., Guangzhou 510663, China

Abstract. The proliferation of malware and its variants have brought great challenges to malware detection. The traditional static analysis methods are complicated and consume a lot of human resource. Moreover, most of the current detection methods mainly focus on the single characteristic of malware. To address the above issues, this paper proposes an Ensemble Learning approach method to detect malwares based on static information. The image feature and entropy features are used separately to train two models. Besides, with the guidance of ensemble learning principle, the two models are combined and obtain better accuracy compared with each of two models. We conduct comprehensive experiments to evaluate the performance of our approach, the results show the effectiveness and efficiency.

Keywords: Ensemble learning · Malware detection · Static information

1 Introduction

In recent years, due to the development of technology and the huge benefits brought by network attacks, the number of malicious software has increased dramatically, and the whole industry has been developing on a large scale. According to kaspersky Security Report 2019, 19.8% of users experienced at least one malware attack in 2018. The ransomware targeted 755,485 individual user computers, including 209,679 enterprise users and 22,440 smes [1]. The 2019 Internet Security Threat Report noted that the use of malware to disrupt commerce and trade increased by 25% compared to 2017, and ransomware attacks on businesses increased by 12%. In addition to the increasing number of attacks, criminals also adopt a variety of new attack strategies, such as malicious E-mail and hijacking legal tools [2]. Therefore, both the number, type and speed of malicious attacks are constantly increasing. In addition, people's life is more and more closely connected with computers, various malware attacks have greatly affected the normal development of personal life and even the society.

Malware is often used to steal confidential information, spread advertising, destroy data, attack servers and other targets, so many businesses and users face a huge security threat. CrowdStrike's Global Threat Report 2020 shows that multiple large attacks

M. Qiu (Ed.): ICA3PP 2020, LNCS 12454, pp. 676–686, 2020.
https://doi.org/10.1007/978-3-030-60248-2_47

in 2019 are among the most profitable ways for cybercriminals. Attackers also try to use sensitive data as a weapon, making victims face more pressure [3]. Over the past five years, there have been a number of malware attacks, including one of the most famous ransomware, SamSam, which has paralyzed medical facilities in Atlanta and the Colorado Department of Transportation. In addition to individuals and businesses, criminals can even pose a serious threat to national security by stealing state secrets, attacking infrastructure to disrupt the normal functioning of society, and undermining other countries' defense systems.

Malware detection methods mainly include two aspects: static analysis and dynamic analysis. Static analysis does not need to run the program to detect the executable file, while dynamic analysis checks its behavior by running the executable program. The two have their own advantages and limitations, and are often complementary to each other. For Windows executable, the static information is extracted from two aspects: binary contents of the executable and the disassembly files of the executable. For Android apps, static features can be extracted from the Android app installer using disassembly. Compared with the method of dynamic analysis, static analysis and detection are faster, cheaper and consume less resources, so this paper will focus on static analysis.

Different from traditional methods that only consider the individual features of malware, this paper extracts multiple features of malware and integrates them to analyze the malicious code from different levels, so as to better detect the malware. Because traditional detection methods are time-consuming, complicated in analysis process and consume a lot of resources, this paper converts the features of malicious code into images and uses convolutional neural network to classify the generated images, including binary machine code images [4] and information entropy images [5]. The process of this method is simple and efficient. In addition, deep learning algorithm can automatically learn the image characteristics of malicious code and benign code, and can be used to process large amounts of data, so it provides the possibility for the application of this method in the actual big data set. The detection model of multi-image feature fusion is divided into three modules: the detection model based on binary machine code image, the detection model based on information entropy image and the fusion model based on ensemble learning.

The main contribution of this paper is as follows:

- A detection model based on binary machine code image is proposed to convert the binary machine code of malware and benign software into RGB image for obtaining more information. The generated binary machine code images are loaded into the convolutional neural network to obtain the classification results.
- A detection model based on information entropy image is proposed, which converts sample data into information entropy image by using information entropy formula, and is used to extract the overall structure characteristics of malware. The convolutional neural network is trained to classify information entropy images.
- A detection model based on ensemble learning is proposed. Two trained convolutional neural networks are integrated as basic learning devices to fuse two image features to improve the accuracy of malware detection.

- We collected data and completed pre-processing, designed and carried out experiments, and used various indicators to evaluate the effect of the model. Experimental results verify the effectiveness of the proposed method.

The reminder of this paper is Structured as follows. We describe the background and related work in Sect. 2. Then we give the design and implementation of our approach in Sect. 3. Section 4 demonstrates some experiments to evaluate the effectiveness and efficiency of our approach. We close our paper with the conclusion in Sect. 5.

2 Related Work

In recent years, the application of polymorphism and deformation technology has led to the fact that malware can hide its true purpose and escape the monitoring of the protection mechanism, which also leads to the continuous increase of the evolution and number of malware. Therefore, the detection time and economic cost increase, and the process is more complicated than before. To solve the above dilemma, experts try to combine machine learning with malware detection. The static analysis method based on machine learning can be divided into the following aspects.

String analysis refers to the extraction of a sequence of characters from an executable file and is the simplest way to obtain the functionality of a program. The information obtained by this method includes application menu item name, domain name, Internet protocol address, attack instruction, registry key, file location modified by the program, etc. String analysis is often combined with other static or dynamic detection methods to improve defects. Ye et al. [6] extracted strings from application program interface (API) calls and semantic sequences that reflect attack intent in 2008 to detect malware. The system includes a parser to extract interpretable strings from each portable executable file (PE file), and a support vector machine (SVM)-based detector. The performance of the system was evaluated on samples collected by Kingsoft anti-virus Laboratory and it performed better than some common anti-virus software.

N-gram [7–9] is one of the most commonly used features of malicious code detection and classification, and is a very important statistical language model. For binary files, n-gram consists of N consecutive bytes. For a file resulting from a disassembly operation, N sequential assembly instructions form an N-gram, in which case only mnemonics are preserved. Although the N-gram method achieves good results, it wastes a lot of computing resources to select a large number of N-gram features, and it is impractical to list all the N-gram features, so the application of n-gram in large data sets is restricted.

One of the most prominent features of a program is the API and its function calls, so this feature is often used to simulate the behavior of a program. In particular, API functions and system calls are closely related to the services provided by the operating system, such as networking, security, file management, and so on. The software has no way to access operating system resources other than through API calls, so this approach provides critical information for rendering malware behavior. For example, Ahmadi et al. [10] analyzed 500,000 malware in 2016, extracted a subset of 794 API function calls, and then calculated the frequency of the subset and built a multi-mode system based on it for malware classification.

Malware attackers often use a variety of code obfuscation techniques to hide the true intention of malware. The two most common methods are compression and encryption, which are mainly aimed at hiding malicious fragments in static analysis. Since files containing compressed or encrypted code usually have higher information entropy than ordinary files, the analysis of information entropy can effectively solve this problem. Sorokin and Jun proposed a method [11] to analyze the information entropy of file structure and detect malicious programs by comparing the similarities and differences of information entropy of unknown files, which has certain detection effect for malicious programs such as encryption and shell.

Compared with the above research work, we propose an ensemble learning approach, which uses mixed features and combined two CNN-based models to achieve high accuracy and fast speed malware detection.

3 The Ensemble Learning Approach

In this section, we will first introduce the design of our approach, and then elaborate the details of our ensemble-learning model.

3.1 The Design

The ensemble learning method proposed in this paper combines the detection model based on binary machine code image and the detection model based on information entropy image. The CNN in these two models is the basic learner needed for our ensemble learning approach. In the fusion model, the basic learner is two convolutional neural networks, which respectively classify malware and benign software by using the extracted features at different levels.

The combination strategies of ensemble learning could be divided into two categories: linear combination and voting combination. Since there are only two basic learners, it is impossible to vote by majority, so the voting method is not suitable for the combination strategy of basic learners. Therefore, the linear grouping method is adopted, which is not conducive to the learning of weighted parameters due to the defects of data noise and overfitting, so the classification probability is simply averaged. For the two existing basic learners, is the probability that x is classified as Y for the specified sample data (0 is the label of benign sample and 1 is the label of malicious sample), then the prediction probability of the integration of the two models is:

$$t = \{1, 2\} f_t(y|x)\, y = \{0, 1\}$$

$$\overline{f}(y|x) = \frac{1}{2}(f_1(y|x) + f_2(y|x))$$

According to this method, the final classification result of sample data can be obtained. The specific fusion process is shown in Fig. 1.

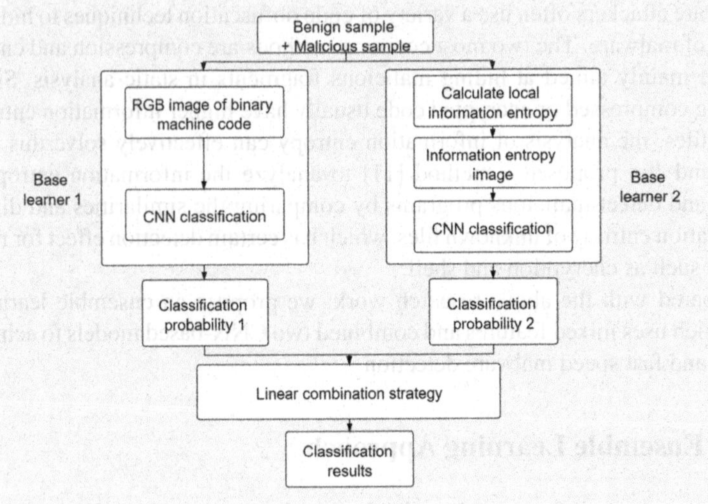

Fig. 1. Ensemble learning model

3.2 Detection Model Based on Binary Image

The technology of transforming binary files into images has been proposed before. For example, Conti et al. transformed the original binary fragments of text files, C++ data structures, image files and audio files into gray images. They proved that the above files could be correctly classified by processing the statistical features of images. The malware and benign software studied in this paper are PE files, which are also binary files. Due to the difference in program instruction sequence between malicious code and benign code, the binary bytes are also different to some extent. Therefore, similar methods can also be applied to the classification of benign and malicious software. Unlike traditional static analysis methods, this method does not extract and analyze features such as strings, function call graphs, and N-gram sequences, so it does not require complex processing and consumes too many resources.

In this paper, the image is generated based on binary machine code and then the detection method is called CNN. However, unlike the previous approach, this article converts binary machine code into AN RGB image. Gray images have a single pixel value, while RGB images have three color channels (R: red, G: green, and B: blue). Compared with gray image, RGB image contains more information, so gray image can be converted into RGB image. RGB images can be obtained by adding color channels to gray images. There are three color channels that can be added. After experimental comparison, it is found that adding green channels is better than red and blue classification. For a pixel point in the RGB image, its gray value is PIX_ARr [K], representing the horizontal coordinate and the vertical coordinate, then R channel value is I, G channel value is PIX_arr [K], and B channel value is J (Fig. 2).

Address	Value (Byte)															
00401000	55	8B	EC	8B	45	0C	53	48	48	56	57	0F	84	FF	01	00
00401010	00	2D	11	01	00	00	0F	84	F5	00	00	00	60	9C	23	D9
00401020	F7	E6	C1	E3	02	90	F7	D6	85	D8	C1	FB	04	1B	FE	F9
00401030	C1	E1	03	90	1B	FB	FC	23	DE	F8	C1	E2	03	F7	DE	C1
00401040	FE	05	2B	D7	F7	EE	FD	F7	EE	90	F7	E7	C1	E2	07	43
00401050	C1	FE	05	F8	F7	EE	C1	F8	00	F7	E7	C1	E0	05	F9	F7
00401060	E8	C1	FA	04	4A	FC	33	FB	F8	F7	D2	33	C8	F7	D8	1B
00401070	F3	C1	E6	04	F7	D0	C1	E8	07	8B	C7	FC	C1	E6	07	0B
00401080	CB	4B	F9	C1	E1	03	F7	D1	90	C1	E7	05	8B	D9	F8	33
00401090	CE	C1	E3	04	85	CE	46	23	FE	FD	F7	E6	F9	33	DF	C1
004010A0	E6	08	23	C8	C1	E1	00	F7	D6	D1	E6	FD	F7	DF	33	D7
004010B0	C1	E9	03	87	F2	C1	E3	07	F7	D8	C1	E1	02	41	F9	C1

Fig. 2. The content of an sample

CNN does not need to pre-process images and extract features like other traditional methods, so the whole classification process becomes simpler. The two significant differences between CNN and ordinary neural network are weight sharing and sparse connection, so the model is more concise and can reduce the number of parameters and the consumption of computing resources, as well as make the efficiency become higher. Using CNN to classify images can also reduce noise and enhance features to be extracted. After converting the binary machine code of malware and benign software into RGB images, it is used as the input of the convolutional neural network to classify the images according to their categories. This CNN will be used as the basic learner of the fusion model in the subsequent steps. The overall model structure is shown in Fig. 3.

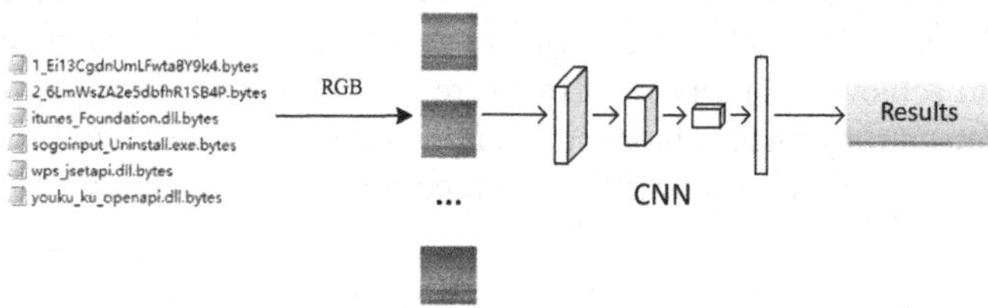

Fig. 3. Detection model based on binary code image

3.3 Detection Model Based on Information Entropy Image

The attackers of malware often use different obfuscation techniques to hide the real purpose of malware, so as to avoid various detection tools and means. Among them, the more common one is the packer technique, namely compression and encryption. As a result, the number of feature codes increases, which seriously reduces the efficiency and accuracy of malware detection and analysis.

In information theory, the entropy of a byte sequence reflects its statistical changes, and the compressed or encrypted code has a higher entropy than the original code.

Lyda et al. analyzed a large number of normal, compressed and encrypted executable files in 2007, and finally observed that the information entropy of the three types of files was 5.09, 6.80 and 7.17 [12]. Therefore, information entropy theory can be used as a means to analyze malicious code.

The specific steps of information entropy image generation are as follows:

1. Binary machine code that reads sample data in bytes. This model USES the same data format as the model mentioned in Sect. 3.2, so the binary machine code is first read into a list per byte, denoted as L, for processing in the next step.
2. Divide data blocks by 16 bytes. The data in L is divided into a data block according to every 16 bytes to calculate the local information entropy. If there are not enough 16 bytes, 0 is added at the end of the file until L can be divided into N complete data blocks.
3. Calculate the local information entropy. For the data blocks obtained in the previous step, the corresponding local entropy is calculated by the following formula, which to some extent measures the structural characteristics of the file. Where I represents the specific byte in the data block, and represents the probability of the occurrence of byte I (that is, the number of byte I in the data block divided by the number of bytes in the data block). The local information entropy calculated by each data block is stored in an array, remembered as Entropy_list [P_i].

$$entropy = -\sum_{i=1}^{16} p_i \log_2 p_i$$

The information entropy images of malicious code and benign code will be used as input of the convolutional neural network, and the two images can be classified by CNN, so as to achieve the purpose of malware detection. The structure of CNN used for information entropy image classification is the same as that used for binary machine code image classification, which will not be repeated in this paper.

4 Experiments

In this section, we will introduce the experimental results of our approach In malware Detection.

4.1 Experimental Setup

This subject USES Python as the development language, USES Keras framework for deep learning experiment, USES Anaconda for Python package management, and USES Spyder as the integrated development environment.

The Malware sample originates from the 2015 Microsoft Malware Classification Challenge data set. It is composed of nine different Malware families with a total number of 1000. Each sample is the content of the binary machine code of the Malware.

Benign samples were downloaded and preprocessed from Tencent software center, containing 980 different normal programs (such as Zoom, Google browser, etc.). In the two CNNS proposed before the training article, the proportion of training set, verification set and test set in the whole sample is 8:1:1 respectively.

This method requires training two neural networks. The first CNN was used to classify binary machine code images, and the input data was RGB images generated by binary machine code. The second CNN is used to classify the information entropy image. The input data is the gray image transformed by the information entropy principle. The two CNNS adopt the same structure. First, a convolution kernel of 10 sizes was used in the first convolutional layer, and tanH was used as the activation function. Then, a maximum pooling layer was set to reduce data redundancy and unnecessary parameters. In the next convolutional layer, 20 convolution kernels of size are selected. The Settings of activation function and pooling layer are the same as before. Then, the Dropout function with the parameter of 0.25 is used to arbitrarily disconnect some neurons to prevent overfitting. And then through a flatten layer, connected to the full-junction layer, the number of neurons is 500, and the activation function is tanh. After passing through a dropout layer with parameter 0.5, a full connection layer containing 2 neurons was finally connected, and Softmax function was used to obtain the classification results.

4.2 The Metrics

In this section, multiple metrics are used to evaluate the experimental results. We first introduce the concept of confusion matrix. For all the samples, TP (True Positive) represented the number of Positive and Positive predictions. FN (False Negative) is the number of true positives and predicted negatives; TN (True Positive) represents the number of negative True and predicted negative; FP (False Positive) represents the number of negative true and Positive predictions. For the classification problem in this topic, the positive sample is the data of malware and the negative sample is the data of benign software.

The evaluation indexes adopted are divided into the following five categories.

1. TPR, also known as recall, is the ratio of the correct amount of malware estimated by the model to the total amount of real malware.

$$TPR = recall = \frac{TP}{TP + FN}$$

2. False positive rate (FPR), that is, how much of the benign software is misjudged as malicious data.

$$\mathrm{FPR} = \frac{\mathrm{FP}}{\mathrm{FP} + TN}$$

3. Accuracy, that is, the proportion of all samples that are correctly classified.

$$accuracy = \frac{TP + TN}{TP + TN + FP + FN}$$

4. Precision, that is, the proportion of the software that is predicted to be malicious that is predicted to be correct.

$$precision = \frac{TP}{TP + FP}$$

5. Generally speaking, the increase in precision is accompanied by the decrease in recall, and the decrease in precision causes the increase in recall. Therefore, in order to balance the two, F1 value was used for comprehensive evaluation.

$$F1 = \frac{2 \times recall \times precision}{recall + precision}$$

4.3 The Experimental Results

We conducted a total of ten experiments to calculate different indicators and take the average. The specific content of each evaluation indicator can be seen in the following table. In the detection experiment based on binary machine code image, RGB image obtained better experimental effect compared with gray image, with the accuracy of 0.902 and F1 value of 0.901, indicating that the model realized the effective discrimination of benign software and malicious software. However, the recall value of this model is relatively low, indicating that some malicious samples are predicted to be benign. For the problem studied in this paper, malicious samples are more harmful when predicted as benign samples, so the recall index of this model is deficient.

In the detection experiment based on information entropy image, the effect of this model is better than that of binary machine code image model. The accuracy was 0.949 and F1 was 0.948, respectively 0.047 higher than the binary code image model. This model has a higher recall value and a lower FPR value, indicating that benign software and malware have lower false positive rates respectively.

The overall effect of the fusion model is better than that of the two single feature models. The accuracy of fusion model is 0.966, 0.064 and 0.017 higher than binary machine code image model and information entropy image model, respectively. F1 value was 0.965, which was also higher than the two single feature models. It can be seen that the recall value of the fusion model is the highest, 0.942, so the false detection rate of malware is low, which is extremely important for the effectiveness of malware detection. Each basic learner has its own limitations, but the fusion model combines the characteristics of different levels of malware, so that the weaknesses of basic learners complement each other and the advantages are combined to better reflect the overall characteristics of malware, so as to reach the best decision (Table 1).

Table 1. The results

The model	Accuracy	Recall	Precision	F1
Binary image + CNN	0.902	0.888	0.917	0.901
Entropy + CNN	0.949	0.930	0.968	0.948
The ensemble model	0.966	0.942	0.990	0.965

5 Conclusion

This paper proposes an ensemble Learning Approach for Malware Detection. The detection model based on binary machine code image is proposed. The binary machine code of malware and benign software is converted into RGB images, and the images are classified using CNN. A detection model based on information entropy image is proposed. Information entropy is used to generate information entropy images of benign software and malware, and image features are extracted by CNN to obtain the classification model. Data were collected and preprocessed, two CNN models were trained and experimental results were obtained. The fusion model is realized by ensemble learning, and the results of single feature model and fusion model are compared and analyzed. Experimental results show the efficiency and effectiveness of this method.

Our future work are two-folds. Firstly, we will evaluate our approach using much more datasets. Secondly, we will expand Our feature sets such as sandbox outputs.

Acknowledgement. The authors gratefully acknowledge The anonymous reviewers for their helpful Suggestions.

References

1. Kaspersky Security Bulletin 2019. Statistics [EB/OL]. https://go.kaspersky.com/rs/802-IJN-240/images/KSB_2019_Statistics_EN.pdf
2. Internet Security Threat Report [EB/OL]. https://docs.broadcom.com/doc/istr-24-2019-e
3. CrowdStrike Global Threat Report 2020 [EB/OL]. https://go.crowdstrike.com/rs/281-OBQ-266/images/Report2020CrowdStrikeGlobalThreatReport.pdf
4. Nataraj, L., Karthikeyan, S., Jacob, G., et al.: Malware images: visualization and automatic classification. In: Proceedings of the 8th International Symposium on Visualization for Cyber Security, pp. 1–7 (2011)
5. Conti, G., Bratus, S.: Voyage of the reverser: a visual study of binary species. Black Hat USA (2010)
6. Jain, S., Meena, Y.K.: Byte level *n*–gram analysis for malware detection. In: Venugopal, K.R., Patnaik, L.M. (eds.) ICIP 2011. CCIS, vol. 157, pp. 51–59. Springer, Heidelberg (2011). https://doi.org/10.1007/978-3-642-22786-8_6
7. Santos, I., Brezo, F., Ugarte-Pedrero, X., et al.: Opcode sequences as representation of executables for data-mining-based unknown malware detection. Inf. Sci. **231**, 64–82 (2013)
8. Hu, X., Shin, K.G., Bhatkar, S., et al.: MutantX-S: scalable malware clustering based on static features. Presented as part of the 2013 USENIX Annual Technical Conference, pp. 187–198 (2013)

9. Yuxin, D., Siyi, Z.: Malware detection based on deep learning algorithm. Neural Comput. Appl. **31**(2), 461–472 (2019)
10. Ahmadi, M., Ulyanov, D., Semenov, S., et al.: Novel feature extraction, selection and fusion for effective malware family classification. In: Proceedings of the Sixth ACM Conference on Data and Application Security and Privacy, pp. 183–194 (2016)
11. Sorokin, I.: Comparing files using structural entropy. J. Comput. Virol. **7**(4), 259 (2011)
12. Lyda, R., Hamrock, J.: Using entropy analysis to find encrypted and packed malware. IEEE Secur. Priv. **5**(2), 40–45 (2007)

Automatic Medical Image Report Generation with Multi-view and Multi-modal Attention Mechanism

Shaokang Yang[1,2], Jianwei Niu[1,2,3](✉), Jiyan Wu[3], and Xuefeng Liu[1,2]

[1] State Key Laboratory of Virtual Reality Technology and Systems, School of Computer Science and Engineering, Beihang University, Beijing 100191, China
niujianwei@buaa.edu.cn

[2] Beijing Advanced Innovation Center for Big Data and Brain Computing (BDBC), Beihang University, Beijing, China

[3] Research Center of Big Data and Computational Intelligence, Hangzhou Innovation Institute of Beihang University, Hangzhou, China

Abstract. Medical image report writing is a time-consuming and knowledge intensive task. However, the existing machine/deep learning models often incur similar reports and inaccurate descriptions. To address these critical issues, we propose a multi-view and multi-modal (MvMM) approach which utilizes various-perspective visual features and medical semantic features to generate diverse and accurate medical reports. First, we design a multi-view encoder with attention to extract visual features from the frontal and lateral viewing angles. Second, we extract medical concepts from the radiology reports which are adopted as semantic features and combined with visual features through a two-layer decoder with attention. Third, we fine-tune the model parameters using self-critical training with a coverage reward to generate more accurate medical concepts. Experimental results show that our method achieves noticeable performance improvements over the baseline approaches and increases CIDEr scores by 0.157.

Keywords: Medical report generation · Medical concepts · Multi-modal attention · Multi-view encoder · Self-critical

1 Introduction

Medical images are widely used in clinical diagnosis and disease treatment, e.g., the consolidation and pneumothorax. Writing report for medical images is time-consuming and requires extensive expertise, and even experienced radiology make mistakes due to excess workload [1]. Therefore, it is desirable to develop an automatic report generation model for medical images. The automated report generation task is presented in Fig. 1, which depicts the medical

Supported by Hangzhou Innovation Institution, Beihang University.

M. Qiu (Ed.): ICA3PP 2020, LNCS 12454, pp. 687–699, 2020.
https://doi.org/10.1007/978-3-030-60248-2_48

images and corresponding radiology report. The medical images are the frontal and lateral views from a patient. The report includes a finding which describes the observations of abnormal/normal phenomena, and an impression/conclusion sentence indicating the most important medical observation.

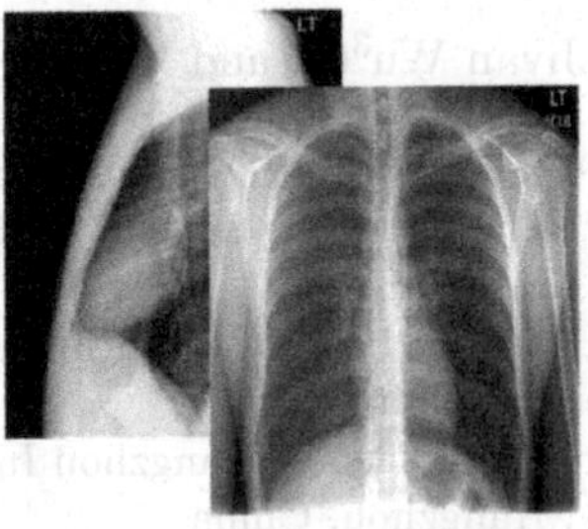

Findings: The cardiomediastinal silhouette is within normal limits for size and contour. The lungs are normally inflated without evidence of focal airspace disease pleural effusion or pneumothorax. Osseous structures are within normal limits for patient age.

Impression: no acute radiographic cardiopulm onary process.

Fig. 1. An example of chest x-ray report. Findings are some sentences that describe the medical observations in details regrading each area of the image. Impression is a conclusive sentence.

This task is similar to the image captioning [2–6] which aims to describe the objects and their relationships in the image using single natural language sentence. However, medical report generation is different from image captioning in the following aspects: 1) Medical reports are featured by long and coherent paragraphs other than a single sentence. 2) Medical images are only subtle changes in some local positions, which makes the training process more difficult and incurs similar sentences in the generated reports. 3) The generated reports require more comprehensive coverage when describing medical-related items, e.g., heart size, bone structure and pneumothorax. To generate highly-accurate medical image reports, recent researches have proposed machine/deep learning based methods. Li et al. [7,8] proposed a retrieval model through reinforcement learning and graph neural networks. Jing et al. [9] and Yuan et al. [10] adopted a hierarchical LSTM decoder with an attention mechanism for long report generation. However, duplicate reports and inexact descriptions are observed in the generated reports of the existing studies [9]. These problems significantly hamper the credibility of automatically generated reports in practical use, since wrong medical-related contents are very misleading.

To address the above issues, we first design a multi-view image encoder to extract visual features. To generate coherent and diverse reports, the medical concepts extracted from the textual report are adopted as semantic features to provide the model with comprehensive information and force model recognizing the subtle changes in local position. Then we design a two-layer LSTM-based multi-modal attention generator to effectively integrate these features. Finally, a coverage reward is included for self-critical training and generating accurate descriptions.

In summary, the main contributions of our work are as follows:

- Introduce medical concepts as the semantic features to generate diverse sentences.
- Develop a multi-view and multi-modal model with the following features: multi-view visual features and semantic features.
- Design a reward term to encourage the model to generate more accurate medical concepts.
- Evaluate the performance in the IU X-ray [11] datasets and experimental results show that we achieve the state-of-the-art performance.

The remainder of this paper is organized as follows. In Sect. 2, we discuss the related works. Section 3 describes our proposed medical report generation method in detail. Section 4 demonstrates the experimental results and the concluding remarks are provided in Sect. 5.

2 Related Work

Visual Captioning with Deep Learning: Visual captioning aims at generating a descriptive sentence for images or videos. Inspired by the success of the encoder-decoder framework in sequence generation [12–17], a lot of visual captioning models [3–6] based on this framework are proposed and have achieved state-of-the-art performance. Recently, due to the introduction of attention mechanisms [6,18–20] and the use of reinforcement learning [21–23], the performance of image captioning has been greatly improved. However, existing image captioning models are not suitable for report generation which requires multiple and visual ground sentences.

Medical Image Report Generation: Deep learning for healthcare has attracted extensive research attention in both industry and academia. Automatic medical image report generation, as a key application in this field, is gaining increasing research interest benefit from the technological advancements in image captioning. Recently, some pioneering works have shown promising results such as Jing et al. [9] and Yuan et al. [10] adopted a hierarchical recurrent decoder with attention mechanism to generate long medical reports, but repetitions are found in their generated reports which hamper the credibility in practical use. Li et al. [7,8] adopted a retrieval-based approach to generate long reports which reduced the flexibility of sentence generation. Inspired by the success of semantic features [3,19,24], we inject the semantic concepts predicted by the concepts extractor into the designed two-layer LSTM-based decoder to generate more diverse and more semantic-coherent reports. Besides, we also designed a coverage reward to further optimize our model to generate more medical concepts.

3 Our Methods

3.1 Overview

To generate a long and semantic-coherent report, we propose a novel multi-view and multi-modal framework. The overall architecture of our framework is shown in Fig. 2, which includes three main components: 1) a multi-view image encoder to extract the visual features from two view images, 2) a medical concepts extractor to predict the concepts from images, 3) a two-layer LSTM-based multi-modal attention generator for the report generation.

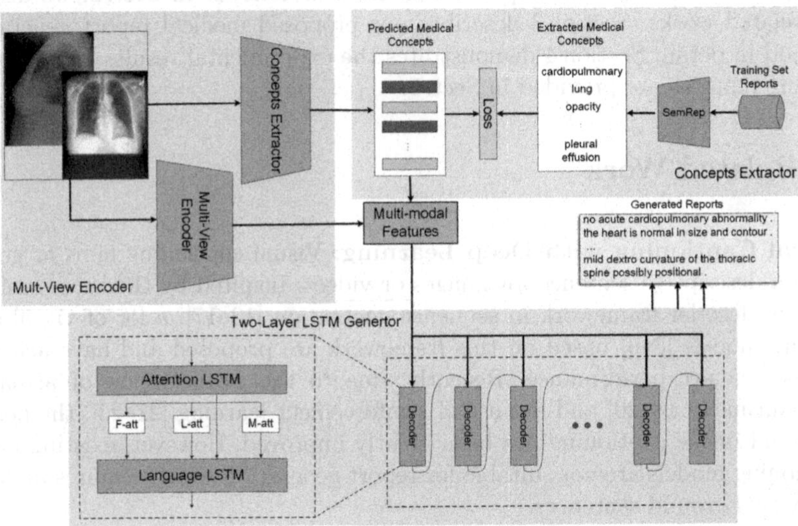

Fig. 2. An overview of our multi-view multi-modal (MvMM) architecture for medical report generation. The architecture consists of three parts: 1) multi-view image encoder to extract visual features from the frontal and lateral view images on the top left. 2) medical concepts extractor to predict the concepts from visual features on the top right. 3) two-layer LSTM with attention generator to generate the final medical reports in the bottom.

3.2 Multi-view Image Encoder

Unlike traditional image captioning in which data consists of pair (I, Y) where I denotes one single image and Y means the corresponding manual annotated description, IU X-rays dataset usually consists of pair $(\{I_f, I_l\}, Y)$ where $\{I_l, I_f\}$ denotes the frontal and lateral view images of the same patient and Y means the corresponding report. However, Jing [9] treated the images of the frontal and lateral view as two independent cases which makes the model hard to train and generate an inconsistent report for the same patient with the two perspective images.

Radiologists wrote the medical reports comprehensively considering the frontal and lateral view images (such as lateral view provides more information about depth, frontal view provides more information about outline and position). To mimic this procedure, we design two separate encoders to extract visual features from the corresponding view images. The frontal and lateral encoders share the same architecture. Specifically, given the frontal and lateral view images $\{I_f, I_l\}$, we first use the multi-view encoder to extract the regional visual features from the corresponding perspective images $\mathbf{V}_p = (v_{p1}, v_{p2}, ...v_{pk}), v_{pi} \in \mathbb{R}^g, p \in \{f, l\}$, and then average them to obtain the corresponding global visual features.

3.3 Medical Concepts Extractor

In previous report generation works, duplicate paragraphs and inexact generated descriptions are the two main hindrances. That is because the pattern among CT images is very similar and the RNN based decoders tend to reproduce the words that frequently occur in the training data. An advance is to furnish decoder with semantic concepts so the decoder obtains the overall perspective of the medical images. In this paper we treat medical concepts that occur in textual reports as the semantic concepts. We employ the SemRep[1] instead of Medical Text Indexer (MTI) which often makes noise to extract those medical concepts that frequently occur in the reports. Then, following common practice (Fang et al. [25], You et al. [19]), we adopt the weakly-supervised approach of Multiple Instance Learning to build a concepts classifier. Due to limited space, please refer to Fang et al. for a detailed explanation. Besides, given the class imbalance of positive and negative, we employ the focal loss [26] to get higher precision.

3.4 Multi-modal Attention Decoder

As illustrated in Fig. 3, we designed a two-layer LSTM-based [20] multi-modal with attention for generator. At each time step, it combines the textual caption, the attentive visual information, and the attentive semantic information as the context to generate the output word.

Visual Attention. To make different descriptions focus on different image regions, we design a visual attention network [6].

$$\mathbf{a}_{v,t} = \boldsymbol{w}_{v,h}^T \tanh(\boldsymbol{W}_v \boldsymbol{V} + (\boldsymbol{W}_g h_t^1)\mathbb{I}^T) \tag{1}$$

$$\boldsymbol{\alpha}_{v,t} = \text{softmax}(\mathbf{a}_{v,t}) \tag{2}$$

where $\mathbb{I}^T \in \mathbb{R}^k$ is a vector with all elements set to 1. $\boldsymbol{W}_v, \boldsymbol{W}_g \in \mathbb{R}^{k \times d}$ and $\boldsymbol{w}_{v,h}^T \in \mathbb{R}^k$ are the parameters to be learned. $\boldsymbol{\alpha}_{v,t} \in \mathbb{R}^k$ is the attention weight of $\boldsymbol{V}$, and the attentive visual features $\hat{\boldsymbol{v}}_{\boldsymbol{t}}$ obtained by:

$$\hat{\boldsymbol{v}}_t = \boldsymbol{V} \boldsymbol{\alpha}_{v,t} \tag{3}$$

[1] https://semrep.nlm.nih.gov/.

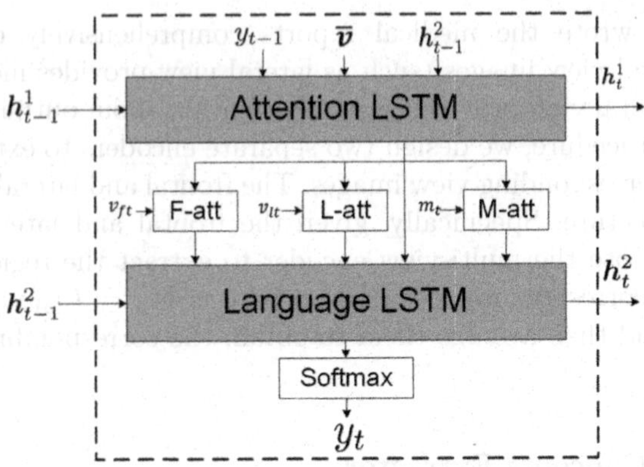

Fig. 3. Illustration of the two-layer LSTM with attention generator.

Notice that we have both frontal $\boldsymbol{V}_f$ and lateral visual features $\boldsymbol{V}_l$, we design two independent visual attention networks following the above methods.

Semantic Attention. As discussed above, since the characteristic of RNN based decoder which tends to reproduce the frequent words in training data and the CT images' pattern similar to each other result in duplicated paragraphs in the generated reports. To overcome such issues, we use the concept extractor to predict medical concepts and serve their word-embedding vectors as the semantic features to aid medical report generation.

In light of the visual attention network, we also design a semantic attention network:

$$\mathbf{a}_{s,t} = \boldsymbol{w}_{s,h}^T \tanh(\boldsymbol{W}_s \boldsymbol{S} + (\boldsymbol{W}_{s,g} \boldsymbol{h}_t) \mathbb{I}^T) \tag{4}$$

$$\boldsymbol{\alpha}_{s,t} = \text{softmax}(\mathbf{a}_{s,t}) \tag{5}$$

where $\boldsymbol{w}_{s,h}^T \in \mathbb{R}^m$ and $\boldsymbol{W}_s, \boldsymbol{W}_{s,g} \in \mathbb{R}^{m \times d}$ are the parameters to be learnt. $\boldsymbol{\alpha}_{s,t} \in \mathbb{R}^m$ is the attention weight of $\boldsymbol{S}$. The attentive semantic features $\boldsymbol{s}_t$ is calculated as:

$$\hat{\boldsymbol{s}}_t = \boldsymbol{S} \boldsymbol{\alpha}_{s,t} \tag{6}$$

Unlike Jing [9] who used two different embedding matrices for words and tags, which increased the risk of model overfitting, we use the same embedding matrix for the medical concepts and report words.

Two-Level LSTM Based Decoder. We have prepared both the visual information and semantic information. Next, we inject the two modal information into the two-layer LSTM generator. To be specific, at each time step t, the input vector of attention LSTM is defined as the concatenation of the previous output of the language LSTM, the mean-pooled visual features from different views $\bar{\boldsymbol{v}}_f = \frac{1}{k} \sum_i \boldsymbol{v}_{i,f}$ and $\bar{\boldsymbol{v}}_l = \frac{1}{k} \sum_i \boldsymbol{v}_{i,l}$ and the embedding of the previously generated word $\boldsymbol{w}_t$:

$$\boldsymbol{x}_t^1 = [h_{t-1}^2, \bar{\boldsymbol{v}}_f, \bar{\boldsymbol{v}}_l, \boldsymbol{w}_t] \tag{7}$$

Once obtaining the output of the attention LSTM h_t^1, we generate the two attentive visual features and the one attentive semantic features as discussed in visual attention and semantic attention sections.

Next, we concatenate the output of the attention LSTM and the two views attentive features and the one attentive semantic features as the input of the language LSTM for medical report generation:

$$\boldsymbol{x}_t^2 = [\hat{\boldsymbol{v}}_{t,f}, \hat{\boldsymbol{v}}_{t,l}, \hat{\boldsymbol{s}}_t, h_t^1] \tag{8}$$

The output h_t^2 of the language LSTM is fed to a softmax layer to predict the next word:

$$w_{t+1} = \text{softmax}(\boldsymbol{W} h_t^2) \tag{9}$$

where the $\boldsymbol{W}$ is the parameter to be learned.

3.5 Self-critical Training with Coverage Reward

To further boost our report generation model and solve the inconsistent between training and inference, we adopt the self-critical strategy [23] to further optimize our model after a pre-training step using cross-entropy. Besides, to encourage our model to generate more correct medical concepts, we additionally design a medical concept coverage reward. Specifically, we use the total medical concepts extracted in Sect. 3 as a medical concept vocabulary. To assess how the generated report covers the medical concepts, we define a soft intersection-over-union (IoU) as the coverage reward:

$$\text{IoU}(y, \hat{y}) = \frac{\boldsymbol{I}(M_p, M_{gt})}{\boldsymbol{U}(M_p, M_{gt})} \tag{10}$$

where M_p and M_{gt} represent the set of medical concept occurring in the generated report and the ground-truth report respectively, $\boldsymbol{I}(\cdot, \cdot)$ denotes the intersection of the input sets, and $\boldsymbol{U}(\cdot, \cdot)$ denotes the union of the input sets. The final reward function is defined as the combination of CIDRr reward and the medical concept coverage reward:

$$R = R^r + \alpha R^c \tag{11}$$

4 Experiments

4.1 Implementation Details

In this section, we describe the implementation details of our proposed model and how to train it.

Medical Concept Extractor. We use semrep to extract the medical concepts that frequently occur in the radiology reports. We filter the extracted medical concepts by minimal 90 and obtain 63 unique medical concepts for satisfactory detection accuracy. Since the medical concepts may contain multiple words such as "enlarged cardiom", we process the reports by jointing those words into one single word with a special token "_" such as "enlarged_cardiom".

Training Details. We set the dimension of hidden states in language LSTM to 1,000, the dimension of hidden units in the attention layer to 512, and the size of the input word embedding to 1,000. We use Adam [27] optimizer for the model training. We first train the model with cross-entropy loss for 80 epochs with an initial learning rate of 5e–4 and then fine-tune model via RL with a fixed learning rate 5e–5 for another 30 epochs. The tradeoff parameter α is set as 10.

4.2 Datasets

IU X-Ray. The Indiana University Chest X-Ray (IU X-Ray) is a public dataset of chest x-ray images paired with their corresponding diagnostic reports. The dataset contains 3,395 radiology reports and 7,470 associated chest x-rays. Each report is associated with a pair of images that are the frontal and lateral views and contains findings, impressions, comparison, and indication sections. We filtered out reports without two complete views and obtained 3,074 reports with multi-view images. We preprocess the reports by tokenizing, converting to lowercases, translate those medical concepts as a new token, and finally obtain 1,990 unique words. We randomly split the dataset by patients into training and testing by a ratio of 8:2. There is no overlap between patients of different sets. Besides, Many previous works [9,28] used the pre-trained model on ImageNet to initialize their model. However, given the fact that the distribution between natural images and medical images is significantly different, the pre-trained model on ImageNet does not satisfy well for the medical image report generation. So we pre-train with the newly proposed ChexPert [29] dataset which is a large scale chest x-ray images for disease classification to obtain better initialized parameters.

4.3 Evaluation Metrics

We use two kinds of metrics for evaluating medical image report generation: 1) automatically evaluates metrics including Cider [30], Bleu [31], Rouge [32] which are widely used in image captioning tasks. 2) human evaluation: we randomly select 100 samples from the testing set for each method and conduct questionnaires with 20 workers. Each query provides corresponding ground truth report and asks the participants to choose the best one matching GT from a given set of reports in terms of language fluency, the correctness of medical concepts, report repeatability. We collect results from 20 participants and compute the average preference percentage for each model.

4.4 Baseline

We compared our proposed model with several state-of-the-art baselines: CNN-RNN [2], Soft-Att [18], CoAtt [9], HRGR [7], KERP [8] and we conduct additional experiments to compare our proposed model with its different variants: 1) the designed multi-view encoder with hierarchical LSTM (MvHL) that is similar to framework in Jing et al. [9] and Yuan et al. [10], 2) the designed multi-view encoder with our proposed two-layer LSTM-based generator (MvTL), 3) the proposed multi-view and multi-modal attention model (MvMM), 4) MvMM^{normal} on normal cases that select from testing set. We re-implemented all of these models and adopted the resnet152 [33] as the image encoder.

Table 1. Automatic evaluation results on IU X-ray Datasets. BLEU-n denotes BLEU score uses up to n-grams. The results in the top 3 sections are reported in KERP [8] with their experimental settings.

Method	CIDER	BLEU-1	BLEU-2	BLEU-3	BLEU-4	ROUGE
CoAtt [9]	0.277	0.455	0.288	0.205	0.154	0.369
HRGR [7]	0.343	0.438	0.298	0.208	0.151	0.322
KERP [8]	0.280	0.482	0.325	0.226	0.162	0.339
CNN-RNN [2]	0.287	0.303	0.201	0.130	0.090	0.321
Soft-Att [18]	0.342	0.302	0.210	0.112	0.101	0.313
MvHL	0.287	0.363	0.234	0.156	0.106	0.327
MvTL	0.373	0.412	0.257	0.175	0.122	0.339
MvMM	**0.500**	**0.443**	**0.306**	**0.215**	**0.145**	**0.374**
MvMM^{normal}	0.675	0.467	0.317	0.223	0.168	0.395

4.5 Automatic Evaluation

Table 1 shows the evaluation results comparison of state-of-the-art methods and our model variants using the standard image captioning evaluation tool[2]. Our proposed model outperforms all baselines (CNN-RNN, Soft-ATT, MvHL and MvTL) on BLEU-1,2,3,4, ROUGE and Cider scores, demonstrating the effectiveness and robustness. From the results of MvHL and MvTL, we can find that our designed two-layer LSTM-based decoder achieves better results regarding the multi-view images than the traditional hierarchical LSTM decoder. From the comparison of MvTL, and MvMM, we can conclude that the performance significantly is improved especially on Cider by introducing the semantic concepts information, which indicates the effectiveness of the proposed MvMM architecture. We also report the performance of our model on normal cases that select from the testing set. When evaluating performance of model on these normal cases, our MvMM significantly improve the performance, which implies that our proposed model further benefit from a disease classifier. Besides, MvMM also achieves equivalent or higher scores compared to CoAtt [9], HRGR [7] and KERP [8] on the same IU X-ray dataset, even though it may not be a fair comparison due to different experimental settings.

[2] https://github.com/tylin/coco-caption.

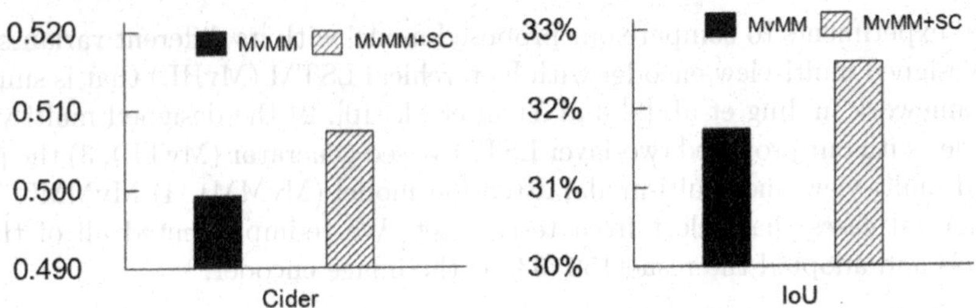

Fig. 4. The performance improvement fine-tuning with coverage reward.

4.6 Self-critical Training Evaluation

To demonstrate the effectiveness of self-critical training with our coverage reward, we further optimize MvMM after a pre-training step using cross-entropy. As shown in Fig. 4, the performance further improves (the cider score improved from 0.50 to 0.51) with the help of our designed coverage reward. Besides, the medical concepts coverage rate (IoU) increases from 31.2% to 32.6%, which indicates the merit of encouraging the proposed model to generate more accurate medical concepts and meanwhile boost the performance.

4.7 Human Evaluation

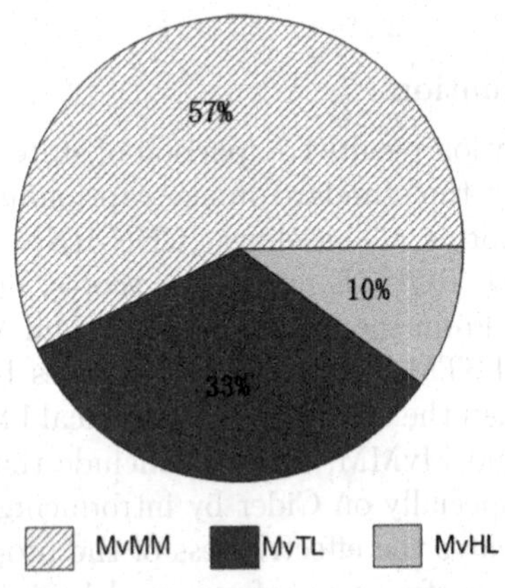

Fig. 5. Illustration of human evaluation about MvMM, MvTL and MvHL.

Figure 5 presents the results of human evaluation in terms of content coverage, specific terminology accuracy, and language fluency. From the results we can conclude that MvMM achieves a higher preference than baselines models, showing that it generates more natural and accurate reports.

4.8 Qualitative Analysis

An illustration of reports generation is shown in Fig. 6. We can find that duplicated reports are found in the MvHL while no duplicate reports are found in reports of MvTL and MvMM. Besides, from the generated reports we can conclude that MvMM can generate more diverse and semantic-coherent reports compared to baseline models, and among the generated reports, MvMM has a higher coverage rate of generating accurate medical concepts.

	GT	MvMM	MvTL	MvHL
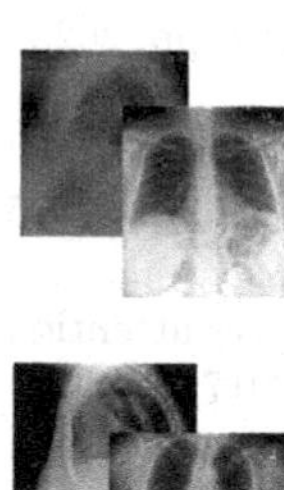	xxxx bilateral effusion. heart size within normal limit. stable position of left subclavian central venous catheter. no focal air space disease. no pneumothorax. mild blunting of the costophrenic xxx bilaterally.	no acute finding. the heart size is within normal limit. there is right paratracheal opacity which may represent adenopathy but or small airspace opacity. no focal airspace disease is seen. no definite pleural effusion is seen. no convincing acute bony finding are identified.	no acute cardiopulmonary abnormality . the cardiomediastinal silhouette is normal in size and contour . the lung are clear without focal airspace disease pleural effusion or pneumothorax . there are mild degenerative change of the thoracic spine .	no acute cardiopulmonary abnormality. the heart is normal in size. the lungs are clear. no focal airspace consolidation. no pneumothorax or pleural effusion.
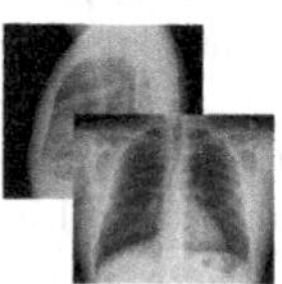	mild cardiomegaly and atherosclerosis. no acute infiltrate. the heart is top normal in size. the mediastinum is stable. the aorta is atherosclerotic. there are mild chronic change without focal consolidation. no pleural effusion is seen.	mild cardiomegaly without acute disease. the heart is mildly enlarged. aorta is tortuous. the lung are hypoinflated with increased retrosternal airspace disease. no pneumothorax or pleural effusion. there are mild degenerative change of the spine.	no acute cardiopulmonary abnormality . the heart is normal in size and contour . there is no pneumothorax or pleural effusion . there are no focal air space opacity . there are degenerative change of the thoracic spine .	no acute cardiopulmonary abnormality. the heart is normal in size. the lungs are clear. no focal airspace consolidation. no pneumothorax or pleural effusion. no pneumothorax or pleural effusion.
	no acute cardiopulmonary disease. the lungs are clear. the heart and pulmonary xxxx appear normal. pleural spaces are clear. the mediastinal contours are normal. bony overlap in the lung apices could obscure a small pulmonary nodule.	no acute cardiopulmonary disease. the lung appear clear. there is a calcified xxxx nodule in the left upper lung zone. this was not well seen on the xxxx xxxx no pleural effusion. there is no evidence of pneumothorax. heart size and mediastinal contour are normal.	no acute cardiopulmonary abnormality . the lung are clear bilaterally . specifically no evidence of focal consolidation on pneumothorax or pleural effusion cardio mediastinal silhouette is unremarkable . visualized osseous structure of the thorax are without acute abnormality.	no acute cardiopulmonary abnormality. the heart is normal in size. the lungs are clear. no focal airspace consolidation. no pneumothorax or pleural effusion.

Fig. 6. Illustration of the report generated by MvMM, MvTL, MvHL. Highlighted phrases are accurate generated sentences or medical concepts.

5 Conclusion

To generate diverse and semantic-coherent medical reports, we proposes a novel multi-view multi-modal attention model. By introducing semantic information into the designed two-layer LSTM-based decoder, we can generate diverse and accurate medical concepts reports. Experimental results show that Ours-with-MC does not only achieve state-of-the-art performance on *IUXray*, but also generates robust reports that include accurate medical concepts and satisfactory human evaluation results.

Acknowledgment. This work has been supported by National Natural Science Foundation of China (61772060, 61976012, 61602024), Qianjiang Postdoctoral Foundation (2020-Y4- A-001), and CERNET Innovation Project (NGII20170315).

References

1. Brady, A., Laoide, R.Ó., McCarthy, P., McDermott, R.: Discrepancy and error in radiology: concepts, causes and consequences. Ulster Med. J. **81**(1), 3 (2012)
2. Vinyals, O., Toshev, A., Bengio, S., Erhan, D.: Show and tell: a neural image caption generator. In: CVPR, June 2015
3. Karpathy, A., Fei-Fei, L.: Deep visual-semantic alignments for generating image descriptions. In: Proceedings of the IEEE Conference on Computer Vision and Pattern Recognition, pp. 3128–3137 (2015)
4. Donahue, J., Anne Hendricks, L., Guadarrama, S., et al.: Long-term recurrent convolutional networks for visual recognition and description. In: CVPR, pp. 2625–2634 (2015)
5. Mao, J., Xu, W., Yang, Y., Wang, J., Yuille, A.L.: Deep captioning with multimodal recurrent neural networks (m-RNN). In: Bengio, Y., LeCun, Y. (eds.) ICLR 2015 (2015). http://arxiv.org/abs/1412.6632
6. Lu, J., Xiong, C., Parikh, D., Socher, R.: Knowing when to look: adaptive attention via a visual sentinel for image captioning. In: CVPR, pp. 375–383 (2017)
7. Li, Y., Liang, X., Hu, Z., Xing, E.P.: Hybrid retrieval-generation reinforced agent for medical image report generation. In: Advances in Neural Information Processing Systems, pp. 1530–1540 (2018)
8. Li, C.Y., Liang, X., Hu, Z., Xing, E.P.: Knowledge-driven encode, retrieve, paraphrase for medical image report generation. In: AAAI 2019, pp. 6666–6673. AAAI Press (2019). https://doi.org/10.1609/aaai.v33i01.33016666
9. Jing, B., Xie, P., Xing, E.P.: On the automatic generation of medical imaging reports. In: ACL 2018, pp. 2577–2586. Association for Computational Linguistics (2018). https://doi.org/10.18653/v1/P18-1240, https://www.aclweb.org/anthology/P18-1240/
10. Yuan, J., Liao, H., Luo, R., Luo, J.: Automatic radiology report generation based on multi-view image fusion and medical concept enrichment. In: Shen, D., et al. (eds.) MICCAI 2019. LNCS, vol. 11769, pp. 721–729. Springer, Cham (2019). https://doi.org/10.1007/978-3-030-32226-7_80
11. Demner-Fushman, D., Kohli, M.D., Rosenman, M.B., et al.: Preparing a collection of radiology examinations for distribution and retrieval. J. Am. Med. Inform. Assoc. **23**(2), 304–310 (2015)
12. Ranzato, M., Chopra, S., Auli, M., Zaremba, W.: Sequence level training with recurrent neural networks. In: ICLR 2016 (2016). http://arxiv.org/abs/1511.06732
13. Wu, Y., Schuster, M., Chen, Z., Le, Q.V., Norouzi, M., Macherey, W.: Google's neural machine translation system: bridging the gap between human and machine translation. CoRR abs/1609.08144 (2016). http://arxiv.org/abs/1609.08144
14. Bengio, S., Vinyals, O., Jaitly, N., Shazeer, N.: Scheduled sampling for sequence prediction with recurrent neural networks. In: Advances in Neural Information Processing Systems, pp. 1171–1179 (2015)
15. Cho, K., Van Merriënboer, B., Gulcehre, C., et al.: Learning phrase representations using RNN encoder-decoder for statistical machine translation. In: EMNLP (2014)
16. Bahdanau, D., Cho, K., Bengio, Y.: Neural machine translation by jointly learning to align and translate. In: ICLR 2015 (2015). http://arxiv.org/abs/1409.0473
17. Sutskever, I., Vinyals, O., Le, Q.V.: Sequence to sequence learning with neural networks. In: NIPS, pp. 3104–3112 (2014)
18. Xu, K., Ba, J., Kiros, R., et al.: Show, attend and tell: neural image caption generation with visual attention. In: ICML, pp. 2048–2057 (2015)

19. You, Q., Jin, H., Wang, Z., et al.: Image captioning with semantic attention. In: CVPR, June 2016
20. Anderson, P., et al.: Bottom-up and top-down attention for image captioning and visual question answering. In: Proceedings of the IEEE Conference on Computer Vision and Pattern Recognition, pp. 6077–6086 (2018)
21. Bahdanau, D., et al.: An actor-critic algorithm for sequence prediction. In: ICLR 2017. OpenReview.net (2017). https://openreview.net/forum?id=SJDaqqveg
22. Tan, B., Hu, Z., Yang, Z., Salakhutdinov, R., Xing, E.P.: Connecting the dots between MLE and RL for sequence generation. In: ICLR 2019 (2019). OpenReview.net. https://openreview.net/forum?id=Syl1pGI9wN
23. Rennie, S.J., Marcheret, E., Mroueh, Y., Ross, J., Goel, V.: Self-critical sequence training for image captioning. In: Proceedings of the IEEE Conference on Computer Vision and Pattern Recognition, pp. 7008–7024 (2017)
24. Liu, F., Ren, X., Liu, Y., Wang, H., Sun, X.: simNet: stepwise image-topic merging network for generating detailed and comprehensive image captions. In: EMNLP 2018, pp. 137–149. Association for Computational Linguistics (2018). https://doi.org/10.18653/v1/d18-1013
25. Fang, H., et al.: From captions to visual concepts and back. In: Proceedings of the IEEE Conference on Computer Vision and Pattern Recognition, pp. 1473–1482 (2015)
26. Lin, T.Y., Goyal, P., Girshick, R., He, K., Dollár, P.: Focal loss for dense object detection. In: ICCV 2017, pp. 2980–2988 (2017)
27. Kingma, D.P., Ba, J.: Adam: a method for stochastic optimization. In: ICLR (2015). http://arxiv.org/abs/1412.6980
28. Xue, Y.: Multimodal recurrent model with attention for automated radiology report generation. In: Frangi, A.F., Schnabel, J.A., Davatzikos, C., Alberola-López, C., Fichtinger, G. (eds.) MICCAI 2018. LNCS, vol. 11070, pp. 457–466. Springer, Cham (2018). https://doi.org/10.1007/978-3-030-00928-1_52
29. Irvin, J., et al.: CheXpert: a large chest radiograph dataset with uncertainty labels and expert comparison. In: AAAI, vol. 33, pp. 590–597 (2019)
30. Vedantam, R., Lawrence Zitnick, C., Parikh, D.: Cider: consensus-based image description evaluation. In: CVPR, pp. 4566–4575 (2015)
31. Papineni, K., Roukos, S., Ward, T., Zhu, W.J.: Bleu: a method for automatic evaluation of machine translation. In: Proceedings of the 40th Annual Meeting on Association for Computational Linguistics, pp. 311–318. Association for Computational Linguistics (2002)
32. Lin, C.Y.: Rouge: a package for automatic evaluation of summaries. In: Text Summarization Branches Out, pp. 74–81 (2004)
33. He, K., Zhang, X., Ren, S., Sun, J.: Deep residual learning for image recognition. In: CVPR, pp. 770–778 (2016)

Poster Paper

PLRS: Personalized Literature Hybrid Recommendation System with Paper Influence

Fanghan Liu, Wenzheng Cai, and Kun Ma(✉)

Shandong Provincial Key Laboratory of Network Based Intelligent Computing, University of Jinan, Jinan 250022, China
{fanghanliu,crainyday}@foxmail.com, ise_mak@ujn.edu.cn

Abstract. The traditional management platform of academic literature depends on the active retrieval of users, and the accuracy of literature retrieval results cannot be guaranteed. In our demonstration, we introduce the personalized literature hybrid recommendation system with paper influence (PLRS for short). The innovation of PLRS is that it can calculate the influence of literature, address the issue of cold start, and actively provide users with high-quality literature in line with users' interests combined with cascade hybrid recommendation algorithm.

Keywords: Cold start · Literature recommendation · Hybrid recommendation algorithm · Keyword extraction

1 Introduction

To save the time that researchers spend searching for literature and help them quickly find the literature they want to view in a large amount of literature [5], we presents a personalized literature hybrid recommendation system with paper influence. The system does not need an active literature search when recommending literature to users. The user portraits are obtained by analyzing the user's behavior and the contact with the third-party account. Then, the initial literature list is screened by the paper influence evaluation model and finally the literature is personalized recommended to users through the cascade-based hybrid recommendation algorithm.

2 Architecture

PLRS is mainly divided into four layers, and its architecture is shown in Fig. 1: information input layer, cold start processing layer, hybrid recommendation model layer, and recommendation list output layer. First of all, users can enter their account password from the information input layer to log in to the system. During the login process, the system will detect whether the user is new or not. If new users, then it enters the cold start processing layer for cold start problem processing. Otherwise, it enters the hybrid recommended model layer. Second,

M. Qiu (Ed.): ICA3PP 2020, LNCS 12454, pp. 703–705, 2020.
https://doi.org/10.1007/978-3-030-60248-2

the cold start processing layer will process the cold start problem of different strategies according to whether the user binds the ORCID ID or not. Then, it obtains the user portrait. Next, the hybrid recommendation model layer uses word2vec word vector model to initially obtain papers with high similarity to user portraits. After filtering through the paper influence assessment model, it combines behavioral data into a multi-level cascading recommendation algorithm for selective recall. Finally, the final recommendation result is displayed to the user through the recommendation list output layer.

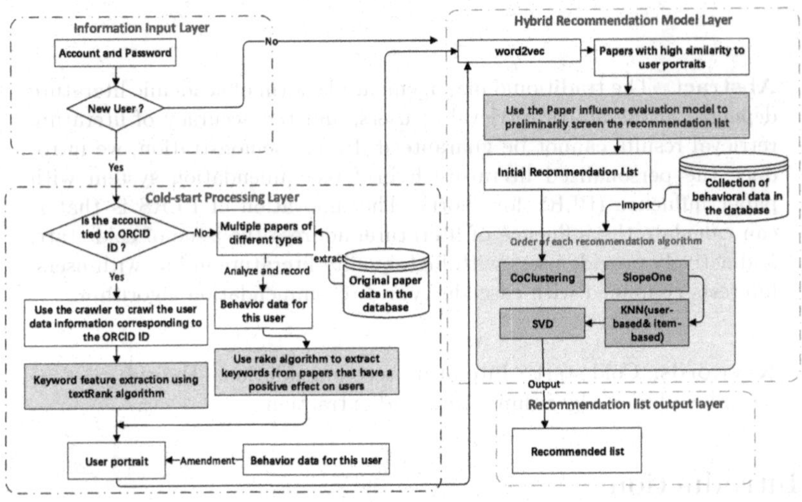

Fig. 1. Architecture of PLRS

We find that there is little difference in the final results of different algorithms for different indicators, so we determined to fuse each algorithm in the form of cascading, and the order of cascading is Coclustering [3], SlopeOne [4], KNN (user-based and item-based) [1], SVD [2]. The hybridization technique is very efficient and tolerant to noise due to the coarse-to-finer nature of the iteration. The candidate results will be selected step by step in this process, and finally the result set with small quantity and high quality will be sent to the output layer of the recommendation list. The cascade process is shown in Fig. 2. Our methods have two advantages. First, the recommendation algorithm we mixed is diverse, and the cascading order is obtained through experimental data. Second, is that the input data of our proposed cascade hybrid algorithm is not the original paper data, but the initial document list obtained through the paper influence evaluation model. This reduces the scope of recommendation compared with the original recommendation data, which helps to shorten the execution time of the algorithm and improve the execution efficiency.

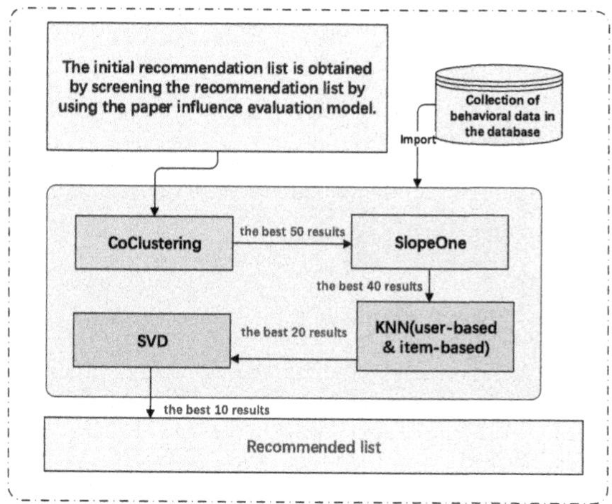

Fig. 2. Cascade hybrid recommendation algorithm

3 Demonstration Scenarios

First of all, we will use slides to introduce the motivation of PLRS and highlight how it works. After that, we will show the demo video of our system to the audience. Finally, we invite them to participate in the interaction by logging in to the system via smartphone or computer.

Acknowledgments. This work was supported by the National Natural Science Foundation of China (61772231), the Shandong Provincial Natural Science Foundation (ZR2017MF025), the Project of Shandong Provincial Social Science Program (18CHLJ39), and the Science and Technology Program of University of Jinan (XKY17348).

References

1. Aggarwal, C.C.: Neighborhood-Based Collaborative Filtering. Recommender Systems, pp. 29–70. Springer, Cham (2016). https://doi.org/10.1007/978-3-319-29659-3_2
2. Baltrunas, L., Ludwig, B., Ricci, F.: Matrix factorization techniques for context aware recommendation. In: Proceedings of the Fifth ACM Conference on Recommender Systems. pp. 301–304 (2011)
3. George, T., Merugu, S.: A scalable collaborative filtering framework based on co-clustering. In: Fifth IEEE International Conference on Data Mining (ICDM'05), p. 4. IEEE (2005)
4. Lemire, D., Maclachlan, A.: Slope one predictors for online rating-based collaborative filtering. In: Proceedings of the 2005 SIAM International Conference on Data Mining, pp. 471–475. SIAM (2005)
5. Ma, K., Yang, B.: A simple scheme for bibliography acquisition using DOI content negotiation proxy. The Electronic Library (2014)

Author Index

Ai, Zhengpeng III-426
Anand Gopalakrishnan, Atul III-601
Anderson, Scott II-415
Andrzejczak, Michal I-661
Asare, Bismark Tei III-580

Ba, Cheikh I-344
Bader, David A. II-157
Baek, Nakhoon II-723
Bai, Jing II-619
Bao, Yungang II-705
Ben Messaoud, Othman I-606
Bian, Haodong II-111
Borowiec, Damian II-492
Bu, Deqing I-47

Cai, Meng-nan II-677
Cai, Shubin III-155
Cai, Wei II-200, II-215
Cai, Wenzheng III-703
Cao, Jian II-587
Cao, Weipeng I-219, I-448, II-352, II-538
Cao, Zhengjia II-215
Cérin, Christophe III-381
Chang, Peng III-494
Chang, Xiaolin I-695, II-619
Chao, Pingfu I-190
Chen, Juan I-92
Chen, Lei I-579, I-681
Chen, Lin III-676
Chen, Shuang II-97
Chen, Xuxin III-537
Chen, Zhijun III-441
Chen, Zhuang II-230
Cheng, Dongxu III-144
Cheng, Qixuan I-528
Cheng, Yongyang II-142, II-646
Chi, Yuanfang II-200
Christopher Victor, Ashish III-601
Chu, Zhaole I-15
Cong, Peijin III-564
Cooke, Jake II-415
Cui, Jiahe II-259

Dai, Chuangchuang III-654
Dai, Hua III-218
Dai, Jialu II-383
Dai, Wenhao III-475
De Giusti, Armando I-262
Deng, Anyuan III-81
Deng, Xiaoge I-495
Ding, Jia III-396
Ding, Jiafeng I-190
Ding, Kai III-3
Ding, Yepeng I-480
Ding, Zhenquan II-398
Dong, Bo I-627
Dong, Changkun II-142
Dong, Runting II-111
Dong, Xiaoyun II-259
Dong, Yong I-92, II-82, II-97
Dowling, Anthony I-3
Dricot, Jean-Michel III-309
Du, Xinlu I-465
Du, Zhihui II-157
Duan, Haihan II-200, II-215
Duan, Huifeng III-367
Duan, Lijuan III-367
Dun, Ming I-232

Eliassen, Frank II-230
Ellinidou, Soultana III-309
Engel, Fabian Herbert I-3

Faiz, Sami II-3
Fan, Dongrui I-61, II-14
Fan, Jianxi II-47
Fan, Kai III-676
Fan, Shuhui I-563
Fan, Weibei II-47
Fan, Xiaopeng I-174
Fang, Junhua I-190

Fang, Xuqi III-537
Fang, Zhengkang III-19
Fei, Haiqiang II-398
Feng, Boqin I-627
Feng, Yujing I-61
Friday, Adrian II-492
Fu, Shaojing III-520
Fu, Xianghua II-432, II-477, II-523, III-340
Fu, Zongkai I-247, II-259

Gai, Keke II-200, III-3, III-19, III-35, III-110, III-282
Gaj, Kris I-661
Gan, Yu I-159, II-82
Gao, Hang III-184
Garraghan, Peter II-492
Ge, Shuxin II-306
Gogniat, Guy III-309
Groß, Julian I-369
Gu, Xiaozhuo III-475
Guan, Hongtao III-609
Guan, Jianbo III-126
Guan, Zhenyu III-144
Gueye, Abdoulaye I-344
Gunturi, Venkata M. V. I-125
Guo, Guibing III-426
Guo, Qiang I-275
Guo, Xiaobing III-654
Guo, Yi I-330
Guo, Yuluo II-111
Guojian, Tang I-719

Hamdi, Wael II-3
Han, Ru III-639
Han, Yetong II-383
Han, Yuejuan II-47
Han, Zhen I-695
Han, Zhu I-695
Hao, Long II-321
Hao, Zhiyu II-398
Harper, Richard II-492
He, Fubao III-297
He, Jianwei II-477, II-523
He, Kai III-65, III-81
He, Yulin II-509
Honan, Reid II-415
Hong, Xiaoguang II-690
Hou, Aiqin III-197
Hou, Biao II-633
Hou, Junteng I-31, I-645
Hu, Jiale III-144
Hu, Jingkun II-157
Hu, Lin II-449
Hu, Mengtao I-386
Hu, Wei I-159, I-330, II-82, II-97
Hu, Wenhui II-603
Hu, Xinrong III-65
Hu, Yuekun I-465
Hu, Ziyue I-174, III-110
Huai, Xu I-579
Huang, Chunxiao III-65
Huang, Feng I-495
Huang, Han I-401
Huang, Hua II-463, III-297, III-355, III-549
Huang, Jiahao III-270
Huang, Jianqiang II-111
Huang, Joshua Zhexue II-509
Huang, Kaixin I-433
Huang, Linpeng I-433
Huang, Qun I-614
Huang, Xiaofu I-465
Huang, Xiaoping II-575
Huang, Xin II-184
Huang, Xinli III-537, III-564
Huang, Yu I-579, II-603
Huang, Zhijian I-563
Hui, Zhao II-142, II-646

Jayanthi, Akhilarka III-601
Ji, Jiawei I-143
Ji, Yimu II-449
Jia, Ziye I-695
Jiang, Congfeng III-381
Jiang, Feng II-142, II-646
Jiang, Jian-guo II-677
Jiang, Linying III-426
Jiang, Peng III-35
Jiang, Shang II-677
Jiang, Shenghong I-143
Jiang, Zoe L. II-126
Jiao, Qiang II-82, II-97
Jin, Hao III-93
Jin, Honghe I-512
Jin, Peipei I-579
Jin, Peiquan I-15, III-623
Jin, Shuyuan III-459

Jin, Xiaolong III-50
Jing, Junchang I-112
Jiwei, Chen I-719

Kandoor, Lakshmi II-184
Kay, Savio II-184
Khedimi, Amina III-381
Köster, Marcel I-369
Krüger, Antonio I-369
Kuang, Xiaoyun III-676

Lagwankar, Ishaan III-592
Lan, Dapeng II-230
Lei, Yongmei I-143
Lewis, Trent W. II-415
Li, Chao III-297
Li, Dawei III-409
Li, Fei I-416
Li, Geng III-324
Li, Haochen III-35
Li, Huiyong I-247
Li, Jianchuan III-623
Li, Jiawei III-218
Li, Jinbao I-305, III-509
Li, Jingjing I-548
Li, Jun I-710
Li, Ling I-47
Li, Lun II-398
Li, Ningwei III-184
Li, Peilong III-93
Li, Peng II-290
Li, Sukun III-663
Li, Wei I-355
Li, Wenming I-61, II-14
Li, Xiangxiang II-32
Li, Xiangxue I-548
Li, Xin I-314
Li, Yi II-14
Li, Yunchun I-232
Li, Yuwen III-282
Li, Yuxiang I-78, I-112
Li, Zhenhao II-82
Li, Zhong III-50
Liao, Chenyang III-270
Liao, Xiaojian I-416
Lin, Chen II-32
Lin, Mufeng II-32
Lin, Yang II-463, III-270
Lin, Zhen III-520
Liu, Bin I-78
Liu, Fanghan III-703
Liu, Feng I-495
Liu, Hongli III-367
Liu, Jianwei III-144, III-324, III-409
Liu, Kaihang II-449
Liu, Lei II-230
Liu, Li I-386
Liu, Liang III-184
Liu, Meiqin III-549
Liu, Qiang II-449
Liu, Shangdong II-449
Liu, Wuji III-197
Liu, Xin II-463
Liu, Xinxin III-549
Liu, Xuefeng I-247, III-687
Liu, Xueyang II-603
Liu, Yanlan II-449
Liu, Yao I-386
Liu, Ye II-383
Liu, Yiyang III-218
Liu, Yizhong III-324, III-409
Liu, Yu I-3, III-297
Liu, Yuan III-426
Liu, Yufei III-537, III-564
Liu, Zhe I-416
Liu, Zihao III-170
Long, Hao I-448
Lu, Fengyuan III-537, III-564
Lu, Jintian III-459
Lu, Ming III-170
Lu, Youyou I-416
Lu, ZhongHua I-290
Lu, Zhonghua III-654
Luan, Hua I-401
Luan, Zerong I-232
Luo, Yan III-93
Luo, Yongping I-15
Luo, Yuchuan III-520
Lv, Huahui III-676
Lv, Xiangyu I-159
Lv, Xingfeng III-509
Lyu, Xukang I-205

Ma, Bingnan I-31
Ma, Dongchao I-465
Ma, Fuhua I-710
Ma, Kun III-703
Ma, Qiangfei III-355

Ma, Xingkong III-609
Mao, Yupeng III-170
Markowitch, Olivier III-309
Maxwell, Thomas II-184
Memmi, Gerard II-274
Meng, Lianxiao II-552
Menouer, Tarek III-381
Miao, Weikai III-231
Ming, Zhong I-219, II-352, II-538, III-155
Mishra, Abhishek I-125
Mišić, Jelena II-619
Mišić, Vojislav II-619
Mu, Lin III-623
Musariri, Manasah I-47

Nagpal, Rahul III-592, III-601
Naiouf, Marcelo I-262
Nana, Laurent III-580
Nie, Feiping II-337
Nie, Yu III-297
Niu, Beifang III-654
Niu, DanMei I-112
Niu, Jianwei I-247, II-259, III-687

Ou, Yan I-61, II-14
Ou, Zhixin I-92
Ouyang, Zhenchao I-247, II-259

Pan, Yu I-305
Park, Seung-Jong II-723
Pei, Songwen II-173
Peng, Jianfei III-184
Peng, Yaqiong II-398
Peng, Zhaohui II-690
Pousa, Adrián I-262

Qian, Haifeng I-548
Qian, Shiyou II-587
Qiao, Yuanhua III-367
Qin, Xiaolin I-314
Qiu, Han II-274
Qiu, Meikang I-159, II-173, II-274, II-463, III-3, III-35, III-297, III-355, III-549, III-654, III-663
Quist-Aphetsi, Kester III-580

Rabbouch, Bochra I-591
Rabbouch, Hana I-591, I-606
Rajapakshe, Chamara II-184
Rajashekar, Vishwas III-592
Ramnath, Sarnath I-125
Ren, Qianqian I-305, I-710
Ren, Shuangyin II-552
Ren, Yi III-126
Rong, Guoping II-32

S N, Durga Prasad III-592
Saâdaoui, Foued I-591, I-606
Sang, Yafei III-494
Sanz, Victoria I-262
Sato, Hiroyuki I-480
Shao, Pengpeng III-170
Sharma, Gaurav III-309
Shen, Chongfei I-61
Shen, Junzhong I-528
Shen, Siqi III-609
Shen, Wei III-197
Shen, Xiaoxian II-245
Shi, Jiakang III-282
Shi, Jiaoli III-65, III-81
Shi, Quanfu II-337
Shi, Yimin II-200
Shu, Jiwu I-416
Si, Chengxiang I-645
Sugizaki, Yukimasa II-365
Sun, Jie III-218
Sun, Qingxiao I-232
Sun, Shibo II-14
Sun, Tao I-495
Sun, Xiaoxiao I-512
Sun, Xudong II-523
Sun, Zhi III-50
Suo, Siliang III-676
Swiatowicz, Frank I-3

Taherkordi, Amir II-230
Takahashi, Daisuke II-365
Tan, Nongdie I-681
Tan, Yusong III-126
Tang, Gaigai II-552
Tang, Minjin I-614
Tang, Shuning II-449
Tang, Xudong III-231
Teng, Meiyan I-314
Teng, Yajun III-475
Tian, Mao III-494
Tian, Mengmeng III-426
Tian, Ze III-639

Tolnai, Alexander John I-3
Tong, Li I-579
Tu, Yaofeng I-433

Valiullin, Timur I-448, II-321

Wan, Shouhong I-15
Wang, Changming III-367
Wang, Danghui II-563, III-639
Wang, Deguang I-528
Wang, Jianwu II-184
Wang, Jihe II-563, II-575
Wang, Jikui II-337
Wang, Li I-275
Wang, Qiang I-448, III-231
Wang, Ruili III-110
Wang, Sa II-705
Wang, Shiyu II-575
Wang, Shupeng I-31, I-645
Wang, Shuxin II-432, II-477, II-523, III-340
Wang, Si-ye II-677
Wang, Tianbo III-251
Wang, Wei I-386
Wang, Weixu II-306
Wang, Xiaoying II-111
Wang, Xinyi I-47
Wang, Xiwen III-170
Wang, Xizhao II-352
Wang, Yan II-47
Wang, Yang I-174
Wang, Yaobin I-47
Wang, Yonghao I-159, I-330, II-97
Wang, Yongjun I-563
Wang, Yu II-603
Wang, Zhe III-126
Wei, Chenghao I-448, II-321
Wei, Guoshuai II-662
Wei, Xiaohui II-245
Wei, Yanzhi II-432, II-477, II-523
Wei, Yihang III-282
Wen, Mei I-528, I-614
Wen, Yuan I-159, II-82
Wen, Yujuan III-564
Wu, Chase Q. I-205, III-197
Wu, Guangjun I-31, I-645
Wu, Haiyu I-627
Wu, Jie I-314
Wu, Jing I-330
Wu, Jiyan III-687
Wu, Kaishun II-383
Wu, Qianhong III-324, III-409
Wu, Quanwang II-662
Wu, Xinxin I-61, II-14
Wu, Yuhao II-538
Wu, Yusheng II-173

Xia, Chunhe III-251
Xia, Jingjing II-47
Xia, Yuanqing I-219
Xiao, Ailing I-465
Xiao, Bowen II-215
Xiao, Wan II-449
Xie, Peidai I-563
Xie, Tao III-520
Xie, Wenhao II-432, II-477, III-340
Xie, Zhongwu II-352
Xiong, Hailing I-681
Xu, Aidong III-676
Xu, Chen III-93
Xu, Chengzhong I-174
Xu, Fang III-81
Xu, Haoran I-563
Xu, Hongzuo I-563
Xu, Jiajie I-190
Xu, Jianqiu III-218
Xu, Jiawei II-587
Xu, Lei III-19
Xu, Liwen I-512
Xu, Wenxing II-14
Xu, Yuan II-705
Xu, Zhengyang II-449
Xu, Zhiwu II-538, III-396
Xue, Guangtao II-587
Xue, Wei I-386

Yan, Ruibo II-142, II-646
Yang, Chenlong I-548
Yang, Geng III-218
Yang, Guang II-690
Yang, Hailong I-232
Yang, Hang III-676
Yang, Jingying III-340
Yang, Lei II-215
Yang, Lin II-552
Yang, NingSheng III-155
Yang, Renyu II-492
Yang, Saqing III-126
Yang, Shaokang III-687

Yang, Wu II-552
Yang, Xueying III-654
Yang, Yang I-47, I-695, II-619
Yang, Yifan II-32
Yang, Zhe I-416
Yang, Zhengguo II-337
Ye, Feng III-170
Ye, Qianwen III-197
Ye, Xianglin I-681
Ye, Xiaochun I-61
Ye, Xuan II-509
Yeung, Gingfung II-492
Yin, Hao III-282
Yin, Jiayuan III-324
Ying, Yaoyao I-433
Yiu, Siu Ming II-126
You, Xin I-232
You, Ziqi III-126
Yu, Dunhui III-441
Yu, Hui III-324, III-409
Yu, Huihua I-579
Yu, Shucheng II-62
Yuan, Baojie II-383
Yuan, Yuan I-92
Yue, Chen III-639
Yue, Hengshan II-245

Zeng, Xiao I-159
Zeng, Yi II-274
Zeng, Yunhui I-275
Zhan, Ke I-290
Zhang, Chaokun II-306
Zhang, Chunyuan I-528, I-614
Zhang, Hanning I-627
Zhang, Jie I-275
Zhang, Jiyong I-448, II-538
Zhang, Jun II-126
Zhang, Junxing II-633
Zhang, Lei I-31, I-645
Zhang, Lu III-639
Zhang, Mengjie III-441
Zhang, Minghui II-603
Zhang, Sen II-157
Zhang, Shengbing II-575
Zhang, Tao II-290
Zhang, Tianwei II-705
Zhang, Wentao III-355
Zhang, Xingsheng III-441
Zhang, Yan-fang II-677
Zhang, Yang I-305
Zhang, Yanlin II-290
Zhang, Yongzheng III-494
Zhang, Yue III-3
Zhang, Yunfang I-92
Zhang, YunQuan I-290
Zhang, Yunyi III-459
Zhang, Zhibo II-184
Zhang, Zhiyong I-78, I-112
Zhang, Zijian III-110
Zhao, Bo-bai II-677
Zhao, Da-ming II-62
Zhao, Hui II-603
Zhao, Jiaxiang II-563
Zhao, Jie III-623
Zhao, Lei I-190
Zhao, PengPeng I-190
Zhao, Qinglin II-306
Zhao, Shuyuan III-494
Zhao, Wenqian II-32
Zhao, Xiaolei I-614
Zhao, Yonglin II-432, II-477, II-523
Zheng, Hongqiang I-330
Zheng, Jianyu II-184
Zheng, Shengan I-433
Zheng, Weiyan I-579
Zhou, Changbao II-245
Zhou, Cong II-538
Zhou, Fangkai III-270
Zhou, Hao I-681
Zhou, Huaifeng III-155
Zhou, Jian-tao II-62
Zhou, Jingren III-623
Zhou, Xiaobo II-306
Zhu, Guanghui I-275
Zhu, Haoran II-619
Zhu, Junwei II-587
Zhu, Liehuang III-19, III-35, III-110, III-282
Zhu, Qingting I-386
Zhu, Weidong II-587
Zhu, Yanmin II-587
Zhu, Zongyao II-587
Zhuang, Chuanzhi III-50
Zhuang, Yuan I-275
Zou, Weidong I-219
Zou, Xingshun I-355
Zou, Yongpan II-383